The Mycota

A Comprehensive Treatise on Fungi as Experimental Systems for Basic and Applied Research

Volume 17

The fungi represent a heterogenous assemblage of eukaryotic microorganisms and have become favored organisms for research at the cellular and molecular level. Such research involvement has been stimulated by interest in the biotechnological application of fungi in processes related to industry, agriculture and ecology. Considering both yeasts and mycelial fungi, THE MYCOTA highlights developments in both basic and applied research and presents an overview of fungal systematics and cell structure. Foremost authorities in research on mycology have been assembled to edit and contribute to the volumes.

All volumes/chapters are systematically reviewed by the series editor(s) and respective volume editor(s).

Soumya Ghosh • Helen N. Onyeaka
Editors

Fungal-Based Cosmetics: Formulation and Usage

Springer

Editors
Soumya Ghosh
Natural and Medical Sciences
Research Center
University of Nizwa
Nizwa, Oman

Helen N. Onyeaka
School of Chemical Engineering
University of Birmingham
Birmingham, UK

ISSN 2945-8048 ISSN 2945-8056 (electronic)
The Mycota
ISBN 978-3-032-09290-8 ISBN 978-3-032-09291-5 (eBook)
https://doi.org/10.1007/978-3-032-09291-5

This Springer imprint is published by the registered company Springer Nature Switzerland AG
The registered company address is: Gewerbestrasse 11, 6330 Cham, Switzerland

Foreword by Ngwogu

In recent years, the cosmetics industry has witnessed a significant shift toward more natural and sustainable ingredients, resulting in a renewed interest in the potential of fungi as sources of bioactive compounds. The present work provides an in-depth exploration of this promising intersection between mycology and cosmetic science. It details with commendable rigor and clarity the diverse applications and benefits of fungal extracts in personal care products.

Comprehensive in its scope, the work surveys the various aspects of fungal-based cosmetics, starting with the exploration of cultivable fungi that offer medicinal properties. It goes on to detail the bioactive compounds derived from these fungi, showcasing their roles as antioxidants, anti-inflammatory agents, and antimicrobial substances. The arrangement of the materials evinces thoughtfulness, with each chapter of the work designed to build on the previous one, leading the reader through a methodological documentation of the efficacy of fungal extracts in skincare and haircare formulations.

Noteworthy sections of work include a discussion on the antiaging properties and the ability of fungal extracts to enhance skin health through emollient, exfoliating, and skin-lightening effects. Additionally, the document addresses the challenges faced in the industrial-scale production of these innovative products, highlighting the need for further research and development.

As the search continues for alternatives to synthetic chemicals in cosmetics, the insights presented in this work are both timely and essential. They jointly constitute a giant stride toward the goal of enhancing the effectiveness of cosmetic formulations and promoting sustainability in product development, and they pave the way for future research work in the field.

The exploration of commercially available fungal-based cosmetics testifies to the growing acceptance and incorporation of these ingredients in the market. The present work is poised to be a reference for researchers, formulators, and anyone interested in the potential of fungi in the world of cosmetics. It encourages a deeper understanding of how fungi can contribute to effective, safe, and sustainable beauty solutions.

Ada Carol Ngwogu
Professor of Medical Microbiology and Public Health,
Head of Department of Medical Microbiology,
Faculty of Basic Clinical Medicine,
College of Medicine and Health Sciences,
Abia State University, Nigeria

Foreword by Gafforov

The cosmetics industry is undergoing a deliberate shift toward sustainable, bio-based innovation. Within this transition, fungi are emerging as versatile sources of high-value ingredients that unite traditional knowledge with modern biotechnology and offer credible alternatives to purely synthetic actives. *Fungal-Based Cosmetics: Formulation and Usage* is a timely contribution that integrates mycology, pharmacognosy, cosmetic chemistry, and product development into a coherent, practice-oriented resource.

The volume traces a clear path from cultivation to finished formulations. It introduces the diversity of cultivable fungi and the bioactive classes they yield and then examines functional activities central to contemporary cosmetic science: antioxidant and anti-inflammatory effects; modulation of melanogenesis (anti-tyrosinase); protection of extracellular-matrix integrity (anti-hyaluronidase, anti-collagenase, anti-elastase); and antimicrobial properties relevant to skin and scalp care. These mechanisms are consistently related to formulation goals and measurable performance endpoints.

A particular strength of the book is its practical scope. The roles of fungal carotenoids, ceramides, and polysaccharides are presented alongside real product categories, offering clear lines of sight from bench to market. Chapters on nanocosmetics, formulation design and compatibility, stability assessment, and scale-up challenges provide a realistic view of what it takes to translate promising fungal actives into reliable, claim-substantiated products.

This work serves multiple audiences. Researchers will find an up-to-date synthesis of mechanisms and materials. Formulators will gain guidance on integrating fungal extracts into emulsions, gels, and anhydrous systems, as well as on compatibility and processing. Industry leaders will appreciate the attention to quality, safety, regulatory expectations, and manufacturing feasibility. By balancing scientific depth with practical instruction, the book functions as both reference and roadmap.

I commend the editors and contributors for assembling a rigorous and forward-looking text. Grounded in evidence and aligned with sustainability principles, *Fungal-Based Cosmetics: Formulation and Usage* will help shape the next generation of skincare, haircare, and personal-care products, rooted in nature and refined by science.

Yusufjon Gafforov
Central Asian Center for Development Studies
New Uzbekistan University,
Tashkent, Uzbekistan

Foreword by Chauhan

The twenty-first century has witnessed a significant shift in consumer demand toward natural, sustainable, and bioactive products. In the cosmetics industry, this trend has driven the exploration of innovative, bio-derived ingredients that promote skin health while meeting aesthetic expectations. Fungi, long valued in pharmaceutical and biotechnological applications, have recently gained attention as a novel source of bioactive compounds for cosmetic formulations.

This book, *Fungal-Based Cosmetics: Formulation and Usage*, represents a pioneering effort to consolidate current knowledge on the role of fungi in cosmetic science. It integrates interdisciplinary research on the biological properties, formulation techniques, and practical applications of fungal-derived compounds, offering a comprehensive resource for professionals and researchers.

The first chapter provides an introductory overview, setting the stage for an in-depth exploration of fungal species and their bioactive metabolites. Subsequent chapters examine the antioxidant, anti-inflammatory, and enzyme-inhibitory properties of these compounds, highlighting their utility in skincare and personal care products. The book further explores the functional and sensory attributes of fungal extracts, including their emollient, exfoliating, and antiaging effects, underscoring their potential in cosmeceutical development.

This book focuses on the practical applications of fungal-derived ingredients, detailing their integration into commercially available products and addressing the transition from laboratory research to market. Additionally, it also explores the biochemical diversity of fungal carotenoids, ceramides, and polysaccharides, emphasizing their contributions to cosmetic efficacy.

The book also addresses emerging innovations, such as the application of nanotechnology in fungal-based cosmetics, alongside practical considerations for formulation, stability, and scalability. These sections provide valuable guidance for cosmetic scientists, formulators, dermatologists, microbiologists, and industry professionals.

In essence, fungal-based cosmetics represent a paradigm shift toward sustainable innovation, with the potential not only to transform the cosmetic industry but also to contribute to environmental stewardship by reducing reliance on synthetic and petrochemical-derived ingredients.

I extend my sincere appreciation to the editors, authors, and contributors whose expertise and dedication have shaped this work. Their efforts highlight the transformative potential of fungi in advancing eco-friendly, science-driven cosmetic solutions.

This volume aims to serve as both a key reference for professionals and an inspiration for researchers to further explore the untapped potential of fungi in redefining modern cosmetic science.

Pankaj Kumar Chauhan
Professor,Faculty of Applied Sciences
and Biotechnology,
Shoolini University, Solan,
Himachal Pradesh, India

Foreword by Al-Harrasi

In a world increasingly drawn to natural, sustainable, and scientifically backed solutions, the cosmetics industry stands at a compelling crossroads. Consumers today not only seek products that enhance beauty but also demand transparency, eco-responsibility, and functional efficacy. Amidst this evolving landscape, fungi—organisms historically revered for their medicinal properties—have emerged as a transformative force in personal care. The book, *Fungal-Based Cosmetics: Formulation and Usage*, is a much-needed and timely contribution to this growing field, blending deep scientific insight with practical application.

Spanning 22 insightful chapters, this volume presents a comprehensive overview of fungi as a novel and potent source of cosmetic ingredients. From the cultivation of medicinal fungi to the isolation of bioactive compounds, the text delves into the diverse functionalities these organisms offer—from antioxidant, anti-inflammatory, and antiaging activities to antimicrobial, emollient, exfoliating, and skin-lightening properties. Notably, chapters on enzymatic inhibition (anti-tyrosinase, anti-hyaluronidase, anti-collagenase, and anti-elastase) illustrate the multi-target potential of fungal-derived ingredients in combating skin aging and supporting dermal health.

What distinguishes this book is its holistic and application-driven approach. Readers are introduced not only to the biochemical and therapeutic potential of fungi but also to formulation strategies, product stability, and commercial applications. Real-world examples—from skincare and haircare to hand sanitizers and soaps—demonstrate that fungal-based cosmetics are no longer futuristic concepts but market-ready realities.

Equally important are the chapters that tackle broader industrial considerations, including challenges in large-scale production, regulatory hurdles, and the emerging intersection of nanotechnology with fungal cosmeceuticals. These insights ground the book in the realities of product development and commercialization while highlighting avenues for future innovation.

The editors and contributing authors deserve commendation for assembling such an authoritative and forward-looking resource. Their work reflects a deep understanding of mycology, cosmetic science, and biotechnology, and it offers researchers, formulators, entrepreneurs, and students a foundation for innovation that is both ethical and evidence-based.

Ahmed Al-Harrasi
Vice Chancellor for Graduate Studies,
Research & External Relations,
Chairman, Natural and Medical Sciences
Research Center,
University of Nizwa, Nizwa, Oman

Series Preface

The Mycota is an encyclopaedic book series published by Springer as a comprehensive treatise spanning the Fungal Kingdom through a focus on fungi as experimental systems for basic and applied research. Articles in this series present introductory information in a comprehensive manner to broaden our general knowledge, along with state-of-the-art information on selected topics in fungal biology. Thus, these articles serve as a timeless reference source for historically relevant discoveries in fungal biology to drive further advances in the field.

The Mycota was founded in 1994 by Karl Esser and Paul Lemke. Since then 15 volumes have been published, with several now in their third edition. The steadily growing interest in *The Mycota*, reflected in particular by the high number of e-book downloads, encouraged Springer to continue this publication with a new international Series Editorial board beginning in 2022.

Historically, the study of fungi originated as a sub-discipline of botany and was a largely descriptive discipline until the early nineteenth century. Subsequent experimental research has propelled the field of mycology, and achievements in the genetics and molecular biology of fungi have benefited studies in the related fields of fungal biochemistry and biophysics, plant pathology, medical mycology, and systematics. Ground-breaking research has been carried out using fungi as recognized by the receipt of nine Nobel Prizes, including the revolutionary 1945 Nobel Prize in Physiology or Medicine for the discovery of penicillin and the foundational 2006 Nobel Prize in Chemistry for establishing the molecular basis of eukaryotic transcription. All such studies expand our knowledge of fungi and of fungal processes and improve our ability to understand, utilize, and control fungi for the benefit of humankind.

Fungi have invaded every conceivable ecological niche. Saprobic forms abound, especially in the decay of organic debris. Pathogenic forms exist with both plant and animal hosts. Fungi even grow on other fungi. They are found in aquatic as well as soil environments, and their spores have been found in the air of extreme environments. Fungi can be variously associated with plants as symbionts in the form of lichens, mycorrhizae, and endophytes and also occur as overt pathogens. Association with animal systems varies; examples include the predaceous fungi that trap nematodes, the micro-fungi that grow in the anaerobic environment of the rumen, and medically important pathogens afflicting humans. Currently, taxonomists have identified about 150,000 species, but it is thought that over 90% of fungal species remain undescribed, with conservative estimates ranging from 2 to 5 million fungal species on earth.

From this perspective, there are new topics in fungal biology to explore, which will expand the current volumes and take them in new directions. Further, there will be new volumes in areas (e.g., Cryptomycota) that were not covered before.

Our understanding of the evolution of fungi is still incomplete and mainly based on species that can be grown in culture. But recent environmental DNA analyses have revealed a highly diverse form of eukaryotic life that branches with the fungi, and thus the resulting and highly diverse clades were named the Cryptomycota. The discovery of novel intermediate forms will redefine the fungal tree of life. Other topics will consider the adaptation of fungi to climate changes, the occurrence of fungal pathogens in the environment, and the dispersal of fungi during global pandemics.

Fungal constituents of the microbiome have received much less attention thus far, yet recent findings from clinical and animal studies clearly establish fungi as a significant component of the oral, gastrointestinal, pulmonary, and skin microbiomes. Finally, the relevance of fungal biology to society is reflected by the increasing number of fungal-related human diseases.

In the history of pharmacology, fungi have always been sources of useful molecules for humans, but the diversity of molecules that can be obtained from fungi is still largely underexplored. However, efforts in fungal genomics provide resources for data-driven genome mining and large-scale comparisons to explore the molecular repertoires produced by fungi. The result will be new compounds with applications in the pharmaceutical and agricultural science industries.

Fungi can also help the entire planet and may, for example, be relevant in specific sectors, such as that linked to pollution from plastic. Fungi produce a wide range of enzymes that have the potential to break down the chemical bonds of plastic polymers, and in this context the potential role of marine fungi in plastic degradation may be of major relevance. Finally, new biomaterials from fungal species may open the door to alternatives to fossil-based materials and thus reduce environmental pollution.

For consistency throughout this series of volumes, the names adopted for major groups of fungi should be followed according to the following paper, which gives an overview of all of the orders in the fungal kingdom: https://pubmed.ncbi.nlm.nih.gov/32660385/

We are grateful to Springer for continuing *The Mycota* series and are especially thankful to all of the Volume Editors in selecting topics and assembling experts from diverse fields of fungal biology.

Sydney, Australia — Dee Carter
Delhi, India — Anuradha Chowdhary
Durham, NC, USA — Joseph Heitman
Bochum, Germany — Ulrich Kück
July 15th, 2022

Volume Preface

The cosmetics industry is undergoing a paradigm shift toward sustainable, safe, and bioactive solutions that align with consumer demand for natural products. Among the vast array of biological resources available, fungi stand out as an exceptional, yet often underexplored, source of valuable compounds with remarkable cosmetic potential. This book, *Fungal-Based Cosmetics: Formulation and Usage*, is dedicated to unveiling the scientific, technological, and commercial dimensions of integrating fungal resources into modern cosmetic and cosmeceutical products.

From ancient medicinal uses of mushrooms to cutting-edge biotechnology in personal care, fungi offer a unique repertoire of bioactive compounds such as, antioxidants, anti-inflammatory agents, antimicrobials, enzymes, pigments, polysaccharides, and structural molecules, that can transform the formulation of skin, hair, and personal hygiene products. In this volume, we present a comprehensive exploration of fungal-based cosmetics, combining foundational knowledge with the latest research findings, industrial applications, and technological innovations.

The book is structured into 21 chapters, beginning with an introduction to the field and progressively delving into the biological properties, functional activities, and formulation aspects of fungal extracts. Readers will gain insights into specific bioactivities such as anti-tyrosinase, anti-hyaluronidase, anti-collagenase, and anti-elastase properties, as well as the role of fungal ingredients in emollient, exfoliation, skin-lightening, and antiaging formulations. Special attention is given to commercially available fungal-based products, highlighting real-world applications in skincare, haircare, soaps, sanitizers, and beyond.

Furthermore, the book addresses critical aspects such as nanotechnology integration, stability studies, large-scale production challenges, and the regulatory and market landscapes for fungal-based cosmetic innovations. Each chapter is crafted to benefit researchers, formulators, industry professionals, and students alike, offering both theoretical understanding and practical guidance.

As the world moves toward a greener and more scientifically informed cosmetics sector, fungi have the potential to become the cornerstone of sustainable beauty solutions. We hope that this book will serve as both an inspiration and a reference point for those committed to advancing cosmetic science through the power of mycological innovation.

Nizwa, Oman Soumya Ghosh

Competing Interests The authors have no competing interests to declare that are relevant to the content of this manuscript.

Contents

1 **Fungal-Based Cosmetics: An Introduction** 1
Soumya Ghosh, Huda Raheem Hashim, Farah K. Mekawy, Arzak. H. Fahmy, Anirban Debnath, Samar Sami Alkafaas, Mustafa Ali Radwan, Arpita Das, Swagata Ghosh, and Helen N. Onyeaka

2 **Exploring Cultivable Fungi for Medicinal Cosmetics** 23
Eunice Ngozi Anaele

3 **Bioactive Compounds of Fungal Extracts Used as Cosmetics Ingredients** 47
E. N. Anaele, F. I. Nwafor, and C. C. Osegbue

4 **Fungi Extracts and Their Use as Antioxidants in Cosmetic Ingredients** 73
Anirban Debnath, Subhendu Bandyopadhyay, and Arpita Das

5 **Anti-inflammatory Activity of Fungal Extracts in Cosmeceutical Formulation** 87
Anushmita Chanda, Hindol Ray, Mehuli Das, Sagarika Paul, Sampurna Saha, Rajib Majumder, and Sanmitra Ghosh

6 **Anti-tyrosinase Activity of Fungal-Based Cosmetics** 111
Ayotunde Adeyemi Adeyanju, Elijah A. Alhassan, Charles Obiora Nwonuma, Abiola F. Olaniran, and Oluwakemi Erinle

7 **Fungal-Based Cosmetics and Their Anti-hyaluronidase Activity** 129
Biswadeep Chaudhuri, Anirban Debnath, and Arpita Das

8 **Anti-collagenase and Anti-elastase Activities of Personal Care Products Made from Fungal Extracts** 143
Bandita Dutta, Sedevino Ltu, Dibyajit Lahiri, Moupriya Nag, Debasmita Bhattacharya, Tanmay Sarkar, Md Osim Aquatar, and Rina Rani Ray

9 **Antimicrobial Activity of Fungal-Based Ingredients in Cosmetics** 161
Pooja Yadav, Tsering Choizom, and Shubham Mahara

10 **Biotransformational Effect of Fungal-Based Cosmetics** 185
Adenike A. Akinsemolu and Helen N. Onyeaka

11 **Emollient, Exfoliation, Skin Lightening, and Anti-aging Properties of Fungal-Based Cosmetics** 199
Rose Daphnee Tchonkouang and Maame Ekua Manful

12 **Fungal-Based Cosmetics: Formulation and Usage** 213
Samar Sami Alkafaas, Bassant E. Abdelmoaty, Amany Bahnsy, Abdullah M. Saif, Sara Samy Elkafas, Wael Hafez, Mthokozisi Dladla, Mohamed T. El-Saadony, and Soumya Ghosh

13 **Commercially Available Cosmetics with Fungal-Based Ingredients 2- Haircare Products** 233
Elijah Kolawole Oladipo, Helen N. Onyeaka, and Stephen Feranmi Adeyemo

14 **Commercially Available Cosmetics with Fungal-Derived Bioactive Compounds: Hand Sanitizer, Washing Soap, and Saponification Products** 241
Tanveen Kaur, Asma Ashraf Khan, and Chandresh Kumari

15 **Fungal Carotenoids and Their Cosmeceutical Effects on Human Bodies** 257
Haidy Nasser, Asmaa M. El-Makawy, Esraa E. Ammar, Shrouk Elshafey, Sohaila Emad, and Heba Bedair

16 **Fungal-Derived Ceramides for the Cosmetics Industry** 277
Roshnai Sen, Anuvab De, Moupriya Nag, Dibyajit Lahiri, Debasmita Bhattacharya, and Tanmay Sarkar

17 **Polysaccharides Obtained from Fungi and Their Usage in Cosmetics** 301
Wan Abd Al Qadr Imad Wan-Mohtar, Nur Asyiqin Zahia-Azizan, Nurul Aqilah Mohd Zaini, and Muhamad Hafiz Abd Rahim

18 **Nanocosmetics and Fungal-Based Cosmetics: Similarities and Differences in Formulation and Usage** 319
Suman Bhandary, Yashaswi Dutta Gupta, and Somtirtha Kool Banerjee

19 **Fungal Food Supplements and Nutraceuticals for a Healthier Life: Techniques and Application** 351
Yashaswi Dutta Gupta, Parijat Roy, Antarika Chowdhury, Sagar Pramanik, and Suman Bhandary

20 Stability of Fungal-Based Cosmeceutical Formulations 387
Adedola S. Adeboye, Tosin Olaniran,
Oluwatoyin O. Onipe, and Helen N. Onyeaka

21 Challenges to the Industrial-Scale Production of Fungal-Based Cosmetics 403
Wilgince Apollon

Index 417

Editors and Contributors

About the Series Editors

Dee Carter is Professor of Microbiology at the University of Sydney. She has a degree in Microbiology and Biochemistry from Otago University, New Zealand, and a PhD from Imperial College, London, UK. After postdoctoral fellowships in Montpellier, France, and UC Berkeley, California, USA, she moved to Australia in 1995 to take up a lectureship at the University of Sydney. Dee's research has encompassed (1) the population genetics and ecology of medically important fungi, including yeast and mold pathogens; (2) responses of fungi to host and antifungal stress, including the production of variant morphological forms; and (3) the transcriptome and proteome response to antifungal therapy, particularly during synergistic interactions between antifungals and natural products. As well as research, Dee teaches microbiology including mycology, epidemiology, and molecular biology at Sydney University.

Anuradha Chowdhary is Professor of Medical Mycology at the Vallabhbhai Patel Chest Institute, University of Delhi, India. She received her medical degree in 1992 and MD (Microbiology) degree in 1996 from the University of Delhi, India, and PhD from the Faculty of Medical Sciences, Radboud University Medical Center, Netherlands. Dr. Chowdhary's research interests include molecular ecology and population genetics of pathogenic fungi, antifungal drug resistance mechanisms, especially *Aspergillus* and *Candida* spp., and the epidemiology of systemic mycoses. She is currently working on molecular epidemiology of *Candida auris*, terbinafine-resistant dermatophytes, and azole-resistant *Aspergillus fumigatus*.

Joseph Heitman is James B. Duke Professor and Chair, Department of Molecular Genetics and Microbiology, Duke University, Durham, NC, USA. His research studies model and pathogenic fungi, addressing fundamental questions of scientific and medical importance. Pioneering studies with Baker's yeast revealed how immunosuppressive natural products interdict signaling cascades via FKBP12-drug complexes, and discovered TOR as a globally conserved nutrient sensor targeted by the immunosuppressive, antiproliferative drug rapamycin. His research discovered unisexual reproduction of pathogenic microbes, with implications for pathogen emergence, how sex generates diversity, and how sex evolved. Dr. Heitman's lab has further developed genetic and genomic approaches elucidating molecular principles of fungal virulence, identifying therapeutic targets, and illustrating convergent evolution of fungal mating-type loci with mammalian, insect, and plant sex chromosomes, defined the calcium-activated protein phosphatase calcineurin as a globally conserved fungal virulence factor, and elucidated functions of RNAi in microbial pathogen genome integrity, hypervirulent outbreak lineages, and drug resistance via epimutation.

Ulrich Kück is a Professor of General and Molecular Botany at the Ruhr-University in Bochum, Germany. He graduated in Biology and Chemistry and has a long-standing experience in the molecular biology of fungi, algae, and plants. His research with fungi has focused on two general aspects: (1) genetic engineering of the secondary metabolism of biotechnically relevant filamentous fungi, including functional genomics and proteomics, and (2) molecular genetic analysis of cellular growth in filamentous fungi, with the focus on sexual development. In particular, he is interested in the function of mating type loci and the involvement of conserved signaling complexes, such as the striatin-interacting phosphatases and kinases (STRIPAK) complex, in the control of cellular and developmental processes.

About the Editors

Soumya Ghosh is currently serving as an Assistant Professor at the Natural and Medical Sciences Research Center, University of Nizwa, Oman (since January 2024), and as a Research Fellow at the University of the Free State, South Africa (since February 2025). He previously worked as a postdoctoral researcher at the University of the Free State, South Africa (2019–2023).

After completing his PhD in Microbiology at Stellenbosch University, South Africa (2011–2015), he pursued his first postdoctoral fellowship at Thompson Rivers University, Canada (2016–2017). Upon returning to India, he briefly served as a visiting faculty in Microbiology at Kuchaman College, Rajasthan. Prior to his doctoral studies, Dr. Ghosh gained international research experience as a Research Fellow at the Biological Center, University of Groningen, The Netherlands (2005), followed by positions at the Centre of Plant Molecular Biology, University of Tübingen, Germany (2005–2008), where he also completed his Biologie-Diplomarbeit (2005–2006), and at the Technical University of Munich, Germany (2008–2010). Dr. Ghosh earned his bachelor's degree in Microbiology from Garware College, Pune, India (1999), and subsequently a master's degree in Zoology from Pune University (2001). His research spans a broad spectrum, ranging from plant biology to microbial diversity, with a focus on bacteria and fungi. In recognition of his scientific contributions, Dr. Ghosh was awarded the C3 scientist rating by the National Research Foundation (NRF), South Africa (January 2024), as well as the Arab Society for Fungal Conservation Outstanding Achievement Award of 2023, Egypt. Additionally, Dr. Ghosh has been enlisted in the Stanford University World top 2% list for the year 2024, 2025 in the subfield of Microbiology.

Helen N. Onyeaka is an Associate Professor at the Department of Chemical Engineering, University of Birmingham, UK, an industrial microbiologist, and a sustainability advocate with over 25 years of experience in food safety, microbiology, and sustainable food systems. She is the Deputy Director of the Birmingham Institute of Sustainability and Climate Action (BISCA) and a Senior Fellow of the Higher Education Academy (SFHEA).

Her research focuses on microbial physiology, foodborne pathogens, predictive microbiology, and sustainable food packaging innovations. Her contributions to applied microbiology and food security have earned her multiple accolades, including the prestigious Basil Jarvis Food Security and Innovation Award. As an educator, Dr. Onyeaka leads postgraduate and undergraduate modules in Food Microbiology, Food Safety, and Chemical Engineering and supervises PhD and MSc research projects. She is also actively involved in professional development training, designing and delivering Continuous Professional Development (CPD) programs for Food Safety Inspection Officers, industry professionals, and international trainees, including specialists from the Saudi Food and Drug Authority. Dr. Onyeaka is engaged in global sustainability initiatives, collaborating with academia, industry, and policymakers to address critical challenges in food systems, climate resilience, and microbiome research.

Contributors

Bassant E. Abdelmoaty Division of Biochemistry, Department of Chemistry, Faculty of Science, Tanta University, Tanta, Egypt

Adedola S. Adeboye African Food Research Network, Gauteng, Pretoria, South Africa

Ayotunde Adeyemi Adeyanju Department of Food Science and Microbiology, College of Pure and Applied Science, Landmark University, Omu-Aran, Kwara State, Nigeria
Centre for Innovative Food Research (CIFR), Department of Biotechnology and Food Technology, Faculty of Science, University of Johannesburg, Johannesburg, Gauteng, South Africa

Stephen Feranmi Adeyemo Division of Medical Artificial Intelligence, Helix Biogen Institute, Ogbomoso, Oyo State, Nigeria

Adenike A. Akinsemolu The Green Institute, Ondo, Nigeria

Elijah A. Alhassan Agricultural and Biosystems Engineering, College of Engineering, Landmark University, Omu-Aran, Kwara State, Nigeria

Samar Sami Alkafaas Molecular Cell Biology Unit, Division of Biochemistry, Department of Chemistry, Faculty of Science, Tanta University, Tanta, Egypt

Esraa E. Ammar Plant Ecology, Botany Department, Faculty of Science, Tanta University, Tanta, Egypt
School of Food and Biological Engineering, Jiangsu University, Zhenjiang, People's Republic of China

E. N. Anaele South-East Zonal Biotechnology Center, University of Nigeria, Nsukka, Nigeria
Department of Microbiology, Faculty of Biological Sciences, University of Nigeria, Nsukka, Nigeria

Eunice Ngozi Anaele South-East Zonal Biotechnology Center, University of Nigeria, Nsukka, Nigeria
Department of Microbiology, Faculty of Biological Sciences, University of Nigeria, Nsukka, Nigeria

Wilgince Apollon Centro de Investigación en Ciencia Aplicada y Tecnología Avanzada (CICATA)-Instituto Politécnico Nacional (IPN), Altamira, Mexico
Department of Agricultural and Food Engineering, Faculty of Agronomy, Autonomous University of Nuevo León, San Nicolás de los Garza, Nuevo León, Mexico

Md Osim Aquatar Department of Environmental Science & Civil Engineering, Jamia Millia Islamia (A Central University, NIRF-3rd), New Delhi, Delhi, India

Amany Bahnsy Division of Biochemistry, Department of Chemistry, Faculty of Science, Tanta University, Tanta, Egypt

Subhendu Bandyopadhyay Department of Biological Sciences, Adamas University, Kolkata, West Bengal, India

Somtirtha Kool Banerjee Department of Microbiology, Lady Brabourne College, Kolkata, West Bengal, India

Heba Bedair Plant Ecology, Botany Department, Faculty of Science, Tanta University, Tanta, Egypt

Suman Bhandary Department of Biological Sciences, School of Life Science and Biotechnology, Adamas University, Kolkata, West Bengal, India

Debasmita Bhattacharya Department of Basic Science and Humanities, Institute of Engineering and Management, Kolkata, Salt Lake Sector, University of Engineering and Management, Kolkata, West Bengal, India
Department of Biotechnology, Institute of Engineering and Management, Kolkata, University of Engineering and Management, Kolkata, West Bengal, India

Anushmita Chanda Department of Biological Sciences, School of Life Science & Biotechnology, Adamas University, Kolkata, West Bengal, India

Biswadeep Chaudhuri Department of Biotechnology, Adamas University, Kolkata, West Bengal, India

Tsering Choizom Department of Food Technology, School of Applied and Life Sciences, Uttaranchal University, Dehradun, Uttarakhand, India

Antarika Chowdhury Department of Biological Sciences, School of Life Science and Biotechnology, Adamas University, Kolkata, West Bengal, India

Arpita Das Department of Biotechnology, Adamas University, Kolkata, West Bengal, India
Department of Health and Social Care, QA Higher Education (QAHE), Birmingham, UK

Mehuli Das Department of Biological Sciences, School of Life Science & Biotechnology, Adamas University, Kolkata, West Bengal, India

Anuvab De Department of Biotechnology, Institute of Engineering and Management, Kolkata, University of Engineering and Management, Kolkata, West Bengal, India

Anirban Debnath Department of Biotechnology, Adamas University, Kolkata, West Bengal, India

Mthokozisi Dladla Human Molecular Biology Unit (School of Biomedical Sciences), Faculty of Health Sciences, University of the Free State, Bloemfontein, South Africa

Bandita Dutta Department of Biotechnology, Maulana Abul Kalam Azad University of Technology, Kolkata, West Bengal, India

Sara Samy Elkafas Production Engineering and Mechanical Design Department, Faculty of Engineering, Menofia University, Menofia, Egypt
Faculty of Control System and Robotics, ITMO University, Saint-Petersburg, Russia
Department of Automation and Control of Technological Processes and Production, Institute of Information Technology and Computer Science, National University of Science and Technology MISIS, Moscow, Russia

Asmaa M. El-Makawy Botany and Microbiology Department, Faculty of Science, Alexandria University, Alexandria, Egypt

Mohamed T. El-Saadony Department of Agricultural Microbiology, Faculty of Agriculture, Zagazig University, Zagazig, Egypt

Shrouk Elshafey Biotechnology branch, Botany Department, Faculty of Science, Tanta University, Tanta, Egypt

Sohaila Emad Biotechnology branch, Botany Department, Faculty of Science, Tanta University, Tanta, Egypt

Oluwakemi Erinle Agricultural and Biosystems Engineering, College of Engineering, Landmark University, Omu-Aran, Kwara State, Nigeria

Arzak. H. Fahmy Faculty of Science, Helwan University, Creative Egyptian Biotechnologists, Giza, Egypt

Sanmitra Ghosh Department of Biological Sciences, School of Life Science & Biotechnology, Adamas University, Kolkata, West Bengal, India

Soumya Ghosh Natural and Medical Sciences Research Center, University of Nizwa, Nizwa, Oman
University of the Free State, Bloemfontein, South Africa

Swagata Ghosh Applied Science Department, Symbiosis Institute of Technology, Symbiosis International Deemed University, Pune, Maharashtra, India

Yashaswi Dutta Gupta Department of Biological Sciences, School of Life Science and Biotechnology, Adamas University, Kolkata, West Bengal, India

Wael Hafez The Emirates Society of Internal Medicine (ESIM), Emirates Medical Association, 46th Street Deira Near International Bowling Centre -, Dubai, United Arab Emirates
Medical Research and Clinical Studies Institute, The National Research Centre, Cairo, Egypt

Huda Raheem Hashim Department of Biology, College of Basic Education, University of Al-Muthanna, Samawah, Iraq
College of Applied Medical Sciences/Department of Pathology, University Of Al-Muthanna, Samawa, Iraq

Tanveen Kaur Faculty of Applied Sciences and Biotechnology, Shoolini University of Biotechnology and Management Sciences, Solan, Himachal Pradesh, India

Asma Ashraf Khan School of Bioengineering and Food Technology, Shoolini Üniversity, Solan, Himachal Pradesh, India

Chandresh Kumari Faculty of Applied Sciences and Biotechnology, Shoolini University of Biotechnology and Management Sciences, Solan, Himachal Pradesh, India

Dibyajit Lahiri Department of Biotechnology, Institute of Engineering and Management, Kolkata, New Town Sector, University of Engineering and Management, Kolkata, West Bengal, India

Sedevino Ltu Department of Biotechnology, Maulana Abul Kalam Azad University of Technology, Kolkata, West Bengal, India

Shubham Mahara Department of Food Technology, School of Applied and Life Sciences, Uttaranchal University, Dehradun, Uttarakhand, India

Rajib Majumder Department of Biotechnology, School of Life Science & Biotechnology, Adamas University, Kolkata, West Bengal, India

Maame Ekua Manful School of Food Science and Environmental Health, Technological University Dublin, Dublin, Ireland
Sustainability and Health Research Hub, Technological University Dublin, Dublin, Ireland

Farah K. Mekawy Faculty of Science, University of Helwan, Giza, Egypt

Moupriya Nag Department of Biotechnology, Institute of Engineering and Management, Kolkata, New Town Sector, University of Engineering and Management, Kolkata, West Bengal, India

Haidy Nasser Botany and Microbiology Department, Faculty of Science, Kafrelsheikh University, Kafrelsheikh, Egypt

F. I. Nwafor Department of Plant Science and Biotechnology, Faculty of Biological Sciences, University of Nigeria, Nsukka, Nigeria

Charles Obiora Nwonuma Department of Biochemistry, College of Pure and Applied Sciences, Landmark University, Omu-Aran, Nigeria

Elijah Kolawole Oladipo Division of Medical Artificial Intelligence, Helix Biogen Institute, Ogbomoso, Oyo State, Nigeria
Department of Chemical Engineering, University of Birmingham, Birmingham, UK
Department of Molecular Biology, Immunology and Bioinformatics, Adeleke University, Ede, Osun State, Nigeria

Abiola F. Olaniran Department of Food Science and Microbiology, College of Pure and Applied Science, Landmark University, Omu-Aran, Kwara State, Nigeria

Tosin Olaniran Department of Consumer and Food Sciences, Faculty of Natural and Agricultural Sciences, University of Pretoria, Hatfield, South Africa

Oluwatoyin O. Onipe Department of Biotechnology and Food Technology, Faculty of Science, University of Johannesburg, South AfricaDoornfontein,

Helen N. Onyeaka School of Chemical Engineering, University of Birmingham, Birmingham, UK

C. C. Osegbue Department of Plant Science and Biotechnology, Faculty of Biological Sciences, University of Nigeria, Nsukka, Nigeria

Sagarika Paul Department of Biological Sciences, School of Life Science & Biotechnology, Adamas University, Kolkata, West Bengal, India

Sagar Pramanik Department of Biological Sciences, School of Life Science and Biotechnology, Adamas University, Kolkata, West Bengal, India

Mustafa Ali Radwan Biochemistry Division, Department of Chemistry, Faculty of Science, Tanta University, Tanta, Egypt

Muhamad Hafiz Abd Rahim Faculty of Food Science and Technology, Universiti Putra Malaysia, Serdang, Selangor, Malaysia

Hindol Ray Department of Biological Sciences, School of Life Science & Biotechnology, Adamas University, Kolkata, West Bengal, India

Rina Rani Ray Department of Biotechnology, Maulana Abul Kalam Azad University of Technology, Kolkata, West Bengal, India

Parijat Roy Department of Biological Sciences, School of Life Science and Biotechnology, Adamas University, Kolkata, West Bengal, India

Sampurna Saha Department of Biological Sciences, School of Life Science & Biotechnology, Adamas University, Kolkata, West Bengal, India

Abdullah M. Saif Division of Biochemistry, Department of Chemistry, Faculty of Science, Tanta University, Tanta, Egypt

Tanmay Sarkar Department of Food Processing Technology, Malda Polytechnic, West Bengal State Council of Technical Education, Government of West Bengal, Malda, West Bengal, India

Roshnai Sen Department of Biotechnology, Institute of Engineering and Management, Kolkata, University of Engineering and Management, Kolkata, West Bengal, India

Rose Daphnee Tchonkouang MED-Mediterranean Institute for Agriculture, Environment and Development, CHANGE-Global Change and Sustainability Institute, Faculty of Sciences, and Technology, Universidade do Algarve, Faro, Portugal

Wan Abd Al Qadr Imad Wan-Mohtar Functional Omics and Bioprocess Development Laboratory, Institute of Biological Sciences, Faculty of Science, Universiti Malaya, Kuala Lumpur, Malaysia

Pooja Yadav School of Health and Allied Sciences, Department of Nutrition and Dietetics, Geeta University, Panipat, Haryana, India

Nur Asyiqin Zahia-Azizan Functional Omics and Bioprocess Development Laboratory, Institute of Biological Sciences, Faculty of Science, Universiti Malaya, Kuala Lumpur, Malaysia

Nurul Aqilah Mohd Zaini Department of Food Sciences, Faculty of Science and Technology, Universiti Kebangsaan Malaysia, Bangi, Selangor, Malaysia

Fungal-Based Cosmetics: An Introduction

1

Soumya Ghosh, Huda Raheem Hashim, Farah K. Mekawy, Arzak. H. Fahmy, Anirban Debnath, Samar Sami Alkafaas, Mustafa Ali Radwan, Arpita Das, Swagata Ghosh, and Helen N. Onyeaka

Abstract

Fungus-based cosmetics are an emerging trend in response to consumer demand for natural, bioactive, and sustainable ingredients. Mushrooms and fungi are rich in bioactive compounds such as polysaccharides, proteins, lipids, phenolic compounds, and terpenoids, which offer dermatological benefits. These compounds exhibit antibacterial, hydrating, anti-inflammatory, and antioxidant properties, making fungi ideal for skincare applications. Additionally, fungal melanin provides a natural alternative to synthetic pigments, addressing concerns over environmental harm and skin irritation caused by artificial colorants. Beyond skincare, fungi are also used in hair care. Fungal-derived oils and enzymes improve scalp health and enhance hair strength and shine. Biopolymers like beta-glucans contribute to skin barrier repair, hydration, and wound healing. Fungal fermentation further increases the bioavailability and efficacy of cosmetic ingredients while reducing the need for artificial additives and preservatives.

Sustainability is a key advantage of fungal-based cosmetics. Cultivating fungi on agro-industrial waste minimizes environmental impact, aligning with the circular economy

S. Ghosh (✉)
Natural and Medical Sciences Research Center, University of Nizwa, Nizwa, Oman

University of the Free State, Bloemfontein, South Africa
e-mail: s.ghosh@unizwa.edu.om

H. R. Hashim
Department of Biology, College of Basic Education, University of Al-Muthanna, Samawah, Iraq

College of Applied Medical Sciences/Department of Pathology, University Of Al-Muthanna, Samawa, Iraq

F. K. Mekawy
Faculty of Science, University of Helwan, Giza, Egypt

A. H. Fahmy
Faculty of Science, Helwan University, Creative Egyptian Biotechnologists, Giza, Egypt

A. Debnath
Department of Biotechnology, Adamas University, Kolkata, West Bengal, India

S. S. Alkafaas
Molecular Cell Biology Unit, Division of Biochemistry, Department of Chemistry, Faculty of Science, Tanta University, Tanta, Egypt

M. A. Radwan
Biochemistry Division, Department of Chemistry, Faculty of Science, Tanta University, Tanta, Egypt

A. Das
Department of Biotechnology, Adamas University, Kolkata, West Bengal, India

Department of Health and Social Care, QA Higher Education (QAHE), Birmingham, UK

S. Ghosh
Applied Science Department, Symbiosis Institute of Technology, Symbiosis International Deemed University, Pune, Maharashtra, India

H. N. Onyeaka
School of Chemical Engineering, University of Birmingham, Birmingham, UK

S. Ghosh, H. N. Onyeaka (eds.), *Fungal-Based Cosmetics: Formulation and Usage*, The Mycota 17,
https://doi.org/10.1007/978-3-032-09291-5_1

and meeting consumer demand for ethical, eco-friendly products. However, challenges remain, including standardizing fungal extracts, navigating regulatory approvals, and ensuring product safety. This book chapter explores the biological properties of fungal compounds, their cosmetic applications, and how fungal biotechnology can transform the beauty industry. It also highlights the challenges and opportunities in this growing field, emphasizing the need for interdisciplinary collaboration among dermatologists, mycologists, and cosmetic scientists to unlock the full potential of fungi in cosmetics.

Keywords

Fungal-derived compounds · Bioactive ingredients · Sustainability · Cosmetic applications

1.1 Introduction

The global cosmetics market, worth tens of billions of dollars, continues to seek innovative products, with a particular focus on natural ingredients (Liu 2022). This demand has led to the widespread availability of natural cosmetics featuring extracts from coconut, jasmine, lemongrass, longan, and many other plants. Mushrooms, revered for thousands of years in Asia for their medicinal properties and once reserved for royalty in China and Japan, are now gaining prominence in the cosmetics industry. These fungi, particularly mushrooms and their derivatives, are being incorporated into cosmetic formulations as highly potent ingredients (Hyde et al. 2010).

Numerous mushrooms have demonstrated therapeutic benefits, particularly in β-glucan production (Kalač 2009). For example, Schizophyllan, derived from *Schizophyllum commune*, contains β-1,6-branched-β-1,3-glucan, which helps reduce inflammation, irritation, and skin damage caused by UV exposure and toxic environments (Kumari et al. 2008). Additionally, extensive research has explored the therapeutic properties of *Agaricus subrufescens* (also known as *A. blazei*) and the protective effects of β-glucans (Angeli et al. 2009). In addition to its anti-mutagenic properties, this fungus has been found to possess anti-clastogenic, anti-genotoxic, and antitumor activities (Mizuno et al. 1990; Luiz et al. 2003; Matuo et al. 2007). Another example is *Lentinula edodes*, which contains the medicinal compound eritadenine. Studies have shown that higher doses of eritadenine result in a dose-dependent reduction in cholesterol levels (Shimada et al. 2002). Furthermore, Ganoderic acids, secondary metabolites produced by *Ganoderma lucidum*, have demonstrated antitumor effects (Dubost et al. 2007; Keypour et al. 2010). Intriguingly, ergothioneine, derived from white button mushrooms (*A. bisporus*), forms a powerful antioxidant, aids in neutralizing free radicals in the human body. Both Ganoderic acids and ergothioneine are recognized as anti-aging agents and are linked to the prevention of cardiovascular diseases and cancer (Sanodiya et al. 2009). Notably, these white button mushrooms (*A. bisporus*) contain ergothioneine 12 times more of this antioxidant than wheat germ (Corey et al. 2009). This makes them a promising additive for enhancing the nutritional value of bread and other food products.

1.1.1 Definition and Scope

The Federal Food, Drug, and Cosmetic Act defines cosmetics as products intended to be rubbed, sprinkled, sprayed, or otherwise applied to the human body, or any part of it, for purposes such as cleansing, beautifying, enhancing attractiveness, or altering appearance (Tian et al. 2008). Cosmetics encompass a wide range of items, including skin care creams, lotions, perfumes, lipsticks, nail polish, eye and facial makeup, hair dyes, sprays, gels, permanent weaves, and even colored contact lenses (Ventura et al. 2008).

Microbial contamination in cosmetic products is a significant concern for the industry due to the potential for financial losses and the risk of turning these products into harmful goods for consumers (Vozenin-Brotons et al. 2001). Most cosmetics are formulated with ingredients that support microbial growth, and the manufacturing process is not sterile. Moreover, storage conditions often provide an optimal environment for microbial proliferation (Wang et al. 2007). During production, cosmetics can become contaminated by environmental microorganisms or through raw materials, which are often water-based and sourced from environments conducive to microbial growth (Ward and Singh 2005). Studies have identified *Penicillium* spp., *Pseudomonas aeruginosa*, *Staphylococcus aureus*, *Escherichia coli*, *Candida albicans*, *Aspergillus fumigatus*, and *A. niger* as common contaminants in cosmetics. While the skin and mucous membranes provide natural defense against microorganisms, contaminated cosmetics can still cause minor trauma and injuries (O'brien et al. 1993). Contaminated products can lead to infections of the skin, mucous membranes, hair, and nails, in addition to compromising the quality of the cosmetics. In severe cases, individuals exposed to microbiologically contaminated items such as mouthwash, creams, and mascara have reported experiencing serious, even life-threatening, illnesses (Ng 2004; Nieto and Chegwin 2008).

1.1.2 Overview of Fungal-Based Cosmetics

Cosmetics, also referred to as personal care products, are designed to cleanse and enhance the skin's appearance (Millikan 2001). Increasing consumer demand for healthier, eco-friendly, and organic options has driven the popularity of cosmetics made from natural or organic materials (Antignac et al. 2011). They are categorized into two main types: topically applied products, such as creams, lotions, and ointments, and nutricosmetics, which are orally consumed and provide similar perceived benefits. Cosmetics must meet essential criteria, including being beneficial to the skin, safe for use, and free from adverse effects (Campana et al. 2004). The growing popularity of natural ingredients in cosmetics, as seen in products like Murad skin care (Murad 2010), highlights a trend toward sustainability and efficacy.

Fungi are increasingly being utilized in cosmeceuticals and nutricosmetics, though there remains significant untapped potential in this area (Cheng et al. 1999). In addition to makeup, skincare products include skin whiteners, anti-aging creams, and skin-reviving solutions aimed at rejuvenation and maintaining wrinkle-free, toned, healthy skin. Many of these products incorporate mushrooms, their derivatives, or fungal-fermented components (Chien et al. 2008). For example, lactic acid, an alpha-hydroxy acid, is used in dermatological preparations for hydrating and smoothing dry, flaky skin. Higher concentrations (over 12%) may serve as chemical peels that remove the skin's outer layer. Research shows that *Rhizopus* strains can produce lactic acid (Fang et al. 2008). Ceramides, known for their rarity, are valuable epidermal moisturizers and are used in products like Elizabeth Arden Ceramide Purifying Toner. Ceramides have been identified in fungi such as *Tuber indicum* (Firenzuoli et al. 2008) and *Phellinus pini* (Firenzuoli et al. 2008), offering potential sustainable sources. Mold lipases might also be bio-transformed to produce ceramides (Gautier et al. 2008). The copolymer chitin-glucan, found in the cell walls of *A. niger*, has moisturizing and anti-aging properties. This component, derived from waste materials of pharmaceutical and food-grade citric acid production, is now used in formulations like KiOsmetine-CG (Golz-Berner and Zastrow 2005; Athalye et al. 2009).

Mushrooms belong to higher fungi phyla like Ascomycota and Basidiomycota, characterized by a stem (stipe), cap (pileus), and gills (lamellae) (Miles and Chang 2004). They are celebrated not only for their culinary benefits—flavor, texture, and nutrition—but also for medicinal properties, such as tonic and pharmacological effects (Wasser 2011). Of the estimated 150,000 mushroom species on Earth, only about 10%, or 15,000, are identified (Mueller and Schmit 2007; Ainsworth 2008). Rich in proteins, vitamins, and minerals, mushrooms are sources of thiamine, β-glucan, selenium, riboflavin, niacin, pantothenic acid, and folic acid (Poucheret et al.

2006; Ahmad et al. 2013); they are associated with vitality, weight management, longevity, and anti-aging benefits (Chang and Wasser 2012; Bashir et al. 2018). Furthermore, they possess diverse bioactivities, including immunomodulatory, antioxidant, anti-inflammatory, antidiabetic, antibacterial, antifungal, antiviral, antitumor, hepatoprotective, and lipid-lowering effects, making them a rich but underutilized reservoir for pharmaceuticals (El Enshasy and Hatti-Kaul 2013; Kalač 2013). Traditionally used in Asia for food, medicine, and cosmetics, species like Shiitake (*Lentinula edodes*), Maitake (*Grifola frondosa*), and Reishi (*G. lucidum*) are now widely recognized (Jikai 2002). Western preferences include Portobello (*A. bisporus*), oyster (*Pleurotus ostreatus*), and tinder fungus (*Fomes fomentarius*) (Zhang et al. 2002). Additionally, polysaccharides such as β-(1–3)-D-glucan, found in *Pleurotus* species, are utilized for their anticancer properties. In Japan, beta-glucans like lentinan and krestin have been extracted from medicinal mushrooms as chemotherapy adjuvants (Ina et al. 2013; Blagodatski et al. 2018). *A. subrufescens* contains β-(1–6)-D-glucan with immune-modulating properties, further demonstrating fungi's potential in enhancing immunity and combating infections (Lourenço et al. 1996; Passi et al. 2002; Pons et al. 2002; Synytsya et al. 2009).

Compounds derived from macrofungi, especially mushrooms—such as ceramides, lentinan, schizophyllan, omega fatty acids (3, 6, 9), carotenoids, and resveratrol—are increasingly used in cosmetics (Hyde et al. 2010; Dos Reis et al. 2022). Many mushroom-derived compounds exhibit strong antioxidant and anti-inflammatory properties, making them suitable for addressing aesthetic concerns like wrinkles, fine lines, uneven tone, and texture. Their nutritional, regenerative, and protective attributes position mushrooms as valuable components in cosmeceuticals and nutricosmetics (Badalyan 2014). Renowned brands incorporate mushroom extracts into products like the Originals Plantidote Mega-Mushroom Treatment (Cheung 2008). Mushroom-derived mycochemicals, such as polysaccharides (lentinan, schizophyllan, PSK, PSP), phenolics, terpenoids, and ergosterols, are extensively studied for skin benefits (Wasser 2010; Chandrasekaran et al. 2012). Ingredients like mushroom-sourced chitosan serve as emulsifiers, antibacterial agents, and skin protectants (Badalyan 2014). Chitin-glucan copolymers, known for moisturizing and anti-aging effects, are used in skin care. *Tremella* extracts treat conditions like neurodermatitis and sclerodermatitis (De Baets and Vandamme 2001a). Furthermore, mushrooms can aid in biotransformation, producing ceramides and lactic acid, which are valuable in cosmetics for hydration and skin renewal (Bowe 2013). Similarly, mushrooms are known to produce chitin-glucan, expanding their potential applications in cosmetics (Jin et al. 2010). Superoxide dismutase, a compound generated by mushrooms, helps mitigate free radical damage and fibrosis, as seen in products like Dermalogica Intensive Moisture Balance. These qualities highlight mushrooms' versatile potential in modern beauty formulations. The cosmetic industry has increasingly focused on fungi due to their unique properties, which make them valuable ingredients in skincare and cosmetic products. While synthetic chemicals and plant-based extracts have traditionally dominated cosmetic formulations, recent research highlights the numerous benefits fungi offer. The growing demand for natural and sustainable products has also encouraged makeup artists to incorporate more traditional, nature-derived ingredients into their practices (De Baets and Vandamme 2001b).

1.2 Types of Fungi Used in Cosmetics

Mushrooms are unequivocally fungi that produce spongy fruiting bodies, especially those with a stalk and a cap. There is an abundance of distinct mushroom species on Earth. They have unquestionably been utilized for food and medicine since ancient times (Sulkowska-Ziaja et al. 2023). Due to their unquestionable significance in human health, nutrition, and illness, as well as their numerous medical characteristics including cancer prevention, antibiotic, antiviral, immune-mediated stimulating properties, and blood lipid-lowering effect, they are undeniably widely used (Sulkowska-Ziaja et al. 2023). Many organizations nationwide have unmistakably conducted comprehensive assessments of fungal species from various sources. Mushrooms have gained an increasing focus in the cosmetic industry due to their diverse range of bioactive compounds, which offer a myriad of benefits for skin health (Purohit et al. 2022). Key species such as Reishi (*Ganoderma*), Shiitake (*Lentinula*), and *Tremella* are highly esteemed for their unique properties. These fungi produce a variety of beneficial compounds including Ganoderic acids, ergosterols, and β-glucans, each of which contributes to their noteworthy anti-aging, antioxidant, and anti-inflammatory properties for cosmeceuticals (Sulkowska-Ziaja et al. 2023).

Medicinal mushrooms like Reishi, Shiitake, and *Tremella* contain an extensive collection of bioactive compounds with proven cosmetic benefits, including anti-inflammatory, antioxidant, and regenerating properties, all of which make them incredibly valuable in cosmetology (Jiang et al. 2021). Furthermore, mushrooms such as Shiitake and Reishi contain notably high levels of antioxidants, allowing them to effectively combat oxidative stress and promote overall skin health with considerable success (Andersone et al. 2023). These properties showcase the significant potential of mushrooms in the development of advanced skincare products. Compounds such as β-glucans and triterpenoids found in these fungi help reduce wrinkles and improve skin elasticity (Mohanta et al. 2022). *Tremella*, often referred to as snow mushroom, is known for its exceptional hydrating properties, making it a popular ingredient in moisturizing products. Extracts from mushrooms like *A. bisporus* and *Boletus edulis* exhibit significant antibacterial properties, which can help in treating skin infections (Li et al. 2023). The benefits of using mushrooms and yeasts in cosmetics are becoming increasingly recognized. Many mushrooms have anti-inflammatory properties that can help reduce skin irritation and redness.

Yeasts, especially *Saccharomyces* and *Candida* species, are also known for their beneficial properties in cosmetics (Alyahyawi et al. 2023). They can improve skin health, reduce pore size, roughness, and hyperpigmentation, as well as provide photoprotection and produce bioactive compounds like β-glucans and triterpenoids. The use of molds, such as *Aspergillus* and *Penicillium*, is crucial in the cosmetics industry, as they offer various benefits like improved efficacy and sustainability (Williams et al. 2023). These molds can produce valuable bioactive compounds that are valuable in modern cosmetic formulations. They are essential for producing enzymes and metabolites that can enhance the effectiveness of cosmetic products (Sulkowska-Ziaja et al. 2023). For example, some *Aspergillus* strains produce enzymes such as amylase and protease, which aid in exfoliating dead skin cells and improving skin texture.

Additionally, the fermentation processes involving *Aspergillus* can result in the production of beneficial compounds such as organic acids and vitamins, which enhance skin hydration and provide antioxidant properties. *Penicillium* molds are well-known for their role in antibiotic production, but they also have applications in cosmetics (Salinas-Solis et al. 2023). Compounds derived from *Penicillium*, such as penicillin, can help in formulating products with antimicrobial properties, making them useful in formulations designed to prevent bacterial growth. The metabolites produced by *Penicillium* can serve as natural preservatives, extending the shelf life of cosmetic products without the need for synthetic preservatives (Kamal et al. 2023). Using molds allows cosmetic formulators to incorporate natural ingredients that can be more appealing to consumers seeking organic or less processed products. The bioactive compounds produced by these molds can enhance the performance of cosmetic products, offering additional benefits such as anti-aging, moisturizing, and skin-soothing properties (Manimaran et al. 2023). The use of molds in cosmetic formulations can contribute to sustainability efforts by reducing the reliance on synthetic chemicals and promoting the use of biotechnologically derived ingredients. Harnessing molds for new formulations can lead to innovative products that stand out in the competitive cosmetics market, appealing to consumers looking for unique and effective solutions (Milovanovic et al. 2024).

Active compounds derived from fungi, including polysaccharides, β-glucans, ergothioneine, peptides, and proteins, are gaining attention for their diverse health benefits and therapeutic potential. These bioactive compounds are integral to the development of functional foods and pharmaceuticals. Fungi such as *Ganoderma* and *Pleurotus* are rich in β-glucans, which exhibit immunomodulatory and anti-inflammatory properties. Studies show that submerged cultivation can enhance their yield significantly, with *G. lingzhi* yielding up to 70.8% glucans. Ergothioneine: This antioxidant, found in various fungi, plays a role in cellular protection and longevity, making it a focus for health span research. Peptides and Proteins: Fungal extracts contain glycoproteins and peptides that have shown potential in treating age-related diseases and enhancing health span in model organisms. Fungal metabolites are linked to anticancer, antidiabetic, and antimicrobial effects, showcasing their versatility in therapeutic applications. The biosynthesis of these compounds is influenced by the cultivation conditions, which can be optimized for higher yields and bioactivity.

1.3 Scientific Basis for Fungal-Based Cosmetics

The cosmetic industry has shown a great deal of interest in the biological activities of fungal compounds because of their many advantageous characteristics. A range of bioactive metabolites produced by fungi can improve cosmetic formulations by providing natural substitutes for artificial components (Mukherjee et al. 2024).

Primary and secondary metabolites are two types of fungal metabolites that have different functions in fungal biology and applications in various domains. While secondary metabolites, frequently low molecular weight molecules, have major industrial and therapeutic relevance, primary metabolites are necessary for growth and development (Behar et al. 2024).

Mushrooms' bioactive compounds enhance antioxidant properties, making them valuable in skin care products for moisturizing, nourishing, and improving skin appearance safely and naturally (Visvanathan et al. 2022).

Free radicals are unstable chemicals that have the potential to harm cells. Antioxidants function by neutralizing these molecules (Zheng et al. 2023). Free radicals may hasten aging and exacerbate skin conditions; therefore, keeping skin healthy depends on avoiding them. In addition, because of its high antioxidant content, mushrooms can shield the skin from oxidative stress brought on by outside influences like pollution and UV rays. To stop premature aging and preserve the integrity of the skin, this protective function is crucial (Venkatakrishna et al. 2023).

Antioxidant components of mushrooms, including *Pleurotus*, *Ganoderm*a, and *Lentinula*, are known to minimize the appearance of wrinkles and fine lines. These antioxidants support the look of more young and bright skin by preventing oxidative damage (Mukherjee et al. 2024).

Moreover, antioxidants contribute to increased skin moisture. They contribute to the maintenance of the skin's moisture barrier and give the appearance of plumper, healthier skin by shielding skin cells from harm (Mukherjee et al. 2024).

Certain mushrooms have the capacity to prevent the synthesis of melanin by reducing tyrosinase activity and melanin content, which may have whitening effects on the skin. Their ability to lessen oxidative stress, which can lead to hyperpigmentation, is frequently associated with their antioxidant qualities (Theeraraksakul et al. 2023).

Cosmetics with anti-inflammatory qualities are essential, especially those meant to enhance skin health. Numerous skin problems, such as redness, irritation, and disorders like eczema or acne, can be brought on by inflammation. Anti-inflammatory drugs can thus be used to help calm the skin and lessen these symptoms (Yin et al. 2024).

As a result, more individuals are realizing the potential therapeutic advantages of fungal terpenes and terpenoids, particularly in skin care formulations meant to address inflammation and skin issues. It is believed that these substances' anti-inflammatory qualities stem from their ability to modify inflammatory pathways. Less redness and irritation, which are common indicators of a variety of skin conditions, could arise from this (Trepa et al. 2024).

Compounds originating from fungus, such as ergothioneine, ceramide, and β-glucans, are becoming more and more valued for their anti-inflammatory qualities, which makes them useful components in cosmetic compositions (Mukherjee et al. 2024).

Fungal cell walls include polysaccharides called β-glucans. They have a reputation for relieving skin irritation and inflammation. Because of this, they work well for redness and irritation of the skin, which are frequent symptoms of eczema and acne. Their relaxing qualities aid in promoting a healthier-looking complexion and calming the skin (Sousa et al. 2023).

Likewise, the lipid molecules known as ceramides are essential to the preservation of the skin barrier. They aid in moisture retention and shield the skin from outside allergens. Ceramides are good for sensitive skin types because they strengthen the skin barrier and reduce symptoms of irritation and dryness (Li et al. 2020).

Moreover, fungi naturally contain ergothioneine, a mega antioxidant. It has been demonstrated that ergothioneine contains anti-inflammatory properties, which may aid in lowering skin irritation and redness. Its soothing benefits are further enhanced by its antioxidant qualities, which shield the skin from oxidative stress (Liu et al. 2023).

Fungal compounds also have the additional purpose of moisturizing effects of *Tremella fuciformis*, a kind of fungus, are thought to be attributed to its high polysaccharide content. These polysaccharides have the ability to create a gel-like coating on the skin that aids in hydration retention. This quality makes it a useful component in skincare formulas, especially in hydration-focused products (Lourith et al. 2021).

A recent study published in Frontiers in Bioengineering and Biotechnology explores the potential of mycelium-based biomaterials for skin wound healing. The research focuses on the mycelia of *G. lucidum* and *P. ostreatus*, which are characterized as self-growing biomaterials suitable for tissue engineering. These mycelia are good prospects for dermal replacements since they have been shown to be biocompatible and effective at promoting tissue healing in vivo. Additionally, the study supports the use of mycelium in wound healing therapies by highlighting its structural similarity to skin collagen. The findings suggest that the healing effects of extracts from *G. lucidum* and *P. ostreatus* are similar to those of existing treatment approaches, with significant increases in cellular metabolic activity and gene expression related to tissue repair processes (Ruggeri et al. 2023).

A 2022 study published in Scientific Reports examines the disparities in skin microbiome and mycobiome between younger and older Korean women, revealing that older women exhibited significantly enhanced alpha-diversity in their bacterial and fungal communities. The identification of specific microbial biomarkers, such as *Cutibacterium*, *Lactobacillus*, and *Malassezia restricta*, elucidates the association between skin aging and microbial composition, underscoring the Y-group's superior microbial diversity and metabolic processes conducive to skin health (Kim et al. 2022).

Current research underscores the escalating adoption of fungal extracts in cosmetic products, spurred by a consumer inclination toward natural ingredients. Fungi, particularly species such as *Tricholoma matsutake* and *P. citrinopileatus*, exhibit bioactive properties that enhance skin health, leading brands like "DR. ANDREW WEIL FOR ORIGINS™" and "ONE LOVE ORGANICS®" to utilize these extracts for their anti-inflammatory and skin-balancing benefits (Esposito et al. 2023).

1.4 Fungal Biotechnology in Cosmetic Formulation

A key component of biotechnology, fermentation has been used for ages to produce food and is currently being used more and more in cosmetics. In order to create sustainable, efficient, and ecologically acceptable active components, microorganisms such as bacteria or yeast must break down organic material (Iram et al. 2023). For instance, kojic acid, a skin-lightening agent, is produced through the fermentation of multiple species of fungi, including *Aspergillus*. This process involves growing the fungus in a nutrient-rich medium under controlled conditions, resulting in the production of kojic acid through complex biosynthetic pathways. The resulting kojic acid is then extracted and purified for use in skincare products to reduce dark spots and hyperpigmentation (GAE Mahmoud et al. 2024).

An article focuses on the production of active compounds for cosmetics using fermentation techniques. It refers to the use of microbes in fermentation to produce bioactive substances such as enzymes, peptides, and antioxidants. The study highlights how crucial fermentation is to producing natural, sustainable, and high-quality components for cosmetic formulas (Gomes et al. 2020).

Significant progress has been made in the extraction and purification of fungal-based chemicals for cosmetic purposes, especially with the use of cutting-edge techniques and green technology. These methods are essential for extracting bioactive substances with a range of advantages, including anti-inflammatory, antibacterial, and antioxidant characteristics (Kumar et al. 2023).

For example, ultrasound-assisted extraction (UAE) has been successfully applied to isolate antioxidant polypeptides from *Saccharomyces cerevisiae*, resulting in substantial scavenging activities against a variety of radicals. Likewise, *Cordyceps militaris* has been exposed to ultrasonic treatments to extract antimicrobial and anticancer polypeptides, exhibiting notable antibacterial properties. These methodologies not only boost the yield but also preserve the bioactivity of the isolated constituents. Furthermore, enzymatic-assisted extraction represents another innovative approach, as demonstrated by the extraction of ACE inhibitory peptides from *A. bisporus* utilizing alkaline protease. Additionally, pulsed electric fields (PEF) have been utilized to facilitate pressure extraction (PE) in *A. bisporus*, leading to a significant enhancement in protein yield and an improved ratio of nucleic acids to proteins. These techniques are advantageous owing to their effectiveness and their capacity to retain the functional characteristics of the extracts (Ahmed et al. 2024).

The isolation and purification of fungal melanin, a pigment with prospective applications in cosmetics, have been investigated employing an array of advanced methodologies. Ultrasound-assisted extraction (UAE), negative pressure cavitation (NPC), microwave-assisted extraction (MWE), and hydrodynamic cavitation extraction (HCE) are among the techniques that have been assessed for their efficacy in extracting melanin from fungal origins. These methodologies provide a compromise between operational efficiency and the retention of the bioactive characteristics of melanin (Suthar et al. 2023).

Carotenoids, including astaxanthin and β-carotene, are of significant interest due to their antioxidant and photoprotective characteristics in cosmetic formulations. The methodologies for extracting these compounds have been refined through the application of avant-garde techniques such as supercritical fluid extraction (SCF) and pressurized liquid extraction (PLE) (Naz et al. 2023).

DNA extraction for quality control: In the arena of maintaining the quality of fungal-derived cosmetic formulation, the methods of DNA extraction and purification play a pivotal role. Approaches that incorporate bead-beating alongside phenol:chloroform:isoamyl alcohol procedures have been demonstrated to markedly elevate DNA yields, which is vital for the molecular identification of bacterial contaminants in cosmetics. These protocols guarantee that cosmetic products are devoid of microbial impurities, thereby augmenting their safety and effectiveness (Michelutti et al. 2020).

Significant biotechnological progress has been made in the field of fungal-based cosmetics, which uses the special qualities of fungus to produce novel and environmentally friendly cosmetic products. The demand for eco-friendly, potent chemicals that may improve skin and hair in several ways is what's driving these breakthroughs (Mago et al. 2023).

Sustainable colorants for beauty products have been shown to be derived from fungi. An environmentally responsible substitute for synthetic dyes, these natural pigments are non-toxic, biodegradable, and light-fast. These pigments are more appealing in skincare products since they fulfill the dual roles of colorant and cosmeceutical benefit, such as antioxidant and UV protection. Large-scale industrial uses of these natural pigments are now feasible due to the biotechnological developments in synthetic biology and microbial culture optimization that have increased their production efficiency (Lyu et al. 2022).

Biotechnological procedures have also made it possible to extract active components from fungus, including resveratrol, hyaluronic acid, and kojic acid. The moisturizing, antioxidant, and skin-lightening qualities of these substances are well-known. Furthermore, fibroblast life extension, increased epidermal suppleness, and cell DNA repair are all facilitated by growth factors and enzymes generated from fungi, which help to heal damaged skin and protect it from UV light (Gomes et al. 2020).

It has been demonstrated that filamentous fungi are particularly useful for industrial uses, such as cosmetics. Research is still being done to improve fungal growing methods in order to increase the yield and efficiency of bioactive component synthesis. Their metabolites have traditionally produced substantial value in medications and cosmetics. With many patents filed for creative use of fungal metabolites in cosmetic formulations, the intellectual property landscape reflects these achievements (Hüttner et al. 2020).

Fungal-based cosmetics have gained significant attention due to their unique bioactive properties and potential benefits for skin and hair care. However, the formulation of these products presents several challenges that need to be addressed to ensure their efficacy and consumer acceptance (Mukherjee et al. 2024).

Creating a stable and visually acceptable emulsion is one of the main problems in the formulation of fungal-based cosmetics. When Tremella fuciformis extract is added directly to an emulsion substrate, for example, the consistency may become quite thick and stringy, with lumps apparent. This is undesirable for application. In order to counteract this, scientists have looked at a number of ways to include the active component. Smaller, imperceptible lumps are produced when the active component is mixed with water before being added to the emulsion substrate. Likewise, combining the active component with ethanol or glycerin can also help with consistency, but the emulsion may still be thick (Woźniak et al. 2023).

Additionally, recent research has shown how crucial it is to optimize the formulation in order to decrease trans-epidermal water loss (TEWL) and improve skin hydration. Over many hours after application, research on a cosmetic product including *Tremella fuciformis* extract showed notable decreases in TEWL and improvements in skin moisture. These findings underscore the potential of fungal-based ingredients in providing long-lasting skin benefits (Woźniak et al. 2023). Furthermore, the quality and safety of cosmetics made from fungi are crucial. Recent research on the safety and efficacy of widely used topical cosmetic treatments found metal contaminants and microbiological contamination in a number of the items. This emphasizes the necessity of strict quality control procedures to guarantee the security of cosmetics made from fungi (Almukainzi et al. 2022).

1.5 Application of Fungal-Based Ingredients in Cosmetic Products

The use of fungal-based ingredients in skincare products, such as moisturizers, anti-aging creams, and serums, is becoming popular due to their unique bioactive properties. Fungi contain a variety of compounds that promote skin health, making them valuable for cosmetic formulations (Salinas-Solis et al. 2023). These compounds include kojic acid, β-glucans, and triterpenoids, which have anti-aging, anti-inflammatory, and antioxidant properties that improve skin hydration and elasticity. For example, an anti-aging face cream containing β-glucan, a fungal-based ingredient, has proven effective in moisturizing, stimulating collagen production, and reducing the signs of aging. Fungal extracts also exhibit significant anti-inflammatory and antimicrobial effects, making them useful for treating skin conditions and preventing pigmentation disorders (Sulkowska-Ziaja et al. 2023) (Table 1.1).

In addition, fungal extracts offer natural alternatives to synthetic preservatives in skincare products. A study has explored the use of *Auricularia polytricha* and *A. bisporus* extracts in a film-forming spray for improving skin hydration, indicating the potential of fungal-based ingredients in products like moisturizers and anti-aging creams (Kamal et al. 2023). *Tremella fuciformis* extract, another fungal ingredient, enhances cosmetic formulations by improving hydration, reducing water loss through the skin, and combating free radicals, making it suitable for use in skincare products (Salinas-Solis et al. 2023). Furthermore, fungal mycelium-based materials are rich in antioxidants and anti-inflammatory compounds, making them beneficial for skincare products such as moisturizers, anti-aging creams, and serums.

The use of fungal-based ingredients in haircare and cosmetic products is increasing due to their unique bioactive properties. Fungi provide a variety of beneficial compounds that enhance the effectiveness of formulations, making them valuable in shampoos, conditioners, and makeup products(Jiang et al. 2021). Fungal ingredients like β-glucans and triterpenoids (e.g., ganoderic acids) exhibit antioxidant and anti-inflammatory properties, which can improve scalp health and hair vitality. Mycelium-derived materials are rich in antioxidants and can be incorporated into hair masks and conditioners for enhanced nourishment (Sulkowska-Ziaja et al. 2023). Fungi produce secondary metabolites with antimicrobial activity, making them suitable for formulations aimed at treating scalp conditions and combating dandruff and other fungal infections. Recombinant proteins from fungi, such as those derived from *G. lucidum*, show promise in reducing melanin synthesis, which could be beneficial in cosmetic products like foundations and concealers (Jiang et al. 2021) (Table 1.1).

Fungal mycelium-based materials are used in cosmetics like shampoos, conditioners, and makeup due to their biodegradability and rich compounds, offering sustainable alternatives in haircare products. The recombinant *Ganoderma* protein exhibits antioxidant and whitening properties, making it suitable for haircare products like shampoos, conditioners, and

Table 1.1 Examples of fungi, their key compounds, and biological functions

Fungal types	Key active compounds	Function	References
Macrofungi (e.g., *Ganoderma lucidum*, *Lentinula edodes*)	β-glucans, chitin, ergosterol, triterpenoids	Antioxidant, anti-aging, skin hydration	Sulkowska-Ziaja et al. (2023)
Filamentous fungi (e.g., *Aspergillus*, *Penicillium*)	Kojic acid, azelaic acid, polyketide pigments	Skin whitening, antimicrobial, colorant	Lin and Xu (2020)
Endophytic and marine fungi	Protocatechualdehyde, carotenoids, flavonoids	Anti-inflammatory, photoprotection	Calado et al. (2021)
Yeasts (e.g., Galactomyces)	Fermentation filtrates, peptides, and amino acids	Barrier improvement, sebum control	Klimek-Szczykutowicz et al. (2024)

makeup due to its bioactive potential (Kamal et al. 2023). Edible and medicinal mushrooms contain bioactive compounds suitable for haircare products like shampoos, conditioners, and masks due to their regenerating, nourishing, and moisturizing properties. Additionally, fungal-based ingredients are utilized in cosmetics like anti-aging products, offering eco-friendly, cost-effective, and innovative solutions for various industries (Milovanovic et al. 2024).

1.6 Safety and Regulatory Aspects

Fungal-derived cosmetics must comply with rigorous safety and regulatory protocols to guarantee their absence of detrimental microorganisms and comply with rigorous safety and regulatory protocols to guarantee their absence of detrimental microorganisms and toxins. To ensure that they are safe for consumer use, these goods are put through rigorous testing for purity, stability, and potential allergenicity, in accordance with rules set by regulatory agencies such as the EMA in Europe and the FDA in the United States.

1.6.1 Safety Assessments of Fungal Ingredients

Fungi have garnered attention in the cosmetics industry for their distinct features, including antibacterial activity and abundant nutrient content (Wu et al. 2016). Ingredients originating from fungi, such as beta-glucans extracted from yeast, are recognized for their moisturizing and anti-inflammatory characteristics (Mirończuk-Chodakowska et al. 2021). Ensuring the safety of fungal elements in cosmetics is of utmost importance. Different techniques, including in vitro and in vivo examinations, are utilized to guarantee its safety for human consumption (Barthe et al. 2021). The European Commission enforces stringent criteria for the safety evaluation of fungal components in cosmetic products, placing significant emphasis on the requirement for comprehensive testing (Andersen 2022). It is necessary to do thorough safety assessments to prevent allergic reactions in consumers, as fungal components may provide allergenic hazards (Alvarez and Boye 2012). Research has demonstrated that cosmetics derived from fungi can effectively treat skin diseases such as acne and eczema. However, it is crucial to verify their safety (Diseases 2006). Consumers usually have a positive opinion of fungal-based cosmetics, but there are ongoing concerns regarding their safety. Therefore, it is crucial for manufacturers to communicate clearly with consumers (Barthe et al. 2021). Conducting toxicological evaluations of fungal extracts is crucial for determining their safety profiles, especially when considering their long-term use (Hintikka et al. 2024). The progress in biotechnology has facilitated the creation of novel substances derived from fungi, with a primary focus on conducting safety evaluations before their commercialization (Meyer et al. 2020). Thorough safety evaluations are crucial to guarantee consumer safety and adherence to regulations when it comes to fungal-based cosmetics, notwithstanding their potential advantages (Manful et al. 2024).

1.6.2 Regulatory Frameworks and Compliance

The distinctive bioactive chemicals found in fungal-based cosmetics have attracted considerable interest, necessitating the implementation of rigorous regulatory frameworks to guarantee their safety and effectiveness. The European Union (EU) and the U.S. Food and Drug Administration (FDA) have implemented extensive regulations on the utilization of fungal components in cosmetic products. Manufacturers must perform thorough safety studies, which include toxicological tests and allergenicity evaluations, before they may market their products according to these guidelines (Ferreira et al. 2022). Adhering to these laws is essential for obtaining market approval, and failure to comply might lead to product recalls or legal proceedings. Moreover, due to the worldwide reach of the cosmetics sector, it is essential to comply with different foreign rules, which might make efforts to ensure compliance more complex (Ferreira et al. 2022). Manufacturers must be knowledgeable about developing regulatory standards and guarantee that their products adhere to the most stringent safety and quality criteria. With the increasing utilization of fungal-based components, it becomes crucial to uphold strict regulatory compliance in order to safeguard consumer well-being and promote advancement in the cosmetics sector (Meyer et al. 2020).

1.6.3 Consumer Perceptions and Acceptance

Consumer attitudes toward fungal-based cosmetics have improved in recent years due to increasing recognition of the advantages of using natural and sustainable ingredients. A significant number of consumers are drawn to the distinctive characteristics of substances produced from fungi, including their anti-inflammatory and antioxidant capabilities, which promote better skin (Rocca et al. 2022). Nevertheless, there was initial doubt because fungi were commonly associated with sickness and rotting. Brands have played a vital role in changing people's views by implementing effective marketing and education strategies that highlight the safety and effectiveness of these chemicals (Case et al. 2022). The growing desire for environmentally friendly and ethically produced goods has also had a substantial impact on the acceptance of cosmetics derived from fungi, in line with the wider consumer movement toward sustainability (Sasounian et al. 2024). However, certain consumers, especially those with sensitive skin or allergies, still exercise caution, emphasizing the significance of openness in the sourcing of ingredients and safety evaluations (Ajayi et al. 2024). In general, although fungal-based cosmetics are becoming more accepted, it is crucial to continue educating consumers and ensuring openness in product information in order to sustain and grow their presence in the market (Case et al. 2022).

1.7 Future Trends and Innovations

1.7.1 Emerging Research and Technologies

The field of fungal-based cosmetics is expected to experience substantial progress in the coming years, propelled by cutting-edge research and pioneering technology. One important trend is the investigation of new fungus species to discover their bioactive chemicals, which provide distinct advantages such as improved skin moisturization, anti-aging effects, and antibacterial capabilities (Mago et al. 2023). Recent developments in biotechnology have made it possible to grow uncommon fungi in well-regulated settings, enhancing the process of extracting powerful components for use in cosmetics (Case et al. 2022). In addition, the incorporation of nanotechnology is anticipated to transform the administration of fungal-derived components, improving their effectiveness and durability in skincare products (Salvioni et al. 2021). The adoption of sustainable production methods, such as fermentation technology, is increasing in response to the rising demand for environmentally friendly and ethically produced cosmetics (Sasounian et al. 2024). The increasing consumer demand for natural and scientifically validated components is expected to drive the advancement of personalized skincare products derived from fungi, which will be customized to meet the specific demands of particular skin types. This is predicted to be a significant innovation in the upcoming years (Goyal and Jerold 2023). The observed patterns suggest a positive outlook for fungal-derived cosmetics, as continuous research is expected to reveal novel uses and advantages.

1.7.2 Potential New Fungal Sources

The exploration of potential novel fungal sources in fungal-based cosmetics is a field that is gaining increasing attention, motivated by the distinct bioactive substances that fungi provide. Recent studies have discovered a number of fungal species, including *C. sinensis* and *G. lucidum*, that show potential in terms of their antioxidant, anti-inflammatory, and skin-rejuvenating effects (Suwannarach et al. 2020). These fungi have a long history of being used in medicine, but their usage in cosmetics is relatively new, offering a new opportunity for ingredient innovation (Wu et al. 2016). Moreover, marine fungi have garnered interest because of their capacity to generate secondary metabolites that possess powerful skincare advantages, such as shielding against UV radiation and maintaining moisture (Vitale et al. 2020). Researchers are investigating the resilience and distinctive biochemical compositions of fungi found in severe conditions, such as *Penicillium antarcticum*, to explore their potential for introducing novel cosmetic capabilities (Varrella et al. 2021). With the cosmetics industry's ongoing pursuit of sustainable and efficient natural ingredients, these emerging fungus sources provide substantial promise for the advancement of cutting-edge skincare solutions (Rocca et al. 2022).

1.7.3 Sustainable and Ethical Considerations

The development of fungal-based cosmetics is increasingly including sustainable and ethical issues in response to consumer desire for environmentally friendly and socially responsible goods. Fungal-derived components, frequently sourced from sustainable materials, offer a hopeful substitute for synthetic compounds and animal-derived substances, thereby aiding in the mitigation of environmental harm (Rocca et al. 2022). Fungi production, particularly through fermentation, generally necessitates less land, water, and energy in comparison to conventional agricultural methods, rendering it a more sustainable alternative (Dai et al. 2021). In addition, it is imperative for companies to prioritize ethical sourcing, which involves adhering to fair trade standards and taking measures to prevent any negative impact on local ecosystems or populations during the harvesting of fungal materials (Lambrechts 2020). Transparency in the supply chain is crucial as it enables consumers to make well-informed decisions on their purchases. Furthermore, the fact that fungal-based cosmetics are biodegradable is in line with the increasing trend of minimizing plastic waste in packaging (Asim et al. 2022). In order to achieve long-term success and gain customer trust, it is crucial for the cosmetics industry to prioritize sustainability and ethical issues while using fungal-based components, as the sector continues to develop and change.

1.7.4 Market Trends and Growth Opportunities

The fungal-based cosmetics industry is growing strongly due to rising consumer desire for skincare products that are natural, sustainable, and effective. An important market trend is the increasing demand for clean beauty products, characterized by consumers' desire for formulations that contain fewer synthetic chemicals and more natural components, including those derived from fungus (Chin et al. 2018). The increasing global emphasis on sustainability is driving the demand for fungal-based cosmetics, as these components are frequently obtained using eco-friendly methods such as fermentation, which result in a reduced environmental footprint (Mishra and Srivastava 2024). Asia and Europe's emerging markets show a particular openness to these products, as they are incorporating the traditional understanding of fungi's medicinal benefits into modern skincare practices (Liu 2022). The anti-aging sector is experiencing substantial expansion, with fungal extracts such as beta-glucans becoming increasingly popular due to their skin-rejuvenating qualities (Mirończuk-Chodakowska et al. 2021). Moreover, advancements in biotechnology and personalized skincare are creating fresh prospects for fungal-derived cosmetics, enabling firms to customize goods according to individual consumer requirements, thereby increasing their market dominance (Gupta et al. 2022). In general, the market for cosmetics made from fungi is set to grow further, with substantial prospects in terms of creating new products and expanding into more regions.

1.7.5 Guidelines for Formulators and Researchers

Although fungi are well-known for their antibacterial activity and abundant nutrient content (Wu et al. 2016), a thorough understanding and knowledge about the unique properties and characteristics of natural ingredients are necessary to formulate effective and safe fungal-based cosmetic products. A previous study (Barthe et al. 2021) reported that different in vitro and ex vivo methods were used to assess the safety of the cosmetic products in terms of genotoxicity, skin sensitization potential, skin and eye irritation, endocrine properties, and dermal absorption. Besides toxicological assessment of fungal formulations, maintaining the stability of the fungal-based products should be kept under consideration. This problem can be mitigated with the implementation of nanotechnology. Nanotechnology can convert natural bioactive compounds into nanoscale products as well as improve their stability, effectivity, and biocompatibility even at low doses (El Enshasy et al. 2018; Dhanjal et al. 2022). However, the potential toxicity of nanoparticles must also be thoroughly assessed to mitigate any potential health and environmental risks. The biosynthesized nanoparticles prepared from plants, bacteria, and fungi are safe to use and cost-effective as compared to the chemical ones (Durán et al. 2011; Dhanjal et al. 2022). Fungi are preferred more over other microbes to produce nanoparticles due to their advantageous growth rates, minimal processing cost, and low energy requirement (Ghosh et al. 2021). Fungal-derived nanoparticles displayed antiviral, antifungal, antibacterial, anticancer, and wound-healing activities which facilitate their incorporation into cosmetic products (Gaikwad et al. 2013; Boroumand Moghaddam et al. 2015). Ag-nanoparticles produced from *P. italicum* displayed prominent DPPH antiradical activity and antibacterial potential against *Staphylococcus aureus* and *E. coli* (Taha et al. 2019). Another Ag-nanoparticles derived from *Fusarium keratoplasticum* showed bacterial growth inhibition potential (Mohamed et al. 2017). Furthermore, the use of fungi in the production of environmentally friendly fungicides and insecticides can contribute to the overall sustainability of the cosmetics industry.

1.7.6 Consumer Education and Engagement Strategies

Although consumers' demand for novel natural fungal-derived cosmetic products is raising sharply in the present days, many people may be unaware of this fact. Therefore, industry professionals must have taken initiatives to educate their consumers about the applications, benefits, safety, and effectiveness of these fungal-derived cosmetics. Advertising, video tutorials, seminars, webinars, educational articles, and interactive workshops can be an effective way to explain the reason for incorporating fungal ingredients in cosmetic products as well as their potential benefits. In addition to this, social media platforms and influencer partnerships can also help to raise awareness among people. By compensating the knowledge gaps, different companies and brands can build trust among consumers and can effectively communicate the ethos and values behind their sustainable offerings, which ultimately helps to fully capitalize on the fungal-based cosmetics marketing (Ding et al. 2022; Liu 2022).

1.8 Vision for the Future of Fungal-Based Cosmetics

As the cosmetics industry continues to evolve, the potential for fungal-based products has captured the attention of researchers and entrepreneurs alike. The unique properties of fungal compounds, combined with the growing demand for natural and sustainable beauty solutions, suggest a promising future for this emerging field.

1.8.1 Long-Term Potential and Market Growth

Presently, the cosmetic industry is becoming one of the most leading and fastest-growing industries throughout the world (Liu 2022). According to recent market analysis, the global natural cosmetics market is expected to grow from 2021 to 2026 with an annual growth rate of approximately 5–7% (Estrela and Abraham 2016; Ratajczak et al. 2023). Its success depends mainly upon the innovation of new products which can meet consumers' expectations (Kumar et al. 2006). Therefore, the incorporation of fungal-based cosmetics into mainstream beauty products can become a promising strategy. A wide range of fungal species produce diverse secondary metabolites, which have prominent antioxidant, antimicrobial, antiaging, regenerative activities. Therefore, applications of fungi or fungal extracts in the cosmetic industry offer several advantages that could drive long-term market growth (Ratajczak et al. 2023). The fungal-based cosmetics are more eco-friendly and safer to use as compared to the synthetic ones and thereby coincide with the increasing consumer demand for natural and sustainable products (Ramli 2015).

1.8.2 Integration into Mainstream Beauty Products

The cosmetics industry is in constant search for natural compounds possessing anti-aging, anti-pigmentation, anti-collagenase, antioxidant, and anti-inflammatory properties to replace them with synthetic chemicals in the cosmetic formulations (Barros et al. 2007; Athalye et al. 2009; Camelini et al. 2014). Natural ingredients are gaining more preference over the synthetic chemicals as they can reduce the deleterious effects of the synthetic ones. A marked increase in the market for cosmeceuticals is observed in recent years as a result of the incorporation of fungi in cosmetic formulations such as creams and lotions. However, integration of fungal-based cosmetics into mainstream beauty products needs to overcome several challenges, such as ensuring product safety, efficacy, and consumer acceptance. Nonetheless, the unique properties of fungal compounds make them attractive candidates for a variety of cosmetic applications (Wu et al. 2016; Mandal and Krishnan 2021). Anti-collagenase, anti-elastase, anti-hyaluronidase, antioxidant, and anti-inflammatory activities of certain fungi facilitate their use in anti-antiaging creams or lotions (Soto et al. 2015; Takahashi et al. 2020; Kim et al. 2021). Beta-glucans present in fungal extracts possess skin-rejuvenating properties (Mirończuk-Chodakowska et al. 2021). Fungal extract of *Phellinus baumii* and *Flammulina velutipes* can inhibit the melanin production and thus are used as skin-whitening agents. Lactic acid and ceramides generated during biotransformation are used as moisturizing agent. Fungal pigments serve as a natural colorant in makeup products. Several fungal species are also used in hair cosmetics formulation due to their antioxidants, vitamin D, iron, selenium, and copper content, which are helpful to improve hair problems (Madnani and Khan 2013). As research and development in this field continues, we can expect to see a growing number of fungal-based cosmetic products entering the mainstream market.

1.8.3 Innovations That Could Shape the Industry

Although fungal-based cosmetics possess a promiscuous future market, it cannot be achieved without the help of scientific advancements and industry innovations (Roth et al. 2023). Fungi exhibit a wide range of health-beneficial properties due to their constituent bioactive compounds. An efficient fungal cultivation and extraction method is needed prior to incorporation into cosmetic products. It can improve the yield and quality of fungal metabolites or bioactive compounds (Meyer et al. 2020; Troiano et al. 2020). It can be possible with the help of biotechnology (Case et al. 2022). Nanotechnology-based incorporation of fungi in cosmetic products helps to improve their effectiveness and stability (Salvioni et al. 2021).

Toxicological assessment of fungal extracts is essential to solve the allergic reactions related concerns of the consumers upon prolonged usage. This ultimately helps to increase the acceptability of the consumers. Another area of innovation is the integration of fungal-based ingredients into novel cosmetic formulations. By combining fungal compounds with other natural or synthetic ingredients, manufacturers could create unique and innovative products that cater to the evolving needs and preferences of consumers (Takahashi et al. 2020; Mandal and Krishnan 2021).

1.9 Conclusion

The development of fungal-based cosmetics and their integration within the beauty and personal care industry underscores an interesting intersection of sustainability and scientific progress while meeting the dynamic and growing consumer demands. This chapter explored the different fungal-derived bioactive molecules including polysaccharides, β-glucans, ceramides, and ergothioneine. These bioactive molecules possess anti-aging, hydrating, antioxidant, and anti-inflammatory properties. When their sustainable production capabilities using agro-industrial waste are combined with their intrinsic properties, it establishes fungi as the ultimate resource for developing eco-friendly and efficient cosmetic products. The introduction of fungal-based product underscores a major transformation in the cosmetics industry. Diverse fungal sources with special bioactive properties including fungi from marine and extreme environments have been revealed by the biotechnology and the development of fermentation approaches. The industry will witness the next wave of innovative products owing to the emergence of nanotechnology, personalized skincare formulations, and extraction methods that are sustainable while reducing the damage to the environment. Recent breakthroughs position fungal-derived compounds as crucial components for the evolution of clean beauty and the development of sustainable cosmetics. The realization of this potential needs more research and developmental efforts. Scientist must prioritize assessing fungi that have been underexploited for their medical and beauty purposes. Researchers must also prioritize the development of advanced extraction techniques and at the same time maintaining the standards of quality products. The collaboration and interdisciplinary research between scientists, dermatologists, and leaders in the industry is key to addressing the challenges related to the safety of the microorganisms used in these products and the challenges around regulatory approvals. The beauty industry is presented with unlimited and transformative possibilities by the advent of fungal-based cosmetics through their innovative approach to sustainable and scientifically validated solutions. By continuously improving research and development and addressing the challenges with scaling up production, the beauty industry will be able to fully leverage the potential of fungi toward achieving sustainability and consumer-oriented innovations. This project redefines beauty criteria and endorses broader global ideas for environmental conservation and scientific progress.

References

Ahmad MF et al (2013) Edible mushrooms as health promoting agent. Adv Sci Focus 1:189–196

Ahmed T, Suzauddula M, Akter K, Hossen M, Islam MN (2024) Green technology for fungal protein extraction – a review. Separations 11:186

Ainsworth GC (2008) Ainsworth & Bisby's dictionary of the fungi. Cabi

Ajayi S, Olaniyi O, Oladoyinbo T, Ajayi N, Olaniyi F (2024) Sustainable sourcing of organic skincare ingredients: a critical analysis of ethical concerns and environmental implications. Asian J Adv Res Reports 18:65–91

Almukainzi M, Alotaibi L, Abdulwahab A, Albukhary N, El Mahdy AM (2022) Quality and safety investigation of commonly used topical cosmetic preparations. Sci Rep 12:18299

Alvarez PA, Boye JI (2012) Food production and processing considerations of allergenic food ingredients: A review. J Allergy 2012:746125

Alyahyawi AR, Khan S, Rafi Z, Singh P, Moheet K, Akasha R, Ahmad S (2023) Exploring Kinnow mandarin's hidden potential: nature's key to antimicrobial and antidiabetic gold nanoparticles (K-AuNPs). Saudi J Biol Sci 30:103782

Andersen KE (2022) Scientific committee on cosmetic products and non-food products intended for consumers

Andersone A, Janceva S, Lauberte L, Ramata-Stunda A, Nikolajeva V, Zaharova N, Rieksts G, Telysheva G (2023) Anti-inflammatory, anti-bacterial, and antifungal activity of Oligomeric Proanthocyanidins and extracts obtained from Lignocellulosic agricultural waste. Molecules 28

Angeli JPF, Ribeiro LR, Camelini CM, De Mendonça MM, Mantovani MS (2009) Evaluation of the antigenotoxicity of polysaccharides and β-glucans from Agaricus blazei, a model study with the single cell gel electrophoresis/Hep G2 assay. J Food Compos Anal 22:699–703

Antignac E, Nohynek GJ, Re T, Clouzeau J, Toutain H (2011) Safety of botanical ingredients in personal care products/cosmetics. Food Chem Toxicol 49:324–341

Asim Z, Shamsi IRA, Wahaj M, Raza A, Abul Hasan S, Siddiqui SA, Aladresi A, Sorooshian S, Seng Teck T (2022) Significance of sustainable packaging: a case-study from a supply chain perspective. Appl Syst Innov 5:117

Athalye SK, Garcia RA, Wen Z (2009) Use of biodiesel-derived crude glycerol for producing Eicosapentaenoic acid (EPA) by the fungus Pythium irregulare. J Agric Food Chem 57:2739–2744

Badalyan SM (2014) Potential of mushroom bioactive molecules to develop healthcare biotech products

Barros L, Ferreira M-J, Queirós B, Ferreira ICFR, Baptista P (2007) Total phenols, ascorbic acid, β-carotene and lycopene in Portuguese wild edible mushrooms and their antioxidant activities. Food Chem 103:413–419

Barthe M, Bavoux C, Finot F, Mouche I, Cuceu-Petrenci-C, Forreryd A, Chérouvrier Hansson A, Johansson H, Lemkine GF, Thénot J-P, Osman-Ponchet H (2021) Safety testing of cosmetic products: overview of established methods and new approach methodologies (NAMs). Cosmetics 8:50

Bashir A, Vaida N, Dar MA (2018) Economic empowerment of women through mushroom cultivation. Int J Res Anal Rev 5:860–864

Behar N, Sharma S, Parihar R, Dubey SK, Mehta S, Pandey V (2024) Role of fungal metabolites in pharmaceuticals, human health, and agriculture. In: Fungal secondary metabolites. Elsevier, pp 519–535

Blagodatski A, Yatsunskaya M, Mikhailova V, Tiasto V, Kagansky A, Katanaev VL (2018) Medicinal mushrooms as an attractive new source of natural compounds for future cancer therapy. Oncotarget 9: 29259

Boroumand Moghaddam A, Namvar F, Moniri M, Md Tahir P, Azizi S, Mohamad R (2015) Nanoparticles biosynthesized by fungi and yeast: A review of their preparation, properties, and medical applications. Molecules 20:16540–16565

Bowe WP (2013) Cosmetic benefits of natural ingredients: mushrooms, feverfew, tea, and wheat complex. J Drugs Dermatol 12:s133–s136

Calado MDL, Silva J, Alves C, Susano P, Santos D, Alves J, Martins A, Gaspar H, Pedrosa R, Campos MJ (2021) Marine endophytic fungi associated with Halopteris scoparia (Linnaeus) Sauvageau as producers of bioactive secondary metabolites with potential dermocosmetic application. PLoS One 16: e0250954

Camelini CM, Rossi MJ, Cardozo FTGS, Gomes A, Sales-Campos C, Giachini AJ (2014) Fungal cultivation and production of polysaccharides. In: Ramawat KG, Mérillon J-M (eds) Polysaccharides: bioactivity and biotechnology. Springer International Publishing, Cham, pp 1–34

Campana F, Zervoudis S, Perdereau B, Gez E, Fourquet A, Badiu C, Tsakiris G, Koulaloglou S (2004) Topical superoxide dismutase reduces post-irradiation breast cancer fibrosis. J Cell Mol Med 8:109–116

Case NT, Berman J, Blehert DS, Cramer RA, Cuomo C, Currie CR, Ene IV, Fisher MC, Fritz-Laylin LK, Gerstein AC, Glass NL et al (2022) The future of fungi: threats and opportunities. G3: Genes|Genomes|Genetics 12:jkac224

Chandrasekaran G, Oh D-S, Shin H-J (2012) Versatile applications of the culinary-medicinal mushroom Mycoleptodonoides aitchisonii (Berk.) Maas G.(higher basidiomycetes): A review. Int J Med Mushrooms 14: 395

Chang S-T, Wasser SP (2012) The role of culinary-medicinal mushrooms on human welfare with a pyramid model for human health. In J Med Mushrooms 14: 95

Cheng MH, Walker TH, Hulbert GJ, Raman DR (1999) Fungal production of eicosapentaenoic and arachidonic acids from industrial waste streams and crude soybean oil. Bioresour Technol 67:101–110

Cheung PC (2008) Mushrooms as functional foods. Wiley

Chien C-C, Tsai M-L, Chen C-C, Chang S-J, Tseng C-H (2008) Effects on tyrosinase activity by the extracts of Ganoderma lucidum and related mushrooms. Mycopathologia 166:117–120

Chin J, Jiang BC, Mufidah I, Persada SF, Noer BA (2018) The investigation of consumers' behavior intention in using green skincare products: a pro-environmental behavior model approach. Sustainability 10:3922

Corey ME, Beelman RB, Seetharman K (2009) Potential for nutritional enrichment of whole-wheat bread with portabella mushroom powder (Agaricus bisporus (JLge) Imbach, Agaricomycetideae). Int J Med Mushrooms 11:157

Dai X, Sharma M, Chen J (2021) Fungi in sustainable food production. Springer

De Baets S, Vandamme E (2001a) Extracellular Tremella polysaccharides: structure, properties and applications. Biotechnol Lett 23:1361–1366

De Baets S, Vandamme EJ (2001b) Extracellular Tremella polysaccharides: structure, properties and applications. Biotechnol Lett 23:1361–1366

Dhanjal DS, Mehra P, Bhardwaj S, Singh R, Sharma P, Nepovimova E, Chopra C, Kuca K (2022) Mycology-nanotechnology interface: applications in medicine and cosmetology. Int J Nanomedicine Volume 17:2505–2533

Ding J, Wu B, Chen L (2022) Application of marine microbial natural products in cosmetics. Front Microbiol 13

Diseases S (2006) Disease control priorities in developing countries – NCBI Bookshelf. Washington (DC): The International Bank for Reconstruction and Development/The World Bank; New York: Oxford University Press

Dos Reis EE, Schenkel PC, Camassola M (2022) Effects of bioactive compounds from Pleurotus mushrooms on COVID-19 risk factors associated with the cardiovascular system. J Integr Med 20:385

Dubost NJ, Beelman RB, Royse DJ (2007) Influence of selected cultural factors and postharvest storage on ergothioneine content of common button mushroom Agaricus bisporus (JLge) Imbach (Agaricomycetideae). Int J Med Mushrooms 9:163

Durán N, Marcato PD, Durán M, Yadav A, Gade A, Rai M (2011) Mechanistic aspects in the biogenic synthesis of extracellular metal nanoparticles by peptides, bacteria, fungi, and plants. Appl Microbiol Biotechnol 90:1609–1624

El Enshasy HA, Hatti-Kaul R (2013) Mushroom immunomodulators: unique molecules with unlimited applications. Trends Biotechnol 31:668–677

El Enshasy HA, El Marzugi NA, Elsayed EA, Ling OM, Malek RA, Kepli AN, Othman NZ, Ramli S (2018) Medical and cosmetic applications of fungal nanotechnology: production, characterization, and bioactivity. In: Prasad R, Kumar V, Kumar M, Wang S (eds) Fungal nanobionics: principles and applications. Springer Singapore, Singapore, pp 21–59

Esposito MM, Patsakos S, Borruso L (2023) The role of the mycobiome in women's health. J Fungi 9:348

Estrela AB, Abraham W-R (2016) Fungal metabolites for the control of biofilm infections. Agriculture 6:37

Fang S-H, Rao YK, Tzeng Y-M (2008) Anti-oxidant and inflammatory mediator's growth inhibitory effects of compounds isolated from Phyllanthus urinaria. J Ethnopharmacol 116:333–340

Ferreira M, Matos A, Couras A, Marto J, Ribeiro H (2022) Overview of cosmetic regulatory frameworks around the world. Cosmetics 9:72

Firenzuoli F, Gori L, Lombardo G (2008) The medicinal mushroom Agaricus blazei Murrill: review of literature and pharmaco-toxicological problems. Evid Based Complement Alternat Med 5:3–15

GAE Mahmoud, A Abdel Shakor, NA Kamal-Eldin, ANA Zohri (2024) Production of kojic acid by aspergillus flavus OL314748 using box-Behnken statistical design and its antibacterial and anticancer applications using molecular docking technique. BMC Microbiol 24:140

Gaikwad S, Ingle A, Gade A, Rai M, Falanga A, Incoronato N, Russo L, Galdiero S, Galdiero M (2013) Antiviral activity of mycosynthesized silver nanoparticles against herpes simplex virus and human parainfluenza virus type 3. Int J Nanomedicine 8: 4303–4314

Gautier S, Xhauflaire-Uhoda E, Gonry P, Piérard G (2008) Chitin–glucan, a natural cell scaffold for skin moisturization and rejuvenation. Int J Cosmet Sci 30: 459–469

Ghosh S, Ahmad R, Zeyaullah M, Khare SK (2021) Microbial nano-factories: synthesis and biomedical applications. Front Chem:9–2021

Golz-Berner K, Zastrow L (2005) Cosmetic preparation containing extracts of Tuberaceae. Google Patents)

Gomes C, Silva AC, Marques AC, Sousa Lobo J, Amaral MH (2020) Biotechnology applied to cosmetics and aesthetic medicines. Cosmetics 7:33

Goyal N, Jerold F (2023) Biocosmetics: technological advances and future outlook. Environ Sci Pollut Res Int 30:25148–25169

Gupta V, Mohapatra S, Mishra H, Farooq U, Kumar K, Ansari MJ, Aldawsari MF, Alalaiwe AS, Mirza MA, Iqbal Z (2022) Nanotechnology in cosmetics and cosmeceuticals – a review of latest advancements. Gels 8:173

Hintikka T, Andersson MA, Lundell T, Marik T, Kredics L, Mikkola R, Andersson MC, Kurnitski J, Salonen H (2024) Toxicity screening of fungal extracts and metabolites, xenobiotic chemicals, and indoor dusts with in vitro and ex vivo bioassay methods. Pathogens 13:217

Hüttner S, Johansson A, Gonçalves Teixeira P, Achterberg P, Nair RB (2020) Recent advances in the intellectual property landscape of filamentous fungi. Fungal Biol Biotechnol 7:16

Hyde K, Bahkali A, Moslem M (2010) Fungi – an unusual source for cosmetics. Fungal Divers 43:1–9

Ina K, Kataoka T, Ando T (2013) The use of lentinan for treating gastric cancer. Anti-Cancer Agents Med Chem 13:681–688

Iram A, Ozcan A, Turhan I, Demirci A (2023) Production of value-added products as food ingredients via microbial fermentation. PRO 11:1715

Jiang C, Ge J, He B, Zeng B (2021) Glycosphingolipids in filamentous fungi: biological roles and potential applications in cosmetics and health foods. Front Microbiol 12:690211

Jikai L (2002) Biologically active substances from mushrooms in Yunnan, China. Heterocycles 57:157–167

Jin J-M, Lee J, Lee Y-W (2010) Characterization of carotenoid biosynthetic genes in the ascomycete Gibberella zeae. FEMS Microbiol Lett 302:197–202

Kalač P (2009) Chemical composition and nutritional value of European species of wild growing mushrooms: a review. Food Chem 113:9–16

Kalač P (2013) A review of chemical composition and nutritional value of wild-growing and cultivated mushrooms. J Sci Food Agric 93:209–218

Kamal A, Saba M, Farooq M (2023) Biocompatible formulations based on mycosynthesized iron oxide nanoparticles: fabrication, characterization, and biological investigation. J Basic Microbiol 63:156–167

Keypour S, Rafati H, Riahi H, Mirzajani F, Moradali MF (2010) Qualitative analysis of ganoderic acids in Ganoderma lucidum from Iran and China by RP-HPLC and electrospray ionisation-mass spectrometry (ESI-MS). Food Chem 119:1704–1708

Kim J-H, An C, Hwang SD, Kim YS (2021) Ceriporia lacerata Mycelium culture medium as a novel anti-aging microbial material for cosmeceutical application. Cosmetics 8:101

Kim H-J, Oh HN, Park T, Kim H, Lee HG, An S, Sul WJ (2022) Aged related human skin microbiome and mycobiome in Korean women. Sci Rep 12:2351

Klimek-Szczykutowicz M, Błońska-Sikora EM, Kulik-Siarek K, Zhussupova A, Wrzosek M (2024) Bioferments and biosurfactants as new products with potential use in the cosmetic industry. Appl Sci 14:3902

Kumar S, Massie C, Dumonceaux MD (2006) Comparative innovative business strategies of major players in cosmetic industry. Ind Manag Data Syst 106:285–306

Kumar M, Barbhai MD, Puranik S, Natta S, Senapathy M, Dhumal S, Singh S, Kumar S, Deshmukh VP, Anitha T (2023) Combination of green extraction techniques and smart solvents for bioactives recovery. TrAC Trends Anal Chem 117286

Kumari M, Survase SA, Singhal RS (2008) Production of schizophyllan using Schizophyllum commune NRCM. Bioresour Technol 99:1036–1043

Lambrechts W (2020) Ethical and sustainable sourcing: toward strategic and holistic sustainable supply chain management. In: Decent work and economic growth. Springer International Publishing, Cham, pp 402–414

Li Q, Fang H, Dang E, Wang G (2020) The role of ceramides in skin homeostasis and inflammatory skin diseases. J Dermatol Sci 97:2–8

Li RS, Xu ZC, Qin DM, Huang JB, Wang MR, Zhao XR, Yang QY, Xiao WL, Wang RR, Li XL (2023) Three new diterpenoids isolated from euphorbia nematocypha hand.-Mazz and their anti-fungal activity. Nat Prod Res, pp 1–10

Lin L, Xu J (2020) Fungal pigments and their roles associated with human health. J Fungi (Basel):6

Liu J-K (2022) Natural products in cosmetics. Nat Products Bioprospecting 12:1–43

Liu H-M, Tang W, Wang X-Y, Jiang J-J, Zhang W, Wang W (2023) Safe and effective antioxidant: the biological mechanism and potential pathways of ergothioneine in the skin. Molecules 28:1648

Lourenço A, Lobo AM, Rodríguez B, Jimeno M-L (1996) Ceramides from the fungus Phellinus pini. Phytochemistry 43:617–620

Lourith N, Pungprom S, Kanlayavattanakul M (2021) Formulation and efficacy evaluation of the safe and efficient moisturizing snow mushroom hand sanitizer. J Cosmet Dermatol 20:554–560

Luiz R, Jordao B, Da Eira A, Ribeiro L, Mantovani M (2003) Mechanism of anticlastogenicity of Agaricus blazei Murill mushroom organic extracts in wild type CHO (K1) and repair deficient (xrs5) cells by chromosome aberration and sister chromatid exchange assays. Mutat Res/Fundam Mol Mech Mutagenes 528:75–79

Lyu X, Lyu Y, Yu H, Chen W, Ye L, Yang R (2022) Biotechnological advances for improving natural pigment production: a state-of-the-art review. Bioresour Bioprocess 9:8

Madnani N, Khan K (2013) Hair cosmetics. Indian J Dermatol Venereol Leprol 79:654–667

Mago P, Sharma R, Hafeez I, Nawaz I, Joshi M, Mehrotra R (2023) Mushroom based cosmeceuticals: an upcoming biotechnology sector. Biosci Biotechnol Res Asia 20:381–394

Mandal S, Krishnan R (2021) Fungi: the budding source for biomaterials. Microbial Biosyst 6:55–65

Manful ME, Ahmed L, Barry-Ryan C (2024) Cosmetic formulations from natural sources: safety considerations and legislative frameworks in the European Union. Cosmetics 11:72

Manimaran K, Yanto DHY, Kamaraj C, Selvaraj K, Pandiaraj S et al (2023) Eco-friendly approaches of mycosynthesized copper oxide nanoparticles (CuONPs) using Pleurotus citrinopileatus mushroom extracts and their biological applications. Environ Res 232:116319

Matuo R, Oliveira R, Silva A, Mantovani M, Ribeiro L (2007) Anticlastogenic activity of aqueous extract of Agaricus blazei in drug-metabolizing cells (HTCs) during cell cycle. Toxicol Mech Methods 17:147–152

Meyer V, Basenko EY, Benz JP, Braus GH, Caddick MX, Csukai M, De Vries RP, Endy D, Frisvad JC, Gunde-Cimerman N, Haarmann T, Hadar Y, Hansen K, Johnson RI, Keller NP, Kraševec N, Mortensen UH, Perez R, Ram AFJ, Record E, Ross P, Shapaval V, Steiniger C, Van Den Brink H, Van Munster J, Yarden O, Wösten HAB (2020) Growing a circular economy with fungal biotechnology: a white paper. Fungal Biol Biotechnol 7:5

Michelutti L, Bulfoni M, Bolzon V, Nencioni E (2020) Preliminary evidence of a molecular detection method to analyze bacterial DNA as a quality indicator in cosmetics. Cosmet 7:54

Miles PG, Chang S-T (2004) Mushrooms: cultivation, nutritional value, medicinal effect, and environmental impact. CRC Press
Millikan LE (2001) Cosmetology, cosmetics, cosmeceuticals: definitions and regulations. Clin Dermatol 19:371–374
Milovanovic I, Zengin G, Maksimovic S, Tadic V (2024) Supercritical carbon-oxide extracts from cultivated and wild-grown Ganoderma lucidum mushroom: differences in ergosterol and ganoderic acids content, antioxidative and enzyme inhibitory properties. Nat Prod Res 38:2522–2528
Mirończuk-Chodakowska I, Kujawowicz K, Witkowska AM (2021) Beta-Glucans from fungi: biological and health-promoting potential in the COVID-19 pandemic era. Nutrients 13:3960
Mishra M, Srivastava D (2024) Harnessing the power of medicinal mushrooms into cosmetics: a fountain of youth from nature
Mizuno T, Inagaki R, Kanao T, Hagiwara T, Nakamura T, Ito H, Shimura K, Sumiya T, Asakura A (1990) Antitumor activity and some properties of water-insoluble hetero-glycans from "Himematsutake," the fruiting body of Agaricus blazei Murill. Agric Biol Chem 54:2897–2905
Mohamed A et al (2017) Enhancing of cotton fabric antibacterial properties by silver nanoparticles synthesized by new Egyptian strain Fusarium Keratoplasticum A1-3. Egypt J Chem 60:63–71
Mohanta YK, Nayak D, Mishra AK, Chakrabartty I, Ray MK, Mohanta TK, Tayung K, Rajaganesh R, Vasanthakumaran M, Muthupandian S, Murugan K, Sharma G, Dahms HU, Hwang JS (2022) Green synthesis of Endolichenic fungi functionalized silver nanoparticles: the role in antimicrobial, anti-cancer, and Mosquitocidal activities. Int J Mol Sci 23
Mueller GM, Schmit JP (2007) Fungal biodiversity: what do we know? What can we predict? Biodivers Conserv 16:1–5
Mukherjee A, Das S, Chakraborty D, Pal N, Das N (2024) Fungi's treasure in cosmeceuticals-a comprehensive chemical approach. S Afr J Bot 166:311–331
Murad H (2010) The water secret: the cellular breakthrough to look and feel 10 years younger. Turner Publishing Company
Naz T, Ullah S, Nazir Y, Li S, Iqbal B, Liu Q, Mohamed H, Song Y (2023) Industrially important fungal carotenoids: advancements in biotechnological production and extraction. J Fungi 9:578
Ng TB (2004) Peptides and proteins from fungi. Peptides 25:1055–1073
Nieto IJ, Chegwin A (2008) Triterpenoids and fatty acids identified in the edible mushroom Pleurotus sajor-caju. J Chil Chem Soc 53:1515–1517
O'brien DJ, Stinson EE, Wessinger EW, Somkuti GA (1993) Production of eicosapentaenoic acid from filamentous fungi utilizing lactose as a primary carbon source. Google Patents
Passi S, De Pità O, Puddu P, Littarru GP (2002) Lipophilic antioxidants in human sebum and aging. Free Radic Res 36:471–477
Pons A, Timmerman P, Leroy Y, Zanetta J-P (2002) Gas-chromatography/mass-spectrometry analysis of human skin constituents as heptafluorobutyrate derivatives with special reference to long-chain bases. J Lipid Res 43:794–804
Poucheret P, Fons F, Rapior S (2006) Biological and pharmacological activity of higher fungi: 20-year retrospective analysis. Cryptogam Mycol 27:311
Purohit A, Sharma R, Shiv Ramakrishnan R, Sharma S, Kumar A, Jain D, Kushwaha HS, Maharjan E (2022) Biogenic synthesis of silver nanoparticles (AgNPs) using aqueous leaf extract of Buchanania lanzan Spreng and evaluation of their antifungal activity against Phytopathogenic fungi. Bioinorg Chem Appl 2022:6825150
Ramli NS (2015) Immigrant entrepreneurs on the world's successful global Brands in the Cosmetic Industry. Procedia Soc Behav Sci 195:113–122
Ratajczak P, Landowska W, Kopciuch D, Paczkowska A, Zaprutko T, Kus K (2023) The growing market for natural cosmetics in Poland: consumer preferences and industry trends. Clin Cosmet Investig Dermatol 16:1877–1892
Rocca R, Acerbi F, Fumagalli L, Taisch M (2022) Sustainability paradigm in the cosmetics industry: state of the art. Cleaner Waste Syst 3:100057
Roth MG, Westrick NM, Baldwin TT (2023) Fungal biotechnology: from yesterday to tomorrow. Front Fungal Biol 4:1135263
Ruggeri M, Miele D, Contardi M, Vigani B, Boselli C, Icaro Cornaglia A, Rossi S, Suarato G, Athanassiou A, Sandri G (2023) Mycelium-based biomaterials as smart devices for skin wound healing. Front Bioeng Biotechnol 11:1225722
Salinas-Solis LJ, Gaytan-Oyarzun JC, Octavio-Aguilar P (2023) Detection of Mitogenic and genotoxic effects of the Turkey tail medicinal mushroom (Trametes versicolor, Agaricomycetes) extracts from Mexico on human lymphocyte cultures. Int J Med Mushrooms 25: 33–41
Salvioni L, Morelli L, Ochoa E, Labra M, Fiandra L, Palugan L, Prosperi D, Colombo M (2021) The emerging role of nanotechnology in skincare. Adv Colloid Interf Sci 293:102437
Sanodiya BS, Thakur GS, Baghel RK, Prasad G, Bisen P (2009) Ganoderma lucidum: a potent pharmacological macrofungus. Curr Pharm Biotechnol 10:717–742
Sasounian R, Martinez RM, Lopes AM, Giarolla J, Rosado C, Magalhães WV, Velasco MVR, Baby AR (2024) Innovative approaches to an eco-friendly cosmetic industry: a review of sustainable ingredients. Clean Technologies 6:176–198
Shimada Y, Morita T, Sugiyama K (2002) Effects of Lentinus edodes on fatty acid and molecular species profiles of phosphatidylcholine in rats fed different

levels of corn oil. Biosci Biotechnol Biochem 66: 1759–1763
Soto ML, Falqué E, Domínguez H (2015) Relevance of natural Phenolics from grape and derivative products in the formulation of cosmetics. Cosmet 2:259
Sousa P, Tavares-Valente D, Amorim M, Azevedo-Silva J, Pintado M, Fernandes J (2023) β-Glucan extracts as high-value multifunctional ingredients for skin health: a review. Carbohydr Polym 121329
Sulkowska-Ziaja K, Trepa M, Olechowska-Jarzab A, Nowak P, Ziaja M, Kala K, Muszynska B (2023) Natural compounds of fungal origin with antimicrobial activity-potential cosmetics applications. Pharmaceuticals (Basel) 16
Suthar M, Dufossé L, Singh SK (2023) The enigmatic world of fungal melanin: a comprehensive review. J Fungi 9:891
Suwannarach N, Kumla J, Sujarit K, Pattananandecha T, Saenjum C, Lumyong S (2020) Natural bioactive compounds from fungi as potential candidates for protease inhibitors and Immunomodulators to apply for coronaviruses. Molecules 25
Synytsya A, Míčková K, Synytsya A, Jablonský I, Spěváček J, Erban V, Kovářiková E, Čopíková J (2009) Glucans from fruit bodies of cultivated mushrooms Pleurotus ostreatus and Pleurotus eryngii: structure and potential prebiotic activity. Carbohydr Polym 76:548–556
Taha ZK, Hawar SN, Sulaiman GM (2019) Extracellular biosynthesis of silver nanoparticles from Penicillium italicum and its antioxidant, antimicrobial and cytotoxicity activities. Biotechnol Lett 41:899–914
Takahashi JA, Barbosa BVR, Martins BDA, P. Guirlanda et al (2020) Use of the versatility of fungal metabolism to meet modern demands for healthy aging, functional foods, and sustainability. J Fungi 6:223
Theeraraksakul K, Jaengwang K, Choowongkomon K, Tabtimmai L (2023) Exploring the biological functions and anti-melanogenesis of phallus indusiatus for mushroom-based cosmetic applications. Cosmetics 10:121
Tian T, Sun Q, Shen J, Zhang T, Gao P, Sun Q (2008) Microbial transformation of polydatin and emodin-8-β-d-glucoside of Polygonum cuspidatum Sieb. et Zucc into resveratrol and emodin respectively by Rhizopus microsporus. World J Microbiol Biotechnol 24: 861–866
Trepa M, Sułkowska-Ziaja K, Kała K, Muszyńska B (2024) Therapeutic potential of fungal terpenes and terpenoids: application in skin diseases. Molecules 29:1183
Troiano D, Orsat V, Dumont MJ (2020) Status of filamentous fungi in integrated biorefineries. Renew Sust Energ Rev 117:109472
Varrella S, Barone G, Tangherlini M, Rastelli E, Dell'anno A, Corinaldesi C (2021) Diversity, ecological role and biotechnological potential of Antarctic marine fungi. J Fungi 7:391
Venkatakrishna K, Gokul T, Kumar KR, Veeramanikandan V, Balaji P (2023) Evaluation of in vitro antioxidant activity of edible basidiomycetes mushroom fungi. J Drug Deliv Ther 13:135–142
Ventura J, Belmares R, Aguilera-Carbo A, Gutiérrez-Sanchez G, Rodríguez-Herrera R, Aguilar CN (2008) Fungal biodegradation of tannins from creosote bush (Larrea tridentata) and tar bush (Fluorensia cernua) for gallic and ellagic acid production. Food Technol Biotechnol 46:213
Visvanathan S, Krishnamoorthy R, Sabesan GS (2022) Fungal cosmetics: mushrooms in beauty care and the new age of natural cosmetics. In: Applied mycology: entrepreneurship with fungi. Springer, pp 1–37
Vitale GA, Coppola D, Palma Esposito F, Buonocore C, Ausuri J, Tortorella E, De Pascale D (2020) Antioxidant molecules from marine fungi: methodologies and perspectives. Antioxidants 9:1183
Vozenin-Brotons M-C, Sivan V, Gault N, Renard C, Geffrotin C, Delanian S, Lefaix J-L, Martin M (2001) Antifibrotic action of Cu/Zn SOD is mediated by TGF-β1 repression and phenotypic reversion of myofibroblasts. Free Radic Biol Med 30:30–42
Wang H, Liu L, Guo Y-X, Dong Y-S, Zhang D-J, Xiu Z-L (2007) Biotransformation of piceid in Polygonum cuspidatum to resveratrol by aspergillus oryzae. Appl Microbiol Biotechnol 75:763–768
Ward OP, Singh A (2005) Omega-3/6 fatty acids: alternative sources of production. Process Biochem 40:3627–3652
Wasser SP (2010) Medicinal mushroom science: history, current status, future trends, and unsolved problems. Int J Med Mushrooms 12:1
Wasser SP (2011) Current findings, future trends, and unsolved problems in studies of medicinal mushrooms. Appl Microbiol Biotechnol 89:1323–1332
Williams LM, Berthon BS, Stoodley IL, Williams EJ, Wood LG (2023) Medicinal mushroom extracts from Hericium coralloides and Trametes versicolor exert differential immunomodulatory effects on immune cells from older adults in vitro. Nutrients 15(9):2227
Woźniak B, Chudzińska J, Szczyglewska P, Nowak I, Feliczak-Guzik A (2023) Optimization of the composition of a cosmetic formulation containing Tremella fuciformis extract (fungi). Cosmetics 10:82
Wu Y, Choi M-H, Li J, Yang H, Shin H-J (2016) Mushroom cosmetics: the present and future. Cosmetics 3:22
Yin Z, Zhang J, Qin J, Guo L, Guo Q, Kang W, Ma C, Chen L (2024) Anti-inflammatory properties of polysaccharides from edible fungi on health-promotion: a review. Front Pharmacol 15:1447677
Zhang Y, Mills GL, Nair MG (2002) Cyclooxygenase inhibitory and antioxidant compounds from the mycelia of the edible mushroom Grifola frondosa. J Agric Food Chem 50:7581–7585
Zheng M, Liu Y, Zhang G, Yang Z, Xu W, Chen Q (2023) The applications and mechanisms of superoxide dismutase in medicine, food, and cosmetics. Antioxidants 12:1675

Exploring Cultivable Fungi for Medicinal Cosmetics

2

Eunice Ngozi Anaele

Abstract

Cultivable fungi are increasingly gaining popularity in medicinal cosmetics due to the associated bioactive fungal compounds laced with numerous health benefits. Fungal compounds used in medicinal cosmetics are safe, ecofriendly, and sustainable. Mushrooms, molds, and yeasts are grown on cheap agricultural and industrial wastes to produce valuable fungal products which are highly beneficial to human health. *Ganoderma*, *Cordyceps*, *Inonotus*, *Agaricus*, *Grifola*, *Lentinula*, *Pleurotus*, *Volvariella*, and *Auricularia* are among the cultivable mushrooms discussed. Ergosterol is known to promote wound healing due to its anti-inflammatory and antibacterial properties. Lactic acid serves as the bioactive agent to reduce acne, remove dead cells, lighten the skin, prevent skin wrinkle, and damage from sunlight while ceramides serve as epidermal hydrating agent and improve skin texture. Extracts of cultivable fungi play key role in inhibition of major enzymes such as tyrosinase, hyaluronidase, elastase, and collagenase which are essential in skin hyperpigmentation and aging. They offer promising effect in the promotion of hair growth and protection against hair breakage and alopecia. Factors affecting growth, development of fruiting body, and production of fungal metabolites as well as the solutions are discussed. Cultivable fungi can provide natural, safe, effective, and sustainable ingredients for medicinal cosmetics.

Keywords

Fungal compounds · Mushrooms · Molds · Yeast · Cultivable fungi · Medicinal cosmetics

2.1 Introduction

Cultivable fungi are gaining much attention not only due to their nutritional value but also their medicinal potential. Mushrooms have been constituents of the human diet for many decades and on the basis of their bioactive components, researchers have recently identified their potential applications in the cosmetics industry. The application of fungal products in medicinal cosmetics is becoming popular because of the increasing demand for natural products, and reduction of harmful, synthetic, and non-biodegradable substances in cosmetics. Although plant products can be used in the production of medicinal cosmetics, bioactive compounds from fungi are preferable as they are laced with economical

E. N. Anaele (✉)
South-East Zonal Biotechnology Center, University of Nigeria, Nsukka, Nigeria

Department of Microbiology, Faculty of Biological Sciences, University of Nigeria, Nsukka, Nigeria
e-mail: eunice.anaele@unn.edu.ng

S. Ghosh, H. N. Onyeaka (eds.), *Fungal-Based Cosmetics: Formulation and Usage*, The Mycota 17,
https://doi.org/10.1007/978-3-032-09291-5_2

production process due to fungal fast growth and easy downstream processing of secondary metabolites (Kumar et al. 2021).

Fungi are known to be a sustainable source of ecofriendly ingredients for medicinal cosmetics. Cultivable fungal products have been applied for treatment of stress and age-related skin disorders. They are also used in the treatment of skin infections due to their antimicrobial properties (Mukherjee et al. 2024; Phasha et al. 2022). They are essentially utilized in cosmetic formulation where the fungal products serve as moisturizing agents and thickeners (polysaccharides). However, some cultivable fungi have not been explored and their products are untapped for application as medicinal cosmetics. This review explores the potential of cultivable fungi in medicinal cosmetics.

2.2 What Are Cultivable Fungi and Medicinal Cosmetics?

Cultivable fungi are fungi that can be nurtured or grown in artificial media, manipulated, and studied to enhance their productivity. Cultivation can be outdoor or indoor environment where the fungal growth conditions are monitored and controlled. Cultivable fungi include mushrooms, molds, and yeasts, which have been known to be cultivated for food and medicine. Following the discovery of numerous bioactive compounds in fungi associated with numerous health benefits, fungi cultivation is currently a treasure (Mukherjee et al. 2024).

Medicinal cosmetics are beauty products that contain bioactive ingredients known to improve skin health or treat a particular skin condition. Such bioactive ingredients include enzymes, vitamins, minerals, phytochemicals, and other biologically active compounds which are incorporated in cream, lotion, ointments, etc. to support or promote healthy skin nail and hair. Medicinal cosmetics include skin moisturizers which prevent skin peeling due to the presence of lactic acid in the cosmetics. Lactic acid serves as the bioactive agent to reduce acne, remove dead cells, lighten the skin, prevent skin wrinkle, and damage from sunlight. In addition, ceramides are also added to cosmetic formulation as epidermal hydrating agent and to improve skin texture (Mukherjee et al. 2024; Mishra and Srivastava 2024).

Another bioactive agent used in medicinal cosmetics is kojic acid, produced mainly by filamentous fungi to promote skin whitening and inhibit melanin production. Fungal melanin could serve as antimicrobial, sunscreen, and antioxidant properties (Suthar et al. 2023). Other fungal extracts or compounds used in medicinal cosmetics have antioxidant, anti-wrinkle, photoprotective, and anti-aging potentials. Additionally, some cultivable fungal bioactive components also prevent hair breakage, treat alopecia, and promote hair development (Wu et al. 2016).

2.3 Role of Cultivable Fungi in Skin and Hair Revitalization

Human skin, the first line of defense, is prone to damage caused by ultraviolet radiation, pollutants, harsh environmental conditions, and aging. Damage to the skin is damage to the body's first line of defense. Natural products from fungi can regulate skin immunity and promote skin vitality. Fungi synthesize abundant bioactive compounds which have played crucial role in skin revitalization. Polysaccharides, phenolic acids, polysaccharopeptides, proteins, free amino acids, glycosides, alkaloids, and triterpenes are fungal compounds for the promotion of antioxidant effects, skin whitening, photoprotective, anti-inflammatory, moisturizing, and stabilization of skin elastin, collagen, and hyaluronic acid (Paterska et al. 2024).

Sterols are natural components found in the cellular membrane, including fungi, which are used in cosmetic preparations. Sterols are known to penetrate the corneal layer of the epidermis, regulating water loss and supplementing the lipid barrier of the skin as well as the epidermal flexibility. Sterols improve skin hydration, prevent premature aging, and are being used in cosmetics as emulsifiers. Ergosterol is known to promote wound healing due to its anti-inflammatory and antibacterial properties. Thus, they are essential in the formulation of anti-acne products and treatment of dry, irritated skin. Ergosterol also promotes regeneration of the skin cells (Mukherjee et al. 2024).

Extracts of cultivable fungi play a key role in inhibition of major enzymes such as tyrosinase, hyaluronidase, elastase, and collagenase which are essential in skin hyperpigmentation and aging. Bioactive compounds from fungi protect the skin from oxidative stress and UV-damage, skin wrinkle reduction, enhancement of skin elasticity, and improvement of skin hydration. They also offer promising effect in the promotion of hair growth and protection against hair breakage or damage. Cultivable fungi provide natural, effective, and sustainable alternative ingredients in the cosmetic industry (Rukhsar et al. 2025).

Sułkowska-Ziaja et al. (2021) analyzed *Ganoderma applanatum*, *Laetiporus sulphureus,* and *Trametes versicolor* on inhibition or tyrosinase and hyaluronidase, assessing their sun protection factor. These mushrooms produced phenolic acid and cinnamic acid, gallic, protocatenhuic, and p-hydroxybenzoic acids. Phenolic acids, sterols, indole derivatives, and kojic acids are content of the mycelial extracts. Triterpenes reported in *Ganoderma* species were responsible for the anticancer or antioxidant effects. In addition, syringic acid was reported in *G. applanatum*, and gentistic acid in *T. versicolor*. The mushroom extracts can absorb ultraviolet rays indicating the potential use as sunscreen agent.

The hair is also prone to stress which might lead to alopecia, baldness, gray hair, dryness, dandruff, and hair aggression. Fungal extracts serve as hair tonic and grooming aid. They also moisturize the sebaceous gland. Cultivable fungi are utilized in the production of hair conditioners and cleansing agents such as shampoo.

Different species of fungi can serve as cosmetic ingredients for skin conditioner, skin protectants, hair conditioners, and cleansing agents (Fig. 2.1).

2.4 Cultivable Fungal Products in Medicinal Cosmetics Are Necessary

The prolonged exposure of the skin to sunlight causes sunburn, thus damaging the skin appearance. The extracellular matrix of the skin is prone to degradation. Melanin, a pigment produced from L-tyrosine and converted to dopaquinone causes skin darkening/hyperpigmentation. The clogging of the skin hair follicle by dead skin cells and oil causes acne, pimple, and skin spot. The hair is also prone to damage as a result of microbial attack or other conditions (Mukherjee et al. 2024). Products from cultivable fungi can be utilized in the formulation of cosmetics for the treatment of such skin and hair conditions (Gupta et al. 2019).

Fungal cosmetic ingredients are in high demand due to their beneficial health potentials including biosafety, biocompatibility, and environmental friendliness. Synthetic products are associated with numerous side effects. The use of plant-based cosmetic ingredients is faced with challenges of prolonged time of development, environmental pollution, deforestation, and climate change. However, cultivable fungi are grown using agricultural and industrial waste for a clean ecosystem, valorizing wastes into beneficial bioactive compounds (Khoo et al. 2022). Production of fungal secondary metabolite is cost-effective, feasible, and sustainable. Cultivable fungi are known to produce enzymes, organic acids, antibiotics, and polysaccharides within a short period, associated with ease of production, and the metabolites have gained wide application in cosmetics. In medicinal cosmetics, cultivable fungi can be used as moisturizing, thickening, and pigmenting agents. The fungal products serve as anti-aging, antioxidant, anti-inflammatory, antimicrobial, skin moisturizing, and skin-whitening agents. Different brands of fungal-based medicinal cosmetics are being marketed for hair and skin care (Mishra and Srivastava 2024; Mukherjee et al. 2024).

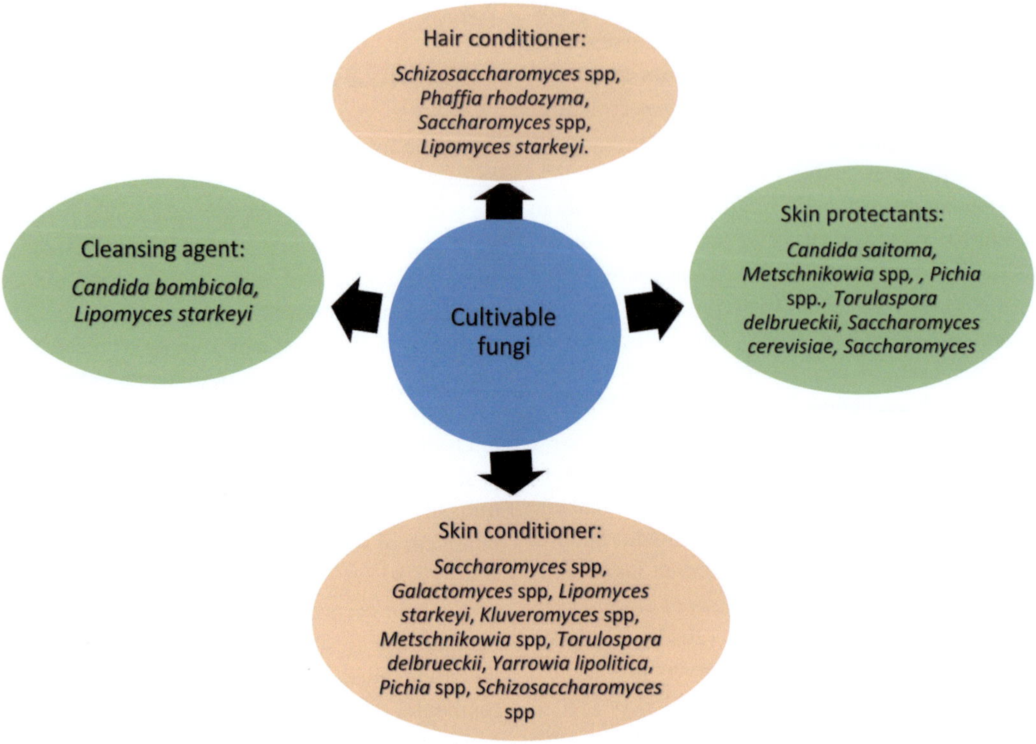

Fig. 2.1 Cultivable yeasts and applications as medicinal cosmetics

The use of fungi in medicine has a long history, leading to the development of an important drug, mevastatin (blood cholesterol-lowering agent and for cardiovascular problem) and antibiotics, penicillin from the fungus, *Penicillium* species (Mukherjee et al. 2024; Das et al. 2024; Phasha et al. 2022). Through intensive research, other drugs have been developed from fungi including ingredients utilized in the formulation of medicinal cosmetics, especially in the era of antimicrobial resistance.

Martins et al. (2023) evaluated the potential of *Butyriboletus regius*, *Ganoderma lucidum*, *Inonotus hispidus*, *Lanmaoa fragrans*, *Pisolithus tinctorius*, *Suillellus luridus*, *Suillellus mendax,* and *Xerocomus subtomentosus* as cosmetic ingredients. *Pisolithus tinctorius* and *Inonotus hispidus* had outstanding phenolic compounds, high anti-hyaluronidase property as well as high superoxide scavenging activity and potent antimicrobial activity especially against Gram-positive bacteria such as *Staphylococcus aureus* and *Enterococcus faecium*. Combination of *I. hispidus* and *P. tinctorius* in cream formulation promoted the antioxidant activity. *Lanmaoa fragrans* was reported to display the best antimicrobial activity against multi-drug-resistant microbes in comparison with other mushrooms. Metabolites from fungi are potential solution to the problem of antimicrobial resistance (Sułkowska-Ziaja et al. 2023).

2.5 Diversity of Cultivable Fungi Used in Medicinal Cosmetics

Nowadays, fungi and their by-products extensively explored for application in cosmetics either as pure compounds or crude extracts for treatment of skin-related conditions (Hyde et al. 2010).

These cultivable fungi used in medicinal cosmetics mostly belong to Ascomycota and Basidiomycota. Mushrooms have gained high popularity in the cosmetics industry; however, other fungi such as mold and yeast are also reported as potential producers of ingredients for medicinal cosmetics. Orally administered products with claims comparable to those of cosmeceuticals have recently been marketed under the names nutricosmetics, nutraceuticals, or cosmeceuticals. Fungal species that are used in cosmeceuticals typically have antioxidant, anti-inflammatory, and antimicrobial properties. These characteristics may aid in defending the skin from oxidative stress brought on by infection, inflammation, and free radicals. Some common fungal species used in cosmeceuticals include *Ganoderma lucidum, Pleurotus ostreatus, Trametes versicolor, Inonotus obliquus, Cordyceps sinensis, Auricularia auricula, Tremella fuciformis, Agaricus blazei, linteus,* and *Grifola frondosa.* These fungi can be used in topical products, like creams and serums, or in oral supplements (Mukherjee et al. 2024).

Among the mushrooms that have exhibited major enzymatic inhibition against elastase, collagenase, tyrosinase, and hyaluronidase are *Pleurotus ostreatus*, *Ganoderma lucidum*, and *Agaricus bisporus*. Hyaluronidase, elastase, collagenase, and tyrosinase are potent enzymes that facilitate skin-aging and hyperpigmentation. Bioactive compounds such as carotenoids, phenolic acids, polysaccharides, and vitamins from cultivable mushrooms protect the skin from UV damage and oxidative stress, wrinkle reduction, and improve skin elasticity and hydration. Mushrooms are also employed in the treatment of inflammation and skin infection due to their potential bioactive constituents (Rukhsar et al. 2025).

Extracts from fungi may be used singly or in combination in the cosmetics formulation including topical creams, serums, lotion, ointments, and facial preparation as cosmetic ingredients (Paterska et al. 2024). *Hypsizygus ulmarius* mycelium, *Ganoderma lucidium,* and *Cordyceps sinensis* were combined in the formulation of a skin relief face mask, a commercially available cosmetics beneficial in soothing, calming, and defending the skin from aging (Taofiq et al. 2016).

2.6 Cultivable Mushrooms Used in Medicinal Cosmetics

2.6.1 *Ganoderma lucidum*

About 80 species of polypore fungi in the genus *Ganoderma* belong to the family Ganodermataceae and majority of these species are found in tropical areas (Richter et al. 2015) and cultivable (Bijalwan et al. 2021). *Ganoderma lucidum* (Curtis) P. Karst. (Reishi or lingzhi mushroom) is a medicinal mushroom with a wide variety of bioactive components, exhibiting nutraceutical and pharmacological properties (Ahmad 2018). Its medicinal application is based on three potentials of *Ganoderma*: (1) no toxicity or adverse effects is associated with the mushroom, (2) no deleterious effect on any particular organ, and (3) improves the normalization of organ function (Valverde et al. 2015). In cancer treatment, *G. lucidum* is utilized either singly or in addition to contemporary medicine to boost the immune system, lessen the harmful effects of chemotherapy, and safeguard cellular DNA by boosting antioxidant capacity in cells (Wasser and Weis 2016). In addition, *Ganoderma* is useful in the treatment of diabetes, gut health maintenance, stimulates probiotics, and reduces hypertension and hypercholesterolemia. The fruit bodies of *Ganoderma* are dried, processed into a powder, and can be added to topical products, including masks, serums, and creams. *Ganoderma lucidum* fruit bodies are abundantly rich in polysaccharides, especially beta-glucans, which have antioxidant, anti-inflammatory, and antimicrobial properties. These properties can help to protect the skin from damage and promote a healthy, youthful appearance. The mushroom extracts stimulate hair follicles to regrow stronger, thicker hair, and inhibits dihydrotestosterone (male hormone that causes hair loss in males).

2.6.2 *Pleurotus ostreatus*

Pleurotus ostreatus (Jacq. ex Fr.) P. Kumm., a member of the family Pleurotaceae and division Basidiomycota, is known as "İstiridye Mantarı" in Türkiye and as "oyster mushroom" around the world (El-Ramady et al. 2022). It is one of the most highly valued edible species with a high general acceptability all around the world. *Pleurotus ostreatus*, the oyster mushroom, tastes like its bivalve namesake as well as copying its shape. It is very similar in texture, too—rather flaccid in comparison to the familiar *Agaricus* species such as field mushrooms. Documented reports have indicated the therapeutic potentials of *Pleurotus* species such as anti-aging, hematological, hepatoprotective, antihypertensive, antioxidative, anticancer, anti-inflammatory, antimicrobial, antiviral, and immunomodulatory functions of the mushroom (Dündar et al. 2024; Gul et al. 2024; Iqbal et al. 2024; Krishna et al. 2023; Muhaxi et al. 2023; Panda et al. 2021). Polysaccharides or pleuran produced by *Pleurotus* are involved in the suppression of skin aging by stimulation of keratinocytes to release zymogen pro-metalloproteinase-9 important in wound healing while the amino acids from the mushroom are essential for skin hydration (Paterska et al. 2024). In another study, Atawong reported the potential of *Pleurotus* as cosmetic ingredient.

2.6.3 *Inonotus* species

Black-brown *Inonotus obliquus* (I. obliquus) (Ach. ex Pers.) popularly known as chaga mushroom, a member of the Hymenochaetaceae family, is a parasitic fungus of the hardwood, birch tree (Miina et al. 2021; Dai 2012; Dai et al. 2009) but can be grown on artificial medium (Chen et al. 2024a). According to Niu et al. (2016) and Balandaykin and Zmitrovich (2015), *Inonotus obliquus* was utilized as a folk remedy in Siberia, Russia, and several other countries. Hippocrates' "Corpus Hippocraticum" is one of the earliest records attesting to the usage of its conks for therapeutic purposes (Szychowski et al. 2018). The fungus was used because of its gastrointestinal, anti-inflammatory, anti-tuberculosis, and antiparasitic qualities (Szychowski et al. 2018). It was also suggested for liver and cardiac conditions. It was most frequently utilized as aqueous macerates, infusions, or inhalations. Up until now, a wide range of characteristics of extracts or compounds obtained from *I. obliquus* have been described in relation to the therapeutic potentials, indicating antiviral, antidiabetic, antioxidant, antiparasitic, immunomodulatory, anti-inflammatory, neuroprotective, anticancer capabilities in vitro, and most recently, antifatigue effects (Ern et al. 2024; Szychowski et al. 2018; Wang et al. 2017). Additionally, *I. obliquus* containing antibacterial soaps were made for external use around Siberia, and chaga infusions were utilized to replace tea around the middle of the twentieth century. Polysaccharides from the mushroom are useful in the formulation of skin cosmetics. The mushroom has shown anti-aging and anti-melanogenic properties for the treatment of hyperpigmentation. Triterpenes produced by the fungus serve as potential anti-alopecia agent. It has been documented that lanosterol, inotodiol, lanost-8,24-diene-3β,21-diol, and trametenolic acid exhibited potential pro-proliferative effect more than the positive control, minoxidil, a known anti-alopecia agent (Elkhateeb et al. 2025). Triterpenes from chaga mushroom are new agent for stimulating hair growth, and have been used in hair shampoo to maintain healthy hair.

Inonotus hispidus (shaggy bracket fungus), a basidiomycete, produces hispidin, a yellow dye with potential application in cosmetics using submerged culture. The addition of a stressor (H_2O_2), light exposure, and caffeic acid/cinnamic/p-coumaric acid improves the production of hispidin. Hispidin synthesis is observed in the mycelium and fruit body of *Inonotus* culture. The product can be used as colorants in soap, body lotion, shampoo, and toothpaste. Colorant from *Inonotus hispidus* had no toxicological properties, however, hispidin colorant fades easily when compared to commercial textile dyes (Bergmann et al. 2022).

2.6.4 *Cordyceps* species

Ophiocordyceps sinensis (*Cordyceps sinensis*; Ophiocordicypetaceae, Ascomycetes) is generally known as the Chinese caterpillar mushroom or Dong-Chong-Xia-Cao (Cheng et al. 2018). *O. sinensis* is known as a medicinal species in Chinese, possessing a wide range of health-promoting and therapeutic properties, including immunomodulatory, antioxidative, anticancer, neuroprotective, and hepatoprotective effects. The rare and highly valuable fruiting body of the *O. sinensis* caterpillar mushroom is produced on an insect larva. Although *O. sinensis* has been used in several commercial cosmetic products, the benefits of *O. sinensis* for cosmetic purposes are still not extensively proven in the literature (Wu et al. 2016). Mycelial fermentation is commonly used for the commercial manufacture of fungal components.

Cordyceps militaris contains numerous bioactive compounds namely cordycepin, adenosin, polysaccharides, ergosterol, mannitol, and active phenolic compounds which have exhibited potential applications in pharmaceutical industry and medicine. The phenolic compounds and carbohydrates extracted from the solid-based residue (SBR or mushroom spent substrate) showed functional characteristics important in cosmetics. These functional properties include antioxidant and anti-tyrosinase properties, photoprotection, cell-proliferative effects as well as low cytotoxicity to fibroblast cell lines (Pintathong et al. 2021). The presence of phenolic compounds is an indicator of excellent antioxidant capacity which could be used in the formulation of cosmetics for anti-aging and anti-wrinkle. Polysaccharides and sugars in SBR extracts can be utilized as cosmetic ingredients for antioxidants while flavonoids exhibit tyrosinase-inhibitory function, indicating potential application as skin-whitening ingredients in cosmetic products. Other compounds detected by Pintathong et al. (2021) in SBR extracts were rutin and quercetin as well as nucleosides, sugars, phospholipids and alkaloids, organic acids, and vitamins, which are promising bioactive materials for the development of ingredients for medicinal cosmetics.

2.6.5 *Auricularia auricula*

Auricularia auricula-judae, commonly referred to as wood ear mushroom or black fungus, has been cultivated in China for over 2100 years because of its culinary and therapeutic qualities, particularly in the case of the Heimuer species (Yao et al. 2018). Additionally, *Auricularia* is a significant dietary fiber source (Cheung 2013). It has been observed that the majority of the carbohydrates in *Auricularia* are indigestible polysaccharides, namely mannans and β-glucans (Bandara et al. 2019). According to Yu et al. (2014) and Ma et al. (2018), compounds derived or extracted from *Auricularia* species have been used in anti-gastrointestinal cancer therapy, including treatments for colorectal, gastric, liver, pancreatic, and esophageal cancers. This suggests that *Auricularia* species may be able to treat disorders related to the stomach. It also has positive impacts on health due to its antitumor, anticoagulant, hypoglycemic, and cholesterol-lowering properties as well as a blood tonic (Yu et al. 2014).

2.6.6 *Trametes versicolor*

Trametes versicolor or *Cariolus versicolor* (Turkey tail mushroom) is a common polypore edible mushroom that may be found all over the world but mainly in Europe, Asia, North America, and United Kingdom (UK), with antioxidant, antimicrobial, antidiabetic, anti-aging, anti-inflammatory, antitumor, hepatoprotective, and immune-boosting potentials. The cultivable fungus contains polysaccharides, polysaccharopeptides, and terpenoids (Mago et al. 2023; Nguyen et al. 2021). *Trametes versicolor* has been widely studied for its potential use in various industries, including the development of bio-products for cosmeceuticals and nutricosmetics (Wasser and

Weis 2016). The versatile mushroom is composed of bioactive compounds that offer numerous health benefits and have shown promising results in skincare and dietary supplements. *Trametes versicolor* extract has gained attention in the cosmeceutical industry due to its antioxidant, anti-inflammatory, and anti-aging properties. Chou et al. (2019) reported that human skin keratinocytes were protected from free-radical-induced inflammation and oxidative stress using enzymatic hydrolysates gotten from the mushroom, *Trametes versicolor* polysaccharopeptides (PSPs). They demonstrated that enzymatic hydrolysates from *T. versicolor* PSPs by 80 U/ml β-1,3-D-glucanase (PSPs-EH80) resulted in enhanced stronger antioxidant and anti-inflammatory effects than PSPs through down-regulation of the extracellular signal-regulated kinase (ERK), P38 mitogen-activated protein kinase, and nuclear factor kappa-light-chain-enhancer of activated β-cells (NF-KB) signaling pathway. PSPs-EH80 was reported to exhibit strong antioxidant and anti-inflammatory activity which can be applied in medicine and cosmetics (Chou et al. 2019).

2.6.7 *Agaricus* species

The species of mushroom known as *Agaricus brasiliensis* or *Agaricus rufotegulis* is also referred to as almond mushroom, almond agaricus, mushroom of the sun, God's mushroom, mushroom of life, royal sun agaricus, jisongrong, or himematsutake (princess matsutake). *A. blazei* is an edible mushroom which has been extensively researched and utilized for its therapeutic and medicinal potentials. *A. blazei* and other species of *Agaricus* are natural sources of antioxidant compounds that possess hepatoprotective properties against liver injury (Yalimaimaiti et al. 2025; Al-Dbass et al. 2012). Ethanolic extract of *A. blazei* exhibited no cytotoxicity at a concentration of 100 μg/mL but showed toxicity on skin cell (decreased keratinocyte cell viability) at a concentration of 10 mg/mL extract. Inhibition of melanin synthesis, elevated collagen synthesis, and increased expression of aquaporin-3 synthase and hyaluronan-2,3 genes at 100 μg/mL of mushroom extract due to presence of the compound, cerebroside D (Paterska et al. 2024). Other bioactive compounds produced by *Agaricus* are cerebroside B, ergosterol, cerevisterol, 5-dihydroergosterol, adenosine, and benzoic acids, which play various roles. Inhibition of melanogenesis and improved collagen biosynthesis was exhibited by 5-dihydroergosterol, while cerebroside B, cerebroside D, and cerevisterol prevented production of nitric oxide in cells. Inhibition of tyrosinase, a key enzyme in the synthesis of melanin, has been reported in extract produced by extracts of *A. blazei* and *A. brasiliensis* (Paterska et al. 2024). *A. bisporus* (portbello) has antimicrobial activities and helps to protect the skin from sunburn (Mago et al. 2023).

2.6.8 *Volvariella volvacea*

Volvariella volvacea, commonly known as paddy straw mushroom, belongs to the family Pluteaceae, order Agaricales, and division Basidiomycota. The mushroom is cultivated in China and Southern Asia for its nutritional and medicinal properties. The fruiting body is rich in terpenes, flavonoids, antioxidant enzymes, phenolics, polypeptides, and sugars (Ali et al. 2024). Polysaccharides extracted from *V. volvacea* mushroom significantly enhanced skin elasticity, skin firmness, and skin moisturization but reduced skin roughness, skin melanin content, and skin scaliness, indicating its high potential as cosmetic ingredient for moisturizing, skin whitening, and anti-aging formulations. The researchers also reported good stability and efficacy of the facial gel-cream preparation containing *V. volvacea* polysaccharide after storage at 4 °C and 45 °C for 3 months (Pintathong et al. 2021; Sangthong et al. 2022).

2.6.9 Other Cultivable Mushrooms Useful in Medicinal Cosmetics

Grifola frondosa (Maitake mushroom), a wood decay basidiomycete, is an edible, nutritious mushroom inhabiting the hardwood species in Europe, Asia, and North America. It possesses high medicinal value such as antidiabetic, antioxidant, anti-inflammatory, antibacterial, antitumor, and hypocholesterolemic activities. The mushroom is produced mainly in Japan, China, and the USA. Maitake mushroom contains beta-glucans and polysaccharides, amino acids, selenium, zinc, vitamins, minerals, and proteins. In cosmetics, *G. frondose* has been used for promotion of hair growth, treatment of dandruff and dry scalp as well as scalp protection.

Schizophyllum commune (gillies or split gill mushroom), an edible basidiomycete fungus, commercially cultivated in Thailand, is explored due to its potential in exhibiting anticancer, anti-inflammatory, antibacterial, antioxidant, anti-aging and immunomodulating effects, skin moisturization and skin conditioning, all due to the beta-glucan polysaccharide, schizophyllan (Debnath et al. 2022; Kumar et al. 2022). It can serve as a thickening agent for food preservation and cosmetics. The mushroom is a white-rot fungus known for its production of essential metabolites (phenolics, ergosterol, and terpenoid compounds) and enzymes that have potential application in agriculture, food, and pharmaceutical industries (Kumar et al. 2022).

Lentinula edodes (Shiitake mushroom) is an edible, white-rot fungus in the family Fagaceae. The mushroom can be cultivated and is rich in bioactive compounds including lentinan, ergothioneine, polysaccharides, phenolic compounds, terpenes, alkaloids, flavonoids, steroids, glycosides, cytosine, and adenosine. Adenosine exhibited superior activity against arbutin in inhibiting melanogenesis signaling pathway, an indication that adenosine could be used as an effective skin-whitening agent. Similarly, ergosterol and cytosine have also been reported to prevent melanogenesis showing their likely potential application as skin-whitening agents. Strong hydrating and antioxidant activities have been attributed to trehalose produced by shiitake mushroom. Research studies have also demonstrated the efficacy of lentinan and ergothioneine in scavenging superoxide anion and free radicals reducing oxidative stress and associated inflammation suggesting the suitability of these compounds in formulating anti-aging cosmetics (Paterska et al. 2024; Mago et al. 2023). Shiitake mushroom is abundant in kojic acid and pantothenic acid which also prevent melanogenesis through tyrosinase inhibition. Extract from shiitake mushroom has no reported toxic effect and can be utilized effectively as cosmetic ingredients.

Phallus indusiatus (bamboo mushroom) exhibited nutraceutical properties (Theeraraksakul et al. 2023) by causing a significant reduction in tyrosinase activity, leading to a decrease in melanin content (anti-melanogenic effect). The mushroom extract can be used as a potential bioactive ingredient in formulating medicinal cosmetics.

Antrodia cinnamomca (Niu-Chang-Chin in Chinese) is a highly valued edible mushroom known for its medicinal potentials including hepatoprotective, antioxidant, anti-inflammatory, anticancer, anti-hypertensive, immune-protective, and neuroprotective effects. It produces bioactive compounds such as terpenoids, antroquinonol, and polysaccharides (Chen et al. 2021). It is a novel ingredient in cosmetics due to its antioxidant effect and tyrosinase inhibition. *A. camphorata* extract also exhibits anti-melanogenesis through tyrosinase inhibition, anti-photoaging, and antioxidant activities (Yang et al. 2024). The mushroom is cultivated by submerged fermentation and wood cultivation.

Phellinus linteus (willo bracket mushroom) is an edible fungus found in Europe, North America, China, Australia, Japan, Korea, and other Asian countries. The mushroom is composed of bioactive compounds such as sterols, phenolic compounds, terpenoids, and polysaccharides, which have been linked to the medicinal values of the mushroom. Antioxidant, anti-aging, hepatoprotective, and immunomodulatory effects are among the reported medicinal values exhibited by the mushroom extracts with potential application in cosmetics industry (Han et al. 2022). Key summary of some cultivable mushrooms is shown in Table 2.1.

Table 2.1 Cultivable mushrooms, bioactive compounds, and growth conditions

Fungi (mushrooms)	Picture	Bioactive compounds produced	Growth media and conditions	References
Ganoderma lucidium		Polysaccharide, proteins, flavonols, phenols, glycosides, triterpenoids, alkaloids	Potato dextrose agar (PDA), grains, saw dust, wood logs, incubation 25–30 °C, pH 4.5–5.5, humidity 60–95%	Paterska et al. (2024), Bijalwan et al. (2021), Subedi et al. (2021), Jayasinghe et al. (2008)
Pleurotus ostreatus		Polysaccharide (pleuran), free amino acids	PDA, wood logs, saw dust, coffee pulp straw, rice bran, and other lignocellulose-rich agricultural wastes, incubation 15–21 ° C, humidity 80–95%	Törős et al. (2022), Belletini et al. (2019), Chae and Ahn (2013)
Inonotus obliquus		Polysaccharides (xylose, rhamnose, mannose, galactose, and glucose), inotodiol, lanosterol, lanost-8,24-diene-3β,21-diol and trametenolic acid	PDA, malt extract agar (MEA), birch trees (in vivo cultivation on living trees)	Elkhateeb et al. (2025), Chen et al. (2024a), Thomas et al. (2020), Miina et al. (2021), Shen et al. (2024)
Cordyceps sinensis		Cordicepin	Sterile rice medium incubates at 9–13 °C for 40–60 days, 1–8 °C for 60–80 days, then 11-16 °C for fruiting body production on mixture of 50–60 kg cereal (brown rice), 10–20 g pupa and 50–60 ml distilled water in 1000 ml polypropylene bottle	Chellapandi and Saranya (2024), Mukherjee et al. (2024), Ha et al. (2024), Nxumalo et al. (2020), Kim et al. (2010)
Auricularia auricula		Melanin	Nutrient-rich agro-wastes: sawdust, rice bran, wheat bran and straw, maize cobs, palm substrates; Rice bran sucrose agar, potato dextrose agar, malt extract agar, sorghum, and paddy grain. Incubate at 20–25 °C, medium pH 5–7 and good air circulation	Regis and Geosel (2024), Akgul et al. (2017), Onyango et al. (2013)
Agaricus species *A. blazei* *A. subrufescens*		Polysaccharide, ergosterol, cerevisterol, cerebroside B and D, Adenosine, Benzoic acid	For cultivation of fruiting bodies, compost can be made from a mixture of the following: wood logs, saw dust, and agricultural residues such as oat, rice, and wheat straw, rice and wheat bran, sugarcane bagasse, tea leaves, reed plants, and animal manure (chicken and horse). The compost can be supplemented with gypsum, urea, ammonium nitrate, and molasses	Huang et al. (2022), Owaid et al. (2017), Alberto (1995)

Trametes versicolor		Polysaccharide-peptides	PDA, MEA incubated in the dark at 25 °C for 8–10 days to produce mycelia. Medium for mother spawn: wheat grain, gypsum, and calcium carbonate. Medium for spawn running and fruiting body: agricultural wastes—sugarcane bagasse and animal manure, 1% gypsum, 1% $CaCO_3$, dark incubation at 25 °C, 85–90% for 15–23 days (spawn running), then at 10-24 °C, humidity 95–100% in sufficient light (fruiting body)	Desisa et al. (2024), Nguyen et al. (2021), Akgul et al. (2017)
Grifola frondosa		Polysaccharides (β-glucans), amino acids, zinc, and selenium	PDA incubated at 25 °C for 21 days (mycelial growth). Spawn: cereal grains—corn, sorghum, rice, barley, moisture content 33–40%, incubation 25 °C for 25 days. Coffee spent ground for mycelial growth. Hardwood sawdust (beech, oak, cherry tree) plus cereal husk or bran (rice, wheat, corn), corn flour, sucrose, and calcium carbonate	Elkhateeb et al. (2025), Cui et al. (2024), Wu et al. (2021), Barreto et al. (2008)
Tricholoma matsutake		β-glucans, D-carnitine, α, α-trehalose, DL malic acid, quinic acid, erucamide	PDA incubated at 25 °C for 5–7 days (mycelial growth). Pine tree root-based mycorrhizal system, temperature 20–25, pH 5–7 and good air circulation (field cultivation). Bed cultivation is possible when the mycelia is cultured in sterilized soil with supplemented nutrients or other treatments (ongoing research)	Hu et al. (2024), Murata et al. (2023), Yamada (2022), Yamanaka et al. (2020)
Volvariella volvacea		Polysaccharides, linoleic acid	Culture media: PDA, MEA. Spawn media: sorghum, wheat, and rye grains, rice straw cuttings, cotton waste, used tea leaves, etc. substrate for fruiting body: rice straw, cotton waste, cassava peel, banana leaves, sugarcane bagasse, oil palm empty fruit bunch, etc. fortified with calcium carbonate, calcium sulphate. Temperature 30–35 °C, relative humidity (RH) 75–80%	Ali et al. (2024), Paterska et al. (2024), Amir et al. (2023), Thiribhuvanamala et al. (2012), Ukoima et al. (2009), Ahiawat and Tewari (2007)

(continued)

Table 2.1 (continued)

Fungi (mushrooms)	Picture	Bioactive compounds produced	Growth media and conditions	References
Lentinula edodes		Ergothioneine, polysaccharides (lentinan), lentinamycin, adenosine, cytosine, trehalose	Substrates for fruiting body cultivation: wood logs, saw dust and agricultural residues (wheat bran and straw, rice bran, sugarcane bagasse), sucrose, calcium carbonate, gypsum, pH 5.5–6.0, moisture 60–65%, incubation 20–24 °C in the dark (ramification), then 12–22 °C, RH 85–90% in dark and finally, exposure to light for fruiting body production	Song et al. (2025), Paterska et al. (2024); Huang et al. (2022), Garcia et al. (2022), Sheng et al. (2021), Zhang (2019)
Schizophyllum commune		Polysaccharide (schizophyllan), phenols	Potato dextrose agar, sawdust, rice bran and straw, wheat bran, etc. incubation in the dark at 25–35 °C, pH 5.5, moisture 65–70%, humidity 50–55%. Solid-state or submerged fermentation	Garcia et al. (2022), Debnath et al. (2022), Kumar et al. (2022), Imtiaj et al. (2008)

2.7 Cultivable Molds Used in Medicinal Cosmetics

Filamentous fungi have been explored for the production of medicinal cosmetics. Species of *Aspergillus*, *Penicillium*, *Fusarium*, *Rhizopus*, *Monascus*, *Curvularia*, *Neurospora*, *Mucor*, *Alternaria*, etc. (Table 2.2) are among cultivable molds isolated from terrestrial, marine environment, or as plant endophyte (Mukherjee et al. 2024; Phasha et al. 2022). Some molds are natural source of bioactive pigments for cosmetic applications. Species of *Monascus* can produce orange, red, and yellow pigments; *Talaromyces* (red pigments), and *Neurospora* (carotenoids), which are potentially useful in cosmetic formulation (Gmoser et al. 2017). *Aspergillus* species are the main producers of the bioactive compounds,

Table 2.2 Cultivable mold-based bioactive compounds and their functions in medicinal cosmetics

Fungi	Bioactive compound	Function in medicinal cosmetics
Aspergillus sp. *A. chevallieri*	Asperisocoumarin A and C. Echinulin and neoechinulin A. Lactic acid	Antioxidant property Echinulin has high antioxidant and UV-photoprotective activity Lactic acid hydrates and smoothens the skin, anti-aging potential
Aspergillus oryzae *A. nidulans* *A. giganteus* *A. parasiticus* *A. flavus*	Penicillins Glucosylceramides Galactosylceramides	Antimicrobial activity Promotion of gut health as prebiotics, and improvement of skin barrier function (Jiang et al. 2021)
Rhizopus sp.	Lactic acid γ-linolenic acid	Skin moisturizer; softens dry, flaky skin; skin-whitening and anti-aging γ-linolenic acid is anti-inflammatory and facilitates healthy skin
Mortierella sp.	Eicosapentaenoic acid and γ-linolenic acid	Omega fatty acid used in anti-aging and promotes healthy skin
Microsporums sp.	Chrysophanol	Skin-whitening agent
Penicillium sp. *P. citrinum*	Penicilloitins A and B. Penicitriketo	Antiaging property Antioxidant activity
P. chrysogenum *P. baculatum* *P. turbatum*	Penicillin	Penicillin inhibits both gram-positive and negative bacteria
Cephalosporium sp. *C. acremonium* *C. caeruleus*	Cephalosporins Fumigacin (helvolic acid)	Antimicrobial activity Fumigacin inhibits gram-negative bacteria
Sporotrichum pruinosum	Produce melanocytic enzymes	Skin-whitening by effective enzymatic degradation of skin melanin instead of inhibition of melanin synthesis (activates depigmentation of skin) (Mohorčič et al. 2007)
Mucor miehei	Extract of protease enzyme	Anti-aging property, gentle exfoliation, and skin whitening
Fusarium sp.	Gallic acid	Elimination of free radicals
Acremonium sp.	Hydroquinonone derivates	Antioxidant activity (Fonseca et al. 2023; Uppala 2015)
Exophiala sp.	Circumdatin 1, Circumdatin C, and Circumdatin G.	UV-A protecting benzodiazepine alkaloid (Zhang et al. 2008)
Paecilomyces sp.	Paecilonic acids A and B.	Anti-aging potential
Talaromyces wortmannii	Fungal crude extract	Anti-inflammatory activity, and antimicrobial activity against *Propionibacterium acnes* (Pretsch et al. 2014)
Eurotium rubrum	Coumarin derivative	Inhibit tyrosinase activity

and *A*. oryzae has GRAS (generally regarded as safe) status.

Aspergillus species grown on corn kernels are known to produce kojic acid. Kojic acid is used in cosmetics due to its depigmentation activity, antibacterial, and antioxidant activities. Kojic acid also inhibits the activity of tyrosinase, an enzyme known to control melanin synthesis responsible for skin color, and its inhibition promotes skin-whitening (Phasha et al. 2022). Tyrosinase inhibition is achieved by blocking the enzyme's active site, thus preventing its binding with the reaction substrate. Notorious tyrosinase inhibitors such as hydroquinone used in cosmetics are associated with disruption of stratum corneum, causing exposure to melanocytes followed by mutagenic effect on mammalian cells due to increased ultraviolet (UV) exposure (Tang et al. 2024; Sułkowska-Ziaja et al. 2021). Reduced superficial protection of the skin exposes the deeper skin cells such as melanocytes to UV radiation and possibly increases the possibility of skin cancer. Hydroquinone also reduces the protective property of the skin, reducing its integrity against physical, and mechanical assault as well as its role in maintaining homeostasis. Such fragile skin is vulnerable to trauma, pathogens, toxins, and UV rays (Owolabi et al. 2020; Watson et al. 2016; Li et al. 1996). Hydroquinone usage has been banned in some countries to reduce the occurrence of skin cancer (Owolabi et al. 2020). The use of 1–2% kojic acid for skin whitening is safe and does not show any allergenic effect (Phasha et al. 2022).

Apart from kojic acid, glycosphingolipid (GSL) is another substance involved in crucial biological processes involving the barrier function of the skin in moisture retention. Glycosphingolipid (GSL) is divided into two: neutral GSL and acidic GSL. Neutral GSL, known as glycosylceramides including glucosylceramide (GlcCer) and galactosylceramide (GalCer), has been commercialized as dry skin dietary supplement or skin moisturizing agent. GlcCer is known for its moisture-holding ability in cosmetics and maintenance of dermal elasticity by elevating dermal fibroblast proliferation and collagen production. It is worth to note that glycosylceramides synthesized by plants contain only GlcCer but lack GalCer (Jiang et al. 2021). In animals, GalCer is restricted to neural tissues, and its extraction for use in cosmetics is not acceptable due to the risk of prion diseases. However, filamentous fungi such as *Aspergillus oryzae*, *A. sojae*, *A. awamori*, *A. nidulans*, *A. versicolor*, and *A. nidurans* have been reported to produce both GlcCer and GalCer as well as production of a new glycosylceramides compound, Aspercerebroside A (anti-inflammatory compound) from *Aspergillus* species (Chen et al. 2024b; Jiang et al. 2021). Glycosylceramides from *Aspergillus* species can promote gut health as prebiotics, and enhance skin barrier function (Jiang et al. 2021).

2.8 Cultivable Yeasts Used in Medicinal Cosmetics

2.8.1 *Saccharomyces cerevisiae*

Antimicrobial peptides from *S. cerevisiae* exhibited antimicrobial, anti-collagenase, antioxidant, and anti-inflammatory effects, an indication of potential cosmetic bioactive ingredients, some of which have been reported safe for cosmetics. The fungal product can be included in formulation of dermal care products and supports wound healing process (Maurício et al. 2024; Costa et al. 2023).

Glucan obtained from *S. cerevisiae* has been reported to be sustainable and exhibit safety profiles for skincare and cosmetics ingredients, contributing to circular economy (Sousa et al. 2024). Beta-glucans from *S. cerevisiae* activate and enhance production of collagen and elastin in normal human dermal fibroblast (Srisukchayakul et al. 2024). Phenol from spent extracts of handcrafted beer (yeast extract and spent hop extract) improved the activity of mitochondria, inhibiting oxidative stress in human keratinocyte cells and could serve as potential ingredient for skin anti-aging cosmetics (Censi et al. 2021).

Yeast extract is mainly produced from waste brewer's yeast, known to be rich in proteins, amino acids, nucleotides, sugars, and trace elements. Yeast extract can be used as cosmetic ingredients (Tao et al. 2022).

Apart from *S. cerevisiae*, other species of cultivable yeasts have produced bioactive extracts with potential applications in cosmetics. These yeasts include *Galactomyces* species, *Yarrowia lipolytica*, *Torulaspora delbrueckii*, *Schizosaccharomyces pombe,* etc. for skin conditioning either as emollient or humectant or miscellaneous.

Biosurfactants produced by *Scheffersomyces shehatae* were investigated by Cedrola et al. (2024) for cosmetic application, indicating the non-irritant, non-toxic, tolerable, and safe cleaning characteristics which are comparable to the cleaning activity of micellar water and 1% SDS solution for foundation products and lipsticks. *Rhodotorulla* species, a UV-mutated marine yeast produced high beta-glucan, which exhibited good wound-healing, anti-aging, antimicrobial, and antioxidant activities which can be applied in cosmetic industry. The identified compound produced by *Rhodotorulla* sp., acid methyl esters used to formulate beta-glucan almond scrub, had good effect on skin, without side effects (Sarkar et al. 2024).

Using *Yarrowia*-based refinery, two species *Yarrowia lipolytica* and *Y. divulgata* are useful in producing bio-emulsifiers and cosmetic ingredients, Eszterbauer & Nemeth (Eszterbauer and Németh 2023) reported the conversion of glycerol to erythritol, bioemulsifiers through fermentation and cell lysates were associated with skin hydrating activity. Other cultivable yeasts are shown in Fig. 2.2.

2.9 Fungal Cultivation Methods

Generally, three cultivation methods are used for production of mushrooms (fruiting bodies). They are bottle culture, bag culture, and out-door bed culture/field culture (Wu et al. 2021). Agricultural and industrial waste are used for fungi cultivation (Table 2.1; Fig. 2.2). Bag culture requires substrate preparation, sterilization, mycelium inoculation, and incubation. Bag culture is mostly used in mushroom cultivation due to the following benefits:

(a) Low cost of plastic bags.
(b) Requirement of small space.
(c) In-door environment is easy to control.
(d) High mushroom yield can be achieved in a shorter time than in field culture.

Apart from mushroom fruiting bodies, fungal mycelia and their metabolites are also in high demand for application in cosmetics. Two major methods used in the cultivation of fungal mycelia and their metabolites are submerged fermentation (SMF) and solid-state fermentation (SSF). It is worth to note that the cultivation of fungi in artificial environmental conditions may affect the fungal composition and activities of fungal enzymes (Jang et al. 2005). Thus, the use of supplements or boosters in fungal cultivation medium to enhance yield of desired fungal product. Such boosters include CaCO3, CaCl2, KH2PO4, NaCl, Na2HPO4 (Thiribhuvanamala et al. 2012), and metabolites from other microbes (Choi et al. 2023). In addition, treatments such as soaking of substrate, adjustment of temperature of incubation, relative humidity, and moisture content also influence fungal performance (Thiribhuvanamala et al. 2012).

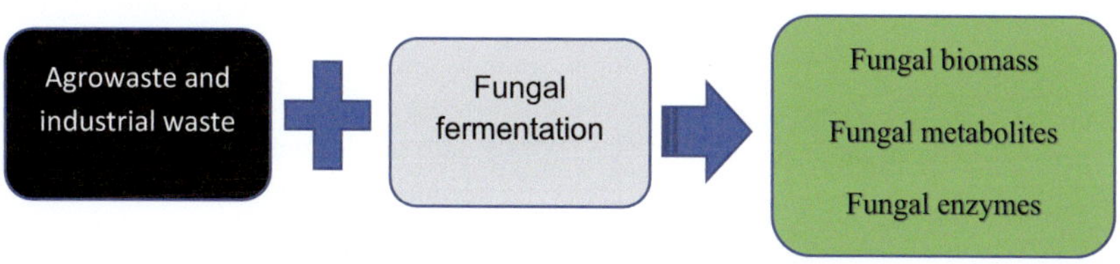

Fig. 2.2 Cultivable fungi valorization of waste into products for medicinal cosmetics

2.10 Submerged Cultivation of Fungi

This involves growing the fungi in broth or liquid medium with or without agitation. The disadvantage of submerged fermentation (SMF) is the existence of the product in dilute form due to the enormous quantity of effluent generated in SMF. It is efficient and gives higher mycelial yield in a shorter time, requires smaller plant space, and supports effective product quality control. Processes such as filtration, centrifugation, ultrafiltration, and concentration are required for cell harvest and concentration of fungal products.

The SMF has been employed in the production of active peptides by *S. cerevisiae* which was incubated for 7 days at 25 °C without agitation, followed by centrifugation to remove yeast cells and sterilization of the supernatant using membrane filters of 0.45 μm and 0.22 μm. The cell-free supernatant underwent ultrafiltration and concentration using special membranes (Maurício et al. 2024).

2.11 Solid-State Fermentation

Solid-state fermentation requires growing the fungi on solid medium/substrate. Cheap and available substrates such as rice bran, wheat bran, and other agricultural/industrial waste which might be supplemented with sucrose, gypsum, or calcium carbonate are used for the cultivation of fungi. Compared to SMF, SSF is less capital intensive and energy requirement, effluent generation is low and fungal product gotten is concentrated (Kumar et al. 2021). The drawback associated with SSF is heat accumulation and contamination *C. militaris* has been cultivated with defatted rice bran fortified with wheat or various cereal grains for the production of cordycepin (Borde and Singh 2023).

Other processes required in the production of fungal products for medicinal cosmetics are summarized in Fig. 2.3 and also documented by Anaele and Nwafor (2025).

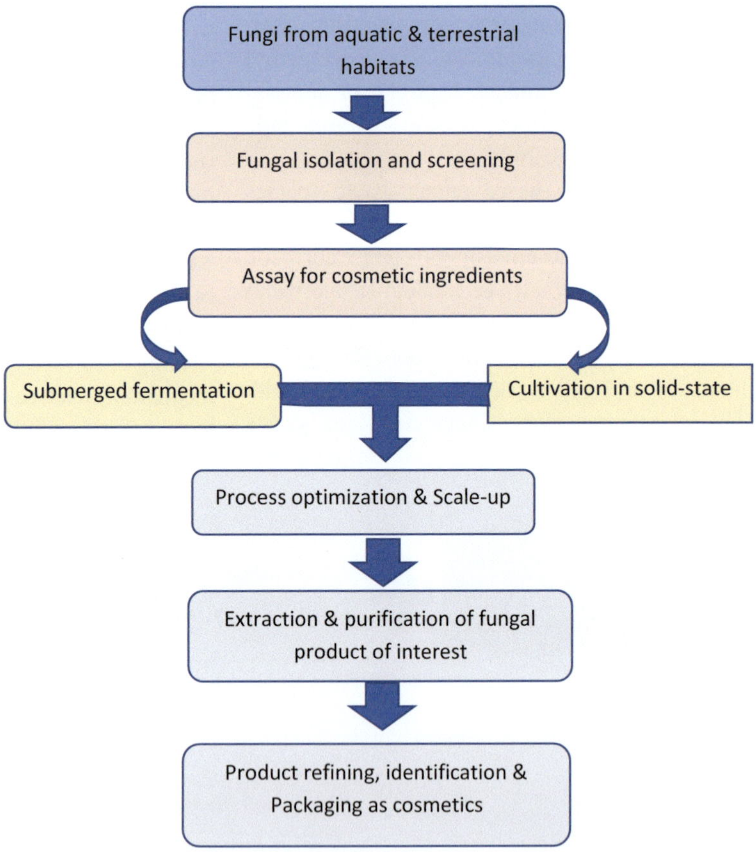

Fig. 2.3 Stages in the production of fungal products for medicinal cosmetics

2.12 Challenges of Fungal Cultivation for Medicinal Cosmetics and Solutions

Fungal cultivation is affected by various environmental factors that affect growth, development of fruiting body, and production of fungal metabolites. With the increased awareness of the benefits of natural cosmetic products, there has been a renewed interest in fungal extracts due to their beneficial bioactivity, cost-effective sustainability, and expansion of the market of fungal products.

There are more than 2000 species of medicinal fungi, each with a particular growth requirement and unique requirement for fruiting body development and production of metabolites. This requires special training and detailed study of each wild fungus with identified therapeutic value. Fungi with potential bioactive compounds are identified from the wild and studied for domestication and optimization of the product of interest (Desisa et al. 2024; Walker et al. 2023; Regis et al. 2024; Jang et al. 2005). Such studies capture the effect of temperature, carbon dioxide, light, nutrient composition of growth medium, and cultivation techniques. Heat treatment has been reported to promote fungal polysaccharide and biomass production (Sengkhamparn and Phonkerd 2014). The fungal cultivation studies require funding, a possible limitation to research on fungal cultivation, development, and commercialization.

Scientific and technical details are required for successful fungal cultivation. A special group of fungi, ectomycorrhizal fungi (EMF), requires special condition for their artificial cultivation. For example, a highly prized mushroom, *Tricholoma matsutake,* is an EMF that has symbiotic relationship with host trees, *Quercus serrata*, *Quercus phillyraeoides*, and *Pinus densiflora* leading to the formation of a fairy ring called "Shiros" around the host plants (Yamanaka et al. 2020). Researchers have explored bed cultivation (without symbiosis) and field cultivation (6 years to grow mushroom without a shiro/no symbiosis) to artificially grow *T. matsutake* fruiting body (Choi et al. 2023; Yamanaka et al. 2020). However, Choi et al. (2023) reported that *Penicillium citreonigrum* isolated from fairy ring (shiros) promoted the growth of *T. matsutake* and its enzyme activity, indicating that the micro-fungus could contribute to the artificial cultivation of the pine mushroom. Fungal cultivation requires special training, thorough research, and funding.

In another study, Desisa et al. (2024) reported the domestication of a popular medicinal fungus, wild turkey tail mushroom (*Trametes versicolor*) from Ethiopian forests using agro-industrial byproducts composed of sugarcane bagasse (80%), horse manure (12%), and poultry manure (8%) with optimal carbon to nitrogen ratio of 31:1 as the best substrate for good mycelial growth, short colonization time, reduced pinhead formation time, and substantial fruiting body yield. The formulation of adequate growth substrate and selection of culture conditions are essential in fungal cultivation. However, the choice of substrate may depend on the cost, and availability of agro-industrial byproduct which differs in worldwide availability, another critical issue in fungal cultivation. Thus, the unavailability of substrates like horse manure, tamarind seed powder, mustard straw, cotton waste, rice straw, and oil palm empty fruit branch known to be suitable for mushroom cultivation in some regions might lead to research to find alternative local substrates for cultivation of mushroom in region where such substrates are not available (Boukary et al. 2024; Umor et al. 2020).

There is no general condition for mycelial growth and fruiting body formation. Thus, research on substrate composition is necessary for each mushroom species. Mushroom cultivation method is developed based on the available local substrate feedstock suitable for each mushroom species. It is also established that a slightly higher nitrogen concentration is required for fruiting body formation than for mycelial growth.

The ratio of cellulose to lignin in substrate is positively correlated to rate of mycelial growth and mushroom yields where presence of high lignin and volatile compounds encourages nematode and mold attacks as reported in *V. volvacea* cultivation (Umor et al. 2020).

The cultivation of fungi has been influenced by the following factors:

(a) Inadequate knowledge of fungi and fungal cultivation.
(b) Lack of trained personnel in fungal cultivation.
(c) Mushroom spawn conservation.
(d) Scarcity of mushroom spawn suppliers.
(e) Inadequate post-harvest technology for fungal packaging and preservation.
(f) Unorganized fungal sector and funding agencies.
(g) Inadequate knowledge of fungal physiology including growth parameters.
(h) Availability of lignocellulosic agro-industrial waste.

Fungal cultivation is affected by the following environmental factors (Boukary et al. 2024; Regis et al. 2024; Walker et al. 2023; Yamada 2022; Jiang et al. 2021; Khan and Chandra 2017):

(a) Media composition.
(b) Temperature.
(c) Humidity.
(d) Concentration of carbon dioxide.
(e) Light.
(f) pH and salinity.
(g) Synergistic association with other microorganisms.

The cultivation of mushroom for production of fruiting body might be influenced by the presence of contaminants, including microbial contamination due to incomplete pasteurization, leading to loss of yield. In addition, the low quality of spawn, insect infestation, low oxygen and high carbon dioxide, fungal and other microbial contaminants might lead to death at the early stage of mushroom cultivation. The growth of the mushroom may be suspended due to light deficiency and low-quality spawn (Umor et al. 2020). It has been documented that fungal strain, substrate chemistry, stages of fungal development, and ecological interactions result in diverse variations of fungal metabolites both in fruiting bodies and mycelia (Berger et al. 2022).

Risks associated with the use of some cultivable fungi as medicinal cosmetics include carcinogenic effects, neurological damage, endocrine disruption, and possible harm to beneficial skin microbiota, rise in contact allergens, and microbial contamination (Mukherjee et al. 2024). Some fungi produce metabolites that can cause adverse health effects in humans. For example, kojic acid has potential bioactivity but there is restriction (1–2%) in its use as cosmetic ingredient due to contact dermatitis, photodamaging effect on the skin, and carcinogenic effect when used on damaged skin (Phasha et al. 2022). Assessment of the safety and toxicity of the fungal bioactive compound should be ascertained before inclusion in cosmetic formulations.

There were challenges associated with regulation and marketing of fungal products used in medicinal cosmetics by different international bodies: the US Food and Drug Administration (FDA), the European Medicines Agency (EMA) in Europe, and Health Canada. In the USA, cosmetic products and ingredients do not require FDA's approval before they are introduced into the market. However, medicinal cosmetics with therapeutic claim are classified as drug and therefore evaluated by the FDA Center for Drug Evaluation and Research (Krause and Tobin 2013). The European Union Cosmetic Regulation (EC No. 1333/2009), which is relevant for natural cosmetic ingredients, considered: (a) requirements for safety testing, (b) ban on animal testing for cosmetic ingredients, and (c) requirement for cosmetic ingredients with restrictions and permissible ingredients (Manful et al. 2024). Health Canada defined cosmetics as "any substance or mixture of substances, manufactured, sold or represented for use in cleansing, improving or altering the complexion, skin, hair or teeth and includes deodorants and perfumes" (Krause and Tobin 2013). Scientific validation and pharmacological studies are necessary to ensure the safety, efficacy, and global standard of medicinal cosmetics.

2.13 Solutions to Poor Mushroom Yield

(a) Spawn production and mushroom cultivation must be developed to suit local context.
(b) The problem of contamination leading to poor fruiting body yield can be addressed by including a pretreatment process (biological or chemical method) during composting for mushroom cultivation.
(c) Organic supplementation such as addition of carbon and nitrogen plus micronutrients to increase the enzymatic activities and enhance formation of fruiting bodies.
(d) Use of bioinoculant. For example, inoculation of *Penicillium citreonigrum* improved mycelial growth of the mushroom, *Tricholoma matsutake* (Choi et al. 2023).
(e) Use of stable mutant mushroom strain as demonstrated in *T. matsutake* bed cultivation where the mutant increased degradation of cellulose and amylase (Yamanaka et al. 2020). Mushroom cultivation can also be improved by inserting multifunctional cellulase genes in the mushroom (*V. volvacea*) through transgenic strain to increase cellulase activities and fruiting body yield (Zhao et al. 2010).
(f) Determination of the safety and toxicology profile of the fungal-based cosmetic ingredient before use.
(g) Establishment of cooperative groups for fungal cultivation to reduce the problem of spawn production and preservation especially in developing countries (Boukary et al. 2024).

2.14 Conclusion

Cultivable fungi are increasingly gaining popularity in medicinal cosmetics due to the associated bioactive fungal compounds laced with numerous health benefits. *Ganoderma*, *Cordyceps*, *Inonotus*, *Agaricus*, *Grifola*, *Lentinula*, *Pleurotus*, *Volvariella*, *Auricularia* (cultivable mushrooms), *Aspergillus* species (mold), and *Saccharomyces* (yeast) are among the fungi with high product potential for medicinal cosmetic ingredients. Ergosterol, lactic acid, and ceramides promote skin health as anti-inflammatory, antibacterial properties, photoprotective effect, or serve as epidermal hydrating agents. Extracts of cultivable fungi play key role in inhibition of major enzymes such as tyrosinase, hyaluronidase, elastase, and collagenase which are essential in skin hyperpigmentation and aging. They offer promising effects in the promotion of hair growth and protection against hair breakage and alopecia. The launching of commercially produced fungal-based cosmetics in the global market is evidence of the detailed research output on fungal-based medicinal cosmetics. Currently, the global market for cultivable fungal-based cosmetics is growing due to the global awareness of adverse effects of synthetic products, and the high health and skincare benefits of fungal-based products. Cultivable fungi can promote public health and the economy of both the developed and developing countries. Most factors affecting growth, development of fruiting body, and production of fungal metabolites as well as the solutions should be properly addressed and fixed. Funding of fungal research and establishment of cultivable fungi cooperative groups should be encouraged, especially in developing countries. Cultivable fungi can provide natural, safe, effective, and sustainable ingredients for medicinal cosmetics.

References

Ahlawat OP, Tewari RP (2007) Cultivation technology of paddy straw mushroom (Volvariella volvacea)

Ahmad MF (2018) Ganoderma lucidum: persuasive biologically active constituents and their health endorsement. Biomed Pharmacother 107:507–519

Akgul H, Sevindik M, Coban C, Alli H, Selamoglu Z (2017) New approaches in traditional and complementary alternative medicine practices: Auricularia auricula and Trametes versicolor. J Tradit Med Clin Natur 6(2):239

Alberto E (1995) Studies on fruitbody production of naturally occurring species of Agaricus on a. bisporus compost. Indian J Mycol Plant Pathol 25(3):266–269

Al-Dbass AM, Al-Daihan SK, Bhat RS (2012) Agaricus blazei Murill as an efficient hepatoprotective and antioxidant agent against CCl4-induced liver injury in rats. Saudi J Biol Sci 19(3):303–309

Ali S, Yousaf N, Usman M, Javed MA, Nawaz M, Ali B et al (2024) Volvariella volvacea (paddy straw mushroom): a mushroom with exceptional medicinal and nutritional properties. Heliyon 10(21):e39747

Amir NF, Mohd-Aris A, Mohamad A, Abdullah S, Yusof FZ, Umor NA (2023) Spawn production and cultivation technology for Volvariella volvacea: a perspective. Food Res 7(s4)

Anaele EN, Nwafor FI (2025) Alkaloids of forest fungi: diversity and therapeutic applications. In: Forest Fungi. Academic Press, pp 343–367

Atawong P, Kriangkrai W (2024) In vitro study on skin-related bioactivity of Pleurotus cystidiosus extract. Doctoral dissertation, Naresuan University

Balandaykin ME, Zmitrovich IV (2015) Review on chaga medicinal mushroom, Inonotus obliquus (higher basidiomycetes): realm of medicinal applications and approaches on estimating its resource potential. Int J Med Mushrooms 17(2):95

Bandara AR, Rapior S, Mortimer PE, Kakumyan P, Hyde KD, Xu J (2019) A review of the polysaccharide, protein and selected nutrient content of Auricularia, and their potential pharmacological value. Mycosphere J 10(1):579–607

Barreto SM, López MV, Levin L (2008) Effect of culture parameters on the production of the edible mushroom Grifola frondosa (maitake) in tropical weathers. World J Microbiol Biotechnol 24:1361–1366

Bellettini MB, Fiorda FA, Maieves HA, Teixeira GL, Ávila S, Hornung PS et al (2019) Factors affecting mushroom Pleurotus spp. Saudi J Biol Sci 26(4): 633–646

Berger RG, Bordewick S, Krahe NK, Ersoy F (2022) Mycelium vs. fruiting bodies of edible fungi – a comparison of metabolites. Microorganisms 10(7):1379

Bergmann P, Takenberg M, Frank C, Zschätzsch M, Werner A, Berger RG, Ersoy F (2022) Cultivation of Inonotus hispidus in stirred tank and wave bag bioreactors to produce the natural colorant hispidin. Fermentation 8(10):541

Bijalwan A, Bahuguna K, Vasishth A, Singh A, Chaudhary S, Dongariyal A et al (2021) Growth performance of Ganoderma lucidum using billet method in Garhwal Himalaya, India. Saudi J Biol Sci 28(5): 2709–2717

Borde M, Singh SK (2023) Enhanced production of cordycepin under solid-state fermentation of Cordyceps militaris by using combinations of grains/ substrates. Braz J Microbiol 54(4):2765–2772

Boukary AA, Olou AB, Piepenbring M, Yorou NS (2024) Mushroom cultivation in tropical Africa: successes, challenges, and opportunities. J Agric Food Res:101264

Cedrola CC, Barbosa FG, Marcelino PRF, da Silva NP, Tavares GD, Barradas TN et al (2024) Unlocking the beauty benefits: exploring biosurfactants from Scheffersomyces shehatae for cosmetics. J Surfactant Deterg 27(5):801–811

Censi R, Vargas Peregrina D, Gigliobianco MR, Lupidi G, Angeloni C, Pruccoli L et al (2021) New antioxidant ingredients from brewery by-products for cosmetic formulations. Cosmet 8(4):96

Chae HJ, Ahn JH (2013) Optimization of rice bran and food waste compost contents in mushroom culture medium to maximize mycelial growth rate and fruit body yield of Pleurotus ostreatus. Int Biodeterior Biodegradation 80:66–70

Chellapandi P, Saranya S (2024) Ophiocordyceps sinensis: a potential caterpillar fungus for the production of bioactive compounds. Expl Res Hypothes Med 9(3):236–249

Chen HY, Cheng KC, Wang HT, Hsieh CW, Lai YJ (2021) Extracts of Antrodia cinnamomea mycelium as a highly potent tyrosinase inhibitor. J Cosmet Dermatol 20(7):2341–2349

Chen HJ, Chen YS, Lin KM, Tsai SW, Liao MJ, Yeh CS, Liu SL (2024a) Comparison of the controlled atmosphere treatment for submerged and solid-state fermentation of Inonotus obliquus. Foods 13(14):2275

Chen YH, Zhu Q, Li J, Yang R, Zhang J, You M et al (2024b) Optimization of fermentation process for new anti-inflammatory Glycosylceramide metabolite from aspergillus sp. Meta 14(2):99

Cheng WY, Wei XQ, Siu KC, Song AX, Wu JY (2018) Cosmetic and skincare benefits of cultivated mycelia from the Chinese caterpillar mushroom, Ophiocordyceps sinensis (ascomycetes). Int J Med Mushrooms 20(7):623

Cheung PCK (2013) Mini-review on edible mushrooms as source of dietary fiber: preparation and health benefit. Food Sci Human Wellness 2:162–166

Choi DH, Han JG, Lee KH, Gi-Hong A (2023) Promotion of Tricholoma matsutake mycelium growth by Penicillium citreonigrum. Mycobiology 51(5):354–359

Chou CH, Tsai MS, Lu HY, Chang CK, Cheng KC, Jhan MH, Hsieh CW (2019) Enzymatic hydrolysates obtained from Trametes versicolor polysaccharopeptides protect human skin keratinocyte against AAPH-induced oxidative stress and inflammatory. J Cosmet Dermatol 18(6):2011–2018

Costa EM, Oliveira AS, Silva S, Ribeiro AB, Pereira CF, Ferreira C et al (2023) Spent yeast waste streams as a sustainable source of bioactive peptides for skin applications. Int J Mol Sci 24(3):2253

Cui FJ, Yang YM, Sun L, Zan XY, Sun WJ, Zeb U (2024) Grifola frondosa polysaccharides: a review on structure/activity, biosynthesis and engineering strategies. Int J Biol Macromol 257:128584

Dai YC (2012) Polypore diversity in China with an annotated checklist of Chinese polypores. Mycoscience 53(1):49–80

Dai YC, Yang ZL, Cui BK, Yu CJ, Zhou LW (2009) Species diversity and utilization of medicinal mushrooms and fungi in China. Int J Med Mushrooms 11(3)

Das S, Das M, Nath R, Nath D, Patra JK, Talukdar AD (2024) Fungal metabolites and their importance in pharmaceutical industry. In: Entrepreneurship with microorganisms. Academic Press, pp 89–120

Debnath S, Hore S, Das P, Saha AK (2022) The evaluation of selected bioactivities of fungal mycelium of a wild Schizophyllum commune Fr. from submerged cultures. J Microbiol Biotechnol Food Sci 12(2): e3250–e3250

Desisa B, Muleta D, Jida M, Dejene T, Goshu A, Negi T, Martin-Pinto P (2024) Domestication of wild-growing Turkey tail mushroom (Trametes versicolor) from Ethiopian forests on augmented agro-industrial byproducts. Mycol Prog 23(1):62

Dündar A, Yalçın P, Arslan N, Acay H, Hatipoğlu A, Boğa M et al (2024) Effect of Pleurotus ostreatus water extract consumption on blood parameters and cytokine values in healthy volunteers. J Am Nutr Assoc 43(8):645–652

Elkhateeb W, Ishihara A, Daba G (2025) Mushrooms natural products cosmetic weapons as anti-hair loss, anti-ageing skin and future prospect. Journal ISSN 2766

El-Ramady H, Törős G, Badgar K, Llanaj X, Hajdú P, El-Mahrouk ME et al (2022) A comparative photographic review on higher plants and macro-fungi: a soil restoration for sustainable production of food and energy. Sustainability 14(12):7104

Ern PTY, Quan TY, Yee FS, Yin ACY (2024) Therapeutic properties of Inonotus obliquus (Chaga mushroom): a review. Mycology 15(2):144–161

Eszterbauer E, Németh Á (2023) Investigations for a Yarrowia-based biorefinery: in vitro proof-of-concept for manufacturing sweetener, cosmetic ingredient, and bioemulsifier. Fermentation 9(9):793

Fonseca S, Amaral MN, Reis CP, Custódio L (2023) Marine natural products as innovative cosmetic ingredients. Mar Drugs 21(3):170

Garcia J, Rodrigues F, Saavedra MJ, Nunes FM, Marques G (2022) Bioactive polysaccharides from medicinal mushrooms: a review on their isolation, structural characteristics and antitumor activity. Food Biosci 49: 101955

Gmoser R, Ferreira JA, Lennartsson PR, Taherzadeh MJ (2017) Filamentous ascomycetes fungi as a source of natural pigments. Fungal Biol Biotechnol 4:1–25

Gul H, Akbar S, Gohar S, Mazhar F (2024) Biochemical analysis and therapeutic potential of extract from mushroom (Pleurotus ostreatus). Int J Appl Clin Res 2(1):84–100

Gupta PL, Rajput M, Oza T, Trivedi U, Sanghvi G (2019) Eminence of microbial products in cosmetic industry. Nat Products Bioprospect 9:267–278

Ha SY, Kim HC, Lim WS, Yang JK (2024) Effect of growth condition on mycelial growth and fruiting body cultivation of Cordyceps militaris wild strain. J Mushroom 22(3):81–86

Han J, Wu T, Jin J, Li Z, Cheng W, Dai X et al (2022) Exosome-like nanovesicles derived from Phellinus linteus inhibit Mical2 expression through cross-kingdom regulation and inhibit ultraviolet-induced skin aging. J Nanobiotechnol 20(1):455

Hu L, Weng J, Wang Z, Huang C, Zhang L (2024) Effect and mechanism of Tricholoma matsutake extract combined with bakuchiol and ergothioneine on UVB-induced skin aging. J Cosmet Dermatol 23(11): 3628–3644

Huang K, El-Seedi HR, Xu B (2022) Critical review on chemical compositions and health-promoting effects of mushroom Agaricus blazei Murill. Curr Res Food Sci 5:2190–2203

Hyde KD, Bahkali AH, Moslem MA (2010) Fungi – an unusual source for cosmetics. Fungal Divers 43:1–9

Imtiaj A, Jayasinghe C, Lee GW, Kim HY, Shim MJ, Rho HS et al (2008) Physicochemical requirement for the vegetative growth of Schizophyllum commune collected from different ecological origins. Mycobiology 36(1):34–39

Iqbal T, Sohaib M, Iqbal S, Rehman H (2024) Exploring therapeutic potential of Pleurotus ostreatus and Agaricus bisporus mushrooms against hyperlipidemia and oxidative stress using animal model. Foods 13(5): 709

Jang Y, Cho SM, June CS, Weon HY, Park JS, Choi SG et al (2005) Changes of enzyme activities and compositions of abnormal fruiting bodies grown under artificial environmental conditions in Pleurotus ostreatus. Mycobiology 33(1):30–34

Jayasinghe C, Imtiaj A, Hur H, Lee GW, Lee TS, Lee UY (2008) Favorable culture conditions for mycelial growth of Korean wild strains in Ganoderma lucidum. Mycobiology 36(1):28–33

Jiang C, Ge J, He B, Zeng B (2021) Glycosphingolipids in filamentous fungi: biological roles and potential applications in cosmetics and health foods. Front Microbiol 12:690211

Khan F, Chandra R (2017) Effect of physiochemical factors on fruiting body formation in mushroom. Int J Pharm Pharm Sci 9:33–36

Khoo SC, Ma NL, Peng WX, Ng KK, Goh MS, Chen HL et al (2022) Valorisation of biomass and diaper waste into a sustainable production of the medical mushroom Lingzhi Ganoderma lucidum. Chemosphere 286: 131477

Kim SY, Shrestha B, Sung GH, Han SK, Sung JM (2010) Optimum conditions for artificial fruiting body formation of Cordyceps cardinalis. Mycobiology 38(2): 133–136

Krause J, Tobin G (2013) Discovery, development, and regulation of natural. In: Using old solutions to new

problems: natural drug discovery in the 21st century, p 1
Krishna KV, Murugan JM, Khan H, Kumar M, Veeramanikandan V, Hatamleh AA et al (2023) Exploring the therapeutic potential of edible Pleurotus mushroom species for oxidative stress and diabetes management. J King Saud Univer Sci 35(9):102926
Kumar V, Ahluwalia V, Saran S, Kumar J, Patel AK, Singhania RR (2021) Recent developments on solid-state fermentation for production of microbial secondary metabolites: challenges and solutions. Bioresour Technol 323:124566
Kumar A, Bharti AK, Bezie Y (2022) Schizophyllum commune: a fungal cell-factory for production of valuable metabolites and enzymes. Bioresources 17(3): 5420–5436
Li Q, Geiselhart L, Mittler JN, Mudzinski SP, Lawrence DA, Freed BM (1996) Inhibition of human T lymphoblast proliferation by hydroquinone. Toxicol Appl Pharmacol 139(2):317–323
Ma Y, Wang C, Zhang Q, Peng X, Feng Y, Meng X (2018) The effects of polysaccharides from Auricularia auricula (Huaier) in adjuvant anti-gastrointestinal cancer therapy: a systematic review and network meta-analysis. Pharmacol Res 132:80–89
Mago P, Sharma R, Hafeez I, Nawaz I, Joshi M, Mehrotra R (2023) Mushroom based cosmeceuticals: an upcoming biotechnology sector. Biosci Biotechnol Res Asia 20(2):381–394
Manful ME, Ahmed L, Barry-Ryan C (2024) Cosmetic formulations from natural sources: safety considerations and legislative frameworks in the European Union. Cosmetics 11(3):72
Maurício EM, Branco P, Araújo ALB, Roma-Rodrigues C, Lima K, Duarte MP et al (2024) Evaluation of biotechnological active peptides secreted by Saccharomyces cerevisiae with potential skin benefits. Antibiotics 13(9):881
Martins T, Machado-Carvalho L, Aires A, Saavedra MJ, Marques G (2023) Antioxidant, antimicrobial and cosmeceutical potential of wild mushroom extracts. Applied Microbiology 3(2):562–579
Miina J, Peltola R, Veteli P, Linnakoski R, Escribano MC, Haveri-Heikkilä J et al (2021) Inoculation success of Inonotus obliquus in living birch (Betula spp.). For Ecol Manag 492:119244
Mishra M, Srivastava D (2024) Harnessing the power of medicinal mushrooms into cosmetics: a fountain of youth from nature. J Indian bot Soc 104(4):200–209
Mohorčič M, Friedrich J, Renimel I, André P, Mandin D, Chaumont JP (2007) Production of melanin bleaching enzyme of fungal origin and its application in cosmetics. Biotechnol Bioprocess Eng 12:200–206
Muhaxi M, Liu F, Ng TB (2023) Structural characterization and in vitro hepatoprotective activity of a novel antioxidant polysaccharide from fruiting bodies of the mushroom Pleurotus ferulae. Int J Biol Macromol 243: 125124
Mukherjee A, Das S, Chakraborty D, Pal N, Das N (2024) Fungi's treasure in cosmeceuticals-a comprehensive chemical approach. S Afr J Bot 166:311–331
Murata H, Yamada A, Ichida H, Nakamura N, Neda H (2023) Biodiversity of Tricholoma matsutake (syn. T. nauseosum) and its related species based on repetitive DNA and genomics. Botany 101(5):138–154
Nguyen BTT, Le VV, Nguyen HTT, Nguyen LT, Tran TTT, Ngo NX (2021) Nutritional requirements for the enhanced mycelial growth and yield performance of Trametes versicolor. J Appl Biol Biotechnol 9(1):1–7
Niu H, Song D, Mu H, Zhang W, Sun F, Duan J (2016) Investigation of three lignin complexes with antioxidant and immunological capacities from Inonotus obliquus. Int J Biol Macromol 86:587–593
Nxumalo W, Elateeq AA, Sun Y (2020) Can Cordyceps cicadae be used as an alternative to Cordyceps militaris and Cordyceps sinensis? –a review. J Ethnopharmacol 257:112879
Onyango BO, Otieno CA, Palapala VA (2013) Effect of wheat bran supplementation with fresh and composted agricultural wastes on the growth of Kenyan native wood ear mushrooms [Auricularia auricula (L. ex hook.) Underw.]. Afr J Biotechnol 12(19)
Owaid MN, Barish A, Ali Shariati M (2017) Cultivation of Agaricus bisporus (button mushroom) and its usages in the biosynthesis of nanoparticles. Open Agric 2(1): 537–543
Owolabi JO, Fabiyi OS, Adelakin LA, Ekwerike MC (2020) Effects of skin lightening cream agents–hydroquinone and kojic acid, on the skin of adult female experimental rats. In: Clinical, cosmetic and investigational dermatology, pp 283–289
Panda MK, Paul M, Singdevsachan SK, Tayung K, Das SK, Thatoi H (2021) Promising anti-cancer therapeutics from mushrooms: current findings and future perceptions. Curr Pharm Biotechnol 22(9):1164–1191
Paterska M, Czerny B, Cielecka-Piontek J (2024) Macrofungal extracts as a source of bioactive compounds for cosmetical anti-aging therapy: a comprehensive review. Nutrients 16(16):2810
Phasha V, Senabe J, Ndzotoyi P, Okole B, Fouche G, Chuturgoon A (2022) Review on the use of kojic acid – a skin-lightening ingredient. Cosmetics 9:64
Pintathong P, Chomnunti P, Sangthong S, Jirarat A, Chaiwut P (2021) The feasibility of utilizing cultured Cordyceps militaris residues in cosmetics: biological activity assessment of their crude extracts. J Fungi 7(11):973
Pretsch A, Nagl M, Schwendinger K, Kreiseder B, Wiederstein M, Pretsch D et al (2014) Antimicrobial and anti-inflammatory activities of endophytic fungi Talaromyces wortmannii extracts against acne-inducing bacteria. PLoS One 9(6):e97929
Regis MAJ, Geösel A, Geösel RM (2024) Cultivation of Auricularia species: a review of the history, health benefits, principles, practices, environmental conditions, research methods, and recent trends. Sydowia 76:21

Richter C, Wittstein K, Kirk PM, Stadler M (2015) An assessment of the taxonomy and chemotaxonomy of Ganoderma. Fungal Divers 71:1–15

Rukhsar S, Usman M, Yousaf N, Murtaza G, Manzoor MA, Azam M (2025) Mushrooms in modern cosmetics: unlocking anti-aging, antioxidant, and therapeutic potential. Arch Dermatol Res 317(1):542

Sangthong S, Pintathong P, Pongsua P, Jirarat A, Chaiwut P (2022) Polysaccharides from Volvariella volvacea mushroom: extraction, biological activities and cosmetic efficacy. J Fungi 8(6):572

Sarkar A, Philip AM, Thakker DP, Bhaskara Rao KV (2024) Extraction and characterization of Beta-Glucan methyl esters Derived from UV-mutated marine yeast and their applications in cosmetics. Thalassas Int J Mar Sci 40(1):721–733

Sengkhamparn N, Phonkerd N (2014) Effects of heat treatment on free radical scavenging capacities and phenolic compounds in Tylopilus alboater wild edible mushrooms. Chiang Mai J Sci 41(5.2):1241–1249

Shen K, Liu Y, Liu L, Khan AW, Normakhamatov N, Wang Z (2024) Characterization, optimization, and scaling-up of submerged Inonotus hispidus mycelial fermentation for enhanced biomass and polysaccharide production. Appl Biochem Biotechnol:1–22

Sheng K, Wang C, Chen B, Kang M, Wang M, Liu K, Wang M (2021) Recent advances in polysaccharides from Lentinus edodes (Berk.): isolation, structures and bioactivities. Food Chem 358:129883

Song X, Shang X, Zhang M, Yu H, Zhang D, Tan Q, Song C (2025) Cultivation methods and biology of Lentinula edodes. Appl Microbiol Biotechnol 109(1):1–15

Sousa P, Tavares-Valente D, Pereira CF, Pinto-Ribeiro I, Azevedo-Silva J, Madureira R et al (2024) Circular economyeast: Saccharomyces cerevisiae as a sustainable source of glucans and its safety for skincare application. Int J Biol Macromol 265:130933

Srisukchayakul P, Werasura T, Teeravet S, Pradapphai P, Kannika A, Muangman T et al (2024) Development of beta-glucan production from microorganisms as active ingredients in cosmeceutical products for skin youthfulness. ScienceAsia 50(2)

Subedi K, Basnet BB, Panday R, Neupane M, Tripathi GR (2021) Optimization of growth conditions and biological activities of nepalese Ganoderma lucidum strain Philippine. Adv Pharmacol Pharm Sci 2021: 4888979

Sułkowska-Ziaja K, Grabowska K, Apola A, Kryczyk-Poprawa A, Muszyńska B (2021) Mycelial culture extracts of selected wood-decay mushrooms as a source of skin-protecting factors. Biotechnol Lett 43: 1051–1061

Sułkowska-Ziaja K, Trepa M, Olechowska-Jarząb A, Nowak P, Ziaja M, Kała K, Muszyńska B (2023) Natural compounds of fungal origin with antimicrobial activity – potential cosmetics applications. Pharmaceuticals 16(9):1200

Suthar M, Dufossé L, Singh SK (2023) The enigmatic world of fungal melanin: a comprehensive review. J Fungi 9(9):891

Szychowski KA, Rybczyńska-Tkaczyk K, Tobiasz J, Yelnytska-Stawasz V, Pomianek T, Gmiński J (2018) Biological and anticancer properties of Inonotus obliquus extracts. Process Biochem 73:180–187

Tang X, Yang T, Yu D, Xiong H, Zhang S (2024) Current insights and future perspectives of ultraviolet radiation (UV) exposure: friends and foes to the skin and beyond the skin. Environ Int 185:108535

Tao Z, Yuan H, Liu M, Liu Q, Zhang S, Liu H et al (2022) Yeast extract: characteristics, production, applications and future perspectives. J Microbiol Biotechnol 33(2): 151

Taofiq O, González-Paramás AM, Martins A, Barreiro MF, Ferreira IC (2016) Mushrooms extracts and compounds in cosmetics, cosmeceuticals and nutricosmetics – a review. Ind Crop Prod 90:38–48

Theeraraksakul K, Jaengwang K, Choowongkomon K, Tabtimmai L (2023) Exploring the biological functions and anti-melanogenesis of phallus indusiatus for mushroom-based cosmetic applications. Cosmetics 10(5):121

Thiribhuvanamala G, Krishnamoorthy S, Manoranjitham K, Praksasm V, Krishnan S (2012) Improved techniques to enhance the yield of paddy straw mushroom (Volvariella volvacea) for commercial cultivation. Afr J Biotechnol 11(64):12740–12748

Thomas PW, Elkhateeb WA, Daba GM (2020) Chaga (Inonotus obliquus): a medical marvel but a conservation dilemma. Sydowia 72:123–130

Törős GH, El-Ramady H, Prokisch J (2022) Edible mushroom of Pleurotus spp.: a case study of oyster mushroom (Pleurotus ostreatus L.). Environ Biodiver Soil Secur 6(2022):51–59

Ukoima HN, Ogbonnaya LO, Arikpo GE, Ikpe FN (2009) Cultivation of mushroom (Volvariella volvacea) on various farm wastes in Obubra local government of Cross River state, Nigeria. Pak J Nutr 8(7):1059–1061

Umor NA, Abdullah S, Mohamad A, Ismail S, Ismail SI, Misran A (2020) Challenges and current state-of-art of the Volvariella volvacea cultivation using agriculture waste: a brief review. In: Advances in waste processing technology, pp 145–156

Uppala L (2015) A review on active ingredients from marine sources used in cosmetics. SOJ Pharm Pharm Sci 2:1–3

Valverde ME, Hernández-Pérez T, Paredes-López O (2015) Edible mushrooms: improving human health and promoting quality life. Int J Microbiol 2015(1): 376387

Walker A, Wannasawang N, Taliam W, Keokanngeun L, Luangharn T, Thongklang N (2023) Optimal conditions for mycelial growth and nutritional values of the Auricularia cornea. Stud Fungi 8(1):0

Wang J, Hu W, Li L, Huang X, Liu Y, Wang D, Teng L (2017) Antidiabetic activities of polysaccharides separated from Inonotus obliquus via the modulation of oxidative stress in mice with streptozotocin-induced diabetes. PLoS One 12(6):e0180476

Wasser SP, Weis AL (2016) Therapeutic effects of substances occurring in higher basidiomycetes mushrooms: a modern perspective. Crit Rev Immunol 19(1):65–96

Watson M, Holman DM, Maguire-Eisen M (2016) Ultraviolet radiation exposure and its impact on skin cancer risk. In: Seminars in oncology nursing, vol 32. WB Saunders, pp 241–254

Wu Y, Choi MH, Li J, Yang H, Shin HJ (2016) Mushroom cosmetics: the present and future. Cosmetics 3(22): 1–13

Wu JY, Siu KC, Geng P (2021) Bioactive ingredients and medicinal values of Grifola frondosa (Maitake). Foods 10(1):95

Yalimaimaiti N, Dong Y, Jia P, Feng X, Luo Y, Hao J et al (2025) A Glycopeptide from Agaricus balchaschensis mitigates cadmium damage in mice. PRO 13(1):168

Yamada A (2022) Cultivation studies of edible ectomycorrhizal mushrooms: successful establishment of ectomycorrhizal associations in vitro and efficient production of fruiting bodies. Mycoscience 63(6): 235–246

Yamanaka T, Yamada A, Furukawa H (2020) Advances in the cultivation of the highly-prized ectomycorrhizal mushroom Tricholoma matsutake. Mycoscience 61(2):49–57

Yang HL, Chen SJ, Yeh JT, Vadivalagan C, Chiu JH, Hseu JH, Hseu YC (2024) The anti-melanogenesis, anti-photoaging, and anti-inflammation of coenzyme Q0, a major quinone derivative from Antrodia camphorata, through antioxidant Nrf2 signaling pathways in UVA/B-irradiated keratinocytes. J Funct Foods 116:106206

Yao FJ, Lu LX, Wang P, Fang M et al (2018) Development of a molecular marker for fruiting body pattern in Auricularia auricula-judae. Mycobiology 46:72–78

Yu J, Sun R, Zhao Z, Wang Y (2014) Auricularia polytricha polysaccharides induce cell cycle arrest and apoptosis in human lung cancer A549 cells. Int J Biol Macromol 68:67–71

Zhang L (2019) Progress in molecular biology and translational science: glycans and glycosaminoglycans as clinical biomarkers and therapeutics-part B, vol 163. Academic Press

Zhang D, Yang X, Kang JS, Choi HD, Son BW (2008) Circumdatin I, a new ultraviolet-a protecting benzodiazepine alkaloid from a marine isolate of the fungus Exophiala. J Antibiot 61(1):40–42

Zhao FY, Lin JF, Zeng XL, Guo LQ, Wang YH, You LR (2010) Improvement in fruiting body yield by introduction of the Ampullaria crossean multi-functional cellulase gene into Volvariella volvacea. Bioresour Technol 101(16):6482–6486

Bioactive Compounds of Fungal Extracts Used as Cosmetics Ingredients

3

E. N. Anaele, F. I. Nwafor, and C. C. Osegbue

Abstract

Fungal extracts are used as cosmetic ingredients due to their numerous potential medicinal properties. Currently, species of fungi beneficially utilized in the production of cosmetics include *Ganoderma lucidum*, *Tremella fuciformis*, *Cordyceps sinensis*, *Inonotus obliquus*, *Grifola frondosa*, *Lentinula edodes*, *Trametes versicolor*, *Polyporus* spp., and fungi of low taxa. Cosmetics produced by incorporation of fungal bio-products include those for hair-beauty and skincare products. Traditionally, these mushrooms are known for their medicinal values being producers of bioactive compounds such as flavonoids, caffeine, choline, ferulic, gallic acid, protocatechuic, p-coumaric, p-hydroxybenzoic, creatine, dithiolthiones, polysaccharides, phytoestrogens, glucosinolates, and prebiotics. These compounds depict notable anti-aging, antioxidant, skin whitening, anti-wrinkle, and moisturizing effects, promoting their utilization in cosmetics production. Despite the various challenges associated with fungal bio-products, their unique constituents make them promising ingredients with a variety of applications in the cosmetic industry. Several species of mushrooms and other fungi have not been described, cultivated, and identified for their possible utilization in the cosmetic industry. Fungal bio-products in cosmetics can revolutionize the cosmetic industry by promoting the use of sustainable, natural, and ecofriendly alternatives for skincare and beauty products.

Keywords

Cosmetic ingredients · Fungal extracts · Fungal-derived compounds · Cosmeceuticals · Nutricosmetics · Fungal enzymes

3.1 Introduction

All cosmetic formulations serve a dual purpose: they are utilized for both transient and extended durations to enhance the aesthetic appeal of the body while also contributing to its physiological well-being to a certain degree, and they exert a psychological influence on observers (Menaa 2014). Cosmetics employed for decorative purposes, such as eyeliners, rouges, mascaras, and facial masks, inherently carry the potential

E. N. Anaele (✉)
South-East Zonal Biotechnology Center, University of Nigeria, Nsukka, Nigeria

Department of Microbiology, Faculty of Biological Sciences, University of Nigeria, Nsukka, Nigeria
e-mail: eunice.anaele@unn.edu.ng

F. I. Nwafor · C. C. Osegbue
Department of Plant Science and Biotechnology, Faculty of Biological Sciences, University of Nigeria, Nsukka, Nigeria

S. Ghosh, H. N. Onyeaka (eds.), *Fungal-Based Cosmetics: Formulation and Usage*, The Mycota 17,
https://doi.org/10.1007/978-3-032-09291-5_3

for both favorable and adverse side effects. Such products may disrupt essential physiological mechanisms, chemically alter certain constituents of the skin (for instance, in the case of bleaching or coloring agents), and potentially lead to their removal or provoke certain allergic reactions (Mukherjee et al. 2024). The scientific investigation of cosmetics necessitates a comprehensive understanding of the chemical properties of ingredients, the interactions between these products and dermal or hair tissues, as well as their implications for overall health. Continuous innovations in formulation methodologies and technological advancements consistently influence the industry, with a focus on enhancing efficacy, ensuring safety, and promoting sustainability. Current trends underscore the significance of biocompatibility, ethical sourcing practices, and environmental considerations, thereby reflecting a broader commitment to consumer health and ecological stewardship (Paterska et al. 2024).

A category of naturally occurring ingredients extensively utilized in cosmetic formulations comprises of plants and fungal

extracts, which represent a substantial reservoir of biologically active compounds that exert significant influences on human epidermis. These extracts may display a diverse array of properties, encompassing both therapeutic applications (in various dermatological conditions, inflamed ailments such as atopic dermatitis, acne, psoriasis) and functions pertinent to dermal care (e.g., astringent, cleansing, antioxidant, antibacterial, moisturizing, regenerative, smoothing, or depigmenting), while offering little or no side effects (Anusiya et al. 2021; Menaa 2014). The study of cosmetics derived from natural origins involves a detailed examination of their biochemical properties, human health effects, and aesthetics. The historical study of cosmetics has garnered significant scholarly attention over the past few decades. This resurgence of interest can, in part, be attributed to the transformative natural movement that has emerged within the cosmetic sector (Mcmullen and Dell'Acqua 2023). Natural origin cosmetics offer a rich array of bioactive compounds that contribute to their efficacy in hair and dermal care. However, integrating these ingredients into cosmetic formulations requires careful consideration of their biochemical properties, extraction methods, and stability. As the industry progresses, there is a growing emphasis on balancing efficacy with ethical and environmental considerations, leading to innovations in both natural and synthetic alternatives (Fig. 3.1).

Demand for bioactive cosmetics incorporated with fungal products has increased significantly, being propelled by the distinctive biochemical properties. Fungi, particularly mushrooms and their derivatives, are increasingly being integrated into cosmetic formulations and are heralded as potent active components within these products (Mago et al. 2023; Wasser and Weis 1999; Hyde et al. 2010). Fungi reported to be used in cosmetic industry are *Ganoderma lucidum, Cordyceps sinensis, Tremella fuciformis, Inonotus obliquus, Grifola frondosa, Trametes versicolor, Lentinula edodes, Polyporus* spp., etc. *Lentinula edodes*, a fungus that produces lentinan, is used to formulate night lotion. Ellagic acid (polyphenol) produced by fungi possesses anticancer, antioxidant, and skin-brightening effects.

In comparison to plant, fungi grow faster and require lesser time to develop bioactive compounds useful in cosmetic sector, to produce orally utilized (nutricosmetics) or topically applied (cosmeceuticals) products. The fungi may be of terrestrial or marine habitat (Agrawal et al. 2018; Bedi et al. 2021). The natural bioactive cosmetic products are in high demand due to the increasing awareness of harmful effects of synthetic products and the pursuit of improved health and beauty. Such skincare products may be composed of single mushroom or its metabolite while other products use two or more mushroom and their derivatives (Hyde et al. 2010).

3.2 Overview of Cosmetics

The term cosmetic originated from the Greek word "Kosmeticos" meaning adorn and preparation (Saraf and Saraf 2022). Cosmetics embody a multifaceted and dynamic domain situated at the

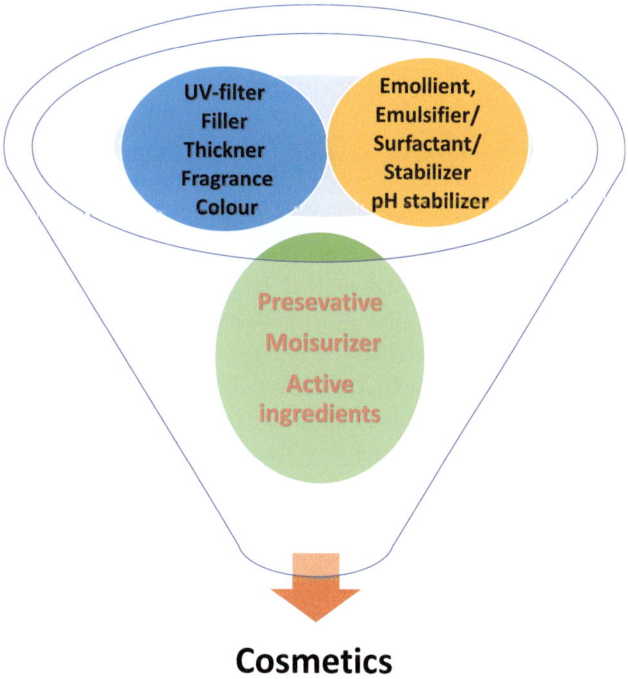

Fig. 3.1 Major ingredients in skincare products

confluence of scientific inquiry, technological advancement, and personal grooming. With historical origins in ancient cultural practices, the sector has undergone significant transformation propelled by advancements in the disciplines of chemistry, dermatology, and materials science. The cosmetics sector is classified into various categories, one of which is skincare, responsible for 36.4% of the global market share in the year 2016. Contemporary cosmetic products include a vast spectrum of items specifically engineered to augment aesthetic appeal, enhance dermal health, and deliver pleasurable sensory experiences. This discipline encompasses various categories such as makeup, skincare, haircare, and fragrance, with each formulation meticulously crafted with distinct objectives and considerations (Mukherjee et al. 2024).

The "active phase" of any cosmetic formulation commences upon its contact with the skin, hair, teeth, or nails, and concludes upon its removal or evaporation. During this active phase, a profound reciprocal interaction occurs, resulting in observable cosmetic modifications on the body (Saraf and Saraf 2022). The cosmetic product functions to safeguard the outermost layer from desiccation, infiltrates beneath the superficial layer, and facilitates the introduction of active compounds into deeper tissues, or adheres superficially to alter the pigmentation or sheen of specific areas. On this note, the origin, source, efficacy, and safety of these ingredients must be taken into consideration (Taofiq et al. 2019). In as much as natural ingredients are considered safer than their synthetic counterparts, there exists a lack of well-defined safety profile, and their potential for eliciting allergic reactions remains uncertain, including the extent of such reactions if they do occur. In addition, the true incidence of adverse effects associated with the utilization of organic ingredients is frequently overlooked or inadequately reported. Consequently, it is imperative for consumers to exercise caution when making use of natural ingredients. In order for products to be designated as "100% organic," they must exclusively comprise ingredients that have been produced through organic methods (Menaa 2014).

3.3 Bioactive Compound with Potential Application in Cosmetics Industry

A bioactive compound is one that affects a living thing, tissue, or cell; this is typically shown in laboratory studies using in vitro or in vivo methods. Following the description by Correia et al. (2012), these compounds or substances can coordinate the essential processes involved in human metabolism, including free radicals scavenging, the inhibition or activation of gene expression, and the functioning of enzymes and receptors. The biochemical activities of these compounds include antibacterial, anticarcinogenic, antimutagenic, antiallergenic, antioxidant, and anti-inflammatory properties, which make them essential. Common sources of bioactive chemicals include plants, animal products, and synthetic materials. A few examples of bioactive substances include polysaccharides, phytoestrogens, terpenoids, glucosinolates, creatine, coenzyme Q, caffeine, choline, and carotenoid flavonoids. These substances have very stable, extremely effective therapeutic characteristics with low potential for toxicity.

However, bioactive compounds in cosmetics offer various benefits for skin and hair care. They help in reducing signs of aging, improving the skin texture, providing protection against environmental damage, promoting the production of collagen synthesis, moisturizing the skin, and even aiding in reducing inflammation or hyperpigmentation. These compounds are often incorporated into skincare products like creams, serums, masks, and hair treatments to enhance their effectiveness in fostering and promoting healthier skin and hair.

3.3.1 Botanical Ingredients in Cosmetics

Plants have been utilized in the realm of cosmetics for numerous centuries owing to their bioactive compounds, inclusive of flavonoids, alkaloids, terpenes, and essential oils. Notable instances encompass *Aloe vera*, recognized for its calming attributes, and green tea extract, which confers antioxidant advantages (Hong et al. 2014). Natural polysaccharide extract of sea buckthorn are known to exhibit strong antioxidant activity with potential application as skin anti-aging substance in cosmetic industry (Wang et al., 2024a). Essential oils are derived from various anatomical segments of plants (such as flowers, leaves, and bark), with essential oils like lavender, rosemary, and tea tree oil being employed for their aromatic and therapeutic characteristics (Mcmullen and Dell'Acqua 2023). Consequently, the study of plant extracts alongside identification of the predominant active phytocompound utilized within cosmetology are of substantial significance, particularly if these natural compounds—whether encapsulated or otherwise—demonstrate enhanced efficacy and reduced toxicity in comparison to conventional chemical formulations employed as pharmaceuticals for the prevention or treatment of ailments (e.g., acne, melasma, dermatitis, melanoma, skin sunburns), nutraceuticals (e.g., anti-inflammatory agents, antioxidants), or cosmeceuticals for mitigation of pathological or physiological aging such as aging of skin cells (Menaa 2014). The recent interest on phyto-based cosmetics has been based on the beneficial effects of such products on skin including the remediation of various skin disorders (Michalak 2023). Ingredients such as shea butter and jojoba oil confer emollient properties, thereby facilitating moisturization and the restoration of the skin barrier. However, cosmetic ingredients derived from plants are associated with challenges including deforestation, environmental pollution, and climate change. The development of such compounds in plants takes prolonged time and is prone to fluctuations in chemical constituents and structure which is influenced by geographical area and season.

3.3.2 Animal-Derived Ingredients

Numerous substances of animal origin are incorporated into beauty care products within the cosmetic sector, forming distinct product category known as zooceuticals (which implies animal origin) (Cristiano and Guagni 2022). In contemporary practice, their usage has diminished relative to historical precedents due to hygienic-sanitary, ethical, and ecological considerations. Additionally, some of these ingredients may provoke irritative or allergic dermatitis. Animal-derived ingredients, including lanolin and beeswax, create occlusive layers that assist in moisture retention while shielding the skin from environmental aggressors. Lanolin, procured from sheep wool, is employed for its emollient properties, offering hydration and establishing a protective barrier on the skin (Mcmullen and Dell'Acqua 2023). Collagen and elastin, both sourced from animal connective tissues, have been incorporated into cosmetics to enhance skin elasticity and diminish the visibility of wrinkles. Conversely, beeswax, derived from honeybees, is esteemed for its moisturizing and emulsifying capabilities, frequently utilized in lip balms and creams. The primary ingredients utilized may include marine collagen, snail slime, and lanolin (Cristiano and Guagni 2022).

3.3.3 Fungal-Derived Ingredients

Diverse compounds endowed with important bio-functions are produced by fungi (Ferreira et al., Ferreira et al. 2010, Ferreira et al. 2009). Most fungal compounds are produced as secondary metabolites with potential application in the cosmetic sector. These secondary metabolites, although not essential for fungal multiplication, their production play beneficial role in protecting the fungus from predators or competitors. Fungal-derived cosmetic ingredients are more advantageous than ingredients from plants and animals. Fungi are easy to cultivate, and the bioactive substance is easy to derive and is not prone to fluctuations in chemical structure. Fungi can also be manipulated with ease to produce the desired distinct ingredient for cosmetic application. With ease in production, fungi have produced non-toxic aromatic compounds and colorants with diverse biological properties, making them suitable for application in cosmetics and other industries (Umashankar and Nygård 2024; Oliveira et al. 2024).

Numerous studies and research papers have revealed the existence of a broad range of bioactive substances, including glycoproteins, vitamins, volatile organic acids, terpenoids, lectins, phenolic compounds, steroids, polysaccharides, and other fungal lipid components (Türsen Uthan et al. 2022; Venturella et al. 2021; Ozen et al. 2019). Numerous mushroom extracts and metabolites exhibit major bio-functions including antimicrobial, antioxidant, antitumor, antiatherogenic, anti-inflammatory, immune-modulatory, and hypoglycemic activities (Sujarit et al., 2021). Edible mushrooms are currently used for cosmeceuticals (topical use) or nutricosmetics (oral use). Fungal-derived cosmetic ingredients are endowed with biological capabilities to improve health and promote beauty (Mukherjee et al. 2024; Taofiq et al. 2016b).

3.4 Fungi Producing Bioactive Compounds Used in Cosmetics

Inonotus obliquus, a medicinal fungus, commonly known as "chaga mushroom," is extensively used in Russia and North Europe due to its acclaimed antitumor, immunomodulatory potentials and other activities (Taofiq et al. 2016a). The anti-tyrosinase action of *Inonotus obliquus* extracts (n-butanol and petroleum ether) was reported by Yan et al. (2014), attributing the bioactivity to trametenolic acid and betulin, known for the ability to lower the body's melanin level and exhibit a noncompetitive form of tyrosinase inhibition.

Tricholoma matsutake or pine mushroom extract possesses anti-aging properties and acts as a tyrosinase inhibitor. Hu et al. (2024) observed that the combination of *T. matsutake* extract with ergothioneine and bakuchiol resulted in a therapeutic effect on skin and prevented direct damage by UV-radiation, reducing the rupture of collagen fiber and ultraviolet-induced inflammation. The report indicates that the combination of the three substances renewed the skin by regulating some biological processes. In addition, Yang et al. (2024) reported the safety and depigmentary effects of a newly detected polysaccharide from *T. matsutake* which suppressed excessive expression of key melanogenic factor by inactivation of c-Jun N-terminal kinase (JNK) resulting in an improved facial appearance indicated by decreased facial hyperpigmentation, minimized melanin index, and enhanced individual topology angle.

Agaricus blazei is rich in bioactive substances with immunological potential for multiple defense purposes (Huang et al. 2022; Gargano et al. 2017). Research have shown that *A. blazei's* principal ingredients, or β-glucans, boost the immunity and act as antitumor agent against liver cancer and myeloma. The bio-potential of *A. blazei* extracts has been demonstrated in a preclinical investigation conducted which revealed inhibition of bacterial growth. Specifically, antibacterial action of *A. blazei* extract is against severe oral infections as well as peritonitis (Lima et al. 2016) and associated implant infections (Wang et al. 2024b; 2024a). *A. blazei* administered orally elicited a reaction specific to β-glucan. Anti-BG antibody generation that results could serve as a gauge for the human immune system's reaction to β-glucan. Anti-allergic effects may arise from *A. blazei's* β-glucans' effect on the immune system. In vitro as well as animal experiments have demonstrated that these actions affect the immune system's Th1/Th2 cell balance including microphage activation (Wang et al. 2024b, 2024a; Huang et al. 2022; Lima et al. 2016).

Another mushroom, *Ganoderma lucidum*, commonly called Reishi, is edible with bitter taste but woody consistency. Miyazaki et al. (Miyazaki and Nishijima 1981) and Mau et al. (2011) documented that 90% of this fungus is water while the remaining 10% consisted of a little amount of protein, fiber, carbohydrate, ash, fat, and a few vitamins and minerals (calcium, potassium, phosphorus, selenium, magnesium, zinc, iron, and copper). The main triterpenic acids detected were Ganoderic acids C2, A, and H while phenolics include p-hydroxybenzoic, protocatechuic, and syringic acids. The fungus is known for its nutritional value, contains triterpenoids and polysaccharides, which are associated with antioxidant, anti-ulcer, anti-inflammatory, antitumorigenic, antimicrobial, anti-tyrosinase, immunomodulatory, and cytotoxic effects (Zhang et al., 2025; Bao et al. 2021; Li et al. 2019; Taofiq et al. 2016b; Taofiq et al. 2017). Most of the pharmacological and health effects have been attributed to the diversity of polysaccharides and high level of germanium present in *Ganoderma* (Sun et al. 2022; Borchers et al. 2019; Ahmad 2018). Recombinant *Ganoderma* proteins have been produced with reports supporting their potency and safety as cosmetic ingredient (Guo et al. 2023).

Pharmaceuticals, nutraceuticals, and nutritional supplements made from *G. lucidum* mycelia, spores, and fruiting bodies are readily accessible in the form of powdered extracts, beverages, oral liquids, teas, chewable tablets, and capsules (Taofiq et al. 2016a; Wu et al. 2016). Polysaccharides, flavonoids, alkaloids, proteins, inositol, mannitol, choline, and vitamins B1, B2, and B6 are among the compounds found in *G. lucidum* (Cör et al. 2018). The fruiting body of *Ganoderma lucidum* consists of 4000 bioactive components, comprising over 140 terpenoids and triterpenes, ergosterols, fatty acids, cerebrosides, proteins, glycoproteins, and trace elements that exhibit potent bioactivities (Bulam et al. 2019a). The varieties of proteins have immunomodulatory effects which could be potentially developed into nutricosmetics or cosmeceuticals (Guo et al. 2023; Li et al. 2019). Being useful for centuries due to its functional metabolites, the cosmeceutical potential of reishi can be valorized to promote circular economy.

Cordyceps sinensis and *C. militaris*, commonly called caterpillar mushroom, are high-value medicinal and edible fungi (Chellapandi and Saranya 2024), and well-studied, explored, and in high demand. Another species, *C. cicadae* used locally for eye treatment, has medicinal and nutritional value similar to the well-known *Cordyceps* but is not well-studied and unexplored (Nxumalo et al. 2020). *Cordyceps* is considered significant in cosmetics because of its wide range of biologically active substances, which boost antioxidant activity and collagen production but reduce inflammation. Even the solid-based residue gotten via solid-state cultivation of *Cordyceps militaris* and preparation of crude extract has potential application in cosmetics (Pintathong et al. 2021). All of the necessary amino acids, proteins, sterols, polysaccharides, nucleosides, and their analogues, as well as macro- and micro-elements (K, Na, Ca, Mg, Fe, Cu, Mn, Zn, Pi, Se, Al, Si, Ni, Sr, Ti, Cr, Ga, V, and Zr) are contents of *Cordyceps*. The polypeptides of *Cordyceps militaris* have been linked to memory and learning enhancement (Yuan et al. 2018). Among the most powerful bioactive components are cordicepic acid, cordicepine (3′-deoxyadenosine), 2′--deoxyadenosine, and several deoxygenated nucleotide analogues (adenosine, dideoxiadenosine, uridine, deoxyuridine), which resemble nucleosides but are absent from other homeopathic remedies. Cordicepine is a potent antioxidant compound known for its adaptogenic properties on the skin, promoting cell regeneration and growth. Beta-glucans, beta-mannans, cyclo-furans, and other polysaccharides are also present in *Cordyceps*. Evidence of antioxidant activity of *Cordyceps* and its potential immunomodulatory and neuroprotective activity, anti-aging, anti-atherosclerosis, anticancer, anti-inflammatory, and hepatoprotective effects have been documented (Ji et al. 2019; Cheng et al. 2018), promoting mushroom extracts utilization in cosmeceutical and skincare products to promote antioxidative, anti-wrinkle, skin-whitening (anti-tyrosinase), and sunscreen effects. In cosmetic formulation, extract of *Cordyceps* is used for its skin-protective and conditioning, antioxidant, emollient, and humectant properties.

Pleurotus species or oyster mushrooms are edible mushrooms, widely cultivated and eaten because of the nutritional composition and positive impact on health (Törős et al. 2022; Galappaththi et al. 2021; Adebayo and Oloke 2017; Carrasco-González et al. 2017). They are useful in green biotechnology (El-Ramady et al. 2022). The mushrooms contain protein, vitamins, essential amino acids, fiber, minerals, carbohydrates, and water (Sałata et al. 2018; Ghosh 2016). The glycoprotein found in *P. citrinopileatus* is used for treating cancer of the blood and the large ergothioneine content is responsible for immunomodulatory and antioxidant effects while the β-glucan contributes to strong immune response through activation of the immune cells for optimal response and elevation of immune cell number (Thi et al. 2023). On the basis of these biochemical constituents, *Pleurotus* species have diverse bio-potentials including hepatoprotective, anti-inflammatory, immunomodulatory, antitumor, antibacterial, radiation protection, and antiviral effects (Törős et al. 2022; Bulam et al., 2019b; Golak-Siwulska et al. 2018; Piska et al. 2017). Thi et al. (2023) extracted β-glucan from *Pleurotus*, determined conditions for its preservation to maintain product stability and recommended refrigeration or storage in the presence of citric acid or NH_4Cl. Studies done on *P. citrinopileatus*, *P. ostreatus*. *P. salmoneostramineus*, *P. ferulae*, and *P. nebrodensis* extracts have revealed antioxidant, anti-aging, UV-protective, and anti-tyrosinase function, decreasing the melanin production (Zerva et al. 2020; Hapsari et al. 2012; Alam et al. 2011a, b; Meng et al. 2011), supporting incorporation of mushroom extracts in oral and topical cosmetic formulation.

Trametes versicolor, commonly known as Turkey tail mushroom, also exhibited potential application in the cosmetic sector. The mushroom has abundant beta-glucans, other polysaccharides, and peptides, showing potential application in inflammation management and immune enhancement (Sun et al. 2022; Akgul et al. 2017; Taofiq et al. 2016a, b). This versatile mushroom possesses a rich composition of bioactive compounds that offer numerous benefits to many aspects of health and have demonstrated promising results in skincare and dietary supplements (Andriani and Yanto 2024). *Trametes versicolor* has the potential to be used as a cosmeceutical because of its bioactive ingredients, which include phenolic compounds, terpenoids, and polysaccharides. Polysaccharides present in this mushroom exhibit moisturizing effects by enhancing water retention in the skin and improving its barrier function. Antioxidant qualities found in phenolic compounds help defend the skin from oxidative stress to lessen aging symptoms. The anti-inflammatory qualities of terpenoids included in *Trametes versicolor* extract have been shown to help calm irritated skin and lessen redness (Wang et al. 2015; 2024c). The mushrooms possess the functional metabolites for potential development of cosmetic products.

Filamentous fungi are also potential sources of biocosmetic ingredients. *Aspergillus niger, Aspergillus oryzae*, *Monascus purpureus*, and *Rhizopus oryzae* produce compounds that have the following effects: anti-aging, whitening, and antioxidant properties. White not fungi, namely: *Geotrichum* species, *Phlebia radiata*, *Trametes hirsute*, *Phanerochaete chrysosporium,* and *Bjerkandera adusta* produce ligninolytic enzymes that are potential source of whitening agent. Fungi that produce mycelia-bound lipase are *Mucor circinelloides*, *Penicillium citrinum*, and especially *Rhizopus oryzae*. Preto et al. (2024) used *R. oryzae* for the production of emollient esters for cosmetics.

Edible fungi such as Ascomycetes, *Monascus*, *Fusarium*, *Penicillium,* and *Neurospora* can serve as natural sources of pigment with bioactivity in food, feed, and cosmetic applications. Ascomycetes, namely, *Talaromyces*, such as *T. purpurogenus,* and *T. atroroseus,* produce red pigment; *Cordyceps unilateralis* produces deep blood-red pigment; *Herpotrichia rhodosticta* produces orange; *Curvularia lunata*; several species of *Drechalera*. Strains of the above species are non-mycotogenic and non-pathogenic to man. *Neurospora* species has GRAS Status, *N. intermedia*, etc. Ascomycetes also produce pigments that have bioactivity, such as anticancer property; *Monascus* pigments also have heat and pH stability during the processing of food (Gmoser et al. 2017).

3.5 Extraction Techniques Used in Cosmetics Production

The resultant quality of cosmetics is mostly dependent on an extraction process that has been carefully planned and carried out. Of course, other crucial elements that impact the final result are the different physicochemical characteristics of the source material, the techniques employed in processing starting materials, the solvent for extraction, the extraction techniques, temperature, pH, and pressure conditions, among others. Fungal bioactive cosmetic ingredients are generally extracted from fungal broth culture and/or the mycelium by utilizing a wide array of organic or aqueous solvents.

There are two categories of extraction techniques: traditional or classical and innovative or modern techniques (Chahal et al., 2024; Sosa-Hernández et al. 2018).

A. **Conventional extraction methods:**
   (i) Maceration
   (ii) Percolation
   (iii) Infusion
   (iv) Decoction

   The standard technique among the aforementioned has been Soxhlet extraction, particularly for materials derived from plants or marine life. However, a significant impediment is the requirement for pre-digestion by acids, which requires time. The need for high-grade solvents, low extract yield, breakdown of heat-sensitive bio-compounds, and prolonged time for extraction, are some of the other drawbacks of the previously listed conventional extraction processes. Resolution to the above challenges led to the development of modern methods of extraction which are regarded as environmentally beneficial and conform to internationally acceptable standard.

B. **Modern extraction methods:**
   (i) Hot continuous extraction (Soxhlet method)
   (ii) Superficial-fluid extraction (SFE)
   (iii) Counter-current extraction (CCE)
   (iv) Enzyme-assisted extraction (EAE)
   (v) Pressurized-liquid extraction (PLE)
   (vi) Microwave-assisted extraction (MAE)
   (vii) Hydro-distillation
   (viii) Steam distillation

3.6 Fungal Bioactive Compound for Cosmetics Industry

Bioactive molecules from fungi include kojic acid, scytonemin, chrysophanol, β-glucans, mycosporine-like amino acids, ergothioneine, chitosans, ceramide, and triterpenoids (ergosterols, ganoderic acid) have potential applications in cosmetics (Mishra and Srivastava 2024; Mukherjee et al. 2024). In cosmetic sector, these fungal metabolites can be used as skin moisturizer, skin whitener, antioxidant, anti-acne, anti-wrinkle, anti-aging, and UV-filter in skincare products. Other applications include use of fungal compounds in toothpaste to prevent plaque formation, and haircare products for strengthening hair, and preventing hair breakage.

Kojic acid is a useful organic acid in cosmetics and healthcare. It functions as a skin-lightening agent by preventing tyrosinase formation and suppressing skin hyperpigmentation (Chib et al. 2023). It enhances the ability of the skincare product in preventing exposure to UV radiation, serving as UV -protector. *Aspergillus* and *Penicillium* are the main fungal genera of kojic-producing strains, with a rising market demand for kojic acid calculated to be worth $39 million by 2026 (de Caldas Felipe et al. 2023; Chib et al. 2023). Endophytic fungi such as *Colletotrichum gloeosporioides* are also promising producers of kojic acid with antimicrobial activity (Nurunnabi et al. 2018). *Aspergillus flavus* isolated from *Annona squamosa* leaves and *Aspergillus oryzae* isolated from *Euphorbia pepils* are potent producers of kojic acid (Calado et al. 2021; Ola et al. 2020; Shehata and Sabry 2020).

Kojic acid is topically applied for skin treatment of melasma spots, light-brown patches, which are hyperpigmentation after an inflammatory response. It protects the skin from ultraviolet radiation, preventing excessive human skin pigmentation by inhibition of tyrosinase, thus hindering melanin production. Kojic acid also promotes the shelf-life of cosmetics. It exhibits antibacterial antifungal, antibiofilm antioxidant, anti-inflammatory, and insecticidal potentials. Potent activities of kojic acid against *Micrococcus luteus*, *Pseudomonas aeruginosa*, *Staphylococcus aureus*, *Fusarium oxysporum*, *Spodoptera frugiperda*, etc. have been documented (Mohamad et al. 2010; Phasha et al. 2022; Sharma et al. 2024; Nurunnabi et al. 2018).

Due to its efficacy in depigmentation, kojic acid is applied widely in the cosmetic sector as an alternative to hydroquinone. In beauty-care products like creams and serums, Kojic acid and its esters are recommended in 1% concentration or less for safety and efficacy (Phasha et al. 2022). Kojic acid is cytotoxic; thus the Cosmeceutical Ingredient Review (CIR) Committee approves the use of kojic acid at concentration not higher than 1%. Kojic acid may cause contact dermatitis, redness, sensitization, and erythema especially when used at concentration above 1% (Phasha et al. 2022; Zilles et al. 2022). The organic acid, kojic acid, is derived from fungi, mainly *Aspergillus* and *Penicillum* and was first isolated in 1907, and identified as 5-hydroxy-2-hydroxymethyl-4H-pyran-4-one. Kojic acid is unstable to heat, light, and oxidative (Zilles et al. 2022). To overcome the challenge of instability, derivatives from kojic such as kojic monooleate, kojic dipalmitate, and the use of nanocarriers have been adopted. The current process of production of kojic acid, gene regulation, is a limitation to commercial production of kojic acid (Chib et al. 2023).

Apart from kojic acid, other fungal-derived compounds inhibiting tyrosinase, and responsible for skin-whitening (brightening) include azelaic acid and 3,4-dihydrox-benzaldehyde. Azelaic acid is produced by a Basidiomycota, *P. ovale*, and it is very effective against hyperpigmentation, when compared to hydroquinone. Apart from tyrosinase activity, azelaic acid has anti-inflammatory, antibacterial, and keratolytic properties.

Vitamin E is another compound of fungal origin with potential application in cosmetics industry. Mushrooms such as *Pleurotus ostreatus*, *Agaricus bisporus*, *Grifola frondosa*, and *Lentinula edodes* are producers of vitamin E. Quintero-cabello et al. (2021) reported various antioxidants produced by Polypopales fungi namely: ergosterol, phenolic compounds β-glucans, ergothioneine, tocopherols, and vitamin C.

Mushrooms produce bioactive compounds such as terperoids, selenium, phenolics, polyphenolics, vitamins, polysaccharides, and volatile organic compounds (Visvanathan et al. 2022). Various genera of mushrooms with potential applications in cosmetics industries include *Ganoderma*, *Pleurotus*, *Tremelia*, *Cordyceps*, *Inonotus*, *Grifola*, *Lentinula*, *Schizophyllum*, *Coriolus*, *Euphorbia*, *Pistacia*, *Trametes*, *Agaricus*, *Phellinus*, and *Antrodia*. Extracts from these mushrooms show potent anti-wrinkle, antioxidant, moisturizing, anti-aging, and skin-whitening properties. They are multifunctional cosmetics components (Sharma et al. 2024).

Vitamin E is known for its significant impact on skin due to its antioxidant, regenerative, UV-radiation protection, moisturizing, anti-aging, and anti-inflammatory effects. Vitamin E is a free-radical scavenger, and free radicals likely harm dermal cells and promote senescence. It also promotes regeneration of worn-out skin cells, reduces the visibility of skin scars, and protects the skin from loss of moisture. Vitamin E acts as UV protection by defending the skin against UV-induced harm (Sułkowska-Ziaja et al. 2023). Thus, vitamin E, a fat-soluble vitamin, is made up of compounds such as tocopherols and tocotrienols and recognized as facilitator of skin regeneration, anti-oxidative stress, and anti-aging.

Chitin is an important fungal cell-wall component that exhibits antiviral, antibacterial, and anti-inflammatory characteristics. In cosmetics, chitin has moisturizing, regenerative, anti-aging, skin firming, and antioxidant effects. It restores skin elasticity and stimulates the synthesis of collagen. Thus, they are useful in the formulation of dry-skin products for maintenance of skin hydration (Park and Kim 2010).

Phenolic acids such as gallic acid exhibit antioxidant, antibacterial, anti-inflammatory, and antiviral effects. They defend the skin against pigmentary disorders (Sharif et al. 2018; Sułkowska-Ziaja et al. 2023).

Ergothioneine is a known metabolite from macrofungi such as *A. bisporus*, *L. edodes,* and *G. frondosa*. It protects skin and safeguards against damage to mitochondrial DNA in the skin keratinocytes and fibroblasts caused by free radicals. Ergothioneine has antimicrobial and anti-inflammatory capabilities and contributes to immune system regulation. It is used in the formulation of anti-aging dermal-care cosmetics (Venturella et al. 2021).

Exopolysaccharides from fungi have different effects such as anti-inflammatory, anti-parasitic, antiviral, antibacterial, antioxidant, antitumor, hypoglycemic, vaso-protective, and hypotensive effects (Chellapandi and Saranya 2024; Sułkowska-Ziaja et al. 2023; Sun et al. 2022; Vlassopoulou et al. 2021). They enhance collagen synthesis, protect the skin from ultraviolet rays, and exhibit moisturizing effect, protecting the skin against aging through their antioxidant potentials (Cheng et al. 2018; Schiano et al. 2021). A fungus, *Schizophyllum commune,* has been reported to produce schizophyllan (-β-glucan), which reduces skin inflammation and itchiness, protecting the skin from ultraviolet rays and other harmful conditions within the environment (Mukherjee et al. 2024). Chitin-glucan, a co-polymer found in fungi such as *Aspergillus niger,* is used to formulate skincare products that exhibit moisturizing and anti-aging effects. Sousa et al. (2024) showed that *Saccharomyces cerevisiae* is a good glucan source which is sustainable and safe for application in the formulation of skincare products, contributing to circular economy. Furthermore, their report indicated no immunomodulatory capacity, cytotoxic effect, genotoxicity, skin sensitization, and no effect on the facial skin normal flora using in vitro model, indicating the promising potential of spent yeast glucan as a component of novel, safe skincare products (Sousa et al. 2024). In another report, UV-mutated marine yeast, *Rhodotorula* sp. DAMB1, produced β-glucan which exhibited potent antioxidant and antimicrobial activity against *Staphylococcus aureus* and other pathogens tested, showing skin-smoothening effect and no side effect or mutated skin infection after using the formulated product, "beta-glucan almond scrub" for a period of 10 weeks (Sarkar et al. 2024).

Selenium is an essential element that accumulates in the body of *Lentinus edodes* and *Agaricus bisporus* which are antitumor agents and used to strengthen bone, the teeth, hair, and nails. It has potential application in the formulation of shampoo (Usman et al. 2021; Ogra et al. 2004).

Terpenoids are terpenes analogues, gotten from 5-carbon isoprene units found in *Ganoderma* species. Terpenoids are wide organic compounds that contain varieties of terpene derivatives, namely ganoderals, ganoderals, ganodermanontriol, ganodermic and ganoderic acid, lanostane, lucidone, and ganodermanondiol (Wu et al. 2016). Terpenoids show anti-infective and immunomodulatory potentials. Organic pigments such as carotenoids (xanthophylls and carotenes) are essential ingredients in the formulation of sun-lotions (Table 3.1).

Glycosphingolipids (GSL) are used as moisturing agents in cosmetics or health products. There are two major types, neutral GSL and acidic GSL (mannosyl inositol phosphoryl ceramide and mannosyl di-inositol phosphoryl ceramide). Neutral GSL is common in plants, animals, and fungi, easy to analyze, exploit, and found beneficial to human health. It has gained huge applications in cosmetics and human health because it is non-toxic and not associated with adverse effects in humans (Tessema et al. 2017),

The neutral GSL is also called glycosylceramide and constitutes glucosylceramide (GlcCer) and/or galactosyl ceramide (GalCer). In nature, GlcCer is found in plants but not GalCer while GalCer is restricted to the neural tissues of animals such as cows. The use of GalCer extracted from cow tissues is prone to safety issues due to the detection of prion, bovine spongiform encephalopathy. Thus, GalCer from animals is not acceptable. However, fungi are known to produce both GlcCer and GalCer (Jiang et al. 2021).

Glycosylceramide is known to enhance the expression of genes involved in skin barrier function, and lipid transport within the keratinocytes. From *Candida utilis* (Torula yeast), the GlcCer is known to enhance dermal fibroblast growth and collagen production for maintenance of dermal elasticity (Fukunaga et al. 2019). GlcCer from *Aspergillus oryzae* and *Saccharomyces cerevisiae* is recognized for its moisture retention effects in cosmetics in Japan (Takahashi et al. 2014). Also, GlcCer from *koji* with *Aspergillus luchensis* and *A. oryzae* elevated occludin (an epidural tight junction protein) expression and ATP-binding cassette subfamily A member 12 (a cellular membrane transporter to enhance ceramide in the keratinocytes). The above report indicates that the cosmetic effect exerted by *koji* was through elevation of ceramide and tight junction in the presence of glycosylceramides (Miyagawa et al. 2019; Jiang et al. 2021).

To increase skin ceramide content, both GalCer and GlcCer are required. It is known that GalCer activates beta-glucocerebrosidase to decompose GlcCer in keratinocytes and elevate ceramide content to improve dry skin and atopic dermatitis (Alessandrini et al. 2004). Moreover, GalCer extracted from cow is not approved for human use due to associated risk of prion. In

Table 3.1 Some bioactive compounds produced by fungi

Bioactive compounds	Fungi species	Activity	References
Saponin	*Ganoderma lucidum* *Plertous ostreatus*	Antioxidant Anticancer	Lee et al. (2019)
Tannin	*Lentinus edodes*	Antimicrobial Antioxidants	Mortimer et al. (2012)
Alkaloids	*Pleurotus ostreatus*	Antimicrobial, anti-inflammatory, antioxidant	Lesa et al. (2022)
Triterpenoids	*Tricholoma matsutake*	Antibacterial, anti-inflammatory	Deepalakshmi and Mirunalini (2011)

cosmetics formulation, the use of only GlcCer which is abundant in plants is deficient since the functional activity of GalCer is missing. Thus, the benefit of GalCer and GlcCer in cosmetics can be feasible through the synthesis of these compounds using microbial fermentation. The use of fungi is advantageous because fungi grow faster than plants and can yield enormous quantity of mycelia for extraction within a short time. Some of the fungi used for fermentation are generally regarded as safe (GRAS), for example, *A. oryzae*, and used in various food fermentations such as sake, shoyu (soybean paste), and *Miso* (soy sauce) for thousands of years. Using biotechnology, fungi can be modified to enhance the production of neutral sphingolipids.

Aspergillus oryzae, *A. sojae*, and *A. awamori* are well-known species of fungi used in varieties of Japanese fermented drinks and foods (Tani et al. 2014; Zhang et al. 2007) that can produce both GlcCer and GalCer. GalCer is known for activation of β-glucocerebrosidase to facilitate the decomposition of GlcCer to ceramide, which is deposited in stratum corneum leading to formation of the epidermal barrier. Apart from its application in cosmetics, glycosylceramide from *A. oryzae* has other beneficial effects, which include:

(a) Improvement of metabolic profiles, and flavor of sake yeast during production of sake (Ferdouse et al. 2018).
(b) When used as prebiotics in obese mice, the glycosylceramide is hydrolyzed by intestinal microbiota and the ratio of intestinal *Blautia coccoides*, a beneficial organism is elevated (Hamajima et al. 2016, 2019).
(c) Alteration of cholesterol metabolism, causing significant reduction in liver cholesterol (Hamajima et al. 2019).

Nowadays, bioactive compounds are used in food, medicine, and cosmetics. These bioactive substances play a variety of roles to sustain human health and revitalize body cells.

3.7 Bioactive Substances as Protective Agents Against DNA Damage

Bioactive chemicals are present in desirable amounts in naturally consumable resources such as plants, vegetables, and cereals. Diets enhanced with these substances can lower the risk of ailments, such as cancer, hypertension, cardiovascular disease, etc. These substances are naturally occurring antibacterial, anti-inflammatory, and anticancerous agents. It also serves as a barrier against DNA damage. The bioactive components found in citrus fruits (lemons, oranges) and fungi have been used to treat conditions including cancer (Yu et al. 2014). It has chemosensitive qualities as well. Natural resources high in antioxidants are highly effective at preventing DNA damage from radiation and other chemicals and processes. Bioactive compounds from fungi have potent anticancer activity (Ma et al. 2018; Ferreira et al. 2010).

3.8 Bioactive Substances as Organic Pigmenting Agents

The substances that are used as coloring agents are mostly carotenoids and flavonoids. Other bioactive substances are also used as natural dyeing agents and food coloring in addition to these. Although anthocyanins have been used for several years to treat illnesses including dysentery, diarrhea, and hypertension, in addition to the therapeutic effects, they are being utilized in the food sector for color impartation (Khoo et al. 2017).

Carotenoids, lycopene, betalains, riboflavin, melanin, and quinones are commonly used fungal pigments (Afroz Toma et al. 2023). *Aspergillus*, *Monascus*, *Cordyceps*, *Fusarium*, *Penicillium*, *Serratia*, *Talaromyces*, etc. are capable of producing these pigments. *Monascus* species produce a red pigment, monascorubramine, which was initially used for the production of red rice and red bean cord but was later applied to other areas due to its potential antimicrobial, anticancer, antioxidant, and cytotoxic properties.

Astaxanthin, a carotenoid with an antioxidant function, can be applied in pharmaceutical, food cosmetics, and other applications. Torularhodin is produced by *Rhodotorula*, *Sporidiobolus*, *Cystofilobasidium*, *Rhodospiridium*, and *Sporobolomyces* and can serve as useful cosmetics, food, and feed ingredient owing to its antioxidant and antimicrobial effects. Torulene produced by *Rhodotorula*, *Sporidiobolus*, *Neurospora*, *Sporobollomyces*, *Dioszegia*, *Cystophilobasdium*, and *Rhodospirillium* is also utilized to improve food, food, and cosmetics due to its antioxidant and antimicrobial properties.

Melanin is a biopolymer produced by *Cyptococcus neoformans*, *Aspergillus fumigatus*, *Sporothrix schenckii*, *Alteromonas nigrifaciens*, etc. and can be utilized as a novel polymer in material engineering as well as serving a protection from unfavorable conditions, making them suitable in food and medicinal use.

3.9 Bioactive Compounds as Agents Against Cancer and Inflammation

Flavonoids are often members of the broad group of phenolic compounds. The six more substances that make up its subclassification are flavones, isoflavones, anthocyanins, flavonols, flavanols, and flavanones. Isoflavones, with their anti-inflammatory qualities, are the most important phenolic chemicals (Venturella et al. 2021). Legumes and soybeans contain a lot of isoflavones. It functions as an antioxidant, antibacterial, and anticancer agent in addition to its anti-inflammatory properties. Daidzein and genistein are the two most prevalent isoflavones. They typically function as phytoestrogens and bind to the mammalian estrogen receptor site through pseudo-hormonal actions. Along with its anti-inflammatory qualities, it also has anticancer properties because it binds to strong estrogens to disrupt their pathways. It is crucial in the battle against all cancers linked to hormones, including breast and prostate cancer (Pejčić et al., 2023). Other bioactive substances with anti-inflammatory and anticancer properties include lycopene. Although flavonoids are produced by plants, studies have revealed the potential of endophytic fungi in flavonoid production as well as its role in inducing flavonoid synthesis in plants (Liu et al. 2022; Wu et al. 2022).

3.10 Bioactive Substances as Antioxidants

The most prevalent and regularly encountered bioactive chemicals are phenol-based compounds, like ascorbic acid and carotenoids. Citrus fruits, Brussels sprouts, tomatoes, etc. are the main foods that contain ascorbic acids. Carotenoids can be found in sweet potatoes, carrots, papayas, etc. They offer the greatest and most abundant supply of antioxidants. However, these fungi are associated with antioxidant activities (Turfan et al. 2020). These substances' primary functions are to guard against free radical damage to cellular membranes and the components that make them up. According to Martillanes et al. (2017), it can slow down microbial activity in certain substances.

3.11 Fungal L-Fucose as Cosmetics Ingredient

For many years, cosmeceutical formulations have used a wide range of fungi, macroalgae, and macroalgal blends because of their moisturizing, skin-conditioning, and viscosity-controlling potentials as well as their intrinsic stability, bioactive qualities, physical makeup, and marine source (Kim et al. 2005). L-fucose is an expensive but rare monosaccharide with high industrial demand due to its value as a useful food and anti-aging skincare products. It is a six-carbon sugar, 6-deoxy-1-galactose, identified with chemical formula; $C_6H_{12}O_5$, and a white crystalline powder that absorbs moisture, with melting point temperature of 150.0–153.0 °C (Wang et al. 2024c, 2024b). It is the only natural laevorotatory sugar beneficial to mammals. L-fucose can be a constituent of polysaccharide from fungi such as *Antrodia malicola* and *Antrodia cinnamomea*. L-fucose acts as an emulsifier and functional food in the food industry. Its biological activities include relief of intestinal disease, antitumor effect, and anti-aging (Wang et al. 2024c).

Due to the high industrial demand for L-fucose, microbial fermentation has been utilized to produce L-fucose and fungi such as *Mortierella alpino* have been successfully used (Wang et al. 2016). L-fucose is used in cosmeceutical products as a skin moisturizing agent. Other medicinal values of L-fucose include (Wang et al. 2024c; 2024b; Yu et al. 2014; Vanhooren and Vandamme 1999):

(a) Prevention of cell colonization of the lungs, an indication of its anticancer function.
(b) Controls white blood cells production, showing its potency in reducing inflammation.
(c) Antigen synthesis for the production of antibody.
(d) Treatment of rheumatoid arthritis.

3.12 Fungal Enzymes Used in Cosmetics

Enzymes from different sources have distinct characteristics and are involved in biotransformation where the end product is used in skincare products. Mushrooms are associated with enzymatic activities (Turfan et al. 2020). High yield of desired product is possible through application of enzyme technology due to enhanced efficacy. The use of enzyme allows for selective catalysis, giving rise to generation of high-purity products during synthesis. The main enzymatic changes produce components with sought-after qualities for the cosmeceutical industry. The basis for choice of modification includes biological effects such as skin-whitening, anti-inflammatory, antimicrobial, and photoprotective effects. A modification can occur via multiple mechanisms, such as transferases catalyzing glycosylation (reverse hydrolysis) or trans-glycosylation, laccases catalyzing β-glucosidases and oligomerization, and esterases (like lipases, feruloyl esterase, or tannase) and proteases catalyzing direct esterification or transesterification (Hong et al. 2014). In cosmeceutical products, the enzyme protease softens or exfoliates the skin by gradual removal of the keratin (desquamated corneocytes). Superoxide dismutase and catalase in skincare products scavenge/remove reactive oxygen species and reduce the dermal aging process. DNA repair enzymes such as photolyases are included in topical skin care formulations to repair DNA damage ascribed to ultraviolet rays (Kuo et al. 2024).

3.13 Beneficial Effects of Fungal Bioactive Compounds

3.13.1 Anti-aging Effect

The process of aging causes all body organs to gradually malfunction and lose some of their reserve capacity. The steady decline in physiological processes that comes with aging makes one more susceptible to chronic illnesses and ultimately to death. Aging is a normal and unavoidable phenomenon. Research on anti-aging focuses on slowing down or even stopping senescence which can either be intrinsic or extrinsic (Kosmadaki and Gilchrest 2004). In summary, the way an individual perceives their aging process is determined by the dynamic interplay of environmental, psychological (both intrinsic and extrinsic), and biological components. The breakdown of the collagen and elastin network appears to be the last pathway connecting all of the aging mechanisms (Khare et al. 2015). Thus, the design of age-renew cosmeceuticals is to retain and/or uphold the self-repairing mechanisms (SRM) of the human body. A crucial role for antioxidants is seen in SRM. A variety of mushroom species are utilized as anti-aging components in topical creams, serums, and face preparations.

3.13.2 Antioxidant Activity

Many studies have examined the antioxidant activity of mushrooms and their different applications in cosmetics. *Volvariella volvacea* and *Lentinula edodes* extracts have the capacity to scavenge free radicals, indicating antioxidant properties (Puttaraju et al. 2006; Cheung 2013). The phenolic chemicals of mushrooms origin are connected with the pharmacological impact, as has been previously established for wine and grapefruits. Moreover, *Lentinula edodes* stimulates the production of glutathione peroxidase (GPx) and superoxidase dismutase (SOD), two antioxidant enzymes that shield the skin from oxidative stress and reverse fibrosis by converting myofibroblasts into fibroblasts. Significant amounts of L-ergothioneine, a thiourea derivative of histidine that can reduce oxidation to the mitochondrial membrane, have been found in *Lentinus edodes, Grifola frondosa, Pleurotus eryngii, Pleurotus ostreatus,* and *portabellas species* (brown *Agaricus bisporus*) (Dubost et al. 2006). Using the lymphocyte proliferation assay, mice were used to evaluate the immunological and antioxidant bioactivities of glucan derived from *Phellinus ribis* (Yuan et al. 2009). Maity et al. (2021) reported that Pleurotus djamar polysaccharide, a water soluble galactoglucan, exhibited antioxidant activity with possible application in functional products or dietary health.

The glucan alkaline extract of somatic hybrid fungus composed of *Pleurotus florida* and *Calocybe indica var.* APK2 exhibited antioxidant effects along with immunological activation of thymocytes, splenocytes, and macrophages. Thus, it should come as no surprise that the inherent antioxidant qualities of *Phellinus linteus* or *Ganoderma lucidum,* whether ingested as food or nutraceutical food, can be transferred in-vivo (Sun et al. 2004).

3.13.3 Anti-wrinkle Activity

One prominent feature of aging skin is the production of wrinkles, attributed to the degradation of type I collagen, a structural protein of the skin (Lupo et al. 2007). Elastase is a metalloproteinase that contributes to the aging of the skin by breaking down elastin and preventing or repairing the creation of wrinkles. However, the breaking down of collagen and elastin causes matrix metalloproteinases (MMP) to occur. It was documented that the mycelial extract from *Tricholoma matsutake* (pine fungus) dramatically reduced MMP levels and, in a dose-dependent manner, lowered elastase activity (Kim et al. 2014).

3.13.4 Moisturizing Effect

Moisturizers lessen lines and wrinkles and the loss of skin hydration. Additionally, moisturizers develop the surface of the skin in terms of its physical qualities and chemical qualities, making it smooth, soft, and moisturized while also increasing the stratum corneum's hydration (Baumann 2016). In cosmetic formulations, traditional moisturizing compounds including hyaluronic acid, pantothenic acid, and 6-palmitoyl-L-ascorbic acid are recognized for their efficient moisturizing properties. When applied topically, the polysaccharide extracted from a hot water extract of a *Tremella* fungus was found to possess a unique property of reducing the impacts of melanin synthesis and lightening skin spots after topical application (Zhang et al. 2013). In another study, Liu and He (2012) reported that 0.05% *Tremella* polysaccharide in cosmetics had a higher moisturizing effect than cosmetic containing 0.02% hyaluronic acid. A similar report of outstanding activity was made on raw *Tremella fuciformis*-generated carboxymethylated polysaccharide when compared to chitosan, ascribing higher moisture retention capacity to *Tremella* polymer (Wang et al. 2015). *Auricularia fuscosuccinea* water extract also displayed water retention capacity comparable to sodium hyaluronate but lesser than *Tremella fuciformis* sporocarp extract (Yao et al. 2018; Yu et al. 2014). *Fomes officinalis* is a type of fungus. Its powder was added to an anhydrous cosmetic formula to improve the look of skin flaws and reduce skin shine. These demonstrated the potential use of mushroom polysaccharides as powerful moisturizing ingredients in cosmetics.

3.14 Skin Whitening

The primary pigment responsible for skin tone, melanin formation also served as a barrier against UV-induced skin irritation (Gupta et al. 2014). Tyrosinase primarily controls two separate oxidation processes in the biosynthetic route for melanin synthesis in different bioforms. In essence, tyrosinase catalyzes the process by which tyrosine is hydroxylated to dihydroxyphenylalanine (DOPA) and then oxidized to DOPAquinone. Depending on the physiological state, DOPAquinone is then further transformed to either pheomelanin (yellow-red pigment) or eumelanin (brown-black pigment). Thus, the most often used method for achieving skin whitening is tyrosinase suppression (Zhao et al. 2022). Mushroom extracts from *Phellinus baumii, Agaricus bisporus, Agaricus hortensis,* and *Ganoderma* spp. displayed tyrosinase inhibition. In B16/F10 melanoma cells, methanol extract from *Phellinus baumii* fruiting bodies demonstrated dose-dependent inhibitory actions of tyrosinase and melanin make up (Lee et al. 2013). In B16 murine melanoma cells, it was also shown that extracts from *Flammulina velutipes* inhibited tyrosinase activity and melanin formation (Nagasaka et al. 2015). *Ganoderma lucidum* showed the greatest tyrosinase inhibition when compared to other basidiomycetes. At the moment, *Ganoderma* extracts are utilized in most face mask cosmetics being marketed.

3.15 Haircare Potential

Hair cosmetics are intended to manipulate the structural properties of hair while also preserving scalp health. Hair conditioners, serums, shampoos, hair styling products (mousses, gels, waxes, and sprays), hair dyes, perming agents, bleaching agents, and straightening products are among the formulations of haircare products (Madnani and Khan 2013). Enhancing scalp hair density is a common goal of prescribed drugs like finasteride and minoxidil (Dias 2015). When paired with cosmetic haircare products, many women who are post-menopaused may attain a more attractive appearance. These products may increase the diameter of the hair fibers in a non-permanent manner. Shampoos are intended for cleaning the scalp and hair. They have been modified to address specific issues including dandruff, androgenic alopecia, as well as variances in hair quality and haircare habits (Tiwari et al. 2022; Usman et al. 2021; Nazir et al. 2012). *Tremella* mushrooms have a hydrophilic agent polysaccharide glucuronoxylomannan (GXM) with anti-inflammatory and wound-healing properties largely used in cosmetology. A little extract from silk film shiitake mushrooms has been added and is patented as a care product for hair. *Ganoderma lucidum*, together with three other botanicals and zinc, was included in another composition that was designed to stimulate human hair growth in males by reducing dihydrotestosterone or prostatic hyperplasia levels (Meehan et al. 2015). Vital nutrients must be given to hair follicles for healthy hair, just like they are for the entire body. Hair difficulties can be brought on by a deficiency of essential nutrients as well as outside variables including severe chemical hair therapy, an unhealthy lifestyle, and protracted illness. Antioxidants (vitamins C, D, E, A, and minerals zinc, copper, selenium) as well as iron (an antioxidant response inducer) are all found in mushrooms and can help to maintain strong, healthy hair as well as prevent dandruff and hair loss (Wu et al. 2016). The cosmetics sector has a lot of promises when it comes to incorporating fungal extracts into haircare products.

3.16 Application of Fungal Compounds in Cosmetics

Cosmetic comes from a Greek word "adorn" which describes the addition of anything ornamental to a person or something. It might include substances that improve the appearance of the body, namely: the skin, hair, nails, lips, mucous membranes, and teeth. Cosmetic substances enhance outward appearance and mask body odor, keeping the skin in good condition. Cosmetics are often formulated for topical applications. In the ancient days, men and women improve their body outlook by application of topical substances to decorate or beautify the body. For example, some people apply ointments and colored soil to their faces and bodies gradually. People even started wearing necklaces and bangles made of baked earth materials rather frequently. Red dye was used to dye hair, and eye shadow was prepared from colorful earth materials like copper ore and lamp black. Nowadays, cosmetics are seen as necessary elements of existence with both oral and topical applications, in addition to psychological impacts. Over the past three to four decades, cosmetics have become very popular, and the application expanded dramatically among male and female gender. Various common formulations of cosmetics include the following products:

- Foundation, used to conceal imperfections, or uneven skin tone. It smoothens the face and may be in the form of liquid, cream, or powder.
- Powder, used to cover-up minor imperfections or blemishes and set the foundation for a matte look.
- Rouge, blush, or blusher is a type of cheek coloring that is used to highlight the color of the cheeks and define the cheekbones. This is available as liquid, cream, and powder.
- Bronzer is a cosmetic product that adds a golden or bronze sheen to skin to give it a little color.
- Mascara is used to lengthen, darken, and thicken lashes. It comes in natural hues.

3.16.1 Application of Fungal Compounds in Cosmeceuticals

Cosmeceuticals are applied on the skin, hair, and nails based on a variety of therapeutic indications like (Nomakhosi and Heidi 2018; Baumann 2016):

(i) Sunscreen, skin lightening or depigmenting.
(ii) Skin moisturization.
(iii) Anti-aging or anti-wrinkle effects.
(iv) Reduction of scar.
(v) Antioxidants and treatment of specific disorders, e.g., acne, melasma, etc.
(vi) Hair breakage arrestor, and hair strengthening.
(vii) Stimulation of hair growth, and improvement of hair texture.

3.16.2 Application of Fungal Compounds in Nutricosmetics

Nutricosmetics are close to cosmeceuticals but are consumed orally as supplements like tablets and pills (Taofiq et al. 2016a).

1. Improves skin hydration and elasticity: certain nutricosmetics contain ingredients that promote the formation of collagen, hyaluronic acid, and elastin, which help to rejuvenate your skin and lessen the appearance of wrinkles, redness, and skin inflammation.
2. Prepares the skin for the sun: Many nutricosmetics contain antioxidants that help protect the skin against UV damage, combat skin ageing, and enhance tanning. However, it is crucial to remember that no nutricosmetic can take place of the use of sunscreen.
3. Strengthens the hair and nails: Biotin, selenium, and zinc are nutrients that help strengthen your hair and they also promote keratin production to give structure and strengthen your nails.

3.17 Fungal Fermentation in Cosmetic Production

The fermentation process has wide applications in various industries, including the production of cosmetics. The procedure involves the conversion of organic substances by fungi under controlled conditions, leading to the production of bio-functional compounds that offer numerous beneficial health impacts (Krzyżostan et al. 2024). Fungal fermentation is an excellent process of generating compounds endowed with bioactivities that have potential application in beauty-care products. Fungi are known to produce a plethora of bioactive compounds which have potential applications in cosmetics industry (Klimek-Szczykutowicz et al. 2024; Pérez-Rivero and López-Gómez 2023; Mago et al. 2023). Fermentation plays an essential role in the manufacture of cosmeceuticals and nutricosmetics.

A variety of cosmeceutical products employ the fermentation process. One significant application is in facial serums and creams that target specific skin concerns such as acne, hyperpigmentation, or aging. Fermented ingredients like plant extracts or botanicals provide nourishment to the skin while improving its texture and appearance. Furthermore, skincare products contain fermented probiotics to support a healthy skin microbiota, which is essential to skin health (Andriani and Yanto 2024).

3.18 Safety and Efficacy of Fungal Bio-product in Cosmetics

Fungal bio-products have gained significant attention in the cosmetic and nutraceutical industries due to their potential beauty- and healthcare benefits (Mishra and Srivastava 2024). These bio-products, obtained from various fungi, provide a range of active compounds endowed with antioxidants, anti-inflammatory, antimicrobial, and moisturizing effects. However, before incorporating fungal bio-products into cosmeceuticals and nutricosmetics, it is very necessary to evaluate their safety profiles to ensure safety and consistency, thereby promoting consumers' well-being (Taofiq et al. 2019). The safety assessment includes (Manful et al. 2024; FDA 2022):

(a) Toxicological examination.
(b) Skin sensitization and irritation test.
(c) Acute and chronic toxicity studies (LD_{50}).
(d) Mutagenicity and genotoxicity assessment.
(e) Good manufacturing practice (GMP).
(f) Labeling requirements, indicating the scientific name of fungal species, the concentration of active ingredients, and potential allergen present.

3.19 Case Studies of Fungal Extracts Used in Cosmetics

Many fungal products have been exploited in cosmetic production and the studies have provided supporting evidence for use of fungi in cosmeceutical and nutraceutical formulations. Case studies have shown that fungal extract or compound improved skin conditions such as eczema, aging, acne, and hyperpigmentation (Williams 2016). Among the outstanding and successful mushroom-cosmetic studies were:

(a) *Ganoderma lucidum* (Reishi mushroom) extract formulation as cosmeceutical (topical agent) is safe for anti-aging and anti-inflammatory effects. However, its safety profile as nutricosmetics has not been ascertained. The mushroom extract also enhanced skin elasticity and decreased wrinkle depth after 8 weeks of application (Thomas et al. 2020; Taofiq et al. 2017).
(b) Extract of *Inonotus obliquus* (Chaga mushroom) with antioxidant bioactivity and incorporated into various cosmeceutical products showed no evidence of significant allergy or toxicity (Fordjour et al. 2023; Park et al. 2023).
(c) *Tremella fuciformis* (snow fungus) extract recognized for its moisturizing and hydrating function, popularly used in Asian skincare products exhibited no significant irritancy or toxicity in the formulation (Ma et al. 2021; Hyde et al. 2010).
(d) *Cordyceps sinensis* and *C. militaris* extracts, known for skin-rejuvenating and anti-aging effects had been used in cosmeceutical formulation but no toxicity or adverse effect was associated with the topical preparations. These mushrooms have been formulated as face cream and gel. These mushrooms possess strong anti-inflammatory property and other bioactivities (Nguyen et al., 2025; Srivastava et al., 2020).

(e) *Agaricus bisporus* (white button fungus) extracts, endowed with anti-inflammatory and antioxidant properties had no significant allergenic or toxicity effect detected when used in cosmetic formulation (Usman et al. 2021).

3.20 Fungal-Based Cosmetics Products in Market

Fungi are recognized as producers of highly valued natural bioactive compounds with applications in the cosmetic sector. The compound hydroxycinnamic acid regulates skin inflammatory disorder, atopic dermatitis, and photo-carcinogenesis (Table 3.2). Varieties of fungi-based cosmetics abound in the market being formulated as skin lotion, skin serum, face cleanser, face mask, face cream, face gel, anti-aging cream, moisturizing cream/lotion/gel, eye mask, hair polish, anti-marks cream, massage gel, soap, shampoo, hair oil, skin scrub, etc. (Srivastava et al. 2020). Oral formulations of fungal-based nutricosmetics also exist in the form of supplements. Displays of such products are available online.

The cosmeceutical product can be prepared from combinations of different mushroom extracts, for example, Shroom-based skincare product, "mycelium glow brightening serum", is an oil serum made from combination of mushroom and vitamin C for skin rejuvenation, even tone and moisturizing effects. The oil serum has an antioxidant property, and contains mushroom extracts such as maitake, chaga, reishi, and cordyceps, in addition to vitamin C. Maitake extract has anti-inflammatory and antioxidant effects serving as skin protection, enhance skin elasticity, and reduce redness. Reishi extract (*Ganoderma lucidum*) contains beta-glycans, which promote skin hydration and thus, is used to formulate skincare products to relieve dry, itchy skin. Chaga mushroom extract is a potent antioxidant for protection of cells from the harmful effects of free radicals. The extract protects the skin and promotes skin repair as well as prevention of inflammation. Cordyceps extract is responsible for protection of collagen and enhance skin elasticity. It also promotes even skin tone.

Most of these fungal-based cosmetics are based on metabolites from mushrooms. It is expected that in the future, products will be developed from the filamentous fungi. It was also noticed that the fungi-based cosmetics are not fully represented in the global market, being manufactured mainly in Asia, Europe, and the USA. Based on the economic potential of mushrooms, it is expected that there will be increased interest in the development of national economy through promotion of fungal research and product development. Valorization of the cosmeceutical and nutraceutical potentials of fungi will boost the economy of both the developing and developed countries.

Table 3.2 Examples of cosmetic products containing fungi and their function

Cosmetic products	Fungi species used	Functions
Vitamega facial moisturizing mask	*Agaricus brasiliensis*	Revitalize and renew skin
Hankook Sansim firming cream (Tan Ryuk SANG)	*Ganoderma lucidum and Pleurotus ostreatus*	Firm skin and vitalized
Root science RS reborn organic face mask	*Inonotus obliquus*	Anti-inflammatory, soothe irritated skin
CV Skinlabs body repair lotion	*Ganoderma lucidum*	Wound-healing and anti-inflammatory
Kose Sekkisei cream	*Cordyceps sinensis*	Moisturizer and suppress melanin production
La prairie advanced marine biology night solution	*Tremella fuciformis*	Moisturizer which nourishes, revitalizes, and hydrates skin
Surkran grape seed lift eye mask	*Tremella polysaccharide*	Improve skin around eyes

3.20.1 Regulation of Fungal-Based Cosmetics Ingredients

The regulation should be done by both national and international regulatory bodies, including the US Government office for the Control of Food and Drug Administration (FDA), Health Canada, and The European Union Cosmetic Regulation (1223/2009). Regulation of fungal products and fungal-based cosmetics is a challenge (Manful et al. 2024; Patel et al., 2022; FDA 2022).

3.20.2 Challenges Associated with Use of Fungal-Based Cosmetics

Some compounds or metabolites produced by fungi are associated with adverse health effects. Thus, safety and toxicology of the bioactive compound from fungi is assessed to ascertain its application in cosmetics. Other challenges include regulatory issues, quality control, product stability, and storage-life (Manful et al. 2024; Lee et al. 2013).

3.20.3 Future Prospects of Fungal Extracts as Cosmetic Ingredient

Bioactive compounds from fungi are mostly unexplored and unexploited. Most cosmetic products in the markets were developed from a few edible mushrooms. The research gap will be filled in the future through increased participation by researchers and collaborative research from both developed and developing countries. It is expected that fungal-based product, with global acceptability and approval, will be developed/produced by scientists from developing countries following good manufacturing practice.

Most research reports consulted during the review study were from developed countries, indicating the paucity of fungal-based cosmetic research in the developing countries where there is an abundance of both mushroom and filamentous fungi. Research increase is expected on bioactive compounds from soil, aquatic, and terrestrial fungi, leading to discovery of potential fungal bioactive new compounds and development of products in the cosmetic sector or other industries. Utilization of advanced techniques in fungal cultivation and extraction of bioactive compounds will enhance discovery, making research less tedious and improve sustainability and market opportunities to meet global demand for fungal products. The challenges associated with product regulation and safety assessment will be reduced or eliminated using modern technology. Fungal extract as cosmetic ingredient will improve the national economy and increase employment opportunities. Safe and natural-based cosmetics will be on the increase, promoting youthful body appearance and human health.

3.21 Conclusion

The use of fungal bioactive compounds as cosmetics ingredients is a positive and desirable development in beauty and biotechnology, and a promising sustainable innovation. Mushrooms and other fungi are source of highly valued compounds with biological functions including anti-aging, antioxidant, skin-whitening, moisturizing, and anti-wrinkle effects. Fungal-based cosmetics have been produced, marketed, and used after assessment of the product safety profile. Yet, many fungi have not been explored. There is paucity of research in the area of fungal bioactive compound as a cosmetic ingredient especially in developing countries, indicating the need for intensive research to promote economic development as well as public health. The unique potential benefits of fungal bioactive compounds as cosmetic ingredients qualify their research despite challenges. It is expected that with advanced technology for fungal cultivation and extraction, good manufacturing practice, and regulatory advancements, bioactive compounds from fungal extract can potentially revolutionize the cosmetics sector through provision of sustainable, safe, natural compounds as alternative ingredients for cosmetics.

References

Patel, J. (2022). Amending the Food, Drug, and Cosmetic Act's Labeling Requirements for Cosmetics. Geo. Wash. L. Rev. Arguendo, 90, 31.

Adebayo EA, Oloke JK (2017) Oyster mushroom (Pleurotus species); a natural functional food. J Microbiol Biotechnol Food Sci 7(3):254–264. https://doi.org/10.15414/Jmbfs.2017/18.7.3.254-264

Afroz Toma M, Rahman MH, Rahman MS, Arif M, Nazir KNH, Dufossé L (2023) Fungal pigments: carotenoids, riboflavin, and polyketides with diverse applications. J Fungi 9(4):454

Agrawal S, Adholeya A, Barrow CJ, Deshmukh SK (2018) Marine fungi: an untapped bioresource for future cosmeceuticals. Phytochem Lett 23:15–20

Ahmad MF (2018) Ganoderma lucidum: persuasive biologically active constituents and their health endorsement. Biomed Pharmacother 107:507–519

Akgul H, Sevindik M, Coban C, Alli H, Selamoglu Z (2017) New approaches in traditional and complementary alternative medicine practices: Auricularia auric ula and Trametes versicolor. J Tradit Med Clin Natur 6(2):239

Alam N, Yoon KN, Lee TS (2011a) Evaluation of the antioxidant and antityrosinase activities of three extracts from Pleurotus nebrodensis fruiting bodies. Afr J Biotechnol 10:2978–2986

Alam N, Yoon KN, Cha YJ, Kim JH, Lee KR, Lee TS (2011b) Appraisal of the antioxidant, phenolic compounds concentration, xanthine oxidase and tyrosinase inhibitory activities of Pleurotus salmoneostramineus. Afr J Agric Res 6:1555–1563

Alessandrini F, Pfister S, Kremmer E, Gerber JK, Ring J, Behrendt H (2004) Alterations of glucosylceramide-β-glucosidase levels in the skin of patients with psoriasis vulgaris. J Invest Dermatol 123(6):1030–1036

Andriani A, Yanto DHY (2024) Fungal applications in skincare products. In: Biomass-based cosmetics: research trends and future outlook, pp 113–132

Anusiya G, Gowthama Prabu U, Yamini NV, Sivarajasekar N, Rambabu K, Bharath G, Banat F (2021) A review of the therapeutic and biological effects of edible and wild mushrooms. Bioengineered 12(2):11239–11268. https://doi.org/10.1080/21655979.2021.2001183

Bao X, Liu C, Fang J, Li X (2021) Structural and immunological studies of a major polysaccharide from spores of Ganoderma lucidum (Fr.) Karst. Carbohydr Res 332(1):67–74

Baumann L (2016) Cosmeceuticals in skin of color. Semin Cutan Med Surg 35(4):233–237

Bedi A, Gupta MK, Conlan XA, Cahill DM, Deshmukh SK (2021) Endophytic and marine fungi are potential source of antioxidants. In: Fungi bio-prospects in sustainable agriculture, environment and nano-technology, pp 23–89

Borchers AT, Stern JS, Hackman RM, Keen CL, Gershwin ME (2019) Mushrooms, tumors, and immunity. Proc Soc Exp Biol Med 221(4):281–293

Bulam S, Üstün NŞ, Pekşen A (2019a) Health benefits of Ganoderma lucidum as a medicinal mushroom. Turk J Agric Food Sci Technol 7:84–93

Bulam S, Üstün NŞ, Pekşen A (2019b) Evaluation of nutritional and medicinal values of edible wild and cultivated Pleurotus ostreatus. Turk J Agric Food Sci Technol 7(12):2054–2061

Calado MDL, Silva J, Alves C, Susano P, Santos D, Alves J et al (2021) Marine endophytic fungi associated with Halopteris scoparia (Linnaeus) Sauvageau as producers of bioactive secondary metabolites with potential dermocosmetic application. PLoS One 16(5):e0250954

Carrasco-González JA, Serna-Saldívar SO, Gutiérrez-Uribe JA (2017) Nutritional composition and nutraceutical properties of the Pleurotus fruiting bodies: potential use as food ingredient. J Food Compos Anal 58: 69–81. https://doi.org/10.1016/J.Jfca.2017.01.016

Chahal S, Sindhu A, Singh A, Sindhu SC (2024) Exploring extraction techniques for medicinal mushroom bioactive compounds: a comprehensive review of advantages and limitations. Pharma Innov J 13(2): 272–280

Chellapandi P, Saranya S (2024) Ophiocordyceps sinensis: a potential caterpillar fungus for the production of bioactive compounds. Explor Res Hypothesis Med 9(3):236–249

Cheng WY, Wei XQ, Siu KC, Song AX, Wu JY (2018) Cosmetic and skincare benefits of cultivated mycelia from the Chinese caterpillar mushroom, Ophiocordyceps sinensis (ascomycetes). Int J Med Mushrooms 20(7):623

Cheung PCK (2013) Mini-review on edible mushrooms as source of dietary fiber: preparation and health benefit. Food Sci Human Wellness 2:162–166

Chib S, Jamwal VL, Kumar V, Gandhi SG, Saran S (2023) Fungal production of kojic acid and its industrial applications. Appl Microbiol Biotechnol 107(7): 2111–2130

Cör D, Knez Ž, Knez Hrnčič M (2018) Antitumour, antimicrobial, antioxidant and antiacetylcholinesterase effect of Ganoderma lucidum terpenoids and polysaccharides: a review. Molecules 23(3):649

Correia RT, Borges KC, Medeiros MF, Genovese MI (2012) Bioactive compounds and phenoliclinked functionality of powdered tropical fruit residues. Food Sci Technol Int 18:539–547

Cristiano L, Guagni M (2022) Zooceuticals and cosmetic ingredients derived from animals. Cosmet 9(1):13. https://doi.org/10.3390/cosmetics9010013

de Caldas Felipe MT et al (2023) Production of kojic acid by aspergillus species: trends and applications. Fungal Biol Rev 45:100313

Deepalakshmi K, Mirunalini S (2011) Therapeutic properties and current medical usage of medicinal

mushroom: Ganoderma lucidum. Int J Pharm Sci Res 2:1922–1929
Dias MF (2015) Hair cosmetics: an overview. Int J Trichology 7:2–15
Dubost NJ, Beelman RB, Peterson D, Royse DJ (2006) Identification and quantification of ergothioneine in cultivated mushrooms by liquid chromatography-mass spectroscopy. Int J Med Mushrooms 8:215–222
El-Ramady H, Abdalla N, Fawzy Z, Badgar K, Llanaj X et al (2022) Green biotechnology of oyster mushroom (Pleurotus ostreatus L.): a sustainable strategy for Myco-remediation and BioFermentation. Sustainability 14:3667
Ferdouse J, Yamamoto Y, Taguchi S, Yoshizaki Y, Takamine K, Kitagaki H (2018) Glycosylceramide modifies the flavor and metabolic characteristics of sake yeast. PeerJ 6:e4768
Ferreira ICFR, Vaz JA, Vasconcelos MA, Martins A (2010) Compounds from wild mushrooms with antitumor potential. Anti-Cancer Agent Med Chem 10:424–436
Ferreira ICFR, Barros L, Abreu RMV (2009) Antioxidants in wild mushrooms. Curr Med Chem 16:1543–1560
Food and Drug Administration (FDA) (2022) Laws regulating cosmetic labelling. In Cosmetics Labeling Guide. U.S. Department of Health and Human Services
Fordjour E, Manful CF, Javed R, Galagedara LW, Cuss CW, Cheema M, Thomas R (2023) Chaga mushroom: a super-fungus with countless facets and untapped potential. Front Pharmacol 14:1273786
Fukunaga S, Wada S, Yamashita M, Morita M, Aoi W, Naito Y, Higashi A (2019) Torula yeast (Candida utilis)-derived glucosylceramide contributes to dermal elasticity in vitro. J Food Biochem 43(7):e12847
Galappaththi MCA, Dauner L, Madawala S, Karunarathna SC (2021) Nutritional and medicinal benefits of oyster (Pleurotus) mushrooms: a review. Fungal Biotechnol 1(2):65–87. https://doi.org/10.5943/Funbiotec/1/2/5
Gargano ML, van Griensven LJ, Isikhuemhen OS, Lindequist U, Venturella G, Wasser SP, Zervakis GI (2017) Medicinal mushrooms: valuable biological resources of high exploitation potential. Plant Biosys-Int J Dealing All Aspect Plant Biol 151(3):548–565
Ghosh K (2016) A review: edible mushrooms as source of dietary fiber and its health effects. J Phys Sci 21:129–137
Gmoser R, Ferreira JA, Lennartsson PR, Taherzadeh MJ (2017) Filamentous ascomycetes fungi as a source of natural pigments. Fungal Biol Biotechnol 4:1–25
Golak-Siwulska I, Kałużewicz A, Spiżewski T, Siwulski M, Sobieralski K (2018) Bioactive compounds and medicinal properties of oyster mushrooms (Pleurotus sp.). Folia Hort 30(2): 191–201. https://doi.org/10.2478/Fhort-2018-0012
Guo ZJ, Liu Y, Yang JY, Jin MY, Mao PW, Zhou XW (2023) Evaluating the application potential of a recombinant Ganoderma protein as bioactive ingredients in cosmetics. Molecules 28(7):3272
Gupta SK, Gautam A, Kumar S (2014) Natural skin whitening agents: a current status. Adv Biol Res (Rennes) 8(6):257–259
Hamajima H, Matsunaga H, Fujikawa A, Sato T, Mitsutake S, Yanagita T et al (2016) Japanese traditional dietary fungus koji *Aspergillus oryzae* functions as a prebiotic for Blautia coccoides through glycosylceramide: Japanese dietary fungus koji is a new prebiotic. Springerplus 5:1–10
Hamajima H, Tanaka M, Miyagawa M, Sakamoto M, Nakamura T, Yanagita T et al (2019) Koji glycosylceramide commonly contained in Japanese traditional fermented foods alters cholesterol metabolism in obese mice. Biosci Biotechnol Biochem 83(8): 1514–1522
Hapsari R, Elya B, Amin J (2012) Formulation and evaluation of antioxidant and tyrosinase inhibitory effect from gel containing the 70% ethanolic Pleurotus ostreatus extract. Int J Med Aromather Plant 2:135–140
Hong YH, Jung EY, Noh DO, Suh HJ (2014) Physiological effects of formulation containing tannase-converted green tea extract on skin care: physical stability, collagenase, elastase, and tyrosinase activities. Integr Med Res 3:25–33
Hu L, Weng J, Wang Z, Huang C, Zhang L (2024) Effect and mechanism of Tricholoma matsutake extract combined with bakuchiol and ergothioneine on UVB-induced skin aging. J Cosmet Dermatol 23(11): 3628–3644
Huang K, El-Seedi HR, Xu B (2022) Critical review on chemical compositions and health-promoting effects of mushroom Agaricus blazei Murill. Curr Res Food Sci 5:2190–2203
Hyde KD, Bahkali AH, Moslem MA (2010) Fungi – an unusual source for cosmetics. Fungal Divers 43:1–9
Ji DB, Ye J, Li CL, Wang YH, Zhao J, Cai SQ (2019) Antiaging effect of Cordyceps sinensis extract. Phytother Res 23(1):116–122
Jiang C, Ge J, He B, Zeng B (2021) Glycosphingolipids in filamentous fungi: biological roles and potential applications in cosmetics and health foods. Front Microbiol 12:690211
Khare N, Khare P, Yadav G (2015) Recent advances in anti-aging. A review. Global J Pharm 9:267–271
Khoo HE, Azlan A, Tan ST, Lim SM (2017) Anthocyanidins and anthocyanins: colored pigments as food, pharmaceutical ingredients, and the potential health benefits. Food Nutr Res 61(1):1361779
Kim HH, Shin CM, Park CH, Kim KH, Cho KH, Eun HC, Chung JH (2005) Eicosapentaenoic acid inhibits UV-induced MMP-1 expression in human dermal fibroblasts. J Lipid Res 46(8):1712–1720
Kim SY, Go KC, Song YS, Jeong YS, Kim EJ, Kim BJ (2014) Extract of the mycelium of T. matsutake inhibits elastase activity and TPA-induced MMP-1 expression in human fibroblasts. Int J Mol Med 34(6):1613–1621

Klimek-Szczykutowicz M, Błońska-Sikora EM, Kulik-Siarek K, Zhussupova A, Wrzosek M (2024) Bioferments and biosurfactants as new products with potential use in the cosmetic industry. Appl Sci 14(9): 3902

Kosmadaki MG, Gilchrest BA (2004) The role of telomeres in skin aging/photoaging. Micron 35:155–159

Krzyżostan M, Wawrzyńczak A, Nowak I (2024) Use of waste from the food industry and applications of the fermentation process to create sustainable cosmetic products: a review. Sustainability 16(7):2757

Kuo CH, Wang HMD, Shieh CJ (2024) Enzymes in biomedical, cosmetic and food application. Catalysts 14(3):162

Lee JS, Shin DB, Lee SM, Kim SH, Lee TS, Jung DC (2013) Melanogenesis inhibitory and antioxidant activities of Phellinus baumii methanol extract. Korean J Mycol 41(2):104–111

Lee HR, Lim HB (2019) Antimutagenic and antioxidative effects of polysaccharides isolated from the water extract of Ganoderma lucidum. J Appl Pharm Sci 9(4):001–007

Lesa KN, Khandaker MU, Iqbal FMR, Sharma R, Islam F, Mitra S, Emran TB (2022) Nutritional value, medicinal importance, and health-promoting effects of dietary mushroom (Pleurotus ostreatus). J Food Qual. https://doi.org/10.1155/2022/2454180

Li LD, Mao PW, Shao KD, Bai XH, Zhou XW (2019) Ganoderma proteins and their potential applications in cosmetics. Appl Microbiol Biotechnol 103:9239–9250

Lima CU, Gris EF, Karnikowski MG (2016) Antimicrobial properties of the mushroom Agaricus blazei–integrative review. Rev Bras 26(6):780–786

Liu H, He L (2012) Comparison of the moisture retention capacity of Tremella polysaccharides and hyaluronic acid. J Anhui Agric Sci 40:13093–13094

Liu XQ, Cheng S, Aroca R, Zou YN, Wu QS (2022) Arbuscular mycorrhizal fungi induce flavonoid synthesis for mitigating oxidative damage of trifoliate orange under water stress. Environ Exp Bot 204:105089

Lupo MP, Cole AL (2007) Cosmeceutical peptides. Dermatol Ther 20(5):343–349

Ma Y, Wang C, Zhang Q, Peng X et al (2018) The effects of polysaccharides from Auricularia auricula (Huaier) in adjuvant anti-gastrointestinal cancer therapy: a systematic review and network meta-analysis. Pharmacol Res 132:80–89

Ma X, Yang M, He Y, Zhai C, Li C (2021) A review on the production, structure, bioactivities and applications of *Tremella* polysaccharides. Int J Immunopathol Pharmacol 35:20587384211000541

Madnani N, Khan K (2013) Hair cosmetics. Indian J Dermatol Venereol Leprosy 79:654–667

Mago P, Sharma R, Hafeez I, Nawaz I, Joshi M, Mehrotra R (2023) Mushroom based cosmeceuticals: an upcoming biotechnology sector. Biosci Biotechnol Res Asia 20(2):381–394

Maity K, Kar E, Maity S, Gantait SK, Das D, Maiti S, Maiti TK, Sikdar SR, Islam SS (2021) Structural characterization and study of immunoenhancing and antioxidant property of a novel polysaccharide isolated from the aqueous extract of a somatic hybrid mushroom of Pleurotus Florida and Calocybe indica variety APK2. Int J Biol Macromol 48:304–310

Manful ME, Ahmed L, Barry-Ryan C (2024) Cosmetic formulations from natural sources: safety considerations and legislative frameworks in the European Union. Cosmetics 11(3):72

Martillanes S, Rocha-Pimienta J, Cabrera-Bañegil M, Martín-Vertedor D, Delgado-Adámez J (2017) Application of phenolic compounds for food preservation: food additive and active packaging. In: Soto-Hernández M, Palma-Tenango M, García-Mateos R (eds) Phenolic compounds: biological activity. IntechOpen, Londres, pp 39–58

Mau JL, Lin HC, Chen CC (2011) Nonvolatile components of several medicinal mushrooms. Food Res Int 34(6):521–526

Mcmullen RL, Dell'Acqua G (2023) History of natural ingredients in cosmetics. Cosmet 10(3):71. https://doi.org/10.3390/cosmetics10030071

Meehan K (2015) Composition to promote hair growth in humans. U.S. Patent US No. 9, 144. 542, 29 September. U.S. Patent and Trademark Office, Washington, DC

Menaa F (2014) Skin anti-aging benefits of phytotherapeutics-based emulsions. Pharm Anal Acta 5(9):10–11. https://doi.org/10.4172/2153-2435.1000e168

Meng TX, Furuta S, Fukamizu S, Yamamoto R, Ishikawa H, Arung ET, Shimizu K, Kondo R (2011) Evaluation of biological activities of extracts from the fruiting body of Pleurotus citrinopileatus for skin cosmetics. J Wood Sci 57:452–458

Michalak M (2023) Plant extracts as skin care and therapeutic agents. Int J Mol Sci 24(20). https://doi.org/10.3390/ijms242015444

Mishra M, Srivastava D (2024) Harnessing the power of medicinal mushrooms into cosmetics: a fountain of youth from nature. J Indian Bot Soc 104(4):200–209

Miyagawa M, Fujikawa A, Nagadome M, Kohama K, Ogami T, Kitamura S, Kitagaki H (2019) Glycosylceramides purified from the Japanese traditional non-pathogenic fungus aspergillus and koji increase the expression of genes involved in tight junctions and ceramide delivery in normal human epidermal keratinocytes. Fermentation 5(2):43

Miyazaki T, Nishijima M (1981) Studies on fungal polysaccharides. XXVII. Structural examination of a water-soluble, antitumor polysaccharide of Ganoderma lucidum. Chem Pharm Bull 29(12):3611–3616

Mohamad R, Mohamed MS, Suhaili N, Salleh MM, Ariff AB (2010) Kojic acid: applications and development of fermentation process for production. Biotechnol Mol Biol Rev 5(2):24–37

Mortimer PE, Karunarathna SC, Li Q, Gui H, Yang X, Yang X, Hyde KD (2012) Prized edible Asian mushrooms: ecology, conservation and sustainability. Fungal Divers 56:31–47

Mukherjee A, Das S, Chakraborty D, Pal N, Das N (2024) Fungi's treasure in cosmeceuticals-a comprehensive chemical approach. S Afr J Bot 166:311–331

Nagasaka R, Ishikawa Y, Inada T, Ohshima T (2015) Depigmenting effect of winter medicinal mushroom Flammulna velutipes (higher basidiomycetes) on melanoma cells. Int J Med Mushrooms 17:511–520

Nazir H, Wang L, Lian G, Zhu S, Zhang Y, Liu Y, Ma G (2012) Multilayered silicone oil droplets of narrow size distribution: preparation and improved deposition on hair. Colloids Surf B Biointerfaces 100:42–49

Nguyen TQ, Van Pham T, Andriana Y, Truong MN (2025) Cordyceps militaris-derived bioactive gels: therapeutic and anti-aging applications dermatology. Gels 11(1):33

Nomakhosi M, Heidi A (2018) Natural options for management of melasma, a review. J Cosmet Laser Ther 20(7–8):470–481. https://doi.org/10.1080/14764172.2018.1427874

Nurunnabi TR, Al-Majmaie S, Nakouti I, Nahar L, Rahman SM, Sohrab MH et al (2018) Antimicrobial activity of kojic acid from endophytic fungus Colletotrichum gloeosporioides isolated from Sonneratia apetala, a mangrove plant of the Sundarbans. Asian Pac J Trop Med 11(5):350–354

Nxumalo W, Elateeq AA, Sun Y (2020) Can Cordyceps cicadae be used as an alternative to Cordyceps militaris and Cordyceps sinensis?–a review. J Ethnopharmacol 257:112879

Ogra Y, Ishiwata K, Ruiz Encinar J, Łobiński R, Suzuki KT (2004) Speciation of selenium in selenium-enriched shiitake mushroom, Lentinula edodes. Anal Bioanal Chem 379:861–866

Ola AR, Sugi Y, Soa CAP, Lerrik RI, Tawa BD (2020) Analysis of production kojic acid from endophytic fungi aspergillus flavus isolated from Annona squamosa leaves using an OSMAC approach. IOP Conf Ser Mater Sci Eng 823:012003

Oliveira LA, Sanches MA, Segundo WO, Santiago PA, Lima RQ, Cortez AC et al (2024) Exploring colorant production by amazonian filamentous fungi: stability and applications. J Basic Microbiol 64(2):2300444

Ozen T, K*ı*z*ı*l D, Yenigun S, Cesur H, Turkekul I (2019) Evaluation of bioactivities, phenolic and metal content of ten wild edible mushrooms from Western Black Sea region of Turkey. Int J Med Mushrooms 21(10):979–994. https://doi.org/10.1615/Intjmedmushrooms.2019031927

Park BK, Kim MM (2010) Applications of chitin and its derivatives in biological medicine. Int J Mol Sci 11(12):5152–5164

Park J, Nguyen TMN, Park HA, Nguyen MTT, Lee NY, Ban SY et al (2023) Protective effects of lanostane triterpenoids from Chaga mushroom in human keratinocytes, HaCaT cells, against inflammatory and oxidative stresses. Int J Mol Sci 24(16):12803

Paterska M, Czerny B, Cielecka-Piontek J (2024) Macrofungal extracts as a source of bioactive compounds for cosmetical anti-aging therapy: a comprehensive review. Nutrients 16(16):2810. https://doi.org/10.3390/nu16162810

Pejčić T, Zeković M, Bumbaširević U, Kalaba M, Vovk I, Bensa M et al (2023) The role of isoflavones in the prevention of breast cancer and prostate cancer. Antioxidants 12(2):368

Pérez-Rivero C, López-Gómez JP (2023) Unlocking the potential of fermentation in cosmetics: a review. Fermentation 9(5):463

Phasha V, Senabe J, Ndzotoyi P, Okole B, Fouche G, Chuturgoon A (2022) Review on the use of kojic acid – a skin-lightening ingredient. Cosmet 9(3):64

Pintathong P, Chomnunti P, Sangthong S, Jirarat A, Chaiwut P (2021) The feasibility of utilizing cultured Cordyceps militaris residues in cosmetics: biological activity assessment of their crude extracts. Journal of Fungi 7(11):973

Piska K, Sułkowska-Ziaja K, Muszyńska B (2017) Edible mushroom Pleurotus ostreatus (oyster mushroom) –its dietary significance and biological activity. Acta Sci Pol Hortorum Cultus 16(1):151–161

Preto AO, Reis WS, Carvalho AKD, Pereira EB (2024) Exploring whole cells of Rhizopus oryzae for efficient synthesis of emollient esters: bioesters for cosmetics. J Braz Chem Soc 35(5):e-20230185

Puttaraju NG, Venkateshaiah SU, Dharmesh SM, Urs SM, Somasundaram R (2006) Antioxidant activity of indigenous edible mushrooms. J Agric Food Chem 54:9764–9772

Quintero-Cabello KP, Lugo-Flores MA, Rivera-Palafox P, Silva-Espinoza BA, González-Aguilar GA, Esqueda M et al (2021) Antioxidant properties and industrial uses of edible polyporales. J Fungi 7(3):196

Sałata A, Lemieszek M, Parzymies M (2018) The nutritional and health properties of an oyster mushroom (*Pleurotus ostreatus* (Jacq. Fr) P. Kumm.). Acta Scientiarum Polonorum Hortorum Cultus 17(2):185–197. https://doi.org/10.24326/Asphc.2018.2.16

Saraf S, Saraf S (2022) Cosmetics – a practical manual, 3rd edn. PharmaMed Press/BSP Books, p 204

Sarkar A, Philip AM, Thakker DP, Bhaskara Rao KV (2024) Extraction and characterization of Beta-glucan methyl esters derived from UV-mutated marine yeast and their applications in cosmetics. Thalassas Int J Mar Sci 40(1):721–733

Schiano I, Raco S, Cestone E, Jesenak M, Rennerova Z, Majtan J (2021) Pleuran-β-Glucan from oyster culinary-medicinal mushroom, Pleurotus ostreatus (Agaricomycetes), soothes and improves skin parameters. Int J Med Mushrooms 23(12):75

Sharif S, Rashid S, Atta A, Irshad A, Riaz M, Shahid M, Mustafa G (2018) Phenolics, tocopherols and fatty acid profiling of wild and commercial mushrooms from Pakistan. J Biol Regul Homeost Agents 32(4):863–867

Sharma S, Singh S, Sarma SJ (2024) Effectiveness of inorganic nitrogen on Kojic acid production from fungi aspergillus sp. BU20S. Korean J Chem Eng 41(4):1163–1172

Shehata RM, Sabry SM (2020) Biotechnological production of Kojic acid synthesized by endophytic fungi, aspergillus oryzae isolated from euphorbia peplis. Egypt Acad J Biol Sci G Microbiol 12(2):35–47

Smithson E, Thompson G (2017) Regulatory considerations for fungal bio-products in cosmetics. *Eur J Dermatol* 27(6):457–464

Sosa-Hernández JE, Escobedo-Avellaneda Z, Iqbal HMN, Welti-Chanes J (2018) State-of-the-art extraction methodologies for bioactive compounds from algal biome to meet bio-economy challenges and opportunities. Molecules 23(11)

Sousa P, Tavares-Valente D, Pereira CF, Pinto-Ribeiro I, Azevedo-Silva J, Madureira R et al (2024) Circular economyeast: Saccharomyces cerevisiae as a sustainable source of glucans and its safety for skincare application. Int J Biol Macromol 265:130933

Srivastava A, Attri BL, Sharma VP (2020) Status report on mushroom based cosmetic products in market. Mushroom Res 29(1):65

Sujarit K, Suwannarach N, Kumla J, Lomthong T (2021) Mushrooms: splendid gifts for the cosmetic industry. Chiang Mai J Sci 48(3):699–725

Sułkowska-Ziaja K, Trepa M, Olechowska-Jarząb A, Nowak P, Ziaja M, Kała K, Muszyńska B (2023) Natural compounds of fungal origin with antimicrobial activity – potential cosmetics applications. Pharmaceuticals 16(9):1200

Sun J, He H, Xie BJ (2004) Novel antioxidant peptides from fermented mushroom Ganoderma lucidum. J Agric Food Chem 52:6646–6652

Sun Y, He H, Wang Q, Yang X, Jiang S, Wang D (2022) A review of development and utilization for edible fungal polysaccharides: extraction, chemical characteristics, and bioactivities. Polymers 14(20):4454

Takahashi K, Izumi K, Nakahata E, Hirata M, Sawada K, Tsuge K et al (2014) Quantitation and structural determination of glucosylceramides contained in sake lees. J Oleo Sci 63(1):15–23

Tani Y, Amaishi Y, Funatsu T, Ito M, Itonori S, Hata Y et al (2014) Structural analysis of cerebrosides from aspergillus fungi: the existence of galactosylceramide in a. oryzae. Biotechnol Lett 36:2507–2513

Taofiq O, González-Paramás AM, Martins A, Barreiro MF, Ferreira IC (2016a) Mushrooms extracts and compounds in cosmetics, cosmeceuticals and nutricosmetics – a review. Ind Crop Prod 90:38–48

Taofiq O, Martins A, Barreiro MF, Ferreira ICFR (2016b) Anti-inflammatory potential of mushroom extracts and isolated metabolites.Trends in Food Science & Technology, 50, 193–210

Taofiq O, Heleno SA, Calhelha RC, Alves MJ, Barros L, González-Paramás AM et al (2017) The potential of Ganoderma lucidum extracts as bioactive ingredients in topical formulations, beyond its nutritional benefits. Food Chem Toxicol 108:139–147

Taofiq O, Rodrigues F, Barros L, Barreiro MF, Ferreira IC, Oliveira MBP (2019) Mushroom ethanolic extracts as cosmeceuticals ingredients: safety and ex vivo skin permeation studies. Food Chem Toxicol 127:228–236

Tessema EN, Gebre-Mariam T, Neubert RH, Wohlrab J (2017) Potential applications of phyto-derived ceramides in improving epidermal barrier function. Skin Pharmacol Physiol 30(3):115–138

Thi TXN, Thi PAN, Le MA, Nguyen HA, Nguyen MH, Thi HPN (2023) Study of the extraction process of the Pleurotus citrinopileatus mushroom and evaluation of the biological activity of the extract. Int J Adv Eng Res Sci 10(04)

Thomas PW, Elkhateeb WA, Daba GM (2020) Chaga (Inonotus obliquus): a medical marvel but a conservation dilemma. Sydowia 72:123–130

Tiwari A, Kumar S, Choudhir G, Singh G, Gangwar U, Sharma V et al (2022) Bioactive metabolites of edible mushrooms efficacious against androgenic alopecia: targeting SRD5A2 using computational approach. J Herbal Med 36:100611

Törős GH, El-Ramady H, Prokisch J (2022) Edible mushroom of Pleurotus spp.: a case study of oyster mushroom (Pleurotus ostreatus L.). Environ Biodiver Soil Secur 6(2022):51–59

Turfan N, Ayan S, Pekşen A, Akın ŞS (2020) Antioxidant enzyme activities of some wild and cultivated edible mushrooms in Turkey. Int J Agric Wildlife Sci 6(2): 202–209. https://doi.org/10.24180/Ijaws.749512

Uthan, E. T., Yamaç, M., & Yıldız, Z. (2022). In vitro prebiotic activity of polysaccharides extracted from edible/medicinal macrofungi species. Mantar Dergisi, 13(1), 15–29.

Umashankar P, Nygård Y (2024) Filamentous fungi as emerging cell factories for the production of aromatic compounds. Fungal Biol Biotechnol 11(1):19

Usman M, Murtaza G, Ditta A (2021) Nutritional, medicinal, and cosmetic value of bioactive compounds in button mushroom (Agaricus bisporus): a review. Appl Sci 11(13):5943

Vanhooren PT, Vandamme EJ (1999) L-Fucose: occurrence, physiological role, chemical, enzymatic and microbial synthesis. Journal of Chemical Technology & Biotechnology: 74(6), 479–497.

Venturella G, Ferraro V, Cirlincione F, Gargano ML (2021) Medicinal mushrooms: bioactive compounds, use and clinical trials. Int J Mol Sci 22:634. https://doi.org/10.3390/Ijms22020634

Visvanathan S, Krishnamoorthy R, Sabesan GS (2022) Fungal cosmetics: mushrooms in beauty care and the new age of natural cosmetics. In: Applied mycology: entrepreneurship with fungi. Springer International Publishing, Cham, pp 1–37

Vlassopoulou M, Yannakoulia M, Pletsa V, Zervakis GI, Kyriacou A (2021) Effects of fungal Beta-Glucans on health – a systematic review of randomized controlled trials. Food Funct 12:3366–3380. https://doi.org/10.1039/D1fo00122a

Wang X, Zhang Z, Zhao M (2015) Carboxymethylation of polysaccharides from Tremella fuciformis for antioxidant and moisture-preserving activities. Int J Biol Macromol 72:526–530

Wang H, Zhang C, Chen H, Yang Q, Zhou X, Gu Z et al (2016) Characterization of an fungal l-fucokinase involved in Mortierella alpina GDP-l-fucose salvage pathway. Glycobiology 26(8):880–887

Wang H, Cheng N, Wu Q, Fang D, Rahman FU, Hao H, Zhang Y (2024) Antioxidant activities of sea buckthorn polysaccharides and their potential application in cosmetic industry. J Dermatol Sci Cosmetic Technol 100023

Wang H, Cheng N, Wu Q, Fang D, Rahman FU, Hao H, Zhang Y (2024a) Antioxidant activities of sea buckthorn polysaccharides and their potential application in cosmetic industry. J Dermatol Sci Cosmet Technol:100023

Wang Q, Wang Y, Liu Y, Yuan K, Lin Y, Qian X et al (2024b) A low-molecular-weight α-glucan from edible fungus Agaricus blazei Murrill activates macrophage TFEB-mediated antibacterial defense to combat implant-associated infection. Carbohydr Polym 346:122659

Wang Y, Zhang X, Tian X, Wang Y, Xing X, Song S (2024c) Research progress on the functions, preparation and detection methods of l-fucose. Food Chem 433:137393

Wasser SP, Weis AL (1999) Therapeutic effects of substances occurring in higher basidiomycetes mushrooms: a modern perspective. Crit Rev Immunol 19(1):65–96

Williams LM (2016) Safety evaluation of fungal bio-products: a regulatory perspective. Regul Toxicol Pharmacol 78:1–10

Wu Y, Choi MH, Li J, Yang H, Shin HJ (2016) Mushroom cosmetics: the present and future. Cosmetics 3(22):1–13

Wu X, Zou K, Liu X, Fu S, Zhang S, Duan Z et al (2022) The novel distribution of intracellular and extracellular flavonoids produced by aspergillus sp. Gbtc 2, an endophytic fungus from Ginkgo biloba root. Front Microbiol 13:972294

Yan ZF, Yang Y, Tian FH, Mao XX, Li Y, Li CT (2014) Inhibitory and acceleratory effects of Inonotus obliquus on Tyrosinase activity and melanin formation in B16 melanoma cells. Evid Based Complement Alternat Med 2014:259836. https://doi.org/10.1155/2014/259836

Yang Y, Lv Z, An Q, Xu D, Sun L, Wang Y et al (2024) Tricholoma matsutake polysaccharides suppress excessive melanogenesis via JNK-mediated pathway: investigation in 8-methoxypsoralen induced B16–F10 melanoma cells and clinical study. Heliyon 10(8)

Yao FJ, Lu LX, Wang P, Fang M et al (2018) Development of a molecular marker for fruiting body pattern in Auricularia auricula-judae. Mycobiology 46:72–78

Yu J, Sun R, Zhao Z, Wang Y (2014) Auricularia polytricha polysaccharides induce cell cycle arrest and apoptosis in human lung cancer A549 cells. Int J Biol Macromol 68:67–71

Yuan G, An L, Sun Y, Xu G, Du P (2018) Improvement of learning and memory induced by Cordyceps polypeptide treatment and the underlying mechanism. Evid Based Complement Alternat Med 2018(1):9419264

Yuan C, Huang X, Cheng L, Bu Y, Liu G, Yi F, Yang Z, Song F (2009) Evaluation of antioxidant and immune activity of Phellinus ribis glucan in mice. Food Chem 115:581–584

Zerva A, Tsafantakis N, Topakas E (2020) Greek basidiomycete wild strains for the production of bioactive compounds and enzymes with applications in cosmetic and biocatalysis industries. In: First international electronic conference on catalysis sciences, vol 10, p 30

Zhang Y, Wang S, Li XM, Cui CM, Feng C, Wang BG (2007) New sphingolipids with a previously unreported 9-methyl-C 20-sphingosine moiety from a marine algous endophytic fungus aspergillus Niger EN-13. Lipids 42:759–764

Zhang K, Meng XY, Sun Y, Guo PY (2013) Preparation of Tremella, Speranskiae tuberculatae and Eriocaulon buergerianum extracts and their performance in cosmetics. Detergent Cosmet 36:28–32

Zhang GP, Pan YM, Ye SM, Lu YC, Fan XJ, Zhang AQ (2025) Bioactive components of Ganoderma lucidum and their efficacy and application in cosmetics. Food Med Homol 2(1)

Zhao W, Yang A, Wang J, Huang D, Deng Y, Zhang X et al (2022) Potential application of natural bioactive compounds as skin-whitening agents: a review. J Cosmet Dermatol 21(12):6669–6687

Zilles JC, Dos Santos FL, Kulkamp-Guerreiro IC, Contri RV (2022) Biological activities and safety data of kojic acid and its derivatives: a review. Exp Dermatol 31(10):1500–1521

Fungi Extracts and Their Use as Antioxidants in Cosmetic Ingredients

4

Anirban Debnath, Subhendu Bandyopadhyay, and Arpita Das

Abstract

Fungi, or mushroom, possess approximately 130 therapeutic properties including antitumor, immunomodulatory, antigenotoxic, antioxidant, anti-inflammatory, antihypertensive, antihyperglycemic, and antimicrobial activities. Fungi can serve as an unlimited source of numerous bioactive compounds such as phenolics, terpenoids, selenium, polysaccharides, vitamins, etc. which are beneficial for skin. In recent days, cosmetic industry constantly searches for natural compounds with antioxidant, antiaging, skin whitening, and moisturizing properties for topical and oral administration. Therefore, diverse fungi with nutritive and medicinal efficiency can serve as potent agent in cosmetic industry as cosmeceuticals and nutricosmetics. This chapter provides overview of some currently used common edible fungi enriched with various valuable bioactive compounds governing anti-hyaluronidase, anti-collagenase, antioxidant, anti-tyrosinase, antibacterial, antiviral, and anti-inflammatory potential that facilitates their usage as natural ingredients for cosmetic formulations.

Keywords

Antioxidant · Cosmetics · Phytocosmetics · Moisturizing · Antimicrobial · Anti-collagenase

4.1 Introduction

Cosmetics, also referred to as personal care products, are any substances which can be rubbed or sprinkled on skin, body, and hair for their cleansing, protecting, and beautifying (Millikan 2001; Taofiq et al. 2016). In the past several decades, cosmetic industry has emerged as a most promising and fastest-growing industries worldwide (Liu 2022). In 2023, the global cosmetic market is expected to reach US$800 billion from US$500 in 2017 with an annual growth rate of 7% (Dhapte-Pawar et al. 2020). Nowadays, in order to meet consumers' increasing demand for healthier, organic, and ecological products, cosmetic industry is looking for natural products with therapeutic potentials to use as raw material in the development of new cosmetic products. Natural products are always preferred over the

A. Debnath
Department of Biotechnology, Adamas University, Kolkata, West Bengal, India

S. Bandyopadhyay
Department of Biological Sciences, Adamas University, Kolkata, West Bengal, India
e-mail: subhendu.bandyopadhyay@adamasuniversity.ac.in

A. Das (✉)
Department of Biotechnology, Adamas University, Kolkata, West Bengal, India

QA Higher Education (QAHE), Birmingham, United Kingdom

S. Ghosh, H. N. Onyeaka (eds.), *Fungal-Based Cosmetics: Formulation and Usage*, The Mycota 17,
https://doi.org/10.1007/978-3-032-09291-5_4

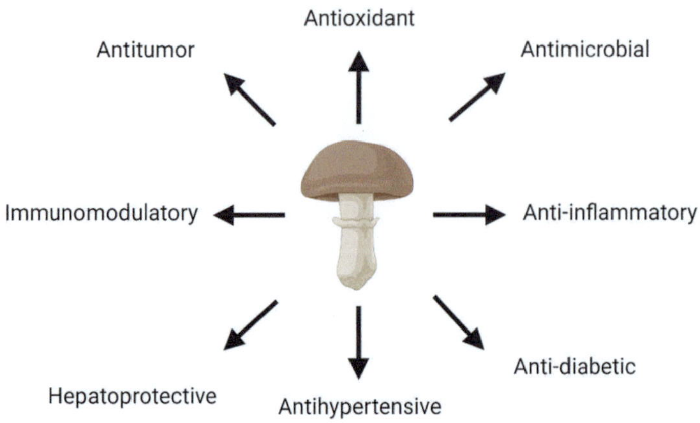

Fig. 4.1 Medicinal properties of fungi

artificial ingredients as they possess lower toxicity, competitive effectiveness unlike the artificial one (Antignac et al. 2011; Alves et al. 2020; Taofiq et al. 2016). Mushroom belongs to phyla Ascomycota and Basidiomycota that are preferred for centuries as they possess unique taste and appearance. They serve an important source of valuable medicines (Wasser 2011; Pan et al. 2015; Rahman and Choudhury 2012). Among the total 150,000 fungi species estimated to found on Earth, only 10% of them (15,000 mushroom species) are known till now (Mueller and Schmit 2007). Only 2000 mushroom species are found safe to be used by human, and approximately 650 species possess therapeutic potentials among the 15,000 known mushroom species (Rai et al. 2005). Mushrooms are an excellent source of selenium, protein, polysaccharides, vitamin B1 (thiamine), vitamin B2 (riboflavin), vitamin B3 (niacin), vitamin B5 (pantothenic acid), vitamin B9 (folic acid), etc. Mushroom helps in maintaining necessary bodily functions and body weight, prevents age-related problems, and promotes long life (Poucheret et al. 2006; Bashir et al. 2014). Recently, a huge amount of attention has been given to numerous mushrooms due to their constituent active compounds which offer several beneficial health effects like immunomodulation, antioxidant, anticancer, anti-inflammatory, antidiabetic, antimicrobial, and hepatoprotective effects as shown in Fig. 4.1 (El Enshasy and Hatti-Kaul 2013). Microchemicals obtained from several mushrooms are enlisted in Table 4.1 along with their bioactivities. The antioxidant and regenerative activities of mushroom facilitate them to be used as cosmetic ingredients such as in cream, lotion, and serum. Kojic acid, a phenolic compound found in several mushrooms, can reduce age spots and discolorations and inhibits melanin production. Mushroom chitosan, which has gel-forming and antibacterial properties, is also used in the cosmetic industry. Lactic acid and ceramides which are the product of biotransformation are used as skin hydrating agent. The major highlight of this chapter is to recapitulate and summing up the potential of some beneficial fungi or their extracts as antioxidant agents and to describe their implication in cosmetic industries (Gao et al. 2004; Bentley 2006).

4.2 Beneficial Constituents of Fungi

Natural sources including fruits, vegetables, herbs, spices, and edible mushrooms are a good source of natural antioxidant that offers a wide range of health benefits. Mushrooms, which are often regarded as mere vegetables, are actually a great source of valuable medicines (Pan et al. 2015; Rahman and Choudhury 2012). Most of

Table 4.1 Fungi extracts with their associated beneficial cosmetic potential

Activity	Fungi	Effects	References
Anti-collagenase	Exopolysaccharide and mycelia extract of *Grifola frondosa*	Reduces or inhibits expression of matrix metalloproteinases to promote collagen synthesis	Bae et al. (2005)
	Phenolics and polysaccharide of *Volvariella volvacea*	Induces collagen production	Ruksiriwanich et al. (2014)
	Tricholoma matsutake mycelial extract	Lowers MMP-1 and MMP-3 at protein-level dose dependently	Kim et al. (2014)
Anti-elastase	*Tricholoma matsutake* mycelial extract	Inhibits elastase m-RNA dose dependently	Kim et al. (2014)
Anti-hyaluronidase	*Pleurotus citrinopileatus*	The extract inhibited hyaluronidase activity	Meng et al. (2012)
	Pleurotus tuber-regium	Prevents the enzyme activity of hyaluronidase	Dandapat and Sinha (2015)
	Trametes lactinea	Its extract in water and in acetone showed declination in hyaluronidase bioactivities	Yahaya and Don (2012)

Table 4.2 Bioactive ingredients of different fungi and their bioactivities

Fungi	Bioactive ingredients	Beneficial properties	References
Abortiporus biennis	Exopolysaccharides	Antioxidant	Long et al. (2019)
Agaricus subrufescens	Polysaccharides	Antioxidant, anti-pigmentation	Taofiq et al. (2019)
Auricularia polytricha	Phenolics, glucans	Anti-hyaluronidase, anti-tyrosinase, anti-inflammatory	Abd Razak et al. (2020)
Calocybe gambosa	Vitamin C	Anti-radiation, stimulating collagen synthesis	Vaz et al. (2011)
Clitocybe aurantiaca	Clitocybin	Procollagen synthesis, antioxidant	Kim et al. (2009)
Cyclocybe aegerita	Ergothioneine, glutathione	Antioxidant	Kalaras et al. (2017)
Flammulina velutipes	Glycosides	Anti-inflammatory, antioxidant	Wu et al. (2016)
Inonotus sanghuang	Chlorogenic acid, quercetin, rutin	Antioxidant	Liu et al. (2017)
Laetiporus sulphureus	Phenolic acids	Anti-hyaluronidase, anti-tyrosinase	Sułkowska-Ziaja et al. (2021)
Macrocybe lobayensis	Phenolic compounds	Antioxidant, antibacterial	Khatua and Acharya (2018)
Pleurotus eryngii	Organic acid, ergosterol	Antioxidant	Souilem et al. (2017)
Polyporus umbellatus	Polysaccharides, steroids, terpenoids	Anti-inflammatory, antibacterial, antioxidant	Bandara et al. (2015)
Suillus bellinii	Cinnamic acid, *p*-hydroxyphenylacetic acids	Antioxidant	Souilem et al. (2017)
Trametes versicolor	Phenolic acids	Antioxidant	Chong et al. (2018)

the edible mushrooms, their fruiting bodies, and mycelia possess numerous nutritive bioactive compounds such as carbohydrates, lipids, vitamins, phenolics, organic acids, monoterpenoid and diterpenoid, proteins, trace elements, etc., which offers immunomodulatory, antimicrobial, antitumor, antihypertensive, hepatoprotective, and antioxidant effects as illustrated in Table 4.2 (Taofiq et al. 2016).

4.2.1 Phenolic Compounds

Mushrooms are enriched with numerous phenolic compounds which possess free radical scavenging or antioxidant activity and therefore exert extensive biological effects (Liu et al. 2013; Vamanu and Nita 2012). Phenolic compounds characterized by one or more hydroxyl groups (-OH) attached directly with an aromatic hydrocarbon ring can be referred to as simple phenols or polyphenols depending on its constituent number of phenol units. Simple phenol includes phenolic acids like kojic acid, chlorogenic acid, benzoic acid, gallic acid, flavonoids, tannins, stilbenes, etc. which are some examples of polyphenols (Kim et al. 2008). Phenolic acid is the major phenolic constituent found in the mushrooms mainly responsible for the antioxidant effects. In addition to this, cinnamic acid, coumaric acids, gallic acid, and rutin are found in wild mushrooms (Barros et al. 2009). Flavonoids such as flavonols, flavones, flavanols, flavanones, anthocyanins, and isoflavonoids serve as potent antioxidant compound (Shahidi 2000; Pinho et al. 2000).

4.2.2 Terpenoids

Terpenoids or terpenes are volatile five-carbon isoprene units found in mushrooms that offer diverse therapeutic potential. It has been reported that 5 monoterpenes, 70 sesquiterpenes, 44 diterpenes, and 166 triterpenes found in mushrooms are investigated for their antitumorigenic, anticarcinogenic, antimicrobial, and neuroprotective properties (Duru and Çayan 2015). *Ganoderma* spp. is well known for its various terpenoid content. Ganoderic acid, ganoderiols, ganodermanontriol, ganodermanondiol, triterpenoids, etc. are found in *Ganoderma lucidum.* Ganoderic acid exerts (a) cytotoxic potential against hepatoma cells in vitro, (b) hepatoprotective activity, (c) antihistamine releasing activity in mast cells of rat, (d) inhibitory effect on farnesyl protein transferase, and (e) angiotensin converting enzyme (Min et al. 1998). Ganoderic acid C and its derivatives prevent the production of cholesterol. Ganoderic acid F promotes protection against atherosclerosis. Triterpenoids are effective antioxidant compound (Morigiwa et al. 1986; Rathee et al. 2012). Ganoderiol and ganodermanotriol exerts good viral inhibition effect. *Ganoderma pfeifferi* exhibits antimicrobial activity against bacteria, virus, and fungus. Sesquiterpene dialdehyde present in *Merulius tremellosus* has antifungal properties. Sesquiterpenes found in *Marasmius alliaceus,* antcin B, antcin K, methylantcinate B, sesquiterpene lactone derived from *Antrodia cinnamomea,* trametenolic acid, ergosterol peroxide, ergosterol, and inotodiol derived from *Inonotus obliquus* has anticarcinogenic potential. Lentinellic acid derived from *Lentinellus omphalodes* has bactericidal activity. Cyathane diterpenes found in *Cyathus africanus*, antacin A and antrolone present in *Antrodia cinnamomea,* and trametenolic acid, ergosterol peroxide, ergosterol, and inotodiol derived from *Inonotus obliquus* possess anti-inflammatory potential (Ganesan et al. 2019; Han et al. 2013; Zhang et al. 2019; Ma et al. 2013). Carotenoids, a type of tetraterpenoid pigment, are a major source of natural antioxidant that comes after polyphenols. It has been extensively used in sun lotions (Urquiaga and Leighton 2000; Jin et al. 2010).

4.2.3 Selenium

Selenium (Se), an essential micronutrient in human, can be present in both inorganic such as sodium selenite or Na_2SeO_3, sodium selenate or Na_2SeO_4, and hydrogen selenide or H_2Se and organic forms including selenonucleic acid, selenopolysaccharide, selenoamino acid, selenoproteins and/or selenoenzymes, etc. Selenium in its inorganic forms is more toxic even at its low concentration; however, in organic form, it showed low toxicity with better bioavailability (Xu et al. 2021). *G. lucidum, L. edodes,* and *P. ostreatus* can take up Se in their bodies, and mycelium and fruiting body can convert it from inorganic to organic form. The mushroom mycelium can be dehydrated and dried into powder form to use in supplement or in functional food. In comparison to the normal mushrooms, selenium-enriched mushrooms possess high antitumorigenic, antimicrobial, anti-inflammatory, and antioxidant potential due to its Se-polysaccharide complexes. $_{\gamma}$-glutamyl-Se-methylselenocysteine and Se-methylselenocysteine (MeSeCys) in selenized garlic have antitumorigenic activity. Selenium improves bone health, strengthens teeth, hair, nails, and extensively applied in shampoo (Xu et al. 2021; McSheehy et al. 2000).

4.2.4 Polysaccharides

Fungal polysaccharides are either homo or heteroglycans based on their monosaccharide composition and can exist as linear and branched glucans. Most of the mushroom polysaccharides belong to the β-glucans (Wasser 2002). Digestive enzymes like trypsinogen, amylase, lactase, and peptidase, which are secreted from the pancreas and brush border of mammals, failed to hydrolyze the β-glycosidic bonds in β-glucans (Van Loo 2006). This nondigestible property of mushroom polysaccharides helps in its likeliness to be used as a prebiotic (Gibson et al. 2004). Mushroom polysaccharides sometimes attached with proteins to form polysaccharide–protein complexes. Besides functioning as a useful prebiotic, mushroom polysaccharides show antitumor, antioxidant, antiviral, antimicrobial, and immune-modulation activity, and some are used in cosmetics products (Lindequist et al. 2005; Thatoi and Singdevsachan 2014). For instance, β-(1-3)-D-glucan found in *Pleurotus* spp. have anticancer properties and are used in cosmetics. Lentinan, schizophyllan, and krestin are some important mushroom polysaccharides which have cytostatic, antitumor, immunomodulating effects (Yao et al. 2011; Bisen et al. 2010).

4.3 Application in Cosmetic Industry

4.3.1 Antiaging Effect with the Help of Antioxidant Activity

Human skin is made up of an upper epidermal layer over an underlying dermis layer. Collagen and elastin are the two main structural proteins of extracellular matrix that provides skin's proper elastic architecture (Kim et al. 2014; Taofiq et al. 2016). Besides these two, hyaluronic acid found in extracellular matrix rejuvenates skin. However, depletion of collagen, elastin, and hyaluronic acid results in skin aging and wrinkle formation (Khare et al. 2015).

Our body has its antioxidant mechanism to neutralize the accumulative effect of free radicals (e.g., reactive oxygen or nitrogen or sulfur species) generated from different biological reactions. However, sometimes body's intrinsic antioxidant pool becomes ineffective in reducing this radical population (Carocho and Ferreira 2013). The exposure of skin to high UV radiation promotes production of reactive oxygen species leading to damage in DNA, inflammation and pigmentation on skin, and increased action of matrix metalloproteases promoting collagen degradation (Masaki 2010). UV radiation also downregulates transforming growth factor beta which facilitates procollagen production and leads to reduced collagen synthesis. Besides elastin and collagen, hyaluronic acid is also a major component of the extracellular matrix, which moisturizes skin and protects skin from oxidative damage (Papakonstantinou et al. 2012). Thus, cosmetic companies looked for natural products with ROS scavenging potential which can functions as inhibitors of extracellular matrix degrading enzymes to delay skin aging while used in antiaging creams or lotions (Soto et al. 2015). Some fungi species with reported cosmetic activity are described in Table 4.1. Numerus reports revealed that several mushroom species possess antioxidant activities mainly as DPPH (2,2-diphenyl-1-picrylhydrazyl), ABTS (2,20-azinobis 3-ethylbenzothiazoline-6-sulfonic acid) radical scavenger, ferric reduction, inhibitor of lipid peroxidation, etc. (Carocho and Ferreira 2013) and therefore have diverse applications in cosmetics.

4.3.1.1 *Agaricus bisporus*

Agaricus bisporus is commonly cultivated throughout the world followed by *Lentinula edodes, Pleurotus* spp., and *Flammulina velutipes*. It is enriched with essential amino acids, proteins, polysaccharides, vitamin C, vitamin D, vitamin B12, folates, minerals, and polyphenols. It is regarded as health promoters as it lacks fat and is often used in dishes due to its beneficial additive and synergistic effects of its bioactive compounds (Aida et al. 2009; Miles and Chang 2004; Ferreira et al. 2009). *Agaricus bisporus* has diverse health promoting properties including antimicrobial, antitumor, anti-inflammatory, anti-aromatase, immunomodulatory, and antioxidant potential (Atila et al. 2017). A report revealed that the ethanolic extract of *Agaricus bisporus* exerts good antioxidant activities in vitro and in vivo due to its phenolic constituents. HPLC also revealed that methanolic extract of *Agaricus bisporus* also has radical scavenging effect (Abah and Abah 2010).

4.3.1.2 *Armillaria mellea*

Armillaria mellea, also called honey mushroom, is used in Asia for treating headache, dizziness, insomnia, palsy, neurasthenia, and infectious diseases (Yang et al. 1984). It is a saprophytic fungus which grows on dead, decaying substances (Erbiai et al. 2021). Methanolic extracts of dried mycelia generated from *Armillaria mellea* culture have been reported to have good antioxidant activity due to its flavonoid and phenol content. Two bioactive polysaccharides (AkP1V-1 and AkPS1V-2) present in alkaline extract of *Armillaria mellea* possess antioxidant activity. AkPS1V-2 possesses greater antioxidant potential in comparison to the AkPS1V-1 (Siu et al. 2016).

4.3.1.3 *Auricularia auricula*

Auricularia auricula belongs to Auriculaceae family are usually ear shaped edible mushrooms. The polysaccharides are mainly responsible for greater antioxidant potential of the mushroom. Mushroom polysaccharides can inhibit lipid peroxidation and exhibit hydroxyl radical scavenging activity in ethanolic, crude, and boiled extracts of *Auricularia auricula* (Sun et al. 2009; Khaskheli et al. 2015; Acharya et al. 2004). The boiled extract is the most potent antioxidant. Freeze dried *Auricularia auricula* fruiting body derived polysaccharides showed higher antioxidant activities in vitro such as ABTS, DPPH radical scavenging activities, ferric reducing ability, and metal ion chelating capacity (Zhang et al. 2022).

4.3.1.4 *Boletus edulis*

Boletus edulis is well known for its exceptional flavor and high nutritional content. It has higher protein content than other mushrooms and higher antioxidant properties (Jaworska et al. 2015). It is enriched with proteins, minerals, and fibers but contains low fat. The fruiting bodies of these mushrooms possess great antioxidant potential and alkaloid content. *Boletus edulis* possesses elevated reducing power as compared to *Agaricus bisporus, Pleurotus eryngii, Coprinus comatus,* and *Pleurotus ostreatus,* but lower than *Pleurotus citrinopileatus* (Ribeiro et al. 2008; Tsai et al. 2007). *Boletus edulis* has higher total phenolic contents governing prominent scavenging activity. It has been reported that *Blastococcus aurantiacus* exhibits greater phenolic content and antioxidant potential as compared to *B. edulis*. In contrast to this, *B. edulis* has better reducing potential and prevents lipid peroxidation. The polysaccharides present in fruiting bodies of *B. edulis* offer good antioxidant activity, chelating activity, and prevention of superoxide radical and hydroxyl radical formation. A good synergistic antioxidant activity was found upon combination action of *B. edulis* and *Marasmius oreades* species mixture (Palacios et al. 2011; Vidović et al. 2010; Vieira et al. 2012).

4.3.1.5 *Ganoderma lucidum*

Ganoderma lucidum is used in Chinese traditional medicine for treating a wide range of health ailments such as decreasing inflammation, regulating immune system, and antiaging properties (Barbieri et al. 2017). It acts as a regulator of different biological activities and also referred to as "the fungus of immortality" as its consumption can increase the lifespan. It grows in decomposed tree bark and root under humid and dim-light conditions (Anwar et al. 2018). Polysaccharides, phenolics, sterols, triterpenoids, proteins, nucleosides, glycopeptides, alkaloids, fatty acids, etc. are responsible for the diverse pharmacological properties of *Ganoderma lucidum* such as anti-inflammatory, antioxidant, anti-tyrosinase, and antibacterial activities (Dong et al. 2019; Ćilerdžić et al. 2018; Yahia et al. 2017). *Ganoderma lucidum* was used in adjuvant therapy in treating hypertension, insomnia, anorexia, and chronic hepatitis. It has been and also used in combination along with chemotherapy, radiotherapy, surgery to mitigate their negative effects in patients' body (Zhong et al. 2019; Suprasert et al. 2014; Zhao et al. 2012) separated four polysaccharides with free radical scavenging properties from *Ganoderma lucidum. Ganoderma applanatum* possesses better antioxidant potential than *Ganoderma lucidum, L. edodes,* and *Trametes versicolor* (Shi et al. 2013; Kozarski et al. 2012).

4.3.1.6 *Grifola frondosa*

Grifola frondosa or "hui-shu-hua" belongs to Grifolaceae family is a Basidiomycetes fungus. Its edible fruiting body is called as Maitake in Japan. Although this mushroom is native to Japan, China, and North America, it is also cultivated throughout the world because of its beneficial health effects (Kodama et al. 2002). This edible mushroom is used as a food-flavoring agent in its dried powder form. The glycoprotein found in its fruiting body showed antitumorigenic, immune-boosting, antihypertensive, antidiabetic, and antiviral effects. The antioxidant property of *Grifola frondosa* by inhibiting DPPH, ABTS, and hydroxyl radical is due to phenols, flavonoids, and polysaccharides content (Yeh et al. 2011; Chen et al. 2012) and is used in treating metabolic disorders (diabetes and nonalcoholic fatty liver disease), inhibiting inflammation, tumor progression, etc. *Grifola frondosa* a hot alkali extract possesses greater antioxidant property over the partially purified polysaccharides (Fan et al. 2011).

4.3.1.7 *Hypsizygus marmoreus*

Hypsizygus marmoreus is a medicinal mushroom species native to Korea, China, Japan, and northern Europe and is enriched with polysaccharides, lectins, enzymes, etc. (Liu et al. 2016; Akavia et al. 2009) which offers antioxidant, anti-inflammatory, anticancer, anti-allergic properties. The fruiting body of this mushroom has a sweet, nutty flavor with crunchy texture. The antioxidant capacity revealed by investigating reactive oxygen species scavenging activity and reducing potential of *H. marmoreus* is mainly due to its constituent bioactive compound, i.e., phenol and polysaccharides. It has been reported that intracellular polysaccharides of *Hypsizygus marmoreus* are a good antioxidant that can improve the adaptive immune system (Chowdhury et al. 2015; Liu et al. 2012).

4.3.1.8 *Lentinus edodes*

Lentinus edodes, the second largest cultivated edible mushroom worldwide (Pettipher 1988), also called shitake mushroom is enriched with vitamin B (B1, B2, B5, B12) and provitamin D2 (Anwar et al. 2018). Bioactive compounds including polysaccharides and ergothioneine present in this mushroom have shown therapeutic application as antioxidant to mitigate the total oxidative stress (Nisar et al. 2017; Zembron-Lacny et al. 2013). *Lentinula edodes* induces antioxidant enzymes (superoxide dismutase, glutathione peroxidase) to protect skin from oxidative stress-induced damage. UV-C radiation elevates the radical scavenging ability of *L. edodes*. Three different polysaccharides isolated from the *Lentinus edodes* have prominent antioxidant efficiency that promotes healthy immune response (Jiang et al. 2010).

4.3.1.9 *Pleurotus ostreatus*

The third most cultivated fungus throughout the world is *Pleurotus ostreatus* (Sánchez 2010). The mycelia and fruiting bodies of *Pleurotus ostreatus* offer strong antioxidant activity due to their bioactive compounds (Anandhi et al. 2013). It is a rich source of polysaccharides, phenolic compounds, flavonoids, steroids, terpenoids, ergothioneine, vitamin C, β-carotene, selenium, etc. Its ethanolic extract bears exogenous antioxidant which prevents oxidative damage and synergizes with body's endogenous antioxidants (Jayakumar et al. 2010; Reis et al. 2012). The methanolic extract of *Pleurotus ostreatus* contains much grounded β-carotene-linoleic acid restraint as compared to other extracts. However, acetone extract shows the highest reducing potential in comparison to other concentrates (Anwar et al. 2018). The hot-water extracts of *P. ostreatus* mycelium possess efficient antioxidant, as well as antimicrobial, and anticancer potential, and therefore can be regarded as a good alternative used in cosmeceuticals (Morris et al. 2017).

4.3.1.10 *Schizophyllum commune*

The wood-decay fungus *Schizophyllum commune,* commonly called as split gill mushroom, belongs to the genus Schizophyllum which is widely distributed throughout the world except Antarctica (Klaus et al. 2011). It grows on rotten woods and dead plants and acts as a pathogen for trees causing white rot (Anwar et al. 2018). *S. commune* is well known for its antioxidant potential due to presence of some antioxidant compounds hydroxybenzoic acid, protocatechuic acid, tocopherol, flavonoid, β-carotene, lycopene and other polysaccharides, and polyphenol components (Mirfat et al. 2010; Klaus et al. 2011; Mayakrishnan et al. 2013). The crude *S. commune* extract functions as an efficient free radical scavenger and thereby increases the shelf life of lipid food products (Yim et al. 2011; Devi et al. 2014).

4.3.2 Skin-Whitening Effect

The melanin pigment that governs skin color also provides protection against UV-induced dermal irritation. Various autoimmune conditions, such as genetic factors, birth control pills, etc., facilitate excess melanin production leading to hyperpigmentation (Lin et al. 2008; Gupta et al. 2014). Overexposure to UV radiation induces more melanin production with increased tyrosinase activity. Tyrosinase catalyzes the hydroxylation of tyrosine to dihydroxyphenylalanine (DOPA) followed by its oxidation to dopaquinone. Dopaquinone is auto-oxidized to convert into dopachrome and finally to either eumelanin or pheomelanin (Meng et al. 2012; Jung et al. 2009).

It has been reported that hydroquinone widely used to treat hyperpigmentation, possess severe side effects such as carcinogenic potential, cell irritants, emission of fish odor from the skin. To combat with the negative effect of these depigmenting agents, cosmetic industry searches for alternative natural products like plants and mushrooms enriched with bioactive chemicals. Reports revealed that phenolic compounds obtained from plants can inhibit the melanin synthesis signaling (Sarkar et al. 2013; Ali et al. 2015), as well as serve as moisturizers, prevent skin aging, and preserve the skin health. *P. ferulae*, *P. nebrodensis*, *P. ostreatus,* and *P. salmoneostramineus* (Alam et al. 2011) have anti-tyrosinase activity. *Phellinus baumii* and *Flammulina velutipes* extracts inhibit melanin production in mice melanoma cells. Presently, *Ganoderma lucidum* extract is used in cosmetic products as it possesses the highest anti-tyrosinase activity among other basidiomycetes (Lee et al. 2013; Chien et al. 2008).

4.3.3 Moisturizing Effect

Tremella polysaccharide exhibits excellent moisturizing potential. The water extract of *Auricularia fuscosuccinea* functions as a good moisturizing agent as sodium hyaluronate, but less effective as compared to sporocarp extract of *Tremella fuciformis*. Fomes officinalis powder was widely used in cosmetic products to reduce skin imperfections (Liu et al. 2012; Siu et al. 2016).

4.3.4 Hair Cosmetics

Hair cosmetics including shampoos, conditioners, serums, sprays, gels, and dyes are used for the betterment of hair quality. Nowadays, hair cosmetics manufacturer are interested in using natural products like mushrooms enriched with vitamin D, antioxidants, iron, selenium, and copper are used to prevent hair problems, hair loss, and dandruff. *Ganoderma lucidum* and several other mushrooms are used to stimulate hair growth in human (Madnani and Khan 2013).

4.4 Conclusion

Consumers' demand for nontoxic natural cosmeceuticals and nutracosmetics is rising sharply over the last few years. Various species of fungi and their extracts obtained from their mycelium and fruiting body bearing a wide range of medicinal properties are used extensively as an essential component in the cosmetic industry for their free radical scavenging, anti-hyperpigmentation, moisturizing, antimicrobial, anti-collagenase, anti-tyrosinase, and anti-inflammatory effects. However, the number of beneficial fungal species identified till now represents a small percentage of the total amount. Thus, more fungal species are needed to be identified and investigated for their medicinal and cosmetic effects. Despite the presence of bioactive ingredients, consumers should consult a doctor before using fungi-derived products. Presently, implication of nano-systems and nano-delivery technologies has improved the efficiency and safety precautions of designing new commercial formulations.

References

Abah SE, Abah G (2010) Antimicrobial and antioxidant potentials of Agaricus bisporus. Adv Biol Res 4(5): 277–282

Abd Razak DL, Jamaluddin A, Abd Rashid NY et al (2020) Assessment of cosmeceutical potentials of selected mushroom fruitbody extracts through evaluation of antioxidant, anti-hyaluronidase and anti-tyrosinase activity. J 3(3):329–342

Acharya K, Samui K, Rai M, Dutta BB, Acharya R (2004) Antioxidant and nitric oxide synthase activation properties of Auricularia auricula. Indian J Exp Biol 42(5):538–540

Aida FMNA, Shuhaimi M, Yazid M, Maaruf AG (2009) Mushroom as a potential source of prebiotics: a review. Trends Food Sci Technol 20(11–12):567–575

Akavia E, Beharav A, Wasser SP, Nevo E (2009) Disposal of agro-industrial by-products by organic cultivation of the culinary and medicinal mushroom Hypsizygus marmoreus. Waste Manag 29(5):1622–1627

Alam N, Yoon KN, Cha YJ et al (2011) Appraisal of the antioxidant, phenolic compounds concentration, xanthine oxidase and tyrosinase inhibitory activities of Pleurotus salmoneostramineus. Afr J Agric Res 6(6): 1555–1563

Ali SA, Choudhary RK, Naaz I, Ali AS (2015) Understanding the challenges of melanogenesis: key role of bioactive compounds in the treatment of hyperpigmentory disorders. J Pigment Dis 2:1–9

Alves A, Sousa E, Kijjoa A, Pinto M (2020) Marine-derived compounds with potential use as cosmeceuticals and nutricosmetics. Molecules 25(11): 2536

Anandhi R, Annadurai T et al (2013) Antihypercholesterolemic and antioxidative effects of an extract of the oyster mushroom, Pleurotus ostreatus, and its major constituent, chrysin, in triton WR-1339-induced hypercholesterolemic rats. J Physiol Biochem 69:313–323

Antignac E, Nohynek GJ, Re T et al (2011) Safety of botanical ingredients in personal care products/ cosmetics. Food Chem Toxicol 49(2):324–341

Anwar H, Hussain G, Mustafa I (2018) Antioxidants from natural sources. In: Antioxidants in foods and its applications, p 3

Atila F, Owaid MN, Shariati MA (2017) The nutritional and medical benefits ofAgaricus bisporus: A review. Journal of Microbiology, Biotechnol Food Sci 7(3):281–286

Bae JT, Sim GS, Lee DH et al (2005) Production of exopolysaccharide from mycelial culture of Grifola frondosa and its inhibitory effect on matrix metalloproteinase-1 expression in UV-irradiated human dermal fibroblasts. FEMS Microbiol Lett 251(2):347–354

Bandara AR, Rapior S, Bhat DJ et al (2015) Polyporus umbellatus, an edible-medicinal cultivated mushroom with multiple developed health-care products as food, medicine and cosmetics: a review. Cryptogam Mycol 36(1):3–42

Barbieri A, Quagliariello V, Del Vecchio V et al (2017) Anticancer and anti-inflammatory properties of Ganoderma lucidum extract effects on melanoma and triple-negative breast cancer treatment. Nutrients 9(3): 210

Barros L, Dueñas M, Ferreira IC et al (2009) Phenolic acids determination by HPLC–DAD–ESI/MS in sixteen different Portuguese wild mushrooms species. Food Chem Toxicol 47(6):1076–1079

Bashir A, Vaida N, Ahmad Dar M (2014) Medicinal importance of mushrooms: a review. Int J Adv Res 2(12):1–4

Bentley R (2006) From miso, sake and shoyu to cosmetics: a century of science for kojic acid. Nat Prod Rep 23(6): 1046–1062

Bisen PS, Baghel RK, Sanodiya BS et al (2010) Lentinus edodes: a macrofungus with pharmacological activities. Curr Med Chem 17(22):2419–2430

Carocho M, Ferreira IC (2013) A review on antioxidants, prooxidants and related controversy: natural and synthetic compounds, screening and analysis methodologies and future perspectives. Food Chem Toxicol 51:15–25

Chen H, Ju Y, Li J, Yu M (2012) Antioxidant activities of polysaccharides from Lentinus edodes and their

significance for disease prevention. Int J Biol Macromol 50(1):214–218
Chien CC, Tsai ML, Chen CC et al (2008) Effects on tyrosinase activity by the extracts of Ganoderma lucidum and related mushrooms. Mycopathologia 166:117–120
Chong Z, Matsuo H, Kuroda M et al (2018) Mushroom extract inhibits ultraviolet B-induced cellular senescence in human keratinocytes. Cytotechnology 70: 1001–1008
Chowdhury MMH, Kubra K, Ahmed SR (2015) Screening of antimicrobial, antioxidant properties and bioactive compounds of some edible mushrooms cultivated in Bangladesh. Ann Clin Microbiol Antimicrob 14(1): 1–6
Ćilerdžić JL, Sofrenić IV, Tešević VV et al (2018) Neuroprotective potential and chemical profile of alternatively cultivated Ganoderma lucidum basidiocarps. Chem Biodivers 15(5):e1800036
Dandapat S, Sinha MP (2015) Antioxidant and anti-inflammatory activity of Pleurotus tuber-regium (Rumph. ex Fr.) singer. Adv Biol Res 9(3):140–145
Devi LS, Dasgupta A, Chakraborty M et al (2014) Chemical composition and antioxidant activity of Schizophyllum commune. Int J Pharm Sci Rev Res 27(2):173–177
Dhapte-Pawar V, Kadam S, Saptarsi S, Kenjale PP (2020) Nanocosmeceuticals: facets and aspects. Future Sci OA 6(10):FSO613
Dong Q, He D, Ni X et al (2019) Comparative study on phenolic compounds, triterpenoids, and antioxidant activity of Ganoderma lucidum affected by different drying methods. J Food Meas Charact 13:3198–3205
Duru ME, Çayan GT (2015) Biologically active terpenoids from mushroom origin: a review. Records Nat Products 9(4):456
El Enshasy HA, Hatti-Kaul R (2013) Mushroom immunomodulators: unique molecules with unlimited applications. Trends Biotechnol 31(12):668–677
Erbiai EH, da Silva LP, Saidi R et al (2021) Chemical composition, bioactive compounds, and antioxidant activity of two wild edible mushrooms Armillaria mellea and Macrolepiota procera from two countries (Morocco and Portugal). Biomolecules 11(4):575
Fan Y, Wu X, Zhang M et al (2011) Physical characteristics and antioxidant effect of polysaccharides extracted by boiling water and enzymolysis from Grifola frondosa. Int J Biol Macromol 48(5):798–803
Ferreira IC, Barros L, Abreu R (2009) Antioxidants in wild mushrooms. Curr Med Chem 16(12):1543–1560
Ganesan N, Baskaran R, Velmurugan BK, Thanh NC (2019) Antrodia cinnamomea – an updated minireview of its bioactive components and biological activity. J Food Biochem 43(8):e12936
Gao JM, Zhang AL, Chen H, Liu JK (2004) Molecular species of ceramides from the ascomycete truffle tuber indicum. Chem Phys Lipids 131(2):205–213
Gibson GR, Probert HM, Van Loo J et al (2004) Dietary modulation of the human colonic microbiota: updating the concept of prebiotics. Nutr Res Rev 17(2):259–275
Gupta SK, Gautam A, Kumar S (2014) Natural skin whitening agents: a current status. Adv Biol Res (Rennes) 8(6):257–259
Han J, Chen Y, Bao L et al (2013) Anti-inflammatory and cytotoxic cyathane diterpenoids from the medicinal fungus Cyathus africanus. Fitoterapia 84:22–31
Jaworska G, Pogoń K, Skrzypczak A, Bernaś E (2015) Composition and antioxidant properties of wild mushrooms boletus edulis and Xerocomus badius prepared for consumption. J Food Sci Technol 52: 7944–7953
Jayakumar T, Thomas PA, Ramesh E, Geraldine P (2010) An extract of the Pleurotus ostreatus mushroom bolsters the glutathione redox system in various organs of aged rats. J Med Food 13(4):771–778
Jiang T, Jahangir MM, Jiang Z et al (2010) Influence of UV-C treatment on antioxidant capacity, antioxidant enzyme activity and texture of postharvest shiitake (Lentinus edodes) mushrooms during storage. Postharvest Biol Technol 56(3):209–215
Jin JM, Lee J, Lee YW (2010) Characterization of carotenoid biosynthetic genes in the ascomycete Gibberella zeae. FEMS Microbiol Lett 302(2):197–202
Jung E, Lee J, Huh S et al (2009) Phloridzin-induced melanogenesis is mediated by the cAMP signaling pathway. Food Chem Toxicol 47(10):2436–2440
Kalaras MD, Richie JP, Calcagnotto A, Beelman RB (2017) Mushrooms: a rich source of the antioxidants ergothioneine and glutathione. Food Chem 233:429–433
Khare N, Khare P, Yadav G (2015) Recent advances in antiaging–a review. Glob J Pharmacol 9:267–271
Khaskheli AA, Talpur FN, Ashraf MA et al (2015) Monitoring the Rhizopus oryzae lipase catalyzed hydrolysis of castor oil by ATR-FTIR spectroscopy. J Mol Catal B Enzym 113:56–61
Khatua S, Acharya K (2018) Functional ingredients and medicinal prospects of ethanol extract from Macrocybe lobayensis. Pharm J 10(6)
Kim MY, Seguin P, Ahn JK (2008) Phenolic compound concentration and antioxidant activities of edible and medicinal mushrooms from Korea. J Agric Food Chem 56(16):7265–7270
Kim YH, Ryoo IJ, Choo SJ et al (2009) Clitocynoic acid, a novel human neutrophil elastase inhibitor from the culture broth of Clitocybe aurantiaca. J Microbiol Biotechnol 19(2):000–000
Kim SY, Go KC, Song YS et al (2014) Extract of the mycelium of T. matsutake inhibits elastase activity and TPA-induced MMP-1 expression in human fibroblasts. Int J Mol Med 34(6):1613–1621
Klaus A, Kozarski M, Niksic M et al (2011) Antioxidative activities and chemical characterization of polysaccharides extracted from the basidiomycete Schizophyllum commune. LWT-Food Sci Technol 44(10):2005–2011
Kodama N, Komuta K, Nanba H (2002) Can maitake MD-fraction aid cancer patients? Altern Med Rev 7(3):236–239
Kozarski M, Klaus A, Nikšić M et al (2012) Antioxidative activities and chemical characterization of

polysaccharide extracts from the widely used mushrooms Ganoderma applanatum, Ganoderma lucidum, Lentinus edodes and Trametes versicolor. J Food Compos Anal 26(1–2):144–153

Lee JS, Shin DB, Lee SM et al (2013) Melanogenesis inhibitory and antioxidant activities of Phellinus baumii methanol extract. Korean J Mycol 41(2): 104–111

Lin JW, Chiang HM, Lin YC, Wen KC (2008) Natural products with skin-whitening effects. J Food Drug Anal 16(2):8

Lindequist U, Niedermeyer TH, Jülich WD (2005) The pharmacological potential of mushrooms. Evid Based Complement Alternat Med 2:285–299

Liu JK (2022) Natural products in cosmetics. Nat Products Bioprospect 12(1):40

Liu H, Wang L, Sun Z et al (2012) Production and antioxidant activity of intracellular polysaccharide by Hypsizigus marmoreus SK-01. Bioresources 7(4)

Liu J, Jia L, Kan J, Jin CH et al (2013) In vitro and in vivo antioxidant activity of ethanolic extract of white button mushroom (Agaricus bisporus). Food Chem Toxicol 51:310–316

Liu M, Meng G, Zhang J et al (2016) Antioxidant and hepatoprotective activities of mycelia selenium polysaccharide by Hypsizigus marmoreus SK-02. Biol Trace Elem Res 172:437–448

Liu K, Xiao X, Wang J et al (2017) Polyphenolic composition and antioxidant, antiproliferative, and antimicrobial activities of mushroom Inonotus sanghuang. LWT-Food Sci Technol 82:154–161

Long Z, Liu H, Li J et al (2019) Preliminary characterization of exopolysaccharides produced by Abortiporus biennis in submerged fermentation. Sains Malays 48(12):2633–2640

Ma L, Chen H, Dong P, Lu X (2013) Anti-inflammatory and anticancer activities of extracts and compounds from the mushroom Inonotus obliquus. Food Chem 139(1–4):503–508

Madnani N, Khan K (2013) Hair cosmetics. Indian J Dermatol Venereol Leprol 79:654

Masaki H (2010) Role of antioxidants in the skin: anti-aging effects. J Dermatol Sci 58(2):85–90

Mayakrishnan V, Abdullah N, Abidin MHZ et al (2013) Investigation of the antioxidative potential of various solvent fractions from fruiting bodies of Schizophyllum commune (Fr.) mushrooms and characterization of phytoconstituents. J Agric Sci 5(6): 58–68

McSheehy S, Yang W, Pannier F et al (2000) Speciation analysis of selenium in garlic by two-dimensional high-performance liquid chromatography with parallel inductively coupled plasma mass spectrometric and electrospray tandem mass spectrometric detection. Anal Chim Acta 421(2):147–153

Meng TX, Zhang CF, Miyamoto T et al (2012) The melanin biosynthesis stimulating compounds isolated from the fruiting bodies of Pleurotus citrinopileatus. J Cosmet Dermatol Sci Appl 2(03):151

Miles PG, Chang ST (2004) Mushrooms: cultivation, nutritional value, medicinal effect, and environmental impact. CRC Press

Millikan LE (2001) Cosmetology, cosmetics, cosmeceuticals: definitions and regulations. Clin Dermatol 19(4):371–374

Min BS, Nakamura N, Miyashiro H et al (1998) Triterpenes from the spores of Ganoderma lucidum and their inhibitory activity against HIV-1 protease. Chem Pharm Bull 46(10):1607–1612

Mirfat AHS, Noorlidah A, Vikineswary S (2010) Scavenging activity of Schizophyllum commune extracts and its correlation to total phenolic content. J Trop Agric Food Sci 38(2):231–238

Morigiwa A, Kitabatake K, Fujimoto Y, Ikekawa N (1986) Angiotensin converting enzyme-inhibitory triterpenes from Ganoderma lucidim. Chem Pharm Bull 34(7): 3025–3028

Morris HJ, Beltrán Y, Llauradó G et al (2017) Mycelia from Pleurotus sp.(oyster mushroom): a new wave of antimicrobials, anticancer and antioxidant bio-ingredients. Int J Phytocosmet Nat Ingredients 4(1):3

Mueller GM, Schmit JP (2007) Fungal biodiversity: what do we know? What can we predict? Biodivers Conserv 16:1–5

Nisar J, Mustafa I, Anwar H et al (2017) Shiitake culinary-medicinal mushroom, Lentinus edodes (Agaricomycetes): a species with antioxidant, immunomodulatory, and hepatoprotective activities in hypercholesterolemic rats. Int J Med Mushrooms 19(11):981

Palacios I, Lozano M, Moro C et al (2011) Antioxidant properties of phenolic compounds occurring in edible mushrooms. Food Chem 128(3):674–678

Pan H, Han Y, Huang J (2015) Purification and identification of a polysaccharide from medicinal mushroom Amauroderma rude with immunomodulatory activity and inhibitory effect on tumor growth. Oncotarget 6(19):17777

Papakonstantinou E, Roth M, Karakiulakis G (2012) Hyaluronic acid: a key molecule in skin aging. Dermatoendocrinol 4(3):253–258

Pettipher GL (1988) Cultivation of the shiitake mushroom (Lentinus edodes) on lignocellulosic waste. J Sci Food Agric 42(3):195–198

Pinho O, Ferreira IMPLVO, Oliveira MBPP, Ferreira MA (2000) Quantification of synthetic phenolic antioxidants in liver pâtés. Food Chem 68(3):353–357

Poucheret P, Fons F, Rapior S (2006) Biological and pharmacological activity of higher fungi: 20-year retrospective analysis. Cryptogam Mycol 27(4):311

Rahman T, Choudhury MBK (2012) Shiitake mushroom: a tool of medicine. Bangladesh J Med Biochem 5(1): 24–32

Rai M, Tidke G, Wasser SP (2005) Therapeutic potential of mushrooms. Nat Product Radiance 4(4):246–257

Rathee S, Rathee D, Rathee D, Kumar V, Rathee P (2012) Mushrooms as therapeutic agents. Rev Bras 22:459–474

Reis FS, Barros L, Martins A, Ferreira IC (2012) Chemical composition and nutritional value of the most widely appreciated cultivated mushrooms: an inter-species comparative study. Food Chem Toxicol 50(2): 191–197

Ribeiro B, Lopes R, Andrade PB et al (2008) Comparative study of phytochemicals and antioxidant potential of wild edible mushroom caps and stipes. Food Chem 110(1):47–56

Ruksiriwanich W, Sirithunyalug J, Boonpisuttinant K, Jantrawut P (2014) Potent in vitro collagen biosynthesis stimulating and antioxidant activities of edible mushroom Volvariella volvacea aqueous extract

Sánchez C (2010) Cultivation of Pleurotus ostreatus and other edible mushrooms. Appl Microbiol Biotechnol 85:1321–1337

Sarkar R, Arora P, Garg KV (2013) Available cosmeceuticals for hyperpigmentation: what is. J Cutan Aesthet Surg 6(1):4–11

Shahidi F (2000) Antioxidants in food and food antioxidants. Food Nahrung 44(3):158–163

Shi M, Zhang Z, Yang Y (2013) Antioxidant and immunoregulatory activity of Ganoderma lucidum polysaccharide (GLP). Carbohydr Polym 95(1):200–206

Siu KC, Xu L, Chen X, Wu JY (2016) Molecular properties and antioxidant activities of polysaccharides isolated from alkaline extract of wild Armillaria ostoyae mushrooms. Carbohydr Polym 137:739–746

Soto ML, Falqué E, Domínguez H (2015) Relevance of natural phenolics from grape and derivative products in the formulation of cosmetics. Cosmetics 2(3):259–276

Souilem F, Fernandes Â, Calhelha RC et al (2017) Wild mushrooms and their mycelia as sources of bioactive compounds: antioxidant, anti-inflammatory and cytotoxic properties. Food Chem 230:40–48

Sułkowska-Ziaja K, Grabowska K, Apola A et al (2021) Mycelial culture extracts of selected wood-decay mushrooms as a source of skin-protecting factors. Biotechnol Lett 43:1051–1061

Sun Y, Liang H, Zhang X et al (2009) Structural elucidation and immunological activity of a polysaccharide from the fruiting body of Armillaria mellea. Bioresour Technol 100(5):1860–1863

Suprasert P, Apichartpiyakul C, Sakonwasun C et al (2014) Clinical characteristics of gynecologic cancer patients who respond to salvage treatment with Lingzhi. Asian Pac J Cancer Prev 15(10):4193–4196

Taofiq O, González-Paramás AM, Martins A et al (2016) Mushrooms extracts and compounds in cosmetics, cosmeceuticals and nutricosmetics – a review. Ind Crop Prod 90:38–48

Taofiq O, Rodrigues F, Barros L et al (2019) Mushroom ethanolic extracts as cosmeceuticals ingredients: safety and ex vivo skin permeation studies. Food Chem Toxicol 127:228–236

Thatoi H, Singdevsachan SK (2014) Diversity, nutritional composition and medicinal potential of Indian mushrooms: a review. Afr J Biotechnol 13(4)

Tsai SY, Tsai HL, Mau JL (2007) Antioxidant properties of Agaricus blazei, Agrocybe cylindracea, and boletus edulis. LWT-Food Sci Technol 40(8):1392–1402

Urquiaga INES, Leighton F (2000) Plant polyphenol antioxidants and oxidative stress. Biol Res 33(2): 55–64

Vamanu E, Nita S (2012) Antioxidant capacity and the correlation with major phenolic compounds, anthocyanin, and tocopherol content in various extracts from the wild edible boletus edulis mushroom. Biomed Res Int

Van Loo J (2006) Inulin-type fructans as prebiotics. Prebiotics: Dev Appl 57:100

Vaz JA, Barros L, Martins A et al (2011) Chemical composition of wild edible mushrooms and antioxidant properties of their water soluble polysaccharidic and ethanolic fractions. Food Chem 126(2):610–616

Vidović SS, Mujić IO, Zeković ZP et al (2010) Antioxidant properties of selected boletus mushrooms. Food Biophysics 5:49–58

Vieira V, Marques A, Barros L et al (2012) Insights in the antioxidant synergistic effects of combined edible mushrooms: phenolic and polysaccharidic extracts of boletus edulis and Marasmius oreades. J Food Nutr Res 51:109–112

Wasser SJAMB (2002) Medicinal mushrooms as a source of antitumor and immunomodulating polysaccharides. Appl Microbiol Biotechnol 60:258–274

Wasser SP (2011) Current findings, future trends, and unsolved problems in studies of medicinal mushrooms. Appl Microbiol Biotechnol 89:1323–1332

Wu Y, Choi MH, Li J et al (2016) Mushroom cosmetics: the present and future. Cosmet 3(3):22

Xu M, Zhu S, Li Y, Xu S, Shi G, Ding Z (2021) Effect of selenium on mushroom growth and metabolism: a review. Trends Food Sci Technol 118:328–340

Yahaya YA, Don MM (2012) Evaluation of Trametes lactinea extracts on the inhibition of hyaluronidase, lipoxygenase and xanthine oxidase activities in vitro. J Phys Sci 23:1–15

Yahia EM, Gutiérrez-Orozco F, Moreno-Pérez MA (2017) Identification of phenolic compounds by liquid chromatography-mass spectrometry in seventeen species of wild mushrooms in Central Mexico and determination of their antioxidant activity and bioactive compounds. Food Chem 226:14–22

Yang J, Yuwu C, Xiaozhang F et al (1984) Chemical constituents of Armillaria mellea mycelium I. Isolation and characterization of armillarin and armillaridin. Planta Med 50(04):288–290

Yao N, White PJ, Alavi S (2011) Impact of β-glucan and other oat flour components on physico-chemical and sensory properties of extruded oat cereals. Int J Food Sci Technol 46(3):651–660

Yeh JY, Hsieh LH, Wu KT, Tsai CF (2011) Antioxidant properties and antioxidant compounds of various extracts from the edible basidiomycete Grifola frondosa (Maitake). Molecules 16(4):3197–3211

Yim HS, Chye FY, Heng PY, Ho CW (2011) Oxidative stability of sunflower oil supplemented with medicinal split gill mushroom, Schizophyllum commune Fr.: Fr. extract during accelerated storage. Int J Med Mushrooms 13(4):357

Zembron-Lacny A, Gajewski M, Naczk M, Siatkowski I (2013) Effect of shiitake (Lentinus edodes) extract on antioxidant and inflammatory response to prolonged eccentric exercise. J Physiol Pharmacol 64(2):249–254

Zhang BB, Guan YY, Hu PF et al (2019) Production of bioactive metabolites by submerged fermentation of the medicinal mushroom Antrodia cinnamomea: recent advances and future development. Crit Rev Biotechnol 39(4):541–554

Zhang Y, Shi Q, Jiang W et al (2022) Comparison of the chemical composition and antioxidant stress ability of polysaccharides from Auricularia auricula under different drying methods. Food Funct 13(5):2938–2951

Zhao H, Zhang Q, Zhao L, Huang X, Wang J, Kang X (2012) Spore powder of Ganoderma lucidum improves cancer-related fatigue in breast cancer patients undergoing endocrine therapy: a pilot clinical trial. Evid Based Complement Alternat Med 2012(1): 809614

Zhong L, Yan P, Lam WC et al (2019) Coriolus versicolor and Ganoderma lucidum related natural products as an adjunct therapy for cancers: a systematic review and meta-analysis of randomized controlled trials. Front Pharmacol 10:703

Anti-inflammatory Activity of Fungal Extracts in Cosmeceutical Formulation

5

Anushmita Chanda, Hindol Ray, Mehuli Das, Sagarika Paul, Sampurna Saha, Rajib Majumder, and Sanmitra Ghosh

Abstract

The extracts from fruitbodies and hyphae have a huge repertoire of biologically active compounds such as phenolic acids, ergosterols, sugar polymers, peptides, proteins, and terpenoids which have anti-inflammatory, anti-aging, antioxidant, anti-tyrosinase, anti-hyaluronidase, antidiabetic, and anti-allergic effect. All these attributes not only make these extracts a perfect candidate to be formulated into cosmeceuticals which are commercially beneficial in two ways: first, it gives a medical benefit to the consumer who uses it; second, it gives an aesthetic benefit by enhancing the appearance of the skin. The antioxidant, antibacterial, and anti-inflammatory effects of these extracts make them ideal for being used as a skincare product for individuals with skin diseases or allergic tendencies. Mushrooms with high chitosan and chitin content have a propensity to increase the growth of keratinocytes and fibroblasts and hence have healing effect. The novel bioactive compounds can be extracted using different solvents and characterized by different advanced techniques like Reverse Phase, Gas Phase or High Pressure Liquid Chromatography, Mass Spectrometry, etc. With increase in the number of data generated by automated processes, the probability of finding new bioactive compounds becomes higher, greater the number of bioactive compounds, the greater the production of new types of products which have a variety of applications.

Keywords

Inflammation · Anti-inflammatory · Bioactive · Chromatography · Oxidative stress · Fungal extracts

5.1 Introduction

5.1.1 Overview of Cosmeceuticals

Cosmeceuticals represent a distinct evolving category of products that bridge the gap between cosmetics and pharmaceuticals. They are defined as commodities with topical application which contain bioactive ingredients designed to improve skin function and health (Goyal et al. 2022).

The authors Anushmita Chanda, Hindol Ray, Mehuli Das, Sagarika Paul, and Sampurna Saha have been equally contributed to this chapter.

A. Chanda · H. Ray · M. Das · S. Paul · S. Saha
S. Ghosh (✉)
Department of Biological Sciences, School of Life Science & Biotechnology, Adamas University, Kolkata, West Bengal, India
e-mail: sanmitra.ghosh@adamasuniversity.ac.in

R. Majumder
Department of Biotechnology, School of Life Science & Biotechnology, Adamas University, Kolkata, West Bengal, India

S. Ghosh, H. N. Onyeaka (eds.), *Fungal-Based Cosmetics: Formulation and Usage*, The Mycota 17,
https://doi.org/10.1007/978-3-032-09291-5_5

Cosmeceuticals are formulated with the intent of providing therapeutic benefits, unlike traditional cosmetics, which are primarily focused on skin beautification(Pandey et al. 2024). Cosmeceuticals are gaining immense popularity, pertaining to the increasing demand of consumers for products that have greater potential other than superficial beauty enhancements. However, the U.S. Food and Drug Administration (FDA) does not legitimately approve the term "cosmeceuticals" (under the Federal Food, Drug, and Cosmetic Act) as a separate category; instead, products are classified as either cosmetics or drugs or a combination of both based on their intended use and claims made by manufacturers (Nutrition 2022). Nevertheless, they must still adhere to labeling regulations and safety standards rendering quality assurance to be a crucial aspect of cosmeceutical production to maintain consumer confidence and product dependability.

The efficacy of cosmeceuticals is largely dependent on the active ingredients such as peptides, antioxidants, botanicals, and vitamins, which can target various skin concerns (Pandey et al. 2024). For instance, antioxidants like vitamin C and E assist to scavenge free radicals, thereby decreasing oxidative stress and averting aging before time. Retinoids, derived from vitamin A, are commonly used in anti-aging formulations due to their ability to promote cell turnover and stimulate collagen production (Goyal et al. 2022). Furthermore, ingredients like hyaluronic acid provide intense hydration, improving skin texture and elasticity. Many cosmeceuticals also incorporate anti-inflammatory agents, which soothe irritated skin and reduce redness in conditions like acne, rosacea, or eczema (Crous et al. 2024). One of the emerging areas of interest in cosmeceuticals is the use of fungi extracts. Historically, fungi have found application in traditional medicine for their healing capabilities. Modern scientific research has begun to reveal the significance of fungi extracts in skincare, particularly their activities against inflammation and oxidative stress (Li et al. 2020; Taofiq et al. 2017; Ziemlewska et al. 2022).

5.1.2 Importance of Anti-inflammatory Properties in Cosmeceutical Formulations

Skin provides the first line of defense against foreign molecules. It not only acts as a boundary between the host and the external environment, but also plays an active role in cutaneous adaptive immunity. Various immune cells like antigen-presenting cells, resident T cells, Langerhans cells, dermal dendritic cells, recruited to dermis and epidermis, together with the non-immune cells, form structures known as skin-associated-lymphoid-tissue (SALT) that is crucial in skin health maintenance and immunomodulation when exposed to antigens (Fernandes et al. 2023; Kabashima et al. 2019).

Inflammatory diseases like acne, eczema, rosacea, and allergic reactions can arise from an imbalance characterized by increased expression of multitude of inflammatory mediators like interleukins (IL), intercellular adhesion molecule-1 (ICAM-1), inducible cyclooxygenase-2 (COX-2), tumor necrosis factor (TNF), nuclear factor-kappa B (NF-κB), and prostaglandin E2 (PGE2). To prevent such outcomes, several anti-inflammatory compounds are often added to cosmeceutical formulations which not only inhibit excess production of inflammatory mediators but also contribute to calming and soothing effects on the skin prevent wrinkle formation and protects the skin from the disastrous effects of the UV rays. Cosmeceuticals with anti-inflammatory agents can be directly applied to the skin with enhanced penetration and absorption and provide stable and non-invasive method for maintaining healthy skin with regular use. Moreover, multi-functional products can be formulated to complement other therapeutic ingredients, and are suitable even for sensitive skin (Mago et al. 2023).

5.2 Fungi as a Natural Production House of Anti-inflammatory Compounds

With the rise in awareness about adverse effects of chemical products together with appreciation for natural alternatives, younger consumers are increasingly seeking "green" products. This shift in consumer purchasing preferences has led the cosmetic brands to focus on various botanical extracts.

Inflammation is a natural process of safeguarding the body that heals infections and injuries, but if the inflammation continues after the injury has healed, it can cause serious conditions including undesirable physiological conditions including several diseases of skin. Numerous bioactive products of fungal origin have been demonstrated to have significant level of anti-inflammatory properties as summarized in Table 5.1 (Antia et al. 2011; Blunt et al. 2018).

For instance, fruiting bodies and mycelia of certain mushrooms that are acknowledged as superfoods are also enriched with bioactive compounds (terpenoids, ceramides, lectins, schizophyllan, carotenoids, phenolics, etc.) that demonstrated appreciable activities against inflammation, microbial infection, oxidative stress, and tyrosinase activities. Some of the most commercially utilized fungi for their extracts to be used in several cosmeceutical products as components of serums, gels, creams, and lotions include *Agaricus bisporus* (Button mushroom), *Cordyceps* sp., *Ganoderma lucidum* (Reishi), *Schizophyllum commune*, *Grifola frondosa* (Maitake), *Lentinula edodes* (Shiitake), *Inonotus obliquus, Phellinus linteus, Cordyceps sinesis*, *Cordyceps militaris,* etc. These agents mainly reduce the concentration of pro-inflammatory mediators (NO, IL-6, IL-1, IL-8, TNF), preventing inflammation (Mago et al. 2023).

Many of the fungal extracts being multi-functional are capable of targeting multiple issues, such as skin moisture, UV rays, oxidative stress, apart from working as agents of anti-inflammation. Antioxidants can scavenge and neutralize the reactive oxygen species produced in response to environmental hassles like UV rays and pollution. Thus, antioxidants play a crucial role in reducing inflammation and protecting cells. Anti-hyaluronidase agents can also aid in preventing inflammation by targeting enzyme hyaluronidase. It breaks down the hyaluronic acid or hyaluronan found in the extracellular matrix of the connective tissues. Hyaluronan is responsible for skin moisture, elasticity, hydration, healing, wound repair, and regulating inflammation. Therefore, hyaluronidase is an effective target in inflammation regulation (Abd Razak et al. 2020). Chronic inflammation is often accompanied by dry, scaly, itchy patches, and lack of moisture. Thus, fungal extracts with moisturizing capacity can also provide relief from inflammation (Kim et al. 2018).

5.3 Mode of Action of Fungal Compounds Against Inflammation

Fungal compounds with anti-inflammatory properties can modulate various biological pathways involved in inflammatory responses, making them valuable in cosmeceutical formulations. Understanding the strategies with which these fungal compounds exhibit their anti-inflammatory effects is crucial for optimizing their use in skincare products.

5.3.1 Cytokine Modulation

One of the primary mechanisms through which fungal compounds exert anti-inflammatory effects is by impeding the synthesis and activity of pro-inflammatory cytokines that play significant roles in inflammatory processes through a complex network of interactions. Certain triterpenoids derived from fungi have been shown to downregulate pro-inflammatory cytokines such as TNF-α, IL-1β, and IL-6, while stimulating the release of anti-inflammatory cytokine IL-10. This shift aids in re-establishment of balance in the immune response, making these compounds potential cosmeceutical agents for inflammatory diseases. For example, recent study shows there is the potential effect of

Table 5.1 Bioactive compounds of fungal origin with anti-inflammatory action used in cosmeceuticals

Name of the fungi	Metabolite obtained	Structure	Mode of action	Reference
Daldinia childiae	Daldinal A	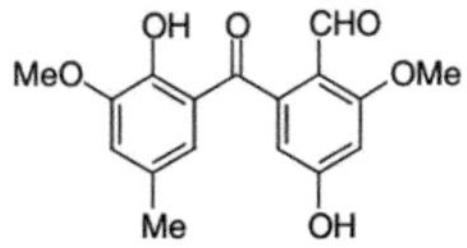	Significantly inhibits the NO generation in LPS-stimulated RAW 264.7 cells	Quang et al. (2006a)
	Daldinal B	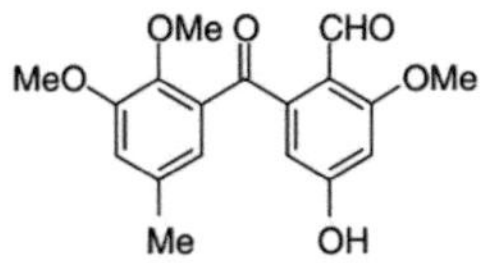		
	Daldinal C	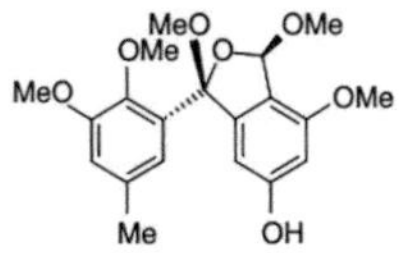		
Agaricus bisporus, Cantharellus cibarius, Lactarius deliciosus	Pyrogallol	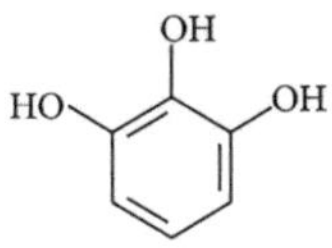	Suppresses NO and transcription of iNOS, IL-1β, and IL-6 mRNAs in response to LPS-induced RAW 264.7 macrophages	Moro et al. (2012)
Trichoderma harzianum	Gliovirin	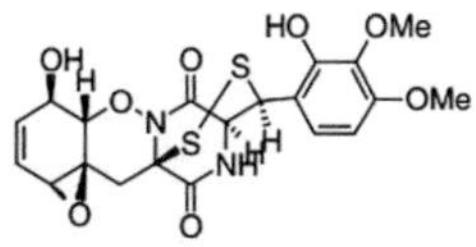	Inhibits the induced TNF-α expression	Rether et al. (2007)
Phomopsis sp.	Phomol	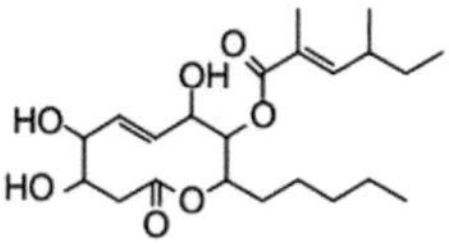	Showed anti-inflammatory activity by suppressing swelling in the mouse ear assay	Weber et al. (2004)
Ascomycete strain A111-95	Galiellalactone	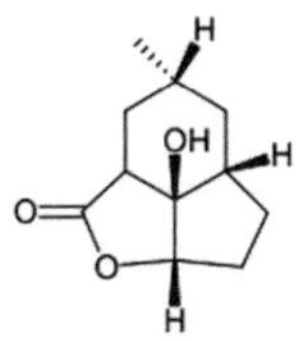	Prevents STAT 3 dimers from binding to their DNA binding sites, suppressing IL-6-mediated SEAP expression	Weidler et al. (2000)
Aspergillus fumigatus	Fumigaclavine C	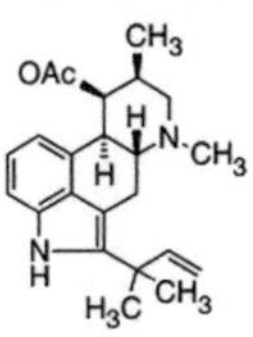	Suppresses the expression of IL-1β, IL-2, IL-12α, IFN-γ, TNF-α	Wu et al. (2005), Zhao et al. (2004)
Albatrellus ovinus	Griofolins	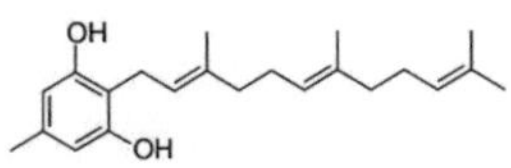	Inhibits the NO production in LPS-induced RAW 264.7 cells. Also suppresses histamine generation from rat peritoneal mast cells	Quang et al. (2006b)

(continued)

Table 5.1 (continued)

Name of the fungi	Metabolite obtained	Structure	Mode of action	Reference
Aspergillus terreus	Lovastatin	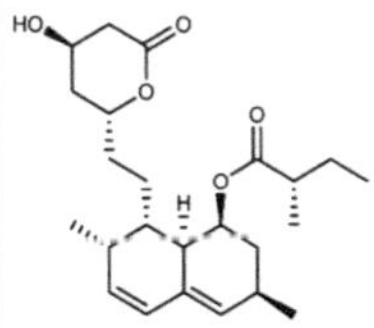	Inhibits NO production in LPS-induced RAW 264.7 cells	Liu et al. (2018b)
Aspergillus unguis	Folipastatin A	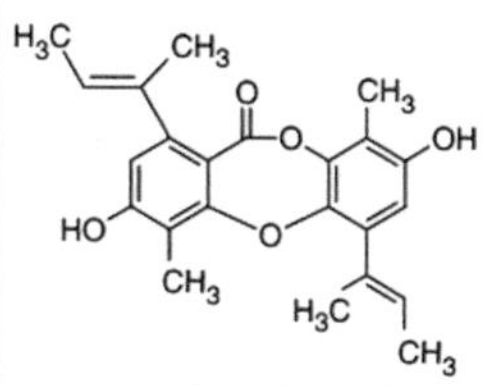	Inhibits the phospholipase A2 from rabbit peritoneal exudate and suppresses the arachidonic acid secretion from rat polymorphonuclear leucocyte	Hamano et al. (1992)
Penicillium sp. *SF-5295*	Viridicatol	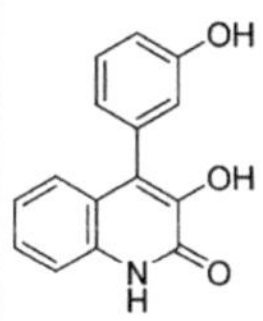	Suppresses the production of NO and prostaglandin E2 in LPS-induced RAW 264.7 and BV2 cells	Ko et al. (2015)
Aspergillus terreus	Luteoride E	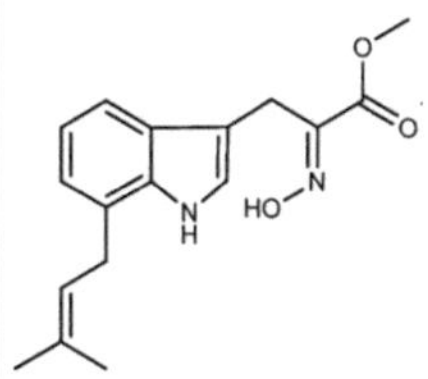	Inhibits the production of NO in LPS-activated RAW264.7 cells	Liu et al. (2018a)
Ganoderma lucidum	Tsugaric acid A	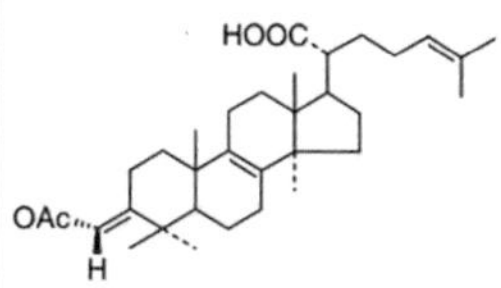	Strongly inhibits SO^{-} ^{2} formation in fMLP/CB-stimulated rat neutrophils	Ko et al. (2008)
Alternaria sp. *SF-5016*	Alternaramide	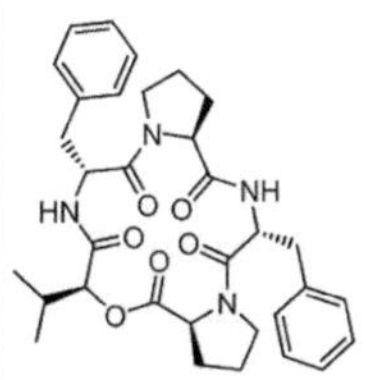	Inhibits the production of NO and prostaglandin E2 in LPS-activated RAW264.7 and BV2 cells	Ko et al. (2015)
Monascus sp.	Citrinin	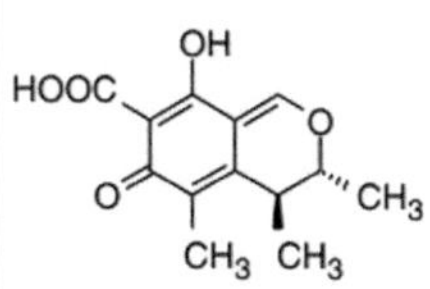	Suppresses LPS/IFN-γ, induced NO in MES-13 cell line with reduced cell viability and nitrite production	Liu et al. (2010)
Penicillium wortmannii	Wortmannin	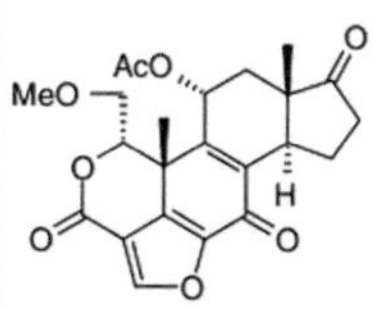	Suppresses the LPS-induced Akt phosphorylation and expression of other members of PI3-kinase superfamily	Brian et al. (1957)

(continued)

Table 5.1 (continued)

Name of the fungi	Metabolite obtained	Structure	Mode of action	Reference
Cladosporium sp., *Trichoderma* sp., *Aspergillus wentii*, *Penicillium islandicum*	Emodin	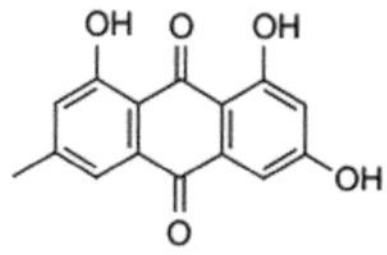	Inhibits expression of IL-1β, IL-6, IL-8, CCL2, NF-kB, and degradation of IkB-α in HUVECs	Izhaki (2002)
Penicillium chrysogenum SCSIO41001	Chrysamide C	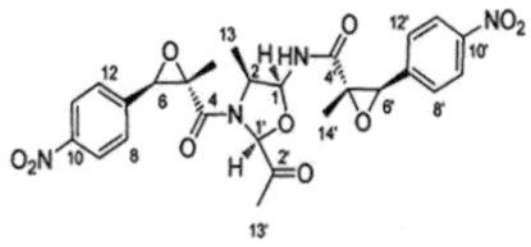	Suppresses the production of IL-17	Chen et al. (2016)
Penicillium paxilli Ma(G)K	Pyrenocine A	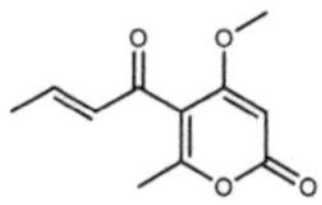	Has a strong anti-inflammatory impact on TNF-α and PGE2 in LPS-activated macrophages	Toledo et al. (2014)
Aspergillus sp. *HKI 0472*	Funalenine	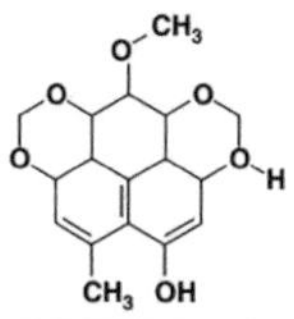	Inhibits enzyme 3α-HSD, COX-1 and COX-2	Wangun and Vigne (2006)
Pholiota squarrosa	Squarrosidine	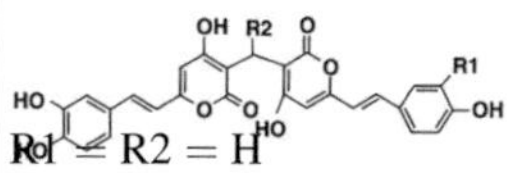 R1 = R2 = H	Inhibits 3α-HSD, COX-1, COX-2 and XO	Wangun and Vigne (2006)
Hypoxylon cohaerens	Cohaerin A	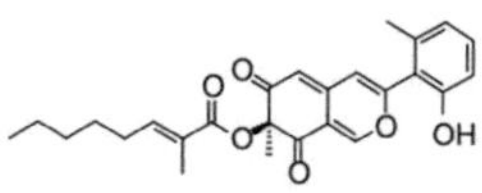	Inhibits NO production	Quang et al. (2004)
	Cohaerin B	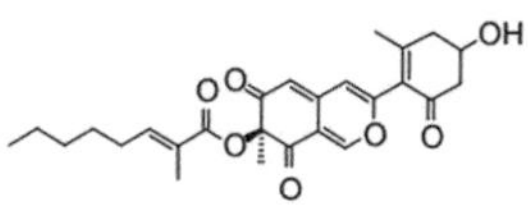		
Hypoxylon fragiforme	Mitorubrinic acid and Mitorubrinol	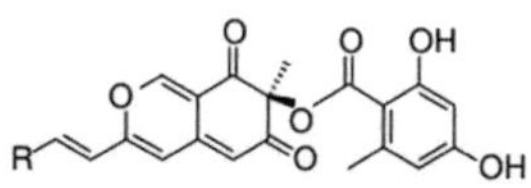 Mitorubrinol R = CH2OH Mitorubrinic acid R = COOH	Inhibits NO production in RAW 264.7 cells	Quang et al. (2006c)
Aspergillus violaceofuscus	Violaceotide A	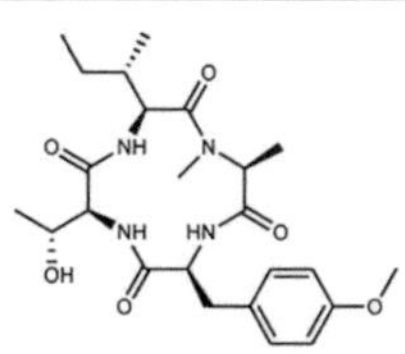	Lowers the expression of IL-10 in LPS-induced THP-1 cells	Liu et al. (2018a)
	Diketopiperazine dimer	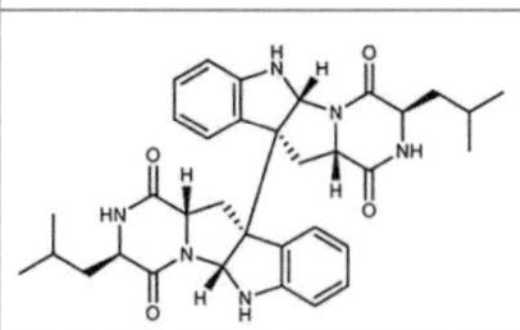		

(continued)

Table 5.1 (continued)

Name of the fungi	Metabolite obtained	Structure	Mode of action	Reference
Penicillium sp. *SF-5629*	Citrinin H1	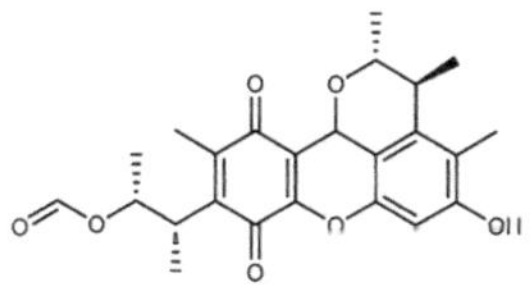	Inhibits the production of NO and PGE2 in BV2 microglia stimulated by LPS	Ngan et al. (2017)
Aspergillus sp. *SF-5044*	Asperlin	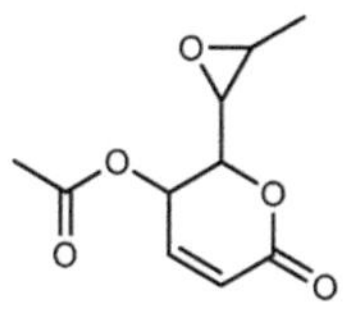	Inhibit the production of NO and PGE2 in LPS-stimulated murine macrophages and RAW 264.7 Reduces the expression of pro-inflammatory cytokines and mediators by increasing heme oxygenase (HO) activity	Lee et al. (2011)
Penicillium thomii KMM 4667	Thomimarine E	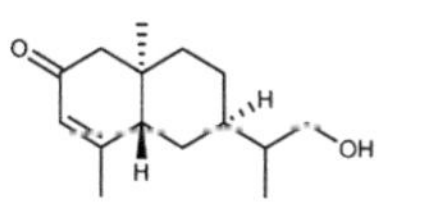	Inhibits the production of NO in LPS-induced RAW 264.7 cells	Afiyatullov et al. (2017)
Penicillium sp.	Brevicompanines E	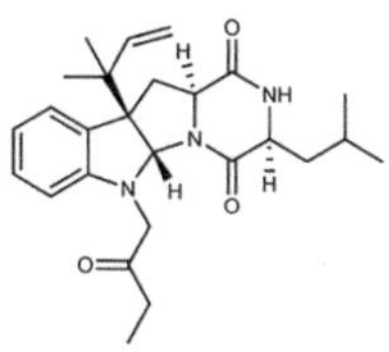	Inhibits NO production in BV2 microglial cells stimulated by LPS. Don't have any cytotoxic effect	Du et al. (2009)
	Brevicompanines H	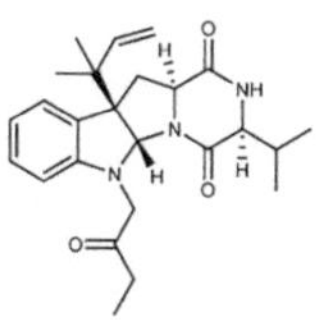		
Fusarium heterosporum CNC-477	Mangicols A and B	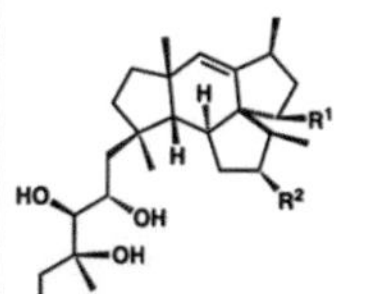 Mangicol A:R1 = OH,R2 = H Mangicol B:R1 = H,R2 = OH	Shows significant anti-inflammatory effects in mouse ear edema stimulated by phorbol myristate acetate	Renner et al. (2000)
Penicillium sp. *SF-5292*	Penicillinolide A	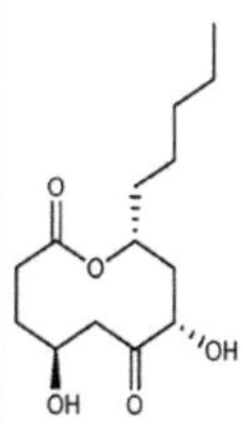	Inhibits the production of NO and PGE2 by lowering the expression of iNOS and COX-2 in macrophages induced by LPS. Also suppresses the transcript level of TNF-α, IL-1β, and IL-6	Lee et al. (2013)
Hypoxylon multiforme	Multiformin D	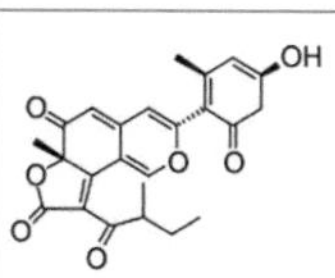	Inhibits NO production in RAW 264.7 cells	Quang et al. (2005)

Cordyceps militaris-based nanoemulsion in reduction of cytokines like TNF-α and IL-1β, which in turn, reduce inflammation (Phull et al. 2022).

5.3.2 Inhibition of NF-κB Signaling Pathway

NF-κB is a transcription factor that mediates immunological responses triggered by pro-inflammatory substances like IL-1, TNF, bacteria, etc. In general, it is confined by IκB, an inhibitory protein, in the cytoplasm. Upon phosphorylation of IκB by IκB kinase-β (IKK-β), NF-κB is released and it further translocates into the nucleus where it induces regulatory genes to produce pro-inflammatory proteins (Liu et al. 2022). Fungal compounds, particularly the terpenoids, can constrain NF-κB activation. By blocking this pathway, these compounds reduce transcription of genes encoding pro-inflammatory mediators, thereby minimizing inflammation (Trepa et al. 2024). For example, the triterpenes found in *Ganoderma lucidum* have been shown to inhibit the addition of phosphate moiety to IκB, leading to retention of NF-κB in cytoplasm and a subsequent decrease in pro-inflammatory cytokine production. This action underscores the potential of fungal compounds in managing chronic inflammatory diseases where these cytokines are overexpressed (Hilliard et al. 2020).

5.3.3 Antioxidant Activity

Oxidative stress is known to exacerbate inflammatory responses. Several fungal metabolites demonstrate significant antioxidant properties, which can counteract oxidative stress by neutralizing free radicals and, subsequently, inflammation and resultant cellular damage. Additionally, antioxidants can modulate signaling pathways involved in inflammation, such as the MAPK pathway, which is activated by oxidative stress. For example, compounds such as 6-pentyl-α-pyrone from *Trichoderma gamsii* and triterpenes and polysaccharides from *Ganoderma lucidum* have shown capacity to scavenge free radicals and reduce oxidative damage in various cell types, further highlighting their potential for anti-inflammatory applications (Lim et al. 2023). These compounds also enhance the activity of endogenous free radical nullifiers, namely glutathione peroxidase and superoxide dismutase (SOD). By suppression of oxidative stress, these extracts help mitigate inflammation and shield skin cells from injury, making them valuable in formulations aimed at preventing and treating inflammatory skin conditions (Seweryn et al. 2021).

5.3.4 Orchestration of Inducible Nitric Oxide Synthase (iNOS/Nitric Oxide (NO) Production

NO is a signaling molecule that, in high concentrations, contributes to inflammation. Some fungal compounds can regulate the production of NO, thereby exerting anti-inflammatory effects (Man et al. 2022). iNOS contributes significantly in inflammation through synthesis of NO from L-arginine. There are two key mechanisms through which NO contribute to inflammation. Firstly, NO enhances cyclooxygenase (COX) activity, prompting elevation of pro-inflammatory prostaglandin expression. Secondly, NO can react with superoxide anions (O_2^-) to generate cytotoxic radical peroxynitrite, which can exacerbate tissue damage. Additionally, the expression of human iNOS is reliant on the activation of the STAT1 pathway, suggesting that targeting this pathway may be a viable strategy for developing iNOS inhibitors. Polysaccharides derived from *Ganoderma lucidum* have been demonstrated to impede the expression of iNOS, the enzyme accountable for NO generation during inflammatory responses and thus impart protection against tissue damage associated with chronic inflammation (Tung et al. 2013).

5.3.5 Suppression of MAPK Pathway

Fungal metabolites can inhibit key components of mitogen-activated protein kinase (MAPK) cascade, leading to downregulation of pro-inflammatory cytokines and chemokines. For example, polysaccharides from *Grifola frondosa* have shown potential of anti-inflammation by inhibiting the MAPK signaling pathways and the activation of Th1, Th2, Th17, and Th22 immune responses (Choi and Choi 2023).

5.4 Methods of Extracting and Detecting Fungal Bioactive Compounds for Use in Cosmeceuticals

5.4.1 Solvent-Based Methods

Extraction solvents like water, ethanol, and methanol are used for the isolation of polar organic compounds, while hexane and carbon tetrachloride can be utilized for the isolation of low polarity compounds from fungi. Polar compound extraction can be either assisted by microwave heat or ultrasonic sound waves which assist cell wall disruption and easy release of intracellular matrix in solvent (Grosso et al. 2015; Kaufmann and Christen 2002). Supercritical fluid extraction extracts non-polar and less polar compounds, with compressed CO_2 gas (Farré et al. 2010; Joana Gil-Chávez et al. 2013). Subcritical water extraction uses pressurized hot water, where both polar and moderately polar substances can be extracted with variation in temperature (Duarte et al. 2014b; Smith 2006). Pressurized liquid extraction uses hexane, ethanol, acetone, or any other solvent in a low volume and isolates target compounds in a short time under high pressure and high temperature (Duarte et al. 2014a).

5.4.2 Chromatographic Techniques

Fungal bioactive compounds are often found in mixtures rendering their separation into individual pure compound a challenging task (Stadler et al. 2014). The most common technique used is reverse phase chromatography that is optimized to separate an assortment of secondary metabolites with functional groups like phenyl, biphenyl, and pentafluoro-phenyl moieties (Kildgaard et al. 2014; Wolfender et al. 2015).

Comprehensive 2D liquid chromatography can carry out orthogonal separation by using high-polarity high-performance liquid chromatography (HPLC) in the first dimension later it uses an absorbent small particle C_{18} in the second dimension (Chen et al. 2012; Dugo et al. 2009). Gas phase chromatography-Mass spectrometry can be used for the separation of volatile organic fungal compounds even in trace amount (as less as 5 ppm) (Lakhdari et al. 2023). Gas chromatography-atmospheric pressure chemical ionization can identify the molecular mass of an organic molecule. Reverse phase chromatography is useful in separating highly polar and ionic compounds with high resolution (Creek et al. 2011; Sørensen et al. 2007). High pH ion chromatography is also used along with ion-pair chromatography (Kvitvang et al. 2014; Magdenoska et al. 2013; van Dam et al. 2002). Although reverse phase chromatography separates a variety of fungal compounds, very small and polar compounds are not very well retained. Therefore to solve this problem hydrophilic interaction liquid chromatography is used.

5.4.3 Mass Spectrometric Analysis

This technique has been optimized to work with liquid chromatography to characterize fungal secondary metabolites without causing significant amount of fragmentation of the products. It involves a soft technique of ionization is carried out by atmospheric pressure chemical ionization, electrospray ionization, and atmospheric pressure photo ionization (Klitgaard et al. 2014; Nielsen et al. 2011; Rasmussen et al. 2014). Libraries such as AntiBase and Dictionary of Natural Products provide the information about the novelty of the compound, mass, composition of the elements in the compound and further comparison of different data can be carried out from these libraries. AntiBase is a library which provides information about a diverse array of fungal compounds. The most commonly used versions of Mass spectrometry are Orbitrap, Time of Flight, and Fourier Transform Ion Cyclotron. The Time of Flight Mass Spectrometers rapidly determine and differentiate multiple compounds with appropriate isotopic ratios. Ample time for separating these compounds is obtained in subsequent runs (Kildgaard et al. 2014). Ion-Trap Mass Spectrometers can perform sequential Mass Spectrometry at a low cost, disadvantage being low mass accuracy. This method can easily identify polymers and peptides of fungal origin. Triple Quadrupole Mass Spectrometry is preferred for sensitive dereplication.

5.4.4 Diode Array Detection

Diode array detection method detects the ultraviolet spectrum of the chromophore of a compound which is then compared with the UV/Vis spectra of diverse molecules present in the database of UV spectrum (Bitzer et al. 2007; de la Cruz et al. 2012; Hansen et al. 2005; Larsen et al. 2005). This method is relatively inexpensive and is carried out after Mass Spectrometric analysis has been carried out since databases like AntiBase and Dictionary of Natural Products do not have a direct data search of the values of the UV/Vis spectra data. Every compound has a unique chromophore system due to which every compound has unique UV spectra. Few examples of such compounds having a unique UV/vis spectra are gamma-naphthopyrones, quinazolines, non-reduced polyketides, and non-ribosomal proteins (Larsen et al. 1999; Nielsen et al. 2009). Since the UV/vis spectrum of these compounds is unique, the data obtained is unique and this helps in the analysis of the structural characteristics of the compounds. Since the structure of the compound is being analyzed, this method reveals the elemental composition of the compounds as well thus helping in identification, screening, and dereplication of the compounds. In this technique, a single compound can have multiple absorption bands, greater the number of absorption bands, more unique the compound is, better the differentiation of the compounds. A good example is mycophenolic acid. Mass Spectrometry has to be carried out since it acts as a confirmatory test for the data obtained, as for example in case of mycophenolic acid which contains 5,7-dihydroxy methylphthalide, there are at least 55 other compounds which have this same compound making the need for mass spectrometry essential. The UV/vis spectra determination helps in identification of the different types of biosynthetic pathways. This data is crucial for easy identification and differentiation of compounds that follow the same biosynthetic pathway (Holm et al. 2014).

5.4.5 Surface Techniques

Different surface techniques have also been useful in detection and identification of fungal compounds. Most of these techniques are surface desorption techniques which ionize and detect the products excreted by the fungi at the surface of the agar medium. There are two main methods that fall into this method of detection: Direct Analysis In Real Time, Desorption Electrospray Ionization Mass Spectrometry as well as the variants which include the soft ionization techniques (Bouslimani et al. 2014; Hsu et al. 2013; Watrous et al. 2012). Desorption Electrospray Ionization Mass Spectrometry and Direct Analysis In Real Time techniques are useful in studying the major structure as their resolution ranges in the millimeters, therefore Desorption Electrospray Ionization Mass Spectrometry is useful in the imprint of droplets of exudates (Figueroa et al. 2014; Gurdak et al. 2014).

In brief, a typical procedure of extracting bioactive compounds from fungi and characterizing them will be as follows:

The procedure for processing the fungal fruitbody is to dry them in an oven at 30 °C followed by powder formation and ethanolic extract preparation. This is followed by use of rotary evaporator to evaporate the solvent in a reduced pressure environment (Heleno et al. 2016). For detection of Ergosterol, the ethanolic extract is again dried and redissolved in methanol followed by filtration with nylon disposable filter. The identification is performed with high-pressure liquid chromatography-ultraviolet detector. The chromatographic separation is conducted using a reverse-phase column operated at 35 °C. The mobile phase used is acetonitrile/methanol at 70:30 v/v and the detection is carried at 280 nm by using ultraviolet detector (Barreira et al. 2014).

5.4.6 Formulation

The base of the cosmetic product, which is a cream, should comprise silicone, chelating agent, pure water, Vitamin E, preservatives, pro-liposomes, lubricant, emollient, emulsifiers, etc. These components in a cream are to be certified to be safe by concerned health surveillance agencies of respective country and by US Food and Drug Administration, Regulation. Considering minimum inhibitory concentration of fungal components, the base cream is normally mixed with around ten times the MIC value of the respective components to form a homogeneous mixture so that bioactivity of all the constituents can be analyzed. Methanolic extract of the final cream is then filtered followed by evaporation using a rotary evaporator. While to analyze phenol content, the dried cream is to be redissolved in water:methanol mixture; for analysis of anti-inflammatory activity and ergosterol content the dried cream can be redissolved in 50% DMSO and methanol respectively. Similarly, different dilutions of DMSO can be used for estimating anti-tyrosinase activity (50%), antioxidant activity (100%) as well as antibacterial activities (5%) in the dried cream formulation of the cosmeceuticals. The one-way Analysis of Variance is applied to analyze the differences between samples (Taofiq et al. 2016).

5.5 Efficacy of Fungal Extracts in Cosmeceutical Formulations

In the ever-evolving world of cosmetic and skincare products, the search for natural and effective remedies has led to the exploration of fungal extracts as a promising cradle of therapeutic compounds. Cosmeceuticals with the ability to escalate the natural process of skin healing have garnered significant attention for their ability to address diverse types of skin concerns, from aging to wound healing apart from beautifying (Geng 2023; Shedoeva et al. 2019). At present, there is a high need for cosmeceuticals that act in skin whitening, moisturizing, sunscreen, anti-inflammatory, anti-wrinkle, anti-aging, skin anti-acne, as well as antioxidation and cell-protection (Alves et al. 2020).

5.5.1 Antioxidant Activities

Fungi are enriched with compounds possessing antioxidant properties that scavenge and deactivate the highly reactive oxygen species, generated due to UV exposure and thus prevent oxidative stress and tissue damage. They can also inhibit lipid peroxidation maintaining skin barrier integrity as well as activate biocatalysts like superoxide dismutase and catalase as weapons of cellular antioxidant protection devices. These include phenols, flavonoids, fatty acids, tocopherols, polysaccharides, etc. (Elkhateeb 2020). Likewise, extract of *Grifola frondosa* being enriched with powerful antioxidants like flavonoids, phenolic compounds, and β-glucans have been incorporated in a large number of serums and sunscreens (Camilleri et al. 2024). Reports also indicated that phenolic content of fungal extracts has a positive correlation with its antioxidative capacity. For instance, research demonstrated that high oxygen scavenging activity in *Agaricus bisporus* extract was attributed to the availability of abundant phenolic acids (gallic and cinnamic acid) (Gąsecka et al. 2018). Another antioxidant compound, trehalose, is found in various mushrooms, such as *Pholiota nameko*, *Lentinula edodes*, *Grifola frondosa*, and *Auricularia auricula*. Polysaccharides and triterpenoids obtained from *Ganoderma amboinense* have also exhibited significant oxidative stress-reducing abilities (Gupta et al. 2019). Marine fungi are the other opulent homes of antioxidants. A novel exopolysaccharide (EPS) named N1 was extracted from the deep-sea fungus *Aspergillus versicolor* N2bc, consisting of mannoglucogalactan and galactofuranose units. Another endophytic strain of *A. versicolor*, SB5 and *Stemphylium lycopersici* (which produces physcion) effectively scavenges hydroxyl and ABTS radicals and reduces NO synthesis (Elawady et al. 2023). Therefore, several fungal extracts have been used in commercially available cosmeceutical formulations.

5.5.2 Anti-inflammatory Activities

Fungi are sources of alkaloids, phenols, steroids, flavonoids, glycosides. and triterpenoids. which contribute to anti-inflammation (Mago et al. 2023). Terpenoids produced by *Pleurotus cornucopiae, Ganoderma lucidum, G. lingzhi, Tricholoma matsutake, Cyathus africanus*, etc. primarily inhibit NO production and modulate inflammatory responses (Dudekula et al. 2020). Fasciculols are sterol-derivatives produced by *Hypholoma lateritium* that inhibit the production of inflammatory mediator, cyclooxygenase-2 (COX-2) and activate Nrf2 pathway, which stimulates the cellular defense mechanisms against inflammation (Ványolós et al. 2020). Studies suggest flavonoids produced by *Sanghuangporus baumii* and polysaccharides produced by *Pholiota nameko, Echinodontium tinctorium,* etc., are potential combatants of inflammation (Abreu et al. 2019; Holm et al. 2014; Wang et al. 2022).

Hyaluronic acid or hyaluronan is an integral component of skin and connective tissues. It is responsible for skin lubrication and moisture. However, it gets broken down upon exposure to UV, pathogen, or hyaluronidase enzyme. Hyaluronidase causes chronic inflammatory response that results in the formation of dry, scaly, itchy skin. Thus, fungal extracts with anti-hyaluronidase and moisturizing properties become essential in cosmeceutical formulations. Studies report *Auricularia polytricha, G. lingzhi, N. lepideus, P. citrinopileatus, P. ostreatus, and S. commune* are able to produce metabolites with moisturizing capacity (Sujarit et al. 2021).

5.5.3 Antimicrobial Activities

The skin is a natural barrier and home simultaneously to numerous microorganisms, mainly bacteria. Opportunistic bacteria can cause serious localized infections and systemic issues like cellulitis, atopic dermatitis, folliculitis, impetigo, furuncle, erysipelas, and psoriasis that lead to skin lesion formation. Antimicrobial ingredients are essential for preventing infections and ensuring cosmetic safety, preferably natural compounds to overcome resistance to traditional antimicrobials (10.13005/bbra/3097, (PDF) Mushrooms: Splendid Gifts for the Cosmetic Industry).

Phenolic acids like 2,4-dihydroxybenzoic, protocatechuic, vanillic, and p-coumaric acids have shown strong activity against bacteria irrespective of gram-character. Isodeoxyhelicobasidin from *V. bombycina* also demonstrated antibacterial effects against several Gram-positive pathogens, including *S. aureus*, methicillin-resistant *S. aureus* (MRSA), quinolone-resistant S. aureus (QRSA), *B. subtilis, S. epidermidis*, and *S. mutans*. ((PDF) Mushrooms: Splendid Gifts for the Cosmetic Industry). Polysaccharides from *Cordyceps militaris, Pleurotus florida,* and *Auricularia* sp. have also been found to possess antibacterial properties (Chang et al. 2019; Rupa et al. 2020; Saraswat et al. 2020). *Lentinula edodes* and other mushrooms like *Boletus, Ganoderma*, and *Lepista* have been reported as effective against most bacteria (Mago et al. 2023).

5.5.4 Anti-tyrosinase Activities

Tyrosinase is crucial in catalysis of the process of melanin production, and its hyperactivity causes skin darkening. UV exposure boosts tyrosinase activity, increasing melanin synthesis. Tyrosinase inhibitors are used in skin-lightening creams where they alter the enzyme's tertiary and quaternary structures reducing melanin production (Sujarit et al. 2021). Many of the fungal phenolic acids and polysaccharides function as tyrosinase inhibitors. Anti-tyrosinase properties have been found to be present in extracts of *Ganoderma applatum, G. lingzhi, Volvariella volvacea*, etc. (Badalyan et al. 2022; Sangthong et al. 2022; Taofiq et al. 2019a, b).

5.5.5 Moisturizing Activities

Certain fungal extracts, especially those rich in polysaccharides like beta-glucans, contribute to skin hydration through the following mechanisms: (a) fungi like *Schizophyllum* containing glucans form a protective film on skin surface, reducing trans-epidermal water loss, (b) fungi like *Tremella* sp. is a potent humectant with rich source of polysaccharides that improves skin's natural moisturizing factor (NMF) production, and (c) ergosterol containing fungal members like *Agaricus* sp. improving skin barrier function by promoting ceramide synthesis (Anil Kumar et al. 2023; Kim et al. 2018; Muthuramalingam et al. 2019; Yang et al. 2021).

5.5.6 Anti-aging Activities

Extracts derived from mushrooms like *Ganoderma lucidum, Cordyceps, Lentinula edodes,* and *Agaricus bisporus* can combat signs of aging. Different mechanisms of action involve inhibiting matrix metalloproteinases (MMPs) that break down collagen and elastin, stimulating collagen and elastin production by fibroblasts and through reduction of advanced glycation end-products (AGEs) formation, which contributes to skin aging (Anil Kumar et al. 2023). Fungal extracts are often multi-functional and are able to address multiple issues simultaneously. This feature facilitates formulating cosmetics with skin healing properties much simpler and easier. Till this date, numerous products are commercially available which contain fungal extracts. At the same time, scientists are attempting to demonstrate novel therapeutic fungal extracts with potential to be utilized in cosmetics. Table 5.2 provides a list of such experiments providing evidence of cosmeceutical potential of fungal extracts.

5.6 Limitations and Regulations on Incorporating Fungal Extracts in Cosmeceutical Formulations

The primary obstacles in incorporating fungal extracts in cosmetics are the risk of fungal allergies, mycotoxins contamination, and lack of standardized fungal extracts. Approximately 10% of general population exhibits fungal allergies and this percentage increases in case of individuals with asthma and hypersensitivity. Fungal extracts are highly complex and show great variability in composition depending upon the extraction, cultivation, and purification process. Ensuring consistent quality and composition of fungal extracts across batches can pose challenge due to dissimilarities in cultivation parameters and extraction procedures. Some fungi may produce toxic metabolites like mycotoxins, so thorough toxicological testing is required to ensure safety for topical application followed by robust clinical studies to substantiate the claimed benefits (Mago et al. 2023). Thus, standardization of a method to obtain functional formulated extracts with long shelf-life is essential for large-scale production (Pfeiffer and Swoboda 2023). Kiki et al. reported some of the undesirable side effects associated with fungi-based cosmetics such as damage to the beneficial microflora of the skin, neurological and endocrine damage and potential carcinogenic effects (Kiki 2023). Fungi-based cosmetics are susceptible to microbial contamination, which may be due to contamination of the crude fungal extracts, insufficient preservative usage, or application after the recommended shelf-life (Kočevar Glavač and Lunder 2018).

There is often a gap in translation of experimental results to commercial applications. For instance, most tyrosinase inhibitors have not been tested for use in topical cosmeceuticals, partly due to the limited availability of many such inhibitors, hindering in vivo testing. While several promising compounds have been identified, there is currently insufficient clinical data to confirm their benefits in cosmetic products

Table 5.2 Bioactive properties of fungal extracts with cosmeceutical application potential

Sl. no.	Mushroom used	Product type	Metabolite	Bioactive property	Toxicity	References
Antioxidative						
1	*Agaricus bisporus*	Extract	Gallic acid, cinnamic acid, p-hydroxybenzoic, p-coumaric	Antioxidant	Yet to be determined	Gąsecka et al. (2018)
2	*Phoma glomerata*	Extract	Peyroisocumarins A and B	Radical scavenging activity	Yet to be determined	Esposito et al. (2021)
3	*Acremonium sp*	Extract	Hydroquinone derivatives	Antioxidant	Yet to be determined	Mukherjee et al. (2024)
4	*Pholiota nameko*	Extract	Trehalose	Reduces oxidative stress	Yet to be determined	Gupta et al. (2019)
5	*Ganoderma amboinense*	Extract	Polysaccharide	Protects against UV-induced damage	Yet to be determined	Zhao et al. (2021)
6	*Aspergillus chevalieri*	Extract	Echinulin and neoechinulin A	Antioxidant	Yet to be determined	da Calado et al. (2021)
7	*Flammulina velutipes*, Pleurotus *florida*, *P. ostreatus*	Extract	Polysaccharide	Antioxidant	Yet to be determined	Saraswat et al. (2020)
8	*Cyclocybe aegerita*, *Boletus edulis*	Extract	Ergothioneine (ERGO), glutathione (GSH)	Reduces oxidative stress by aiding cellular defense systems	Yet to be determined	Mukherjee et al. (2024)
9	*Auricularia auricula-judae (Bull.)*	Extract	Polysaccharides	Antioxidant	No cytotoxicity was observed	Choi et al. (2019)
10	*Pleurotus eryngii*	Extract	Cinnamic acid, ergosterol, p-Coumaric acid, p-hydroxyphenylacetic acid	Reduces oxidative stress	Non-cytotoxic	Sujarit et al. (2021)
Anti-inflammatory						
1	*Cordyceps militaris*	Nanoemulsion	Cordycepin	Anti-inflammatory	Non-toxic	Rupa et al. (2020)
2	*Phallus indusiatus*	Extract	phenols	Reduces the levels of NO	Non-toxic	Theeraraksakul et al. (2023)
3.	*Agaricus blazei Murrill*	Semi solid cream	Tocopherols, fatty acids, organic acids, sugars	Suppress inflammatory responses, antioxidant, skin hydration and elasticity	Non-toxic	Taofiq et al. (2019a)

4	*Agaricus bisporus*	Creams	Phenolic compounds	Suppress the production of pro-inflammatory mediators, antioxidant	Non-toxic	Usman et al. (2021)
5.	*Cyathus africanus*	Extract	Neocyathin	Reduces NO production and suppresses expression of inducible NO synthase protein.	No significant cytotoxicity	Wei et al. (2018)
6	*Macrolepiota procera*	Extract	Lanostane triterpenoids	Inhibits NO production	Slight toxicity to murine RAW 264.7 macrophages	Chen et al. (2018)
7	*Phellinus igniarius*	Extract	Nepetidin, Betulic acid	Suppressed the activity of NF-κB	Not yet demonstrated	Jiang et al. (2018)
8	*Pholiota nameko*	Extract	β-D-glucan	Inhibits the release of inflammatory mediators, downregulated the NF-κB signaling pathway	Non-toxic	Abreu et al. (2019)
9	*Hypholoma lateritium*	Extract	Fasciculols	Inhibits COX-2 and activates of Nrf2	Non-toxic	Ványolós et al. (2020)
10	*Echinodontium tinctorium*	Extract	Polysaccharides	Inhibits NO and TNF-α production	Non-toxic	
Antimicrobial						
1	*Cordyceps militaris*	Nano emulsion	Cordycepin acid, adenosine, unsaturated fatty acids, polysaccharides	Antibacterial *E. coli* and *S. aureus*, action similar to neomycin 30	Non-toxic	Rupa et al. (2020)
2	*Pleurotus florida*	Extract	Polysaccharides	Antibacterial specific against *P. aeruginosa*	Yet to be determined	Saraswat et al. (2020)
3	*Auricularia* sp.	Extract	Chitosan	Antibacterial effect against *S. aureus* and *E. coli*	Yet to be determined	Chang et al. (2019)
4	*Taiwanofungus camphoratus*	Extract	Not determined	Potent antifungal property against *A. fumigatus* and *C. albicans*. Antibacterial property against *S. aureus* and *Lactobacillus* spp.	Yet to be determined	Kim et al. (2022)
	Tricholosporum goniospermum	Extract	Not specified	Antibacterial against *E. coli, B. cereus,* and *P. aeruginosa.* Antifungal against *A. crocatum* and *A. gypseum*	Not yet determined	Angelini et al. (2020)
Anti-tyrosinase						
1	*Volvariella volvacea*	Gel cream	Polysaccharides	Anti-tyrosinase	Non-toxic	Kočevar Glavač and Lunder (2018)
2	*Phallus indusiatus*	Extract	Phenols, flavonoids (chalcone, resveratrol, coumarin), 5-(hydroxymethyl)-2-furfural	Anti-tyrosinase	No cytotoxicity was observed	Theeraraksakul et al. (2023)

(continued)

Table 5.2 (continued)

Sl. no.	Mushroom used	Product type	Metabolite	Bioactive property	Toxicity	References
3	*Ganoderma applanatum, Laetiporus sulphureus, Trametes versicolor*	Extract	Kojic acid, Sterols, phenolic acids	Anti-tyrosinase	Not yet demonstrated	Sułkowska-Ziaja et al. (2021)
4	*Ganoderma lingzhi*	Semi-solid cream	Protocatechuic and syringic acids	Inhibits tyrosinase	Non-toxic	Taofiq et al. (2019b)
5	*Ganoderma lucidum*	Extract	β-glucan	Significant anti-tyrosinase potential	Not yet determined	Badalyan et al. (2022)
6	*Myrothecium sp*	Extract	6-n-pentyl-pyrone	Inhibits tyrosinase	Not yet determined	Agrawal et al. (2018)
7	*Tricholosporum goniospermum*	extract	Not specified	Inhibits tyrosinase	Not yet determined	Angelini et al. (2020)

(Agrawal et al. 2019). These extracts are also difficult to purify leading to toxicity in case of large-scale production. In vivo results often do not correspond with the in vitro data. Other factors contributing to this issue include the fact that researchers are more focused in identifying novel therapeutic extracts, or other beneficial properties of known compounds, but much of the extensive knowledge about the compounds remains unknown (Yao and Xu 2022). A review by Zhang et al. emphasizes that connections between the functionalities and structural properties of fungal sugar polymers can only be fully understood once their purified structures are determined. A thorough exploration of the structure-function correlation of anti-inflammatory, antioxidant, wound-recovery, moisturizing, anti-aging, and skin tone lightening effects of the polysaccharides will guide researchers in developing modified polysaccharides with improved bioactive properties (Zhang et al. 2022).

Cosmetic products containing fungal extracts must adhere to strict labelling requirements, including listing all ingredients and making no misleading claims about the product's effects. Moreover, manufacturers of cosmeceuticals with fungal extracts must abide by GMP guidelines to safeguard product quality, safety, and consistency. After marketing safety of the cosmeceutical usage is regularly monitored by the regulatory agencies and any adverse effect found is to be immediately reported (Abd Razak et al. 2020; Wu et al. 2016). Fungal extracts must be approved for use in cosmetics by regulatory bodies like the FDA and EU. Novel ingredients require extensive safety data to gain approval. Regulations for fungal-derived cosmeceuticals vary globally. In the USA, the FDA considers fungal products safe and categorizes botanical ingredients as food additives or supplements. In Europe, European Medicines Agency (EMA) differentiates between medications and cosmetics, defining cosmetics as products used externally for cosmetic purposes. The EU's Cosmetic Products Regulation (EU Regulation 1223/2009), effective from July 11, 2013, enforces stringent safety standards and mandates detailed safety reporting for cosmetics. In Canada, the term "cosmeceutical" is not used. Instead, products are classified as either cosmetics or drugs based on their claims and composition. Canadian regulations, governed by the Consumer Packaging and Labeling Act and the Food and Drugs Act, emphasize safety, packaging, labeling, and advertising. The Canadian cosmetic regulation system primarily focuses on the safety of chemicals and products, accurate labeling, and mandatory reporting of all cosmetic items. (Mukherjee et al. 2024).

Despite these challenges, the use of fungal extracts in cosmeceuticals is increasing due to their potential benefits and consumer demand for natural ingredients. Careful attention to quality control, safety, and regulatory compliance is crucial for successful incorporation of these ingredients in cosmetic products.

5.7 Enhancing Penetration and Efficacy Through Advanced Delivery Systems

Strong anti-inflammatory effects have been shown by fungus-derived bioactive substances, including peptides, terpenoids, and polyphenols. However, difficulties relating to bioavailability, stability, and targeted distribution to specific tissues or cell types frequently hamper their successful therapeutic application. Advanced delivery methods provide novel solutions to these issues. Modern technology is utilized in the development of these systems to enable accurate drug distribution to the targeted site of action, enhancing therapeutic efficacy and reducing non-specific accumulation inside the body. They are therefore more effective than traditional drug delivery methods. Usually, their composition consists of nanomaterials or tiny devices with multifunctional constituents that are biocompatible, biodegradable, and have a high viscoelasticity, which leads to a longer half-life in circulation (Ezike et al. 2023). Some of the advanced drug delivery approaches are discussed below that have the potential to significantly enhance the penetration and efficacy of fungal bioactive compounds with anti-inflammatory effects:

Liposome-based drug delivery systems are regarded as effective and innovative nano-delivery systems as they can successfully distribute a diverse variety of biologically active substances (Hofheinz et al. 2005). In practical terms, liposomes work as carriers for many drugs, considerably increasing their therapeutic efficacy. They accomplish this via altering drug absorption, extending drug biological half-life, decreasing metabolism, and lowering possible toxicity (Andra et al. 2022). Liposomes are truly unique in their intrinsic capacity to distinguish between hydrophilic and hydrophobic substances and dissolve them. This property, together with their biocompatibility and biodegradability, makes liposomes an excellent drug delivery vehicle (Çağdaş et al. 2014). Apart from their pharmaceutical applications, liposomes also serve as an effective carrier for cosmetic formulations, further highlighting their diverse potential in the field of drug delivery and cosmetics (Bhupathyraaj et al. 2013).

5.8 Conclusion

Fungi are seen as prospective biofactories for fabrication of anti-inflammatory compounds that may serve as a safe substitute for their chemical counterparts with fewer negative effects when used in medicinal or cosmeceutical products. Macrofungi or mushrooms are highly valued for their rich repertoire of bioactive compounds with therapeutic qualities, which are increasingly sought after for their applications in cosmeceutical products (Badalyan et al. 2022; He et al. 2012; Wang and Marcone 2011). Mushroom growth being seasonal and variable with weather changes, their constant availability for research and production purpose is compromised. Moreover, identification of exact biomolecule with therapeutic significance is difficult as the anti-inflammatory activities have mostly been identified in crude extract with several metabolites together. Thus, for better understanding of the anti-inflammatory processes and any adverse effects of these compounds, it is imperative to isolate and identify the active metabolites. Furthermore, the assurance of the effectiveness and purity of mushroom products intended for medicinal use is hampered by the lack of established testing techniques (Elsayed et al. 2014).

To sum up, in order to firmly establish fungi as important repertoire for the synthesis of anti-inflammatory cosmeceuticals, more research is required to solve these issues. The identification of novel species for use in cosmeceuticals and organic cosmetics will be made easier by advancements in the fields of fungal biology, proteomics, metabolomics, genomics, systems pharmacology, biotechnology, and molecular cosmetology, as well as a more detailed exploration of the molecular mechanisms underlying the therapeutic and cosmetic activities of fungi.

References

Abd Razak DL, Jamaluddin A, Abd Rashid NY, Sani NA, Abdul Manan M (2020) Assessment of cosmeceutical potentials of selected mushroom Fruitbody extracts through evaluation of antioxidant, anti-hyaluronidase and anti-tyrosinase activity. J 3(3). https://doi.org/10.3390/j3030026

Abreu H, Simas FF, Smiderle FR, Sovrani V, Dallazen JL, Maria-Ferreira D, Werner MF, Cordeiro LMC, Iacomini M (2019) Gelling functional property, anti-inflammatory and antinociceptive bioactivities of β-D-glucan from the edible mushroom Pholiota nameko. Int J Biol Macromol 122:1128–1135. https://doi.org/10.1016/j.ijbiomac.2018.09.062

Afiyatullov S, Leshchenko EV, Sobolevskaya MP, Antonov AS, Denisenko VA, Popov RS, Khudyakova YV, Kirichuk NN et al (2017) New Thomimarine E from marine isolate of the fungus Penicillium thomii. Chem Nat Compd 53(2):290–294. https://doi.org/10.1007/s10600-017-1972-9

Agrawal S, Adholeya A, Barrow CJ, Deshmukh SK (2018) Marine fungi: an untapped bioresource for future cosmeceuticals. Phytochem Lett 23:15–20. https://doi.org/10.1016/j.phytol.2017.11.003

Agrawal S, Deshmukh SK, Barrow CJ (2019) Marine fungi as a potential source of future cosmeceuticals. In: Satyanarayana T, Deshmukh SK, Deshpande MV (eds) Advancing frontiers in mycology & mycotechnology: basic and applied aspects of fungi. Springer, pp 627–669. https://doi.org/10.1007/978-981-13-9349-5_25

Alves A, Sousa E, Kijjoa A, Pinto M (2020) Marine-derived compounds with potential use as cosmeceuticals and Nutricosmetics. Molecules (Basel, Switzerland) 25(11):2536. https://doi.org/10.3390/molecules25112536

Andra VVSNL, Pammi SVN, Bhatraju LVKP, Ruddaraju LK (2022) A comprehensive review on novel liposomal methodologies, commercial formulations, clinical trials and patents. BioNanoScience 12(1): 274–291. https://doi.org/10.1007/s12668-022-00941-x

Angelini P, Venanzoni R, Angeles Flores G, Tirillini B, Orlando G, Recinella L, Chiavaroli A, Brunetti L, Leone S, Di Simone SC, Ciferri MC, Zengin G, Ak G, Menghini L, Ferrante C (2020) Evaluation of antioxidant, antimicrobial and Tyrosinase inhibitory activities of extracts from Tricholosporum goniospermum, an edible wild mushroom. Antibiotics 9(8). https://doi.org/10.3390/antibiotics9080513

Anil Kumar NV, Quispe C, Herrera-Bravo J, Herrera Belén L, Loren P, Salazar LA, Silva V, Erdogan Orhan I, Senol Deniz FS, Nemli E, Capanoglu E, Olatunde A et al (2023) Potential of mushrooms bioactive for the treatment of skin diseases and disorders. J Food Biochem 2023(1):5915769. https://doi.org/10.1155/2023/5915769

Antia BS, Aree T, Kasettrathat C, Wiyakrutta S, Ekpa OD, Ekpe UJ, Mahidol C, Ruchirawat S, Kittakoop P (2011) Itaconic acid derivatives and diketopiperazine from the marine-derived fungus aspergillus aculeatus CRI322-03. Phytochemistry 72(8):816–820. https://doi.org/10.1016/j.phytochem.2011.02.013

Badalyan SM, Barkhudaryan A, Rapior S (2022) Medicinal macrofungi as cosmeceuticals: a review. Int J Med Mushrooms 24(4):1–13. https://doi.org/10.1615/IntJMedMushrooms.2022043124

Barreira JCM, Oliveira MBPP, Ferreira ICFR (2014) Development of a novel methodology for the analysis of Ergosterol in mushrooms. Food Anal Methods 7(1): 217–223. https://doi.org/10.1007/s12161-013-9621-9

Bhupathyraaj M, Maheswaran A, Brindha P, Mullaicharam AR, Masilamani K (2013) Liposomal drug delivery systems – a review. Int J Pharm Sci Rev Res 23(1):295–301

Bitzer J, Köpcke B, Stadler M, Hellwig V, Ju Y-M, Seip S, Henkel T (2007) Accelerated dereplication of natural products, supported by reference libraries. Chimia *61*(6):6. https://doi.org/10.2533/chimia.2007.332

Blunt W et al (2018) Marine natural products. Nat Prod Rep 35(1):8–53. https://doi.org/10.1039/C7NP00052A

Bouslimani A, Sanchez LM, Garg N, Dorrestein PC (2014) Mass spectrometry of natural products: current, emerging and future technologies. Nat Prod Rep 31(6): 718–729. https://doi.org/10.1039/C4NP00044G

Brian PW, Curtis PJ, Hemming HG, Norris GLF (1957) Wortmannin, an antibiotic produced by *Penicillium wortmanni*. Trans Br Mycol Soc 40(3):365–IN3. https://doi.org/10.1016/S0007-1536(57)80033-

Çağdaş M, Sezer AD, Bucak S, Çağdaş M, Sezer AD, Bucak S (2014) Liposomes as potential drug carrier systems for drug delivery. In: Application of nanotechnology in drug delivery. IntechOpen. https://doi.org/10.5772/58459

Camilleri E, Blundell R, Baral B, Karpiński TM, Aruci E, Atrooz OM (2024) Unveiling the full spectrum of maitake mushrooms: a comprehensive review of their medicinal, therapeutic, nutraceutical, and cosmetic potential. Heliyon 10(9):e30254. https://doi.org/10.1016/j.heliyon.2024.e30254

Chang AKT, Frias RR, Alvarez LV, Bigol UG, Guzman JPMD (2019) Comparative antibacterial activity of commercial chitosan and chitosan extracted from *Auricularia* sp. Biocatal Agric Biotechnol 17:189–195. https://doi.org/10.1016/j.bcab.2018.11.016

Chen X, Cao Y, Lv D, Zhu Z, Zhang J, Chai Y (2012) Comprehensive two-dimensional HepG2/cell membrane chromatography/monolithic column/time-of-flight mass spectrometry system for screening anti-tumor components from herbal medicines. J Chromatogr A 1242:67–74. https://doi.org/10.1016/j.chroma.2012.04.034

Chen S, Wang J, Lin X, Zhao B, Wei X, Li G, Kaliaperumal K, Liao S, Yang B, Zhou X, Liu J, Xu S, Liu Y (2016) Chrysamides A-C, three dimeric Nitrophenyl trans-Epoxyamides produced by the Deep-Sea-derived fungus Penicillium chrysogenum SCSIO41001. Org Lett 18(15):3650–3653. https://doi.org/10.1021/acs.orglett.6b01699

Chen H-P, Zhao Z-Z, Li Z-H, Huang Y, Zhang S-B, Tang Y, Yao J-N, Chen L, Isaka M, Feng T, Liu J-K (2018) Anti-proliferative and anti-inflammatory Lanostane triterpenoids from the polish edible mushroom Macrolepiota procera. J Agric Food Chem 66(12):3146–3154. https://doi.org/10.1021/acs.jafc.8b00287

Choi E-J, Choi JK (2023) Extracts of Grifola frondosa inhibit the MAPK signaling pathways involved in keratinocyte inflammation and ameliorate atopic dermatitis. Nutr Res Pract 17(6):1056–1069. https://doi.org/10.4162/nrp.2023.17.6.1056

Choi Y-J, Park I-S, Kim M-H, Kwon B, Choo Y-M, Jeong S-I, Yu K-Y, Choe CH, Kim J (2019) The medicinal mushroom Auricularia auricula-judae (bull.) extract has antioxidant activity and promotes procollagen biosynthesis in HaCaT cells. Nat Prod Res 33(22): 3283–3286. https://doi.org/10.1080/14786419.2018.1468332

Creek DJ, Jankevics A, Breitling R, Watson DG, Barrett MP, Burgess KEV (2011) Toward global metabolomics analysis with hydrophilic interaction liquid chromatography–mass spectrometry: improved metabolite identification by retention time prediction. Anal Chem 83(22):8703–8710. https://doi.org/10.1021/ac2021823

Crous C, Pretorius J, Petzer A (2024) Overview of popular cosmeceuticals in dermatology. Skin Health Disease 4(2):e340. https://doi.org/10.1002/ski2.340

da Calado M et al (2021) Marine endophytic fungi associated with Halopteris scoparia (Linnaeus) Sauvageau as producers of bioactive secondary metabolites with potential dermocosmetic application. PLoS One 16(5):e0250954. https://doi.org/10.1371/journal.pone.0250954

de la Cruz M, Martín J, González-Menéndez V, Pérez-Victoria I, Moreno C, Tormo JR, El Aouad N, Guarro J, Vicente F, Reyes F, Bills GF (2012) Chemical and physical modulation of antibiotic activity in Emericella species. Chem Biodivers 9(6):1095–1113. https://doi.org/10.1002/cbdv.201100362

Du L, Yang X, Zhu T, Wang F, Xiao X, Park H, Gu Q (2009) Diketopiperazine alkaloids from a deep ocean sediment derived fungus Penicillium sp. Chem Pharm Bull 57(8):873–876. https://doi.org/10.1248/cpb.57.873

Duarte K, Justino CIL, Gomes AM, Rocha-Santos T, Duarte AC (2014a) Chapter 4 – Green analytical methodologies for preparation of extracts and analysis of bioactive compounds. In: Rocha-Santos T, Duarte AC (eds) Comprehensive analytical chemistry, vol 65. Elsevier, pp 59–78. https://doi.org/10.1016/B978-0-444-63359-0.00004-5

Duarte K, Justino CIL, Pereira R, Freitas AC, Gomes AM, Duarte AC, Rocha-Santos TAP (2014b) Green analytical methodologies for the discovery of bioactive compounds from marine sources. Trends Environ Anal Chem 3–4:43–52. https://doi.org/10.1016/j.teac.2014.11.001

Dudekula UT, Doriya K, Devarai SK (2020) A critical review on submerged production of mushroom and their bioactive metabolites. 3 Biotech 10(8):337. https://doi.org/10.1007/s13205-020-02333-y

Dugo P, Cacciola F, Donato P, Airado-Rodríguez D, Herrero M, Mondello L (2009) Comprehensive two-dimensional liquid chromatography to quantify polyphenols in red wines. J Chromatogr A 1216(44):7483–7487. https://doi.org/10.1016/j.chroma.2009.04.001

Elawady ME, Hamed AA, Alsallami WM, Gabr EZ, Abdel-Monem MO, Hassan MG (2023) Bioactive metabolite from Endophytic aspergillus versicolor SB5 with anti-acetylcholinesterase, anti-inflammatory and antioxidant activities: in vitro and in Silico studies. Microorganisms 11(4):1062. https://doi.org/10.3390/microorganisms11041062

Elkhateeb WA (2020) What medicinal mushroom can do?

Elsayed EA, El Enshasy H, Wadaan MAM, Aziz R (2014) Mushrooms: a potential natural source of anti-inflammatory compounds for medical applications. Mediat Inflamm 2014:805841. https://doi.org/10.1155/2014/805841

Esposito R, Ruocco N, Viel T, Federico S, Zupo V, Costantini M (2021) Sponges and their symbionts as a source of valuable compounds in cosmeceutical field. Mar Drugs 19(8):444. https://doi.org/10.3390/md19080444

Ezike TC, Okpala US, Onoja UL, Nwike CP, Ezeako EC, Okpara OJ, Okoroafor CC, Eze SC, Kalu OL, Odoh EC, Nwadike UG, Ogbodo JO, Umeh BU, Ossai EC, Nwanguma BC (2023) Advances in drug delivery systems, challenges and future directions. Heliyon 9(6):e17488. https://doi.org/10.1016/j.heliyon.2023.e17488

Farré M, Pérez S, Gonçalves C, Alpendurada MF, Barceló D (2010) Green analytical chemistry in the determination of organic pollutants in the aquatic environment. TrAC Trends Anal Chem 29(11):1347–1362. https://doi.org/10.1016/j.trac.2010.07.016

Fernandes A, Rodrigues PM, Pintado M, Tavaria FK (2023) A systematic review of natural products for skin applications: targeting inflammation, wound healing, and photo-aging. Phytomed Int J Phytother Phytopharmacol 115:154824. https://doi.org/10.1016/j.phymed.2023.154824

Figueroa M, Jarmusch AK, Raja HA, El-Elimat T, Kavanaugh JS, Horswill AR, Cooks RG, Cech NB, Oberlies NH (2014) Polyhydroxyanthraquinones as quorum sensing inhibitors from the Guttates of Penicillium restrictum and their analysis by desorption electrospray ionization mass spectrometry. J Nat Prod 77(6):1351–1358. https://doi.org/10.1021/np5000704

Gąsecka M, Magdziak Z, Siwulski M, Mleczek M (2018) Profile of phenolic and organic acids, antioxidant properties and ergosterol content in cultivated and wild growing species of Agaricus. Eur Food Res Technol 244(2):259–268. https://doi.org/10.1007/s00217-017-2952-9

Geng X (2023) The application of nanotechnology in natural cosmeceuticals. Theor Nat Sci 3:110–116. https://doi.org/10.54254/2753-8818/3/20220204

Goyal A, Sharma A, Kaur J, Kumari S, Garg M, Sindhu RK, Rahman MH, Akhtar MF, Tagde P, Najda A, Banach-Albińska B, Masternak K, Alanazi IS, Mohamed HRH, El-Kott AF, Shah M, Germoush MO, Al-Malky HS, Abukhuwayjah SH et al (2022) Bioactive-based cosmeceuticals: an update on emerging trends. Molecules (Basel, Switzerland) 27(3):828. https://doi.org/10.3390/molecules27030828

Grosso C, Valentão P, Ferreres F, Andrade PB (2015) Alternative and efficient extraction methods for marine-derived compounds. Mar Drugs 13(5). https://doi.org/10.3390/md13053182

Gupta PL, Rajput M, Oza T, Trivedi U, Sanghvi G (2019) Eminence of microbial products in cosmetic industry. Nat Products Bioprospect 9(4):267–278. https://doi.org/10.1007/s13659-019-0215-0

Gurdak E, Green FM, Rakowska PD, Seah MP, Salter TL, Gilmore IS (2014) VAMAS Interlaboratory study for desorption electrospray ionization mass spectrometry (DESI MS) intensity repeatability and Constancy. Anal Chem 86(19):9603–9611. https://doi.org/10.1021/ac502075t

Hamano K, Kinoshita-Okami M, Hemmi A, Sato A, Hisamoto M, Matsuda K, Yoda K, Haruyama H, Hosoya T, Tanzawa K (1992) Folipastatin, a new depsidone compound from aspergillus unguis as an inhibitor of phospholipase A2. Taxonomy, fermentation, isolation, structure determination and biological properties. J Antibiot 45(8):1195–1201. https://doi.org/10.7164/antibiotics.45.1195

Hansen ME, Smedsgaard J, Larsen TO (2005) X-hitting: an algorithm for novelty detection and Dereplication by UV spectra of complex mixtures of natural products. Anal Chem 77(21):6805–6817. https://doi.org/10.1021/ac040191e

He J-Z, Ru Q-M, Dong D-D, Sun P-L (2012) Chemical characteristics and antioxidant properties of crude water soluble polysaccharides from four common edible mushrooms. Molecules (Basel, Switzerland) 17(4): 4373–4387. https://doi.org/10.3390/molecules17044373

Heleno SA, Diz P, Prieto MA, Barros L, Rodrigues A, Barreiro MF, Ferreira ICFR (2016) Optimization of ultrasound-assisted extraction to obtain mycosterols from *Agaricus bisporus* L. by response surface methodology and comparison with conventional Soxhlet extraction. Food Chem 197:1054–1063. https://doi.org/10.1016/j.foodchem.2015.11.108

Hilliard A, Mendonca P, Soliman KFA (2020) Involvement of NFκB and MAPK signaling pathways in the preventive effects of Ganoderma lucidum on the inflammation of BV-2 microglial cells induced by LPS. J Neuroimmunol 345. https://doi.org/10.1016/j.jneuroim.2020.577269

Hofheinz R-D, Gnad-Vogt SU, Beyer U, Hochhaus A (2005) Liposomal encapsulated anti-cancer drugs. Anti-Cancer Drugs 16(7):691–707. https://doi.org/10.1097/01.cad.0000167902.53039.5a

Holm DK, Petersen LM, Klitgaard A, Knudsen PB, Jarczynska ZD, Nielsen KF, Gotfredsen CH, Larsen TO, Mortensen UH (2014) Molecular and chemical characterization of the biosynthesis of the 6-MSA-derived Meroterpenoid Yanuthone D in *Aspergillus niger*. Chem Biol 21(4):519–529. https://doi.org/10.1016/j.chembiol.2014.01.013

Hsu C-C, ElNaggar MS, Peng Y, Fang J, Sanchez LM, Mascuch SJ, Møller KA, Alazzeh EK, Pikula J, Quinn RA, Zeng Y, Wolfe BE, Dutton RJ, Gerwick L, Zhang L, Liu X, Månsson M, Dorrestein PC (2013) Real-time metabolomics on living microorganisms using ambient electrospray ionization flow-probe. Anal Chem 85(15):7014–7018. https://doi.org/10.1021/ac401613x

Izhaki I (2002) Emodin – a secondary metabolite with multiple ecological functions in higher plants. New Phytol 155(2):205–217. https://doi.org/10.1046/j.1469-8137.2002.00459.x

Jiang Z, Jin M, Zhou W, Li R, Zhao Y, Jin X, Li G (2018) Anti-inflammatory activity of chemical constituents isolated from the willow bracket medicinal mushroom Phellinus igniarius (Agaricomycetes). Int J Med Mushrooms 20(2):119–128. https://doi.org/10.1615/IntJMedMushrooms.2018025536

Joana Gil-Chávez G, Villa JA, Fernando Ayala-Zavala J, Basilio Heredia J, Sepulveda D, Yahia EM, González-Aguilar GA (2013) Technologies for extraction and production of bioactive compounds to be used as nutraceuticals and food ingredients: an overview. Compr Rev Food Sci Food Saf 12(1):5–23. https://doi.org/10.1111/1541-4337.12005

Kabashima K, Honda T, Ginhoux F, Egawa G (2019) The immunological anatomy of the skin. Nat Rev Immunol 19(1):19–30. https://doi.org/10.1038/s41577-018-0084-5

Kaufmann B, Christen P (2002) Recent extraction techniques for natural products: microwave-assisted extraction and pressurised solvent extraction. Phytochem Anal 13(2):105–113. https://doi.org/10.1002/pca.631

Kiki MJ (2023) Biopigments of microbial origin and their application in the cosmetic industry. Cosmetics 10(2). https://doi.org/10.3390/cosmetics10020047

Kildgaard S, Mansson M, Dosen I, Klitgaard A, Frisvad JC, Larsen TO, Nielsen KF (2014) Accurate Dereplication of bioactive secondary metabolites from marine-derived fungi by UHPLC-DAD-QTOFMS and a MS/HRMS library. Mar Drugs 12(6). https://doi.org/10.3390/md12063681

Kim YA, Kim DH, Park CB, Park TS, Park BJ (2018) Anti-inflammatory and skin-moisturizing effects of a flavonoid glycoside extracted from the aquatic plant Nymphoides indica in human keratinocytes. Molecules (Basel, Switzerland) 23(9):2342. https://doi.org/10.3390/molecules23092342

Kim JH, Tam CC, Chan KL, Mahoney N, Cheng LW, Friedman M, Land KM (2022) Antimicrobial efficacy of edible mushroom extracts: assessment of fungal resistance. Appl Sci 12(9). https://doi.org/10.3390/app12094591

Klitgaard A, Iversen A, Andersen MR, Larsen TO, Frisvad JC, Nielsen KF (2014) Aggressive dereplication using UHPLC–DAD–QTOF: screening extracts for up to 3000 fungal secondary metabolites. Anal Bioanal Chem 406(7):1933–1943. https://doi.org/10.1007/s00216-013-7582-x

Ko H-H, Hung C-F, Wang J-P, Lin C-N (2008) Antiinflammatory triterpenoids and steroids from Ganoderma lucidum and G. tsugae. Phytochemistry 69(1):234–239. https://doi.org/10.1016/j.phytochem.2007.06.008

Ko W, Sohn J, Kim Y-C, Oh H (2015) Viridicatol from marine-derived fungal strain Penicillium sp. SF-5295 exerts anti-inflammatory effects through inhibiting NF-κB signaling pathway on lipopolysaccharide-induced RAW264.7 and BV2 cells. Nat Prod Sci 21: 240. https://doi.org/10.20307/nps.2015.21.4.240

Kočevar Glavač N, Lunder M (2018) Preservative efficacy of selected antimicrobials of natural origin in a cosmetic emulsion. Int J Cosmet Sci. https://doi.org/10.1111/ics.12461

Kvitvang HFN, Kristiansen KA, Bruheim P (2014) Assessment of capillary anion exchange ion chromatography tandem mass spectrometry for the quantitative profiling of the phosphometabolome and organic acids in biological extracts. J Chromatogr A 1370:70–79. https://doi.org/10.1016/j.chroma.2014.10.029

Lakhdari W, Benyahia I, Bouhenna MM, Bendif H, Khelafi H, Bachir H, Ladjal A, Hammi H, Mouhoubi D, Khelil H, Alomar TS, AlMasoud N, Boufafa N, Boufahja F, Dehliz A (2023) Exploration and evaluation of secondary metabolites from Trichoderma harzianum: GC-MS analysis, phytochemical profiling, antifungal and antioxidant activity assessment. Molecules 28(13):13. https://doi.org/10.3390/molecules28135025

Larsen TO, Franzyk H, Jensen SR (1999) UV-guided isolation of Verrucines a and B, novel Quinazolines from Penicillium verrucosum structurally related to Anacine from Penicillium aurantiogriseum. J Nat Prod 62(11):1578–1580. https://doi.org/10.1021/np990251p

Larsen TO, Petersen BO, Duus JØ, Sørensen D, Frisvad JC, Hansen ME (2005) Discovery of new natural products by application of X-hitting, a novel algorithm for automated comparison of full UV spectra, combined with structural determination by NMR spectroscopy. J Nat Prod 68(6):871–874. https://doi.org/10.1021/np040248s

Lee D-S, Jeong G-S, Li B, Lee SU, Oh H, Kim Y-C (2011) Asperlin from the marine-derived fungus aspergillus sp. SF-5044 exerts anti-inflammatory effects through heme oxygenase-1 expression in murine macrophages. J Pharmacol Sci 116(3):283–295. https://doi.org/10.1254/jphs.10219fp

Lee D-S, Ko W, Quang TH, Kim K-S, Sohn JH, Jang J-H, Ahn JS, Kim Y-C, Oh H (2013) Penicillinolide a: a new anti-inflammatory metabolite from the marine fungus Penicillium sp. SF-5292. Mar Drugs 11(11): 4510–4526. https://doi.org/10.3390/md11114510

Li L-Q, Song A-X, Yin J-Y, Siu K-C, Wong W-T, Wu J-Y (2020) Anti-inflammation activity of exopolysaccharides produced by a medicinal fungus Cordyceps sinensis Cs-HK1 in cell and animal models. Int J Biol Macromol 149:1042–1050. https://doi.org/10.1016/j.ijbiomac.2020.02.022

Lim JS, Hong J-H, Lee DY, Li X, Lee DE, Choi JU, Lee KY, Kim KH, Cho Y-C (2023) 6-Pentyl-α-Pyrone from Trichoderma gamsii exert antioxidant and anti-inflammatory properties in lipopolysaccharide-stimulated mouse macrophages. Antioxidants 12(12): 12. https://doi.org/10.3390/antiox12122028

Liu B-H, Chi J-Y, Hsiao Y-W, Tsai K-D, Lee Y-J, Lin C-C, Hsu S-C, Yang S-M, Lin T-H (2010) The fungal metabolite, citrinin, inhibits lipopolysaccharide/interferon-γ-induced nitric oxide production in glomerular mesangial cells. Int Immunopharmacol 10(12): 1608–1615. https://doi.org/10.1016/j.intimp.2010.09.017

Liu J, Gu B, Yang L, Yang F, Lin H (2018a) New anti-inflammatory Cyclopeptides from a sponge-derived fungus aspergillus violaceofuscus. Front Chem 6:226. https://doi.org/10.3389/fchem.2018.00226

Liu M, Sun W, Wang J, He Y, Zhang J, Li F, Qi C, Zhu H, Xue Y, Hu Z, Zhang Y (2018b) Bioactive secondary metabolites from the marine-associated fungus aspergillus terreus. Bioorg Chem 80:525–530

Liu D, Zhong Z, Karin M (2022) NF-κB: a double-edged sword controlling inflammation. Biomedicine 10(6). https://doi.org/10.3390/biomedicines10061250

Magdenoska O, Martinussen J, Thykaer J, Nielsen KF (2013) Dispersive solid phase extraction combined with ion-pair ultra high-performance liquid chromatography tandem mass spectrometry for quantification of nucleotides in *Lactococcus lactis*. Anal Biochem 440(2):166–177. https://doi.org/10.1016/j.ab.2013.05.023

Mago P, Sharma R, Hafeez I, Nawaz I, Joshi M, Mehrotra R (2023) Mushroom based cosmeceuticals: an upcoming biotechnology sector. Biosci Biotechnol Res Asia 20(2):381–394. https://doi.org/10.13005/bbra/3097

Man M-Q, Wakefield JS, Mauro TM, Elias PM (2022) Regulatory role of nitric oxide in cutaneous inflammation. Inflammation 45(3):949–964. https://doi.org/10.1007/s10753-021-01615-8

Moro C, Palacios I, Lozano M, D'Arrigo M, Guillamón E, Villares A, Martínez JA, García-Lafuente A (2012) Anti-inflammatory activity of methanolic extracts from edible mushrooms in LPS activated RAW 264.7 macrophages. Food Chem 130(2):350–355. https://doi.org/10.1016/j.foodchem.2011.07.049

Mukherjee A, Das S, Chakraborty D, Pal N, Das N (2024) Fungi's treasure in cosmeceuticals-a comprehensive chemical approach. S Afr J Bot 166:311–331. https://doi.org/10.1016/j.sajb.2024.01.036

Muthuramalingam K, Choi SI, Hyun C, Kim YM, Cho M (2019) β-Glucan-based wet dressing for cutaneous wound healing. Adv Wound Care 8(4):125–135. https://doi.org/10.1089/wound.2018.0843

Ngan NTT, Quang TH, Kim K-W, Kim HJ, Sohn JH, Kang DG, Lee HS, Kim Y-C, Oh H (2017) Anti-inflammatory effects of secondary metabolites isolated from the marine-derived fungal strain Penicillium sp. SF-5629. Arch Pharm Res 40(3):328–337. https://doi.org/10.1007/s12272-017-0890-5

Nielsen KF, Mogensen JM, Johansen M, Larsen TO, Frisvad JC (2009) Review of secondary metabolites and mycotoxins from the aspergillus Niger group. Anal Bioanal Chem 395(5):1225–1242. https://doi.org/10.1007/s00216-009-3081-5

Nielsen KF, Månsson M, Rank C, Frisvad JC, Larsen TO (2011) Dereplication of microbial natural products by LC-DAD-TOFMS. J Nat Prod 74(11):2338–2348. https://doi.org/10.1021/np200254t

Nutrition, C. for F. S. and A (2022) Cosmeceutical. *FDA*. https://www.fda.gov/cosmetics/cosmetics-labeling-claims/cosmeceutical

Pandey A, Jatana GK, Sonthalia S (2024) Cosmeceuticals. In *StatPearls*. StatPearls Publishing. http://www.ncbi.nlm.nih.gov/books/NBK544223/

Pfeiffer S, Swoboda I (2023) Problems encountered using fungal extracts as test solutions for fungal allergy diagnosis. J Fungi (Basel, Switzerland) 9(10):957. https://doi.org/10.3390/jof9100957

Phull A-R, Ahmed M, Park H-J (2022) Cordyceps militaris as a bio functional food source: pharmacological potential, anti-inflammatory actions and related molecular mechanisms. Microorganisms 10(2). https://doi.org/10.3390/microorganisms10020405

Quang DN, Hashimoto T, Stadler M, Asakawa Y (2004) New Azaphilones from the inedible mushroom Hypoxylon rubiginosum. J Nat Prod 67(7):1152–1155. https://doi.org/10.1021/np0400631

Quang DN, Hashimoto T, Stadler M, Radulović N, Asakawa Y (2005) Antimicrobial azaphilones from the fungus Hypoxylon multiforme. Planta Med 71(11):1058–1062. https://doi.org/10.1055/s-2005-873129

Quang DN, Harinantenaina L, Nishizawa T, Hashimoto T, Kohchi C, Soma G-I, Asakawa Y (2006a) Inhibitory activity of nitric oxide production in RAW 264.7 cells of daldinals A–C from the fungus Daldinia childiae and other metabolites isolated from inedible mushrooms. J Nat Med 60(4):303–307. https://doi.org/10.1007/s11418-006-0010-1

Quang DN, Hashimoto T, Arakawa Y, Kohchi C, Nishizawa T, Soma G-I, Asakawa Y (2006b) Grifolin derivatives from Albatrellus caeruleoporus, new inhibitors of nitric oxide production in RAW 264.7 cells. Bioorg Med Chem 14(1):164–168. https://doi.org/10.1016/j.bmc.2005.08.005

Quang DN, Hashimoto T, Asakawa Y (2006c) Inedible mushrooms: a good source of biologically active substances. Chem Record (New York, NY) 6(2):79–99. https://doi.org/10.1002/tcr.20074

Rasmussen BB, Nielsen KF, Machado H, Melchiorsen J, Gram L, Sonnenschein EC (2014) Global and phylogenetic distribution of quorum sensing signals, acyl Homoserine lactones, in the family of Vibrionaceae. Mar Drugs 12(11):11. https://doi.org/10.3390/md12115527

Renner MK, Jensen PR, Fenical W (2000) Mangicols: structures and biosynthesis of a new class of sesterterpene polyols from a marine fungus of the genus Fusarium. J Org Chem 65(16):4843–4852. https://doi.org/10.1021/jo000081h

Rether J, Serwe A, Anke T, Erkel G (2007) Inhibition of inducible tumor necrosis factor-alpha expression by the fungal epipolythiodiketopiperazine gliovirin. Biol Chem 388(6):627–637. https://doi.org/10.1515/BC.2007.066

Rupa EJ, Li JF, Arif MH, Yaxi H, Puja AM, Chan AJ, Hoang V-A, Kaliraj L, Yang DC, Kang SC (2020) Cordyceps militaris fungus extracts-mediated Nanoemulsion for improvement antioxidant, antimicrobial, and anti-inflammatory activities. Molecules (Basel, Switzerland) 25(23):5733. https://doi.org/10.3390/molecules25235733

Sangthong S, Pintathong P, Pongsua P, Jirarat A, Chaiwut P (2022) Polysaccharides from Volvariella volvacea mushroom: extraction, biological activities and cosmetic efficacy. J Fungi (Basel, Switzerland) 8(6):572. https://doi.org/10.3390/jof8060572

Saraswat A, Mathur P, Sanyal D (2020) Assessing cosmeceuticals properties of some macrofungi for improved skincare. Int J Pharm Pharm Sci:15–19. https://doi.org/10.22159/ijpps.2020v12i2.36074

Seweryn E, Ziała A, Gamian A (2021) Health-promoting of polysaccharides extracted from Ganoderma lucidum. Nutrients 13(8). https://doi.org/10.3390/nu13082725

Shedoeva A, Leavesley D, Upton Z, Fan C (2019) Wound healing and the use of medicinal plants. Evid Based Complement Alternat Med 2019(1):2684108. https://doi.org/10.1155/2019/2684108

Smith RM (2006) Superheated water: the ultimate green solvent for separation science. Anal Bioanal Chem 385(3):419–421. https://doi.org/10.1007/s00216-006-0437-y

Sørensen JL, Nielsen KF, Thrane U (2007) Analysis of Moniliformin in maize plants using hydrophilic interaction chromatography. J Agric Food Chem 55(24):9764–9768. https://doi.org/10.1021/jf0715875

Stadler M, Læssøe T, Fournier J, Decock C, Schmieschek B, Tichy H-V, Peršoh D (2014) A polyphasic taxonomy of Daldinia (Xylariaceae). Stud Mycol 77(1):1–143. https://doi.org/10.3114/sim0016

Sujarit K, Suwannarach N, Kumla J, Lomthong T (2021) Mushrooms: splendid gifts for the cosmetic industry. Chiang Mai J Sci:699–725

Sułkowska-Ziaja K, Grabowska K, Apola A, Kryczyk-Poprawa A, Muszyńska B (2021) Mycelial culture extracts of selected wood-decay mushrooms as a source of skin-protecting factors. Biotechnol Lett 43(5):1051–1061. https://doi.org/10.1007/s10529-021-03095-0

Taofiq O, Heleno SA, Calhelha RC, Alves MJ, Barros L, Barreiro MF, González-Paramás AM, Ferreira ICFR (2016) Development of mushroom-based cosmeceutical formulations with anti-inflammatory, anti-Tyrosinase, antioxidant, and antibacterial properties. Molecules 21(10). https://doi.org/10.3390/molecules21101372

Taofiq O, Heleno SA, Calhelha RC, Alves MJ, Barros L, González-Paramás AM, Barreiro MF, Ferreira ICFR (2017) The potential of Ganoderma lucidum extracts as bioactive ingredients in topical formulations, beyond its nutritional benefits. Food Chem Toxicol 108 (Pt A):139–147. https://doi.org/10.1016/j.fct.2017.07.051

Taofiq O, Rodrigues F, Barros L, Barreiro MF, Ferreira ICFR, Oliveira MBPP (2019a) Mushroom ethanolic extracts as cosmeceuticals ingredients: safety and ex vivo skin permeation studies. Food Chem Toxicol 127:228–236. https://doi.org/10.1016/j.fct.2019.03.045

Taofiq O, Rodrigues F, Barros L, Peralta RM, Barreiro MF, Ferreira ICFR, Oliveira MBPP (2019b) Agaricus blazei Murrill from Brazil: an ingredient for nutraceutical and cosmeceutical applications. Food Funct 10(2):565–572. https://doi.org/10.1039/c8fo02461h

Theeraraksakul K, Jaengwang K, Choowongkomon K, Tabtimmai L (2023) Exploring the biological functions and anti-Melanogenesis of phallus indusiatus for mushroom-based cosmetic applications. Cosmetics 10(5). https://doi.org/10.3390/cosmetics10050121

Toledo TR, Dejani NN, Monnazzi LGS, Kossuga MH, Berlinck RGS, Sette LD, Medeiros AI (2014) Potent anti-inflammatory activity of Pyrenocine a isolated from the marine-derived fungus Penicillium paxilli

Ma(G)K. Mediat Inflamm 2014(1):767061. https://doi.org/10.1155/2014/767061
Trepa M, Sułkowska-Ziaja K, Kała K, Muszyńska B (2024) Therapeutic potential of fungal terpenes and Terpenoids: application in skin diseases. Molecules 29(5). https://doi.org/10.3390/molecules29051183
Tung NT, Cuong TD, Hung TM, Lee JH, Woo MH, Choi JS, Kim J, Ryu SH, Min BS (2013) Inhibitory effect on NO production of triterpenes from the fruiting bodies of *Ganoderma lucidum*. Bioorg Med Chem Lett 23(5): 1428–1432. https://doi.org/10.1016/j.bmcl.2012.12.066
Usman M, Murtaza G, Ditta A (2021) Nutritional, medicinal, and cosmetic value of bioactive compounds in button mushroom (Agaricus bisporus): a review. Appl Sci 11(13):13. https://doi.org/10.3390/app11135943
van Dam JC, Eman MR, Frank J, Lange HC, van Dedem GWK, Heijnen SJ (2002) Analysis of glycolytic intermediates in *Saccharomyces cerevisiae* using anion exchange chromatography and electrospray ionization with tandem mass spectrometric detection. Anal Chim Acta 460(2):209–218. https://doi.org/10.1016/S0003-2670(02)00240-4
Ványolós A, Muszyńska B, Chuluunbaatar B, Gdula-Argasińska J, Kała K, Hohmann J (2020) Extracts and steroids from the edible mushroom Hypholoma lateritium exhibit anti-inflammatory properties by inhibition of COX-2 and activation of Nrf2. Chem Biodivers 17(9):e2000391. https://doi.org/10.1002/cbdv.202000391
Wang S, Marcone MF (2011) The biochemistry and biological properties of the world's most expensive underground edible mushroom: truffles. Food Res Int 44(9):2567–2581. https://doi.org/10.1016/j.foodres.2011.06.008
Wang S, Liu Z, Wang X, Liu R, Zou L (2022) Mushrooms do produce flavonoids: metabolite profiling and transcriptome analysis of flavonoid synthesis in the medicinal mushroom Sanghuangporus baumii. J Fungi 8(6). https://doi.org/10.3390/jof8060582
Wangun K, Vignie H (2006) Isolation, structure elucidation and evaluation of anti-inflammatory and anti-infectious activities of fungal metabolites. https://www.semanticscholar.org/paper/Isolation%2C-structure-elucidation-and-evaluation-of-Wangun-Vigne/c6270f1032911ca43f5405c8e942f1fb976400ab
Watrous J, Roach P, Alexandrov T, Heath BS, Yang JY, Kersten RD, van der Voort M, Pogliano K, Gross H, Raaijmakers JM, Moore BS, Laskin J, Bandeira N, Dorrestein PC (2012) Mass spectral molecular networking of living microbial colonies. Proc Natl Acad Sci 109(26):E1743–E1752. https://doi.org/10.1073/pnas.1203689109
Weber D, Sterner O, Anke T, Gorzalczancy S, Martino V, Acevedo C (2004) Phomol, a new antiinflammatory metabolite from an endophyte of the medicinal plant Erythrina crista-galli. J Antibiot 57(9):559–563. https://doi.org/10.7164/antibiotics.57.559
Wei J, Guo W-H, Cao C-Y, Kou R-W, Xu Y-Z, Górecki M, Di Bari L, Pescitelli G, Gao J-M (2018) Polyoxygenated cyathane diterpenoids from the mushroom Cyathus africanus, and their neurotrophic and anti-neuroinflammatory activities. Sci Rep 8(1):2175. https://doi.org/10.1038/s41598-018-20472-4
Weidler M, Rether J, Anke T, Erkel G (2000) Inhibition of interleukin-6 signaling by galiellalactone. FEBS Lett 484(1):1–6. https://doi.org/10.1016/S0014-5793(00)02115-3
Wolfender J-L, Marti G, Thomas A, Bertrand S (2015) Current approaches and challenges for the metabolite profiling of complex natural extracts. J Chromatogr A 1382:136–164. https://doi.org/10.1016/j.chroma.2014.10.091
Wu X-F, Fei M-J, Shu R-G, Tan R-X, Xu Q (2005) Fumigaclavine C, an fungal metabolite, improves experimental colitis in mice via downregulating Th1 cytokine production and matrix metalloproteinase activity. Int Immunopharmacol 5(10):1543–1553. https://doi.org/10.1016/j.intimp.2005.04.014
Wu Y, Choi M-H, Li J, Yang H, Shin H-J (2016) Mushroom cosmetics: the present and future. Cosmetics 3(3). https://doi.org/10.3390/cosmetics3030022
Yang M, Zhang Z, He Y, Li C, Wang J, Ma X (2021) Study on the structure characterization and moisturizing effect of *Tremella* polysaccharide fermented from GCMCC5.39. Food Sci Human Wellness 10(4):471–479. https://doi.org/10.1016/j.fshw.2021.04.009
Yao Y, Xu B (2022) Skin health promoting effects of natural polysaccharides and their potential application in the cosmetic industry. Polysaccharides 3(4):Article 4. https://doi.org/10.3390/polysaccharides3040048
Zhang T, Guo Q, Xin Y, Liu Y (2022) Comprehensive review in moisture retention mechanism of polysaccharides from algae, plants, bacteria and fungus. Arab J Chem 15(10):104163. https://doi.org/10.1016/j.arabjc.2022.104163
Zhao Y, Liu J, Wang J, Wang L, Yin H, Tan R, Xu Q (2004) Fumigaclavine C improves concanavalin A-induced liver injury in mice mainly via inhibiting TNF-alpha production and lymphocyte adhesion to extracellular matrices. J Pharm Pharmacol 56(6): 775–782. https://doi.org/10.1211/0022357023592
Zhao S, Lei M, Xu H, He H, Suvorov A, Wang J, Qiu J, Zhou Q, Yang J, Chen L (2021) The normal cell proliferation and wound healing effect of polysaccharides from *Ganoderma amboinense*. Food Sci Human Wellness 10(4):508–513. https://doi.org/10.1016/j.fshw.2021.04.013
Ziemlewska A, Wójciak M, Mroziak-Lal K, Zagórska-Dziok M, Bujak T, Nizioł-Łukaszewska Z, Szczepanek D, Sowa I (2022) Assessment of cosmetic properties and safety of use of model washing gels with Reishi, Maitake and lion's mane extracts. Molecules (Basel, Switzerland) 27(16):5090. https://doi.org/10.3390/molecules27165090

Anti-tyrosinase Activity of Fungal-Based Cosmetics

6

Ayotunde Adeyemi Adeyanju, Elijah A. Alhassan, Charles Obiora Nwonuma, Abiola F. Olaniran, and Oluwakemi Erinle

Abstract

The fascinating world of fungal-based cosmetics and their potential as anti-tyrosinase agents are explained in this chapter. The cosmetic industry is interested in inhibiting tyrosinase, an enzyme involved in melanin formation, to fight hyperpigmentation disorders. Fungi, known for their diverse secondary metabolites, have demonstrated promising bioactive compounds that may contribute to skin-lightening properties. The importance of tyrosinase inhibition in skincare is understood in this chapter. Common tyrosinase inhibitors were discussed: inhibitors produced from microorganisms, aldehydes and their derivatives, and plant-based substances (particularly polyphenols). The relationship between fungus-based inhibitors' anti-tyrosinase activities and skin hyperpigmentation was also presented. To solve the problem of hyperpigmentation, the chapter explains how fungi-derived inhibitors may be crucial in controlling the creation of melanin. A practical case of Kojic acid in cosmetics was discussed. This chapter advances our understanding of natural cosmetic chemicals, but they also open the door for the creation of novel, potent products that address skin pigmentation issues.

Keywords

Fungi-based cosmetics · Kojic acid · Tyrosinase · Hyperpigmentation · Melanin · Eumelanin · Human tyrosinase · Polyphenols

6.1 Introduction

The cosmetics industry has over the years witnessed remarkable breakthroughs that stemmed from human ingenuity to manufacture products that enhance beauty and make one feel good. Natural and synthetic-based materials are being engineered to form the arrays of cosmetic products on display in the market. The primary

A. A. Adeyanju (✉)
Department of Food Science and Microbiology, College of Pure and Applied Science, Landmark University, Omu-Aran, Kwara State, Nigeria

Centre for Innovative Food Research (CIFR), Department of Biotechnology and Food Technology, Faculty of Science, University of Johannesburg, Johannesburg, Gauteng, South Africa
e-mail: adeyanju.adeyemi@lmu.edu.ng

E. A. Alhassan · O. Erinle
Agricultural and Biosystems Engineering, College of Engineering, Landmark University, Omu-Aran, Kwara State, Nigeria

C. O. Nwonuma
Department of Biochemistry, College of Pure and Applied Sciences, Landmark University, Omu-Aran, Nigeria

A. F. Olaniran
Department of Food Science and Microbiology, College of Pure and Applied Science, Landmark University, Omu-Aran, Kwara State, Nigeria

S. Ghosh, H. N. Onyeaka (eds.), *Fungal-Based Cosmetics: Formulation and Usage*, The Mycota 17,
https://doi.org/10.1007/978-3-032-09291-5_6

function of cosmetic products is to purify, perfume, safeguard, and modify the appearance of the exterior body parts of human. Water, preservatives, emulsifiers, moisturizers, thickeners, colors, fragrances, emollients, and pH stabilizers constitute the basic ingredients present in most cosmetics. The chemical compounds of these ingredients are paramount in cosmetic formulation to minimize risk to human health. Thus, the strict regulations of these constantly evolve multi-billion-dollar industries. The damaging impact an unhealthy product can have on the well-being of the consumers and the environment cannot be overemphasized. In recent years, the rising interest in herbal, natural, and organic products has fueled rapid growth in the cosmetics industry, paving the way for further expansion centered on safety and sustainability (Kumar 2005).

Tyrosinase is a multifunctional metalloenzyme containing binuclear copper ions and is widely present in various organisms. It plays a critical role in enzymatic browning and melanogenesis (Kilimnik and Dembitsky 2016). As a rate-limiting enzyme in melanin metabolism (Garcia-Jimenez et al. 2017), it catalyzes the oxidation of tyrosine to dihydroxyphenylalanine (DOPA) and subsequently converts DOPA to DOPA quinone (Wuttisin et al. 2017). In mammals, this process contributes to post-inflammatory hyperpigmentation, leading to darkened skin areas and melasma, an undesirable skin discoloration. Hyperpigmentation is characterized by an increase in melanin production and accumulation or a higher number of melanocytes. Melanocytes produced dark colored pigments called melanin. Browning of fruits and vegetables, and skin disorder resulting from excessive production of melanin are traceable to Tyrosinase (Zolghadri et al. 2019). Therefore, adopting tyrosinase inhibitors is critical for remediating these disorders. Over the years, researchers have found different arrays of potent inhibitors for cosmetic, medical, agricultural, food, and environmental purposes. Some plants, animals, and microorganisms have been isolated and purified to obtain tyrosinases. As a result, natural tyrosinase inhibitors have gained significant attention in the cosmetic industry due to their safety advantages.

Melanogenesis is the process that produces melanin by the melanocytes, primarily the reason for pigmentation. When it comes to shielding the skin from UV damage and skin photo-carcinogenesis, skin pigmentation and melanogenesis are the most potent photoprotective factors. In humans, depigmentation and abnormal melanin loss can pose a severe threat to facial aesthetics and dermatology (Brenner and Hearing 2008). On the contrary, various skin conditions, including cervical poikiloderma, acanthosis nigricans, periorbital hyperpigmentation, lentigines, melasma, and neurodegenerative disorders related to Parkinson's disease, as well as an elevated risk of skin cancer, are characterized by excessive melanin synthesis and pigment accumulation (Dogra and Sarangal 2014; Nouveau et al. 2016). Enzymes such as tyrosinase and other tyrosinase-related proteins (TYRP1 and TYRP2) play a crucial part in melanin synthesis, despite the complex melanogenesis process characterized by many chemical and enzymatic reactions.

In the skin, whitening agents are displayed at several stages of melanin production through melanogenesis. Some whitening agents prevent the enzyme from functioning or hinder the movement of melanosomes (pigment granules) from melanocytes to neighboring keratinocytes. Agents that limit melanin formation and accumulation in the skin must be encouraged in cosmetic products because of the detrimental effects this enzyme depicts. They are generally termed anti-tyrosinase because of the inhibiting properties that impede this enzyme functionality. Inhibitors are used as browning-inhibiting chemical in agriculture and food industries and depigmentation agent in cosmetic and pharmaceuticals. They are mostly synthesized using different screening techniques from natural, semi-synthetic, and synthetic materials (Roh et al. 2004). Plants-based extracts have been synthesized with potent characteristics to inhibit tyrosinase activities in the food and cosmetic industries (Kasrati et al. 2015; Zhang et al. 2014), vegetables (Umesha and Naidu 2015), and fruits (Abdennacer et al. 2015).

Despite the array of inhibitors available, only a handful are suitable as skin whitening agents and for clinical use. Prominently used whitening agents (active substances) in cosmetic products are kojic acid, hydroquinone, arbutin, and azelaic acid (Özer et al. 2007; Petit and Pierard 2003; Wuttisin et al. 2017). Similarly, *Glycyrrhiza glabra* Linneva (Leguminosae) or *Morus alba* L. (Moraceae) and many other plant extracts effectively impede the production of melanin and have been used as whitening agents with minimal negative effects (Bernard and Berthon 2000; Özer et al. 2007). Citrus peel (*Citrus unshiu*) has been demonstrated to exhibit antioxidant and anti-tyrosinase activities (Kim et al. 2008b). Nobiletin, the primary active compound in citrus peel, acts as a tyrosinase inhibitor (Sasaki and Yoshizaki 2002). There have been reports of the whitening effect of antioxidants like ascorbic acid and its derivatives, suggesting it is possible to avoid melanogenesis through reduction in reactive oxygen species (ROS) levels in melanocytes (Balaguer et al. 2008). Because tyrosinase inhibition can reduce synthesis of melanin, thereby causing whitening, the anti-tyrosinase properties of mushrooms have also been harnessed for cosmetic purpose (Baek et al. 2015). Nonetheless, among the various tyrosinase sources, *Agaricus bisporus* mushroom tyrosinase is a substantial and affordable tyrosinase source that exhibits high homology and similarity to human tyrosinase (Zhou et al. 2017). Owing to these advantageous qualities, mushroom tyrosinase's structural, functional, and biochemical features have up to now been thoroughly investigated as a model system for enzyme-catalyzed reactions, melanogenic investigations, and structural studies of enzyme inhibitors (Ioniță et al. 2014; Zhou et al. 2017). However, most of these studies have employed extracts from mushrooms; the anti-tyrosinase properties of supernatants from mycelium culture have received less attention. It's interesting to note that a study highlighted the antioxidant activity of *Phellinus igniarius* broths without mycelium (Lung et al. 2010). *Lactobacillus rhamnosus* spent culture supernatant could be a desirable source for cosmetics with anti-tyrosinase and antioxidant properties (Tsai et al. 2013). These findings imply the possibility of the inclusion of antioxidant and anti-tyrosinase active ingredients in mycelium culture media. The research confirmed that a necessary step toward the development of effective natural antioxidants and tyrosinase inhibitors would be to examine the antioxidant and anti-tyrosinase activity from a huge quantity extract of mycelium culture. Premised on the ease of producing mycelium inexpensively in comparison to farming mushrooms and the capacity to grow controllable extensive products than cultivating mushroom.

Broadly, this chapter elucidates on the anti-tyrosinase activity of fungal-based cosmetics with clear focus on tyrosinase and some common inhibitors, anti-tyrosinase mechanism, methods of determining anti-tyrosinase activity, anti-tyrosinase activity of fungal-based cosmetics with practical case studies of some fungal-based cosmetics and their anti-tyrosinase activity.

6.2 Tyrosinase and Some Common Inhibitors

Tyrosinase (EC 1.14.18.1) is a copper-dependent, multifunctional enzyme with mixed-function oxidase activity, present in plants, animals, and microorganisms (Chen et al. 2016). It enhances the hydroxylation of monophenols and the oxidation of o-diphenols to o-quinones, which can then undergo polymerization to form brown or black pigments (Baber et al. 2023; Yuan et al. 2020). Tyrosinase can exist in three forms: deoxy, oxy, and met (Panzella and Napolitano 2019; Xie et al. 2003). The most common type of tyrosinase is the mushroom tyrosinase, having molecular weight (120 kDa), and it has two H subunits and two L subunits with molecular weights of 43 kDa and 13 kDa, respectively, and has two active sites. Melanin, the primary pigment determining skin color, is produced in response to UV radiation exposure. Excessive exposure to UV radiation stimulates elevated melanin synthesis in the skin, facilitated by heightened tyrosinase enzyme activity. Within the melanin biosynthesis pathway, tyrosinase helps during the conversion of tyrosine into dihydroxyphenylalanine (DOPA) and subsequently oxidize it to dopaquinone (Meng et al. 2012; Moreiras et al. 2021). Following its formation, dopaquinone undergoes autooxidation to dopachrome and is ultimately transformed into eumelanin, a brown-black pigment, facilitated by dopachrome tautomerase (Zhu et al. 2022). On the other hand, under the influence of cysteine or glutathione, dopaquinone has the potential to transform into cysteinyl DOPA, leading to the creation of pheomelanin, a pigment with yellow-red hues (Jung et al. 2009) as in Fig. 6.1.

Human tyrosinase, a type-3 copper metalloenzyme belonging to the tyrosinase family (EC 1.14.18.1), is found in enzymes across fungi, plants, mammals, and bacteria. The glycoprotein is composed of 529 amino acids, including an 18-residue N-terminal signal sequence and 6 to 7 N-glycosylation sites, with an approximate molecular weight of 67 kDa. It also has 2 active sites, magnetically coupled copper centers, an aquo (hydroxo)-bridged ligand in the reactive state of the human tyrosinase which is harmonized by six histidine residues (H180, H202, H211 for CuA, H363, H367, H390 for CuB), greatly preserved by catechol oxidases, tyrosinases, and hemocyanins (Mermer and Demirci 2023; Roulier et al. 2020). The human tyrosinase physiologically performs two basic functions: monophenolase (o-hydroxylation of L-tyrosine to L-DOPA) and then diphenolase activity using a molecular oxygen (oxidation of L-DOPA in dopaquinone) (Li et al. 2023; Parvez et al. 2007), even though the enzyme can oxidize a large array of monophenolic and diphenolic substrates. This two-step oxidation process starts up the melanin pigments synthesis, which then progresses non-enzymatically (Chang 2012; Şöhretoğlu et al. 2018). During melanogenesis in human skin, in addition to human tyrosinase enzyme (hsTYR) two other enzymes also contribute to the process: tyrosinase-related proteins 1 and 2 (hsTYRP1 and hsTYRP2), even though the function of hsTYRP1 is still not clear but is related to the protection of hsTYR. The melanogenesis process yields a dark brown-black eumelanin polymer and a yellow to red pheomelanin polymer, however, based on the presence or absence of nucleophilic cysteines (Lee et al. 2016). The eumelanin is basically 5,6-dihydroxyindole-2-carboxylic acids with the indole rings attached to the pyrrole counterparts, while the pheomelanin is a benzothiazine and benzothiazole unit. In human skin, eyes, and hair bulb, melanogenesis occurs in the dendritic cells (melanocytes), whereas, at subcellular layer, melanin is produced in the melanosomes (a special membrane-bound organelles). Melanin shields the skin from UV damage and generates free radicals. The hsTYR serves a central role in the melanogenesis mechanism by catalytically promoting the rate-limiting first two reaction steps in the sequence. Therefore, to curb or suppress the melanogenesis (melanin synthesis), there will be a need for tyrosinase inhibitors (Roulier et al. 2020).

Tyrosinase inhibitor, which also functions as a melanin inhibitor, interferes with melanin production by blocking the upstream signal transduction of tyrosinase (i.e., downregulation of MC1R and MITF expression) or downstream movement (i.e., the melanosome transfer or abrasion of the epidermis resulting in melanin loss) (Li et al. 2021). Furthermore, tyrosinase inhibitor can also perform the following functions: (1) act as a reducing agent to chemically reduce dopaquinone, (2) scavenge o-Dopaquinone, (3) can act as a substrate with good affinity to chelate enzyme producing different products, (4) non-specific or specific enzyme deactivators, e.g., alkaline and acid, and (5) a true inhibitor that binds to enzyme and restricts its activity (Fernandes and Kerkar 2017). Furthermore, tyrosinase inhibitors remove copper, in order to retard the binding activity of the substrate (Tang and Chen 2015). Over the years, tyrosinase inhibitors have been used frequently in cosmetics and medicine; however, it has been a major cause of hyperpigmentation (Xie et al. 2003). To evaluate the inhibition ability of these inhibitors, mushroom tyrosinase has been likened to human tyrosinase since they belong to the same family, and since it can be tested in vitro. However, using mushroom tyrosinase as a model could be deceptive, as studies have shown that

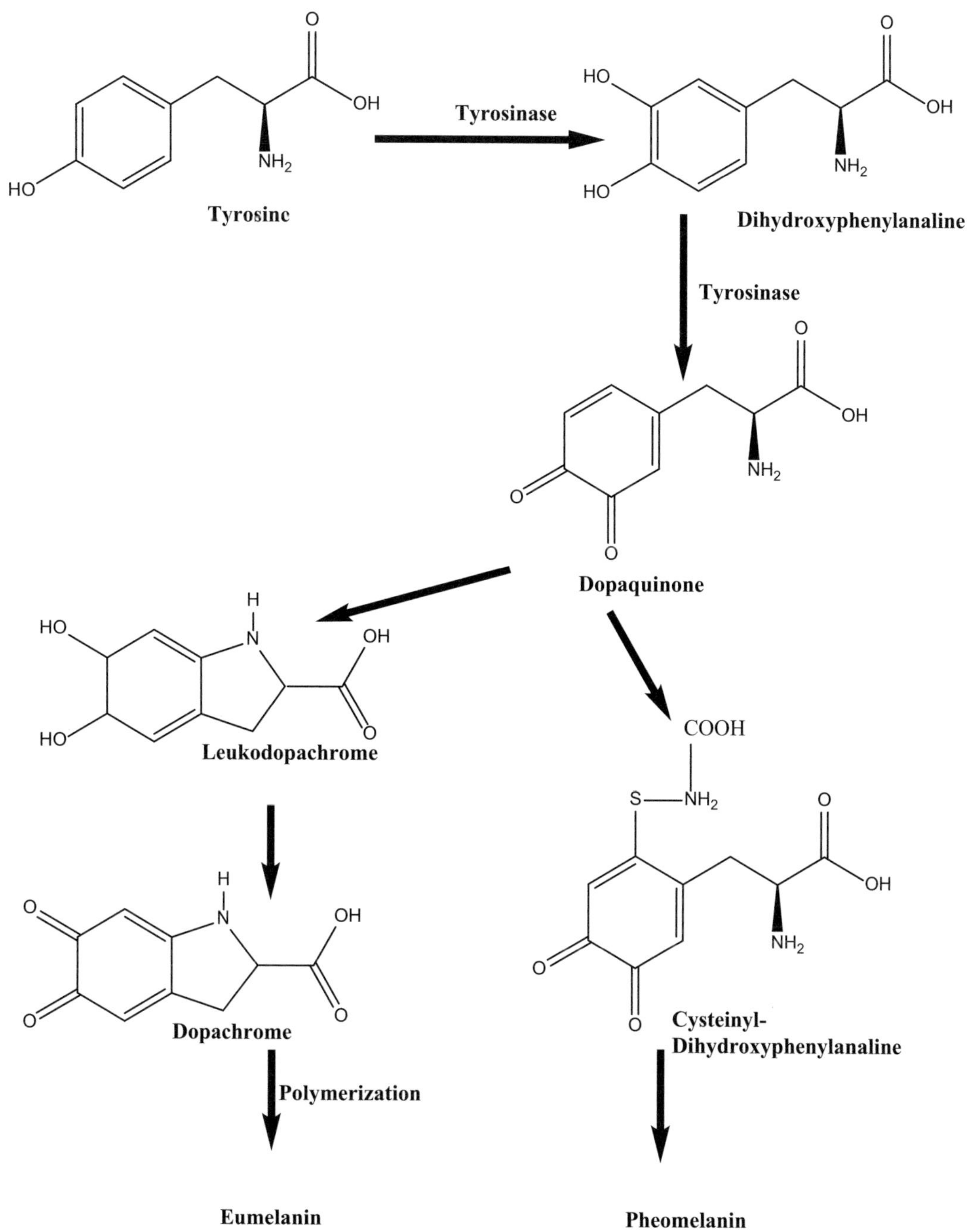

Fig. 6.1 Melanin biosynthesis: the role of tyrosinase

mushroom tyrosinase inhibition is significantly less than human tyrosinase, possibly due to the different associated enzymes or other related factors. For instance, using kojic acid, the inhibition of mushroom tyrosinase is about IC_{50} 6μM, while the inhibition of human tyrosinase is about IC_{50} 500μM. Therefore, it is necessary to evaluate tyrosinase inhibitors in cosmetic industries with human models, as this is still lacking in the literature (Mann et al. 2018; Roulier et al. 2020).

Some of the standard tyrosinase inhibitors frequently used in cosmetics are: kojic acid, hydroquinone, arbutin, ellagic acid, azelaic acid, L-Ascorbic acid, etc. (Yu et al. 2019). Even though these inhibitors are very effective, they may cause some adverse effects like dermatitis, skin irritation, depigmentation, and sometimes cancer (Zhang et al. 2021; Zolghadri et al. 2019). In addition, Kojic acid can cause liver toxicity while hydroquinone can affect the immune response. Basically, tyrosinase inhibitors are divided into two: natural and synthetic. Some of the synthetic tyrosinase inhibitors reported in the literature include N-hydroxy-N′--phenylthiourea, N-cyclopentyl-N-nitrosohydroxylamine, sildenafil methyl ether, 2,5-disubstituted-oxadiazoles, oxazolones, tetraketones, 4,4′-dihydroxybiphenyl, 4-hydroxybenzaldehyde derivatives, polyphenolic curcumin derivatives, etc. (Masum et al. 2019). Even though these synthetic inhibitors have high inhibition potential, their harmful side effects and cost make researchers and cosmetic industries to sought after naturally occurring (Bonesi et al. 2018; Zhang et al. 2021). The natural tyrosinase inhibitors can be categorized into 3: (1) plant polyphenols, (2) aldehyde and other derivatives, and (3) inhibitors from fungi (Parvez et al. 2007).

6.2.1 Natural Tyrosinase Inhibitors

6.2.1.1 Plant-Based Tyrosinase Inhibitors (Polyphenols)

Polyphenols are benzene ring or phenolic ring biologically active compound found abundantly in plants, which is the reason for the color in flowers. Some polyphenols have been identified as potent or weak tyrosinase inhibitors. Polyphenols are generally classified into four main groups: phenolic acids, flavonoids, stilbenes, and lignans (Abbas et al. 2017; Yu et al. 2019).

6.2.1.1.1 Flavonoids

Flavonoids are among the most utilized polyphenol derivatives with phenols and pyrane rings, which are found in the flowers, leaves, bark, and seeds of most plants. Flavonoids protect against UV damage, pathogens, and other micro or macro-organisms (Şöhretoğlu et al. 2018). It is one of the largest parts of the developed plant-based tyrosinase inhibitors (Parvez et al. 2007). Basically, flavonoids are divided into the following groups: flavonols (e.g., quercetin, kaempferol, etc.), flavones (e.g., luteolin, apigenin, etc.), flavanones (e.g., naringenin), isoflavones (e.g., daidzein), anthocyanins (e.g., cyanidin), and flavanols (e.g., catechins) (Obaid et al. 2021; Zhang et al. 2021). The differences between various flavonoids are attributed to the positioning of substituents (like hydroxy, glycosides, methoxy, etc.) and the conjugation of the A- and B-rings (Şöhretoğlu et al. 2018). The structure of the flavonoids makes it possible for it to function as the substrates and the tyrosinase inhibitor (Jakimiuk et al. 2022). Flavonoids like quercetin, kaempferol, and morin play an inhibitory role against tyrosinase, while others like rhamnetin and catechin function as a substrate and tyrosinase inhibitor as a cofactor like in catechin or as a scavenger of free radicals like in rhamnetin (Masum et al. 2019; Xie et al. 2003). Some of the natural plants that have been explored in extracting this flavonoid are *Trifolium nigrescens* Subsp., mung bean (*Vigna radiatae* L.), calamondin peel, *Morus yunnanensis*, *Maackia faurie*, *Potentilla bifurca*, roots of *Morus lhou*, *Artocapus altilis*, *Pulsatilla cernua*, etc. Studies on the flavone derivative (7,8,4-trihydroxyflavone) indicate that it noncompetitively inhibits tyrosinase diphenolase activity ($IC_{50} = 10.31\,\mu M$). The enzyme features a single binding site, and thermodynamic analysis reveals that the binding process is primarily governed by hydrogen bonding and van der Waals interactions. The hydrogen bonding occurred at the His244 residues and Met280 of active sites (Zolghadri et al. 2019). Flavonols reversibly inhibit tyrosinase activity through a multi-kinetic mechanism, binding to a single site on the enzyme via hydrogen bonds and van der Waals interactions and functioning competitively (Lee et al. 2016). Aside from flavonols causing conformational changes on the enzyme, they also inhibit L-DOPA by chelating copper due to the free 3-hydroxyl group (Obaid et al. 2021). Moreover, some flavonols, such as quercetin, act as cofactors and do not inhibit monophenolase activity, while galangin inhibits monophenolase activity without acting as a cofactor. Kaempferol neither functions as a cofactor nor inhibits monophenolase activity, but quercetin, galangin, and kaempferol all inhibit diphenolase activity (Kim et al. 2006; Masum et al. 2019). Quercetin derivatives happened to be a vital tyrosinase inhibitor (IC_{50},1.9 Mm) among all the flavonols (Jakimiuk et al. 2022). Isoflavone like 6,7,4-trihydroxyisoflavone has been discovered as an important inhibitor that inhibits tyrosinase monophenolase activity (IC_{50}, 9.2 mM) compared to others like daidzen and genistein, due to the C-6 and C-7 hydroxyl groups (Obaid et al. 2021). However, some isoflavones like 7,8,4′--trihydroxyisoflavone irreversibly inhibit both monophenolase and diphenolase enzymes. In addition, some isoflavones can act as potent suicide substrates of tyrosinase (Chen et al. 2016). Thus, it is essential to emphasize that the hydroxyl groups at the C7 and C8 positions can completely transform the inhibitory mechanism of isoflavones, shifting it from a reversible competitive inhibition to an irreversible suicide inhibition (Ng et al. 2009; Zolghadri et al. 2019). Flavanones like hesperetin reversibly and competitively inhibit tyrosinase, by disrupting the hydrophobic interactions. It also chelates copper ion harmonizing with histidine residues like HIS61, HIS85, and HIS259 domicile in the active sites of the enzyme. Flavanols like catechin exhibit high inhibition (IC_{50}, 30.2 mg/ml) of tyrosinase activity (Xie et al. 2003). Proanthocyanidins inhibit both monophenolase and diphenolase activity in a mixed competitive manner, by chelating between the B-ring hydroxyl group and the dicopper ions of the enzymes (Yu et al. 2019). However, some flavanols act as important alternative (Obaid et al. 2021).

6.2.1.1.2 Phenolic Acids

Phenolic acids are categorized into 2: hydroxycinnamic acids (e.g., caffeic acid, ferulic acid, coumaric acid, etc.) and hydroxybenzoic acids (protocatechuic acid, gallic acid, etc.) are used to deactivate tyrosinase activity (Panzella and Napolitano 2019; Zolghadri et al. 2019). Phenolic acids have been extracted from the following plants: *Hypericum laricifolium* Juss, *Phellinus linteus*, ginseng leaves, *Breynia officinalis, Spiranthes sinensis*, green coffee beans, etc. Phenolic acids act as mixed-type, reversible inhibitors of diphenolase activity. For example, 3-phenylbenzoic acid inhibits monophenolase activity in a noncompetitive manner (IC_{50} = 6.97 mM) and inhibits diphenolase activity in a mixed-type fashion (IC_{50} = 36.3 mM) (Masum et al. 2019; Zolghadri et al. 2019).

6.2.1.1.3 Stilbenes

For stilbenes, resveratrol is the most common. Some of the natural plant sources: *Pleurotus ferulae, Morus alba*, Carignan grape juice, *vitis viniferae caulis*, *Artocarpus gomezianus*, etc. Oxyresveratrol from *Morus alba* and *Cudrania cochinchinensis* inhibits tyrosinase by 0.1 mM and 2.33 Mm, respectively. Resveratrol is unable to inhibit diphenolase activity but suppresses L-tyrosine oxidation. Preincubation of resveratrol with tyrosinase for approximately 30 min has been shown to inhibit both monophenolase and diphenolase activities of tyrosinase (Satooka and Kubo 2012). However, adding L-cysteine to resveratrol reduced the inhibition potential, indicating suicide inhibition mechanism (Lee et al. 2016).

6.2.1.1.4 Lignans

Lignans are diverse and have a complex structure with three primary precursors (Tang and Chen 2015). The natural plant sources of lignans are the exocarp of *Marrubium cylleneum*, *Crataegus pinnatifida*, *Pinellia ternate*, etc. They can inhibit tyrosinase activity (Zolghadri et al. 2019).

6.2.1.2 Aldehydes and Other Derivatives-Based Tyrosinase Inhibitors

Aldehydes such as trans-cinnamaldehyde, 2-hydroxy-4-methoxybenzaldehyde, cuminaldehyde, anisaldehyde, benzaldehyde, cinnamic acid, etc., can deactivate the activity of tyrosinase by its relationship with nucleophilic groups like amino, hydroxy, and sulfhydryl groups (Chen et al. 2003). Electron-donating groups like methoxy and isopropyl at the para position of cumin-aldehyde enhance the stability of the Schiff base at the enzyme's active site via an inductive effect. Aldehydes like benzaldehyde derivatives are viable alternative inhibitor of L-DOPA oxidation. However, some aromatic aldehydes like (2E)-alkenals, cumic acid, etc., are noncompetitive inhibitors of tyrosinase (Jiménez et al. 2001; Parvez et al. 2007).

6.2.1.3 Microorganism-Based Tyrosinase Inhibitors

Microorganisms like fungi and bacteria are very active inhibitors of tyrosinase activity, especially fungi. Fungal inhibitors such as azelaic acid (1,7-heptanedicarboxylic acid, produced by yeast) and kojic acid (5-hydroxy-2-(hydroxymethyl)-gamma-pyrone, a metabolite from *Aspergillus niger* and Penicillium) are effective at chelating transition metal ions and scavenging free radicals (Baber et al. 2023; Vasantha et al. 2014). The tyrosinase inhibition mechanism is described in the following section (Fernandes and Kerkar 2017; Parvez et al. 2007).

6.3 Skin Hyperpigmentation and Anti-tyrosinase Mechanism of Fungi-Based Inhibitors

Skin hyperpigmentation can arise from various factors, such as autoimmune disorders, exposure to UV radiation, hormonal fluctuations, genetic predisposition, hormonal therapy, contraceptive use, and reactions to medications. This results in an excessive release of melanin by melanocytes, giving rise to hyperpigmentation (Thawabteh et al. 2023). Melanin synthesis is controlled through signal transduction pathways that govern the mRNA expression of tyrosinase and tyrosinase-related proteins (TRP1 and TRP2) (Chang 2012). Melanin production is activated by a mix of hormonal and chemical factors, with the cAMP-mediated pathway being the most crucial (Meng et al. 2012). cAMP activation stimulates the production of micropthalmia-associated transcription factor (MITF) via cAMP-dependent protein kinase A (PKA) and the CREB transcription factor. MITF controls melanogenesis by interacting with the promoter regions of tyrosinase-related proteins (TRP-1 and TRP-2), which results in the enhanced expression of the tyrosinase enzyme responsible for melanin synthesis (Park et al. 2011). Chemicals that suppress MITF expression have the potential to serve as inhibitors throughout the melanogenesis cycle.

Hyperpigmentation conditions are often perceived as undesirable skin irregularities, characterized by increased darkening, light to dark brown spots, uneven grey patches on areas like the face, neck, and torso, and the presence of pale to dark brown spots on the skin. In response to these concerns, researchers have delved into the exploration of novel ingredients, particularly those of natural origin, as potential and effective solutions. The depigmentation process works by interfering with key steps in the melanogenic pathway or melanosome transfer, reducing melanin production and ultimately preventing melanogenesis. Despite the effectiveness of hydroquinone, a commonly used treatment for hyperpigmentation disorders, concerns about its safety have been raised due to reported carcinogenic potential, exogenous ochronosis, skin irritation, and an unpleasant fishy odor (Sarkar et al. 2013). Alternatives such as corticosteroids and kojic acid, while effective, may induce long-term adverse effects. In the pursuit of safer options, researchers are exploring bioactive compounds from alternative sources.

Fungi serve as a major source of bioactive compounds with potentials in agriculture, food, and pharmaceutical industries, producing antibiotics, enzymes, enzyme inhibitors, and growth promoters. Among them, certain fungi, particularly those of the Aspergillus genus, have been identified for their anti-tyrosinase activity. As shown in Table 6.1, various Aspergillus species produce compounds with tyrosinase inhibitory properties. Notably, the well-known

Table 6.1 Some of the compounds identified in fungi and their anti-tyrosinase mechanism

Source	Compound	Structure	Mechanism of action	References
Aspergillus spp.	6-Methoxy 7,4′ Dihydroxyiso	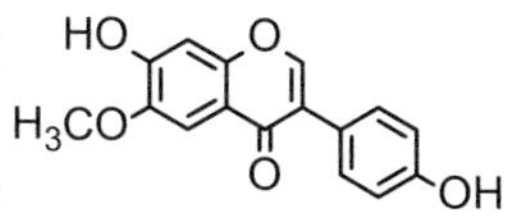	Inhibition of monophenolase activity through competitive mechanisms.	Chang (2012)
Aspergillus spp.	7,4′ Dihydroxyisoflavone	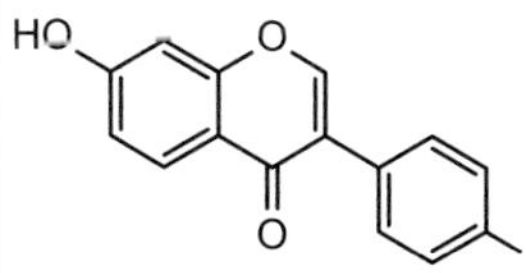	Inhibition of monophenolase activity through competitive mechanisms.	Chang (2012)
Aspergillus spp.	6,7,4′ Trihydroxyisoflavone	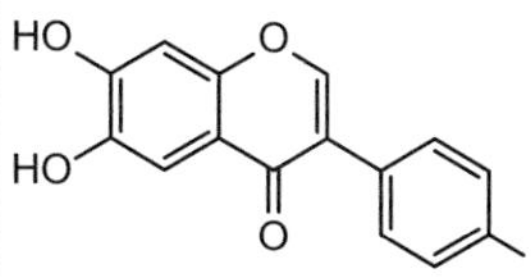	Inhibition of monophenolase activity through competitive mechanisms.	Chang (2012)
Aspergillus spp.	Kojic acid	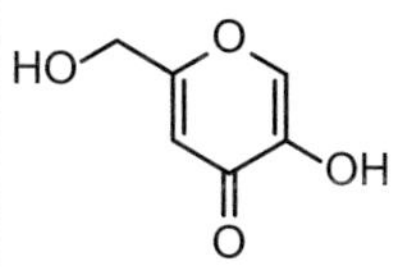	Bind copper at its active site through chelation.	Vasantha et al. (2014)
Aspergillus spp.	7,8, 4′ Trihydroxyisoflavone	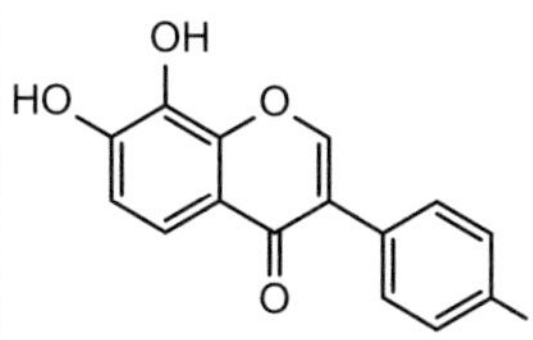	Suppression of monophenolase and diphenolase activities.	Chang (2012)
Aspergillus spp.	Terrein	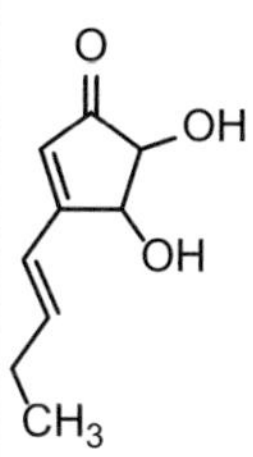	Reducing MITF expression by stimulating ERK activity, Suppression of MITF promoter activity.	Kim et al. (2007, 2008a, b)
Mushroom	Caffeic acid	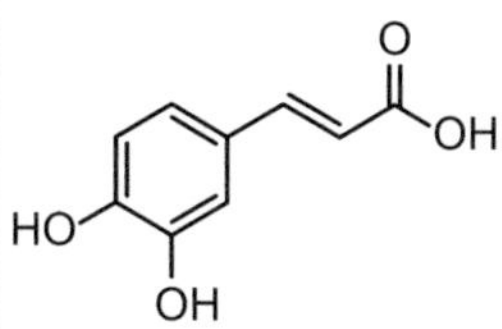	Inhibition of tyrosinase protein expression.	Chaiprasongsuk et al. (2016)
Mushroom	Ferulic acid	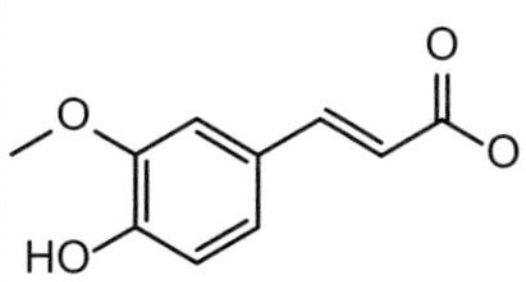	Inhibition of tyrosinase protein expression.	Chaiprasongsuk et al. (2016)
Mushroom	p-Coumaric acid	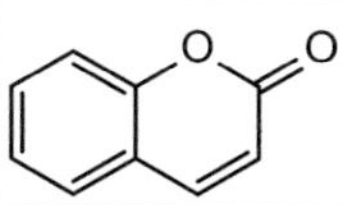	Decreased mRNA expression of MITF and tyrosinase.	Jun et al. (2012)

(continued)

Table 6.1 (continued)

Source	Compound	Structure	Mechanism of action	References
Mushroom	Gallic acid	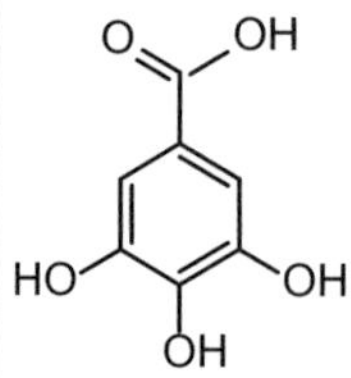	Decreases melanin synthesis by down-regulating MITF.	Su et al. (2013)

tyrosinase inhibitor kojic acid (5-hydroxy-2-(hydroxymethyl)-gamma-pyrone) has been reported in *A. albus*, *A. candidus*, *A. niger*, and *Penicillium* species. Beyond its tyrosinase inhibition, kojic acid is also recognized for its chelating properties and ability to scavenge free radicals (Fernandes and Kerkar 2017).

The *Aspergillus genus* not only synthesizes the renowned tyrosinase inhibitor, kojic acid, but also produces a variety of compounds exhibiting anti-tyrosinase activity. As an example, *Aspergillus niger* generates metallothioneins, potent tyrosinase inhibitors known for their strong affinity in chelating copper (Goetghebeur and Kermasha 1996). Another noteworthy compound, Decumbenone A, identified in *P. decumbens* and *A. sulphureus*, functions as a melanin formation inhibitor. A recently discovered decaline derivative, decumbenone C from Aspergillus, exhibits cytotoxicity against human melanoma cells with an IC_{50} of 0.9μM (Zhuravleva et al. 2012). Terrein, extracted for the first time from *A. terreus*, demonstrates inhibitory effects on melanin synthesis by downregulating MITF through the induction of ERK activity and inhibiting MITF promoter activity (Kim et al. 2007, 2008a).

Thriving under challenging conditions of pH, temperature, salinity, and nutrient levels, marine fungi are emerging as promising reservoirs of innovative bioactive compounds. Exploring the realm of marine-derived fungi has uncovered a spectrum of compounds, both familiar and novel, showcasing inhibitory activity against tyrosinase. *Alternaria* sp., for instance, has produced two kojic acid derivatives and phomaligol A, all exhibiting tyrosinase inhibitory properties. Similarly, *Myrothecium* sp. MFA 58 has yielded 6-n-pentyl-α-pyrone and myrothenone A, surpassing the efficacy of kojic acid (Li et al. 2005). *Botrytis* sp. has contributed a pyrone derivative with anti-tyrosinase activity (Zhang et al. 2007). Additionally, *Pestalotiopsis* sp. Z233, isolated from algae, has demonstrated the production of two sesquiterpene compounds with tyrosinase inhibitory activity induced by abiotic stress from $CuCl_2$. These findings underscore the potential of marine fungi as valuable sources of novel compounds with anti-tyrosinase properties (Wu et al. 2013).

Apart from marine fungi, diverse fungal groups demonstrate anti-tyrosinase properties. Azelaic acid, synthesized by the yeast Pityrosporum ovale, competes with L-tyrosine for the α-carboxylate binding site, thereby inhibiting tyrosinase. Additionally, yeasts produce cytosolic proteins called metallothioneins, recognized for their specific binding of substantial amounts of heavy metal ions and high cysteine content (Fernandes and Kerkar 2017). Entomopathogenic fungi, like *Paecilomyces gunnii*, present potential bioactive compounds. Three novel polyphenolic tyrosinase inhibitors: paecilomycones A, B, and C, have been isolated, exhibiting IC_{50} values of 110, 170, and 140μM, respectively. These compounds compete for the enzyme's active site, with inhibition influenced by the number of hydroxyl groups (Lu et al. 2014).

Pleurotus species rank among the most extensively cultivated edible mushrooms worldwide, valued for their substantial nutritional content and recognized for displaying medicinal attributes, including anti-inflammatory effects, antioxidant properties, immunomodulation, and antimicrobial activity. The observed medicinal characteristics have instilled optimism among scientists regarding their potential incorporation as cosmetic ingredients. In research conducted by Meng et al. (2012), investigators explored extracts obtained from the fruiting bodies of *P. citrinopileatus* Singer for their potential to hinder tyrosinase activity and melanin production. Using kojic acid as a benchmark, the findings demonstrated a noteworthy 100% inhibition of tyrosinase activity at a concentration of 100μg/mL. Simultaneously, the n-hexane-, ethyl acetate-, and n-butanol-soluble extracts exhibited inhibitory activities of up to 28.8, 27.4, and 41.0%, respectively, under identical conditions. The depigmentation mechanism of these extracts involves disrupting signals in the melanogenesis pathway. Exposure of B16 melanoma cells to *P. citrinopileatus* extract, particularly the n-hexane (50μg/mL), diethyl ether (75μg/mL), and ethyl acetate (100μg/mL) fractions, resulted in melanin inhibition of 63.1, 64.0, and 58.8%, respectively.

Various ethanolic mycelium extracts were investigated for their ability to inhibit tyrosinase (Park et al. 2015). The results showed the significant tyrosinase inhibition potential of *Inonotus mikadoi* extract, reaching up to 46% at a concentration of 10 mg/mL. Additionally, *Coriolus versicolor* and *Fomitopsis* sp. demonstrated tyrosinase inhibitory activities of 26.3 and 23.9%, respectively, at the same concentration. The researchers suggested that extracts with notable tyrosinase inhibition effects could be explored further for potential incorporation as functional additives in cosmetics.

Inonotus obliquus, commonly referred to as "chaga mushroom," is a medicinal fungus valued for its presumed anti-tumor and immunomodulatory properties (Taofiq et al. 2016). Yan et al. (2014) examined the tyrosinase inhibitory effects of *Inonotus obliquus* extracts obtained using petroleum ether and n-butanol. At 10μg/mL, both extracts showed tyrosinase inhibition with IC_{50} values of 3.81 and 6.32μg/mL, respectively. Betulin and trametenolic acid reduced melanin content and showed noncompetitive tyrosinase inhibition. Betulin demonstrated significant tyrosinase inhibitory activity with an IC_{50} of 5.13 μM, surpassing the efficacy of the positive control, Kojic acid (6.43 μM). These findings indicate the potential for further exploration of these compounds as cosmetic ingredients for addressing hyperpigmentation.

Several phenolic compounds and their derivatives, as detailed in Table 6.1, have been recognized for their capacity to hinder tyrosinase activity. In a recent study, Chaiprasongsuk et al. (2016) explored the impact of polyphenols namely caffeic acid (CA), ferulic acid (FA), quercetin (QU), rutin (RU), and avobenzone (AV), on anti-melanogenesis following UVA radiation exposure to B16F10 melanoma cells. Evaluating melanin content and tyrosinase activity inhibition resulted in IC_{30} values, ranking the compounds as QU $>$ RU $\approx$ CA $\approx$ AV $>$ FA. Particularly, CA demonstrated a more pronounced tyrosinase inhibition effect compared to FA. These findings align with Thangboonjit et al. (2014) who reported IC_{30} values for tyrosinase inhibition by p-coumaric acid, caffeic acid, and ferulic acid as 22.86, 43.09, and 51.85μM, respectively. The mechanism of these compounds in inhibiting melanogenesis is linked to their ability to suppress tyrosinase protein expression in UVA-irradiated B16F10 cells.

6.4 Fungal-Based Cosmetics: Practical Case of Kojic Acid

UV light from the sun activates tyrosinase activity by the oxidation of copper ion present in the enzyme, in contrast, kojic acid tends to inhibit the oxidation of the copper ion and prevents the initiation of tyrosinase. Inhibiting tyrosinase activity prevents melanin production (Saeedi et al. 2019). Hyperpigmentation disorders, such as post-inflammatory hyperpigmentation, melasma, solar lentigines, and freckles, result from excessive melanin production or an increased melanin content (Plensdorf et al. 2017). Factors such as sun exposure, hormonal changes, aging, and inflammation can trigger melanin production (Lee et al. 2016). Anti-hyperpigmentation agents are currently in high demand to counter the global health challenges associated with hyperpigmentation. Hence, research should concentrate on discovering natural products with no toxicity and drug resistance (Di Petrillo et al. 2016). To control tyrosinase-dependent melanogenesis is an effective basis for anti-melanoma therapy and an effective decline of melanocyte biosynthesis of melanin to address skin disorder hyperpigmentation (Ferro et al. 2018; Mughal et al. 2022). Most skin-lightening products associated with pharmaceuticals, cosmeceuticals, and agriculture function primarily through tyrosinase activity inhibition (Hassan et al. 2023). Endophytic fungi are a source of bioactive metabolites with potential for skin depigmentation. Furthermore, endophytic fungi are known to produce certain metabolites of high medicinal importance but exclusively compared to plants such as vinca alkaloids like vincristine and vinblastine, paclitaxel, and camptothecin which have long been identified as plant-derived compounds with remarkable anticancer efficacy (Zhao et al. 2011).

Kojic acid (KA), a biologically active compound isolated from the endophytic fungus *Aspergillus flavus*, is commonly incorporated into cosmetic formulations such as creams, lotions, and soaps. Its inclusion is due to its antityrosinase properties, which help counteract the stimulatory effects of ultraviolet light on tyrosinase activity, thereby reducing skin hyperpigmentation (Ola et al. 2019). Kojic acid is found in fungi like *A. flavus*, *A. oryzae*, *A. tamarii*, and *A. parasiticus*, and can also be produced from Asian foods like soy sauce and rice wine, serving as a primer for the fungus (Ariff et al. 1996; Wakisaka et al. 1998). KA is also classified as an organic acid that is obtained from fungi during aerobic fermentation (Saeedi et al. 2019). Microphthalmia-associated transcription factor (MITF) upregulates tyrosinase gene transcription, leading to increased tyrosinase protein expression, enhanced enzyme activity, and the subsequent stimulation of melanocyte pigmentation, proliferation, and survival. Transcription factors control gene expression by binding to specific gene sequences. Ultraviolet radiation, metal ions, and free radicals stimulate tyrosinase gene transcription, while kojic acid and its esters inhibit melanin formation in B16F1 melanoma cells (Fig. 6.2) (Jiang et al. 2011; Lajis et al. 2012).

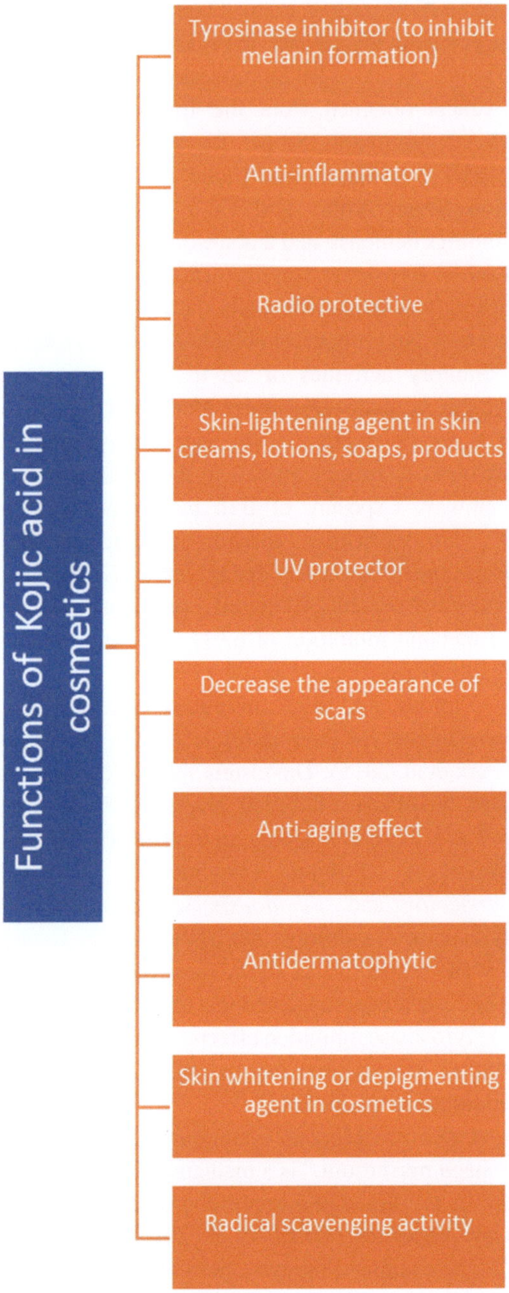

Fig. 6.2 The usefulness of kojic acid in cosmetic formulations

KA (5-hydroxy-2-hydroxymethyl-4-pyrone) promotes skin lightening at 1–4% concentration by inhibiting melanin production in melanocytes through tyrosinase (Lim 1999). Some studies evaluate the depigmentation mechanism and safety (Phasha et al. 2022). They recommended a KA concentration of 1% or less for topical use, as it is effective and safe at this level. A preclinical study showed KA at 1% was effective in cream therapy for a photo-hypersensitive patient within 6 months (Saeedi et al. 2019). Oral administration of KA in rat was found to be absorbed rapidly in rats but was slowly absorbed and distributed when given transdermally (Burnett et al. 2010). Additionally, fungi from different genera *Aspergillus* sp. (Vasantha et al. 2014), *Trichoderma* sp. (Tsuchiya et al. 2008), *Paecilomyces* sp. (Lu et al. 2014), *Phellinus linteus* (Kang et al. 2004), *Daedalea dickinsii* (Morimura et al. 2007), *Dictyophora indusiate* (Sharma et al. 2004) along with a liquid culture of *Neolentinus lepideus* (Ishihara et al. 2018) have been reported as a source of novel tyrosinase inhibitor by producing bioactive compounds. Similarly, marine fungi species, such as *Myrothecium* sp., isolated from algae, also produce tyrosinase inhibitors (X. Li et al. 2005) and *Pestalotiopsis* sp. Z233 (Wu et al. 2013). These novel fungi anti-tyrosinase sources are not yet commercialized.

6.5 Conclusions

In conclusion, a more comprehensive understanding regarding tyrosinase inhibitors' potential in skincare is provided in this chapter. Natural inhibitors, which include polyphenols produced from plants, aldehydes, and microorganisms, are classified in a way that makes it easier to understand the various sources of these bioactive substances. The diverse secondary metabolites produced by fungi present a promising source of bioactive compounds that exhibit the potential to regulate melanin synthesis. The relationship between fungal-based inhibitors' anti-tyrosinase processes and skin hyperpigmentation provides insight into possible treatment approaches for uneven skin tone-related cosmetic issues. The real-world application of compounds produced from fungi in cosmetics is demonstrated by the case study using Kojic acid, which highlights the function of these compounds in formulations intended to reduce hyperpigmentation. As the journey into fungal-based cosmetics continues, further research and development will be crucial for unlocking the full potential of these natural agents. Collaboration between researchers, cosmetic scientists, and industry professionals will play a pivotal role in translating these scientific discoveries into effective and safe skincare solutions. The future holds exciting possibilities for fungal-based cosmetics, offering a harmonious blend of nature's richness and modern skincare innovation.

References

Abbas M, Saeed F, Anjum FM, Afzaal M, Tufail T, Bashir MS, Ishtiaq A, Hussain S, Suleria HAR (2017) Natural polyphenols: an overview. Int J Food Prop 20(8): 1689–1699. https://doi.org/10.1080/10942912.2016.1220393

Abdennacer B, Karim M, Yassine M, Nesrine R, Mouna D, Mohamed B (2015) Determination of phytochemicals and antioxidant activity of methanol extracts obtained from the fruit and leaves of Tunisian Lycium intricatum Boiss. Food Chem 174:577–584. https://doi.org/10.1016/J.FOODCHEM.2014.11.114

Ariff AB, Salleh MS, Ghani B, Hassan MA, Rusul G, Karim MIA (1996) Aeration and yeast extract requirements for kojic acid production by aspergillus flavus link. Enzym Microb Technol 19(7):545–550. https://doi.org/10.1016/S0141-0229(96)00065-8

Baber MA, Crist CM, Devolve NL, Patrone JD (2023) Tyrosinase inhibitors: a perspective. Molecules 28(15):5762. https://doi.org/10.3390/MOLECULES28155762

Baek SH, Nam IJ, Kwak HS, Kim KC, Lee SH (2015) Cellular anti-melanogenic effects of a Euryale ferox seed extract ethyl acetate fraction via the lysosomal degradation machinery. Int J Mol 16(5):9217–9235. https://doi.org/10.3390/IJMS16059217

Balaguer A, Chisvert A, Salvador A (2008) Environmentally friendly LC for the simultaneous determination of ascorbic acid and its derivatives in skin-whitening cosmetics. J Sep Sci 31(2):229–236. https://doi.org/10.1002/JSSC.200700414

Bernard P, Berthon JY (2000) Resveratrol: an original mechanism on tyrosinase inhibition. Int J Cosmet Sci 22(3):219–226. https://doi.org/10.1046/J.1467-2494.2000.00019.X

Bonesi M, Xiao J, Tundis R, Aiello F, Sicari V, Loizzo MR (2018) Advances in the Tyrosinase inhibitors from plant source. Curr Med Chem 26(18):3279–3299.

https://doi.org/10.2174/0929867325666180522091311

Brenner M, Hearing VJ (2008) The protective role of melanin against UV damage in human skin. Photochem Photobiol 84(3):539–549. https://doi.org/10.1111/J.1751-1097.2007.00226.X

Burnett CL, Bergfeld WF, Belsito DV, Hill RA, Klaassen CD, Liebler DC, Marks JG, Shank RC, Slaga TJ, Snyder PW, Andersen FA (2010) Final report of the safety assessment of Kojic acid as used in cosmetics. Int J Toxicol 29(6). https://doi.org/10.1177/1091581810385956

Chaiprasongsuk A, Onkoksoong T, Pluemsamran T, Limsaengurai S, Panich U (2016) Photoprotection by dietary phenolics against melanogenesis induced by UVA through Nrf2-dependent antioxidant responses. Redox Biol 8:79–90. https://doi.org/10.1016/J.REDOX.2015.12.006

Chang TS (2012) Natural Melanogenesis inhibitors acting through the down-regulation of Tyrosinase activity. Materials 5(9):1661–1685. https://doi.org/10.3390/MA5091661

Chen QX, Song KK, Wang Q, Huang H (2003) Inhibitory effects on mushroom Tyrosinase by some Alkylbenzaldehydes. J Enzyme Inhib Med Chem 18(6):491–496. https://doi.org/10.1080/14756360310001613094

Chen J, Yu X, Huang Y (2016) Inhibitory mechanisms of glabridin on tyrosinase. Spectrochim Acta A Mol Biomol Spectrosc 168:111–117. https://doi.org/10.1016/J.SAA.2016.06.008

Di Petrillo A, González-Paramás AM, Era B, Medda R, Pintus F, Santos-Buelga C, Fais A (2016) Tyrosinase inhibition and antioxidant properties of Asphodelus microcarpus extracts. BMC Complement Altern Med 16(1):1–9. https://doi.org/10.1186/S12906-016-1442-0/TABLES/4

Dogra S, Sarangal R (2014) Pigmentary disorders: an insight. Pigment Int 1(1):5. https://doi.org/10.4103/2349-5847.135429

Fernandes MS, Kerkar S (2017) Microorganisms as a source of tyrosinase inhibitors: a review. Ann Microbiol 67(4):343–358. https://doi.org/10.1007/S13213-017-1261-7

Ferro S, Deri B, Germanò MP, Gitto R, Ielo L, Buemi MR, Certo G, Vittorio S, Rapisarda A, Pazy Y, Fishman A, De Luca L (2018) Targeting Tyrosinase: development and structural insights of novel inhibitors bearing Arylpiperidine and Arylpiperazine fragments. J Med Chem 61(9):3908–3917

Garcia-Jimenez A, Teruel-Puche JA, Garcia-Ruiz PA, Berna J, Rodríguez-López JN, Tudela J, Garcia-Canovas F (2017) Action of 2,2′,4,4′--tetrahydroxybenzophenone in the biosynthesis pathway of melanin. Int J Biol Macromol 98:622–629. https://doi.org/10.1016/J.IJBIOMAC.2017.02.032

Goetghebeur M, Kermasha S (1996) Inhibition of polyphenol oxidase by copper-metallothionein from aspergillus Niger. Phytochemistry 42(4):935–940. https://doi.org/10.1016/0031-9422(96)86993-7

Hassan M, Shahzadi S, Kloczkowski A et al (2023) Tyrosinase inhibitors naturally present in plants and synthetic modifications of these natural products as anti-melanogenic agents: a review. Molecules 28(1):378. https://doi.org/10.3390/MOLECULES28010378

Ioniță E, Aprodu I, Stănciuc N, Râpeanu G, Bahrim G (2014) Advances in structure–function relationships of tyrosinase from Agaricus bisporus – investigation on heat-induced conformational changes. Food Chem 156:129–136. https://doi.org/10.1016/J.FOODCHEM.2014.01.089

Ishihara A, Ide Y, Bito T, Ube N, Endo N, Sotome K, Maekawa N, Ueno K, Nakagiri A (2018) Novel tyrosinase inhibitors from liquid culture of Neolentinus lepideus. Biosci Biotechnol Biochem 82(1):22–30. https://doi.org/10.1080/09168451.2017.1415125

Jakimiuk K, Sari S, Milewski R, Supuran CT, Şöhretoğlu D, Tomczyk M (2022) Flavonoids as tyrosinase inhibitors in in silico and in vitro models: basic framework of SAR using a statistical modelling approach. J Enzyme Inhib Med Chem 37(1):421–430. https://doi.org/10.1080/14756366.2021.2014832

Jiang Z, Li S, Liu Y, Deng P, Huang J, He G (2011) Sesamin induces melanogenesis by microphthalmia-associated transcription factor and tyrosinase up-regulation via cAMP signaling pathway. Acta Biochim Biophys Sin 43(10):763–770. https://doi.org/10.1093/ABBS/GMR078

Jiménez M, Chazarra S, Escribano J, Cabanes J, García-Carmona F (2001) Competitive inhibition of mushroom tyrosinase by 4-substituted benzaldehydes. J Agric Food Chem 49(8):4060–4063. https://doi.org/10.1021/JF010194H/ASSET/IMAGES/LARGE/JF010194HF00004.JPEG

Jun H et al (2012) p-Coumaric acid inhibition of CREB phosphorylation reduces cellular melanogenesis. Eur Food Res Technol 235(6):1207–1211. https://doi.org/10.1007/S00217-012-1830-8/METRICS

Jung E, Lee J, Huh S, Lee J, Kim YS, Kim G, Park D (2009) Phloridzin-induced melanogenesis is mediated by the cAMP signaling pathway. Food Chem Toxicol 47(10):2436–2440. https://doi.org/10.1016/J.FCT.2009.06.039

Kang HS, Jin HC, Won KC, Jong CP, Jae SC (2004) A sphingolipid and tyrosinase inhibitors from the fruiting body of Phellinus linteus. Arch Pharm Res 27(7):742–750. https://doi.org/10.1007/BF02980143/METRICS

Kasrati A, Alaoui Jamali C, Bekkouche K, Wohlmuth H, Leach D, Abbad A (2015) Comparative evaluation of antioxidant and insecticidal properties of essential oils from five Moroccan aromatic herbs. J Food Sci Technol 52(4):2312–2319. https://doi.org/10.1007/S13197-014-1284-Z/METRICS

Kilimnik A, Dembitsky VM (2016) Anti-melanoma agents derived from fungal species. Mathews J Pharm Sci 1(1):1–16

Kim D, Park J, Kim J, Han C, Yoon J, Kim N, Seo J, Lee C (2006) Flavonoids as mushroom tyrosinase inhibitors: a fluorescence quenching study. J Agric Food Chem 54(3):935–941. https://doi.org/10.1021/JF0521855/ASSET/IMAGES/LARGE/JF0521855F00005.JPEG

Kim DS, Cho HJ, Lee HK, Lee WH, Park ES, Youn SW, Park KC (2007) Terrein, a fungal metabolite, inhibits the epidermal proliferation of skin equivalents. J Dermatol Sci 46(1):65–68. https://doi.org/10.1016/j.jdermsci.2006.11.011

Kim DS, Lee HK, Park SH, Lee S, Ryoo IJ, Kim WG, Yoo ID, Na JI, Kwon SB, Park KC (2008a) Terrein inhibits keratinocyte proliferation via ERK inactivation and G2/Mcell cycle arrest. Exp Dermatol 17(4):312–317. https://doi.org/10.1111/J.1600-0625.2007.00646.X

Kim SS, Lee JA, Kim JY, Lee NH, Hyun CG (2008b) Citrus Peel wastes as functional materials for cosmeceuticals. J Appl Biol Chem 51(1):7–12. https://doi.org/10.3839/JABC.2008.002

Kumar S (2005) Exploratory analysis of global cosmetic industry: major players, technology and market trends. Technovation 25(11):1263–1272. https://doi.org/10.1016/J.TECHNOVATION.2004.07.003

Lajis AFB, Hamid M, Ariff AB (2012) Depigmenting effect of kojic acid esters in hyperpigmented B16F1 melanoma cells. J Biomed Biotechnol 2012. https://doi.org/10.1155/2012/952452

Lee SY, Baek N, Nam TG (2016) Natural, semisynthetic and synthetic tyrosinase inhibitors. J Enzyme Inhib Med Chem 31(1):1–13. https://doi.org/10.3109/14756366.2015.1004058

Li X, Kim MK, Lee U, Kim SK, Kang JS, Choi HD, Son BW (2005) Myrothenones a and B, Cyclopentenone derivatives with Tyrosinase inhibitory activity from the marine-derived fungus Myrothecium sp. Chem Pharm Bull 53(4):453–455. https://doi.org/10.1248/CPB.53.453

Li J, Feng L, Liu L, Wang F, Ouyang L, Zhang L, Hu X, Wang G (2021) Recent advances in the design and discovery of synthetic tyrosinase inhibitors. Eur J Med Chem 224:113744. https://doi.org/10.1016/J.EJMECH.2021.113744

Li J, Li C, Peng X, Li S, Liu B, Chu C (2023) Recent discovery of tyrosinase inhibitors in traditional Chinese medicines and screening methods. J Ethnopharmacol 303:115951. https://doi.org/10.1016/J.JEP.2022.115951

Lim JTE (1999) Treatment of Melasma using Kojic acid in a gel containing hydroquinone and glycolic acid. Dermatologic Surg 25(4):282–284. https://doi.org/10.1046/J.1524-4725.1999.08236.X

Lu R, Liu X, Gao S, Zhang W, Peng F, Hu F, Huang B, Chen L, Bao G, Li C, Li Z (2014) New tyrosinase inhibitors from paecilomyces gunnii. J Agric Food Chem 62(49):11917–11923

Lung MY, Tsai JC, Huang PC (2010) Antioxidant properties of edible basidiomycete Phellinus igniarius in submerged cultures. J Food Sci 75(1):E18–E24. https://doi.org/10.1111/J.1750-3841.2009.01384.X

Mann T, Gerwat W, Batzer J, Eggers K, Scherner C, Wenck H, Stäb F, Hearing VJ, Röhm KH, Kolbe L (2018) Inhibition of human Tyrosinase requires molecular motifs distinctively different from mushroom Tyrosinase. J Invest Dermatol 138(7):1601–1608. https://doi.org/10.1016/J.JID.2018.01.019

Masum MN, Yamauchi K, Mitsunaga T (2019) Tyrosinase inhibitors from natural and synthetic sources as skin-lightening agents. Rev Agric Sci 7:41–58. https://doi.org/10.7831/RAS.7.41

Meng T-X, Zhang C-F, Miyamoto T, Ishikawa H, Shimizu K, Ohga S, Kondo R (2012) The melanin biosynthesis stimulating compounds isolated from the fruiting bodies of *Pleurotus citrinopileatus*. J Cosmetics Dermatol Sci Appl 02(03):151–157. https://doi.org/10.4236/JCDSA.2012.23030

Mermer A, Demirci S (2023) Recent advances in triazoles as tyrosinase inhibitors. Eur J Med Chem 259:115655. https://doi.org/10.1016/J.EJMECH.2023.115655

Moreiras H, Seabra MC, Barral DC (2021) Melanin transfer in the epidermis: the pursuit of skin pigmentation control mechanisms. Int J Mol Sci 22(9):4466. https://doi.org/10.3390/IJMS22094466

Morimura K, Yamazaki C, Hattori Y, Makabe H, Kamo T, Hirota M (2007) A Tyrosinase inhibitor, Daedalin a, from mycelial culture of Daedalea dickinsii. Biosci Biotechnol Biochem 71(11):2837–2840. https://doi.org/10.1271/BBB.70266

Mughal EU, Ashraf J, Hussein EM, Nazir Y, Alwuthaynani AS, Naeem N, Sadiq A, Alsantali RI, Ahmed SA (2022) Design, synthesis, and structural characterization of Thioflavones and Thioflavonols as potential Tyrosinase inhibitors: in vitro and in Silico studies. ACS Omega. https://doi.org/10.1021/ACSOMEGA.2C01841/ASSET/IMAGES/MEDIUM/AO2C01841_M003.GIF

Ng LT, Ko HH, Lu TM (2009) Potential antioxidants and tyrosinase inhibitors from synthetic polyphenolic deoxybenzoins. Bioorg Med Chem 17(13):4360–4366. https://doi.org/10.1016/J.BMC.2009.05.019

Nouveau S, Agrawal D, Kohli M, Bernerd F, Misra N, Nayak C (2016) Skin hyperpigmentation in Indian population: insights and best practice. Indian J Dermatol 61(5):487. https://doi.org/10.4103/0019-5154.190103

Obaid RJ, Mughal EU, Naeem N, Sadiq A, Alsantali RI, Jassas RS, Moussa Z, Ahmed SA (2021) Natural and synthetic flavonoid derivatives as new potential tyrosinase inhibitors: a systematic review. RSC Adv 11(36):22159–22198. https://doi.org/10.1039/D1RA03196A

Ola ARB, Metboki G, Lay CS, Sugi Y, De Rozari P, Darmakusuma D, Hakim EH (2019) Single production of Kojic acid by aspergillus flavus and the revision of Flufuran. Molecules 24(22):4200. https://doi.org/10.3390/MOLECULES24224200

Özer Ö, Mutlu B, Kivçak B (2007) Antityrosinase activity of some plant extracts and formulations containing

Ellagic acid. Pharm Biol 45(6):519–524. https://doi.org/10.1080/13880200701446746

Panzella L, Napolitano A (2019) Natural and bioinspired phenolic compounds as Tyrosinase inhibitors for the treatment of skin hyperpigmentation: recent advances. Cosmetics 6(4):57. https://doi.org/10.3390/COSMETICS6040057

Park SY, Jin ML, Kim YH, Kim Y, Lee SJ (2011) Aromatic-turmerone inhibits α-MSH and IBMX-induced melanogenesis by inactivating CREB and MITF signaling pathways. Arch Dermatol Res 303(10):737–744. https://doi.org/10.1007/S00403-011-1155-7/METRICS

Park KM, Kwon KM, Lee SH (2015) Evaluation of the antioxidant activities and Tyrosinase inhibitory property from mycelium culture extracts. Evid Based Complement Alternat Med 2015. https://doi.org/10.1155/2015/616298

Parvez S, Kang M, Chung HS, Bae H (2007) Naturally occurring tyrosinase inhibitors: mechanism and applications in skin health, cosmetics and agriculture industries. Phytother Res 21(9):805–816. https://doi.org/10.1002/PTR.2184

Petit L, Pierard GE (2003) Skin-lightening products revisited. Int J Cosmet Sci 25(4):169–181

Phasha V, Senabe J, Ndzotoyi P, Okole B, Fouche G, Chuturgoon A (2022) Review on the use of Kojic acid – a skin-lightening ingredient. Cosmet 9(3):64. https://doi.org/10.3390/COSMETICS9030064

Plensdorf S, Livieratos M, Dada N (2017) Pigmentation disorders: diagnosis and management. Am Fam Physician 96(12):797–804

Roh JS, Han JY, Kim JH, Hwang JK (2004) Inhibitory effects of active compounds isolated from safflower (Carthamus tinctorius L.) seeds for Melanogenesis. Biol Pharm Bull 27(12):1976–1978. https://doi.org/10.1248/BPB.27.1976

Roulier B, Pérès B, Haudecoeur R (2020) Advances in the design of genuine human tyrosinase inhibitors for targeting melanogenesis and related pigmentations. J Med Chem 63(22):13428–13443

Saeedi M, Eslamifar M, Khezri K (2019) Kojic acid applications in cosmetic and pharmaceutical preparations. Biomed Pharmacother 110:582–593. https://doi.org/10.1016/J.BIOPHA.2018.12.006

Sarkar R, Arora P, Garg KV (2013) Cosmeceuticals for hyperpigmentation: what is available? J Cutan Aesthet Surg 6(1):4. https://doi.org/10.4103/0974-2077.110089

Sasaki K, Yoshizaki F (2002) Nobiletin as a Tyrosinase inhibitor from the Peel of citrus fruit. Biol Pharm Bull 25(6):806–808. https://doi.org/10.1248/BPB.25.806

Satooka H, Kubo I (2012) Resveratrol as a kcat type inhibitor for tyrosinase: potentiated melanogenesisinhibitor. Bioorganic med chem 20(2):1090–1099. https://doi.org/10.1016/j.bmc.2011.11.030

Sharma VK, Choi J, Sharma N, Choi M, Seo SY (2004) In vitro anti-tyrosinase activity of 5-(hydroxymethyl)-2-furfural isolated from Dictyophora indusiata. Phytother Res 18(10):841–844. https://doi.org/10.1002/PTR.1428

Şöhretoğlu D, Sari S, Barut B, Özel A (2018) Tyrosinase inhibition by some flavonoids: inhibitory activity, mechanism by in vitro and in silico studies. Bioorg Chem 81:168–174. https://doi.org/10.1016/J.BIOORG.2018.08.020

Su TR, Lin JJ, Tsai CC, Huang TK, Yang ZY, Wu MO, Zheng YQ, Su CC, Wu YJ (2013) Inhibition of Melanogenesis by Gallic acid: possible involvement of the PI3K/Akt, MEK/ERK and Wnt/β-catenin signaling pathways in B16F10 cells. Int J Mol Sci 14(10): 20443–20458. https://doi.org/10.3390/IJMS141020443

Tang HC, Chen YC (2015) Identification of tyrosinase inhibitors from traditional Chinese medicines for the management of hyperpigmentation. Springerplus 4(1): 1–18. https://doi.org/10.1186/S40064-015-0956-0/FIGURES/19

Taofiq O, Martins A, Barreiro MF, Ferreira ICFR (2016) Anti-inflammatory potential of mushroom extracts and isolated metabolites. Trends Food Sci Technol 50:193–210. https://doi.org/10.1016/J.TIFS.2016.02.005

Thangboonjit W, Limsaeng-u-rai S, Pluemsamran T, Panich U (2014) Comparative evaluation of Antityrosinase and antioxidant activities of dietary Phenolics and their activities in melanoma cells exposed to UVA. Siriraj Med J 66(1)

Thawabteh AM, Jibreen A, Karaman D, Thawabteh A, Karaman R (2023) Skin pigmentation types,causes and treatment—a review. Molecules 28(12):4839. https://doi.org/10.3390/molecules28124839

Tsai CC, Chan CF, Huang WY, Lin JS, Chan P, Liu HY, Lin YS (2013) Applications of lactobacillus rhamnosus spent culture supernatant in cosmetic Antioxidation, whitening and moisture retention applications. Molecules 18(11):14161–14171. https://doi.org/10.3390/MOLECULES181114161

Tsuchiya T, Yamada K, Minoura K, Miyamoto K, Usami Y, Kobayashi T, Hamada-Sato N, Imada C, Tsujibo H (2008) Purification and determination of the chemical structure of the Tyrosinase inhibitor produced by Trichoderma viride strain H1-7 from a marine environment. Biol Pharm Bull 31(8): 1618–1620. https://doi.org/10.1248/BPB.31.1618

Umesha SS, Naidu KA (2015) Antioxidants and antioxidant enzymes status of rats fed on n-3 PUFA rich garden cress (Lepidium Sativum L) seed oil and its blended oils. J Food Sci Technol 52(4):1993–2002. https://doi.org/10.1007/S13197-013-1196-3/METRICS

Vasantha KY, Murugesh CS, Sattur AP (2014) A tyrosinase inhibitor from aspergillus Niger. J Food Sci Technol 51(10):2877–2880. https://doi.org/10.1007/S13197-014-1395-6/METRICS

Wakisaka Y, Segawa T, Imamura K, Sakiyama T, Nakanishi K (1998) Development of a cylindrical apparatus for membrane-surface liquid culture and production of kojic acid using aspergillus oryzae

NRRL484. J Ferment Bioeng 85(5):488–494. https://doi.org/10.1016/S0922-338X(98)80067-6

Wu B, Wu X, Sun M, Li M (2013) Two novel Tyrosinase inhibitory Sesquiterpenes induced by CuCl2 from a marine-derived fungus Pestalotiopsis sp. Z233. Mar Drugs 11(8):2713–2721. https://doi.org/10.3390/MD11082713

Wuttisin N, Boonmak J, Thaipitak V (2017) Anti-tyrosinase activity of orange peel extract and cosmetic formulation. Int Food Res J 24(5):2128–2132

Xie LP, Chen QX, Huang H, Wang HZ, Zhang RQ (2003) Inhibitory effects of some flavonoids on the activity of mushroom tyrosinase. Biochemistry (Moscow) 68(4): 487–491. https://doi.org/10.1023/A:1023620501702/METRICS

Yan ZF, Yang Y, Tian FH, Mao XX, Li Y, Li CT (2014) Inhibitory and acceleratory effects of Inonotus obliquus on tyrosinase activity and melanin formation in b16 melanoma cells. Evid Based Complement Alternat Med 2014. https://doi.org/10.1155/2014/259836

Yu Q, Fan L, Duan Z (2019) Five individual polyphenols as tyrosinase inhibitors: inhibitory activity, synergistic effect, action mechanism, and molecular docking. Food Chem 297:124910. https://doi.org/10.1016/J.FOODCHEM.2019.05.184

Yuan Y, Jin W, Nazir Y, Fercher C, Blaskovich MAT, Cooper MA, Barnard RT, Ziora ZM (2020) Tyrosinase inhibitors as potential antibacterial agents. Eur J Med Chem 187:111892. https://doi.org/10.1016/J.EJMECH.2019.111892

Zhang D, Li X, Kang JS, Choi HD, Son BW (2007) A new α-Pyrone derivative, 6-[(E)-Hept-α-pyrone], with Tyrosinase inhibitory activity from. Notes 28(5):887

Zhang YC, Li G, Jiang C, Yang B, Yang HJ, Xu HY, Huang LQ (2014) Tissue-specific distribution of Ginsenosides in different aged ginseng and antioxidant activity of ginseng leaf. Molecules 19(11): 17381–17399. https://doi.org/10.3390/MOLECULES191117381

Zhang X et al (2021) Recent advance in the discovery of tyrosinase inhibitors from natural sources via separation methods. J Enzyme Inhib Med Chem 36(1): 2104–2117. https://doi.org/10.1080/14756366.2021.1983559

Zhao J, Shan T, Mou Y, Zhou L (2011) Plant-derived bioactive compounds produced by Endophytic fungi. Mini-Rev Med Chem 11(2):159–168. https://doi.org/10.2174/138955711794519492

Zhou L, Liu W, Zou L, Xiong Z, Hu X, Chen J (2017) Aggregation and conformational change of mushroom (Agaricus bisporus) polyphenoloxidase subjected to thermal treatment. Food Chem 214:423–431. https://doi.org/10.1016/J.FOODCHEM.2016.07.041

Zhu YZ, Chen K, Chen YL, Zhang C, Xie YY, Hider RC, Zhou T (2022) Design and synthesis of novel stilbene-hydroxypyridinone hybrids as tyrosinase inhibitors and their application in the anti-browning of freshly-cut apples. Food Chem 385:132730. https://doi.org/10.1016/J.FOODCHEM.2022.132730

Zhuravleva OI, Afiyatullov SS, Vishchuk OS, Denisenko VA, Slinkina NN, Smetanina OF (2012) Decumbenone C, a new cytotoxic decaline derivative from the marine fungus aspergillus sulphureus KMM 4640. Arch Pharm Res 35(10):1757–1762. https://doi.org/10.1007/S12272-012-1007-9/METRICS

Zolghadri S, Bahrami A, Hassan Khan MT, Munoz-Munoz J, Garcia-Molina F, Garcia-Canovas F, Saboury AA (2019) A comprehensive review on tyrosinase inhibitors. J Enzyme Inhib Med Chem 34(1): 279–309. https://doi.org/10.1080/14756366.2018.1545767

7 Fungal-Based Cosmetics and Their Anti-hyaluronidase Activity

Biswadeep Chaudhuri, Anirban Debnath, and Arpita Das

Abstract

The worldwide cosmetic industry, esteemed billions of US dollars, is consistently looking for innovative, natural, and low-cost raw resources. This demand in modern days has prompted the far and wide accessibility of normal cosmetic items from various plants like coconut, jasmine, lemongrass, and longan in stores around the world. Recently, mushrooms have been revered as products that improve health in Asia. In nations like China and Japan, their advantages were once so exceptionally regarded that they were held for eminence in the field of healthcare cosmetics. Fungi, especially mushrooms and their derivatives, are increasingly being used in cosmetics to promote their potency as active ingredients. Future developments for fungal-based cosmetics appear quite promise in terms of expansion and innovation. Technological developments in extraction and formulation, together with the investigation of diverse fungus species and the creation of customized skincare products, are propelling this field's advancement. The emphasis on sustainability, scientific confirmation, and regenerating capabilities further adds to the allure of fungal cosmetics. With the development of science and technology, the extraction techniques to obtain strong and stable fungal components for cosmetic products are getting simpler and more efficient. The potential uses of these components are growing as research into various fungus species reveals novel bioactive substances with special advantages for skin health.

The trend toward personalized skincare is fueling the development of customized products that precisely target specific skin concerns of specific patient requirements. This specific area also has high industrial values that can provide a wide range of personalized cosmetics for almost every skin type and requirement. The focus on sustainability is highly relevant, considering that fungi can be cultivated with minimal environmental impact compared to traditional plant sources. This eco-conscious aspect makes fungal-based cosmetics an appealing option for consumers. Moreover, ongoing scientific validation ensures that fungal ingredients meet safety and efficacy standards, thus contributing to building consumer trust as well as product safety for future use. Fungal-based products are well-positioned to play a significant role as the cosmetic industry continues to evolve. They offer effective, environmentally friendly, and tailored

B. Chaudhuri · A. Debnath
Department of Biotechnology, Adamas University, Kolkata, West Bengal, India

A. Das (✉)
Department of Biotechnology, Adamas University, Kolkata, West Bengal, India

QA Higher Education (QAHE), London, UK

S. Ghosh, H. N. Onyeaka (eds.), *Fungal-Based Cosmetics: Formulation and Usage*, The Mycota 17,
https://doi.org/10.1007/978-3-032-09291-5_7

skincare solutions that cater to a diverse and discerning customer base. With ongoing advancements and increasing consumer interest in novel products, the future of fungal cosmetics appears to be both innovative and impactful for the future generation to come in a safe and eco-friendly way.

Keyword

Anti-aging · Hyaluronidase · Fungal polysaccharides · Skincare · Tyrosinase inhibitors

7.1 Introduction

Over the last 10 years, the cosmetics industry has grown at some of the fastest rates. The cosmetics industry is always looking for high-quality organic ingredients with relevant bioactive properties, like anti-aging, anti-wrinkle, anti-pigmentation, anti-collagenase, antioxidant, and anti-inflammatory capabilities, in order to create successful formulations (Angeli et al. 2009; Athalye et al. 2009; Barros et al. 2007a). Due to the distinct biological and chemical properties of fungi, there has been a notable increase in the market for cosmeceuticals made from them in recent years. Sporobores from both wild and farmed mushrooms are the source of various nutraceuticals that, when taken orally, can be utilized as nutricosmetics. In addition to their terrestrial counterparts, marine fungi produced over a thousand metabolites, some of which are being studied as possible cosmeceuticals and may have therapeutic applications because of their unique biological properties (Angeli et al. 2009; Barros et al. 2007b). Many skincare products, such as those that are anti-aging, antioxidant-rich, skin-rejuvenating, and haircare products, contain fungi. Numerous fungi are used in the biological transformation process, and the resultant compounds, such as ceramides and lactic acid, may find use in skincare products. To produce a variety of skincare products, the cosmetics industry has long relied on plant-based elements, synthetic substances, and minerals. Regardless, the rising interest with beauty care products got from parasites implies an eminent change in skincare procedure, bridling the unmistakable and strong bioactive substances present in growths. This paper digs into the conceivable outcomes of parasitic-based beauty care products, with a particular spotlight on their advantages for the skin, especially in the domain of hostile to maturing, and the rising interest encompassing these parts inside the excellence area (Campana et al. 2004; Attard et al. 2015).

Cosmeceuticals are topical creams that contain certain bioactive ingredients that mimic the actions of medications by enhancing skin-related processes. In the worldwide cosmeceutical industry, organic materials are progressively taking the place of synthetic ones because of their biocompatibility, safety, and environmental friendliness. The ability to preserve the potency of active substances, guarantee their distribution in a physiologically active form, and achieve a biological impact through defined processes are three critical criteria that determine the success of cosmeceutical manufacture (Angeli et al. 2009; Attard et al. 2015; Bakkali et al. 2008). Success in this industry also depends on guaranteeing cost, safety, and the integration of all these factors. Many cosmetic goods, such as creams, lotions, and ointments, contain phytochemicals, enzymes, vitamins, minerals, and other active ingredients (Bakkali et al. 2008; Bitencourt et al. 2018).

Such bioactive compounds can come from a number of natural sources including bacteria, fungi, and plants. Numerous plants, bacteria, and fungi have uses, especially when it comes to promoting human health and strong, glossy skin, hair, and nails at the cellular level (Angeli et al. 2009; Athalye et al. 2009; Barros et al. 2007a). Since prehistoric area, medicinal herbs and selective microorganisms have been the main producers of naturally occurring organic compounds. In this context, the production of cosmetic compounds derived from plants has several obstacles, including deforestation, contamination of the environment, and climate change, which can be avoided by the use of microorganisms in the near future (Brunner

2005; Burčová et al. 2018; Campalani et al. 2020).

The recent resurgence of a number of previously identified bacterial chemicals makes the discovery of new structures more challenging. It's always looking for new and complementary bio assets, especially ones like mushrooms that can be generated in large quantities. Many new natural compounds have been discovered and isolated from marine fungus species in the past few decades. Lipids, alkaloids, terpenoids, and chemicals derived from shikimate, polyketides, and non-ribosomal peptides (NRPs) are some examples of these molecules (Bitencourt et al. 2018; Chhouk et al. 2018).

7.2 Hyaluronidase and Skin Aging

Hyaluronidase is an enzyme that breaks down hyaluronic acid, which is an important part of the skin's extracellular matrix that keeps it hydrated, supple, and shaped. The breakdown of hyaluronic acid causes the skin to lose moisture, which in turn causes wrinkles and a loss of firmness. Thus, halting hyaluronidase action fills in as a technique for forestalling skin maturing and protecting an energetic appearance (Angeli et al. 2009; Chouaibi et al. 2019).

7.2.1 Degradation of Hyaluronic Acid

Hyaluronidase breaks down hyaluronic acid, causing a decrease in the skin's moisture levels, resulting in dryness, reduced elasticity, and increased susceptibility to fine lines and wrinkles (Angeli et al. 2009; Chhouk et al. 2018).

7.2.2 Collagen and Elastin Breakdown

The presence of hyaluronic acid is vital for safeguarding the collagen and elastin fibers, which are responsible for the skin's structure and flexibility. When hyaluronic acid is broken down by hyaluronidase, these fibers become more prone to harm, leading to the aging of the skin (Angeli et al. 2009; Ciarlini et al. 2017).

7.2.3 Impact on Skin Volume

Hyaluronic corrosive (HA) is a normally occurring substance of healthy skin that assumes an essential part in keeping up with skin volume and hydration. It contributes to the extracellular matrix's moisture retention, giving the skin a firm and plump appearance. The natural production of HA decreases as we get older, causing our skin to lose volume and elasticity, which in turn contributes to wrinkles and sagging. Skin volume can be significantly affected by the application of HA, whether through supplements, injectable, or topical products. By recharging the skin's HA levels, these medicines can re-establish dampness, further develop versatility, and improve the skin's general totality (Gao et al. 2004; Bubalo et al. 2018). Skin HA items, similar to serums and creams, give surface hydration, while injectable HA fillers offer a quicker and designated expansion in volume, especially in regions inclined to drooping, like the cheeks and under the eyes (Golz-berner and Zastrow 2005; Gautier et al. 2008). In general, HA is esteemed in skincare for its capacity to draw in and hold water, giving a brief lift in skin volume and a more energetic appearance. Normal utilization of HA can assist with alleviating the noticeable impacts of maturing, keeping up with skin's perfection and volume after some time. HA plays a role in maintaining the skin's fullness and volume. By reducing this volume, hyaluronidase creates the sunken appearance commonly observed in aging skin, especially in areas like the cheeks and under the eyes (Gutiérrez et al. 2002; Jin et al. 2010).

7.2.4 Exacerbation of External Aging Factors

Various external factors like UV radiation, pollution, and smoking can heighten the effects of hyaluronidase, speeding up the breakdown of hyaluronic acid and further contributing to skin aging (Angeli et al. 2009; Chouaibi et al. 2019).

7.3 Anti-hyaluronidase Activity of Fungal Compounds

7.3.1 Polysaccharides

Beta-glucans (fungal polysaccharides), are well known for their capacity to safeguard the outer layer of human skin, making them exceptionally useful in skincare items (Kakumyan and Matsui 2009; Kalač 2009). One of their essential capabilities is to obstruct the activity of hyaluronidase, a catalyst liable for separating hyaluronic corrosive in the skin. Since hyaluronic corrosive is pivotal for holding skin dampness, flexibility, and in general young appearance, beta-glucans assist with keeping up with these characteristics by forestalling hyaluronic corrosive debasement, in this way offering hostile to maturing benefits and adding to a more energetic look (Kim et al. 2006; Keypour et al. 2010). The restraint of hyaluronidase movement by beta-glucans upholds the skin's inherent ability to trap dampness and maintain its primary strength, which thus decreases the perceivability of scarcely discernible differences and kinks, advancing smoother and firmer skin. Also, beta-glucans contain cancer prevention agent properties, further reinforcing their defensive job by protecting the skin against natural stressors, like UV radiation and contamination (Krištofíková et al. 1991; Kumari et al. 2008). The combination of these effects makes polysaccharides derived from fungi, like beta-glucans, a useful component in skincare products meant to keep skin looking young and prevent aging signs (Lourenço et al. 1996; Lemann 2007).

7.3.2 Phenolic Compounds

Phenolic compounds, which are significant in skincare because of their cell reinforcement properties, are known to be delivered by certain organisms (Luiz et al. 2003; Machado et al. 2005). These phenolic compounds are extremely effective at eliminating free radicals, protecting the skin from oxidative stress, a major contributor to premature aging (Margolin and Klibanov 1987; Mantzouridou and Tsimidou 2008). By removing these harmful free radicals, the phenolic helps to preserve the skin's primary uprightness and reduces the risk of damage that can cause signs of aging like wrinkles, barely discernible differences, and a lack of flexibility (Matuo et al. 2007; Angeli et al. 2009). Despite their disease avoidance specialist limits, these parasitic construed phenolic compounds furthermore show the ability to control the activity of impetuses like hyaluronidase (Mizuno et al. 1990; Menoli et al. 2001). Hyaluronidase is obligated for isolating hyaluronic destructive, an essential piece of the skin's extracellular grid that stays aware of hydration, strength, and in everyday skin prosperity (Mizuno et al. 1999). These phenolic intensifiers protect the degrees of hyaluronic corrosive in the skin by restraining hyaluronidase, accordingly supporting the skin's normal dampness maintenance and flexibility (Ng 2004; Mohorčič et al. 2007). This shielding is basic for keeping a youthful appearance and hindering the debasement of skin structure long term. Due to their dual antioxidant and enzyme inhibition functions, these fungal phenolic compounds provide the skin with comprehensive protection (Obrien et al. 1993; Nieto and Carolina Chegwin 2008). Other than the way that they defend the skin from oxidative damage achieved by normal elements like UV radiation and defilement, they also thwart enzymatic corruption of major skin parts. As an outcome of this, the worth of these mixtures in skincare items that battle the noticeable indications of maturing and advance strong, sound skin is developing. Their ordinary start areas of strength for and influences make them a drawing in choice for those attempting to overhaul their skincare routine with convincing, nature-deduced trimmings (Park and Hwang 1999; Passi et al. 2002).

7.3.3 Triterpenoids

Ganoderma lucidum, a notable medicinal fungus, produces compounds called triterpenoids, which are progressively perceived for their likely advantages in skincare, especially in enemy of maturing plans (Pons et al. 2002; Angeli et al. 2009). These triterpenoids have potent anti-inflammatory properties that can help soothe irritated skin, reduce redness, and stop chronic inflammation from deteriorating the structural integrity of the skin (Angeli et al. 2009; Sadruddin and Arora 2009). Furthermore, these mixtures have been displayed to restrain the movement of hyaluronidase, a compound that separates hyaluronic corrosive in the skin (Angeli et al. 2009). Hyaluronic acid is an essential part of the extracellular matrix of the skin that keeps the skin moist, supple, and plump. *Ganoderma lucidum* triterpenoids preserve hyaluronic acid levels by inhibiting hyaluronidase. This keeps the skin hydrated and reduces the appearance of fine lines and wrinkles. This double activity of calming and hostile to hyaluronidase properties makes triterpenoids important in battling the indications of maturing (Sandewicz et al. 2003; Sanodiya et al. 2009). Besides, the joining of *Ganoderma lucidum* separates into skincare items offers a characteristic and successful way to deal with improving skin wellbeing and appearance (Shimada et al. 2002; Angeli et al. 2009). These concentrates support the skin's normal guard components as well as add to a younger and energetic coloring (Lourenço et al. 1996; Chouaibi et al. 2019). As examination keeps on investigating the maximum capacity of parasitic determined intensifies in beauty care products, *Ganoderma lucidum* stands apart as a promising fixing in the improvement of hostile to maturing skincare arrangements, joining customary restorative information with present-day skincare science. In synopsis, the triterpenoids created by *Ganoderma lucidum* hold huge commitment in the domain of hostile to maturing skincare, because of their capacity to decrease irritation and safeguard the skin's fundamental parts, offering a characteristic answer for keeping up with young and sound skin (Tian et al. 2008; Synytsya et al. 2009).

7.4 Fungi as Potential Resources

Despite not falling under the categories of plants or animals, fungi, which are members of a separate kingdom of organisms, possess characteristics that are extremely beneficial for cosmetic purposes (Bakkali et al. 2008; Angeli et al. 2009). They can flourish in different conditions and have created ways of getting by in brutal circumstances, making them a significant wellspring of bioactive mixtures. Dissimilar to plants, growths don't rely upon photosynthesis for supplements; all things considered, they get them through deterioration, beneficial interaction, or parasitism. This remarkable natural cycle empowers organisms to deliver a wide cluster of optional metabolites, like polysaccharides, peptides, proteins, and phenolic compounds, a large number of which have shown possible in skincare. Further, due to their antioxidant properties, fungal-derived phenolic compounds can further shield the skin from oxidative stress, a major cause of skin aging. Oxidative pressure, coming about because of free extremists, causes the disintegration of collagen and elastin, significant proteins for protecting skin solidness and adaptability. By checking free revolutionaries, these parasitic mixtures might decelerate the maturing system and empower a more energetic look (Vozenin-Brotons et al. 2001; Ventura et al. 2008).

Due to their well-known skin-soothing, moisturizing, and anti-inflammatory properties, fungal biology's production of polysaccharides like beta-glucans is fascinating. These polysaccharides can make defensive layers on the skin, holding dampness and further developing the skin's obstruction capability. Certain organisms additionally produce intensifies like triterpenoids and phenolics with strong cell reinforcement properties, shielding the skin from oxidative pressure and ecological harm (Ventura et al. 2008; Vysotskaya et al. 2009). Cosmetics made from fungi are especially interesting for their potential anti-aging properties, particularly for inhibiting hyaluronidase, an enzyme that breaks down hyaluronic acid in the skin. Although hyaluronic acid is necessary for maintaining skin's structure, elasticity, and hydration, its degradation by increased hyaluronidase activity can cause dryness, wrinkles, and sagging (Wang et al. 2007; Angeli et al. 2009). Contagious mixtures show potential in restraining hyaluronidase movement, in this way saving hyaluronic corrosive levels in the skin. For instance, the ability of beta-glucans found in fungi like *Ganoderma lucidum* (Reishi) and Trametes versicolor to inhibit hyaluronidase has been the subject of research. This effectively aids in maintaining skin hydration and reduces the appearance of fine lines and wrinkles (Yabuta 1924; Ward and Singh 2005).

Skin, being the biggest organ in the human body, is vital for both infection/pathogen resistance and stress management related to weather conditions. Research on common skin disorders such as pigmentation, cancer, microbial infections, wound healing, ageing, and UV-induced wrinkles has led to the development of better skincare products. Fungi are found in a lot of skincare products, including those that are anti-aging, high in antioxidants, skin-rejuvenating, and haircare products (Barros et al. 2007a; Angeli et al. 2009). Numerous fungi are involved in the biological transformation process, and the substances that are produced—like lactic acid and ceramides—may be used in skincare products. Schizophyllum fungi produce Schizophyllan can reduce skin inflammation as they contain 1,6-branched-1,3-glucan, which reduces inflammation of the skin, itchiness, and UV damage. Rhizopus strains can be used to produce lactic acid. In dermatological cosmetic products, it is used to soften and moisturize dry, flaky skin (Lourenço et al. 1996; Yang et al. 2009).

7.5 Fungal-Based Ingredients in Modern Skincare Products

The skincare business is progressively recognizing the advantages of involving contagious-based parts in skincare items. To meet the growing demand for natural and effective skincare products, brands are beginning to incorporate extracts from various fungi, taking advantage of their bioactive properties (Ng 2004; Zhang et al. 2007). *Tremella fuciformis*, otherwise called Snow Mushroom, is a famous parasitic component with a background marked by utilization in Chinese medication. It is profoundly esteemed for its ability to keep up with skin dampness (Zulfakar et al. 2007; Hyde et al. 2010). It is a great natural substitute for hyaluronic acid in moisturizing products because studies have demonstrated that it can retain up to 500 times its weight in water. Snow mushroom extract is now a key component of a variety of serums, creams, and masks that aim to provide deep hydration and plumping benefits as a result. *Cordyceps sinensis*, which is known for its anti-inflammatory and antioxidant properties, is another fungal component that is frequently used. Cordyceps extracts are frequently used in skincare products to soothe irritated skin, reduce redness, and protect the skin from environmental stressors (Lourenço et al. 1996; Angeli et al. 2009). This makes it an important expansion to definitions focusing on delicate or maturing skin. Moreover, *Ganoderma lucidum* (Reishi), otherwise called the "Mushroom of Everlasting status," has acquired ubiquity for its enemy of maturing and skin-mitigating properties. Triterpenoids and polysaccharides are abundant in reishi extracts, which contribute to the plant's ability to reduce inflammation, increase skin hydration, and guard against oxidative damage. As a result, luxury skincare brands that emphasize anti-aging and rejuvenation frequently use Reishi (Mohorčič et al. 2007; Angeli et al. 2009).

In addition, *Aspergillus niger*'s cell walls include chitin-glucan, a form of copolymer (Margolin and Klibanov 1987; Angeli et al. 2009). Because chitin-glucan bond has strong moisturizing properties on skin that can assist to counteract some signs of skin ageing and wrinkles, it is anticipated to be employed in skin-moisturizing and anti-aging formulations. Lentinula edodes, which includes lentinan, is used and marketed in the form of night lotion. Another kind of polyphenol found in fungi is called ellagic acid, and it has anti-oxidant, anti-cancer, and skin-whitening properties. Gallic acid, a phenolic acid found in pineapples, oak bark, green tea, grapes, strawberries, bananas, and lemons, has well-established antioxidant, anti-inflammatory, antimicrobial, and detoxifying qualities that can further be another important natural resource (Wang et al. 2007; Zulfakar et al. 2007).

7.6 Chitin-Glucan Co-polymer

Chitin-glucan is a kind of co-polymer found predominantly in the cell walls of life forms, particularly *Aspergillus niger*. This compound is gathering thought for its probably use in skincare things in view of its important immersing properties, which can help with directing signs of developing (Ng 2004; Angeli et al. 2009). Chitin-glucan's ability to retain moisture makes it a promising fixing in foe of developing and soaking subtleties, adding to better, more lively looking skin. Despite chitin-glucan, night treatments much of the time incorporate Lentinula edodes, for the most part known as shiitake mushrooms, which contain lentinan (Sandewicz et al. 2003; Athalye et al. 2009). Lentinan is known for its worthwhile properties in skin wellbeing the board, adding another layer of feasibility to these things. Ellagic destructive, a polyphenol present in various life forms, offers an extent of skincare benefits, including cell support, anticancer, and skin-lighting up influences. Its phone support properties help with protecting the skin from oxidative tension, while its actual limit in skin lighting up makes it a pursued fixing in things highlighted achieving an an even skin tone (Menoli et al. 2001; Keypour et al. 2010).

7.7 Gallic Acid from Fungi

Gallic acid (naturally occurring phenolic compound) is another compound with interesting properties suitable for skincare products. Known for its disease anticipation specialist, quieting, antimicrobial, and detoxifying properties, gallic destructive adds to the overall prosperity and presence of the skin. Fungi, including specific types of mushrooms, are referred to deliver gallic corrosive as an optional metabolite (Angeli et al. 2009; Bitencourt et al. 2018). Due to its potent antioxidant activity, which enables it to scavenge free radicals and shield cells from oxidative stress, this compound has been the subject of extensive research. This property is especially useful for preventing cellular damage, which can cause long-term illnesses and aging of skin that particularly interests modern cosmetic industries (Zulfakar et al. 2007; Vysotskaya et al. 2009). Its broad extent of effects makes it an adaptable fixing in various subtleties. Tyrosinase inhibitors are remarkable for their ability to back off the skin, making them significant in things expected to additionally foster appearance. Eliminates from Ganoderma species, for example, have shown tyrosinase inhibitory properties (Yang et al. 2009). These concentrates are at this point being utilized in facial covering drugs open accessible, highlighting their reasonability in propelling a more splendid, even more even composition. As a rule, the solidification of these infectious decided strengthens in skincare things displays the capacity of ordinary trimmings in further developing skin prosperity and appearance. From immersing and threatening to developing benefits to cell support and shading further creating influences, these trimmings offer an extensive method for managing skincare (Campana et al. 2004; Angeli et al. 2009) (Fig. 7.1).

7.8 Fungal Cosmetics Patents

There are several US patents on fungal cosmetics. A few notable initial patents are listed below:

(a) **US Patent 6,843,995**: The invention is a cosmetic gel preparation featuring an active complex procured from aqueous truffle extracts for the formation of a gel. Such invention, along with its cosmetic values, also has several medical aspects such as enhancing immune system stimulation and skin regeneration, promoting a healthier and more balanced skin integrity.
(b) **US Patent 6,645,502 B2**: A skin rejuvenating mushroom-based cream that is

Fig. 7.1 Cosmetics containing mushrooms or their extracts. (1) Aquamella Cream (*Tremella fuciformis*). (2) Origins Plantidote Mega-Mushroom Treatment. (3) Vitamega shampoo (*Agaricus subrufescens*). (4) Alqvimia Eternal Youth (*Schizophyllum commune*). (5) Grape Seed Lift Eye Mask (*Tremella polysaccharide*). (6) Aveeno anti-aging cream (*Ganoderma lucidum* and *Lentinula edodes*). (7) Skin Hydrating Gel (*Tremella polysaccharide*). (8) La Prairie Advanced skin rejuvenating cream (*Tremella fuciformis*). (9) Sulwhasoo Hydroaidskin rejuvenating cream (*Schizophyllum commune*). (10) Yves Saint Laurent Temps Majeur Elixir De Nui skin repairing and soothing cream (*Ganoderma lucidum*). (11) Murad's Sleep Reform Rapid Skin Rejuvenating Serum (*Schizophyllan*). (12) Kose Sekkiseiskin hydrating and brightening gel (*Cordyceps sinensis*). These are some of the well-known fungal cosmetic products available in the worldwide market. (Hyde et al. 2010)

having pigmented anhydrous composition. This product helps in reducing the shiny appearance of oily skin and also improves the overall integrity and healthy appearance of the skin surface (Angeli et al. 2009; Attard et al. 2015).

7.9 Advancements in Fungal Isolation, Extraction, and Formulation Techniques

One of the most exciting developments in fungal cosmetics are developments in extraction and formulation technologies (Barros et al. 2007a). In the past, bioactive compounds from fungi were frequently extracted using ineffective and inconsistent methods. Nonetheless, late developments are essentially working on the proficiency and security of these mixtures, making them more powerful for restorative applications. Procedures like high-level hydrolysis, enzymatic cycles, and supercritical liquid extraction are presently empowering more exact and powerful confinement of important parasitic mixtures (Bakkali et al. 2008; Bitencourt et al. 2018). These upgraded techniques vow to build the power and dependability of contagious-based fixings, prompting more predictable and compelling skincare items. With these progressions in extraction advancements, there is likewise a more prominent potential for making imaginative corrective details. It is possible to develop new products by incorporating fungal extracts into a wide variety of cosmetic bases, such as creams, serums, masks, and gels. This adaptability considers the making of multi-practical items that can address different skin issues, including against maturing, hydration, skin inflammation treatment, and skin fix. As contagious concentrates become more flexible in various definitions, the extent of parasitic beauty care products is probably going to expand, contacting a more extensive crowd and meeting a different exhibit of skincare needs (Ward and Singh 2005).

7.10 Exploration of Diverse Fungal Species

The future improvement of cosmetic products will be highly influenced by the findings of a more useful fungal species in this area. Generally, organisms utilized in beauty care/cosmetic products have been restricted to a few deep-rooted species, similar to specific sorts of mushrooms (Synytsya et al. 2009; Tian et al. 2008). On the other hand, ongoing research is demonstrating the potential of lesser-known fungi that possess distinct properties. These new-found contagious species contain a scope of bio-active mixtures that could offer inventive advantages for skin well-being. For example, certain growths produce compounds with strong cell reinforcement, mitigating, and antimicrobial properties, which can be very profitable in skincare. The investigation of these less natural parasites might bring about the distinguishing proof of new dynamic fixings with explicit benefits, for example, further developed skin boundary fix, upgraded collagen creation, or designated treatment for skin conditions. The cosmetics industry can develop more specialized and effective products that address a wider range of skin issues if it embraces diversity (Campana et al. 2004; Angeli et al. 2009).

7.11 Personalized Skincare Solutions

The eventual fate of cosmetic beauty care products is supposed to be impacted by customized skincare, an arising pattern that is acquiring importance. Skincare products can now be tailored to specific skin types and concerns as knowledge about fungal ingredients grows. Customized skincare involves evaluating a singular's skin profile and picking items containing fixings that best location their specific prerequisites. In the domain of contagious cosmetic products, this includes creating custom-made definitions utilizing the particular properties of various parasitic concentrates (Cheng et al. 1999; Chien et al. 2008; Corey et al. 2009). For instance, an item could be planned with a specific sort of parasitic concentrate eminent for its mitigating properties to take care of people with delicate skin, or with an alternate concentrate known for its saturating impacts for those with dry skin. This customized approach lifts the adequacy of skincare items and offers shoppers arrangements that are more lined up with their singular requirements (Angeli et al. 2009; Attard et al. 2015).

7.12 Sustainability and Environmental Impact

Sustainability is becoming increasingly important in the future of fungal cosmetics. As consumers lean toward eco-friendly and ethical products, the sustainable nature of fungal-based ingredients offers a significant advantage(Dubost et al. 2006, 2007). Fungi can be cultivated with minimal environmental impact compared to traditional plant sources. They require less water, fewer pesticides, and can be grown in controlled settings, which helps to reduce the overall carbon footprint of production. This aligns with the broader industry trend toward greener practices. By incorporating fungal ingredients, cosmetic brands can appeal to environmentally conscious consumers and demonstrate their commitment to sustainability. Additionally, using fungal by-products and waste materials in cosmetics supports a circular economy by promoting resource reuse and minimizing waste (Angeli et al. 2009; Bitencourt et al. 2018).

7.13 Scientific Studies and Validation

The future success of fungal cosmetics hinges on demonstrating their benefits through scientific validation. To gain consumer trust, it is essential to prove the safety and effectiveness of fungal extracts with rigorous clinical studies. Such research will provide the evidence needed to support the claims of fungal-based ingredients and differentiate them from other market offerings (Wasser 2017; Yabuta 1924). Additionally, scientific validation is crucial for securing regulatory approvals and meeting industry standards. As fungal-based ingredients become more prevalent, adhering to safety and effectiveness criteria will be vital for their broader acceptance. Establishing strong clinical evidence and obtaining regulatory endorsements will enhance the credibility of fungal cosmetics, encouraging both manufacturers and consumers to adopt these innovative products (Ey et al. 2007; Fang et al. 2008; Angeli et al. 2009).

7.14 Innovation and Efficiency of the Product

Like all upcoming fields, innovation has to be efficient and effective in fungal-based cosmetic industries. Scientists and industrialists need to hand and hand for better communication and product development with appropriate market research and product development understanding of the needs of the masses. Continuous advancement of fungal cosmetics will be driven by innovation (Ferreira et al. 2013; Wasser 2017). By blending fungal extracts with peptides, vitamins, and plant extracts, sophisticated formulations with enhanced benefits can be developed. For instance, combining fungal extracts with hyaluronic acid could increase hydration levels, while integrating them with peptides might enhance anti-aging properties (Angeli et al. 2009; Bitencourt et al. 2018). Consumers seeking skincare solutions with multiple benefits are likely to be drawn to such inventive formulations. Furthermore, introducing fungal extracts into new product types such as emulsions, micellar waters, and beauty supplements will expand the range of choices for consumers. This flexibility in product development will allow fungal cosmetics to appeal to a wider audience and address various skincare needs (Sandewicz et al. 2003; Barros et al. 2007a). Fungal extracts have gained recognition for their ability to regenerate and heal, and this will continue to be a major focus in the development of fungal-based cosmetics. Fungi produce a variety of bioactive compounds that aid in skin repair, reduce inflammation, and promote cellular regeneration (Sandewicz et al. 2003; Synytsya et al. 2009). These qualities make fungal extracts highly valuable for creating products that target skin health and tackle specific issues like acne scars, sun damage, or aging signs. As interest in regenerative skincare increases, there will be a movement to produce products that utilize these beneficial properties for precise skin repair and rejuvenation. Ongoing exploration of the regenerative capabilities of fungi is likely to uncover new approaches for harnessing their potential, resulting in a wider array of products aimed at enhancing skin health and appearance. This expansion will cater to a range of skincare needs, encompassing everything from repairing damage to promoting overall skin vitality (Campana et al. 2004; Barros et al. 2007a).

7.15 Future Scope and Conclusion

Cosmetics derived from fungi have a lot of promise, but in order to fully take advantage of their benefits, there are challenges that need to be overcome. Maintainable obtaining and development of parasites remains as a significant obstacle (Firenzuoli et al. 2008; Wasser 2017). Numerous organisms used in beauty care products are assembled from normal territories, prompting likely issues of overharvesting and damage to the climate. To ensure a consistent and sustainable supply of these valuable ingredients, it is essential to develop sustainable cultivation methods like fungal fermentation. Normalizing parasitic concentrates represents another test. Fungi's bioactive compounds can vary a lot depending on the species, growth conditions, and extraction methods (Angeli et al. 2009; Bitencourt et al. 2018). This fluctuation can affect the adequacy and security of the last superficial items. Subsequently, laying out normalized conventions for the extraction and plan of parasitic fixings is pivotal for keeping up with item quality and customer certainty. In spite of these difficulties, what's to come possibilities of parasitic-based beauty care products seem hopeful. Continuous investigation of the bioactive mixtures created by parasites is probably going to uncover new fixings offering extraordinary advantages for the skin (Barros et al. 2007a; Athalye et al. 2009). Furthermore, headways in biotechnology and maturation cycles could bring about more reasonable and proficient creation strategies, making contagious-based beauty care products more open to a more extensive crowd. The pattern toward normal and supportable magnificence items likewise forecasts well for the extension of contagious-based beauty care products. Purchasers are progressively looking for items that are successful while additionally being harmless to the ecosystem and morally obtained. Fungi are well-suited to meet this demand due to their potential for novel skincare applications and extensive history in traditional medicine.

So, there is tremendous potential and a brilliant future ahead for contagious cosmetic products. The progression of extraction and definition advancements, investigation into different contagious species, and the rise of customized skincare arrangements all add to the improvement of this groundbreaking area. The focus on maintainability, logical/scientific approval, and regenerative properties adds to the engaging quality of these fungal-based beauty care products. As the business advances, such products are ready to assume a critical part, giving proficient, harmless to the ecosystem, and customized skincare choices to take special care of a diverse and discerning consumer base in the near future.

Acknowledgments We duly acknowledge Honourable Chancellor, Adamas University, Kolkata, India for providing us the infrastructure and facilities. This research work is financially supported by Seed fund for university-sponsored research project (Ref No. AU/REG/2019-20/12/008), under Adamas University, India.

Conflict of Interest The authors report no conflict of interest. The authors alone are responsible for the content and writing of the paper.

References

Angeli JPF, Ribeiro LR, Camelini CM, de Mendonça MM, Mantovani MS (2009) Evaluation of the antigenotoxicity of polysaccharides and β-glucans from *Agaricus blazei*, a model study with the single cell gel electrophoresis/Hep G2 assay. J Food Compos Anal 22:699–703

Athalye SK, Garcia RA, Wen Z (2009) Use of biodiesel-derived crude glycerol for producing eicosapentaenoic acid (EPA) by the fungus *Pythium irregulare*. J Agric Food Chem 57:2739–2744

Attard TM, McElroy CR, Rezende CA, Polikarpov I, Clark JH, Hunt AJ (2015) Sugarcane waste as a valuable source of lipophilic molecules. Ind Crops Prod 76: 95–103

Bakkali F, Averbeck S, Averbeck D, Idaomar M (2008) Biological effects of essential oils–a review. Food Chem Toxicol 46(2):446–475

Barros L, Ferreira M-J, Queirós B, Ferreira ICFR, Baptista P (2007a) Total phenols, ascorbic acid, β-carotene and lycopene in Portuguese wild edible mushrooms and their antioxidant activities. Food Chem 103:413–419

Barros L, Baptista P, Estevinho LM, Ferreira ICFR (2007b) Effect of fruiting body maturity stage on chemical composition and antimicrobial activity of *Lactarius* sp. mushrooms. J Agric Food Chem 55: 8766–8771

Bitencourt RG, Ferreira NJ, Oliveira AL, Cabral FA, Meirelles AJ (2018) High pressure phase equilibrium of the crude green coffee oil–CO2–ethanol system and the oil bioactive compounds. J Supercrit Fluids 133: 49–57

Brunner G (2005) Supercritical fluids: technology and application to food processing. J Food Eng 67(1–2): 21–33

Bubalo MC, Vidović S, Redovniković IR, Jokić S (2018) New perspective in extraction of plant biologically active compounds by green solvents. Food Bioprod Process 109:52–73

Burčová Z, Kreps F, Greifová M, Jablonský M, Ház A, Schmidt Š, Šurina I (2018) Antibacterial and antifungal activity of phytosterols and methyl dehydroabietate of Norway spruce bark extracts. J Biotechnol 282:18–24

Campalani C, Chioggia F, Amadio E, Gallo M, Rizzolio F, Selva M, Perosa A (2020) Supercritical CO2 extraction of natural antibacterials from low value weeds and agro-waste. J CO_2 Util 40:101198

Campana F, Zervoudis S, Perdereau B, Gez E, Fourquet A, Badiu C, Tsakiris G, Koulaloglou S (2004) Topical superoxide dismutase reduces post-irradiation breast cancer fibrosis. J Cell Mol Med 8:109–116

Cheng MH, Walker TH, Hulbert GJ, Raman DR (1999) Fungal production of eicosapentaenoic and arachidonic acids from industrial waste streams and crude soybean oil. BioresourTechnol 67:101–110

Chhouk K, Kanda H, Goto M (2018) Efficacy of supercritical carbon dioxide integrated hydrothermal extraction of Khmer medicinal plants with potential pharmaceutical activity. J Environ Chem Eng 6(2): 2944–2956

Chien C-C, Tsai M-L, Chen C-C, Chang S-J, Tseng C-H (2008) Effects on tyrosinase activity by the extracts of *Ganoderma lucidum* and related mushrooms. Mycopathologia 166:117–120

Chouaibi M, Rezig L, Hamdi S, Ferrari G (2019) Chemical characteristics and compositions of red pepper seed oils extracted by different methods. Ind Crop Prod 128:363–370

Ciarlini JJ, Marangoni A, Bolzan A (2017) Selectivity of supercritical CO2 extraction and atmospheric pressure techniques for the major volatile compounds of Eugenia involucrata leaves from southern Brazil. Food Bioprod Process 106:29–34

Corey ME, Beelman RB, Seetharman K (2009) Potential for nutritional enrichment of whole-wheat bread with portabella mushroom powder (*Agaricus bisporus* (J. Lge) Imbach, Agaricomycetideae). Int J Med Mushrooms 11:157–166

Dubost NJ, Beelman RB, Peterson D, Royse DJ (2006) Identification and quantification of ergothioneine in cultivated mushrooms by liquid chromatography-mass spectroscopy. Int J Med Mushrooms 8:215–222

Dubost NJ, Beelman RB, Royse DJ (2007) Influence of selected cultural factors and postharvest storage on ergothioneine content of common button mushroom *Agaricus bisporus* (J. Lge) Imbach (Agaricomycetideae). Int J Med Mushrooms 9:163–176

Ey J, Schömig E, Taubert D (2007) Dietary sources and antioxidant effects of ergothioneine. J Agric Food Chem 55:6466–6474

Fang S-H, Rao YK, Tzeng Y-M (2008) Anti-oxidant and inflammatory mediator's growth inhibitory effects of compounds isolated from *Phyllanthus urinaria*. J Ethnopharmacol 116:333–340

Ferreira JA, Lennartsson PR, Edebo L, Taherzadeh MJ (2013) Zygomycetes-based biorefinery: Presentstatus and future prospects. Bioresour Technol 135:523–532

Firenzuoli F, Gori L, Lombardo G (2008) The medicinal mushroom *Agaricus blazei* Murrill: review of literature and pharmaco-toxicological problems. Evid Based Complement Altern Med 5:3–15

Gao J-M, Zhang A-L, Chen H, Liu J-K (2004) Molecular species of ceramides from the ascomycete truffle *Tuber indicum*. Chem Phys Lipids 131:205–213

Gautier S, Xhauflaire-Uhoda E, Gonry P, Piérard GE (2008) Chitin-glucan, a natural cell scaffold for skin moisturization and rejuvenation. Int J Cosmet Sci 30: 459–469

Golz-berner K, Zastrow L (2005) Cosmetic preparation containing extracts of Tuberaceae. United States Patent 6843995

Gutiérrez A, Del Río JC, Martínez-Íñigo MJ, Martínez MJ, Martínez ÁT (2002) Production of new unsaturated lipids during wood decay by ligninolytic basidiomycetes. Appl Environ Microbiol 68:1344–1350

Hyde KD, Bahkali AH, Moslem MA (2010) Fungi – an unusual source for cosmetics. Fungal Divers 43:1–9

Jin J-M, Lee J, Lee Y-W (2010) Characterization of carotenoid biosynthetic genes in the ascomycete *Gibberellazeae*. FEMS Microbiol Lett 302:197–202

Kakumyan P, Matsui K (2009) Characterization of volatile compounds in *Astraeus* spp. Biosci Biotechnol Biochem 73:2742–2745

Kalač P (2009) Chemical composition and nutritional value of European species of wild growing mushrooms: a review. Food Chem 113:9–16

Keypour S, Rafati H, Riahi H, Mirzajani F, Moradali MF (2010) Qualitative analysis of ganoderic acids in *Ganoderma lucidum* from Iran and China by RP-HPLC and electrospray ionisation-mass spectrometry (ESI-MS). Food Chem 119:1704–1708

Kim HH, Cho S, Lee S, Kim KH, Cho KH, Eun HC, Chung JH (2006) Photoprotective and anti-skin-aging effects of eicosapentaenoic acid in human skin in vivo. J Lipid Res 47:921–930

Krištofíková L, Rosenberg M, Vlnová A, Šajbidor J, Čertík M (1991) Selection of *Rhizopus* strains for l (+)-lactic acid and γ-linolenic acid production. Folia Microbiol 36:451–455

Kumari M, Survase SA, Singhal RS (2008) Production of schizophyllan using *Schizophyllum commune* NRCM. Bioresour Technol 99:1036–1043

Lemann P (2007) Mushrooms in cosmetics. In: SpecialChem cosmetics and personal care, innovations and solutions

Lourenço A, Lobo AM, Rodríguez B, Jimeno M-L (1996) Ceramides from the fungus *Phellinus pini*. Phytochemistry 43:617–620

Luiz RC, Jordao BQ, da Eira AF, Ribeiro LR, Mantovani MS (2003) Mechanism of anticlastogenicity of *Agaricus blazei* Murill mushroom organic extracts in wild type CHO (K-1) and repair deficient (xrs5) cells by chromosome aberration and sister chromatid exchange assays. Mutat Res Fundam Mol Mech Mutagen 528:75–79

Machado MP, Rodrigues E, Terezan AP, Ribeiro UR, Mantovani MS (2005) Cytotoxicity, genotoxicity and antimutagenicity of hexane extracts of *Agaricus blazei* determined in vitro by the comet assay and CHO/HGPRT gene mutation assay. Toxicol In Vitro 19:533–539

Mantzouridou F, Tsimidou MZ (2008) Lycopene formation in *Blakesleatrispora*. Chemical aspects of a bioprocess. Trends Food Sci Technol 19:363–371

Margolin AL, Klibanov AM (1987) Peptide-synthesis catalyzed by lipases in anhydrous organic-solvents. J Am Chem Soc 109:3802–3804

Matuo R, Oliveira RJ, Silva AF, Mantovani MS, Ribeiro LR (2007) Anticlastogenic activity of aqueous extract of *Agaricus blazei* in drug-metabolizing cells (HTCs) during cell cycle. Toxicol Mech Meth 17:147–152

Menoli R, Mantovani MS, Ribeiro LR, Speit G, Jordao BQ (2001) Antimutagenic effects of the mushroom *Agaricus blazei* Murrill extracts on V79 cells. Mutat Res Genet Toxicol Environ Mutagen 496:5–13

Mizuno T, Inagaki R, Kanao T, Hagiwara T, Nakamura T, Ito H, Shimura K, Sumiya T, Asakura A (1990) Studies on the host-mediated antitumor polysaccharides.14. Antitumor-activity and some properties of water-insoluble hetero-glycans from himematsutake, the fruiting body of *Agaricus-blazei* Murill. Agric Biol Chem 54:2897–2905

Mizuno M, Minato K, Ito H, Kawade M, Terai H, Tsuchida H (1999) Anti-tumor polysaccharide from the mycelium of liquid-cultured Agaricus blazei mill. Biochem Mol Biol Int 47:707–714

Mohorčič M, Friedrich J, Renimel I, André P, Mandin D, Chaumont J-P (2007) Production of melanin bleaching enzyme of fungal origin and its application in cosmetics. Biotechnol Bioprocess Eng 12(3):200–206

Ng TB (2004) Peptides and proteins from fungi. Peptides 25:1055–1073

Nieto IJ, Carolina Chegwin A (2008) Triterpenoids and fatty acids identified in the edible mushroom *Pleurotus sajor-cajú*. J Chil Chem Soc 53:1515–1517

Obrien DJ, Stinson EE, Wessinger EW, Somkuti GA (1993) Production of eicosapentaenoic acid from filamentous fungi utilizing lactose as a primary carbon source. United States Patent 5246842

Park S-S, Hwang S-M (1999) Purification and characterization of iron-containing superoxide dismutase from *Lentinus edodes*. J MicrobiolBiotechnol 9:854–860

Passi S, De Pità O, Puddu P, Littarru GP (2002) Lipophilic antioxidants in human sebum and aging. Free Radic Res 36:471–477

Pons A, Timmerman P, Leroy Y, Zanetta J-P (2002) Gas-chromatography/mass-spectrometry analysis of human skin constituents as heptafluorobutyrate derivatives with special reference to long-chain bases. J Lipid Res 43:794–804

Sadruddin S, Arora R (2009) Resveratrol: biologic and therapeutic implications. J Cardiometab Syndr 4:102–106

Sandewicz IM, Russ JG, Zhu VX (2003) Anhydrous cosmetic compositions containing mushroom extract United States Patent 6645502

Sanodiya BS, Thakur GS, Baghel RK, Prasad GBKS, Bisen PS (2009) Ganoderma lucidum: a potent pharmacological macrofungus. Curr Pharm Biotechnol 10: 717–742

Shimada Y, Morita T, Sugiyama K (2002) Effects of *Lentinus edodes* on fatty acid and molecular species profiles of phosphatidylcholine in rats fed different levels of corn oilBioscience. BiotechnolBiochem 66: 1759–1763

Synytsya A, Míčková K, Synytsya A, Jablonský I, Spěváček J, Erban V, Kováříková E, Čopíková J (2009) Glucans from fruit bodies of cultivated mushrooms *Pleurotus ostreatus* and *Pleurotus eryngii*: structure and potential prebiotic activity. CarbohydrPolym 76:548–556

Tian T, Sun Q, Shen J, Zhang T, Gao P, Sun Q (2008) Microbial transformation of polydatin and emodin-8-β-d-glucoside of *Polygonum cuspidatum* Sieb. et Zucc into resveratrol and emodin respectively by *Rhizopus microsporus*. World J Microbiol Biotechnol 24: 861–866

Ventura J, Belmares R, Aguilera-Carbo A, Gutiérrez-Sanchez G, Rodríguez-Herrera R, Aguilar CN (2008) Fungal biodegradation of tannins from creosote bush (*Larrea tridentata*) and tar bush (*Fluorensiacernua*) for gallic and ellagic acid production. Food Technol Biotechnol 46:213–217

Vozenin-Brotons MC, Sivan V, Gault N, Renard C, Geffrotin C, Delanian S, Lefaix JL, Martin M (2001) Antifibrotic action of Cu/Zn SOD is mediated by TGF-beta1 repression and phenotypic reversion of myofibroblasts. Free Radic Biolological Med 30:30–42

Vysotskaya MR, Maslova GV, Petrova VA et al (2009) Electrochemical recovery of chitin-glucan complex from *Pleurotus ostreatus* basidial fungus and properties of the product. Russ J Appl Chem 82: 1390–1395

Wang H, Liu L, Guo Y-X, Dong Y-S, Zhang D-J, Xiu Z-L (2007) Biotransformation of piceid in *Polygonum cuspidatum* to resveratrol by *aspergillus oryzae*. Appl Microbiol Biotechnol 75:763–768

Ward OP, Singh A (2005) Omega-3/6 fatty acids: alternative sources of production. Process Biochem 40:3627–3652

Wasser SP (2017) Medicinal mushrooms in human clinical studies. Part I. Anticancer, oncoimmunological, and immunomodulatory activities: a review. Int J Med Mushrooms 19(4):279

Yabuta T (1924) The constitution of kojic acid, a gamma-pyrone derivative formed by *aspergillus oryzae* from carbohydrates. J Chem Soc 125:575–587

Yang FQ, Feng K, Zhao J, Li SP (2009) Analysis of sterols and fatty acids in natural and cultured Cordyceps by one-step derivatization followed with gas chromatography-mass spectrometry. J Pharm Biomed Anal 49:1172–1178

Zhang ZY, Jin B, Kelly JM (2007) Production of lactic acid from renewable materials by *Rhizopus* fungi. Biochem Eng J 35:251–263

Zulfakar MH, Edwards M, Heard CM (2007) Is there a role for topically delivered eicosapentaenoic acid in the treatment of psoriasis? Eur J Dermatol 17:284–291

Anti-collagenase and Anti-elastase Activities of Personal Care Products Made from Fungal Extracts

8

Bandita Dutta, Sedevino Ltu, Dibyajit Lahiri, Moupriya Nag, Debasmita Bhattacharya, Tanmay Sarkar, Md Osim Aquatar, and Rina Rani Ray

Abstract

In modern civilization, skin ageing has become one of the most prominent dermatological issues. In human existence, it becomes a complicated and inexorable process. The extracellular matrix is comprised of proteoglycans, including elastin, collagen, metalloproteinase (MMP) and fibronectin. Collagen and elastin are the most prevalent compounds present in the ECM matrix, serving the role of adherence factor, connective tissues, cellular conditions for growth and morphogenesis. The change of morphology, cell migration and tissue desorption has become significant in the remodulation of tissue during the degeneration of proteins, including cell growth, embryonic development and disease activity. Elastin and collagenase belong to the protease family and they control elastin and collagen breakdown along with some other ECM protein breakdown whilst enduring skin ageing. Photoaging due to UV radiation and solar exposure is another important aspect of skin ageing. It induces reactive oxygen species (ROS) formation that contributes to the reduction of skin elasticity. Therefore, demand for anti-collagenase and anti-elastase compounds has risen higher in order to prevent naïve skin ageing.

Keywords

Collagenase · Anti-aging · Fungal compounds · Natural therapeutics

B. Dutta · S. Ltu · R. R. Ray (✉)
Department of Biotechnology, Maulana Abul Kalam Azad University of Technology, Kolkata, West Bengal, India

D. Lahiri (✉) · M. Nag
Department of Biotechnology, Institute of Engineering and Management, Kolkata, New Town Sector, University of Engineering and Management, Kolkata, West Bengal, India
e-mail: dibyajit.lahiri@uem.edu.in

D. Bhattacharya
Department of Basic Science and Humanities, Institute of Engineering and Management, Kolkata, Salt Lake Sector, University of Engineering and Management, Kolkata, West Bengal, India

T. Sarkar
Department of Food Processing Technology, Malda Polytechnic, West Bengal State Council of Technical Education, Government of West Bengal, Malda, West Bengal, India

M. O. Aquatar
Department of Environmental Science & Civil Engineering, Jamia Millia Islamia (A Central University, NIRF-3rd), New Delhi, Delhi, India

8.1 Introduction

Skin aging is a natural phenomenon which occurs as a result of the loss of skin elasticity. However, photoaging is another aspect of skin ageing due to the exposure to solar radiation. Physical orientation to the skin due to the UV exposure instigates

S. Ghosh, H. N. Onyeaka (eds.), *Fungal-Based Cosmetics: Formulation and Usage*, The Mycota 17,
https://doi.org/10.1007/978-3-032-09291-5_8

reactive oxygen species (ROS), lipid peroxides as well as several other enzymes formation that potentially induce the alteration of connective tissues. ROS helps to reduce skin elasticity whereas lipid peroxide metabolizes to the secondary products that impair extracellular matrix (ECM). They both are involved in diseases like cancer, diabetes and arthritis. ROS is involved in the biological system including metabolic pathways and gives rise to several reactive species including nitric oxide (NO) and superoxide (O_2^-) (Benaiges et al. 1998).

The overproduction of ROS may induce the transition of some redox-active metal ions like copper (II) and iron (II). These two metal ions produce critical oxidative stress and enable tissue damage through the destruction of cellular DNA, carbohydrate, lipid, and protein content. (Kaur et al. 2006). O_2 has a small plasma half-life while superoxide dismutase (SOD) usually breaks down O_2^- into H_2O_2 in order to generate novel SOD mimetics (Arct and Pytkowska 2008). Collagen is present in around 80% of the total skin dry weight providing the skin tensile strength. Elastin fibre meshwork contains 2–4% ECM and glycoaminoglycans (GAGs), which provide skin elasticity and skin hydration (Jenkins 2002). Fibroblasts develop collagen, elastin and GAGs, which are significantly associated with photoaging and enable apparent changes in skin including pigmentation, skin tenderness and wrinkles. ROS is able to induce the development of proteinases that modify matrix metalloproteinases (MMPs) as well as serine proteases of the ECM matrix (Y.-J. Kim et al. 2004). Collagenases and gelatinases are the examples of zinc-containing transmembrane endopeptidases MNPs. Metalloproteinases like collagenase-2 (MMP-8) may break various molecules present within the cell, such as aggrecan, laminin, fibronectin, elastin and gelatine, in addition to collagen. The enzyme collagenase is able to break down the X-gly link in collagen including synthetic peptides with the structure -Pro-X-Gly-Pro, wherein X represents any respective amino acid with a blocked amino end terminus. ECM is also degraded by the collagenase of bacterium *Clostridium histolyticum* (ChC). In both physiological and in vitro circumstances, this bacteria-mediated collagenase hydrolyses triple-helical collagen utilizing synthetic peptides as substrates (Raffetto and Khalil 2008; Van Wart and Steinbrink 1981).

Serine proteases, one type of elastase, that implicate another proteolytic mechanism in the destruction of the ECM. Elastase is a protease of the family chymotrypsin and is principally accountable for elastin breakdown, a protein present in the ECM. Elastin protein is essential for contributing elastic property to lungs, arteries, skin and ligaments due to its distinctive elastic recoiling capabilities. Elastases have the capacity to cleave elastin including a diverse substrate portfolio that enables in cleaving fibronectin, collagen and other ECM proteins. In normal circumstances, elastase activity is required after injury to cleave non-native proteins in the ECM amid phagocytosis by neutrophils, allowing tissue healing. With the purpose of overseeing the ageing process, elastase and collagenase inhibitors need to be discovered in order to prevent skin ageing and loss of skin elasticity.

8.2 Personal Care Products

Personal care products or cosmetics encompass products that are administered to the body's exterior parts, which include the skin, hair, nails, lips, external genital and anal regions, in addition to the teeth and mucous lining of the mouth, for the main purpose of cleansing, protection, healing and maintenance. The largest organ in the human body is the skin and it is exposed to varied external elements like environment pollutions, UV radiation and lifestyle choices which contributes to the decline of collagen and elastin. The presence of these skin's matrix viz collagen and elastin significantly contributes in the development of unwanted skin problems like premature ageing, pigmentations, wrinkles, dull and unhealthy skin. Collagen, which is the protein that gives strength and structure to the skin dermis and elastin which provides elasticity and suppleness to the skin works in collaboration to maintain texture and fine appearance of overall skin health.

Decrease in the amount of elastin and collagen can potentially bring about signs of ageing. Skin ageing has two distinct categories, intrinsic ageing and extrinsic ageing, these factors establish to describe the phenomenon of skin ageing. Ageing of skin involves elemental factors including predisposition as well as external elements, such as UV radiation, pollution and choice of lifestyle. All these factors collectively contribute to breakdown of collagen and elastin. As a consequence the dermis gradually becomes thinner over time leading to loss of firmness and wrinkles that are typically associated with ageing. The explanation behind the diminished level of collagen and elastin is due to the inhibiting activity of the enzyme collagenase and elastase respectively. These enzymes degrade the vital function of elastin and collagen whose sole purpose is to keep the skin a youthful and a firm appearance. These enzymes slowly increase and become active with the increase in age (intrinsic) and enzymatically degraded collagen and elastin, hence it is mainly attributed to the overtime occurring gradual decline in skin elasticity. Photoaging (extrinsic), which is sometimes referred to as the reverse phenomenon, takes place, due to prolonged exposure to solar radiation. Reactive oxygen species (ROS) portray an instrumental part in regulating connective tissue architecture, which results in the skin developing noticeable changes after prolonged exposure to ultraviolet (UV) radiation.

In both dermatology and skin care, there is importance placed on collagenase activity or collagenase inhibition. This concept revolves around reducing the activity of collagen enzymes called matrix metalloproteinases (MMPs). Anti-collagenase agents help protect these proteins by inhibiting these enzymes. This reinforcement helps maintain a framework within the skin promoting elasticity, firmness and lessening the manifestation of ageing process such as fine lines and wrinkles. Integrating skincare products and therapeutic treatments with anti-collagenase and anti-elastase properties into dermatological routines is an effective approach to combat the ageing process from its root ensuring the long-lasting preservation of collagen and elastin, for a youthful and resilient skin integrity. Therefore, reduced dermal thickness, collagen density, and the production and replacement of crucial structural proteins all contribute to skin ageing (Halla et al. 2018).

8.3 Collagenase

The enzyme collagenase breaks the peptide bond present in collagen, a structural protein found abundantly in the ECM of human connective tissue. Collagen seems to be an important systemic protein found in skin tendon. It is present in the ECM of human connective tissue and is a fibrous protein that is abundant in humans. The connective tissue in the bones, cartilage, ligaments, tendons, and skin is where it is most commonly found (Wang 2021).

The fundamental structure of collagen is made up of three amino acids, namely, glycine (33%), proline, and hydroxyproline (22%). These three amino acids structurally form a triplex helix, made up of triple chains. Collagen is known to be a significant protein, abundantly generated in a human being. Each alpha chain roughly contains 1014 amino acids and possesses a molecular weight of about 100 kDa. A left-handed helix formed by these chains is secondary in structure and has three amino acids per turn. A hard structure (tertiary structure) is created when the chains are triple helixed around one another. The quaternary structure of collagen, or the super helix, is one example of this. Due to hydrogen bonds at intramolecular level, formed by glycine in the neighbouring chains, the structure of collagen is extremely durable. To create a tight and stable framework, the chains are bent around one another thereby forming triple helix (León-López et al. 2019).

The disintegration of triple-helical interstitial collagens is essential for the growth of embryos, organ morphological development, remodelling of tissues and restoration, and tissue regeneration. Diseases caused by abnormal collagenolysis include arthritis, malignancy, atherosclerosis, aneurysms, and fibrosis. Collagenases from the matrix metalloproteinase (MMP) family begin the process in vertebrates (Chung et al. 2004).

The types, amounts, and occasionally the molecular makeup of collagen expressed vary based on the type of tissue. Among all five common types of collagen, the collagen types I, II and III are known to be the most prevalent extracellular matrix (ECM) proteins. The primary enzyme involved in collagen breakdown is collagenase and is a part of an enzyme family, matrix metalloproteinase (MMP). The interactions between cells and the ECM are critical for maintaining structure and function of tissue. (Donahue et al. 2006) (Table 8.1).

Despite being widely dispersed throughout the body, collagen is subject to a natural process that will inevitably cause it to gradually deteriorate. UV radiation acting as a powerful catalyst in initiating ROS resulting in reduced collagen levels, may be a result of natural ageing. Other factors like hormonal changes and ultraviolet (UV) radiation may influence the skin and other body parts and initiate reactive oxygen species (ROS) production, which causes oxidative stress. This oxidative harm to the skin's lipids, DNA and proteins caused by ROS is a direct cause of skin ageing. The MAP-kinase pathway is another method that ROS might indirectly stimulate the creation of MMPs. Thereby leading to manifestations of common dermatological conditions, including premature ageing, UVB-induced erythema (sunburn), impairment of the skin barrier function, and elevated sebum secretion (Badalyan et al. 2022; Masaki 2010; Fisher et al. 2002; Sim et al. 2007).

The matrix metalloproteinases (MMPs), a family of proteinases, act on the extracellular matrix (ECM) at a neutral pH and have the capability to destroy any component of the extracellular matrix collectively (Gross and Lapiere 1962).

Collagenase, stromelysins, gelatinase, membrane-type MMPs (MT-MMPs), and others are among the five subgroups of MMPs, a category of zinc-dependent endopeptidases which cleave non-matrix proteins and one or several other components of the extracellular matrix extracellular proteinases, based on the substrate they metabolize (Pientaweeratch et al. 2016).

The collagenase enzyme, classified as a metalloproteinase, possesses the capacity to enzymatically degrade various molecules, including aggrecan, elastin, fibronectin, gelatin, laminin, and collagen. Consequently, interventions of collagenase inhibitory activity involving agents, namely anti-collagenase, capable of obstructing collagenase activity are hypothesized to yield advantageous outcomes, potentially contributing to the preservation of skin health by impeding the degradation of the dermal matrix and contributing to undesired wrinkles and premature skin ageing also known as photoaging (Ghimeray et al. 2015) (Table 8.2).

8.4 Elastase

Elastic fibres, which are made of the protein elastin, differ from collagenous fibres in size, pattern, and chemical structure. They lack the homogenous components found in collagen unit fibrils. In electron micrographs, they seem to have a uniform nucleus encircled by tiny fibrils, while at other instances they appear to be entirely made of compact amorphous material. It is unknown if there are two components or if both are different variations of the same material. The strands of fibre vary in diameter and frequently branched and reconnect to create large networks in free connective tissue under the light microscope. They give the tissue a yellow colour when present in large concentrations. Elastic ligaments have very coarse fibres that are placed in parallel bundles. Elastin is found in the walls of arteries as sheets or membranes punctured by perforations of varied sizes. Elastic fibres are particularly resistant to hot water, strong alkaline substances, and additionally to proteolytic enzyme trypsin digestion. They may be digested by elastase, an enzyme found in the pancreas. Chemical study reveals that elastin, like collagen, is high in glycine and proline, but it varies in its high valine

Table 8.1 Representation of common types of collagen belonging to different family

Collagen	Chain	Allocation
Type I	[α1(I)]2α2(I)	skin, teeth, bone tendon, vascular ligature, ligament and organs
Type II	[α1(II)]3	cartilages
Type III	[α1(III)]3	skin, blood vessels and muscle
Type IV	[α1(IV)]2α2 (IV) [α3(IV)]2α4 (IV) [α5(IV)]2α6 (IV)	basement membrane and basal lamina.
Type V	α1(V), α2(V), α3(V)	Rhabdomyosarcoma, the placenta, the skin, the chorioamnion, and pericellular/stromal in most connective tissue

Table 8.2 Types of MMPs and their mode of activities

Subgroup	Type	MMP	Collagen substrates
Collagenase	I	MMP-1(interstitial collagenase)	I, II, III, VII, X Gellatin
	II	MMP-8 (neutrophile collagenase)	I, II, III, VII, VIII, X Aggrecan, Gellatin
	III	MMP-13 (collagenase 3)	I, II, III, IV, IX, X Gelatin
Gelatinase	A	MMP-2 (gelatinase A)	Gelatin, Col I, II, III, IV, VII, X
	B	MMP-9(gelatinase B)	Gelatin, Col IV, V
Stromelysins	I	MMP-3 (Stromelysin-1)	Col II, IV, IX, X, XI, gelatin
	II	MMP-10(Stromelysin-2)	Col IV, laminin, fibronectin, elastin
	III	MMP-11(Stromelysin-3)	Col IV. fibronectin, laminin, aggrecan
Matrilysins	I	MMP-7 (Matrilysin-1)	Fibronectin, laminin, Col IV, gelatin
	II	MMP-26 (Matrilysin-2)	Fibrinogen, fibronectin, gelatin
MT-MMP	(MT1-MMP)	MMP-14	Gelatin, fibronectin, laminin
	(MT2-MMP)	MMP-15	
	(MT3-MMP)	MMP-16	
	(MT4-MMP)	MMP-17	Fibrinogen, fibrin
	(MT5-MMP)	MMP-24	Gelatin, fibronectin, laminin
	(MT6-MMP)	MMP-25	Gelatin
Others	(macrophage metalloelastase)	MMP-12	Elastin, fibronectin, Col IV
		MMP-19	Aggrecan, elastin, fibrillin, Col IV. gelatin
	Enamelysin	MMP-20	Aggrecan
	XMMP	MMP-21	
		MMP-23	Gelatin, casein, fibronectin
	CMMP	MMP-27	Unknown
	Epilysin	MMP-28	

concentration and the presence of an uncommon amino acid, desmosine. Elastic fibres, as the name indicates, are extremely distensible and, when ruptured rebound like rubber bands. Changes in this characteristic, as well as a decrease in their quantity, are considered to contribute to the loss of suppleness of the skin and blood-vessel walls in old age.

8.4.1 Elastase 1

A serine protease elastase (EC 3.4.21.36) is able to stimulate the breakdown of the carboxyl groups of the small hydrophobic amino acids including alanine, valine and glycine. The principle function of elastase is to cleave the elastin protein, which retains the elastic property of the connective tissue. Total of eight genes are reported to encode elastase enzymes, among which four come under the chymotrypsin category (Cotten 2020).

8.4.2 Elastase 2

Neutrophil elastase is a chymotryptic serine protease, monomeric, 218-amino acid (25 kDa), that is produced via promyelocytes and promonocytes during the initial stages of production of primary granules. During differentiation it is primarily formed as a proenzyme before being finally stored in its active state at azurophilic granules (Bouma and Thrasher 2011). Neutrophil elastase is able to break down the outer layer of the Gram-negative bacteria's cell wall such as *E. coli*. The pancreatic abundance of exocrine is used to indirectly measure the fecal elastase-1. Among the commercial two reagents, the monoclonal antibody is able to detect pancreatic elastase-1 while the other polyclonal antisera detects the whole elastase. Though both reagent-specific assays are able to measure total content of elastase, they are not able to detect enzyme activity. Both the reagents detect different epitopes as nonhuman elastase is specific towards polyclonal elastase reagent. Chronic pancreatitis (CP) patients receiving supplements with porcine enzyme are more vulnerable as they lack specificity of the polyclonal reagent against human elastase. The watery stool or overgrowth of the small bowel bacteria may lead to the false positive result for stool elastase.

8.5 Function of Collagenase and Elastase

The ECM consists of proteoglycans intricately intertwined with matrix metalloproteinases (MMPs), encompassing components like collagen, elastin and fibronectin. It exhibits a dynamic and intricate structural architecture. MMPs function collaboratively to enzymatically cleave distinct ECM constituents (Mott and Werb 2004; Oikarinen 2004).

It is noteworthy that the substrate specificity of MMPs is selective, with individual MMPs demonstrating varying affinities for particular ECM components. Most MMPs, if not all of them, are capable of cleaving ECM components such as elastin, laminin, aggrecan or fibronectin (Loffek et al. 2011).

Specifically, collagenases (MMP-1, MMP-8, MMP-13), MMP-2, and MT1-MMP are the principal enzymatic effectors held accountable for the deterioration of fibrillar collagen types, for example, collagen I, II and III (Page-McCaw et al. 2007).

Roughly three-quarters of the way down the N-terminal end of the collagen molecule, there exists a precise site where proteolytic cleavage occurs. Following this cleavage event, the resultant fragments undergo rapid denaturation, resulting in the formation of gelatin. Subsequently, gelatin can be subjected to enzymatic degradation by an extensive collection of proteases, including but not strictly confined to MMP-2 and MMP-9, both of which are classified as gelatinases. Additionally, other MMPs such as stromelysins, matrilysins, and macrophage metalloelastase demonstrate the capacity to engage in proteolysis of the collagen IV component found within the basement membrane (Visse and Nagase 2003).

In the realm of collagenase enzymes, interstitial collagenase (MMP-1), collagenase-3 (MMP-13) and neutrophil collagenase (MMP-8) represent distinct variants. Collagenase enzymatically severs X-glycoside bond of the collagen, as well as synthetic peptides featuring the -Pro-X-Gly-Pro- amino acid sequence, wherein X can represent a broad range of amino acids, with the caveat that the amino terminus must be sterically hindered (Van Wart and Steinbrink 1981).

These enzymes share the remarkable characteristic of simultaneously degrading the three alpha chains of type I, type II and type III collagens, leading to the formation of fragments approximately one-quarter and three-quarters the original size of collagen molecule. Subsequent to this first cleavage event, the triple helical structure of the collagen undergoes disruption, causing the two resultant collagen fragments to become unstable at physiological temperatures. This instability renders the polypeptide chains susceptible to further degradation by other proteases. Furthermore, it is noteworthy that these collagenase enzymes also exhibit the ability to cleave type X collagen and various other collagen types (Shingleton et al. 1996).

In both healthy and diseased connective tissue, collagen turnover is started by the enzymes known as collagenases.

The regulation of MMP functions occur on many scales. Growth factors and cytokines regulate the stimulation of gene expression in many ways. Since MMPs are so significant, they must be meticulously controlled at many points: (i) stimulus via cytokines and growth factors; (ii) proMMP's production and release; (iii) proenzymes activation; and (iv) suppression of activated enzymes.

Scleroprotein elastin is the key protein that aids the connecting tissue in maintaining the mechanical properties along with collagen. The prime protein which can hydrolyse this scleroprotein elastin are elastases, which break down the peptide bonds. Elastase belongs to a group of enzymes known as serine proteases. Being a glycoprotein it has a polypeptide chain of 218 amino acids. The acinar cells of the pancreas produce pancreatic elastase enzyme, which is resistant to breakdown by the intestinal enzymes including neutrophil elastase and pancreatic elastase 1 (Protease E) and 2. These three enzymes function as a synthetic substrate on succinyl-trialanine-p-nitroanilide as amidase and esterase. Proelastase produced by the mammalian pancreas is activated by trypsin to produce pancreatic elastase. Once proelastase is produced, it is maintained in the zymogen granules until it reaches the duodenum, where conversion of proelastase to elastase takes place. This conversion occurs by the tryptic fragmentation of a peptide bond, transforming the inactive state of the precursor molecule into its active form. Elastase is released by the neutrophils in order to kill the microorganisms and destroy the host tissue in the course of inflammation. Being a member of serine protease, the elastase has a unique attribute of getting strongly adhered towards the DNA and getting confined within Neutrophil Extracellular Traps (NETs). Elastases are known to be involved in the physiopathology of atherosclerosis, cystic fibrosis, pulmonary emphysema, inflammation and infections. They participate physiologically in actions including phagocytosis, neutrophil migration, and tissue remodelling (Brilan 2022).

8.6 Importance of Anti-collagenase and Anti-elastase

Two distinct categories—intrinsic ageing and extrinsic ageing—have been established to describe the phenomenon of skin ageing.

Intrinsic skin ageing: Natural ageing, also known as intrinsic skin ageing, is mainly attributed to the overtime occurring gradual decline in skin elasticity. When skin age, the amount of collagen declines slowly with each passing year which leads to weakening of the skin dermis layer and elasticity decline. The reduction in the synthesis of collagen resulting from diminished mechanical tension (stress or stretching) with ageing is manifested by a decrease in the transcriptional activity of genes encoding specific types of collagen (interstitial collagens). This phenomenon exerts an influence to enzymes participating in post-translational modification of the procollagen peptides, potentially leading to modification in both quality and quantity of collagen yield. Concurrently, it

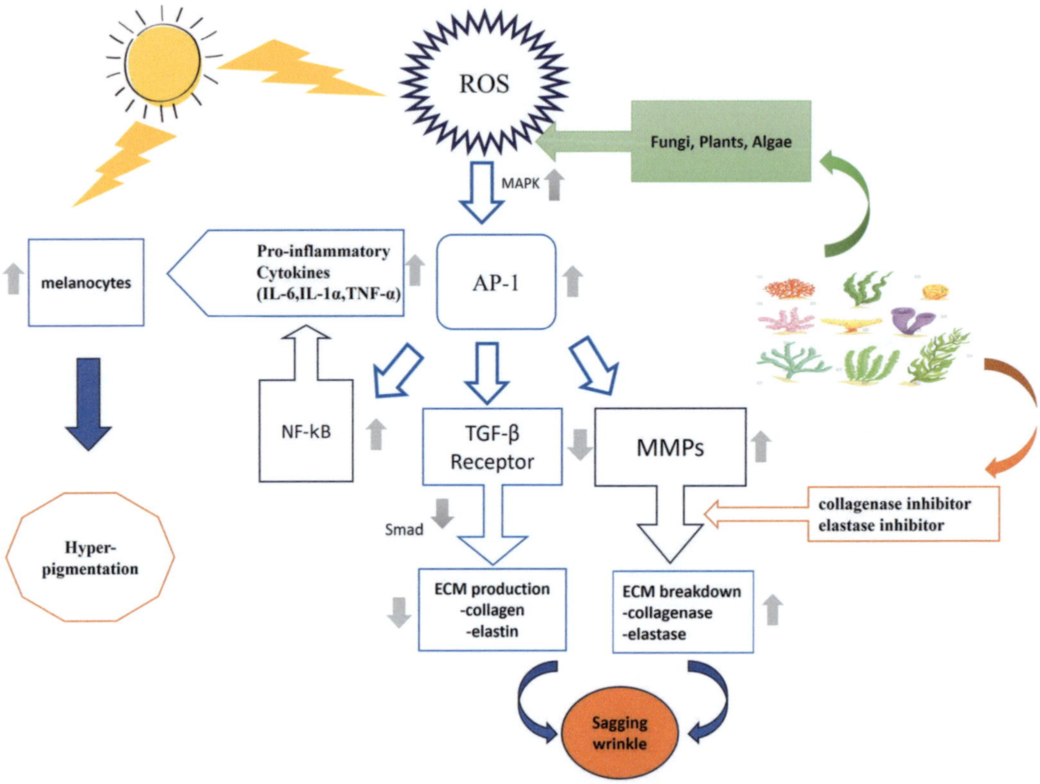

Fig. 8.1 Ultraviolet radiation, ROS generation and collagen and elastase breakdown stimulating ageing

promotes an upregulation in the production of matrix metalloproteinases (MMPs) enzymes, which play a crucial role in the enzymatic breakdown of collagen (Varani et al. 2006; Lambert et al. 2001).

Extrinsic skin ageing: Photoaging, which is sometimes known as reverse phenomenon, is primarily caused due to prolonged exposure to solar radiation. Reactive oxygen species (ROS) portray an instrumental part in the regulation of connective tissue architecture, which results in the skin developing noticeable changes after prolonged exposure to ultraviolet (UV) radiation (Thring et al. 2009) (Fig. 8.1).

A key factor in preserving skin integrity, firmness, and elasticity is collagen, which predominates as a structural protein in the extracellular matrix. A great part of the skin is made up of collagen, which is composed of 10–15% collagen type III and 85–90% collagen type I (Wang 2021).

However, certain collagenase enzymes, most notably matrix metalloproteinases (MMPs), can degrade collagen, which plays a significant part of the skin ageing process; as a result, it weakens collagen present in the dermis or under layer of the skin and causes wrinkles to appear accompanied by a loss of skin suppleness as a result of ageing skin. With potential implications for dermatological research and therapeutic approaches aimed at preserving youthful skin characteristics, anti-collagenase interventions are of utmost importance in reducing this process (Srisuksomwong et al. 2023).

8.7 Collagenase and Elastase Inhibitory Activities

8.7.1 Anti-elastase and Anti-collagenase Enzymes from Plant

Pitaya (*Hylocereus* spp.), also known as dragon fruit, has recently emerged as an indispensable component of the expanding niche market for exotic fruits as well as within the local markets of demanding producer nations like Malaysia, Vietnam, Mexico, Colombia, Nicaragua and Costa Rica. The vine cactus species earned global recognition for being a decorative plant and a fruit crop with great commercial potential. It belongs to the plant family Cactaceae and its fruit has been considered to be the most visually appealing, embodying a bright red scaly-coated skin on the exterior and yellow or red pulp containing tiny, uniformly distributed black seeds on the interior. Pitayas are classified into three types: ones with red coated skin and red flesh (*Hylocereus polyrhizus*), some with white coated skin and yellow flesh (*Selenicereus megalathus*), other ones with red coated skin and white flesh (*Hylocereus undatus*) (Nerd and Mizrahi 1996).

According to the anti-collagenase activity assay, red pitaya peels at 1000 μg/ml concentration are able to reduce collagenase around 96.92 ± 0.02%, compared to the ascorbic acid as a standard with reduction of 97.97 ± 0.18%. With an IC50 of (16.280.14) g/ml, red pitaya peel extract inhibited collagenase, whereas ascorbic acid inhibited collagenase with an IC50 of (7.670.11) g/ml. This observation revealed that red pitaya peel extract may inhibit collagenase enzymes as well as conventional ascorbic acid, indicating its anti-ageing efficacy.

Collagen, which is derived from procollagen, is the essential and most important molecular component responsible for the formation of human skin. Protein collagen is widely present in every part of the human connective tissues. Dermal fibroblasts generate procollagen in response to transforming growth factor- (TGF-) as well as activator protein-1 (AP-1), both control collagen development and decomposition, respectively. UV exposure stimulates MMP enzymes released by fibroblasts, keratinocytes, and some other cells to increase collagen degradation by AP-1 while decreasing collagen synthesis (Fisher et al. 1997; Varani et al. 2000).

Therefore, a result, connective tissues are susceptible to degradation during photoaging (Chung et al. 2001). With maturation collagen content decreases by around 1% every year. However, as older individuals have higher amounts of MMPs, the rate of collagen reduction is significantly higher in them (Seo and Chung 2006).

MMPs, which are zinc-enriched proteinases, commonly referred to as interstitial collagenase, start with the decomposition of the most prevalent interstitial collagens of collagen types I, II and III in the dermis. MMP-2 directs breaking of type I, II, III, IV and VII collagens, wherein type IV and VII are abundant in the dermal-epidermal junction (DEJ). In addition to MMPs, elastin is digested by elastase enzyme, a different kind of the skin interstitial fibre. Reducing both of these anatomical components of skin affects the integrity of the skin and suppleness, which inevitably promotes premature ageing of the skin and development of wrinkles (Chanvorachote et al. 2009).

The peel of red pitaya has polyphenol compound, supplemented with hydroxyl groups that are able to react with the collagenase enzyme side chain carrying the functional groups. The conformational alteration due to the hydrophobic interaction between the benzene rings in polyphenols leads to collagenase impairment (Madhan et al. 2007).

Collagenase having Zn ion at the active site plays an important role in interacting with the inhibitor. Therefore, this active site may get adhered to by polyphenols via Zn ion and initiate inhibition of enzyme from substrate breakdown which could contribute to the extract of red pitaya peel as a good collagenase inhibitory compound (Bigg et al. 1994; McDonald et al. 1996).

The extract of red pitaya peel is an effective elastase enzyme inhibitor and is used to assess its potential to help in skin anti-ageing attributes. It was observed that red pitaya peel extract is able to reduce 87.62 ± 0.05% at its highest concentration 1000 μg/ml.

Anti-elastase activity has been categorized into four groups: inactive (IC50 > 100.00 g/ml), weakly active (IC50: 50.01–100.00 g/ml), moderately active (IC50: 15.01–50.00 g/ml) and very active (IC5015.00 g/ml). The extract was shown to inhibit elastase with a moderately active IC50 of 29.830.21 g/ml.

For all biological being aging is an unavoidable phenomenon that occurs incessantly in the course of life span. The damage of the cell mitochondria is dependent on the ROS production, which is directly proportional to the cell longevity (Dua and Srivastava 2016). Several inherent and external factors regulate the complex biological process of skin ageing. Extrinsic variables are mostly induced by extended exposure to sunlight, whereas intrinsic factors are primarily caused by the passage of time. (Assaf et al. 2017). The skin ageing gets stimulated by the ultraviolet (UV) radiation that produces peroxyl free radicals, which disintegrate into malondialdehyde (MDA). MDA integrates with collagen and reduces the elasticity of the skin and water water-holding property of the skin, leading to skin wrinkling (Kim et al. 2007). With the increase of elastase activity several health complications including cystic fibrosis, psoriasis, rheumatoid arthritis, chronic obstructive airway disease (COPD), impaired wound healing, and premature ageing syndrome also get increased. Elastase chiefly cleaves the amino acid valine which resides in collagen and elastin fibrils and affects the inflammation (Siedle et al. 2003, 2007; Klaas et al. 2002).

Biological actions of plants are regulated by the presence of phytochemicals. According to earlier research (Wu et al. 2006), the significant percentage inhibition of elastase demonstrated by the extract of red pitaya peel might be attributed to varied possession of many bioactives including flavonoids, vitamin C, betacyanins and polyphenols. In reality, vitamin C is considered the most powerful antioxidant, because of its ability to combat ROS to slow down the process of ageing while simultaneously protecting the skin's collagen and elastin fibres. The flavonoid and polyphenol chemicals include hydroxyl groups that are effective at establishing linkage with carboxyl groups of serine amino acids at active sites of elastase enzyme to modify the action mechanism of enzyme. As a result, elastase is no longer able to break the peptide bonds, which considerably aids in the prevention of losing skin elasticity and development of wrinkles (Iván et al. 2009). Furthermore, topical anti-ageing targeted therapies that are used for restoring skin tenderness make use of popularly recognized compounds possessing superior scavenging free radical qualities such as plant-derived extract having pharmacological property, antioxidant and vitamins dietary supplements. Primarily, since antioxidants possess the ability to scavenge ROS, they act as anti-ageing substances, leaving the skin healthy. As living systems can maintain ROS equilibrium in cells, the human skin has barriers against UV rays by a sophisticated antioxidant defence mechanism that includes endogenous and exogenous antioxidants (Sander et al. 2002). Therefore, topical usage of skin treatments containing plant extracts with polyphenol-rich content will provide protection of the skin from the detrimental consequences of ultraviolet radiation, which causes early ageing. It was studied that plant by-products are rich in bioactive products.

Several studies found that byproducts of plant viz coconut testa, cocoa pod, and rice bran all have anti-elastase actions, with percentages of inhibition ranging from 6.80 ± 3.16%, 2.54 ± 0.04%, and 9.16 ± 0.54% respectively (Jamaluddin et al. 2016; Abdul Karim et al. 2014). These comparisons clearly revealed that red pitaya peels have substantially better elastase inhibitory characteristics and may be used as an effective slow-ageing ingredient in the formulation of plant-derived cosmetic products and other medicaments. Thus, the extract of red pitaya peel with its anti-elastase action was equivalent to that of ascorbic acid, a well-known antioxidant used in beauty goods, with just a little difference between the two.

In another study, anti-collagenase and anti-elastase properties were observed from different plant extracts. Extracts from different plants showed varying results indicating the presence of anti-elastase action with inhibitory behaviour on the following extracts; white tea-89%, cleavers-58%, burdock root-51%, bladderwrack-50%, anise and angelica-32%. Anti-collagenase action was also detected in the extracts of white tea-87%, green tea-47%, rose tincture-41%, and lavender-31%, exhibiting the highest inhibitory activity. However, some plant extracts inhibited both collagenase (C) and elastase (E) namely; white tea (C = 87%, E = 89%) > cleavers (C = 7%, E = 58%) > rose tincture (C = 41%, E = 22%) > green tea (C = 47%, E = 10%) > rose aqueous (C = 26%, E = 24%) > angelica (C = 17%, E = 32%) > anise (C = 6%, E = 32%) > pomegranate (C = 11%, E = 15%) (Thring et al. 2009).

In vitro enzymatic tests that imitate collagen and elastin fibre breakdown were tested on the extracts of *P. candolleivar mirifica* (PMM) and *C. grandis* fruit juice (CGJ). PMM strongly suppressed elastase action showing an IC50 of 143 ± 4.78 μg/mL, and its anti-collagenase action showed equivalent to the positive control, Epigallocatechin gallate (EGCG). The elastase inhibitory action of PMM was substantially more elevated than CGJ ($P < 0.001$) on the basis of same concentration, but the collagenase inhibitory action of both extracts showed equivalent result. Therefore, *P. candollei var. mirifica* root extract may help reduce the skin ageing process and may potentially be utilized in anti-ageing cosmetic products (Chattuwatthana and Okello 2015).

Citrus fruits, *Citrus unshiu*, incorporate a wide range of phytochemicals, comprising carotenoids, vitamins, coumarins, phenolic compounds, minerals, pectins and limonoids, all of which play important roles in human health (Jayaprakasha and Patil 2007).

A study that looked at anti-collagenase versus anti-elastase properties of *C. unshiu* at different growth cycles found that inhibitory activity of elastase was higher in the fruit's flesh (13–29%) compared to the fruit peels (1–10%) during all stages of the fruit maturation. When only the flesh of fruits was taken into account, the immature green stages were more active than the colour-changing and mature stages. The inhibitory action of elastase in the leaf versus flower extracts was extremely low (0.6–4%). While in the collagenase inhibition experiment, during all stages of fruit-maturity, the inhibitory activity in the food was greater (9.7–106.6%) than in the peel (0–8.9%). The immature green stage (85%) displayed the highest collagenase inhibition. A dilution concentration of the extract was used to examine the immature green stage extract in greater detail. Additionally, the concentration of the extract had a significant impact on the anti-collagenase activity. In both the enzyme assays, the immature green phase showed an increased inhibition action, despite the fact that elastase and collagenase had somewhat different inhibitory activities (Eun et al. 2020).

8.7.2 Anti-elastase and Anti-collagenase Enzymes from Algae

Due to the ever-growing demand on to generate safe goods through eco-friendly bioprocessing, algae has become an abundant and sustainable source for generating new bio-based products (Michalak and Chojnacka 2014; Yun et al. 2015). On beaches along many coasts around the world, beach-cast seaweeds have been documented. These seaweeds are an accumulation of sizable amounts of algae that have been pulled up from the ocean floor or coastal platforms (Holden et al. 2018; White and White 2020).

With all the valuable biomass given the numerous uses for seaweeds and it's metabolites, squandering without making the most out of nature given valuable is ludicrous. Empirical evidence has shown that biomass gathered from beach casting can be used to produce biochemicals, feed, food, fertilizer, and fuel (Milledge and Harvey 2016; Masri et al. 2018).

Eight extracts out of 30 exhibited 50% inhibition at 250 μg/mL, two of them inhibited collagenase with high potency (IC50 values of 41.04 μg/mL and 59.34 μg/mL, respectively). The two extracts inhibit collagenase more effectively and similarly to the positive control EDTA (IC50 = 59.26 μg/mL). Results obtained showed potent scope for beach-casted seaweed biomass as an ingredient to skin treatment bio-products delivering a compatible anti-ageing effect.

Some forms of red macroalgae and brown are utilized in cosmetic products considering their high levels of amino acids, vitamins, minerals, lipids and carbohydrates as well as the existence of other physiologically active chemicals. Additionally, Agar, carrageenan, and alginates are phycocolloids made from macroalgae or seaweeds (Borowitzka 2013; Sanghvi and Lo 2010). *Ascophyllum nodosum, Alaria esculenta, Chondrus crispus, Laminaria longicruris, L. saccharina, L. digitata, Mastocarpus stellatus, Ulva lactuca,* and other *Porphyra* species are some common Macroalgae used in cosmetics. Algae often respond to a variety of stress circumstances in their natural environment by producing a diverse array of defensive bioactive compounds. Several of these bioactive compounds are appreciated to possess some therapeutic property for skincare and are widely utilized in cosmeceutical products to protect against ageing, oxidative stress and UV radiation as well as in smoothing, whitening and moisturizing agent in a variety of cosmetic products (Ariede et al. 2017).

The brown macroalgae *Macrocystis pyrifera* are rich in two antioxidant chemicals like phlorotannins, phloroeckol, and tetrameric phloroglucinol. These phlorotannins may have antioxidant and anti-diabetic properties, which might assist in reducing skin ageing. Algae extracts of *Chaeloceros* sp., *Chlorococcum* sp., *Monodus* sp. and *Thalassiosira* sp. can be utilized to boost collagen stimulation in anti-aging products(Leyton et al. 2016).

8.7.3 Anti-elastase and Anti-collagenase Enzymes from Fungi

The crude fungal extract is able to inhibit the activities of the certain enzymes which are mainly associated with breaking down extracellular matrix components in the dermis (hyaluronidase, collagenase and elastase) and formation of melanin (tyrosinase). It was observed that the crude fungal extract can suppress the collagenase action at a concentration of 200 μg/mL.

The extracts derived from *A. chevalieri* isolates 13 and 20, *Diaporthe* sp., *Exilicaulis, Penicillium sect.* and *S. gracilariae* showed a stronger inhibition strength were examined at low doses, yielding IC50 values of 54.3 μg/mL, 65.6 μg/mL, 98.6 μg/mL, 107.0 μg/mL and 112.3 μg/mL respectively.

However, extracts of *A. arundinis, Fusarium* incarnatum-equiseti complex and *N. oryzae* had no influence on collagenase activity (Calado et al. 2021).

Fermentation utilizing these fungi that are filamentous additionally enhanced the inhibitory action of elastase across all fermented rice bran substances. Preventing elastase activity might be implemented to prevent early onset of skin aging (Kim et al. 2007). A type of extracellular matrix polymer, elastin affects skin elasticity, and a proteinase enzyme elastase has the ability to digest elastin, which may promote ageing (Nar et al. 2001). In single fermentation culture, the fermented extract of *M. purpureus* had the highest elastase inhibition action (23.22%), followed by *A. niger* (12.80%). According to the observation, active metabolites of the fermentation may induce the elastase inhibitory effect. The inhibition of *Aspergillus niger* and *Monascus purpureus* combined-culture fermented extracts has higher activity than single culture fermented extracts, indicating that the enzymes produced by fungal association produce a spectrum of metabolic products that collectively assist in inhibiting elastase function.

Fermentation	Sample	Elastase inhibition activity (%)
Single cultured	*Monascus purpereus* *Aspergillus niger*	23.22 ± 0.10 12.80 ± 0.10
Mixed cultured	*Monascus purpereus + Aspergillus niger*	27.66 ± 0.43

8.8 Anti-elastase and Anti-collagenase Activities in Cosmetic Formulations

Any substance with a supple activity that is applied or sprayed on various parts of the body with the purpose of cleansing, beautifying, making them appear more attractive, protecting them, or maintaining their condition is referred to as a "cosmetic product" (Luque De Castro 2011). Demand for product consumption that is viewed as healthier, organic, and environmentally friendly is driving an increase in the demand for cosmetics that contain clean ingredients today (Antignac et al. 2011).

Given the alterations in the release of hormones that occur with age, ageing of the skin is triggered by an inherent, inevitable, or cellular systems impacting every body organs and not merely the skin, resulting in a drastic degradation of collagen, degeneration of elastic fibre, dryness and the development of wrinkled skin and laxatives (Papakonstantinou et al. 2012). The elastase, collagenase, and MMP-1 enzyme inhibitors can therefore be possible cosmetic components for skin ageing therapy by restoring skin suppleness, increasing moisture content, stimulating the synthesis of collagen, and having a skin-lightening effect. As a result, the skincare manufacturers are looking for clean natural substances, extracts or supplements that can slow down the ageing process. Natural phenolic compounds are intriguing candidates for the manufacture in the cosmetic sector of anti-aging creams or lotions due to reports that they have ROS-scavenging properties (Soto et al. 2015).

For centuries, mushrooms have been valued for their naturally emerging bioactive properties, they are now explored as possible cosmetics ingredients. Many mushrooms and their compounds have been demonstrated to provide skin and hair benefits. Active components like polyphenolics, phenolics, selenium, terpenoids, vitamins, polysaccharides, and other volatile chemical compounds offer outstanding anti-ageing, antioxidant, skin brightening, anti-wrinkle, and skin moistening effects, making these ingredients perfect choices for cosmeceutical products. (Y. Wu et al. 2016). Mushrooms are extremely abundant in protein, minerals and vitamins, as well as -β-glucan, thiamine, selenium, niacin, riboflavin, folic acid and panthothenic acid (Ahmad et al. 2013). Elastase is a type of enzyme that breaks down elastin, an important protein responsible for skin elasticity. This contributes to skin ageing and wrinkle development, whereas matrix metalloproteinases (MMPs) are enzymes that break down both collagen and elastin. The extract of *Tricholoma matsutake* (pine mushroom) mycelium significantly decreases the elastase action and the MMP dose-dependent levels (Kim et al. 2014). These extracts might be a potential biomaterial for cosmetic anti-wrinkle treatment. The reproductive structure of *Dictyophora indusiata* also called veiled lady mushroom was analysed in methanol extract to check for its potential in inhibiting cholinesterase activity, wrinkle reduction, and inhibiting melanin production. This inhibition activity exhibited by the extract against collagenase and elastase shows comparable result as the positive control epigallocatechin gallate (EGCG), indicating a promising natural anti-cholinesterase and skin care agent (Sin et al. 2013). The constituents that make up haircare products include shampoos, serums, conditioners, straightening products, hair styling products (sprays, waxes, gels, and mousses), hair colours, perming agents, and bleaching agents (Madnani and Khan 2013). Shampoos may be used to eliminate excess sebum from the hair, conditioners can restore shine, and style products can make hair more manageable. Depending on the requirements of the patient, each of these medications comes in a variety of formulations (Draelos 1991). A haircare cosmetic product has been developed and patented, composed of a fine silk film infused with shiitake mushroom extract. Additionally, a different formula designed to stimulate hair growth in humans was introduced, containing *Ganoderma lucidum* in combination with three additional plant ingredients and zinc. This combination may encourage hair growth in men by reducing levels of dihydrotestosterone or

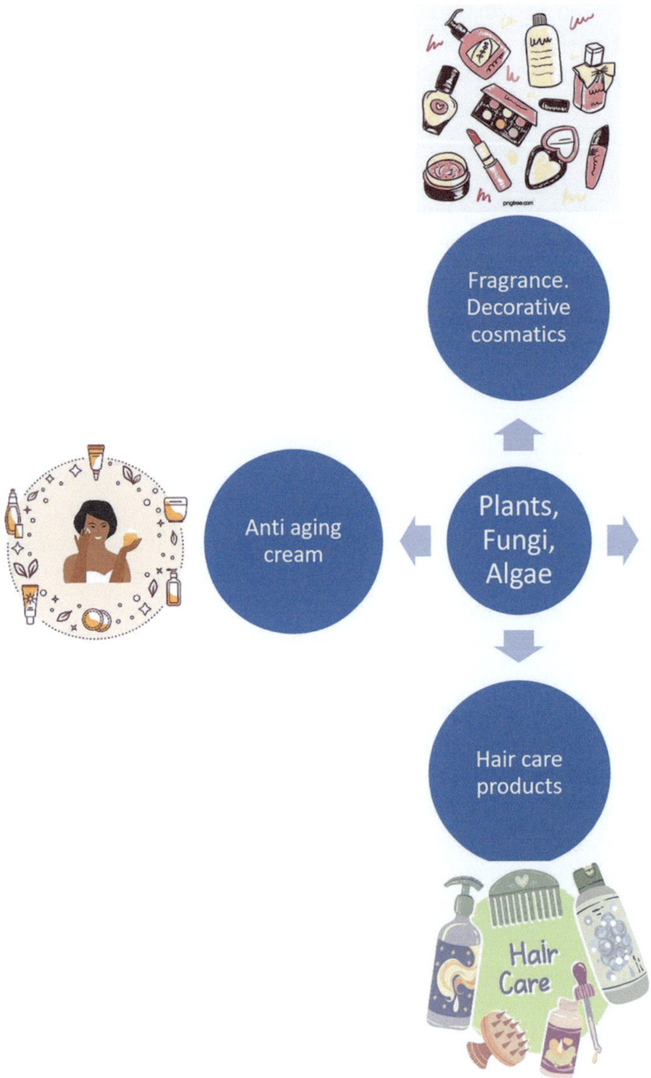

Fig. 8.2 Personal care products developed from plants, fungi and algae

addressing prostate enlargement. Haircare requires as much care as skincare to maintain healthy hair follicles. However, due to external modification from harsh chemical treatment (colour, perm, straightening), environment pollutant exposure, unhealthy lifestyle can damage hair. Mushrooms are excellent source of antioxidants, vitamin D, and minerals like iron, copper, and selenium. They pose helpful for a healthy, strong hair, avoiding dandruff and hair loss control. Therefore, cosmetic sector has a lot of opportunity for infusing mushrooms and their extracts into haircare products (Fig. 8.2).

8.9 Pharmacological and Cosmeceutical Potential of Anti-collagenase and Anti-elastase

Cosmetics are divided into two distinct groups: one being cosmeceuticals-topical use and another being nutricosmetic-oral use (Hyde et al. 2010).

Cosmeceuticals combine cosmetics and medications. They are topically applied treatments such as ointments, creams, and lotions that consist of biologically active chemicals that have medicinal drug-like properties.

Cosmeceuticals refer to goods that are used topically, whereas nutricosmetics are therapies or nutritional supplementation of nutrients having similar claimed effects but are taken orally (so-called inner beauty). The main advantage is its anti-ageing properties, which are believed to diminish wrinkles by neutralizing free radicals generated from prolonged sun exposure.

There are several standards for cosmetics items, but the most important is that they must be harmless, have zero consequences, and should contribute a favourable impact on the skin. Products comprising of natural ingredients gained a considerable attention in the field of cosmetics lately. Many fungi-derived products are employed in both nutricosmetics and cosmeceuticals, with promising future development (Hyde et al. 2010) (Anunciato and da Rocha Filho 2012).

Some common elements in nutricosmetics are carotenoids, different vitamins, mushroom extracts (such as polyphenolic isoflavones), polyphenols, glycopolyglycans, micronutrients, polyunsaturated fatty acids, additional mushroom-based components, and amino acids. (Barel et al. 2014).

A variety of goods are utilized in the production of nutritional supplements. For example, *Grifola frondosa* fruit body powder is used in Perricone MD's Maitake Mushroom Extract. This includes active glycoprotein, which is thought to aid with cardiovascular health and weight loss. Mushroom Wisdom Inc. produces health-promoting supplements derived from the same fungus (e.g., Grifron Maitake Caplet). Super Tremella Tablets, beneficial for bone and skin health, are also produced by the same company (Hyde et al. 2010).

8.10 Conclusions

Overall, this study highlighted the anti-collagenase and anti-elastase properties of enzymes extracted from several plants, algae and fungi. These enzymes have importance in pharmaceutical and cosmetical industry for helping the development of medicinal and skin care products in order to prevent premature ageing. This study additionally emphasizes the diversity of fungi, algae and plants and their bioactive compounds that stimulate anti-collagenase and anti-elastase activities. Being the natural source of bioactive compounds with biological activities, these approaches have been found to be clean, safe, less expensive, and a potent replacement to the many artificial skin anti-ageing chemicals which have been utilized in the current industry.

References

Abdul Karim A, Azlan A, Ismail A, Hashim P, Abd Gani SS, Zainudin BH, Abdullah NA (2014) Phenolic composition, antioxidant, anti-wrinkles and tyrosinase inhibitory activities of cocoa pod extract. BMC Complement Altern Med 14(1):381. https://doi.org/10.1186/1472-6882-14-381

Ahmad MF, Ahmad FA et al (2013) Edible mushrooms as health promoting agent. Adv Sci Focus 1(3):189–196. https://doi.org/10.1166/asfo.2013.1034

Antignac E, Nohynek GJ, Re T, Clouzeau J, Toutain H (2011) Safety of botanical ingredients in personal care products/cosmetics. Food Chem Toxicol 49(2):324–341. https://doi.org/10.1016/j.fct.2010.11.022

Anunciato TP, da Rocha Filho PA (2012) Carotenoids and polyphenols in nutricosmetics, nutraceuticals, and cosmeceuticals. J Cosmet Dermatol 11(1):51–54. https://doi.org/10.1111/j.1473-2165.2011.00600.x

Arct J, Pytkowska K (2008) Flavonoids as components of biologically active cosmeceuticals. Clin Dermatol 26(4):347–357. https://doi.org/10.1016/j.clindermatol.2008.01.004

Ariede MB, Candido TM, Jacome ALM, Velasco MVR, De Carvalho JCM, Baby AR (2017) Cosmetic attributes of algae – a review. Algal Res 25:483–487. https://doi.org/10.1016/j.algal.2017.05.019

Assaf H, Adly MA, Hussein MR (2017) Aging and intrinsic aging: pathogenesis and manifestations. In: Farage MA, Miller KW, Maibach HI (eds) Textbook of aging skin. Springer, Berlin Heidelberg, pp 783–794. https://doi.org/10.1007/978-3-662-47398-6_13

Badalyan SM, Barkhudaryan A, Rapior S (2022) Medicinal macrofungi as cosmeceuticals: a review. Int J Med Mushrooms 24(4):1–13. https://doi.org/10.1615/IntJMedMushrooms.2022043124

Barel AO, Paye M, Maibach HI (2014) Oral cosmetics: a general overview. In: Handbook of cosmetic science and technology. CRC Press, pp 587–596. https://doi.org/10.1201/b16716-56

Benaiges A, Marcet P, Armengol R, Betes C, Gironés E (1998) Study of the refirming effect of a plant complex. Int J Cosmet Sci 20(4):223–233. https://doi.org/10.1046/j.1467-2494.1998.176608.x

Bigg HF, Clark IM, Cawston TE (1994) Fragments of human fibroblast collagenase: interaction with metalloproteinase inhibitors and substrates. Biochim Biophys Acta Protein Struct Mol Enzymol 1208(1): 157–165. https://doi.org/10.1016/0167-4838(94)90173-2

Borowitzka MA (2013) High-value products from microalgae – their development and commercialisation. J Appl Phycol 25(3):743–756. https://doi.org/10.1007/s10811-013-9983-9

Bouma G, Thrasher A (2011) Disorders of phagocyte function. In: Blood and bone marrow pathology. Elsevier, pp 263–272. https://doi.org/10.1016/B978-0-7020-3147-2.00017-1

Brilan S (2022) Role of elastases in human body. Enz Eng 11:185

Calado MDL, Silva J, Alves C, Susano P, Santos D, Alves J, Martins A, Gaspar H, Pedrosa R, Campos MJ (2021) Marine endophytic fungi associated with Halopteris scoparia (Linnaeus) Sauvageau as producers of bioactive secondary metabolites with potential dermocosmetic application. PLoS One 16(5):e0250954. https://doi.org/10.1371/journal.pone.0250954

Chanvorachote P, Pongrakhananon V, Luanpitpong S, Chanvorachote B, Wannachaiyasit S, Nimmannit U (2009) Type I pro-collagen promoting and anti-collagenase activities of Phyllanthus emblica extract in mouse fibroblasts. J Cosmet Sci 60(4):395–403

Chattuwatthana T, Okello E (2015) Anti-collagenase, anti-elastase and antioxidant activities of Pueraria candollei var. mirifica root extract and Coccinia grandis fruit juice extract: an in vitro study. Eur J Med Plants 5(4): 318–327. https://doi.org/10.9734/EJMP/2015/14129

Chung JH, Seo JY, Choi HR, Lee MK, Youn CS, Rhie G, Cho KH, Kim KH, Park KC, Eun HC (2001) Modulation of skin collagen metabolism in aged and Photoaged human skin in vivo. J Invest Dermatol 117(5):1218–1224. https://doi.org/10.1046/j.0022-202x.2001.01544.x

Chung L, Dinakarpandian D, Yoshida N, Lauer-Fields JL, Fields GB, Visse R, Nagase H (2004) Collagenase unwinds triple-helical collagen prior to peptide bond hydrolysis. EMBO J 23(15):3020–3030. https://doi.org/10.1038/sj.emboj.7600318

Cotten SW (2020) Evaluation of exocrine pancreatic function. In: Contemporary practice in clinical chemistry. Elsevier, pp 573–585. https://doi.org/10.1016/B978-0-12-815499-1.00033-8

Donahue TR, Hiatt JR, Busuttil RW (2006) Collagenase and surgical disease. Hernia J Hernias Abdominal Wall Surgery 10(6):478–485. https://doi.org/10.1007/s10029-006-0146-7

Draelos ZK (1991) Hair cosmetics. Dermatol Clin 9(1): 19–27

Dua D, Srivastava NS (2016) A study on antioxidant and anti-aging properties of few medicinal plants. Int J Pharm Pharm Sci 8:344–347

Eun C-H, Kang M-S, Kim I-J (2020) Elastase/collagenase inhibition compositions of citrus unshiu and its association with phenolic content and anti-oxidant activity. Appl Sci 10(14):4838. https://doi.org/10.3390/app10144838

Fisher GJ, Wang Z, Datta SC, Varani J, Kang S, Voorhees JJ (1997) Pathophysiology of premature skin aging induced by ultraviolet light. N Engl J Med 337(20): 1419–1429. https://doi.org/10.1056/NEJM199711133372003

Fisher GJ, Kang S, Varani J, Bata-Csorgo Z, Wan Y, Datta S, Voorhees JJ (2002) Mechanisms of photoaging and chronological skin aging. Arch Dermatol 138(11):1462–1470. https://doi.org/10.1001/archderm.138.11.1462

Ghimeray AK, Jung US, Lee HY, Kim YH, Ryu EK, Chang MS (2015) In vitro antioxidant, collagenase inhibition, and in vivo anti-wrinkle effects of combined formulation containing Punica granatum, Ginkgo biloba, Ficus carica, and Morus alba fruits extract. Clin Cosmet Investig Dermatol 8:389–396. https://doi.org/10.2147/CCID.S80906

Gross J, Lapiere CM (1962) Collagenolytic activity in amphibian tissues: a tissue culture assay. Proc Natl Acad Sci USA 48(6):1014–1022. https://doi.org/10.1073/pnas.48.6.1014

Halla N, Fernandes I, Heleno S, Costa P, Boucherit-Otmani Z, Boucherit K, Rodrigues A, Ferreira I, Barreiro M (2018) Cosmetics preservation: a review on present strategies. Molecules 23(7):1571. https://doi.org/10.3390/molecules23071571

Holden JJ, MacNeill SK, Juanes F, Dudas SE (2018) Beach-cast deposition and commercial harvesting of a non-indigenous alga, Mazzaella japonica: implications for macrofauna communities in Baynes sound, British Columbia. Estuar Coast Shelf Sci 210:162–171. https://doi.org/10.1016/j.ecss.2018.06.008

Hyde KD, Bahkali AH, Moslem MA (2010) Fungi – an unusual source for cosmetics. Fungal Divers 43(1): 1–9. https://doi.org/10.1007/s13225-010-0043-3

Iván G, Szabadka Z, Ördög R, Grolmusz V, Náray-Szabó G (2009) Four spatial points that define enzyme families. Biochem Biophys Res Commun 383(4): 417–420. https://doi.org/10.1016/j.bbrc.2009.04.022

Jamaluddin A et al (2016) Effects of solid state fermentation by monascus purpureus on phenolic content and biological activities of coconut testa and rice bran. J Teknol 78(11–2). https://doi.org/10.11113/jt.v78.9939

Jayaprakasha GK, Patil BS (2007) In vitro evaluation of the antioxidant activities in fruit extracts from citron and blood orange. Food Chem 101(1):410–418. https://doi.org/10.1016/j.foodchem.2005.12.038

Jenkins G (2002) Molecular mechanisms of skin ageing. Mech Ageing Dev 123(7):801–810. https://doi.org/10.1016/S0047-6374(01)00425-0

Kaur G, Jabbar Z, Athar M, Alam MS (2006) Punica granatum (pomegranate) flower extract possesses potent antioxidant activity and abrogates Fe-NTA induced hepatotoxicity in mice. Food Chem Toxicol

44(7):984–993. https://doi.org/10.1016/j.fct.2005.12.001

Kim Y-J, Uyama H, Kobayashi S (2004) Inhibition effects of (+)-catechin–aldehyde polycondensates on proteinases causing proteolytic degradation of extracellular matrix. Biochem Biophys Res Commun 320(1):256–261. https://doi.org/10.1016/j.bbrc.2004.05.163

Kim YH, Kim KS, Han CS, Yang HC, Park SH, Ko KI, Lee SH, Kim KH, Lee NH, Kim JM, Son K-H (2007) Inhibitory effects of natural plants of Jeju Island on elastase and MMP-1 expression. J Cosmet Sci 58(1): 19–33

Kim SY, Go KC, Song YS, Jeong YS, Kim EJ, Kim BJ (2014) Extract of the mycelium of T. matsutake inhibits elastase activity and TPA-induced MMP-1 expression in human fibroblasts. Int J Mol Med 34(6):1613–1621. https://doi.org/10.3892/ijmm.2014.1969

Klaas CA, Wagner G, Laufer S, Sosa S, Della Loggia R, Bomme U, Pahl HL, Merfort I (2002) Studies on the anti-inflammatory activity of phytopharmaceuticals prepared from arnica flowers. Planta Med 68(5): 385–391. https://doi.org/10.1055/s-2002-32067

Lambert CA, Colige AC, Lapière CM, Nusgens BV (2001) Coordinated regulation of procollagens I and III and their post-translational enzymes by dissipation of mechanical tension in human dermal fibroblasts. Eur J Cell Biol 80(7):479–485. https://doi.org/10.1078/0171-9335-00181

León-López A, Morales-Peñaloza A, Martínez-Juárez VM, Vargas-Torres A, Zeugolis DI, Aguirre-Álvarez G (2019) Hydrolyzed collagen – sources and applications. Molecules 24(22):4031. https://doi.org/10.3390/molecules24224031

Leyton A, Pezoa-Conte R, Barriga A, Buschmann AH, Mäki-Arvela P, Mikkola J-P, Lienqueo ME (2016) Identification and efficient extraction method of phlorotannins from the brown seaweed Macrocystis pyrifera using an orthogonal experimental design. Algal Res 16:201–208. https://doi.org/10.1016/j.algal.2016.03.019

Loffek S, Schilling O, Franzke C-W (2011) Biological role of matrix metalloproteinases: a critical balance. Eur Respir J 38(1):191–208. https://doi.org/10.1183/09031936.00146510

Luque De Castro MD (2011) Cosmetobolomics as an incipient '-omics' with high analytical involvement. TrAC Trends Anal Chem 30(9):1365–1371. https://doi.org/10.1016/j.trac.2011.04.013

Madhan B, Krishnamoorthy G, Rao JR, Nair BU (2007) Role of green tea polyphenols in the inhibition of collagenolytic activity by collagenase. Int J Biol Macromol 41(1):16–22. https://doi.org/10.1016/j.ijbiomac.2006.11.013

Madnani N, Khan K (2013) Hair cosmetics. Indian J Dermatol Venereol Leprol 79(5):654–667. https://doi.org/10.4103/0378-6323.116734

Masaki H (2010) Role of antioxidants in the skin: anti-aging effects. J Dermatol Sci 58(2):85–90. https://doi.org/10.1016/j.jdermsci.2010.03.003

Masri MA, Younes S, Haack M, Qoura F, Mehlmer N, Brück T (2018) A seagrass-based biorefinery for generation of single-cell oils for biofuel and Oleochemical production. Energ Technol 6(6):1026–1038. https://doi.org/10.1002/ente.201700604

McDonald M, Mila I, Scalbert A (1996) Precipitation of metal ions by plant polyphenols: optimal conditions and origin of precipitation. J Agric Food Chem 44(2): 599–606. https://doi.org/10.1021/jf950459q

Michalak I, Chojnacka K (2014) Algal extracts: technology and advances. Eng Life Sci 14(6):581–591. https://doi.org/10.1002/elsc.201400139

Milledge J, Harvey P (2016) Golden tides: problem or Golden opportunity? The valorisation of Sargassum from beach inundations. J Mar Sci Eng 4(3):60. https://doi.org/10.3390/jmse4030060

Mott JD, Werb Z (2004) Regulation of matrix biology by matrix metalloproteinases. Curr Opin Cell Biol 16(5): 558–564. https://doi.org/10.1016/j.ceb.2004.07.010

Nar H, Werle K, Bauer MM, Dollinger H, Jung B (2001) Crystal structure of human macrophage elastase (MMP-12) in complex with a hydroxamic acid inhibitor. J Mol Biol 312(4):743–751. https://doi.org/10.1006/jmbi.2001.4953

Nerd A, Mizrahi Y (1996) Reproductive biology of cactus fruit crops. In: Janick J (ed) Horticultural reviews, 1st edn. Wiley, pp 321–346. https://doi.org/10.1002/9780470650608.ch7

Oikarinen A (2004) Connective tissue and aging. Int J Cosmet Sci 26(2):107–107. https://doi.org/10.1111/j.1467-2494.2004.213_6.x

Page-McCaw A, Ewald AJ, Werb Z (2007) Matrix metalloproteinases and the regulation of tissue remodelling. Nat Rev Mol Cell Biol 8(3):221–233. https://doi.org/10.1038/nrm2125

Papakonstantinou E, Roth M, Karakiulakis G (2012) Hyaluronic acid: a key molecule in skin aging. Dermatoendocrinol 4(3):253–258. https://doi.org/10.4161/derm.21923

Pientaweeratch S, Panapisal V, Tansirikongkol A (2016) Antioxidant, anti-collagenase and anti-elastase activities of *Phyllanthus emblica*, *Manilkara zapota* and silymarin: an *in vitro* comparative study for anti-aging applications. Pharm Biol 54(9):1865–1872. https://doi.org/10.3109/13880209.2015.1133658

Raffetto JD, Khalil RA (2008) Matrix metalloproteinases and their inhibitors in vascular remodeling and vascular disease. Biochem Pharmacol 75(2):346–359. https://doi.org/10.1016/j.bcp.2007.07.004

Sander CS, Chang H, Salzmann S, Müller CSL, Ekanayake-Mudiyanselage S, Elsner P, Thiele JJ (2002) Photoaging is associated with protein oxidation in human skin in vivo. J Invest Dermatol 118(4): 618–625. https://doi.org/10.1046/j.1523-1747.2002.01708.x

Sanghvi AM, Lo YM (2010) Present and potential industrial applications of macro- and microalgae. Recent Pat Food Nutr Agric 2(3):187–194

Seo JY, Chung JH (2006) Thermal aging: a new concept of skin aging. J Dermatol Sci Suppl 2(1):S13–S22. https://doi.org/10.1016/j.descs.2006.08.002

Shingleton WD, Hodges DJ, Brick P, Cawston TE (1996) Collagenase: a key enzyme in collagen turnover. Biochem Cell Biol 74(6):759–775. https://doi.org/10.1139/o96-083

Siedle B, Gustavsson L, Johansson S, Murillo R, Castro V, Bohlin L, Merfort I (2003) The effect of sesquiterpene lactones on the release of human neutrophil elastase. Biochem Pharmacol 65(5):897–903. https://doi.org/10.1016/s0006-2952(02)01652-0

Siedle B, Hrenn A, Merfort I (2007) Natural compounds as inhibitors of human neutrophil elastase. Planta Med 73(5):401–420. https://doi.org/10.1055/s-2007-967183

Sim G-S, Lee B-C, Cho HS, Lee JW, Kim J-H, Lee D-H, Kim J-H, Pyo H-B, Moon DC, Oh K-W, Yun YP, Hong JT (2007) Structure activity relationship of antioxidative property of flavonoids and inhibitory effect on matrix metalloproteinase activity in UVA-irradiated human dermal fibroblast. Arch Pharm Res 30(3):290–298. https://doi.org/10.1007/BF02977608

Sin D-B et al (2013) Antioxidant and anti-inflammatory activities of fruiting bodies of Dyctiophora indusiata. Journal of Mushroom 11(4):269–277. https://doi.org/10.14480/JM.2013.11.4.269

Soto M, Falqué E, Domínguez H (2015) Relevance of natural Phenolics from grape and derivative products in the formulation of cosmetics. Cosmetics 2(3):259–276. https://doi.org/10.3390/cosmetics2030259

Srisuksomwong P, Kaenhin L, Mungmai L (2023) Collagenase and Tyrosinase inhibitory activities and stability of facial cream formulation containing cashew leaf extract. Cosmet 10(1):17. https://doi.org/10.3390/cosmetics10010017

Thring TS, Hili P, Naughton DP (2009) Anti-collagenase, anti-elastase and anti-oxidant activities of extracts from 21 plants. BMC Complement Altern Med 9(1):27. https://doi.org/10.1186/1472-6882-9-27

Van Wart HE, Steinbrink DR (1981) A continuous spectrophotometric assay for clostridium histolyticum collagenase. Anal Biochem 113(2):356–365. https://doi.org/10.1016/0003-2697(81)90089-0

Varani J, Warner RL, Gharaee-Kermani M, Phan SH, Kang S, Chung J, Wang Z, Datta SC, Fisher GJ, Voorhees JJ (2000) Vitamin a antagonizes decreased cell growth and elevated collagen-degrading matrix Metalloproteinases and stimulates collagen accumulation in naturally aged human Skin1. J Invest Dermatol 114(3):480–486. https://doi.org/10.1046/j.1523-1747.2000.00902.x

Varani J, Dame MK, Rittie L, Fligiel SEG, Kang S, Fisher GJ, Voorhees JJ (2006) Decreased collagen production in chronologically aged skin: roles of age-dependent alteration in fibroblast function and defective mechanical stimulation. Am J Pathol 168(6):1861–1868. https://doi.org/10.2353/ajpath.2006.051302

Visse R, Nagase H (2003) Matrix Metalloproteinases and tissue inhibitors of Metalloproteinases: structure, function, and biochemistry. Circ Res 92(8):827–839. https://doi.org/10.1161/01.RES.0000070112.80711.3D

Wang H (2021) A review of the effects of collagen treatment in clinical studies. Polymers 13(22):3868. https://doi.org/10.3390/polym13223868

White LN, White WL (2020) Seaweed utilisation in New Zealand. Bot Mar 63(4):303–313. https://doi.org/10.1515/bot-2019-0089

Wu L, Hsu H-W, Chen Y-C, Chiu C-C, Lin Y-I, Ho JA (2006) Antioxidant and antiproliferative activities of red pitaya. Food Chem 95(2):319–327. https://doi.org/10.1016/j.foodchem.2005.01.002

Wu Y, Choi M-H, Li J, Yang H, Shin H-J (2016) Mushroom cosmetics: the present and future. Cosmet 3(3):22. https://doi.org/10.3390/cosmetics3030022

Yun EJ, Choi I-G, Kim KH (2015) Red macroalgae as a sustainable resource for bio-based products. Trends Biotechnol 33(5):247–249. https://doi.org/10.1016/j.tibtech.2015.02.006

Antimicrobial Activity of Fungal-Based Ingredients in Cosmetics

9

Pooja Yadav, Tsering Choizom, and Shubham Mahara

Abstract

Fungal-based ingredients are gaining popularity in cosmetic industries due to their natural antimicrobial properties and skin-friendly benefits. These ingredients include chitosan, β-glucans, mycocins, and fungal enzymes, which help to protect the cosmetic products from harmful bacteria and fungi while also supporting our skin health. They work by disrupting the microbial cell membranes, preventing biofilm formation and even boost the skin natural defense mechanism. From acne-fighting skin care to dandruff shampoos and free from preservatives, fungal derivatives are directing the future toward safer and more sustainable cosmetics. While challenges related to large-scale production and regulatory approvals and ongoing researches continue to find solutions for full potential in beauty and personal care products.

Keywords

Antimicrobial · Cosmetics · Fungi · Stability · UV protection

P. Yadav (✉)
School of Health and Allied Sciences, Department of Nutrition and Dietetics, Geeta University, Panipat, Haryana, India

T. Choizom · S. Mahara
Department of Food Technology, School of Applied and Life Sciences, Uttaranchal University, Dehradun, Uttarakhand, India

9.1 Introduction

According to the Cambridge Dictionary of English, "substances that you put on your face or body that are intended to improve your appearance" are considered cosmetics. Under European Commission (EC) regulation No.1223/2009, cosmetics are defined that "any substance or mixture intended to be placed in contact with the external parts of the human body (epidermis, hair system, nails, lips, and external genital organs) or with the teeth and the mucous membranes of the oral cavity with a view exclusively or mainly cleaning them, perfuming them, changing their appearance, protecting them, keeping them in good condition, or correcting body odours." This definition is more comprehensive cosmetics and pharmaceuticals are combined to form the term "cosmeceutical," which refers to items that are applied topically as cosmetics but contain certain compounds that can enhance biological function and produce better skin (Balboa et al. 2015). In 1962, Reed coined the term "cosmeceutical," which Kligman later reinterpreted in 1993 as a category that combines cosmetics and medications with a medicinal purpose. These days, this definition encompasses a more extensive and accurate set of actions that assert a bioactivity effect (Aranaz et al. 2018; Morganti and Coltelli 2019).

Europe, including the UK, leads the globe in cosmetics production with €78.6 billion, next to the USA with €67.2 billion. The European

S. Ghosh, H. N. Onyeaka (eds.), *Fungal-Based Cosmetics: Formulation and Usage*, The Mycota 17,
https://doi.org/10.1007/978-3-032-09291-5_9

cosmetics supply chain is also essential contributing €29.19 billion to the gross value added (GVA) overall, with France and Germany contributing the most. It also creates 2 million direct, indirect, and induced jobs. The richest countries, like Norway and Switzerland, invest over €225 per person yearly on cosmetics, whereas European consumers spend an average of €135 per person yearly (Association CETPC 2019).

In 2020, the beauty and cosmetics sector needs to adapt to the evolving demands of a discerning and volatile consumer base. Finding a balance between the "chemistry of the natural" and "cosmetic" items was (and still is) the largest challenge. Certain trends and related industries emerge in this dynamic environment, showing positive indications of recovery. According to the sample interviewed, the upcoming keywords for the cosmetics sector are "sustainable growth" (18.9% in 2020 compared to 13.2% in 2018), "natural/organic" (10.9%), "care" (7.8%), "ethics" (7.5%), "e-commerce" (7.1%), "social beauty" (7.0%), "personalization" (6.7%), and "safety" (6.3%) (Dini and Laneri 2021).

Concerns about the environment, culture, or religion; the need to adhere to certain personal beliefs (vegetarianism or veganism); and health or safety considerations are just a few of the difficulties that the cosmetics business, which produces both cosmetics and cosmeceuticals, is currently confronting. The increasing demand for "green products," or those that customers believe have a smaller environmental impact than conventional items, is one of the market's current trends (Liobikienė and Bernatonienė 2017).

The increasing consumer interest in natural products has led to the use of extracts from aromatic, herbal, and medicinal plants as active ingredients in cosmeceuticals and nutricosmetics formulations. They contain phenolic acids, polyphenols, triterpenes, stilbenes, flavonoids, steroids, steroidal saponins, carotenoids, sterols, fatty acids, sugars, polysaccharides, peptides, and a number of other physiologically active substances (Dini 2018) whose characteristics and degree vary according to the pedoclimatic conditions and agricultural methods (Laneri et al. 2020; Dini et al. 2021) also from algae, mushrooms, and plant-based by products, bioactive extracts were obtained (Laneri et al. 2019; Dini et al. 2020; Juliano and Magrini 2018) and also by the technology of plant cell culture (Laneri et al. 2021). Significant biosynthetic capacity in the kingdom of fungi produces substances with intricate chemical structures and strong biological activity. Fungi are saprophytic creatures that are closely biochemically linked to the makeup of the substrate they live on. Their biosynthetic capacities are greatly determined by this fact (Deacon 2005).

The primary and secondary metabolites are the two main categories of biogenetic compounds that come from fungi (Kohlmünzer 1998). The chemicals known as primary metabolites which include sugar alcohols, quaternary amine bases, polysaccharides, disaccharides, and monosaccharides are produced during the primary metabolism of sugars. Secondary metabolites are substances that can be further classified as follows: substances originating from the metabolism of fatty acids (polyacetylenes); substances derived from the metabolism of active acetate (polyketides, isoprenoids, and sterols); substances derived biogenetically from shikimic acid (phenols and phenolic acids); substances formed from the transformation of amino acids (amines, toxic amines, alkaloids, and peptides); and substances formed from the transformation of aromatic amino acids (ergot alkaloids) (Kohlmünzer 1998).

Along with nitrogen-containing substances like urea, amino acids, peptides, proteins, lectins, amines, alkaloids, indole derivatives, vitamins, purine compounds, isoxazole derivatives, and phenoxazine derivatives, fungal metabolites also include non-nitrogenous compounds like carbohydrates, lipids, polyacetylenes, polyketides, isoprenoids, sterols, organic acids, and phenolic compounds (Deacon 2005; Kohlmünzer 1998). These compounds can be substituted for commonly used antibiotics because fungi produce a variety of secondary metabolites with bacteriostatic or fungistatic activity. In addition, fungi accumulate compounds with antiviral activity, making them a promising source of these compounds (Suay et al. 2000).

New technologies like metabolomics could be very helpful in the future for the identification and characterization of new natural antibacterial compounds. In the future, this information may be crucial for developing new treatment strategies. Most natural substances, including those derived from fungi, lack adequate therapeutic efficacy when used alone, but when combined with conventional antibiotics, they may have a synergistic effect that reduces side effects and enhances pharmacokinetic and pharmacodynamic characteristics (Jakubczyk and Dussart 2020). Advances in separation techniques, fermentation technologies, and structure determination techniques have made additional microorganisms attractive as possible sources of new bioactive secondary metabolites (Jakubczyk and Dussart 2020; Agrawal and Verma 2021).

9.2 Antimicrobial Properties of Fungi

Many studies have been conducted on the nutritional composition, flavor, and nutritional value of specific compounds and extracts from mushrooms as cosmetic components. Key biological characteristics of several mushroom extracts and their secondary metabolites include hypoglycemic, immunomodulatory, anticancer, antibacterial, anti-inflammatory, antioxidant, and antiatherogenic actions. Polysaccharides (PS) from fungi, such as chitin and chitosan, have specific properties suitable for cosmeceutical uses, including UV protection, cleaning, conditioning, smoothing, moisturizing, and antioxidant. Many fungi have been found to contain β-(1–3)-d-glucan, which is used in cosmetics like *Pleurotus* species and is one of the fungal-based antimicrobial substances in cosmetics. Dermatologists and scholars who are interested in cosmetics may find that mushrooms and their components can be used in the cosmetics industry. Secondary metabolites found in mushrooms include polyketides, peptides, and terpenoids. Terpenoids, which have anti-infective and immune-modulating qualities, are chemical substances that are present in mushrooms like *Ganoderma* spp. Cosmetics, especially sun creams, contain carotenoids, an organic pigment. When it comes to Gram-positive bacteria, peptides like plectasin have antibacterial qualities. Many fungi create polyketides, which are naturally occurring pigments that include melanins, anthraquinones, hydroxyanthraquinones, azaphilones, quinones, and naphthoquinones. A great source of naturally occurring compounds with bioactivity is slime molds, which are protists that eat food particles and rotting vegetation.

Numerous secondary metabolites, such as terpenoids, peptides, and polyketides, are found in mushrooms. Chemicals called terpenoids, which are present in mushrooms like *Ganoderma* spp., have the ability to modulate the immune system and prevent infections. Sun creams and other cosmetics include carotenoids, an organic pigment. Peptides with antibacterial qualities against Gram-positive bacteria include plectasin. Natural pigments called polyketides, which include melanins, anthraquinones, hydroxyanthraquinones, azaphilones, quinones, and naphthoquinones, are made by a variety of fungi. It has come to light that slime molds, which are protists that eat food particles and decaying vegetation, are great sources of naturally occurring compounds with bioactivity (Table 9.1).

9.3 Fungal-Based Ingredients in Cosmetics

Fungal-based ingredients are becoming more and more popular in cosmetics because of their skin-friendly qualities, bioactivity, and natural origin. Beneficial compounds can be found in common fungi like *Ganoderma lucidum*, *Aspergillus*, *Trametes versicolor*, and *Saccharomyces*.

9.3.1 Slime Molds as a Source of Cosmetic Ingredients

Slime molds are protists that go through two stages of development. They act like fungus in one stage and like protozoa (amoeba) in another. They consume decomposing vegetation, other

Table 9.1 Fungal secondary metabolites

Secondary metabolites	Antimicrobial properties	References
Terpenoids	Terpenoids are chemical compounds that are present in mushrooms such as *Ganoderma* spp. that are generated from five-carbon isoprene units. They have anti-infective and immune-modulating qualities. Mushrooms contain organic pigments called carotenoids, which are utilized in cosmetics, particularly sun creams	Gao et al. (2003), Jeong et al. (2008)
Peptides	The defensin family includes the peptide plectasin, which is made by fungi and exhibits antibacterial properties against gram-positive bacteria such as *Streptococcus aureus* and *S. pneumoniae.* It is sensitive to a number of gram-positive bacteria, such as *streptococcus*, *staphylococcus*, *enterococcus*, *Corynebacterium*, and *bacillus*, and it has 40 amino acids	Mygind et al. (2005)
Polyketides	Fungal polyketides, including melanins, anthraquinones, hydroxyanthraquinones, azaphilones, quinones, and naphthoquinones, are naturally occurring pigments produced by various fungi, including filamentous ascomycete genera	Lin and Xu (2022), Malik et al. (2012)

microorganisms, and food particles while in the protozoan phase. These microbes fall into two categories: cellular slime molds (Dictyostelium) and acellular slime molds (Myxogastria) (Willey et al. 2014).

Slime molds have recently gained recognition as excellent sources of naturally occurring chemicals with bioactivity (Dembitsky et al. 2005). One of the well-known groups of acellular slime molds that generates a lot of metabolites is Myxomycetes (Dembitsky et al. 2005). Fruiting bodies were used to directly extract these metabolites. Therefore, the production of a large number of fruiting bodies can be achieved through the in vitro cultivation of Myxomycetes. In vitro cultured species of Myxomycetes are primarily found in the genera Didymium and Physarum (Tafakori 2021) (Table 9.2).

9.3.2 Mushroom Extracts and Individual Compounds Used as Components in Cosmetics

People have consumed mushrooms since ancient times due to their flavor, nutritional content, and nutritional value. The occurrence of a wide range of bioactive metabolites, including phenolic compounds, terpenoids, polysaccharides, lectins, steroids, glycoproteins, and other lipid components, has been extensively studied. Numerous mushroom extracts and their secondary metabolites have been reported to have important biological qualities, such as hypoglycemic, immunomodulatory, anticancer, antibacterial, anti-inflammatory, antioxidant, and antiatherogenic activities (Taofiq et al. 2016a). The ingredients for the fungal extraction were first evaluated on the parameters given in Fig. 9.1.

However, the anti-hyaluronidase, anti-collagenase, anti-elastase, and antityrosinase characteristics of certain compounds and mushroom extracts have received comparatively little published investigation. Even less well-known, these investigations and the reported bioactivities give dermatologists and scholars with a passion in cosmetics hope for the future of the use of mushrooms and their components in the industry (Taofiq et al. 2016b) (Table 9.3).

9.3.3 Fungal-Based Antimicrobial Compounds in Cosmetics

Polysaccharides (PS) have certain properties that make them appropriate for cosmeceutical uses including protection, collagen stimulation, or hyaluronic acid synthesis, as well as cosmetic uses like emulsifying ability, thickening, and gel formation (Guillerme et al. 2017).

Table 9.2 Slime molds as a source of cosmetic ingredients

Fungi	Cosmetic ingredients	Characteristics	References
Slime mold	Pigments	Naturally occurring colorants, slime molds are usually red, orange, and yellow in hue. Due to their distinct chemical structures, these pigments are necessary for personal hygiene goods like soap as well as cosmetics like lipsticks and nail polish	Amberg and Fogarassy (2019), Basch et al. (2006)
	Plasmodial pigments	These pigments have demonstrated remarkable cytotoxic and antibiological activity, strong UV-visible light absorption capabilities, and the ability to bind metals to provide strong protection for the weak in natural plasmodia, according to early exploratory studies	Tafakori (2021), Basch et al. (2006), Steglich (1989)
	Pigments of dormant sclerotia	Occasionally brown or black, *Fuligo septica* develops dormant sclerotia and yellow plasmodia in old mucosal veins. There is a correlation between color and sclerotia vitality; 60% of yellow sclerotia reproduce	Krzywda et al. (2008)
	Pigments of fruiting bodies	Fruiting body colors are produced by the oxidative dimerization of tryptophan and are members of the carbazole alkaloids class. These substances are well-known for having potent anticancer effects. Because of their antibacterial, anti-inflammatory, and anticancer qualities, indolocarbazoles, lactones, and heterocyclic compounds are also significant pigments in cutaneous cosmetics	Du and Ryan (2015), Chambers et al. (2021), Sartori et al. (2021)
	Spore pigments	*Reticularia lycoperdon*, a member of the order Liceales, possesses an additional ethanol-extractable yellow-green pigment, but up until now, melanin has been the only pigment found in the spores of the investigated species from the acellular orders Physarales and Stemonitales (only *Stemonitis herbatica*)	Sartori et al. (2021), Loganathan and Kalyanasundaram (1999), Loganathan et al. (1989), Płonka and Rakoczy (1997), Rakoczy and Panz (1994)
	Lipids	Lipids and their derivatives are essential components of dermatological chemicals and cosmetics nanoparticles. Investigating the presence of sterols, esters, fatty acids, phospholipids, di- and triglycerides, and slime molds	Dembitsky et al. (2005), Stoyneva-Gärtner et al. (2022a)
	Fatty acids	Both cellular and acellular slime molds include fatty acids, which are essential raw materials used in industrial cosmetics as softeners, emulsifiers, and other products	Dembitsky et al. (2005), Yang et al. (2020)
	Steroids	Cell membranes contain secondary triterpenoid metabolites called sterols, which are crucial for regulating the membranes' permeability, fluidity, and stability. Additionally, sterols are antioxidants with anti-inflammatory, anti-carcinogenic, and antibacterial qualities that also affect how enzymes, receptors, and channels work	Stoyneva-Gärtner et al. (2022a)

(continued)

Table 9.2 (continued)

Fungi	Cosmetic ingredients	Characteristics	References
	Esters and lactones	Esters are ring-shaped, vital emollients and solubilizers that are frequently utilized as flavoring and odorants in personal care cosmetics	Korn et al. (1965), Stoyneva-Gärtner et al. (2022b)
	Protein	For the manufacturing of cosmetics and the fluidity of membranes, fatty-acid desaturases such as Delta-5 and Delta-6 are essential. A secondary metabolite with antibacterial properties was produced by *Dictyostelium discoideum*, the first slime mold to identify a Delta-5 desaturase	Czumaj and Śledziński (2020), Saito and Ochiai (1999), Murata et al. (2016)

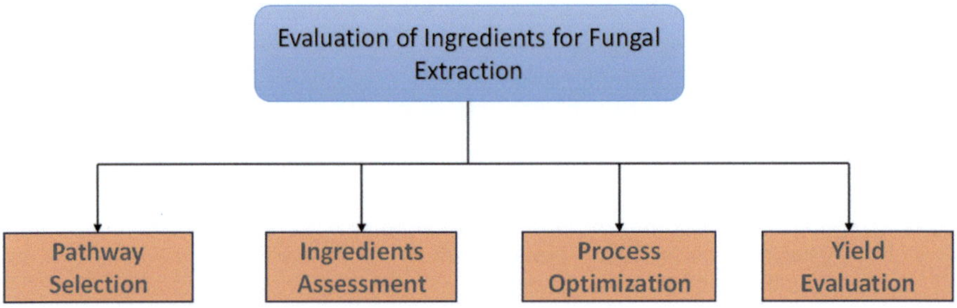

Fig. 9.1 Evaluation of ingredients for fungal extracts

Table 9.3 Mushroom extracts with reported cosmeceutical potential

Mushroom species	Bioactive	Effects	References
Agaris bisporus	2-amino-3H-phenoxszin-3-one	At 0.5 M, 1 M, and 2 M suppressed melanin production by 80%, 54.1% and 39.4%, respectively	Miyake et al. (2010)
Agaris brasiliensis	Extract	Tyrosinase activity was decreased by methanolic extract at 10 mg/mL up to 49.82 ± 0.51% compared to 48.95 ± 0.35% for kojic acid	Huang et al. (2014)
Coriolus versicolor	Extract	Tyrosinase activity was reduced up to 26.3 ± 8.3% at 10 mg/mL	Park et al. (2015)
Ganoderma luciderm	Extract	The extract reduced tyrosinase activity by up to 80% at 1 mg/mL	Chien et al. (2008)
Lentinus edode	Extract	Tyrosinase enzyme was inhibited by acetonic, methanolic, and hot water extracts at 0.125–1.0 mg/mL by 11.94–54.22, 15.12–54.61, and 3.09–47.32%, respectively	Yoon KiNam et al. (2011)
Pleurotus ostreatus	Extract	Tyrosinase activity was suppressed by acetone, methanol, and hot water extracts at 0.125–1.0 mg/mL by 11.37–52.05%, 11.36–59.56%, and 9.60–49.60%, respectively	Alam et al. (2010)
Tricholoma matsutake	Extract	It reduced MMP-3 expression and inhibited elastase activity by 81.4 ± 3.92% at 100 g/mL	Kim et al. (2014)

Chitin and chitosan are polysaccharides found in a wide variety of marine organisms, such as crabs and mollusks. While chitin, a polymer of glucosamide and N-acetylglucosamide, is insoluble in aqueous solution, chitosan, the N-deacetylated form of chitin produced by chemical or enzymatic deacetylation of chitin, is soluble in acidic aqueous solutions. These compounds have a variety of applications, including UV protection, cleansing, conditioning, smoothing, moisturizing, and antioxidant (Aranaz et al. 2018). Polysaccharides, such as glucans, chitin, and its derivative chitosan, are vital components of the fungal cell wall. The majority of polysaccharides found in mushrooms are either homoglycans or heteroglycans, depending on the kind of monosaccharide molecule present. They can form peptidoglycans or polysaccharide–protein complexes by joining with other proteins. The anticancer polysaccharides, particularly β-(1–3)-d-glucan, have been discovered in many fungi and are utilized in cosmetics, such as *Pleurotus* species (Synytsya et al. 2009). Japan was the first to isolate β-(1–3)-d-glucan, which has a triple helical structure, from the fruiting body of *Lentinus edodes* in the late 1960s (Ikekawa 2001). Since then, many groups have identified β-(1-6)-d-glucan from *Agaricus subrufescens*, lentinan from *Lentinus edodes*, protein-bound PSK from *Coriolus versicolor*, and schizophyllan from *Schizophyllum commune* (Angeli et al. 2009; Hobbs 2005; Yao et al. 2011).

Mushroom cell walls are made of chitin, which is a complex structure made up of poly-N acetyl-1,4-D-glucosamine subunits connected by (1–4) glycosidic linkages. Despite having hygroscopic and sorptive properties, this chemical is insoluble in water. Interestingly, chitin exhibits antiviral, anti-inflammatory, and antibacterial properties (Ifuku 2014). Chitin is used in cosmetics and is mostly valued for its hydrating properties. By restoring skin elasticity, producing antioxidant effects, and promoting collagen synthesis, it provides antiaging, firming, and regeneration advantages. It effectively maintains the proper moisture levels and is especially noticeable in formulations designed for dry and aged skin (Park and Kim 2010). Chitosan has antibacterial properties because it contains amino sugars. Notably, arthropods including crabs, shrimp, squid, and crayfish have shells that contain chitosan in addition to fungus (No et al. 2002). Chitosan has a broad range of antibacterial action and works well against some Gram-positive and Gram-negative bacteria as well as fungus. Gram-positive bacteria, such as *Listeria monocytogenes, Bacillus megaterium, B. cereus, S. aureus, Lactobacillus plantarum, L. brevis,* and *L. bulgaris*, have been found to be more susceptible to its effects. Although it exhibits activity against Gram-negative bacteria, including *Vibrio parahaemolyticus, P. fluorescens, S. typhimurium,* and *Escherichia coli*, its potency is rather low (Coma et al. 2003; Kumar 2000).

A polymer of D-glucose monomers joined by β-glycosidic linkages is called β-glucan. Numerous investigations have revealed that the primary sources of β-glucan include plants, algae, bacteria, fungi, and yeast (Du et al. 2014). Although yeast is generally thought to be the finest source of β-glucan, the quantity and variety of β-glucans (Zhu et al. 2016), the properties and composition of β-glucans derived from yeast differ depending on the species (Nakov et al. 2019). In terms of cosmetics, β-glucan can improve skin immunity by activating Langerhans cells, which are found in the epithelium of the epidermis and mucosal tissues. Additionally, it increases the synthesis of hyaluronic acid, collagen, and elastin by means of a moisturizing agent and pattern recognition receptors (PRRs) on Langerhans cells (Calvani et al. 2020). According to these research findings, β-glucan is frequently utilized as the primary component of numerous skin care products (Kogan et al. 2005). These substances' cross-links are very advantageous for the health of skin cells (Petravić-Tominac et al. 2010).

9.3.4 Fungal Enzymes with Antimicrobial Properties

Natural proteins called enzymes are made of a lengthy chain of amino acids joined by peptide bonds. Several multistep procedures known as post-translational modification occur after the formation of the primary amino acid chain, modifying the lengthy chain to take on its three-dimensional form. Different enzymes contain different amino acid sequences and types, which are clearly related to the stability and activity of the enzyme. Only a small number of amino acids in (2–4) are referred to as the active site, however not all enzyme molecules participate in the catalytic activity (Gurung et al. 2013). Since fungal enzymes are frequently more stable (at high temperatures and severe pH) than enzymes derived from plants and animals, they are becoming more and more important in agriculture, industry, and human health. Fungal enzymes aid in the simplification of raw material processing and are utilized in the production of food, drinks, confections, textiles, and leather. Marine fungi that live on wood could be a source of exoenzymes.

9.3.5 Antibacterial Activity of Substances of Fungal Origin

The antibacterial activity of individual compounds and particular extracts derived from fungal fruiting bodies has been demonstrated in numerous research investigations. It is thought that the existence of fungal fruiting bodies with these characteristics results from defense mechanisms that fungi have developed to help them survive in the environment. A range of naturally occurring chemicals with antibacterial activity against pathogenic organisms are gaining more attention as the problem of bacterial resistance to current antibiotics intensifies Pharmaceuticals 2023,16,12,003 of 32. Notably, the antibiotic chemicals parasol, which were discovered from *Sparassis crispain* in 1920, were among the first compounds with antibacterial action (Sharma et al. 2022). Numerous chemicals having antibacterial properties are known to be produced by fungi (Agrawal and Verma 2021). Antibiotics are substances that prevent bacteria from growing and dividing. Microbiologist Selman Waksman found two antibiotics, streptomycin and neomycin, and came up with the term "antibiotic." These compounds are dual in nature, producing a wide range of activity while also demonstrating selective efficacy against specific bacterial strains (Clardy et al. 2009). Among the most significant antibiotic classes produced by fungi are penicillins, cephalosporins, fusidans, fusafungin, and fumigacin (helvolic acid). Figure 9.2 illustrates the various extraction approaches for *Phallus indusiatus*, highlighting the solvents, compounds obtained, antioxidant activity levels, and assays used for evaluation.

9.3.5.1 Penicillin

Penicillins are members of the β-lactam group of antibiotics, which have a thiazolidine ring paired with a β-lactam ring in their chemical structure. The way penicillins and other β-lactam antibiotics work is by attaching themselves to penicillin-binding proteins (PBPs) and blocking their function. Because they only affect cells where peptidoglycan synthesis takes place, antibiotics in this group have very low levels of general and organ toxicity against human cells. The commercial production of penicillins involves specific species, including *Aspergillus persicinum, Aspergillus flavus, Aspergillus giganteus, Aspergillus nidulans, Aspergillus oryzae,* and *Aspergillus parasiticus (*Bush and Bradford 2016*;* Miller 2002*).*

9.3.5.2 Cephalosporin

In 1948, Giuseppe Brotzu at the very beginning isolated the precursors of cephalosporin from the culture of *Cephalosporium acremonium (*Bo 2000*;* Fernandez et al. 2021*).* Cephalosporins and β-lactam antibiotics work similarly. Strains of *C. acremonium* and *Paecilomyces persicinius* are used to produce these chemicals commercially. Based on their molecular makeup, cephalosporins can be categorized into multiple subgroups (P1–P5) (Sartori et al. 2021). Like all β-lactam antibiotics, cephalosporins hinder the development of a full cell wall by blocking the links that bind peptidoglycan (murein) subunits together. They block the activities of bacterial enzymes by forming covalent bonds with their active sites, specifically carboxypeptidase and transpeptidase. As a result, they prevent the

Fig. 9.2 Extraction pathway in *Phallus indusiatus* extracts

formation of bacterial cell walls. Both Gram-positive and Gram-negative bacteria, as well as infections caused by other bacteria, are treated by cephalosporins. These antibacterial drugs are used to treat infections brought on by a variety of microorganisms, including *E. coli* and *Staphylococcus aureus* (Khan et al. 2019). Cephalosporins can be used to treat microbially induced skin disorders. They have been used in dermatology to treat skin disorders such as folliculitis and surgical site infections (Harrison and Bratcher 2008).

9.3.5.3 Fusidans

Fusidic acid is one of the most well-known fusidans. The synthesis of proteins by Gram-positive bacteria is inhibited by fusidic acid. Fusidic acid was first isolated from *Fusidium coccineum* in 1962 and later recovered from *Isaria kogana* and *Mucor ramannianus* (Cordes 2007; Schöfer and Simonsen 2010).

Nowadays, fusafungine is extracted using biotechnological techniques from species including *Mortierella ramanniana, Fusidium coccophilum,* and *Calcarisporium arbuscula* (Cordes 2007). As a peptide antibiotic, fusafungine exhibits bacteriostatic activity against a wide range of harmful microorganisms. It exhibits independent anti-inflammatory effects in addition to its antibacterial qualities. Inhibiting pro-inflammatory cytokines, increasing NK cell activity, and inducing lymphocytes to generate IL-2 are the likely mechanisms of action (Levy et al. 1995).

9.3.5.4 Isoprenoids

The broad collection of secondary metabolites known as isoprenoid compounds is present in Basidiomycota. These substances are inherently connected to the biogenetic process that develops "active isoprene" by starting with active acetate and moving through mevalonic acid. Monoterpenes, sesquiterpenes, diterpenes, triterpenes, tetraterpenes, and steroids are produced as a result of the latter's further modifications (Savidov et al. 2018; Ninkuu et al. 2021).

9.4 Historical Use of Fungi in Beauty Products

Many cultures and eras have used fungus for skincare and beauty, and ancient methods acknowledged its therapeutic and aesthetic benefits long before contemporary science did. Fungi have been used in cosmetics for centuries across cultures, valued for skin health, healing, and natural coloring. Yeast, reishi, chaga, fermented rice fungi, and birch polypore have served roles from anti-aging and blemish reduction to inflammation relief, while various fungi also provided eco-friendly natural dyes (Fig. 9.3). An outline of the historical applications of fungi in beauty is shown below:

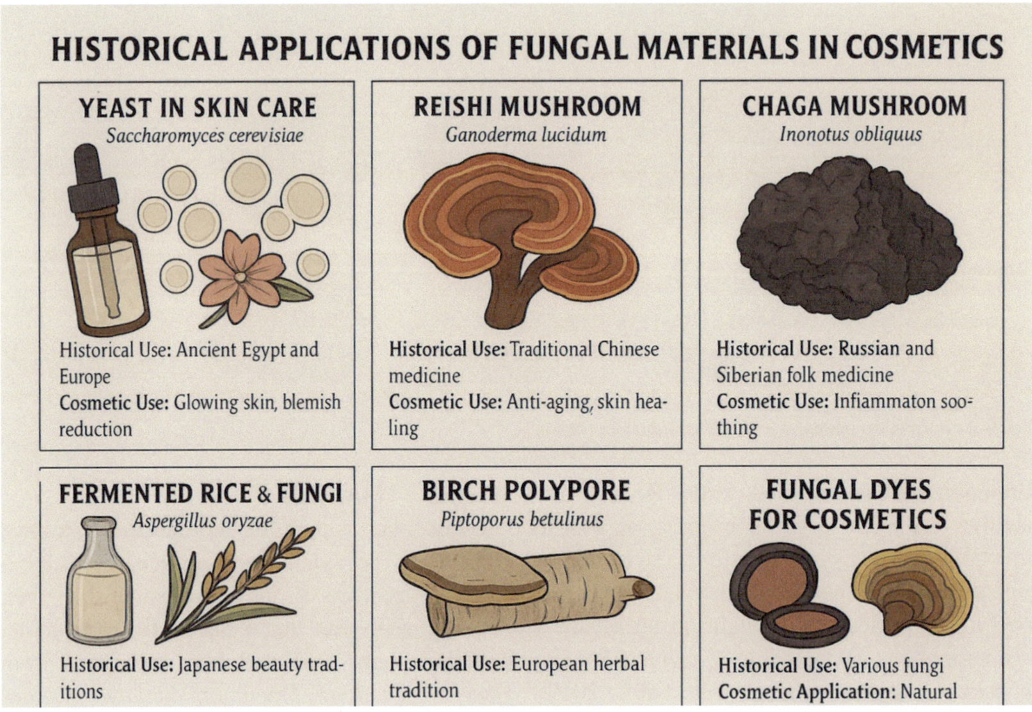

Fig. 9.3 Examples of historical applications of fungal materials in cosmetics

9.4.1 Traditional Chinese Medicine (TCM)

Since more than 2000 years ago, fungi such as *Tremella fuciformis* and Reishi (*Ganoderma lucidum*) have been utilized. Reishi was known as the "Mushroom of Immortality" and was thought to promote youthful skin, longevity, and vitality. Yang Guifei, a famously beautiful imperial concubine, utilized tremella, also called the "beauty mushroom," as a natural moisturizer and anti-aging cream. People ate these fungi as teas, soups, or topically as poultices (Wasser 2002).

9.4.2 Japanese and Korean Beauty Rituals

Because of its skin-brightening and detoxifying qualities, shiitake mushrooms (*Lentinula edodes*) were utilized in traditional Japanese beauty regimens. Traditionally, shiitake was mashed into pastes or used in face steams because it was rich in kojic acid, a natural skin lightener. Because of their anti-inflammatory and skin-evening properties, these fungi are still used in serums, masks, and moisturizers in modern K-beauty (Kaur 2021).

9.4.3 Ayurvedic Practices

In Ayurveda, some fungi, such as *Cordyceps sinensis*, have long been used to support skin vitality and stamina, albeit they are not as common as herbs. *Cordyceps* was frequently employed in tonics that were thought to improve skin tone and texture and restore prana, or life energy (Tuli et al. 2013).

9.4.4 Apothecaries in Europe

Fungi like yeast (*Saccharomyces cerevisiae*) were utilized in basic skincare procedures in medieval Europe to cure skin irritations and improve complexion. Because of its high vitamin B content and skin-soothing qualities, yeast was included to tonics and face masks.

9.5 Growing Popularity of Fungal-Based Ingredients

Both consumer demand for natural alternatives and scientific validation have contributed to a discernible increase in the use of fungal-based ingredients in skincare and wellness products. Long utilized in traditional medical systems such as Ayurveda and Traditional Chinese Medicine, mushrooms are now being acknowledged by contemporary research for their diverse range of bioactive chemicals. These include antioxidants, polyphenols, triterpenes, polysaccharides (particularly β-glucans), and polysaccharides, which provide anti-inflammatory, hydration, and skin barrier repair effects (Qin et al. 2023).

The "snow mushroom," *Tremella fuciformis*, is one of the most commonly utilized species. Because of its exceptional water-binding properties, its polysaccharides function as a natural humectant, much like hyaluronic acid. Numerous cosmetic studies and formulations emphasize its capacity to hydrate the skin, improve elasticity, and create a plumping effect without leaving the skin feeling oily (Raut and Adhikari 2021). Another well-known species is *Ganoderma lucidum*, sometimes known as reishi, which has long been used in traditional medicine and is now used in cosmetics because of its calming and antioxidant qualities. It helps counteract the effects of environmental stresses, promotes skin regeneration, and lessens redness (Wasser 2011).

In herbal medicine, *Cordyceps sinensis* is used to boost resilience and energy. It is also being investigated for topical applications for its anti-fatigue, rejuvenating, and tone-evening properties on dull or weary skin (Hui et al. 2005). Consumers looking for natural substitutes will find these components appealing, but they also give manufacturers sustainable, eco-friendly active ingredients that complement the trend toward biotechnological innovation in cosmetic product creation.

9.6 Cosmetic Antimicrobial Agents

Cosmetic antimicrobial agents are those substances which are used in the cosmetics to inhibit the growth of microorganisms such as bacteria, fungi, and yeast. These agents help to preserve the product, extend shelf life, and protect users from infections.

9.6.1 Potential Application of Fungi with Antibacterial Activity in Cosmetics

One of the most common fungal source materials in East Asian traditional medicine, including TCM, is *Ganoderma lucidum* (Paterson 2006).

Additional research has verified that *G. lucidum* acetone extract has antibacterial properties, primarily against Gram-negative *K. pneumoniae* bacteria. Furthermore, when *G. lucidum* extracts were combined with antimicrobial drugs such ampicillin, cefazolin, oxytetracycline, and chloramphenicol, a synergistic interaction was noted. When cefazolin was used against *B. subtilis* and *Klebsiella oxytoca*, this synergy was very noticeable (Yoon et al. 1994). The effectiveness of several extracts, such as water, ethanol, methanol, and acetone, against the aminoglycoside antibiotic gentamicin sulfate has been noteworthy. This efficiency holds true for a variety of bacterial species: *B. subtilis, S. typhimurium, P. aeruginosa, K. pneumoniae, S. aureus*, and *E. coli* (Quereshi et al. 2010).

9.6.2 Potential Application of Fungi with Antifungal Activity in Cosmetics

Researchers are now looking at natural substances with antifungal qualities because of worries about the toxicity of synthetic azole derivatives and polyene antibiotics. The hunt for alternate resources is prompted by the growing resistance of *Candida* species to conventional antifungal medications. Both low-molecular-weight chemicals like terpenes (sesquiterpenes), steroids, and organic acids, as well as high-molecular-weight compounds like proteins and peptides, are responsible for the antifungal activity of substances made by fungus (Choudhary et al. 2022).

9.6.3 Potential Application of Fungi with Antiviral Activity in Cosmetics

The antiviral effects of fungal-derived compounds are usually through inhibition of viral enzymes, disruption of the production of nucleic acids, or via indirect immune stimulatory effects. As many chemical compounds proven to be antiviral have been registered as a drug, continuous intensive research is being undertaken to search for active natural substances (fungal, for example) and report scientifically on the antiviral activity of both fruiting body extracts and isolated compounds (Linnakoski et al. 2018). For example, triterpenes ganoderiol (Ganodermanontriol) and Ganodermic acid, *G. lucidum*-derived, have activity against HIV-1 (Galappaththi et al. 2022). The antiviral activity in the aqueous–methanol extracts isolated from the fruiting bodies of *L. sulphureus* was found to be due to the inhibition of HIV reverse transcriptase, which is essential to the transcription process and thus the suppression of virus replication. To our knowledge, the antiviral effect reported in the extracts is thought to be due to immune modulatory polysaccharides present therein (Kac and Pohleven 2005).

The ethanol extract of *Phellinus baumii* fruiting bodies was used to separate polyphenols. Polyphenols identified as hispidin, hypholomine B, inoscavin A, davallialactone, and pelligridin D were subjected to scatter spectroscopy. All of these demonstrated a reduction in the virus-induced cytopathic effect (CPE) and an inhibitory effect on the activity of the influenza virus's neuraminidase (H1N1, H5N1, and H3N2). An enzyme called virus neuraminidase breaks down the cell membrane of an infected cell, allowing viruses to leave the cell. Neuraminidase's strong affinity for salic acid of membrane receptors makes it possible for viruses to bind to the cell membrane and enter the cell (Hwang et al. 2015).

9.7 Cosmetic Applications of Antimicrobial Fungal-Derived Compounds

The potential advantages of antimicrobial fungal-derived chemicals in treating and preventing microbiological illnesses have drawn a lot of interest in a variety of industries, including cosmetics. These compounds are used in a variety of cosmetic formulations, especially in skincare and haircare products (Taofiq et al. 2016b; Hyde et al. 2010; Wu et al. 2016; Bowe and Pugliese 2014).

Recent research has thoroughly examined the antimicrobial qualities of mushrooms, demonstrating their potential as important sources of organic substances with antioxidant and antibacterial effects. Eiamthaworn et al.'s study concentrated on *Cordyceps militaris* extracts and assessed their biological activity against skin. The preservation properties of bioactive substances obtained from mushrooms are among their most important practical uses. The antimicrobial properties of fungus metabolites provide a strong natural alternative to traditional synthetic preservatives, extending the shelf life of cosmetic formulations (Taofiq et al. 2016c).

Results show that extracts from *Cordyceps militaris* have strong antibacterial activity against a range of skin pathogens, including pathogenic bacteria. This research highlights mushrooms' potential as a natural source of antimicrobial compounds, which could lead to the creation of treatments for skin infections (Eiamthaworn et al. 2022).

Conversely, Taofiq et al. outline the creation of cosmeceutical formulations based on mushrooms. These mixtures are intended to have both antibacterial and anti-inflammatory qualities. Furthermore, formulations designed for sensitive cutaneous diseases or acne-prone individuals can benefit from the use of mushroom metabolites. Their antibacterial properties can be used to successfully control the growth of germs linked to acne. Furthermore, including fungal metabolites with antimicrobial properties into sun protection formulations is a clever way to make use of them. These products provide a holistic approach to skin health by combining antibacterial qualities, cutaneous hydration preservation, and sun protection (Shankar and Sharma 2022) antioxidant properties, and antityrosinase. The study highlights the potential of mushroom chemicals to treat several skin-related issues at once, demonstrating their multifunctional nature. According to this study, mushrooms can be used as a variety of ingredients in skin care products, providing advantages that go beyond their traditional antibacterial properties. Studies indicate that substances derived from mushrooms might contribute to the development of successful skin care products, especially in the fight against skin infections, the reduction of inflammation, and the provision of antioxidant support (Taofiq et al. 2016c).

9.7.1 Potential Practical Applications

The preservation properties of bioactive substances obtained from mushrooms are among their most important practical uses. The antibacterial properties of fungus metabolites provide a strong natural alternative to traditional synthetic preservatives, extending the shelf life of cosmetic compositions (Paul and Joshi 2022). Additionally, the use of mushroom metabolites can be seen in formulations designed for sensitive or acne-prone skin problems. They can successfully control the growth of microorganisms linked to acne pathology by using their antibacterial properties (Shankar and Sharma 2022). Because of its medicinal and preservation qualities, fungus-based compounds are being used more and more in cosmetics. Here are a few noteworthy applications of antimicrobial fungal-derived compounds:

9.7.1.1 Natural Preservatives

Fungal substitutes are being investigated because synthetic preservatives, such as parabens, have come under fire for possible health hazards. Derived from the cell walls of fungi, chitosan is a non-toxic, biodegradable preservative that works well against a variety of spoiling organisms, such as *Candida albicans* and *Escherichia coli (*Kulka and Sionkowska 2023*)*. Its cationic composition allows it to adhere to negatively charged microbial surfaces, boosting its potency.

9.7.1.2 Anti-acne Formulations

The Gram-positive bacterium *Cutibacterium acnes*, formerly known as *Propionibacterium acnes*, is frequently the cause of acne. A natural substitute for harsh chemical therapies like benzoyl peroxide, fungus metabolites from *Trichoderma* species have demonstrated promise in suppressing this infection (Sułkowska-Ziaja et al. 2023).

9.7.1.3 Antifungal Skin Care

Trichophyton species and *Malassezia fungi*, respectively, are the causes of athlete's foot and dandruff, and fungal-derived antifungals can treat these conditions. These infections can be targeted by polyene-like chemicals derived from fungi without significantly altering the skin's natural microbiota (Kaur et al. 2021).

9.7.1.4 Products for Wound Healing

Chitin and chitosan help wounds heal by promoting tissue regeneration and providing antimicrobial protection. Creams and ointments made for sensitive or injured skin benefit greatly from these qualities (Rinaudo 2006).

9.7.1.5 Antioxidant and Anti-aging Agents

Pleurotus species (mushrooms) contain ergothioneine, a fungus metabolite that is a strong antioxidant that combats free radicals and shields skin from oxidative stress and early aging (Cheah and Halliwell 2012). Similarly, by preventing the breakdown of collagen, terpenoids from *Ganoderma lucidum*, or *reishi* mushrooms, have demonstrated potential in minimizing wrinkles (Dewanjee et al. 2024).

9.7.1.6 Moisturizing and Barrier Support

Beta-glucans, which are derived from fungus such as *Saccharomyces cerevisiae*, improve the skin barrier and replicate the moisturizing effects of hyaluronic acid. According to studies, they work better than other synthetic humectants at enhancing moisture retention (Lee et al. 2024).

9.7.1.7 Natural Pigments

Fungal pigments have gained attention due to the need for clean-label colorants. For example, red and yellow pigments produced by *Monascus* species serve as antioxidants and are a safer substitute for artificial dyes that cause skin discomfort (Dufossé 2018).

9.7.1.8 Exfoliation and Skin Renewal

Aspergillus oryzae proteases and other fungal enzymes gently break down dead skin cells, giving sensitive skin types a natural exfoliation option (Kumar et al. 2023).

9.7.1.9 UV Protection

Fungi like *Aureobasidium pullulans* have melanin that absorbs UV rays, which makes it a promising bio-based sunscreen ingredient with less negative environmental effects than chemical filters.

9.7.2 Microbiological Factors Causing Skin Diseases

The root cause of many skin conditions, including fungal infections and acne, is pathogenic microorganisms. Metabolites from mushrooms with strong antibacterial and antifungal properties have encouraging potential for treating these infections. They have the ability to specifically target and eliminate harmful microorganisms while also promoting skin health when added to topical formulations. Model bacteria like *S. aureus* are prevalent in the field of bacterially induced skin diseases. This species frequently causes anything from cellulitis to boils and impetigo, as well as more serious symptoms like methicillin-resistant *S. aureus* (MRSA) infections. Research investigators are focused on understanding the virulence factors and mechanisms that enable *S. aureus* to colonize and invade the skin in order to create effective antibiotic treatments (Turner et al. 2019).

Furthermore, *Streptococcus pyogenes*, also known as Group A *Streptococcus* (GAS), is a cause of conditions like impetigo, erysipelas, and cellulitis and needs careful examination to fully understand its pathogenic complexities (Avire et al. 2021). Likewise, *P. acnes* is prominent as the causative agent of acne vulgaris, a common dermatological issue. The function of *P. acnes* in the development of acne, its interaction with the immune system, and possible treatment approaches that target this bacterium have all been the subject of scientific investigation (Platsidaki and Dessinioti 2018).

Trichophyton species are common among dermatophytes that attack the integument and are responsible for a variety of fungal skin illnesses, including *Tinea corporis*, *Tinea pedis* (athlete's foot), and *Tinea cruris* (jock itch). Parallel disorders like athlete's foot are caused by a different group of fungi that includes *Microsporum* and *Epidermophyton*. In order to combat this variety of microorganisms, a thorough investigation of metabolites generated from mushrooms that possess strong antibacterial and antifungal properties highlights a promising path for developing treatment strategies for various skin illnesses (Wang et al. 2021).

9.7.3 Selected Compounds of Fungal Origin with Cosmetic Properties

Ergothioneine is a notable example of an exogenous amino acid that is produced from thiourea-histidine. High water solubility and stability at elevated temperatures are characteristics of this chemical. Macrophages are its primary source, and their ability to chelate metal ions aids in preventing the body from accumulating reactive oxygen species. Ergothioneine also assists in regulating immune system function and possesses antibacterial and anti-inflammatory actions (Taofiq et al. 2016b; Bowe and Pugliese 2014). The fungus-derived ingredient ergothioneine is widely used in cosmetic products. The skin is protected from the damaging effects of free radicals by its protective qualities against fibroblasts' and keratinocytes' damage to the DNA of mitochondria. Preparations that fight aging are their main use. Ergothioneine is a commonly used resource in the field of cosmetics that is extracted from *A. bisporus*, *G. frondosa*, and *L. edodes* (Hyde et al. 2010).

The complex structure of poly-N acetyl-1,4-D-glucosamine subunits connected by (1–4) glycosidic linkages makes up chitin, which is what gives mushrooms their cell walls. Hygroscopic and sorptive properties are displayed by this chemical, which is insoluble in water. It is noteworthy that chitin exhibits anti-inflammatory, antiviral, and antibacterial qualities (Ifuku 2014).

Chitin is used in cosmetics and is mostly valued for its hydrating properties. By restoring skin elasticity, producing antioxidant effects, and promoting collagen synthesis, it provides antiaging, firming, and regeneration advantages. It effectively maintains the proper moisture levels and is especially noticeable in formulations designed for dry and aged skin (Park and Kim 2010).

Polysaccharides are common in both filamentous and macromycete fungi, and their chemical structures dictate their biological activity. Fungal polysaccharides, which are mostly represented by glucans, can also have galactan or mannan structures. The arrangement of the polysaccharide molecule and the kind of glycosidic link (or) have a major influence on its biological activity. Glucans, which have fundamental biological function, are particularly noteworthy. These polysaccharides frequently have significant molecular weights between 100 and 1000 kDa. The linear structures of the most active polysaccharides, which lack lengthy side chains, promote improved solubility and absorption. Usually, there are three bonds in the main chain, and branching happens through six bonds (Ifuku 2014). Polysaccharides derived from mushrooms are used in cosmetic preparations because of their antioxidant qualities, which include promoting the production of collagen, shielding skin cells from ultraviolet light, exhibiting anti-inflammatory and antimicrobial properties, and having moisturizing effects (Hosseinnejad and Jafari 2016).

A group of closely similar substances known as tocopherols and tocotrienols are collectively referred to as vitamin E. They share a structure that includes an isoprene side chain and a chromanol. This vitamin is known for its capacity to scavenge free radicals, which is believed to protect against degenerative diseases like cancer and heart disease. Additionally, vitamin E demonstrates effectiveness in thwarting the growth of a variety of microorganisms. This beneficial quality of vitamin E adds to its possible function in strengthening the body's defenses against infections and supporting the health of the immune system. In the realm of cosmetology, vitamin E holds significant sway due to its favorable impacts on the skin, encompassing antioxidant, moisturizing, regenerative, UV radiation shielding, anti-inflammatory, and anti-aging properties (Fiume et al. 2018). One essential tactic for skin lightening is to inhibit tyrosinase, the enzyme that forms melanin. Many substances used for skin brightening modify tyrosinase activity (Sharif et al. 2018). Among the substances obtained from fungi that have skin-whitening properties are 3,4-dihydroxybenzaldehyde, azelaic acid, and kojic acid (Taofiq et al. 2016c). Kojic acid, which functions similarly to hydroquinone, is an essential inhibitor of melanogenesis. It operates by blocking tyrosinase activity while further delivering antibacterial and antioxidant characteristics. This chemical is largely generated by fungus species belonging to the *Aspergillus* and *Penicillium* genera (Chang 2009).

9.7.4 Commercially Available Cosmetics with Cosmetic Properties

Commercially available cosmetics are products designed primarily to enhance appearance, improve skin condition, and promote personal hygiene. These include a wide range of items like moisturizers, foundations, lipsticks, mascaras, sunscreens, and anti-aging creams. Modern cosmetics are more than just aesthetic enhancers; thanks to advancements in dermatological research and cosmetic chemistry, many now offer functional benefits often referred to as "cosmeceuticals." These cosmetic properties include anti-aging effects, UV protection, acne treatment, hydration, and skin brightening. Because of their dual role, cosmetics are essential in daily skincare routines.

9.7.4.1 Skin Hydration and Moisturizers

Hyaluronic acid (HA): HA's superior moisture-retention qualities make it a common ingredient in cosmetics. The effectiveness of HA varies according to its molecular weight; smaller molecular weights penetrate deeper for anti-aging effects, while greater molecular weights are effective for hydration. Serums and moisturizers frequently include HA (Kim et al. 2018).

***Aloe Vera* extracts:** Because of its calming and moisturizing qualities, *aloe vera* is employed in cosmetic compositions. *Aloe vera* extract compositions have been demonstrated in studies to increase skin moisture through a humectant mechanism (Fonseca-Santos et al. 2015).

9.7.4.2 Anti-aging and Skin Rejuvenation

Resveratrol: An antioxidant and anti-inflammatory polyphenolic molecule that is present in berries and grapes, resveratrol is used in cosmetics. It assists in defending the skin from oxidative stress and photoaging (Liu 2022).

Cocoa-derived theobromine has been used in anti-aging treatments because it is effective at preventing UV-induced skin damage and increasing skin suppleness (Scapagnini et al. 2014).

9.7.4.3 UV Protections and Sunscreen

Nanoparticles of zinc oxide and titanium dioxide are frequently found in sunscreens due to their capacity to scatter and reflect ultraviolet light. When formulated at the nano-scale, they offer better protection from the sun while having less opacity than their bulk counterparts (Maghraby et al. 2024).

MAAs, or mycosporine-like amino acids: Sunscreens contain algae-derived MAAs because of their ability to absorb UV rays and provide protection against UV-induced skin damage (Gupta et al. 2019).

9.7.4.4 Hair Care and Condition

Fibroin and sericin: These two types of silk proteins are employed in hair care products because of their ability to moisturize and condition hair. They aid in preventing brittleness and enhancing the smoothness and shine of hair.

Algal Oils: High in omega-3 fatty acids, oils made from microalgae are used in hair care products to nourish hair follicles, lessen dandruff, and encourage healthy hair development (Gupta et al. 2019).

9.7.4.5 Anti-pollutant and Skin Barrier Protection

***Calendula Officinalis* Extract:** Because of its anti-inflammatory and skin barrier-repairing qualities, calendula extract is utilized in cosmetics. According to studies, it can improve skin hydration and barrier function by lowering trans-epidermal water loss and reducing skin erythema (Scapagnini et al. 2014).

9.8 Challenges of Incorporating Fungal-Based Ingredients into Cosmetic Formulation

Fungal-based compounds have bioactive qualities that make them promising for use in cosmetic compositions. Nonetheless, several obstacles need to be overcome to guarantee the product's effectiveness, safety, and customer acceptance. Here is a thorough explanation of these difficulties backed up by academic sources:

9.8.1 Shelf-Life and Stability

Polysaccharides and phenolic acids, which are chemicals generated from fungi, can be sensitive to light, pH, and temperature. Careful formulation techniques are needed to guarantee the stability of these bioactives throughout the product's shelf life. For example, research has demonstrated that adding mushroom extracts to cosmetic creams can preserve their bioactivity; nevertheless, careful process optimization is required to avoid degradation (Taofiq et al. 2016c).

9.8.2 Standardization and Uniformity

Inconsistencies in the concentration of active chemicals may result from natural variability in fungal species, growing circumstances, and extraction techniques. Achieving batch-to-batch uniformity, which is essential for both product performance and regulatory compliance, is made more difficult by this unpredictability.

9.8.3 Sensory Attributes

Certain fungi can give cosmetics unfavorable textures, hues, or smells. These sensory attributes may have an impact on consumer acceptance, requiring formulation modifications or the use of masking agents to provide a satisfying user experience (Wu et al. 2016).

9.8.4 Compatibility of Formulations

Interactions between fungus extracts and other cosmetic components may result in instability or decreased effectiveness. For instance, careful formulation is needed when including *Tremella fuciformis* extract to preserve microbiological purity and guarantee compatibility with other ingredients (Woźniak et al. 2023).

9.8.5 The Dangers of Microbial Contamination

Microbial contamination of fungus-based components might jeopardize product safety and shelf life. The health concerns posed by contaminants such as *Aspergillus flavus* and *Candida* species highlight the necessity of strict quality control procedures during production and storage.

9.8.6 Perception of the Consumer

For some consumers, the word "fungus" may evoke negative associations with mold or illnesses. To dispel doubt and encourage acceptance, it is crucial to inform customers about the advantages and security of chemicals derived from fungi.

9.9 Regulatory Requirements and Safety Consideration

In sectors like food, chemicals, medicines, and consumer goods, regulatory requirements and safety considerations are essential to ensuring legal compliance, environmental protection, and public health. Minimum requirements for worker safety, manufacturing procedures, product quality, and cleanliness are established by these regulations. Following these rules safeguards customers, employees, and businesses' reputations. In a competitive market, it is critical to be abreast of safety procedures and regulatory developments.

9.9.1 Regulation and Licensing of Cosmeceutical

The pharmaceutical registration procedure shouldn't be as time-consuming as the drug registration procedure. Good Clinical Practices (GCP) must be applied during the evaluation in order to provide pertinent clinical reports regarding its therapeutic efficacy. It should be noted that this product should only be used when homeostasis is maintained or when moderate skin diseases are present. Cosmetics are not meant to solve serious or enduring problems with the skin and hair. The therapeutic efficacy date of the Cosmos article must be supplied along with toxicity studies. The toxicity studies must suggest that there is low chance of adverse effects for treating minor skin and hair issues. Typically, in Western nations, cosmetics are registered under the label "Over The Counter (OTC) product". The idea that a substance can only be classified as a medication or cosmetic based on its pharmacological activity is up for legal discussion. Minoxidil, for instance, can be utilized (Epstein 2009).

9.9.2 Safety Concerns of Cosmetics

Although many people throughout the world view cosmetics as a blessing, it is important to recognize the risks and toxicity of these products. Given that many people use cosmetics, safety should be guaranteed and included into the products during the production process. Additionally, excessive use of cosmetics can result in toxicity and long-term adverse effects, some of which are permanent.

9.9.2.1 Safety Concern

- Products related to cosmetics and personal hygiene may raise the following safety issues: eye infections that are severe.
- Ocular irritation and scrapes.
- Product contaminants made outside of the USA.
- Allergy or intolerance to ingredients included in the product.
- Bacterial dispersion on the skin's surface.
- Fire risks associated with aerosol products.

A certain amount of time cannot be exceeded by the cosmetics' effect. Numerous regulatory agencies, including the US FDA, recommend washing off cosmetics before bed (Bhuvaneswari and Chandan 2018).

The following steps must be taken to insure safer use of the cosmetic:

The first step in comprehending the safety of cosmetics is to be aware of the substances they contain. This can be accomplished by looking through databases such as skin deep cosmetics by EWG, which includes information on ingredient lists for more than 69,000 distinct cosmetic goods. These databases allow for searches to be conducted by manufacturer, product name, ingredient, and even brand name. A hazard key score, which ranges from 1 to 10, can be used to display the hazard nature of the components in a particular cosmetic product. The product's safety and the danger key score are adversely correlated. Certain database programs, such as the California Safe Cosmetics Program Product Database (CSCP), offer information about the connection between an ingredient and a particular illness or condition.

9.10 Future Prospect and Research Opportunities

To maximize the extraction and formulation of fungal-based antimicrobial agents used in cosmetics, more study is required. The development of biotechnology-based techniques like genetic engineering and fungal fermentation may increase the stability and effectiveness of these components and encourage the broad usage of antimicrobial agents in the cosmetics sector.

9.11 Conclusion

An extensive review of fungal-based antimicrobial agents used in cosmetics is provided in this chapter. Consumer demand for cosmetics and cosmeceuticals is rising, particularly for the prevention and amelioration of indications of skin aging. The process of aging is complicated and involves both hereditary and environmental variables. The primary issues are typically oxidative stress and the resulting changes to the extracellular matrix. In order to alter these molecular mechanisms and create cosmeceuticals, the industry is currently searching for bioactive substances with particular activities. The cosmetics industry faces a number of challenges, including the growing demand from consumers for cosmetics and cosmeceuticals from businesses that address their environmental concerns, disclose the sources of ingredients, offer products free of ingredients derived from animals, or guarantee that specific cultural and religious concerns are taken into consideration. Fungal-based active ingredients have a promising antimicrobial effect in cosmetics, with promise to protect both the product integrity and skin health. Fungal active ingredients can combat pathogenic microbes as well as support the body's microbiome, making them ideal components in skincare and personal care formulations. With increasing awareness for natural and sustainable ingredients, the development of fungi in cosmetics is expected to continue exponentially, providing new and innovative approaches for beauty.

References

Agrawal K, Verma P (2021) Fungal metabolites: a recent trend and its potential biotechnological applications. In: New and future developments in microbial biotechnology and bioengineering. Elsevier, pp 1–14

Alam N et al (2010) Antioxidant activities and tyrosinase inhibitory effects of different extracts from Pleurotus ostreatus fruiting bodies. Mycobiology 38(4):295–301

Amberg N, Fogarassy C (2019) Green consumer behavior in the cosmetics market. Resources 8(3):137

Angeli JPF et al (2009) β-Glucan extracted from the medicinal mushroom Agaricus blazei prevents the genotoxic effects of benzo [a] pyrene in the human hepatoma cell line HepG2. Arch Toxicol 83(1):81–86

Aranaz I et al (2018) Cosmetics and cosmeceutical applications of chitin, chitosan and their derivatives. Polymers 10(2):213

Association CETPC (2019) Socio-economic contribution of the european cosmetics industry

Avire NJ, Whiley H, Ross K (2021) A review of streptococcus pyogenes: public health risk factors, prevention and control. Pathogens 10(2):248

Balboa EM et al (2015) Cosmetics from marine sources. In: Springer handbook of marine biotechnology, pp 1015–1042

Basch E et al (2006) Marigold (Calendula officinalis L.) an evidence-based systematic review by the natural standard research collaboration. J Herb Pharmacother 6(3–4):135–159

Bhuvaneswari V, Chandan R (2018) Review on safety assessment of a cosmetic product. Int J Curr Pharm Res 10(3):1–6

Bo G (2000) Giuseppe Brotzu and the discovery of cephalosporins. Clin Microbiol Infect 6:6–8

Bowe WP, Pugliese S (2014) Cosmetic benefits of natural ingredients. J Drugs Dermatol 13(9):1021–1025

Bush K, Bradford PA (2016) β-Lactams and β-lactamase inhibitors: an overview. Cold Spring Harb Perspect Med 6(8):a025247

Calvani F et al (2020) Innovation in skin regeneration: novel approaches with β-1, 3/1, 6-Glucan-based treatments. Recent Progress Mater 2(2):1–1

Chambers GE, Sayan AE, Brown RC (2021) The synthesis of biologically active indolocarbazole natural products. Nat Prod Rep 38(10):1794–1820

Chang T-S (2009) An updated review of tyrosinase inhibitors. Int J Mol Sci 10(6):2440–2475

Cheah IK, Halliwell B (2012) Ergothioneine; antioxidant potential, physiological function and role in disease. Biochim et Biophys Acta (BBA)-Mol Basis Dis 1822(5):784–793

Chien C-C et al (2008) Effects on tyrosinase activity by the extracts of Ganoderma lucidum and related mushrooms. Mycopathologia 166:117–120

Choudhary M et al (2022) Antifungal metabolites, their novel sources, and targets to combat drug resistance. Front Microbiol 13:p. 1061603

Clardy J, Fischbach MA, Currie CR (2009) The natural history of antibiotics. Curr Biol 19(11):R437–R441

Coma V, Deschamps A, Martial-Gros A (2003) Bioactive packaging materials from edible chitosan polymer—antimicrobial activity assessment on dairy-related contaminants. J Food Sci 68(9):2788–2792

Cordes MG (2007) Fusidic acid

Czumaj A, Śledziński T (2020) Biological role of unsaturated fatty acid desaturases in health and disease. Nutrients 12(2):356

Deacon J (2005) Introduction: the fungi and fungal activities. Fungal Biol:1–15

Dembitsky VM et al (2005) Secondary metabolites of slime molds (myxomycetes). Phytochemistry 66(7): 747–769

Dewanjee D et al (2024) Ganoderma in skin. In: Ganoderma: cultivation, chemistry, and medicinal applications, vol 2, p 79

Dini I (2018) Spices and herbs as therapeutic foods. In: Food quality: balancing health and disease. Elsevier, pp 433–469

Dini I, Laneri S (2021) The new challenge of green cosmetics: natural food ingredients for cosmetic formulations. Molecules 26(13):3921

Dini I et al (2020) An environmentally friendly practice used in olive cultivation capable of increasing commercial interest in waste products from oil processing. Antioxidants 9(6):466

Dini I et al (2021) Trichoderma strains and metabolites selectively increase the production of volatile organic compounds (VOCs) in olive trees. Metabolites 11(4): 213

Du Y-L, Ryan KS (2015) Expansion of bisindole biosynthetic pathways by combinatorial construction. ACS Synth Biol 4(6):682–688

Du B, Bian Z, Xu B (2014) Skin health promotion effects of natural beta-glucan derived from cereals and microorganisms: a review. Phytother Res 28(2): 159–166

Dufossé L (2018) Microbial pigments from bacteria, yeasts, fungi, and microalgae for the food and feed industries. In: Natural and artificial flavoring agents and food dyes. Elsevier, pp 113–132

Eiamthaworn K et al (2022) Efficacy of Cordyceps militaris extracts against some skin pathogenic bacteria and antioxidant activity. J Fungi 8(4):327

Epstein H (2009) Cosmeceutical vehicles. Clin Dermatol 27(5):453–460

Fernandez J, Jimenez-Rodriguez TW, Blanca-Lopez N (2021) Classifying cephalosporins: from generation to cross-reactivity. Curr Opin Allergy Clin Immunol 21(4):346–354

Fiume MM et al (2018) Safety assessment of tocopherols and tocotrienols as used in cosmetics. Int J Toxicol 37(2_suppl):61S–94S

Fonseca-Santos B, Corrêa MA, Chorilli M (2015) Sustainability, natural and organic cosmetics: consumer, products, efficacy, toxicological and regulatory considerations. Braz J Pharm Sci 51(1):17–26

Galappaththi MC et al (2022) A review of Ganoderma triterpenoids and their bioactivities. Biomolecules 13(1):24

Gao Y et al (2003) Effects of Ganopoly®(a ganoderma lucidum polysaccharide extract) on the immune functions in advanced-stage cancer patients. Immunol Investig 32(3):201–215

Guillerme J-B, Couteau C, Coiffard L (2017) Applications for marine resources in cosmetics. Cosmetics 4(3):35

Gupta PL et al (2019) Eminence of microbial products in cosmetic industry. Nat Products Bioprospect 9:267–278

Gurung N et al (2013) A broader view: microbial enzymes and their relevance in industries, medicine, and beyond. Biomed Res Int 2013(1):329121

Harrison CJ, Bratcher D (2008) Cephalosporins: a review. Pediatr Rev 29(8):264–273

Hobbs C (2005) The chemistry, nutritional value, immunopharmacology, and safety of the traditional food of medicinal split-gill fugus Schizophyllum commune Fr.: Fr.(Schizophyllaceae). A literature review. Int J Med Mushrooms 7(1&2):127

Hosseinnejad M, Jafari SM (2016) Evaluation of different factors affecting antimicrobial properties of chitosan. Int J Biol Macromol 85:467–475

Huang H-C et al (2014) Inhibition of melanogenesis in murine melanoma cells by Agaricus brasiliensis methanol extract and anti-reactive oxygen species (ROS) activity. Afr J Microbiol Res 8(6):519–524

Hui KP-Y, Sit W-H, Wan JM-F (2005) Induction of S phase cell arrest and caspase activation by polysaccharide peptide isolated from Coriolus versicolor enhanced the cell cycle dependent activity and apoptotic cell death of doxorubicin and etoposide, but not cytarabine in HL-60 cells. Oncol Rep 14(1):145–155

Hwang BS et al (2015) Anti-influenza activities of polyphenols from the medicinal mushroom Phellinus baumii. Bioorg Med Chem Lett 25(16):3256–3260

Hyde K, Bahkali A, Moslem M (2010) Fungi – an unusual source for cosmetics. Fungal Divers 43:1–9

Ifuku S (2014) Chitin and chitosan nanofibers: preparation and chemical modifications. Molecules 19(11): 18367–18380

Ikekawa T (2001) Beneficial effects of edible and medicinal mushrooms on health care. Int J Med Mushrooms 3(4):2

Jakubczyk D, Dussart F (2020) Selected fungal natural products with antimicrobial properties. Molecules 25(4):911

Jeong YT et al (2008) Ganoderma applanatum: a promising mushroom for antitumor and immunomodulating activity. Phytother Res 22(5): 614–619

Juliano C, Magrini GA (2018) Cosmetic functional ingredients from botanical sources for anti-pollution skincare products. Cosmetics 5(1):19

Kac J, Pohleven F (2005) Screening of selected wood-damaging fungi for the HIV-1 reverse transcriptase inhibitors. Acta Pharma 55:69–79

Kaur R (2021) Chemopreventive effects of mushrooms on breast cancer. Alabama Agricultural and Mechanical University

Kaur N et al (2021) A review on antifungal efficiency of plant extracts entrenched polysaccharide-based nanohydrogels. Nutrients 13(6):2055

Khan DA et al (2019) Cephalosporin allergy: current understanding and future challenges. J Allergy Clin Immunol Pract 7(7):2105–2114

Kim SY et al (2014) Extract of the mycelium of T. matsutake inhibits elastase activity and TPA-induced MMP-1 expression in human fibroblasts. Int J Mol Med 34(6):1613–1621

Kim H et al (2018) Seeking better topical delivery technologies of moisturizing agents for enhanced skin moisturization. Expert Opin Drug Deliv 15(1):17–31

Kogan G et al (2005) Antioxidant properties of yeast (1→3)-β-D-glucan studied by electron paramagnetic resonance spectroscopy and its activity in the adjuvant arthritis. Carbohydr Polym 61(1):18–28

Kohlmünzer S (1998) Farmakognozja: podręcznik dla studentów farmacji. Państwowy Zakład Wydawnictw Lekarskich

Korn ED, Greenblatt CL, Lees AM (1965) Synthesis of unsaturated fatty acids in the slime mold Physarum polycephalum and the zooflagellates Leishmania tarentolae, Trypanosoma lewisi, and Crithidia sp.: a comparative study. J Lipid Res 6(1):43–50

Krzywda A et al (2008) Sclerotia of the acellular (true) slime mould Fuligo septica as a model to study melanization and anabiosis. Cell Mol Biol Lett 13(1): 130–143

Kulka K, Sionkowska A (2023) Chitosan based materials in cosmetic applications: a review. Molecules 28(4): 1817

Kumar MNR (2000) A review of chitin and chitosan applications. React Funct Polym 46(1):1–27

Kumar A et al (2023) Industrial applications of fungal lipases: a review. Front Microbiol 14:1142536

Laneri S et al (2019) Dosage of bioactive molecules in the nutricosmeceutical Helix aspersa muller mucus and formulation of new cosmetic cream with moisturizing effect. Nat Prod Commun 14(8):1934578X19868606

Laneri S et al (2020) Aloe barbadensis: a plant of nutricosmetic interest. Nat Prod Commun 15(7): 1934578X20932744

Laneri S et al (2021) Plant cell culture extract of Cirsium eriophorum with skin pore refiner activity by modulating sebum production and inflammatory response. Phytother Res 35(1):530–540

Lee H-N et al (2024) Effect of modification methods on the physical properties and immunomodulatory activity of particulate β-glucan. Food Sci Biotechnol 33(7): 1615–1621

Levy D et al (1995) Alkali cation transport through liposomes by the antimicrobial fusafungine and its constitutive enniatins. Biochem Pharmacol 50(12): 2105–2107

Lin L, Xu J (2022) Production of fungal pigments: molecular processes and their applications. Journal of Fungi 9(1):44

Linnakoski R et al (2018) Antiviral agents from fungi: diversity, mechanisms and potential applications. Front Microbiol 9:2325

Liobikienė G, Bernatonienė J (2017) Why determinants of green purchase cannot be treated equally? The case of green cosmetics: literature review. J Clean Prod 162: 109–120

Liu J-K (2022) Natural products in cosmetics. Nat Products Bioprospect 12(1):40

Loganathan P, Kalyanasundaram I (1999) The melanin of the myxomycete Stemonitis herbatica. Acta Protozool 38(2)

Loganathan P, Paramasivan P, Kalyanasundaram I (1989) Melanin as the spore wall pigment of some myxomycetes. Mycol Res 92(3):286–292

Maghraby YR et al (2024) Overview of Nanocosmetics with emphasis on those incorporating natural extracts. ACS Omega 9(34):36001–36022

Malik K, Tokkas J, Goyal S (2012) Microbial pigments: a review. Int J Microbial Res Technol 1(4):361–365

Miller EL (2002) The penicillins: a review and update. J Midwifery Womens Health 47(6):426–434

Miyake M et al (2010) Inhibitory effects of 2-Amino-3 H-phenoxazin-3-one on the melanogenesis of murine B16 melanoma cell line. Biosci Biotechnol Biochem 74(4):753–758

Morganti P, Coltelli M-B (2019) A new carrier for advanced cosmeceuticals. Cosmetics 6(1):10

Murata C et al (2016) Synthesis and SAR of 4-methyl-5-pentylbenzene-1, 3-diol (MPBD), produced by Dictyostelium discoideum. Bioorg Med Chem Lett 26(5):1428–1433

Mygind PH et al (2005) Plectasin is a peptide antibiotic with therapeutic potential from a saprophytic fungus. Nature 437(7061):975–980

Nakov G et al (2019) Beta glucans in biscuits enriched with barley flour made with different sweeteners

Ninkuu V et al (2021) Biochemistry of terpenes and recent advances in plant protection. Int J Mol Sci 22(11):5710

No HK et al (2002) Antibacterial activity of chitosans and chitosan oligomers with different molecular weights. Int J Food Microbiol 74(1–2):65–72
Park BK, Kim M-M (2010) Applications of chitin and its derivatives in biological medicine. Int J Mol Sci 11(12):5152–5164
Park KM, Kwon KM, Lee SH (2015) Evaluation of the antioxidant activities and tyrosinase inhibitory property from mycelium culture extracts. Evid Based Complement Alternat Med 2015(1):616298
Paterson RRM (2006) Ganoderma–a therapeutic fungal biofactory. Phytochemistry 67(18):1985–2001
Paul S, Joshi S (2022) Industrial perspectives of fungi. In: Industrial microbiology and biotechnology. Springer, pp 81–105
Petravić-Tominac V et al (2010) Biological effects of yeast β-glucans. Agric Conspec Sci 75(4):149–158
Platsidaki E, Dessinioti C (2018) Recent advances in understanding Propionibacterium acnes (Cutibacterium acnes) in acne. F1000Res 7:1953
Płonka P, Rakoczy L (1997) Usefulness of the EPR method in the research on slime moulds. In: Application of the in vitro cultures in plant physiology. The Franciszek Górski Department of Plant Physiology, Polish Academy of Science, pp 175–180
Qin P et al (2023) Extraction and utilization of active substances from edible fungi substrate and residue: a review. Food Chem 398:133872
Quereshi S, Pandey A, Sandhu S (2010) Evaluation of antibacterial activity of different Ganoderma lucidum extracts. J Sci Res 3:9–13
Rakoczy L, Panz T (1994) Melanin revealed in spores of the true slime moulds using the electron spin resonance method. Acta Protozool 33(4)
Raut JK, Adhikari MK (2021) Mushroom: a true super food. In: Comprehensive insights in vegetables of Nepal. Nepal Academy of Science & Technology, Khumaltar
Rinaudo M (2006) Chitin and chitosan: properties and applications. Prog Polym Sci 31(7):603–632
Saito T, Ochiai H (1999) Identification of Δ5-fatty acid desaturase from the cellular slime mold Dictyostelium discoideum. Eur J Biochem 265(2):809–814
Sartori SK, Diaz MAN, Diaz-Muñoz G (2021) Lactones: classification, synthesis, biological activities, and industrial applications. Tetrahedron 84:132001
Savidov N et al (2018) Highly oxygenated isoprenoid lipids derived from fungi and fungal endophytes: origin and biological activities. Steroids 140:114–124
Scapagnini G et al (2014) Cocoa bioactive compounds: significance and potential for the maintenance of skin health. Nutrients 6(8):3202–3213
Schöfer H, Simonsen L (2010) Fusidic acid in dermatology: an updated review. Eur J Dermatol 20(1):6–15
Shankar A, Sharma KK (2022) Fungal secondary metabolites in food and pharmaceuticals in the era of multi-omics. Appl Microbiol Biotechnol 106(9): 3465–3488
Sharif S et al (2018) Phenolics, tocopherols and fatty acid profiling of wild and commercial mushrooms from Pakistan. J Biol Regul Homeost Agents 32(4):863–867
Sharma N et al (2022) Medicinal, nutritional, and nutraceutical potential of Sparassis crispa s. lat.: a review. IMA Fungus 13(1):8
Steglich W (1989) Slime moulds (Myxomycetes) as a source of new biologically active metabolites. Pure Appl Chem 61(3):281–288
Stoyneva-Gärtner M, Uzunov B, Gärtner G (2022a) Aeroterrestrial and extremophilic microalgae as promising sources for lipids and lipid nanoparticles in dermal cosmetics. Cosmetics 9(1):11
Stoyneva-Gärtner M et al (2022b) Potential of slime molds as a novel source for the cosmetics industry. Cosmet 10(1):3
Suay I et al (2000) Screening of basidiomycetes for antimicrobial activities. Antonie Van Leeuwenhoek 78: 129–140
Sułkowska-Ziaja K et al (2023) Natural compounds of fungal origin with antimicrobial activity – potential cosmetics applications. Pharmaceuticals 16(9):1200
Synytsya A et al (2009) Glucans from fruit bodies of cultivated mushrooms Pleurotus ostreatus and Pleurotus eryngii: structure and potential prebiotic activity. Carbohydr Polym 76(4):548–556
Tafakori V (2021) Slime molds as a valuable source of antimicrobial agents. AMB Express 11(1):92
Taofiq O et al (2016a) Anti-inflammatory potential of mushroom extracts and isolated metabolites. Trends Food Sci Technol 50:193–210
Taofiq O et al (2016b) Mushrooms extracts and compounds in cosmetics, cosmeceuticals and nutricosmetics – a review. Ind Crop Prod 90:38–48
Taofiq O et al (2016c) Development of mushroom-based cosmeceutical formulations with anti-inflammatory, anti-tyrosinase, antioxidant, and antibacterial properties. Molecules 21(10):1372
Tuli HS et al (2013) Cordycepin: a bioactive metabolite with therapeutic potential. Life Sci 93(23):863–869
Turner NA et al (2019) Methicillin-resistant Staphylococcus aureus: an overview of basic and clinical research. Nat Rev Microbiol 17(4):203–218
Wang R et al (2021) Invasive dermatophyte infection: a systematic review. Mycoses 64(4):340–348
Wasser S (2002) Medicinal mushrooms as a source of antitumor and immunomodulating polysaccharides. Appl Microbiol Biotechnol 60:258–274
Wasser SP (2011) Current findings, future trends, and unsolved problems in studies of medicinal mushrooms. Appl Microbiol Biotechnol 89:1323–1332
Willey JM, Sherwood LM, Woolverton CJ (2014) Prescott's microbiology. McGraw-Hill
Woźniak B et al (2023) Optimization of the composition of a cosmetic formulation containing Tremella fuciformis extract (fungi). Cosmet 10(3):82
Wu Y et al (2016) Mushroom cosmetics: the present and future. Cosmet 3(3):22

Yang M, Zhou M, Song L (2020) A review of fatty acids influencing skin condition. J Cosmet Dermatol 19(12): 3199–3204

Yao N, White PJ, Alavi S (2011) Impact of β-glucan and other oat flour components on physico-chemical and sensory properties of extruded oat cereals. Int J Food Sci Technol 46(3):651–660

Yoon KiNam YK et al (2011) Detection of phenolic compounds concentration and evaluation of antioxidant and antityrosinase activities of various extracts from the fruiting bodies of Lentinus edodes

Yoon SY et al (1994) Antimicrobial activity of Ganoderma lucidum extract alone and in combination with some antibiotics. Arch Pharm Res 17:438–442

Zhu F, Du B, Xu B (2016) A critical review on production and industrial applications of beta-glucans. Food Hydrocoll 52:275–288

10 Biotransformational Effect of Fungal-Based Cosmetics

Adenike A. Akinsemolu and Helen N. Onyeaka

Abstract

The cosmetics industry is harnessing microbial biotransformation through the adoption of fungi-based cosmetics. Fungi-based cosmetics offer similar benefits to their chemical and more traditional alternatives, including antioxidant, anti-aging, anti-inflammatory, and anti-wrinkling properties, which justify their application in the production of facial creams, sunscreen, body lotions, skin cleansers, shampoo, conditioner, and other skincare and haircare products. This article explores fungi-based cosmetics, with emphasis on the various species of fungi used in their production, the properties of each that are harnessed for various personal care products, and the implications of the cosmetic industry's adoption of the cosmetics on its sustainability. This is achieved through a compilation and synthesis of existing literature on fungi-based cosmetics, an evaluation of the various fungi-based cosmetics in the market, and an evaluation of the economic and commercial viability of the cosmetics using industry data and projections.

A. A. Akinsemolu (✉)
The Green Institute, Ondo, Nigeria
e-mail: a.akinsemolu@bham.ac.uk

H. N. Onyeaka
School of Chemical Engineering, University of Birmingham, Birmingham, UK

Keywords

Fungal-based cosmetics · Biotransformation · Skin health · Sustainable beauty · Cosmetic microbiology · Environmental impact · Mycology · Biotechnology · Dermatology · Eco-friendly products

10.1 Introduction

10.1.1 Background

The use of cosmetics is universal and present in every society on earth. For millennia, past and current civilizations have used fragrances and perfumes, oils and lotions, and substances such as clay that alter their appearance. Essentially, cosmetics are all substances that are applied to or placed on an external part of the human body to clean, scent, change, or enhance appearance, eliminate or correct body odor, protect, or maintain the body part in good condition (Lopez-Galindo and Viseras 2004). They protect skin health, enhance beauty, and enable users to maintain hygiene. Products that make up the cosmetics industry include a wide range of personal care products used on different parts of the body such as moisturizers, perfumes, hair shampoos and conditioners, shower gels, sunscreens, creams, and deodorants (Srinivasulu et al. 2022). The extensive range of cosmetics available across the world and their constantly high demand are

S. Ghosh, H. N. Onyeaka (eds.), *Fungal-Based Cosmetics: Formulation and Usage*, The Mycota 17,
https://doi.org/10.1007/978-3-032-09291-5_10

responsible for the size of the cosmetics industry, which has recorded positive growth every year for the past two decades and generates revenues of almost $129 billion every year (Petruzzi 2024a, b). The universality of cosmetic use among individuals and communities from all over the world means that innumerable personal care products are manufactured, purchased, used, and disposed of every day. All these elements of cosmetic use present the challenge of environmental exploitation. They cause the exploitation of environmental resources to provide raw materials, require resources such as water and energy for their production, necessitate the further exploitation of natural resources to produce packaging, and their use emits chemicals, plastic, and gases into the environment, water, and the soil (Cubas and Bianchet 2022; Arshad et al. 2020). These detrimental effects of the cosmetics industry on the environment have led to calls for the introduction of sustainable practices in the production, use, and disposal of cosmetics. The calls have been occasioned by the rise of sustainable beauty products that provide all the benefits of traditional cosmetics, but their production, use, and disposal cause minimal or no damage to the environment and its ability to sustain current and future generations. The industry has integrated elements of sustainability in every phase of the cosmetics lifecycle, beginning from the design phase to the post-use phase in which empty packaging, unused cosmetics, and fragments of used cosmetics are disposed of (Martins and Marto 2023).

One of the approaches that the cosmetics industry has adopted to drive sustainability is the use of microorganisms in the production of products. Fungi, a group of eukaryotic microorganisms, are one of the types of microorganisms whose natural abilities are being harnessed to produce natural and sustainable cosmetics. These applications leverage the natural bioactive molecules produced by different species of fungi, adapting them for various cosmetic purposes such as moisturizing the skin, enhancing its anti-aging ability, and preventing inflammation (Mukherjee et al. 2024). The past few decades have seen a rise in the demand for fungi-based cosmetics owing to their biological and chemical properties that enable the production of high-quality, organic, and comprehensive products with multiple benefits (Mukherjee et al. 2024). The media has noted this trend of the growing popularity of fungi-based cosmetics, driven by the discovery and popularization of their benefits, particularly when incorporated in skincare products (Peters 2023; Mago et al. 2023).

10.1.2 Importance of Fungal-Based Cosmetics

The benefits of fungi-based cosmetics, which are driving the growth in the demand for the products, highlight the importance of integrating microorganisms in the production, packaging, and disposal of cosmetics. For instance, fungi have natural bioactive compounds, including vitamins, polyphenols, and carotenoids, which provide natural antioxidant properties that are harnessed for use in the formulation of anti-aging creams and oils that prevent wrinkles (Mago et al. 2023). Similarly, fungi have antimicrobial properties and are used to eliminate harmful microorganisms, making them particularly suitable for the formulation of creams for the prevention and treatment of skin conditions caused by harmful microorganisms (Mago et al. 2023). Third, some fungi such as some species of mushrooms interfere in the production of inflammatory mediators, effectively serving as anti-inflammatory agents. Consequently, they are used in the production of anti-inflammatory creams and oils (Mago et al. 2023). These and other fungi-based cosmetics leverage the natural abilities of fungi, eliminating the need for chemicals that serve similar purposes. In addition to reducing the use of harmful chemicals whose manufacture and disposal harm the environment, fungi-based cosmetics are natural, making them biodegradable (Kržišnik and Gonçalves 2023). Consequently, they cause significantly less pollution compared to cosmetics made of synthetic ingredients. Further, fungal-based cosmetics are highly effective to produce, requiring less energy than their synthetic alternatives while producing less waste across the various stages of their lifecycle (Kržišnik and Gonçalves 2023). Together, these benefits of fungi-based cosmetics designate them as a potential revolutionary shift in the beauty industry from chemical-based personal care products to cosmetics made from fungi and other microorganisms.

10.1.3 Scope of the Article

The growing interest in fungi-based cosmetics and their applications in dermatology warrants a comprehensive evaluation of the products' biotransformational effect, which is the objective of this article. In this article, we discuss the integration of fungi and their natural biological mechanisms in the production of eco-friendly cosmetics. Having introduced fungi-based cosmetics and their transformative role in the cosmetics industry, the second chapter of this article explores the properties and bioactive compounds of the microorganisms that make them suitable for various applications in the industry before delving into the specific biotransformation processes adapted for the production of personal care products. In the third chapter, we discuss the applications of these processes in the production of specific personal-care products using case studies of actual fungal-based cosmetics that are currently under production and use across the world. The fourth chapter delves into the environmental and economic impact of the adaptation of fungi for applications in the cosmetics industry. The chapter discusses some environmental benefits of fungi-based cosmetics over other cosmetics and links them to sustainability. The economic benefits derived from the improved sustainability as well as cost-effectiveness of the use of fungi over other material in the production of cosmetics are discussed in this chapter as well. The final, and fifth, chapter highlights some of the technical and regulatory challenges experienced in the development and efforts to mainstream fungi-based cosmetics further. The chapter concludes by exploring some of the emerging trends in the adoption of fungi-based cosmetics and using them to identify potential areas of research in the topic.

10.2 The Science Behind Fungal-Based Cosmetics

10.2.1 Mycology and Fungal Diversity

Different species of fungi are used in various phases of the cosmetics lifecycle, particularly the production, use, and disposal in the post-use phase. Mushrooms, particularly, have many uses in the cosmetics industry. One commonly used fungal species is *Schizophyllum commune,* a mushroom with anti-inflammatory compounds that reduce skin itchiness and inflammation and provide soothing effects after contact with UV rays (Mukherjee et al. 2024). The species belongs to the genus *Schizophyllum* and is typically found on rotting wood (Gupta et al. 2021). Molds, too, have been adapted for the production of personal care products. For instance, *Aspergillus niger,* a type of mold that belongs to the genus *Aspergillus*, has a natural compound in its cells that is used in the formulation of skin moisturizers and anti-aging products (Mukherjee et al. 2024). *Lentinula edodes*, a type of mushroom known popularly as the Shiitake mushroom, has been adapted for applications in the cosmetic industry as well. The mushroom, which belongs to the genus *Lentinula*, is used in the production of night creams due to its high concentration of lentinan (Mukherjee et al. 2024). Lentinan has two primary applications in dermatology. It inhibits the production of inflammatory mediators, reducing inflammation, and reduces oxidative stress, which slows down the rate of skin aging (Zi et al. 2020). *Ganoderma*, a genus of another species of mushrooms, has metabolites with antioxidant and anti-inflammatory properties, which have led to the adaptation of the fungi for creams that treat burns and promote skin healing, skin lightening products, and in the treatment of dermatitis (Yin et al. 2019). Fungi are also used in the production of antimicrobial topical creams used to treat dermatological problems caused by pathogens. Filamentous fungi, particularly those

Table 10.1 Examples of fungi used in the production of cosmetics

Genus/species	Properties	Applications in the cosmetics industry	Reference
Ganoderma	Antioxidant and anti-inflammatory	Skin healing creams, burn salves, and skin lightening creams	Yin et al. (2019)
Lentinula edodes	Antioxidant, antimicrobial, and anti-inflammatory	Night creams	Zi et al. (2020)
Aspergillus Niger	Antimicrobial, antioxidant, and anti-inflammatory	The production of chitin-glucan, which is used in moisturizers and anti-aging products	Mukherjee et al. (2024)
Schizophyllum commune	Anti-inflammatory	Dermatological and skin healing creams	Mukherjee et al. (2024)
Fusidium, Aspergillus, Penicillium, and *Cephalosporium*	Antimicrobial	The production of antimicrobial skin creams	Sułkowska-Ziaja et al. (2023)

that belong to the genera *Fusidium, Aspergillus, Penicillium,* and *Cephalosporium,* are used widely in the cosmetics industry (Sułkowska-Ziaja et al. 2023). They are key raw materials in the production of antimicrobial compounds that are infused with creams to promote skin health (Table 10.1).

The growing use of fungi and compounds produced from the microorganisms in the cosmetics industry relies on biotransformation processes that transform organic compounds in fungi into compounds and forms that are beneficial to the skin and can be adapted for skin care products, haircare products, and other cosmetics. These processes, and the compounds on which they rely, are part of the unique and inherent properties of fungi that make the microorganisms suitable for cosmetics. The primary property of fungi that allows their use in cosmetics is their production of bioactive natural compounds or metabolites. Fungi produce bioactive secondary metabolites that have been adapted for applications across various industries, particularly the pharmaceutical industry (Bills and Gloer 2016). Mushrooms, for instance, contain an array of bioactive metabolites that make them suitable for use in the production of moisturizers. Some contain phenolic and polyphenolic compounds such as kojic acid, which inhibits the production of melanin and is used as a key ingredient in skin-lightening creams, serums, and lotions while others contain polysaccharides such as Lentinan, which is used as an anti-inflammatory and antibacterial agent in creams and lotions (Wu et al. 2016; Sułkowska-Ziaja et al. 2023). In addition to producing bioactive metabolites that are adapted for applications in the production of cosmetics, fungi can produce and accumulate compounds with antimicrobial activity (Sułkowska-Ziaja et al. 2023). This ability is used in the production of creams that prevent or fight skin infections caused by pathogens. These properties of fungi designate the microorganisms as alternative sources of some of the chemical compounds used in the manufacture of cosmetics.

10.2.2 Biotransformation Processes

Biotransformation refers to the use of biological systems to cause structural changes in chemical compounds (Sultana 2018). The process modifies the functional groups of substrates and transforms them into natural products. Therefore, it is favored as a replacement for the chemical synthesis of compounds, which yields by-products that harm the environment. Since biotransformation relies on natural biological processes, it is primarily carried out by biological organisms such as microorganisms. In the cosmetics industry, biotransformation has the potential to replace the traditional and formerly conventional methods of production, which relied on chemicals to produce the chemical compounds used in cosmetics, yielding more efficient, sustainable, and cost-

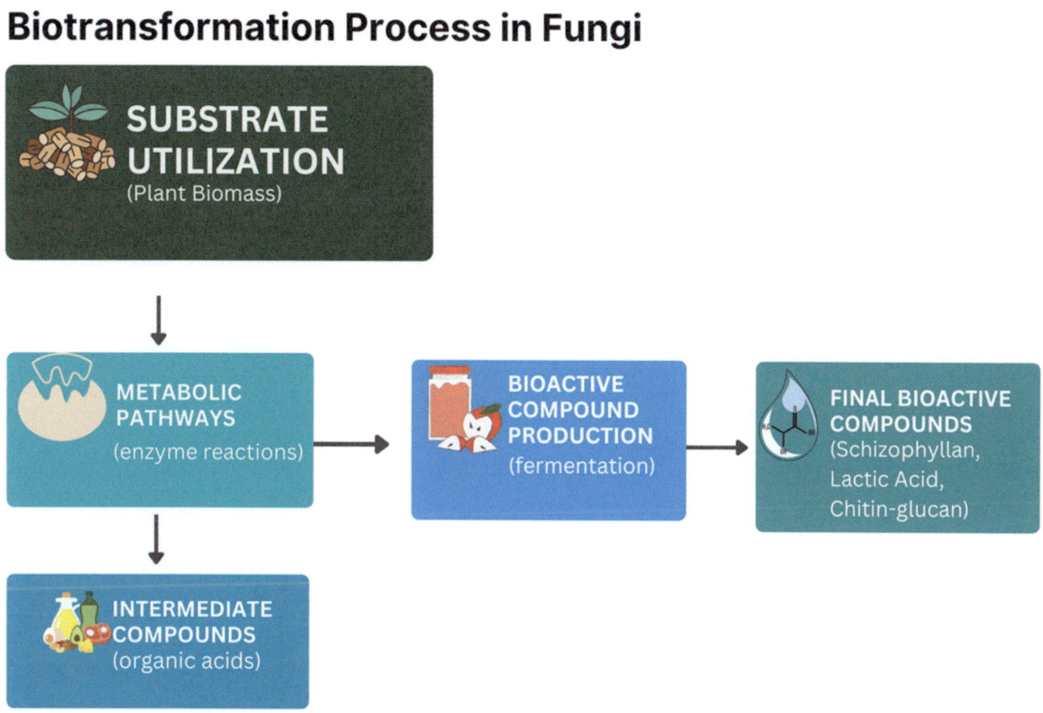

Fig. 10.1 Biotransformation process of fungi

effective production processes. Further, cosmetics produced from microorganisms such as fungi through biotransformation are biodegradable, eliminating one of the primary challenges associated with conventional chemical-based personal care products (Manan et al. 2021). Consequently, biotransformation using fungi has the potential to revolutionize the cosmetics industry by replacing chemicals used in personal care products as well as the chemical synthesis processes used in their production (see Fig. 10.1). Compared to chemical synthesis and transformation of compounds, the biotransformation process is more economically viable. Biotransformation is more commercially feasible than its alternative chemical synthesis approaches (Sultana 2018). Microbial biotransformation is particularly cost-effective since microorganisms generate the metabolites needed for the process naturally, eliminating the need for extraneous cofactors in their generation (Shanu-Wilson et al. 2020). Besides, some of the metabolites needed for chemical processes such as the production of cosmetics are not easily synthesized using unnatural means, making microorganisms the most effective approach for their synthesis. The efficiency, commercial viability, and eco-friendly nature of fungi-based cosmetics establish biotransformation as an integral process in the cosmetics industry.

10.2.3 Bioactive Compounds in Fungi

Different species of fungi possess or produce bioactive compounds with cosmetic benefits that adapt them for applications in the cosmetics industry. The compounds fall under different categories of ingredients used in cosmetics and personal care products, including minerals, phytochemicals, vitamins, and enzymes that are used in the synthesis of end products such as ceramides that are incorporated into creams, lotions, oils and ointments, and hygiene products. One key bioactive compound found in fungi is Schizophyllan, a type of glucan produced by

Table 10.2 Examples of fungal substrates used in the cosmetics industry

Species/genus	Substrate/ bioactive compound	Application in the cosmetics industry	Reference
Agaricus bisporus	2-amino-3H-phenoxazin-3-one	The production of skin-lightening creams	Hashem and Attia (2023), Usman et al. (2021)
Aspergillus japonicus	Ellagic acid	The production of topical creams, skin moisturizers, and skin lightening creams	Gulsunoglu-Konuskan and Karbancioglu-Guler (2021), Mukherjee et al. (2024)
Aspergillus Niger	Chitin-glucan	The production of anti-aging creams and ointments	Araújo et al. (2023) and Mukherjee et al. (2024)
Schizophyllum commune	Schizophyllan	The production of skin healing and restoration creams	Mukherjee et al. (2024)

Schizophyllum commune (Mukherjee et al. 2024). The compound is used in dermatology in the formulation of creams and ointments with multiple cosmetic benefits, including reducing itchiness and inflammation, protecting, and restoring skin that has been damaged by exposure to UV rays. Lactic acid, another bioactive compound produced by fungi such as *Rhizopus oryzae*, has various cosmetic benefits, including skin moisturizing, softening, lightening, and rejuvenation and the reduction of acne (Abedi and Hashemi 2020; Mukherjee et al. 2024). Chitin-glucan, a third bioactive compound found in the walls of the cells of fungi such as *Aspergillus Niger*, combines the bioactive properties of the two compounds of which it is comprised, which include antimicrobial, antioxidant, and anti-inflammatory qualities (Araújo et al. 2023; Mukherjee et al. 2024). The compound is used in moisturizers and anti-aging ointments. In addition to fortifying personal care products with the three qualities, chitin-glucan is biodegradable, making the cosmetics formulated using the compound eco-friendly and sustainable (Araújo et al. 2023). Filamentous fungal species such as *Aspergillus niger* and *Rhizopus oligosporus* produce phenolic compounds, a group of bioactive compounds that have a range of applications in the cosmetics industry (Dulf and Vodnar 2017). For instance, Ellagic acid, a bioactive compound produced by *Aspergillus japonicus,* is a polyphenol with anti-inflammatory, anti-oxidant, and anti-microbial properties as well as the ability to eliminate toxins from the skin (Gulsunoglu-Konuskan and Karbancioglu-Guler 2021). These cosmetic benefits have been adapted for topical creams, skin moisturizers, and skin lightening creams (Mukherjee et al. 2024). Other bioactive compounds with cosmetic benefits that adapt them for the production of personal care products include phenoxazines and amide derivatives. Phenoxazines include compounds such as 2-Amino-3H-phenoxazin-3-one, which is synthesized by *Agaricus bisporus* and inhibits the production of melanin, making it suitable for skin lightening creams (Hashem and Attia 2023; Usman et al. 2021). Finally, amide derivatives are synthesized by fungi and used in the manufacture of skin cleansers (Hashem and Attia 2023; Burnett et al. 2017) (Table 10.2).

10.3 Applications of Fungal-Based Cosmetics

10.3.1 Skin Care

The skin serves as a barrier that protects all tissues and organs under it from irritation, infection, and mechanical stress. One of the ways through which the skin protects tissues and organs in the body is by regulating the loss of moisture from the body by establishing a barrier between them and atmospheric conditions. If the barrier is compromised, more water than necessary is lost by the body and tissue and organs are susceptible to harmful chemicals and pathogens from the environment (Ryosuke et al. 2021). Moisturizing the skin is essential to maintaining the integrity and mechanical strength of this barrier and preventing significant moisture loss through the skin. The skin has a natural mechanism for moisturization, which pulls water from deep epidermal layers to the stratum corneum, a membrane that contains cholesterol, fatty acids, and ceramides that ensure hydration by creating a water barrier. The water is then lost to evaporation at a rate that is determined by the effectiveness of the above mechanism. Moisturizers work by supplementing this barrier. They increase the water content of the stratum corneum, form a barrier that prevents further moisture loss, and cover any fissures through which water would be lost (Purnamawati et al. 2017). These mechanisms create a protective film on the skin that provides a smoothing effect, enhances water retention, and reduces trans-epidermal water loss (Purnamawati et al. 2017). Some bioactive compounds produced or naturally occurring in some fungal species have properties that can be harnessed for use in moisturizers. For instance, polysaccharides extracted from *Tremella fuciformis*, a species of mushrooms, preserve water in the stratum corneum, maintaining the skin's moisture content while protecting its physical appearance and integrity by keeping it smooth, soft, and free from wrinkles (Mukherjee et al. 2024; Wu et al. 2016). Other bioactive compounds synthesized by fungi and used in moisturizers include chitin-glucagon, lactic acid, and ceramides.

In addition to moisturizers, fungal extracts are incorporated in anti-aging creams and lotions to maintain or restore skin firmness, reduce or slow down the appearance of wrinkles, and enhance overall radiance. Skin aging is the result of two sets of factors namely intrinsic and extrinsic factors. An example of the intrinsic mechanisms responsible for the aging of the skin and its subsequent manifestation through wrinkles, dry skin, and surface lines is the presence of free radicals in the body, whose reactions induce aging in tissues and cells, including skin cells and tissues (Chaudhary et al. 2020). Other intrinsic causes of skin aging include genetics and somatic capacity while the extrinsic factors responsible for skin aging include exposure to the sun and other environmental influences such as pollution, the lifestyle adopted by an individual, and their nutrition (Krutmann et al. 2021). Fungal extracts address skin aging caused by both sets of factors by repairing the skin barrier, maintaining overall skin health, and countering the mechanisms that exacerbate aging. Extracts such as lentinan, for instance, induce the activity of the antioxidants that address the oxidative stress responsible for the loss of collagen and skin elasticity (Luo et al. 2024). Oxidative stress results from the presence of excessive free radicals in the body. Other fungal extracts prevent wrinkles and address other signs of aging by reducing oxidative damage to the mitochondria, whose functional and mechanical decline are associated with skin aging (Stout and Birch-Machin 2019). In addition to addressing the oxidative stress responsible for the signs of aging, fungal extracts prevent wrinkles and restore good skin elasticity by addressing the two primary causes of wrinkles, which are the degradation of elastin by elastase and a decline in the concentration of structural proteins in the middle layer of the skin (Luo et al. 2024). Fungal extracts such as Clitocybin A, which is produced by the fungi species *Hygrophoropsis aurantiaca,* inhibit the production and activity of elastase, which impairs skin elasticity, addressing one of the main symptoms of skin aging (Luo et al. 2024). Similarly, polyphenols, which are extracted from filamentous fungi, counter the effects of exposure to environmental conditions such as UV rays on the dermis and epidermis, which restores the mechanisms that synthesize collagen and improve skin elasticity (Bjørklund et al. 2022). Incorporating these and other fungal extracts in anti-aging creams addresses both the causes and symptoms of skin aging.

In addition to moisturizing and slowing down the aging process of the skin, fungi-based cosmetics also restore damaged skin to health. Fungal extracts such as carotenoids, which are produced from fungal species such as *Mucoraceae*, have restorative effects on the skin (Naz et al. 2023). They repair skin barriers that have been damaged by environmental factors such as exposure to UV exposure, which has

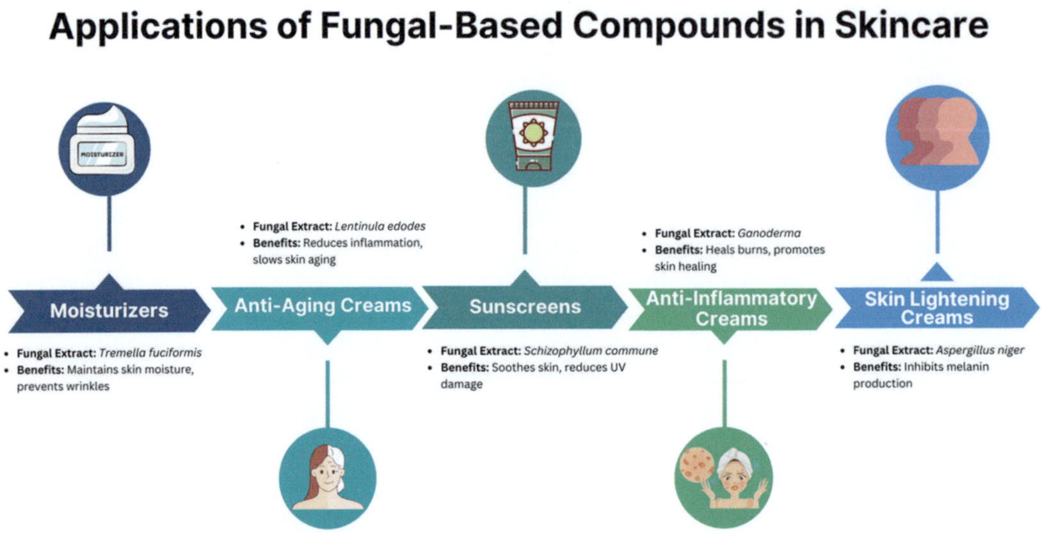

Fig. 10.2 Applications of fungal-based compounds in skincare

been linked to skin dryness and loss of elasticity. Some of the mechanisms used by extracts such as carotenoids in skin repair include oxygen scavenging and the inhibition of the formation of reactive oxygen species (ROS), which impair natural skin repair mechanisms (Bakac et al. 2023). Finally, some fungal extracts have anti-inflammatory properties that heal the skin from inflammation and repair itchy, dry, and irritated skin. Triterpenoids, for instance, aid in the healing and restoration of skin that is damaged by inflammatory conditions and allergic reactions such as rashes through their natural anti-inflammatory and antihistamine properties (Kumar and Quispe 2023). Ultimately, various fungal extracts have antioxidants, anti-inflammatory, antimicrobial, healing, and restorative properties that adapt them for the production of skincare products (See Fig. 10.2). Some popular fungal-based skincare brands and products include Shiseido, a Japanese cosmetics company that incorporates *reishi* mushrooms in products such as its Ultimate Power Infusing Concentrate, Origins, an American cosmetics company that uses *reishi* and *chaga* mushrooms in products such as its skin relief facial cleanser, and Dermatologica, an American producer of personal care products, which uses kojic acid derived from fungi in skin care products for hyperpigmentation (Poulter 2023). Other popular skincare products and brands that contain fungal extracts include ALUMIER MD—ALUMINEYE, a luxury eye cream that contains extracts derived from the *Tremella mushroom*, DR ANDREW WEIL FOR ORIGINS, a soothing treatment lotion that contains *coprinus*, *reishi*, and *chaga* mushrooms, VOLITION, a hydrating serum that contains extracts of *Tremella fuciformis*, and REN—EVERCALM ULTRA COMFORTING RESCUE MASK, a hydrating, moisturizing, and healing mast that contains extracts of *albarellus ovinus* (Littleton 2020).

10.3.2 Hair Care

In addition to their adaptation for skincare products such as moisturizers, skin lightening creams, anti-aging and anti-wrinkle creams, and sunscreen to protect the skin against UV rays or restore damaged skin, fungi have been adapted for the development of haircare products that improve hair appearance, strength, and overall health. One of the fungi-derived ingredients that is common in haircare products is collagen, which is produced from fungi such as *Rhizopus niveus* (Mizuta et al. 2023). Collagen is incorporated into haircare products such as moisturizers and shampoo to provide the amino acids needed for the nourishment of hair bulbs to moisturize and stimulate hair growth (Jadach et al. 2024). Chitin-glucans have also been adapted for haircare products. The polymers, which are extracted from the cell walls of various fungal species, are used together with collagen to give hair more volume, increase its thickness, and strengthen hair to either address or prevent damage (Gupta and Rajput 2019). Chitosan is also used to promote hair growth. Third, the hydrating and moisturizing benefits responsible for the integration of fungal extracts in skin care products are also responsible for the addition of the extracts to shampoos and other hair products. Mushrooms such as *Agaricus subrufescens* are used to hydrate hair, keeping it moisturized for extended periods while enhancing its shine and softening it (Hyde and Bahkali 2010). Similarly, *Ganoderma lucidum, Lentinula edodes,* and *Mucor miehei,* which are common ingredients in antioxidant and anti-aging skin creams, are used to nourish the hair and strengthen it to prevent hair loss (Hyde and Bahkali 2010). Finally, Tremella, a type of mushroom with a hydrophilic agent known for its anti-inflammatory properties, is used to keep the scalp moisturized and hair hydrated, reducing the onset and severity of conditions such as dandruff and alopecia (Wu et al. 2016). Some examples of haircare products that have fungal extracts include Tela Beauty Organic's Encore, Texture, and Styling Cream, a product of an American cosmetics company, which is promoted as a styling paste that protects the hair from the effects of the harsh sun (Tela Beauty Organics 2024). The cream has extracts of *Ganoderma lucidum*, popularly known as the Reishi mushroom, which has antioxidant properties (Ziemlewska et al. 2022). Another popular haircare product that is formulated using fungal extracts is Symhair Thermo, a product of the German chemical company Symrise, which is manufactured from a mushroom-based ingredient and used to protect the hair from the effects of the heat during styling (Ferrer 2023).

10.3.3 Makeup

The application of fungal extracts in the cosmetics industry is not limited to products that protect and keep the skin healthy. Some fungi-based products are used to enhance appearance. For instance, Kojic acid, a fungal extract produced through the fermentation of fungal species such as *Aspergillus* and *Penicillium,* is used as an agent of skin lightening (Phasha et al. 2022). Extracts of kojic acid have been used in various make-up products such as the Neostrata dark spot corrector, a product of American beauty brand Neostrata, which reduces discoloration and brightens skin (Neostrata 2024). Fungal substrates such as kojic acid have fundamental advantages over chemical color cosmetics, including biodegradability upon disposal, low cost of production, and sustainability.

10.4 Environmental and Economic Impact

10.4.1 Sustainability of Fungal-Based Cosmetics

The advancements in cosmetic microbiology that have enabled the above applications of fungal substrates and bioactive compounds as ingredients in personal care products are beneficial not only for the properties that improve skin and hair health but also for their implication on the sustainability of the cosmetics industry. Some of the traditional ingredients used in the cosmetics industry are harmful to the environment. One of the harmful ingredients of cosmetics is plastic. According to a recent study, cosmetics can contain over fifty thousand particles of microplastics and nanoplastics per gram (Aristizabal et al. 2024). Further, every one use of personal care products can release close to three hundred thousand microbeads of microplastics into drainage and sewerage systems (Aristizabal et al. 2024). Micro and nanoplastics are not biodegradable and they absorb contaminants from the environment, which they transfer to aquatic ecosystems upon the disposal of wastewater containing the plastic particles and other contaminants. In the terrestrial environment, microplastics affect soil microbiota and structure, impairing its ability to sustain the growth of trees and food. In both terrestrial and aquatic environments, micro and nanoplastics are consumed by animals, affecting bodily functions and mechanisms such as reproduction and digestion (Lamichhane et al. 2023). The ripple effects of these effects on animals and microorganisms on the environment are felt across ecosystems. In addition to microorganisms, other cosmetic ingredients, particularly chemicals, are environmental pollutants that affect both the environment and the health of the organisms and microorganisms that inhabit it. Chemicals such as asbestos, lead, mercury, butylated hydroxyanisole (BHA), polyethylene glycols (PEGs), and others that make up the active ingredients of cosmetics and personal care products are considered chemical pollutants (Mukherjee et al. 2024). Their effects on the environment include toxicity to animals, plants, and microorganisms and the degeneration of soil quality, impairing its ability to support agriculture (Juliano and Magrini 2017; Chen and Li 2022). The use of fungal extracts in place of the chemical ingredients in cosmetics reduces the overall negative environmental impact of the products since the extracts are biodegradable and form part of the natural microbiome of the environment in which runoff water from the use of cosmetics and other domestic activities is directed for remediation or disposal. Further, the use of fungal extracts as replacements for similar chemicals or chemicals that have similar properties reduces energy usage and associated environmental impacts. For instance, the production of carotenoids using fungi reduces the use of raw materials such as plants, the amount and cost of energy needed for the process, and the use of chemicals (Naz et al. 2023). The reduced energy usage reduces emissions from the combustion of non-renewable energy sources while the reduced need for the use of chemicals in the extraction and production of cosmetics reduces chemical pollution significantly, enhancing the overall sustainability of the cosmetics industry, by reducing both wastage and the carbon footprint of the industry.

10.4.2 Economic Advantages

In addition to their sustainability, fungi-based cosmetics are relatively more cost-effective compared to cosmetics produced using conventional or traditional ingredients. The cost-effectiveness of producing fungi-based cosmetics stems from production efficiency, the reduced cost of raw materials, less energy needs, and technological advancements in biotechnology. The bioactive compounds produced by fungi that are used in the production of cosmetics are either naturally occurring in microorganisms or produced through biological mechanisms that utilize natural sources of energy. For mushrooms, for instance, natural sources of energy include dead wood, which is a sustainable source of energy compared to the sources of energy that would be used in the production of alternative chemical substrates. Further, fungal-based substrates are biodegradable, which reduces the cost of waste treatment, expanding their environmental and economic benefits beyond production and usage to disposal. Third, technological advances in microbiology and biotechnology have made the production of fungal substrates cost-effective. For instance, metabolic and genetic engineering techniques are used in the culturing of microorganisms and production of substrates and compounds using the engineered microorganisms (Naz et al. 2023). Further, genetic engineering can also enhance the desired traits in microorganisms (Rafeeq et al. 2023). In cosmetic production, this could include increasing their capacity to produce the substrates and bioactive compounds needed for use as ingredients in cosmetics. These advancements reduce the cost of producing fungal-based cosmetics while enhancing the efficiency of the production of the substrates used in the process, enhancing the overall economic and commercial viability of fungal-based cosmetics. This economic viability is matched by current and future potential for market growth owing to the rising demand for sustainable beauty products, which is characterized by an increasing prioritization of organic, sustainable, and vegan ingredients by the consumers of the products of the cosmetics industry (Petruzzi 2024a, b). Ultimately, the demand for fungal-based cosmetics is projected to keep rising as the demand for sustainable beauty products grows.

10.5 Challenges and Future Prospects

10.5.1 Technical and Regulatory Challenges

Despite the growing interest in sustainable cosmetics and the rising demand for fungal-based cosmetics, their production faces some technical and regulatory challenges. First, the cultivability of some species of fungi or their substrates requires tight control of environmental conditions and demands throughput cultivation conditions to optimize production (Cairns et al. 2021). The production of *Aspergillus Niger*, for instance, requires high-throughput assays, which are difficult to achieve when using submerged fermentation to produce substrates from the fungi. Second, some of the production processes used in the production of fungal substrates are prone to failure due to limitations in equipment and production processes. The production of chitosan, for instance, often fails at the aeration and coagulation stages of production, limiting the capacity for the mass production of the substrate from fungi (Elsoud et al. 2023). Third, fungi-based cosmetics face the universal challenge of regulation that is reported across all products produced by or using microorganisms. The use of microorganisms such as fungi raises the issue of safety, which leads to the placement of strict and significant regulations for products produced using the microorganisms (Molitorisová et al. 2021). Consequently, fungi-based cosmetic ingredients face relatively lengthy and extensive approval processes. However, innumerable products have passed these approval processes, laying a foundation for the approval and mainstreaming of other fungi-based ingredients in the cosmetics industry.

10.5.2 Future Research Directions

Future research in fungi-based cosmetics can expand the current pool of fungal species used in the production of cosmetics to include other species with antioxidant, anti-inflammatory, and anti-aging properties. Such research efforts can focus on fungi with antifungal properties, whose substrates and bioactive compounds can be used in skin creams to prevent and treat fungal dermatological conditions and hair products that protect the scalp from or treat dandruff. This aligns with emerging trends in fungal biotechnology for cosmetics, which include the adaptation of mushrooms for the production of cosmetic ingredients, the use of waste such as plant biomass in the production of substrates such as chitosan, and the use of biotechnological advancements, particularly genetic engineering, to enhance the production of fungal substrates and extracts with applications in the cosmetics industry.

10.6 Conclusion

Concisely, the cosmetics industry is one of the fastest-growing industries globally. Like most industries, the cosmetics industry faces the challenge of sustainability, particularly due to its reliance on pollutants such as microplastics and other pollutants as ingredients. One of the current trends in the industry is the adoption of fungal-based cosmetics, which solve the problem of pollution from cosmetic use while providing skin and hair-friendly products that are cheap to produce, economically and commercially viable, and less harmful to the environment. Consequently, the past decade has seen the rapid growth of the production and penetration of cosmetics with fungal substrates such as chitosan, mushroom extracts, and other fungal substrates as ingredients. Although the regulation process for these products is lengthy and extensive, advancements in biotechnology promise to discover and adapt more fungi for the production of cosmetics.

References

Abedi E, Hashemi S (2020) Lactic acid production–producing microorganisms and substrates sources-state of art. Heliyon 6(10)

Araújo D, Rodrigues T et al (2023) Chitin-glucan complex hydrogels: physical-chemical characterization, stability, in vitro drug permeation, and biological assessment in primary cells. Polymers 15(4):791. https://doi.org/10.3390/polym15040791

Aristizabal M, Jiménez-Orrego K et al (2024) Microplastics in dermatology: potential effects on skin homeostasis. J Cosmet Dermatol 23(3):766–772. https://doi.org/10.1111/jocd.16167

Arshad H, Mehmood M et al (2020) Evaluation of heavy metals in cosmetic products and their health risk assessment. Saudi Pharm J 28(7):779–790

Bakac E, Percin E et al (2023) A narrative review: the effect and importance of carotenoids on aging and aging-related diseases. Int J Mol Sci 24(20):15199. https://doi.org/10.3390/ijms242015199

Bills G, Gloer J (2016) Biologically active secondary metabolites from the fungi. Microbiol Spectrum 4(6):10–1128. https://doi.org/10.1128/microbiolspec.funk-0009-2016

Bjørklund G, Shanaida M et al (2022) Natural compounds and products from an anti-aging perspective. Molecules 27(20):7084

Burnett C, Heldreth B et al (2017) Safety assessment of amino acid alkyl amides as used in cosmetics. Int J Toxicol 36(1_suppl):17S–56S. https://doi.org/10.1177/1091581816686048

Cairns T, Barthel L et al (2021) Something old, something new: challenges and developments in aspergillus Niger biotechnology. Essays Biochem 65(2):213–224

Chaudhary M, Khan A et al (2020) Skin ageing: pathophysiology and current market treatment approaches. Curr Aging Sci 13(1):22–30

Chen X, Li X (2022) The impact of hazardous substances in cosmetics, and treatment measures. IOP Conf Ser Earth Environ Sci 1011(1):012024. https://doi.org/10.1088/1755-1315/1011/1/012024

Cubas A, Bianchet R (2022) Plastics and microplastic in the cosmetic industry: aggregating sustainable actions aimed at alignment and interaction with UN sustainable development goals. Polymers 14(21):4576

Dulf F, Vodnar D (2017) Phenolic compounds, flavonoids, lipids and antioxidant potential of apricot (Prunus armeniaca L.) pomace fermented by two filamentous fungal strains in solid state system. Chem Cent J 11:1–10

Elsoud M, Mohamed S et al (2023) Characterization and optimization of chitosan production by aspergillus terreus. Arab J Sci Eng 48(1):93–106. https://doi.org/10.1007/s13369-022-07163-z

Fela Beauty Organics (2024) Encore, Texture and Style Paste. https://telabeautyorganics.com/products/encore. Accessed 15 July 2024

Ferrer B (2023) Mycocosmetic era: Shroom boom in skincare and cosmetics introduces new fungi applications. https://www.personalcareinsights.com/news/mycocosmetic-era-shroom-boom-in-skincare-and-cosmetics-introduces-new-fungi-applications.html. Accessed 15 July 2024

Gulsunoglu-Konuskan Z, Karbancioglu-Guler F (2021) Development of a bioprocess for production of ellagic acid from chestnut (Castanea sativa mill.) waste by fermentation with aspergillus spp. Food Biosci 42:101058. https://doi.org/10.1016/j.fbio.2021.101058

Gupta P, Rajput M (2019) Eminence of microbial products in cosmetic industry. Nat Products Bioprospect 9:267–278. https://doi.org/10.1007/s13659-019-0215-0

Gupta P, Gupta H, Poluri KM (2021) Biomedical applications of polysaccharide-based nanocomposites from fungal origin. In: Microbial and natural macromolecules, pp 233–272. https://doi.org/10.1016/B978-0-12-820084-1.00010-7

Hashem A, Attia M (2023) Bioactive compounds and biomedical applications of endophytic fungi: a recent review. Microb Cell Fact 22(1):107

Hyde K, Bahkali A (2010) Fungi—an unusual source for cosmetics. Fungal Divers 43:1–9. https://doi.org/10.1007/s13225-010-0043-3

Jadach B, Mielcarek Z et al (2024) Use of collagen in cosmetic products. Curr Issues Mol Biol 46(3):2043–2070. https://doi.org/10.3390/cimb46030132

Juliano C, Magrini G (2017) Cosmetic ingredients as emerging pollutants of environmental and health concern. A mini-review. Cosmetics 4(2):11. https://doi.org/10.3390/cosmetics4020011

Krutmann J, Schikowski T et al (2021) Environmentally-induced (extrinsic) skin aging: Exposomal factors and underlying mechanisms. J Invest Dermatol 141(4):1096–1103. https://doi.org/10.1016/j.jid.2020.12.011

Kržišnik D, Gonçalves J (2023) Environmentally conscious technologies using fungi in a climate-changing world. Earth 4(1):69–77. https://doi.org/10.3390/earth4010005

Kumar N, Quispe C (2023) Potential of mushrooms bioactive for the treatment of skin diseases and disorders. J Food Biochem 2023(1):5915769. https://doi.org/10.1155/2023/5915769

Lamichhane G, Acharya A et al (2023) Microplastics in environment: global concern, challenges, and controlling measures. Int J Environ Sci Technol 20(4):4673–4694. https://doi.org/10.1007/s13762-022-04261-1

Littleton S (2020) Spike your skincare with these 'magic' mushroom products. https://www.dazeddigital.com/

beauty/article/47739/1/skincare-mushroom-products-alumier-lush-moon-juice-ren. Accessed 15 July 2024

Lopez-Galindo A, Viseras C (2004) Pharmaceutical and cosmetic applications of clays. In: Interface science and technology, vol 1. Elsevier, pp 267–289. https://doi.org/10.1016/S1573-4285(04)80044-9

Luo J, Ganesan K et al (2024) Unlocking the power: new insights into the anti-aging properties of mushrooms. J Fungi 10(3):215. https://doi.org/10.3390/jof10030215

Mago P, Sharma R et al (2023) Mushroom based cosmeceuticals: an upcoming biotechnology sector. Biosci Biotechnol Res Asia 20(2):381–394

Manan S, Ullah M et al (2021) Synthesis and applications of fungal mycelium-based advanced functional materials. J Bioresour Bioprod 6(1):1–10. https://doi.org/10.1016/j.jobab.2021.01.001

Martins A, Marto J (2023) A sustainable life cycle for cosmetics: from design and development to post-use phase. Sustain Chem Pharm 35:101178. https://doi.org/10.1016/j.scp.2023.101178

Mizuta S, Matsuyama Y et al (2023) Characterization of collagen in salmon nasal cartilage: enzymatic solubilization with fungal acid protease. Fish Sci 89(4): 507–513. https://doi.org/10.1007/s12562-023-01697-3

Molitorisová A, Monaco A et al (2021) An analysis of the regulatory framework applicable to products obtained from mushroom and mycelium. SSRN 3955899. https://doi.org/10.2139/ssrn.3955899

Mukherjee A, Das S et al (2024) Fungi's treasure in cosmeceuticals-a comprehensive chemical approach. S Afr J Bot 166:311–331. https://doi.org/10.1016/j.sajb.2024.01.036

Naz T, Ullah S et al (2023) Industrially important fungal carotenoids: advancements in biotechnological production and extraction. J Fungi 9(5):578. https://doi.org/10.3390/jof9050578

Neostrata (2024) Dark Spot Corrector. https://www.neostrata.com/products/dark-spot-corrector/F30162XB.html. Accessed 15 July 2024

Peters S (2023) Mushroom Skincare Is Huge Right Now. https://www.elle.com/beauty/g42695414/mushroom-skincare/. Accessed 15 July 2024

Petruzzi D (2024a) Global demand for green feature in cosmetics products as of 2019. https://www.statista.com/statistics/803595/global-demand-for-natural-organic-environmental-friendly-cosmetics/. Accessed 15 July 2024

Petruzzi D (2024b) Cosmetics industry – statistics & facts. https://www.statista.com/topics/3137/cosmetics-industry/. Accessed 15 July 2024

Phasha V, Senabe J et al (2022) Review on the use of kojic acid – a skin-lightening ingredient. Cosmet 9(3):64. https://doi.org/10.3390/cosmetics9030064

Poulter S (2023) Sales of cosmetics that contain fungi are mushrooming! Upmarket brands favour unusual 'ingredient' in their latest beauty products. https://www.dailymail.co.uk/news/article-11668669/Cosmetics-contain-fungi-mushrooming-Brands-favour-unusual-ingredient-products.html. Accessed 15 July 2024

Purnamawati S, Indrastuti N et al (2017) The role of moisturizers in addressing various kinds of dermatitis: a review. Clin Med Res 15(3–4):75–87

Rafeeq H, Afsheen N et al (2023) Genetically engineered microorganisms for environmental remediation. Chemosphere 310:136751. https://doi.org/10.1016/j.chemosphere.2022.136751

Ryosuke O, Yoshie S et al (2021) The association between activity levels and skin moisturising function in adults. Dermatol Reports 13(1)

Shanu-Wilson J, Evans L et al (2020) Biotransformation: impact and application of metabolism in drug discovery. ACS Med Chem Lett 11(11):2087–2107

Srinivasulu M, Chandra M et al (2022) Personal care products – fragrances, cosmetics, and sunscreens – in the environment. In: Environmental micropollutants, pp 131–149. https://doi.org/10.1016/B978-0-323-90555-8.00015-5

Stout R, Birch-Machin M (2019) Mitochondria's role in skin ageing. Biology 8(2):29

Sułkowska-Ziaja K, Trepa M et al (2023) Natural compounds of fungal origin with antimicrobial activity – potential cosmetics applications. Pharmaceuticals 16(9):1200

Sultana N (2018) Microbial biotransformation of bioactive and clinically useful steroids and some salient features of steroids and biotransformation. Steroids 136:76–92. https://doi.org/10.1016/j.steroids.2018.01.007

Usman M, Murtaza G et al (2021) Nutritional, medicinal, and cosmetic value of bioactive compounds in button mushroom (Agaricus bisporus): a review. Appl Sci 11(13):5943. https://doi.org/10.3390/app11135943

Wu Y, Choi M et al (2016) Mushroom cosmetics: the present and future. Cosmet 3(3):22. https://doi.org/10.3390/cosmetics3030022

Yin Z, Yang B et al (2019) Preventive and therapeutic effect of Ganoderma (Lingzhi) on skin diseases and care. In: Ganoderma and health: pharmacology and clinical application, pp 311–321. https://doi.org/10.1007/978-981-32-9421-9_14

Zi Y, Jiang B et al (2020) Lentinan inhibits oxidative stress and inflammatory cytokine production induced by benzo (a) pyrene in human keratinocytes. J Cosmet Dermatol 19(2):502–507. https://doi.org/10.1111/jocd.13005

Ziemlewska A, Wójciak M et al (2022) Assessment of cosmetic properties and safety of use of model washing gels with Reishi, Maitake and lion's mane extracts. Molecules 27(16):5090. https://doi.org/10.3390/cosmetics9030064

11 Emollient, Exfoliation, Skin Lightening, and Anti-aging Properties of Fungal-Based Cosmetics

Rose Daphnee Tchonkouang and Maame Ekua Manful

Abstract

In the last few decades, incorporating bio-based resources typically derived from plants, animals, and microorganisms into cosmetics products has received increasing attention. Fungal species have proven to be a valuable group of organisms with several industrial applications. Fungal-based cosmetics have emerged as a promising frontier in skincare because they are eco-friendly, cost-effective, and a good source of natural bioactive compounds, offering a spectrum of skin benefits. This chapter discusses the emollient, exfoliation, skin-lightening, and anti-aging properties of fungal species and fungal-derived compounds.

Keywords

Fungi · Cosmetics · Anti-aging · Exfoliation · Skin-lightening · Hyperpigmentation

R. D. Tchonkouang (✉)
MED-Mediterranean Institute for Agriculture, Environment and Development, CHANGE-Global Change and Sustainability Institute, Faculty of Sciences, and Technology, Universidade do Algarve, Faro, Portugal

M. E. Manful
School of Food Science and Environmental Health, Technological University Dublin, Dublin, Ireland

Sustainability and Health Research Hub, Technological University Dublin, Dublin, Ireland

11.1 Introduction

The cosmetic industry has evolved in response to shifting consumer preferences. The industry has seen an increased awareness of more environmentally friendly, sustainable, and animal-friendly sourcing of cosmetic ingredients (Martins and Marto 2023). Fungi are one of the renewable and regenerative resources that have been earmarked as a source of natural bioactive extracts for cosmetic applications. These bioactive substances, including polysaccharides, glycoproteins, phenolic compounds, terpenoids, and lipid components, have been studied for their role in promoting skin health (Niego et al. 2023).

Fungi have been traditionally utilized for medicinal uses due to their rich phytochemical content, whose biological properties have been studied and are an ongoing area of research. Fungal extracts (e.g., compounds from *Ganoderma lucidum*) have been shown to exhibit antimicrobial, anti-cancer, and antioxidant properties, which makes them potentially suitable candidates for effective cosmetic usage (Shamsaei et al. 2021). With a short growth life-cycle, prolific growth rate, minimal cultivation inputs and growth requirements, and low dependence on fertilizers and agrochemicals, fungal species are a relatively economical source of ingredients for cosmetic applications. Mushrooms are known to be the first fungal species used in cosmetic formulations. Some examples of mushrooms

S. Ghosh, H. N. Onyeaka (eds.), *Fungal-Based Cosmetics: Formulation and Usage*, The Mycota 17,
https://doi.org/10.1007/978-3-032-09291-5_11

include *Lentinula edodes* and *Ganoderma* species, which are commonly utilized as fungal sources in skincare products (Hyde et al. 2010; Niego et al. 2023).

Skincare is an important consumer routine due to the potential to improve the quality of the skin complexion (Rodan et al. 2016). Furthermore, the practice of skincare aids in the management of chronic, acute, or inflammatory skin disorders such as acne vulgaris, dry skin, uneven skin pigmentation, peeling skin (desquamation), flaky skin (xeroderma), and wrinkling, among others (Fischer et al. 2010; Del Rosso 2013a). Cosmetics infused with dermatologically active agents for skin-related therapeutic purposes play a huge role in managing a healthy skin (Dreno et al. 2023). Moreover, the course and biological origin of such effective agents in cosmetics has risen as one of the major consumer expectations on cosmetic labels (Liu 2022).

Fungal-based cosmetics are one of the emerging sectors in response to consumer need for effective natural cosmetic agents (Devi et al. 2020). The attributes consumers look out for in topical products include, but not limited to anti-aging, conditioning, emollient, anti-hyperpigmentation, exfoliation, and skin-lightening. Hence, the chapter aims to discuss the utilization of fungal-based compounds in the formulation of skincare products. The various potential biological properties and roles that fungal-derived cosmetics play in meeting consumer needs for anti-aging, emollient, skin lightening, and exfoliation are discussed.

11.2 Fungi in Cosmetics

Fungi are one of the living organisms with abundant biological species on the planet with an estimated 5.1 million species in existence (Gupta et al. 2019). Among the fungal ecosystem, mushrooms are the most prolific sources of phytochemicals known to possess pharmacological properties (Gupta et al. 2019). In the cosmetic sector, the species commonly used include *Agaricus subrufescens,* Shiitake (*Lentinula edodes*), and Reishi mushrooms (Hyde et al. 2010; Wu et al. 2016). Table 11.1 summarizes some of the commonly used fungal extracts in cosmetics with their associated bioactive properties with potential cosmetic effects.

The main fungal compounds reported in the literature include both hydrophobic and hydrophilic compounds. These compounds possess chemical structures that enable their usability in cosmetic formulations such as solubility, emulsification, humectability, and hydration properties.

The commonly isolated fungal compounds include polyphenols and phenolic acids, organic acids (lactic acid), fatty acids (ceramides), and polysaccharides (chitosan) (Gupta et al. 2019).

Chitosan is known to possess antimicrobial properties and has been an active ingredient employed in mouthwashes (Aranaz et al. 2018). Studies by Chang et al. (2019) revealed that chitosan derived from *Auricularia* spp. exerted antimicrobial effects on *Escherichia coli* and *Staphylococcus aureus*. Fungal species such as *Lentinula edodes*, *Aspergillus niger, Termitomyces titanicus*, and *Trichodema reesei* are known to possess chitosan-rich mycelium and are potential biological sources for chitosan recovery (Muñoz et al. 2015; John Kasongo et al. 2020; Huq et al. 2022). Chitosan is found in mycelial cell walls and its extraction requires the breakdown of the cell wall. Chitosan extraction is commonly achieved with chemical-based methods such as acid and alkaline hydrolysis to degrade cell walls and disassociate chitosan from alkali-soluble macromolecules such as proteins, fatty acids, and carbohydrates (Alimi et al. 2023). Other non-alkali treatments and greener alternatives such as the use of ionic liquids, enzymatic methods, deep eutectic solvents, ultrasound, and microwave-assisted extraction have been reviewed by Pellis et al. (2022).

Ceramides are sphingolipids that supplement the human epidermal lipids and contribute to skin barrier maintenance. The lipids found in the human *stratum corneum* contain about 30–40% ceramides, the deficiency thereof can lead to a myriad of skin disorders (Kahraman et al. 2019). Ceramides in topical products are an important ingredient that functions as skin and hair conditioning agents, and are one of the important

Table 11.1 Fungal species, bioactive agents, and biological properties relevant for cosmetic applications

Name of fungi	Bioactive agent	Cosmetological effect	References
Tremella fuciformis	Polysaccharides (mainly acid heteropolysaccharides, neutral heteropolysaccharides, cell wall polysaccharides, extracellular polysaccharides, and acid oligosaccharides)	Anti-aging, photoprotective, wound healing, and enhancement of skin's protective barrier.	Mineroff and Jagdeo (2023)
Polyporus umbellatus	Polysaccharides, steroids, ergosterol, terpenoids	Anti-inflammatory, antibacterial, antioxidant properties	Badalyan et al. (2022)
maitake (Grifola frondosa)	Glycoproteins; glycans	Wound healing, antioxidant activities, maintenance of skin barrier strength; collagen synthesis	Yang et al. (2019) and Bajza et al. (2020)
Hypsizygus ulmarium	Polysaccharides, polyphenols	Antioxidant effects. Polysaccharides offer skin hydration benefits	Thimmaraju and Govindan (2022)
Lentinula edodes	Phenolic compounds Polysaccharides (e.g., lentinan)	Anti-aging, anti-inflammatory, antimicrobial, anti-tyrosinase properties	Taofiq et al. (2016) and Zi et al. (2020)
Trametes versicolor	Protein SIRT1, phenolic acids, polysaccharopeptides (PSPs)	Anti-tyrosinase, antihyaluronidase, antioxidative, anti-inflammatory effects	Jhan et al. (2016), Chou et al. (2019) and Badalyan et al. (2022)
Inonotus obliquus	Terpernoids (betulic acid and betulin)	Anti-tyrosinase effects, antioxidant, anti-aging, photo-protective properties	Yun et al. (2011), Lee and Cha (2019) and Wontcheu Fotso et al. (2023)

manufactured cosmetic ingredients. Current research has therefore sought to obtain natural sources of ceramides from fungal mycelium. Glycoceramides also known as gylcosphingolipds found in fungi cell membranes are involved in the organism's biological processes such as growth and stress response (Ryuta et al. 2016; Santos et al. 2020). *Aspergillus* spp. is one of the abundant sources of glycosphingolipids (Jiang et al. 2021). Lipid extracts derived from fungi are potential sources of glycosphingolipids (Singh and Del Poeta 2016) and are commonly extracted with solvents such as chloroform-methanol mixtures or petroleum ether. This is followed by a separation of the sphingolipid fraction with an akali-hydrolysis process. Ceramides are then isolated via silicic acid column chromatography and purified using thin-layer chromatography with silica gel (Rupčić and Marić 1998).

Polyphenol extraction from fungi has been increasingly studied due to their potential biological properties including anti-inflammatory and antioxidant properties. Polyphenols from fungi mycelia have been predominantly obtained through the use of organic solvents such as ethanol and petroleum ether. These solvents have been associated with drawbacks including low extraction yields, lengthier extraction periods, and flammability of solvents (Dantas et al. 2022). Various techniques have therefore been employed in combination with organic solvents to increase extraction yields, and these include ultrasound, microwave-assisted extraction, and supercritical fluid extraction (Zheng et al. 2021). Ongoing research has focused on greener alternatives to the commonly employed organic solvents. In particular, the utilization of deep eutectic solvents for the maximal recovery of polyphenols from fungal fruiting bodies is being investigated. In a study by Chu et al. (2022), *Antrodia cinnamomea* ethanolic extracts derived with 70% ethanol yielded the maximum phenolic content of 10.95 mg/g. The phenolic content was found to double (21.83 mg/g) when the extraction was performed using an equal combination of choline chloride and malonic acid (deep eutectic solvent). Zheng et al. (2021) found that the use of deep eutectic solvents (equal parts of choline chloride and malic acid) yielded the highest phenolic content (6.37 mg/g) compared to those extracted with 60% ethanol (4.20 mg/g). Some of the common phenolic compounds extracted from fungi include xanthones, tyrosol, flavonols, and flavonoids (Dantas et al. 2022).

Certain fungal species are prolific producers of organic acids which are useful cosmetic agents. Organic acids such as citric acid, gluconic acid, and malic acid are derived from fungi belonging to the *Aspergillus* genera whereas lactic acid and fumaric acids have been derived from those from the *Rhizopus* genera (Liaud et al. 2014). The fermentative action of filamentous fungi such as *Rhizopus oryzae* and *Rhizopus arrihzus* has been exploited for the production of lactic acid (Ojo and de Smidt 2023). These fungi rely on carbon sources and convert hexose or pentose-rich substrates into lactic acid through the glucose metabolic pathway (Maas et al. 2008). The utilization of food processing wastes or side streams has been of particular interest with various materials such as potato, pineapple and corn waste, and wheat straw having been studied (Jin et al. 2005). The isolation and optimal recovery of pure lactic acid from fungal fermentation broth or substrates is an area with ongoing developments. Research has investigated the potential of in situ recovery of lactic acid from fermentation bioreactors through the immobilization of fungal cells in fibrous material and polyvinyl alcohol cryogel (Spiricheva et al. 2007).

11.3 Emollient Properties

Compounds conferring emollient properties in cosmetics form an oil seal-like barrier on the *stratum corneum* which reduces skin dehydration, thus softening the skin by enhancing hydration (Liu 2022). Emollient compounds help retain skin moisture and elasticity. The major class of emollients includes lipids, biopolymers, and organic acids. Fungal-derived compounds such as ceramides and lactic acid have been shown to exert dermal emollient action. A study found that the skin hydration and desquamation properties of females aged 50 years and older improved significantly after a 2-week application of 15% lactic acid and ceramide skin lotion (Draelos et al. 2020). In another study, a commercial emollient containing 5% urea, ceramide, and lactate significantly improved the skin barrier function and hydration of women aged 60 years and above after 28-day application (Danby et al. 2016). Further clinical research is required to generate additional insights on the validate the in vivo emollient properties of ceramides and lactates derived from fungal species.

Research has found that extracts from mushrooms such as *Tremella fuciformis* (snow mushroom) contain polysaccharides that offer similar skin benefits as hyaluronic acid (Visvanathan et al. 2022). One of the most often utilized ingredients in cosmetic formulations is hyaluronic acid (HA). It is a polysaccharide that is a part of the glycosaminoglycan family and is composed of N-acetylglucosamine and D-glucuronic acid disaccharide units. High molecular weight HA, when applied to the skin, hyaluronic acid creates a viscoelastic film that moisturizes the skin. The primary action of the high molecular weight hyaluronic acid polymer is to form films, which inhibit the skin's ability to evaporate water, producing an occlusive effect (Juncan et al. 2021). Due to its ability to imitate the effects of hyaluronic acid, the extract derived from the mushroom serves as a body moisturizer. When applied topically, snow mushroom extracts effectively provide the skin with ample hydration and moisturization, resulting in a soft and supple appearance. This, in turn, produces a plumping effect that contributes to a more youthful appearance of the skin (Visvanathan et al. 2022).

Certain emollients have been commonly referred to as natural moisturizing factors (NMF) (Hon et al. 2018). These NMFs are endogenous compounds that attract water molecules to mature terminal keratinocytes (corneocytes) and therefore hydrate the skin. NMFs naturally make up about 10% of corneocytes and contribute to the maintenance of the osmotic pressure within corneocytes which reduces the inflow of or the loss of excessive extracellular moisture (Gunnarsson et al. 2021). NMFs are derivative products from the degradation of filaggrin, a structural epidermal protein, composed of low molecular weight compounds (Harding et al. 2013). The supplementation of the skin's natural NMFs with topical products has been a proposed remedy for various skin disorders. NMFs present in fungal mycelium such as lactic acid, ceramides, and amino acids have been reviewed for their potential to address skin ailments such as atopic dermatitis and xerosis (Hon et al. 2018).

11.4 Exfoliants from Fungi

Exfoliants are substances that eliminate the dead cell layers of the skin and are utilized to address various skin conditions such as aging, photodamage, acne, and dry skin, among others (Banchhor and Saraf 2008). Skin exfoliation leads to an improvement of the overall texture and radiance of the skin by aiding in the elimination of dead skin cells that accumulate on its surface (Packianathan and Kandasamy 2011). Cosmetic exfoliation can be done through mechanical exfoliation and chemical exfoliation. Mechanical exfoliation involves using microdermabrasion or scrubs with small particles such as walnut shells, apricot pits, and salt crystals to remove dead skin cells and loosen the outer corneocytes of the *stratum corneum* (Behalpade and Gajbhiye 2022). Chemical exfoliation methods include the use of acids (such as alpha-hydroxy acids and beta-hydroxy acids), non-acids (e.g., N-acetylglucosamine), enzymes (e.g., papain and bromelain), or other agents (Banchhor and Saraf 2008; Small and Green 2012; David Troncoso et al. 2022). Another example is fungal-derived oxalic acid which has exfoliating properties (Palmieri et al. 2019; Sułkowska-Ziaja et al. 2023).

Fungi are capable of producing proteases that can be applied in topical exfoliants (Dhevagi et al. 2021). *Aspergillus*, *Penicillium*, *Rhizopus*, *Mucor*, *Humicola*, *Thermoascus*, and *Thermomyces* have been employed to produce proteases (de Souza et al. 2015). Although exfoliation is a recognized technique for removing melanin and lightening the skin, the application of leave-on products containing exfoliating ingredients such as retinol and glycolic acid can result in serious side effects like burning and stinging, skin irritation, peeling, and dryness (Gold and Biron 2011). In contrast, enzyme-based exfoliating ingredients are generally more gentle and safer and can be used for sensitive skin types that are less tolerant to acid (Packianathan and Kandasamy 2011). Several proteolytic enzymes have been tested as exfoliants, including papain, bromelain, proteases from pumpkin, proteases from *Bacillus subtilis*, trypsin, subtilisin, and keratinases. These proteolytic enzymes possess the ability to mimic the natural desquamation of the skin, thereby serving as exfoliants (Trevisol et al. 2022). Desquamation refers to the gradual shedding of dead skin cells from the outer layer of the skin (*stratum corneum*) (Has 2018). The principle behind enzymatic exfoliation is the enzymes' (primarily proteases) capacity to break down skin proteins, primarily keratin, and desmosomes, which are adhesion-specialized cell structures that keep cells linked in the *stratum corneum* (Trevisol et al. 2022). Proteases are grouped into six categories based on their catalytic domain: serine, cysteine, aspartic, threonine, glutamic, and metalloproteases. The first three categories are thought to play a role in the process of desquamation (Del Rosso 2013b).

Due to their action in facilitating desquamation, proteases accelerate skin regeneration, cleanse deeply, and enhance the absorption of cosmetic products. Enzymatic exfoliation is suitable for pigmentation, acne spots, oily skin, and rough skin (Behalpade and Gajbhiye 2022). Impaired or faulty desquamation manifests as visible scales and flakes on the surface of the skin (Del Rosso 2013b). A fungus-derived topical protease found in Lipotec's Actizyme advanced botanical component, an efficient cosmetic product sold in South Africa, functions like a crucial epidermal protein involved in exfoliation (Gimenez et al. 2018). Fungi offer several benefits for enzyme production, including cost-effectiveness, high productivity, faster production, and easy modification of enzymes. Additionally, the enzymes are produced extracellularly, which requires simple purification methods (El-Gendi et al. 2022). Fungal proteases are advantageous because using fungi to produce enzymes is considered safe, as they are classified as GRAS (generally recognized as safe) (de Souza et al. 2015).

Maintaining the stability of enzymes in cosmetic formulations poses a significant challenge (Pinto et al. 2011). The stability and activity of enzymes relies on factors such as temperature, pH level, the presence of activators and inhibitors, the specificity toward certain substrates, the enzyme concentration, and the specific protease activity (Uçak and Afreen 2022; Bialkowski 2022). The stability of protease produced by *Penicillium bilaiae* was tested at different temperatures (20–70 °C) and pH levels (3–9). The protease was highly stable at 25 °C, retaining 100% of its activity even after 2 h of incubation, whereas at 70 °C, it was inactivated, losing 90% of its initial activity after 10 min of incubation. The enzyme exhibited higher stability under acidic pH values between 5.5 and 6.5 (Ben Mefteh et al. 2019). It is suggested to use proteolytic enzymes that are active between 20 and 35 °C to prevent loss of activity during application on the skin. Using warm water or other products before applying the enzyme can enhance the effectiveness of enzymatic exfoliation by matching the skin temperature to the enzyme's temperature requirements (Trevisol et al. 2022). Enzyme encapsulation technologies provide a means to enhance stability by effectively bypassing the detrimental effects of protease degradation and denaturation caused by the high water content present in lotions and creams (Del Rosso 2013b). Therefore, it is crucial to ensure that the formulation of the carrier product enables the enzymes to successfully conduct the necessary reactions on the skin.

11.5 Skin-Lightening Properties

Hyperpigmentation is a common issue in the fields of dermatology and cosmetology that affects people of any gender or ethnicity and finding effective solutions to control it is a significant challenge (Badalyan et al. 2022). Skin pigmentation can increase due to various factors such as aging, hormonal changes, imbalances of the endocrine system, inflammation, and exposure to UV and infrared radiation, leading to conditions like lentigines, melasma, post-inflammatory hyperpigmentation, and actinic (sun) damage (Juhasz and Levin 2018). Melanin, a pigment that is responsible for skin color, is produced by melanocytes through a process called melanogenesis (Schlessinger et al. 2023). This is a complex process that starts with the enzymatic conversion of L-tyrosine or L-dihydroxyphenylalanine (DOPA) by tyrosinase (a copper-dependent enzyme), also known as polyphenol oxidase. This first step is the rate-limiting step of melanogenesis (Gold and Biron 2011; Boo 2020). After maturation, the pigment is packaged in melanosomes and migrates to keratinocytes (Gold and Biron 2011).

Skin hyperpigmentation results from tyrosinase accumulation, abnormally high melanin production, or abnormal distribution of the pigment (Saeedi et al. 2019). Exposure to UV triggers the production of melanin in humans as a defense mechanism to shield skin cells from the damaging effects of UV rays (Ullah et al. 2019). There are various options available for treating hyperpigmentation, including cosmetic products, topical and oral drugs, chemical peels, and light-based treatments (Moolla and Miller-Monthrope 2022). Hydroquinone, azelaic acid, 4-butyl resorcinol, and arbutin are substances that inhibit an enzyme called tyrosinase, which plays a key role in the production of melanin. Other agents that can lighten the skin have different ways of working, such as reducing the production of tyrosinase (tretinoin, glucosamine, retinol, retinaldehyde, N-acetyl glucosamine), speeding up the turnover of skin cells (vitamin C, vitamin E, thioctic acid, retinoids, lactic acid, glycolic acid, salicylic acid, liquiritin, gluconolactone), preventing the transfer of melanin to keratinocytes within the skin (linoleic acid, niacinamide), reducing inflammation (niacinamide, soy milk), and scavenging free radicals (glycyrrhetinic acid, topical steroids, ubiquinone, bakuchiol) (Resende et al. 2022). Kojic acid, azelaic acid, and p-coumaric are some of the fungal compounds that have skin-whitening properties (Badalyan et al. 2022; Sułkowska-Ziaja et al. 2023).

Kojic acid (5-hydroxy-2-hydroxymethyl-4-pyrone) is a natural substance produced from the secondary metabolism of fungal species during the aerobic fermentation of carbohydrates (Rasmey and Abdel-Kareem 2021). It is commonly used in skincare products to lighten the skin and prevent browning (Faig et al. 2017; Chib et al. 2023). Kojic acid effectively inhibits the activity of the enzyme tyrosinase by binding to copper, an important cofactor in the melanogenesis pathway (Masub and Khachemoune 2022). Due to its ability to inhibit tyrosinase, kojic acid protects the skin from UV rays, impedes melanin synthesis, and reduces hyperpigmentation (Phasha et al. 2022). Kojic acid and its related compounds have various uses in skincare and dental care products. They serve as antioxidants, anti-proliferative agents, antimicrobial, and anti-inflammatory agents (Saeedi et al. 2019; Sarkar 2022; Sułkowska-Ziaja et al. 2023). Currently, the industrial production of kojic acid is limited to filamentous fungi, specifically *Aspergillus* and *Penicillium* species. Among these, *A. oryzae* is the primary producer of Kojic acid (Chib et al. 2023). Kojic acid is currently produced in industry as a byproduct during the fermentation of malting rice in sake production in Japan (Lassfolk et al. 2019). The Personal Care Products Council conducted a survey on the use of kojic acid and found that it is commonly used in various personal care products, with concentrations ranging from 0.1% to 2%. The highest concentration is typically found in facial and neck creams, lotions, and powders (Burnett et al. 2010). Studies have shown that using products with up to 2% kojic acid on the skin does not have any negative effects (Felipe et al. 2023). The current data on the safety of kojic acid in dermatology also suggests that a 2% concentration in cosmeceuticals is safe, implying that a limit of 2% may be appropriate (Phasha et al. 2022). The prevailing side effect associated with kojic acid is irritant contact dermatitis (Grimes 2009), a condition that has been observed in some individuals utilizing cosmetic products or medicinal creams infused with 1% kojic acid, as evidenced by various case studies (Burnett et al. 2010).

Azelaic acid is another substance that can inhibit tyrosinase activity, therefore inhibiting epidermal melanin production (Schulte et al. 2015). It is a naturally occurring and straight-chain saturated (containing 9 atoms) dicarboxylic acid (Todea et al. 2021) that is produced by *Pityrosporum ovale* (a fungal species found on normal skin also known as *Malassezia fur fur*), among others (Grobel and Murphy 2018; Sułkowska-Ziaja et al. 2023). Azelaic acid selectively inhibits hyperactive or abnormal melanocytes, making it useful to treat abnormal pigmentation disorders such as lentigo maligna melanoma (Wang et al. 2021). It prevents melanocytes from functioning by interfering with DNA synthesis and inhibiting mitochondrial oxidoreductase activation (Searle et al. 2022). In vitro studies revealed that azelaic acid interferes with the activity of abnormal melanocytes without affecting normal melanocytes (Yu and Kim 2010). It may also lighten the skin by neutralizing free radicals due to its antioxidant properties (Searle et al. 2022). Additionally, it exhibits keratolytic, anti-inflammatory, and antibacterial properties (Sułkowska-Ziaja et al. 2023). Topical azelaic acid has been utilized in the treatment of melasma, *acne rosacea*, *acne vulgaris*, and post-inflammatory hyperpigmentation (Choulis 2014). The administration of azelaic acid in the form of a 20% cream is currently approved for the dermatological treatment of pigmentation (Wang et al. 2021). A randomized, double-blinded study that was conducted across multiple centers assessed the effectiveness, safety, and tolerability of azelaic acid cream (20%) in comparison to a cream containing no active ingredients (vehicle-only) for treating facial hyperpigmentation in patients with dark skin. Following a 24-week treatment, it was found that azelaic acid led to a significantly greater reduction in pigment intensity (Xing et al. 2022). Kircik (2011) demonstrated the safety and efficacy of utilizing a 15% concentration of azelaic acid in treating post-inflammatory hyperpigmentation caused by acne. Some potential side effects of azelaic acid include dryness, irritation, burning, transient erythema, and pruritus (Moolla and Miller-Monthrope 2022).

The use of fungi-derived compounds such as p-Coumaric acid provides a potential solution to hyperpigmentation by reducing oxidative stress and inflammation, and acting as anti-aging and skin-whitening agent (Badalyan et al. 2022). Edible mushrooms are the fruiting structures formed by species of macroscopic filamentous fungi (i.e., macrofungi), including those from Ascomycota and Basidiomycota phylum (Ghorai 2022). P-Coumaric acid (4-hydroxycinnamic acid) is a specialized metabolite found in various edible plants and fungi, especially in edible mushrooms (Abazari et al. 2021; Combes et al. 2021). Barros (2008) detected p-Coumaric acid in three different edible mushroom species (*Agaricus arvensis*, *Agaricus silvicola*, and *Lepista nuda*). In addition, Toledo et al. (2016) reported the presence of p-Coumaric acid in *Ramaria patagonica*. This phytochemical was identified as a potent and selective inhibitor of the enzyme tyrosinase (Boo 2019). Because it has a similar chemical structure to tyrosine (the substrate of tyrosinase), it can potentially hinder melanin production by competitively binding to active sites on the tyrosinase enzyme, preventing substrate access (An et al. 2010). It was previously demonstrated that 3 μM of p-coumaric acid reduced tyrosinase activity by 50% (Boo 2019). A 1.5% p-coumaric acid cream was applied twice daily to the forearm of individuals to examine its effect on human pigmented skin in vivo. The 1.5% cream caused a significant decrease in the melanin index of UV-induced pigmented skin after 8 weeks, leading to a lighter skin complexion (Seo et al. 2011). P-coumaric acid inhibits human tyrosine much more effectively than arbutin or kojic acid. It not only impedes cellular melanogenesis but also preserves the viability of human epidermal melanocytes when exposed to UV, acting as an antiphotoaging agent (Song et al. 2011). Park and Kim (2020) reported no significant cytotoxicity of p-coumaric acid-peptide conjugate on SK-MEL-2 melanoma cells. p-Coumaric acid did not show significant cytotoxicity in the effective concentration range that inhibits cellular melanogenesis. Furthermore, no toxic effects were observed in animal and human studies (Boo 2019). Adverse skin reactions observed after application of p-coumaric acid cream include scaling and mild development of papules and pruritus (Seo et al. 2011).

11.6 Anti-aging Properties

The mechanism of anti-aging effects produced by cosmetic compounds can be grouped into four actions: hydration, quenching of free radicals, UV filtration, and cell repair mechanisms (Li 2015). Extracts from mushrooms can prevent wrinkles, fine lines, sunspots, and other aging symptoms. This is because various fungal molecules possess antioxidant and regenerative properties (Angelini et al. 2022). Reishi mushroom is a popular ingredient in skincare products due to its ability to prevent redness, fine wrinkles, and premature aging while also purifying and moisturizing the skin. The high beta-glucan content helps to tone and soften the complexion of the skin. In addition, this mushroom's capacity to scavenge free radicals prevents skin damage from UV rays and pollution, as well as age-related skin issues. The mushroom extract, which is high in antioxidants, can aid in lowering oxidative stress brought about by free radicals. Its ability to scavenge free radicals makes it a strong natural anti-aging ingredient. Additionally, the extract can maintain tone, firmness, and youthful appearance of the skin (Visvanathan et al. 2022).

Fungal compounds that hydrate skin include biopolymers such as hyaluronic acid. Minimum research has focused on the recovery of hyaluronic acid from filamentous fungi (Shikina et al. 2022). Hyaluronic acid serves as a humectant due to its anionic strength which attracts water molecules to the skin thereby maintaining skin hydration (Walker et al. 2024). Skin hydration maintains the elasticity of the skin, the absence of which leads to wrinkle formation. Cosmetic products containing *Tremella fuciformis* extract, which acts as a natural alternative to hyaluronic acid, exhibited a 12.4% reduction in transepidermal water loss (TEWL) compared to formulations without this active ingredient. Moreover, these products effectively maintained sufficient hydration levels within the epidermis (Woźniak et al. 2023). Chitin-glucan, a copolymer present in several mushroom cell walls, possesses effective moisturizing qualities, has the potential to combat skin aging, and can therefore be utilized in skincare formulations for moisturization and anti-aging purposes (Wu et al. 2016).

Lactic acid in cosmetics performs an exfoliating function by removing dead cells in the *stratum corneum*. This action paves the way for the development of new skin cells (Murphrey et al. 2024). Lactic acid can be produced via the action of various fungi such as species of the *Rhizopus* and *Saccharomyces* genera (Zhang et al. 2007; Song et al. 2022). Keratinocytes undergo a self-destruction mechanism as they age and move up the epidermis forming a dead tissue within the *stratum corneum* (Murphrey et al. 2024). Therefore, the facilitation of their removal from the skin is beneficial for skin health and maintenance. Furthermore, exfoliation is known to stimulate cell regeneration and to boost the formation of collagen (Sharma et al. 2008).

11.7 Conclusion

The current trend in cosmetics is characterized by a focus on eco-friendly manufacturing techniques and the natural sourcing of raw materials. On a global scale, there is a heightened demand for incorporating natural ingredients in cosmetic formulations. As a result, natural products now play a pivotal role in the advancement of cosmetics as an increasing number of individuals are becoming conscious of the importance of adopting safe, healthy, and efficient skincare routines. There has been a significant focus on extracting secondary metabolites from numerous fungi sources. This approach has proven to be highly valuable as fungi provide a rich source of natural products for the cosmetic industry.

The use of fungi in cosmetics is gaining popularity, and several fungi-based natural products are already being used as active ingredients in cosmetic products for their anti-aging, anti-wrinkle, antioxidant, skin-lightening, emollient, anti-tyrosinase, anti-hyperpigmentation, and exfoliating properties. These products have therapeutic potential that goes beyond cosmetics and can effectively treat a range of dermatological conditions. Fungal-derived compounds such as p-coumaric acid, azelaic acid, and kojic acid have been successfully used in the treatment of abnormal skin pigmentation thanks to their depigmenting potency. Proteolytic enzymes from fungi are used in skincare products as enzyme exfoliants that eliminate dead skin cells, resulting in more radiant, uniform, and smoother skin. Compounds derived from fungi, like ceramides and lactic acid, have been found to possess emollient properties. Antioxidants obtained from fungal species, including polyphenols, serve as powerful free radical scavengers, effectively preventing and treating both intrinsic and extrinsic skin aging. Additionally, certain fungal compounds such as hyaluronic acid can effectively hydrate the skin to produce an anti-aging effect. Yet, limited research has been conducted on the extraction of hyaluronic acid from filamentous fungi.

Despite the enormous diversity of fungi that exists in nature, only a select few have been identified and utilized in the cosmetic industry with mushrooms being one of the most commonly used fungi in the cosmetic industry. More research and exploration of fungal species will help in discovering new species and fungal-derived substances that can be utilized in the formulation of skincare products. The potential for the use of fungi in the cosmetic industry is vast and holds great promise for future growth. Consequently, ongoing efforts aimed at establishing industry-wide regulatory standards to guarantee the safety and effectiveness of fungal-based cosmetic products are necessary.

References

Abazari MF, Nasiri N, Karizi SZ et al (2021) An updated review of various medicinal applications of p-Co umaric acid: from antioxidative and anti-inflammatory properties to effects on cell cycle and proliferation. Mini-Rev Med Chem 21:2187–2201. https://doi.org/10.2174/1389557521666210114163024

Alimi BA, Pathania S, Wilson J et al (2023) Extraction, quantification, characterization, and application in food packaging of chitin and chitosan from mushrooms: a review. Int J Biol Macromol 237:124195–124195. https://doi.org/10.1016/j.ijbiomac.2023.124195

An SM, Koh J-S, Boo YC (2010) p-coumaric acid not only inhibits human tyrosinase activity in vitro but also melanogenesis in cells exposed to UVB. Phytother Res 24:1175–1180. https://doi.org/10.1002/ptr.3095

Angelini P, Girometta CE, Venanzoni R, Bertuzzi G (2022) Anti-aging properties of medicinal mushrooms in systemic aesthetic medicine. In: Arya A, Rusevska K (eds) Biology, cultivation and applications of mushrooms. Springer, Singapore, pp 185–202

Aranaz I, Acosta N, Civera C et al (2018) Cosmetics and cosmeceutical applications of chitin, chitosan and their derivatives. Polymers (Basel) 10:213. https://doi.org/10.3390/polym10020213

Badalyan SM, Barkhudaryan A, Rapior S (2022) Medicinal macrofungi as cosmeceuticals: a review. IJM 24(4): 1–13. https://doi.org/10.1615/IntJMedMushrooms.2022043124

Bajza Á, Kocsis D, Berezvai O et al (2020) Verification of P-glycoprotein function at the dermal barrier in diffusion cells and dynamic "skin-on-a-chip" microfluidic device. Pharmaceutics 12:804. https://doi.org/10.3390/pharmaceutics12090804

Banchhor M, Saraf S (2008) PHCOG REV.: review article potentiality of papain as an antiaging agent in cosmetic formulation. Pharmacogn Rev 2:266–270

Behalpade S, Gajbhiye S (2022) Review article: skin care with exfoliation process. Int J Curr Sci 12:372

Ben Mefteh F, Frikha F, Daoud A et al (2019) Response surface methodology optimization of an acidic protease produced by *Penicillium bilaiae* Isolate TDPEF30, a newly recovered endophytic fungus from healthy roots of date palm trees (*Phoenix dactylifera* L.). Microorganisms 7:74. https://doi.org/10.3390/microorganisms7030074

Bialkowski K (2022) Proteolytic activity of cosmetic enzyme peel products. Acta Biochim Pol 69:895–899. https://doi.org/10.18388/abp.2020_6485

Boo YC (2019) p-coumaric acid as an active ingredient in cosmetics: a review focusing on its antimelanogenic effects. Antioxidants 8:275. https://doi.org/10.3390/antiox8080275

Boo YC (2020) Up- or downregulation of melanin synthesis using amino acids, peptides, and their analogs. Biomedicine 8:322. https://doi.org/10.3390/biomedicines8090322

Burnett CL, Bergfeld WF, Belsito DV et al (2010) Final report of the safety assessment of Kojic acid as used in cosmetics. Int J Toxicol 29:244S–2473S. https://doi.org/10.1177/1091581810385956

Chang AKT, Frias RR, Alvarez LV et al (2019) Comparative antibacterial activity of commercial chitosan and chitosan extracted from Auricularia sp. Biocatal Agric Biotechnol 17:189–195. https://doi.org/10.1016/j.bcab.2018.11.016

Chib S, Jamwal VL, Kumar V et al (2023) Fungal production of kojic acid and its industrial applications. Appl Microbiol Biotechnol 107:2111–2130. https://doi.org/10.1007/s00253-023-12451-1

Chou C-H, Tsai M-S, Lu H-Y et al (2019) Enzymatic hydrolysates obtained from Trametes versicolor polysaccharopeptides protect human skin keratinocyte against AAPH-induced oxidative stress and inflammatory. J Cosmet Dermatol 18:2011–2018. https://doi.org/10.1111/jocd.12959

Choulis NH (2014) Chapter 14—dermatological drugs, topical agents, and cosmetics. In: Aronson JK (ed) Side effects of drugs annual. Elsevier, pp 273–294

Chu J, Ming Y, Cui Q et al (2022) Efficient extraction and antioxidant activity of polyphenols from *Antrodia cinnamomea*. BMC Biotechnol 22:9. https://doi.org/10.1186/s12896-022-00739-5

Combes J, Imatoukene N, Couvreur J et al (2021) Intensification of p-coumaric acid heterologous production using extractive biphasic fermentation. Bioresour Technol 337:125436. https://doi.org/10.1016/j.biortech.2021.125436

Danby SG, Brown K, Higgs-Bayliss T et al (2016) The effect of an emollient containing urea, ceramide NP, and lactate on skin barrier structure and function in older people with dry skin. Skin Pharmacol Physiol 29:135–147. https://doi.org/10.1159/000445955

Dantas SBS, Moraes GKA, Araujo AR, Chapla VM (2022) Phenolic compounds and bioactive extract produced by endophytic fungus *Coriolopsis rigida*. Nat Prod Res 37:2037–2042. https://doi.org/10.1080/14786419.2022.2115492

David Troncoso F, Alberto Sánchez D, Luján Ferreira M (2022) Production of plant proteases and new biotechnological applications: an updated review. ChemistryOpen 11:e202200017. https://doi.org/10.1002/open.202200017

de Barros LB (2008) Caracterización química y propiedades bioactivas de hongos silvestres portugeses comestibles. In: Chemical characterization and bioactive properties of portuguese wild edible mushrooms. https://doi.org/10.14201/gredos.22597

de Souza PM, Bittencourt ML de A, Caprara CC et al (2015) A biotechnology perspective of fungal proteases. Braz J Microbiol 46:337–346. https://doi.org/10.1590/S1517-838246220140359

Del Rosso JQ (2013a) The role of skin care as an integral component in the management of acne vulgaris: part 1: the importance of cleanser and moisturizer ingredients, design, and product selection. J Clin Aesthet Dermatol 6:19–27

Del Rosso JQ (2013b) Application of protease technology in dermatology. J Clin Aesthet Dermatol 6:14–22

Devi R, Kaur T, Guleria G et al (2020) Chapter 9—fungal secondary metabolites and their biotechnological applications for human health. In: Rastegari AA, Yadav AN, Yadav N (eds) New and future developments in microbial biotechnology and bioengineering. Elsevier, pp 147–161

Dhevagi P, Ramya A, Priyatharshini S et al (2021) Industrially important fungal enzymes: productions and applications. In: Yadav AN (ed) Recent trends in mycological research: volume 2: environmental and industrial perspective. Springer, Cham, pp 263–309

Draelos ZD, Hall S, Munsick C (2020) A 14-day controlled study assessing qualitative improvement with 15% lactic acid and ceramides in skin moisturization and desquamation. J Clin Aesthet Dermatol 13:E54–E58

Dreno B, Khosrotehrani K, Silva G et al (2023) The role of dermocosmetics in the management of cancer-related skin toxicities: international expert consensus. Support Care Cancer 31(12):672. https://doi.org/10.1007/s00520-023-08116-4

El-Gendi H, Saleh AK, Badierah R et al (2022) A comprehensive insight into fungal enzymes: structure, classification, and their role in mankind's challenges. J Fungi 8:23. https://doi.org/10.3390/jof8010023

Faig JJ, Moretti A, Joseph LB et al (2017) Biodegradable kojic acid-based polymers: controlled delivery of bioactives for melanogenesis inhibition. Biomacromolecules 18:363–373. https://doi.org/10.1021/acs.biomac.6b01353

Felipe MT de C, Barbosa R do N, Bezerra JDP, Souza-Motta CM de (2023) Production of kojic acid by *Aspergillus species*: trends and applications. Fungal Biol Rev 45:100313. https://doi.org/10.1016/j.fbr.2023.100313

Fischer T, Perosino E, Poli F et al (2010) Chemical peels in aesthetic dermatology: an update 2009. J Eur Acad Dermatol Venereol 24:281–292. https://doi.org/10.1111/j.1468-3083.2009.03409.x

Ghorai S (2022) Edible mushroom: a potent producer of industrial enzymes. JSR 66:148–166. https://doi.org/10.37398/JSR.2022.660116

Gimenez A, Laporta O, Ollagnier M et al (2018) Natural mushroom enzyme for mild exfoliation. S Afr Pharm Cosmet Rev 45:38–39. https://doi.org/10.10520/EJC-e4625e73b

Gold MH, Biron J (2011) Efficacy of a novel hydroquinone-free skin-brightening cream in patients with melasma. J Cosmet Dermatol 10:189–196. https://doi.org/10.1111/j.1473-2165.2011.00573.x

Grimes PE (2009) Management of hyperpigmentation in darker racial ethnic groups. Semin Cutan Med Surg 28:77–85. https://doi.org/10.1016/j.sder.2009.04.001

Grobel H, Murphy SA (2018) Chapter 77—acne vulgaris and acne rosacea. In: Rakel D (ed) Integrative medicine, 4th edn. Elsevier, pp 759–770.e5

Gunnarsson M, Mojumdar EH, Topgaard D, Sparr E (2021) Extraction of natural moisturizing factor from the stratum corneum and its implication on skin molecular mobility. J Colloid Interface Sci 604:480–491. https://doi.org/10.1016/j.jcis.2021.07.012

Gupta PL, Rajput M, Oza T et al (2019) Eminence of microbial products in cosmetic industry. Nat Prod Bioprospect 9:267–278. https://doi.org/10.1007/s13659-019-0215-0

Harding CR, Aho S, Bosko CA (2013) Filaggrin—revisited. Int J Cosmet Sci 35:412–423. https://doi.org/10.1111/ics.12049

Has C (2018) Peeling skin disorders: a paradigm for skin desquamation. J Invest Dermatol 138:1689–1691. https://doi.org/10.1016/j.jid.2018.05.020

Hon KL, Kung JSC, Ng WGG, Leung TF (2018) Emollient treatment of atopic dermatitis: latest evidence and clinical considerations. Drugs Context 7:212530. https://doi.org/10.7573/dic.212530

Huq T, Khan A, Brown D et al (2022) Sources, production and commercial applications of fungal chitosan: a review. J Bioresour Bioprod 7:85–98. https://doi.org/10.1016/j.jobab.2022.01.002

Hyde KD, Bahkali AH, Moslem MA (2010) Fungi—an unusual source for cosmetics. Fungal Divers 43:1–9. https://doi.org/10.1007/s13225-010-0043-3

Jhan M-H, Yeh C-H, Tsai C-C et al (2016) Enhancing the antioxidant ability of trametes versicolor polysaccharopeptides by an enzymatic hydrolysis process. Molecules 21:1215. https://doi.org/10.3390/molecules21091215

Jiang C, Ge J, He B, Zeng B (2021) Glycosphingolipids in filamentous fungi: biological roles and potential applications in cosmetics and health foods. Front Microbiol 12:690211. https://doi.org/10.3389/fmicb.2021.690211

Jin B, Yin P, Ma Y, Zhao L (2005) Production of lactic acid and fungal biomass by Rhizopus fungi from food processing waste streams. J Ind Microbiol Biotechnol 32:678–686. https://doi.org/10.1007/s10295-005-0045-4

John Kasongo K, Tubadi DJ, Bampole LD et al (2020) Extraction and characterization of chitin and chitosan from *Termitomyces titanicus*. SN Appl Sci 2:1–8. https://doi.org/10.1007/s42452-020-2186-5

Juhasz MLW, Levin MK (2018) The role of systemic treatments for skin lightening. J Cosmet Dermatol 17:1144–1157. https://doi.org/10.1111/jocd.12747

Juncan AM, Moisă DG, Santini A et al (2021) Advantages of hyaluronic acid and its combination with other bioactive ingredients in cosmeceuticals. Molecules 26:4429. https://doi.org/10.3390/molecules26154429

Kahraman E, Kaykın M, Bektay HŞ, Güngör S (2019) Recent advances on topical application of ceramides to restore barrier function of skin. Cosmetics 6:52. https://doi.org/10.3390/COSMETICS6030052

Kircik LH (2011) Efficacy and safety of azelaic acid (AzA) gel 15% in the treatment of post-inflammatory hyperpigmentation and acne: a 16-week, baseline-controlled study. J Drugs Dermatol 10:586–590

Lassfolk R, Suonpää A, Birikh K, Leino R (2019) Chemoenzymatic three-step conversion of glucose to kojic acid. Chem Commun 55:14737–14740. https://doi.org/10.1039/C9CC07405H

Lee EJ, Cha HJ (2019) *Inonotus obliquus* extract as an inhibitor of α-MSH-induced melanogenesis in B16F10 mouse melanoma cells. Cosmetics 6:9. https://doi.org/10.3390/cosmetics6010009

Li X (2015) Anti-aging cosmetics and its efficacy assessment methods. IOP Conf Ser: Mater Sci Eng 87: 012043. https://doi.org/10.1088/1757-899X/87/1/012043

Liaud N, Giniés C, Navarro D et al (2014) Exploring fungal biodiversity: organic acid production by 66 strains of filamentous fungi. Fungal Biol Biotechnol 1:1. https://doi.org/10.1186/s40694-014-0001-z

Liu J-K (2022) Natural products in cosmetics. Nat Prod Bioprospect 12:40. https://doi.org/10.1007/s13659-022-00363-y

Maas RHW, Springer J, Eggink G, Weusthuis RA (2008) Xylose metabolism in the fungus *Rhizopus oryzae*: effect of growth and respiration on l(+)-lactic acid production. J Ind Microbiol Biotechnol 35:569–578. https://doi.org/10.1007/s10295-008-0318-9

Martins AM, Marto JM (2023) A sustainable life cycle for cosmetics: from design and development to post-use phase. Sustain Chem Pharm 35:101178. https://doi.org/10.1016/j.scp.2023.101178

Masub N, Khachemoune A (2022) Cosmetic skin lightening use and side effects. J Dermatolog Treat 33:1287–1292. https://doi.org/10.1080/09546634.2020.1845597

Mineroff J, Jagdeo J (2023) The potential cutaneous benefits of *Tremella fuciformis*. Arch Dermatol Res 315:1883–1886. https://doi.org/10.1007/s00403-023-02550-4

Moolla S, Miller-Monthrope Y (2022) Dermatology: how to manage facial hyperpigmentation in skin of colour. Drugs Context 11:2021-11–2. https://doi.org/10.7573/dic.2021-11-2

Muñoz G, Valencia C, Valderruten N et al (2015) Extraction of chitosan from *Aspergillus niger* mycelium and synthesis of hydrogels for controlled release of betahistine. React Funct Polym 91–92:1–10. https://doi.org/10.1016/j.reactfunctpolym.2015.03.008

Murphrey MB, Miao JH, Zito PM (2024) Histology, stratum corneum. In: StatPearls. StatPearls Publishing, Treasure Island

Niego AGT, Lambert C, Mortimer P et al (2023) The contribution of fungi to the global economy. Fungal Divers 121:95–137. https://doi.org/10.1007/s13225-023-00520-9

Ojo AO, de Smidt O (2023) Lactic acid: a comprehensive review of production to purification. Processes 11:688. https://doi.org/10.3390/pr11030688

Packianathan N, Kandasamy R (2011) Skin care with herbal exfoliants. Funct Plant Sci Biotechnol 5(Special Issue 1):94–97

Palmieri F, Estoppey A, House GL et al (2019) Chapter two—oxalic acid, a molecule at the crossroads of bacterial-fungal interactions. In: Gadd GM, Sariaslani S (eds) Advances in applied microbiology. Academic, pp 49–77

Park K-Y, Kim J (2020) Synthesis and biological evaluation of the anti-melanogenesis effect of coumaric and caffeic acid-conjugated peptides in human melanocytes. Front Pharmacol 11:922

Pellis A, Guebitz GM, Nyanhongo GS (2022) Chitosan: sources, processing and modification techniques. Gels 8:5–25. https://doi.org/10.3390/gels8070393

Phasha V, Senabe J, Ndzotoyi P et al (2022) Review on the use of kojic acid—a skin-lightening ingredient. Cosmetics 9:64. https://doi.org/10.3390/cosmetics9030064

Pinto CAS de O, Lopes PS, Sarruf FD et al (2011) Comparative study of the stability of free and modified papain incorporated in topical formulations. Braz J Pharm Sci 47:751–760. https://doi.org/10.1590/S1984-82502011000400012

Rasmey A-H, Abdel-Kareem M (2021) Optimization of culture conditions for kojic acid production in surface fermentation by *Aspergillus oryzae* isolated from wheat grains. Bull Pharm Sci Assiut Univ 44:201–211. https://doi.org/10.21608/bfsa.2021.174146

Resende DISP, Ferreira MS, Lobo JMS et al (2022) Skin depigmenting agents in anti-aging cosmetics: a medicinal perspective on emerging ingredients. Appl Sci 12: 775. https://doi.org/10.3390/app12020775

Rodan K, Fields K, Majewski G, Falla T (2016) Skincare bootcamp: the evolving role of skincare. Plast Reconstr Surg Glob Open 4:e1152. https://doi.org/10.1097/GOX.0000000000001152

Rupčić J, Marić V (1998) Isolation and chemical composition of the ceramide of the *Candida lipolytica* yeast. Chem Phys Lipids 91:153–161. https://doi.org/10.1016/S0009-3084(97)00106-0

Ryuta T, Yasushi T, Miwa K et al (2016) Relationship between the glycosphingolipids and phospholipids synthesis and the mycelial growth in *Neurospora crassa*. Trace Nutr Res 33:13–20

Saeedi M, Eslamifar M, Khezri K (2019) Kojic acid applications in cosmetic and pharmaceutical preparations. Biomed Pharmacother 110:582–593. https://doi.org/10.1016/j.biopha.2018.12.006

Santos FC, Marquês JT, Bento-Oliveira A, de Almeida RFM (2020) Sphingolipid-enriched domains in fungi. FEBS Lett 594:3698–3718. https://doi.org/10.1002/1873-3468.13986

Sarkar D (2022) Intensification of Tropical Cyclones due to climate change: a case study from Bay of Bengal (2001–2020). https://doi.org/10.13140/RG.2.2.22584.62728

Schlessinger DI, Anoruo M, Schlessinger J (2023) Biochemistry, melanin. In: StatPearls. StatPearls Publishing, Treasure Island

Schulte BC, Wu W, Rosen T (2015) Azelaic acid: evidence-based update on mechanism of action and clinical application. J Drugs Dermatol 14:964–968

Searle T, Ali FR, Al-Niaimi F (2022) The versatility of azelaic acid in dermatology. J Dermatol Treat 33:722–732. https://doi.org/10.1080/09546634.2020.1800579

Seo YK, Kim SJ, Boo YC et al (2011) Effects of p-coumaric acid on erythema and pigmentation of human skin exposed to ultraviolet radiation. Clin Exp Dermatol 36:260–266. https://doi.org/10.1111/j.1365-2230.2010.03983.x

Shamsaei S, Getso M, Ahmadikia K et al (2021) Recent findings on the role of fungal products in the treatment of cancer. Clin Transl Oncol 23:197–204. https://doi.org/10.1007/s12094-020-02428-1

Sharma SR, Poddar R, Sen P et al (2008) Effect of vitamin C on collagen biosynthesis and degree of birefringence in polarization sensitive optical coherence tomography (PS-OCT). AJB 7:2049–2054. https://doi.org/10.5897/AJB07.486

Shikina EV, Kovalevsky RA, Shirkovskaya AI, Toukach PV (2022) Prospective bacterial and fungal sources of hyaluronic acid: a review. Comput Struct Biotechnol J 20:6214–6236. https://doi.org/10.1016/j.csbj.2022.11.013

Singh A, Del Poeta M (2016) Sphingolipidomics: an important mechanistic tool for studying fungal pathogens. Front Microbiol 7:1–14. https://doi.org/10.3389/fmicb.2016.00501

Small R, Green B (2012) 24—skin care products. In: Usatine RP, Pfenninger JL, Stulberg DL, Small R (eds) Dermatologic and cosmetic procedures in office practice. W.B. Saunders, Philadelphia, pp 286–297

Song K, An SM, Kim M et al (2011) Comparison of the antimelanogenic effects of p-coumaric acid and its methyl ester and their skin permeabilities. J Dermatol Sci 63:17–22. https://doi.org/10.1016/j.jdermsci.2011.03.012

Song L, Yang D, Liu R et al (2022) Microbial production of lactic acid from food waste: latest advances, limits, and perspectives. Bioresour Technol 345:126052. https://doi.org/10.1016/j.biortech.2021.126052

Spiricheva OV, Sen'ko OV, Veremeenko DV, Efremenko EN (2007) Lactic acid production by immobilized cells of the fungus *Rhizopus oryzae* with simultaneous product extraction. Theor Found Chem Eng 41:150–153. https://doi.org/10.1134/S0040579507020078

Sułkowska-Ziaja K, Trepa M, Olechowska-Jarząb A et al (2023) Natural compounds of fungal origin with antimicrobial activity—potential cosmetics applications. Pharmaceuticals 16:1200. https://doi.org/10.3390/ph16091200

Taofiq O, Heleno SA, Calhelha RC et al (2016) Development of mushroom-based cosmeceutical formulations with anti-inflammatory, anti-tyrosinase, antioxidant, and antibacterial properties. Molecules 21:1372. https://doi.org/10.3390/molecules21101372

Thimmaraju A, Govindan S (2022) Novel studies of characterization, antioxidant, anticoagulant and anticancer activity of purified polysaccharide from *Hypsizygus ulmarius* mushroom. Bioact Carbohydr Diet Fibre 27:100308. https://doi.org/10.1016/j.bcdf.2022.100308

Todea A, Deganutti C, Spennato M et al (2021) Azelaic acid: a bio-based building block for biodegradable polymers. Polymers 13:4091. https://doi.org/10.3390/polym13234091

Toledo C, Barroetaveña C, Fernandes Â et al (2016) Chemical and antioxidant properties of wild edible mushrooms from native Nothofagus spp. forest, Argentina. Molecules 21:1201. https://doi.org/10.3390/molecules21091201

Trevisol TC, Henriques RO, Souza AJA, Furigo A Jr (2022) An overview of the use of proteolytic enzymes as exfoliating agents. J Cosmet Dermatol 21:3300–3307. https://doi.org/10.1111/jocd.14673

Uçak İ, Afreen M (2022) Chapter 13—enzymes. In: Galanakis CM (ed) Nutraceutical and functional food components (2nd). Academic, pp 537–571

Ullah S, Park C, Ikram M et al (2019) Tyrosinase inhibition and anti-melanin generation effect of cinnamamide analogues. Bioorg Chem 87:43–55. https://doi.org/10.1016/j.bioorg.2019.03.001

Visvanathan S, Krishnamoorthy R, Sabesan GS (2022) Fungal cosmetics: mushrooms in beauty care and the new age of natural cosmetics. In: Shukla AC (ed) Applied mycology: entrepreneurship with fungi. Springer, Cham, pp 1–37

Walker K, Basehore BM, Goyal A, Zito PM (2024) Hyaluronic acid. In: StatPearls. StatPearls Publishing, Treasure Island

Wang Z, Xiang H, Dong P et al (2021) Pegylated azelaic acid: synthesis, tyrosinase inhibitory activity, antibacterial activity and cytotoxic studies. J Mol Struct 1224:129234. https://doi.org/10.1016/j.molstruc.2020.129234

Wontcheu Fotso YA, Ghazi S, Belkaid A et al (2023) Extraction, chemical composition, antiradical capacity, and photoprotective effect of *Inonotus obliquus* from Eastern Canada. Nutraceuticals 3:380–402. https://doi.org/10.3390/nutraceuticals3030029

Woźniak B, Chudzińska J, Szczyglewska P et al (2023) Optimization of the composition of a cosmetic formulation containing *Tremella fuciformis* extract (fungi). Cosmetics 10:82. https://doi.org/10.3390/cosmetics10030082

Wu Y, Choi M-H, Li J et al (2016) Mushroom cosmetics: the present and future. Cosmetics 3:22. https://doi.org/10.3390/cosmetics3030022

Xing X, Dan Y, Xu Z, Xiang L (2022) Implications of oxidative stress in the pathogenesis and treatment of hyperpigmentation disorders. Oxidative Med Cell

Longev 2022:e7881717. https://doi.org/10.1155/2022/7881717
Yang D, Zhou Z, Zhang L (2019) An overview of fungal glycan-based therapeutics. Prog Mol Biol Transl Sci 163:135–163. https://doi.org/10.1016/bs.pmbts.2019.02.001
Yu JS, Kim AK (2010) Effect of combination of taurine and azelaic acid on antimelanogenesis in murine melanoma cells. J Biomed Sci 17:S45. https://doi.org/10.1186/1423-0127-17-S1-S45
Yun JS, Pahk JW, Lee JS et al (2011) *Inonotus obliquus* protects against oxidative stress-induced apoptosis and premature senescence. Mol Cell 31:423–429. https://doi.org/10.1007/s10059-011-0256-7
Zhang ZY, Jin B, Kelly JM (2007) Production of lactic acid from renewable materials by Rhizopus fungi. Biochem Eng J 35:251–263. https://doi.org/10.1016/j.bej.2007.01.028
Zheng N, Ming Y, Chu J et al (2021) Optimization of extraction process and the antioxidant activity of phenolics from *Sanghuangporus baumii*. Molecules 26:3850. https://doi.org/10.3390/molecules26133850
Zi Y, Jiang B, He C, Liu L (2020) Lentinan inhibits oxidative stress and inflammatory cytokine production induced by benzo(a)pyrene in human keratinocytes. J Cosmet Dermatol 19:502–507. https://doi.org/10.1111/jocd.13005

Fungal-Based Cosmetics: Formulation and Usage 12

Samar Sami Alkafaas, Bassant E. Abdelmoaty, Amany Bahnsy, Abdullah M. Saif, Sara Samy Elkafas, Wael Hafez, Mthokozisi Dladla, Mohamed T. El-Saadony, and Soumya Ghosh

Abstract

A growing trend in the skincare and cosmetics industry is the use of fungi's beneficial properties, like antioxidants and anti-inflammatory agents, to create cosmetics (Sułkowska-Ziaja, Trepa et al. 2023). This chapter looks at how fungal-based cosmetics are made, used, and applied. We start off by talking about the opportunities and difficulties associated with using fungi in cosmetics. Subsequently, the examination explores distinct elements and extracts from a range of mushrooms, emphasizing their possible role in hair, skincare, and general health.

The transformative effects of fungal-based cosmetics on skin health are highlighted in this chapter. These effects range from improving moisture retention to offering anti-aging benefits like lifting and firming. We also looked into the intriguing skin-whitening properties of extracts from fungi. Examined is the diverse range of uses for fungal-based cosmetics in haircare, demonstrating their ability to revitalize and enhance hair health while addressing several issues.

In conclusion, this chapter offers an extensive overview of fungal-based cosmetics, addressing the nuances of formulation, the

S. S. Alkafaas (✉)
Molecular Cell Biology Unit, Division of Biochemistry, Department of Chemistry, Faculty of Science, Tanta University, Tanta, Egypt
e-mail: samar.alkafas@science.tanta.edu.eg

B. E. Abdelmoaty · A. Bahnsy · A. M. Saif
Division of Biochemistry, Department of Chemistry, Faculty of Science, Tanta University, Tanta, Egypt

S. S. Elkafas
Production Engineering and Mechanical Design Department, Faculty of Engineering, Menofia University, Menofia, Egypt

Faculty of Control System and Robotics, ITMO University, Saint-Petersburg, Russia

Department of Automation and Control of Technological Processes and Production, Institute of Information Technology and Computer Science, National University of Science and Technology MISIS, Leninsky Prospekt 4, 119049 Moscow, Russia

W. Hafez
The Emirates Society of Internal Medicine (ESIM), Emirates Medical Association, 46th Street Deira Near International Bowling Centre -, Dubai, United Arab Emirates

Medical Research and Clinical Studies Institute, The National Research Centre, Cairo, Egypt

M. Dladla
Human Molecular Biology Unit (School of Biomedical Sciences), Faculty of Health Sciences, University of the Free State, Bloemfontein, South Africa

M. T. El-Saadony
Department of Agricultural Microbiology, Faculty of Agriculture, Zagazig University, Zagazig, Egypt

S. Ghosh
Natural and Medical Sciences Research Center, University of Nizwa, Nizwa, Oman

University of the Free State, Bloemfontein, South Africa

S. Ghosh, H. N. Onyeaka (eds.), *Fungal-Based Cosmetics: Formulation and Usage*, The Mycota 17,
https://doi.org/10.1007/978-3-032-09291-5_12

distinct qualities of mushroom compounds, and the variety of effects that can be achieved in the areas of skincare and haircare. In the beauty industry, fungus-based cosmetics are viewed as exciting new developments that will unite modern beauty science with the marvels of nature. These goods provide an enticing combination of cutting-edge solutions and natural benefits as the beauty business develops.

Keywords

Cosmetics · Mushrooms · Moisture retention

12.1 Introduction

A novel paradigm known as "fungus-based cosmetics" has emerged in the rapidly changing field of cosmetic science as a result of the search for environmentally friendly and naturally inspired beauty products (Wu et al. 2016). Cosmetics represent a compelling fusion of innovation, sustainability, and biological elegance by utilizing the power of fungi, an often-underestimated kingdom of life. Fungal-based formulations show the promise of nature's pharmacy for skincare and beauty enhancement as the beauty industry works to appeal to environmentally conscious consumers (Martins and Marto 2023; El Enshasy et al. 2018).

Fungi, with their rich biodiversity and unique biochemical profiles, offer a treasure trove of bioactive compounds that can elevate cosmetics to new heights (Abo Nahas et al. 2021). From moisturizers and serums to foundations and color cosmetics (Wu et al. 2016; Kalra et al. 2020), the versatility of fungal-derived ingredients provides formulators with an array of options to create products that not only deliver performance but also champion environmental responsibility (Sułkowska-Ziaja et al. 2023; Vitale et al. 2020).

This exploration of Fungal-Based Cosmetics navigates the complex process of its formulation using fungal-derived ingredients, shedding light on the scientific principles behind each choice. They also delve into the intricate world of fungal species, their bioactive components, and the artistry of formulating cosmetics that are both effective and ecologically sensitive. By embracing fungi, the beauty industry can pave the way for sustainable practices that go beyond mere aesthetics and foster a harmonious relationship between personal care and environmental stewardship.

12.2 Formulation of Fungal-Based Cosmetics

Chen et al. state that both topical application and ingestion of the extract are safe. It affects skin hydration by creating an occlusive layer on the skin's surface and, when added to skin care formulations, binds to water in the skin more successfully than hyaluronic acid. To make the base emulsion, the following steps must be completed.

Each raw ingredient required for the recipe was measured and weighed (Abendrot et al. 2021).

- To prepare the aqueous phase, continuous stirring at 500 rpm was required after the addition of sodium gluconate to water in a beaker on a magnetic stirrer.
- A blend, or glycerin and xanthan gum, was created.
- After adding the blend to the beaker containing water, the mixture was stirred for approximately 2 min until it became transparent and homogenous.
- A separate beaker was used to hold the emulsifier and emollient for the oil phase.
- Both the water-filled beaker and the oil-phase beaker were heated to approximately 75–80 °C.
- The aqueous phase is first filled with the melted oil phase, and then a preservative.
- The mixed phases were agitated for 2 min at a speed of 700–800 rpm using a magnetic stirrer, the emulsion was manually stirred to a temperature of 20–25 °C for 2 min at a speed of 9500 rpm.

- After adding tocopherol, the pH was adjusted to 5.0–5.5.
- The stability of the emulsion was confirmed using a centrifuge test. The process was weighing two grams of the emulsion into an Eppendorf-style tube, mounting it on a centrifuge rotor that was rotating, and modifying the following parameters: The rotor's range of RPM is 5.5.
- Centrifugation period: 2 min.

12.2.1 Optimizing the Emulsion Medium's Addition of Tremella Fungi Extract

- The following protocol was used to treat the emulsion substrate before adding the active ingredient in a weighed quantity and Tremella fungi extract (Evonik) at a concentration of 0.1 weight percent.
- After combining the emulsion substrate with an equal amount of water, the active ingredient was added in a weighed quantity.
- The emulsion substrate was mixed with weighed amounts of glycerin and the active ingredient.
- The emulsion substrate was supplemented with a weighed quantity of the active ingredient and an equivalent amount of ethanol.
- The active ingredients were weighed and combined with 200 times more water using a magnetic stirrer. Subsequently, the mixture was added to the emulsion substrate and agitated at 500 rpm for approximately 15 min.
- After the active ingredient was weighed and combined with three times the amount of water, it was shaken at 500 rpm for approximately half an hour using a magnetic stirrer before being added to the emulsion substrate.
- The weighed amount of active substance from the emulsion substrate preparation stage was combined with water and agitated at 500 rpm/min for approximately 30 min using a magnetic stirrer. The remaining steps for making the emulsion substrate were completed after adding sodium gluconate and mixing the mixture.
- The active ingredient was weighed and combined with sodium gluconate, which was obtained from the preparation step, to create the emulsion substrate. After adding the blend, the remaining steps in the emulsion substrate preparation process were completed while the active ingredients were added and stirred with a magnetic stirrer for approximately 30 min at 500 rpm (Woźniak et al. 2023).

12.3 Individual Compounds and Mushroom Extracts as Ingredients in Cosmetics

For centuries, precious mushrooms have been explored as potential ingredients, in addition to being a valuable source of bioactive substances for humans (Wu et al. 2016). Since they are more effective than their synthetic counterparts and have fewer hazardous side effects, natural components are continuously sought after by the cosmetics industry (Taofiq et al. 2016a). For many years, mushrooms have become a significant component of food. Currently, they are making their way into nutricosmetics and cosmeceuticals, which are both forms of cosmetics. The ingredients used as examples include phenolics, polysaccharides, polyphenolics, terpenoids, selenium, vitamins, and volatile organic compounds. These substances are excellent choices for cosmetic products because of their moisturizing, anti-aging, anti-wrinkle, and skin-whitening properties (Wu et al. 2016). Extrinsic processes are linked to UV radiation exposure, which results in the formation of free radical species and is intrinsic or natural. The two primary causes of skin aging are hormonal variations that occur with aging and mechanisms that affect the body's organs and skin (Papakonstantinou et al. 2012). Reactive oxygen species (ROS) induce transcription factor activator protein-1 (AP-1), which promotes degradation of collagen and elastin by upregulating matrix metalloproteinases (MMPs) (Leem 2015). Environmental stress caused by reactive nitrogen species (RNS) and ROS leads to inflammation. It also increases the expression

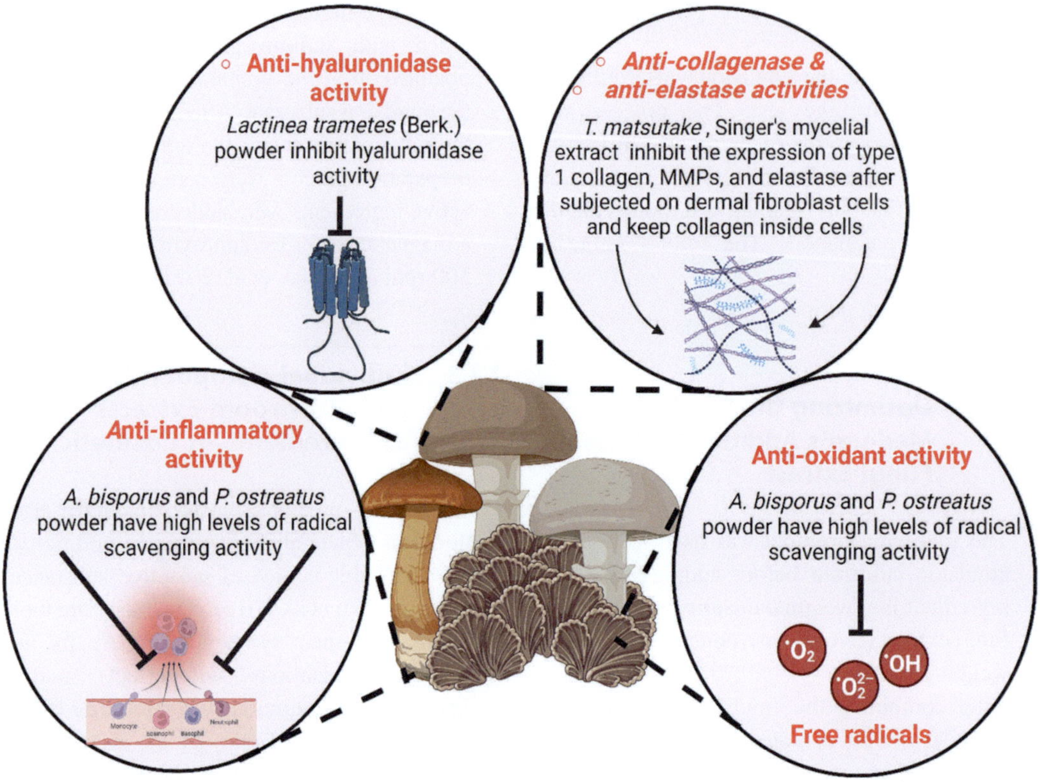

Fig. 12.1 Biological activities of fungi. (The figure was drawn by using Biorender https://www.biorender.com/)

of enzymes (as collagenase and elastase) and the secretion of inflammatory mediators. This reduced the elasticity and tensile strength of the skin (Tanuja et al. 2015). Additionally, as a result of rising bacterial drug resistance across a broad clinical range, the frequency of skin and soft tissue infections is increasing in the antimicrobial domain (Gündoğdu et al. 2016). Any material with a soft action that is applied or sprinkled on the different body parts of a human being with the express purpose of cleaning, beautifying, amplifying attractiveness, safeguarding them, and maintaining their health is referred to as a "cosmetic product" (Fig. 12.1).

12.3.1 Activity of Antioxidant

Reactive oxygen species (ROS) accelerate the aging process of the skin by causing intracellular and extracellular oxidative stress, which also leads to wrinkles and abnormal pigmentation. When talking about skin aging, UV exposure is usually brought up because it causes cells to produce more ROS. Antioxidants are a useful tool for halting photoaging-induced skin aging. The processes that the body uses to produce and eliminate ROS are described in this review. The effects of reactive oxygen species (ROS) produced in the skin and how they change the skin are also covered (Masaki 2010). A summary of representative antioxidant effects on skin is also provided, with an emphasis on skin aging. The two species with the highest reduction power and radical scavenging activity were *A. bisporus* and *P. ostreatus*. These results corroborate the findings of Reis et al. (2012), who found that the most potent DPPH radical scavenging activity was present in *A. bisporus* methanolic extracts obtained during maceration (Heleno et al. 2015). In the process of creating cosmeceutical designs, identifying extracts or bioactive metabolites with antioxidant activity is essential. These substances scavenge reactive oxygen species (ROS) and inhibit the enzymes tyrosinase and matrix metalloproteinase, which degrade collagen and cause hyperpigmentation (Masaki 2010). Numerous biological processes occur within the body's tissues and cells, all essential for optimum operation. These reactions frequently result in the production of free radicals, which are unpaired electrons. Among them are reactive sulfur species (RSS), reactive nitrogen species (RNS), and free radicals (ROS). Normally, the body uses glutathione peroxidase and other naturally occurring antioxidants to balance the production and neutralization of ROS. However, excessive ROS generation usually depletes it, resulting in oxidative stress in the cells (Ferreira et al. 2010). Many different biological processes occur within the body's tissues and cells, all of which are essential for peak performance. Oftentimes, these reactions result in the production of free radicals—unpaired electrons. Some instances are reactive sulfur species (RSS), reactive nitrogen species (RNS), and free radicals (ROS). Using glutathione peroxidase and other naturally occurring antioxidants, the body normally regulates the production and neutralization of ROS. However, it is often exhausted by excessive ROS production, resulting in oxidative stress to the cells (Carocho and Ferreira 2013; Alkafaas et al. 2023).

12.3.2 Anti-inflammatory Activity (Types of Fungi That Have Anti-inflammatory Activity)

An injury's natural physiological response is inflammation, which usually presents as pain, heat, redness, swelling, and loss of function. It causes the expression of cyclooxygenase-2 (COX-2), which can lead to cancer and inflammatory illnesses. Excess production of inflammatory mediators, including interleukins (IL 1β, IL-6, IL-8), tumor necrosis factor (TNF-α), 5-lipoxygenase (5-LOX), inducible nitric oxide synthase (iNOS), nuclear factor-κ B (NF-κB), and intercellular adhesion molecule-1 (ICAM-1), can lead to inflammatory diseases and cancer (Taofiq et al. 2016b). Studies have shown the anti-inflammatory potential of mushrooms and their isolated constituents, such as polysaccharides, terpenes, phenolic compounds, sterols, fatty acids, polysaccharide–protein complexes, and other bioactive metabolites, based on their ability to reduce the synthesis of inflammatory mediators. Reduced levels of inflammatory mediators, such as NO, as well as other inflammatory mediators released from inflammatory cells, such as interleukin (IL) 1β, IL-6, IL-8, TNF-α, and PGE2, have been associated with anti-inflammatory mechanisms. COX-2, TNF-α, iNOS, and IL-1β are just a few of the pro-inflammatory enzymes and cytokines whose expression is regulated by the transcription factor NF-κB. Thus, the key to reducing inflammation is to identify natural inhibitors of one or more NF-κB pathway steps (Taofiq et al. 2016b). Atopic dermatitis (AD), a chronic inflammatory skin disease, is caused by several physiological and environmental factors. Usually, it is accompanied by redness, rashes, and intense itching (Park et al. 2015a). According to recent reports, 1–2% of adults and 10–15% of children are affected by this illness. Even though the exact physiological mechanism underlying the disease is still unknown, research has shown that the overproduction of pro-inflammatory cells, such as macrophages, is the primary cause of the overproduction of pro-inflammatory mediators, such as TNF-α, IL-1β, IL-6, and IL-8 (Wu et al. 2011).

12.3.3 Anti-tyrosinase Action

The main enzyme responsible for melanogenesis in mammals and unintended enzymatic browning of fruit after harvesting is tyrosinase. This led researchers to search for potent tyrosinase inhibitors for application in the food and cosmetic industries. Despite recent advances in our understanding of tyrosinase inhibitors derived from synthetic and plant sources, there is a dearth of research on inhibitors of microbial origin. As a result, by reviewing tyrosinase inhibitors made by microbes, this article offers an updated database of tyrosinase inhibitors from microbial sources

(Fernandes and Kerkar 2017). Melanin is the primary pigment that determines a person's skin tone. The constant exposure of human skin to UV radiation modifies the skin's structure and function. After extended UV exposure, the body produces more tyrosinase, which causes the skin to produce too much melanin. An enzyme called tyrosinase regulates how quickly the process of making melanin proceeds. Because it converts tyrosine into dopaquinone, it is also referred to as oxidizing dihydroxyphenylalanine (DOPA). Many signal transduction pathways that regulate the mRNA expression of tyrosinase and tyrosinase-related proteins (TRP1 and TRP2) are responsible for the increased production of melanin (Chang 2012).

The most common hormonal and chemical mediator for the initiation of melanin biosynthesis is the cAMP-mediated pathway (Meng et al. 2012). The transcription factors' activation by cAMP microphthalmia-associated transcription factor (MITF) is upregulated in response to cAMP-response element binding protein (CREB) and cAMP-dependent protein kinase A (PKA). Melanin-producing enzyme tyrosinase is expressed when MITF binds to the promoter region of two tyrosinase-related proteins, TRP-1 and TRP-2. This system regulates the complex procedures that go into melanogenesis (Vachtenheim and Borovanský 2010). Researchers are looking for novel substances or combinations of substances to treat hyperpigmentation disorders, which are undesirable abnormalities typically characterized by the appearance of pale brown to dark brown spots on the skin; irregular grey patches on the face, neck, and trunk; and darker skin. Natural ingredients could be helpful in the treatment of these ailments. Depigmentation is caused by the inhibition of one or more steps in the melanogenic pathway or melanosome transfer because these actions reduce the release of melanin and ultimately prevent melanogenesis (D'Mello et al. 2016). Although hydroquinone is used to treat hyperpigmentation disorders, safety concerns are raised. These consist of the exogenous ochronosis effect, possible carcinogenicity, cell irritants, and an unpleasant fishy smell that is exuded by the skin (Sarkar et al. 2013). Other studies have looked at the use of corticosteroids and kojic acid, which have been shown to have adverse local or systemic effects after prolonged exposure despite their high efficacy. Because of the adverse effects of the aforementioned depigmenting agents, the hunt has been replaced. Natural products, such as those made from marine algae, rhizomes, mushrooms, or plants, that are rich in bioactive chemicals and can be used to treat a variety of disorders, have gained importance as research topics. Many compounds, mostly phenolics, including arbutin, resveratrol, ellagic acid, and gentilic acid, have been found in plants. These compounds have been shown to obstruct one or more stages of the melanogenic pathway (Ali et al. 2015). In addition to inhibiting melanogenesis, these phytochemicals have been shown to have anti-aging properties, function as moisturizers, and provide skin support and stability, all of which contribute to the maintenance of healthy skin. Table 12.1 lists studies that show how individual bioactive metabolites and mushroom extracts inhibit tyrosinase.

12.3.4 The Activity of Anti-hyaluronidase

The extracellular matrix of the skin provides mechanical and structural support while preserving the integrity of the skin. One of the main causes of skin aging is the disintegration of the skin matrix. Structural proteins like collagen and elastin are necessary for good skin, but they are not enough to support a healthy skin matrix. The skin also requires the proper elements to promote dermal migration, proliferation, and regeneration (Ito 2014). According to Papakonstantinou et al., (2012), hyaluronic acid (HA) is a naturally occurring polymer that is derived from glucose and has several functions in the skin, including increased viscosity, skin rejuvenation, and moisture retention (Papakonstantinou et al. 2012). The amount of HA in prokaryotic and eukaryotic cells is equal. The erectile tissues of the penis, lung, aorta, heart valves, umbilical cord, and vitreous of the eye are the next most common organs in

Table 12.1 Extracts from mushrooms with possible cosmeceutical uses

Mushroom species	Species bioactivity	Species effect	References
Agaricus bisporus	Anti-tyrosinase	The amount of melanin that was produced was inhibited by 80% at 0.5μM, 54.1% at 1μM, and 39.4% at 2μM.	Miyake et al. (2010)
Coriolus versicolor	Anti-tyrosinase	The amount of tyrosinase activity was reduced to 26.3 ± 8.3% at 10 mg/mL.	Park et al. (2015b)
Ganoderma lucidum	Anti-tyrosinase	Tyrosinase activity was up to 80% inhibited by the extract at 1 mg/mL.	Chien et al. (2008a)
Inonotus mikadoi	Anti-tyrosinase	It demonstrated inhibition of tyrosinase activity up to 46.0 ± 7.5% at 10 mg/mL.	Park et al. (2015b)
Inonotus obliquus	Anti-tyrosinase	When tyrosinase inhibitory activity was measured at 10 μg/mL in petroleum ether and n-butanol extract, the IC50 values were 3.81 and 6.32 μg/mL, respectively.	Yan et al. (2014)
Coprinus comatus	Anti-inflammatory, Antioxidant	-------	Mattila et al. (2001)
Grifola frondosa	Anti-collagenase	It reduced MMP-1 mRNA expression by 61.1% at 250μg/mL.	Bae et al. (2010)
Volvariella volvacea	Anti-collagenase	Collagen biosynthesis is stimulated by sonicated aqueous extracts up to 146.77 ± 13.20%.	Taofiq et al. (2016a)
Tricholoma matsutake	Anti-collagenase, anti-elastase	It reduced MMP-3 expression and 81.4 ± 3.92% inhibited elastase activity at 100μg/mL.	Kim et al. (2014)
Pleurotus citrinopileatus Singer	Anti-hyaluronidase	The extract reduced hyaluronidase activity by 9.7%, 10.8%, and 25.4% at 2.0, 1.10, and 4.1 mg/mL, respectively.	Meng et al. (2011)
Pleurotus tuber-regium	Anti-hyaluronidase	Activity of hyaluronidase was at 100μg/mL, the aqueous and methanolic extracts both significantly decreased, by 22.19 and 3.94%, respectively.	Dandapat and Sinha (2015)
Trametes lactinea	Anti-hyaluronidase	The inhibition of hyaluronidase activity was found in both acetonic and aqueous extracts, at 88.3 ± 0.14% and 88.6 ± 0.11%, respectively.	Yahaya and Don (2012)
Agaricus bisporus	Antimicrobial	–	De Silva et al. (2012)

MMP-1 and MMP-3 are the names of the respective matrix metalloproteinases

humans, following the skin. HA is a non-sulfated glycosaminoglycan (GAG) consisting of recurrent polymeric disaccharides linked by a glucuronidic β (1→3) bond between N-acetyl-d-glucosamine and d-glucuronic acid. Hydrogen bonds to form a stable tertiary structure in aqueous solutions. Despite having a straightforward composition without branching points or changes in its sugar composition, HA demonstrates a broad range of physicochemical characteristics. HA polymers can assume a huge range of forms and configurations depending on factors such as pH, salt concentration, size, and related positive ions. Unlike other GAGs, HA is not covalently bound to a protein core, but it can aggregate with proteoglycans. Because HA contains a lot of water, solutions have a high viscosity even at low concentrations (Papakonstantinou et al. 2012). Hyaluronic acid levels decrease with age, dehydrating the skin and impairing its capacity for self-repair and renewal. Among the products available in the cosmetic industry are creams and lotions for topical HA application; however, they face significant challenges because of their tendency to cause an inflammatory reaction. *Pleurotus citrinopileatus*, commonly known as the golden oyster mushroom, is edible.

Antigenotoxicity of it's angiotensin-converting enzyme (ACE) inhibition, antioxidant, anticancer, antihyperlipidemic, and anti-atopic dermatitis properties have all been shown in a number of studies (Meng et al. 2011). Because it has been demonstrated that these extracts inhibit hyaluronidase activity by 9.6%, 10.7%, and 25.4%, respectively, at concentrations of 2.0, 1.1, and 4.1 mg/mL, they are essential components of cosmetic products. *Lactinea trametes* (Berk) Sac is a macrofungus that belongs to the Polyporaceae family and can inhibit hyaluronidase (Yahaya and Don 2012) when the HA substrate is present. Two different extraction solvents, each containing 100μg/mL, were used to measure activity. At 88.6 ± 0.11%, the aqueous extract showed the highest level of hyaluronidase inhibition. The acetone extract had the second-highest level of inhibition, at 88.3 ± 0.14%. The authors were unable to determine a correlation between the extract's phenolic content and the inhibition of hyaluronidase activity, despite the fact that the positive control, apigenin, inhibited hyaluronidase activity by up to 87.4 ± 0.03%. Table 12.1 lists the studies that show how particular bioactive metabolites and extracts from mushrooms inhibit hyaluronidase.

12.3.5 Inhibitors of Collagen and Elastin Actions

The connective tissue between the dermis beneath the epidermis, which makes up human skin, keeps the two layers firmly connected. Dermal fibroblasts produce the two structural proteins, collagen and elastin, that make up the extracellular matrix (ECM). These proteins are necessary for the skin's protective function (Kim et al. 2014). Connective tissues, including the skin, cartilage, lungs, aorta, and elastic ligaments, are made of an ECM protein called elastin.

Elastase is a metalloproteinase that inhibits elastin expression. Prolonged exposure to elastases can result in wrinkles and a decrease in the skin's resilience, despite elastin's high resistance to proteolytic degradation (Onar et al. 2012). Interest in screening natural matrices, such as plants, mushrooms, rhizomes, and marine algae, for the production of active cosmetic ingredients that can lessen wrinkles in the skin and slow down the aging process has increased due to age-related increases in elastase activity. UV radiation-induced collagen degradation is the primary cause of aging because collagen is the most significant component of the skin's extracellular matrix and is responsible for restoring the elasticity, flexibility, and strength of the skin. When collagen, elastin, and hyaluronic acid levels in the extracellular matrix decrease with age, the skin loses its elasticity and strength, which is manifested as wrinkles (Ndlovu et al. 2013). The extracellular matrix, which is linked to a variety of physiological and pathological conditions, such as bone resorption, wound healing, carcinogenesis, and inflammatory and vascular diseases, is hydrolyzed by matrix metalloproteinases (MMPs) and zinc-dependent endopeptidases (Thomas et al. 2014). The main membrane-anchored metalloproteinase that degrades collagen is MMP-1. Tissue inhibitors of metalloproteinases (TIMPs) are naturally occurring inhibitors that have been demonstrated to reduce MMP expression. This is an undesirable effect that further protects the extracellular matrix. Skin aging is mostly caused by excessive production of reactive oxygen species (ROS) resulting from exposure to ultraviolet (UV) radiation. These ROS (reactive oxygen species) cause mitogen-activated protein kinases to become active, which in turn causes activator protein factor 1 (AP-1). This leads to an overabundance of MMP expression, which degrades collagen and wrinkles the skin. Many investigations are being carried out to identify naturally occurring AP-1 inhibitors that can be used as cosmeceutical ingredients to inhibit MMP expression (Thomas et al. 2014). In this work, the anti-elastase and anti-collagenase properties of the mycelial extract of T. matsutake (S. Ito & S. Imai) Singer were examined. Dermal fibroblast cells were exposed to concentrations ranging from 1 to 100μg/mL for 72 h. Type 1 collagen, MMPs, and elastase expression were then measured. The extract showed a dose-dependent reduction in the protein expression of MMP-1, MMP-3, and elastase mRNA of up to 81.4 ± 3.92% at 100μg/mL, while the positive control, phophoramidon, showed a decreasing effect of up to 89.6 ± 7.74% at 10μM. The results showed that T. matsutake extract has anti-aging potential by halting the extracellular matrix's degradation, despite its inability to inhibit collagen expression (Kim et al. 2014).

12.3.6 Anti-microbial Activity

Non-pathogenic bacteria like fungus, some species of Streptococcus, and Staphylococcus aureus are always colonizing the skin. Environmental elements including temperature, humidity, and sebum secretion affect the distribution and density of skin microflora in addition to an individual's age (Elsner 2006). Inflammatory skin diseases such as psoriasis, furuncle, carbuncle abscess, erysipelas, cellulitis, folliculitis, and atopic dermatitis have been linked to these microorganisms (Alsterholm et al. 2010). Atopic dermatitis is a common skin condition linked to increased skin colonization by bacteria like Staphylococcus aureus. This usually makes the illness worse (Salah and Faergemann 2015). The cosmetic industry is constantly searching for new, natural bioactive compounds to replace synthetic antimicrobial agents because these microbes also develop resistance to common topical antimicrobials. Inflammatory skin conditions have long been treated externally with medicinal plant pastes and infusions (Nesy and Mathew 2014). Numerous studies have demonstrated the antimicrobial properties of mushrooms in addition to their bioactive components. It has been observed that the bioactive components of mushroom extracts possess antimicrobial properties (Alves et al. 2012). Research has shown that the antimicrobial qualities of mushrooms, both edible and inedible, inhibit the growth of pathogenic microorganisms (Reid et al. 2016). Lentinula edodes was the most intriguing species because it proved to be efficient against both gram-positive and gram-negative bacteria. The species listed below belong to the genera Lepista, Ganoderma, and Boletus. Secondary metabolites called phenolic compounds can be found in mushrooms and other natural sources. They provide defense against insects, bacteria, viruses, and UV radiation. Studies have indicated that these substances exhibit antibacterial characteristics by dissolving the invasive pathogen's membrane and cell wall, ultimately resulting in the pathogen's elimination (Ribeiro et al. 2015). 2,4-dihydroxybenzoic, protocatechuic, vanillic, and p-coumaric acids were among the phenolic acids that showed interesting anti-gram-positive and anti-gram-negative characteristics, according to Alves et al. (2013). Moreover, studies have demonstrated the antimicrobial properties of furulic, caffeic, ellagic, syringic, and chlorogenic acids (Alves et al. 2013).

12.4 Cosmetics, Cosmeceuticals, and Nutricosmetics (Skincare)

A "cosmetic product" is defined as any mildly active material that is applied topically by rubbing or misting to various body parts with the specific intent of cleaning, beautifying, enhancing their attractiveness, safeguarding, and preserving their state. The tens of billions of US dollars global cosmetics market is always looking for better natural raw materials that can either support new bioactivities or compete with artificial synthetic counterparts. Because of the strength and effectiveness of mushrooms, Asians have been using them as medicine for thousands of years. For instance, they were set aside for royalties in China and Japan (Taofiq et al. 2016a).

Cosmeceuticals are a kind of hybrid medication that combine components of pharmaceutical and cosmetic industries to enhance appearance. Despite being used in cosmetics, they contain ingredients that have an impact on the skin's biological processes. Products under the cosmeceutical category include both human and animal treatments. Veterinary cosmeceutical products can be broadly classified into two categories: shampoos and anti-ectoparasites. The most well-known brand in the former category in Western markets is probably mera-pets (Merapet Inc., Ahmedabad, India). Human cosmeceuticals are mostly anti-aging products, with some moisturizing and anti-acne options as well. Among the well-known products are SensiClear (Mission Scientific Skincare Inc., Gold River, CA, USA), La Roche (F. Hoffmann-La Roche Ltd., Basel, Switzerland), Nu-Derm (Obagi Medical Products Inc., Irvine, CA, USA), and Bliss (Hut.com Ltd., Cheshire, UK). These days, nearly all of them include products made with extracts from mushrooms (Wu et al. 2016). These days, it's common to use natural skin care ingredients like dairy products, minerals, phytonutrients, microbial metabolites, and portions of animal protein. Extracts from coconut, jasmine, lemon grass, longan, and other medicinal plants have been used to treat a wide range of skin conditions. These extracts are widely used in many countries, including Africa, where they are crucial for treating a variety of skin conditions. They have been the subject of extensive research. Cosmetics that contain pure compounds or raw materials derived from fungi, particularly mushrooms and their products, are being used to treat a variety of skin-related conditions (Taofiq et al. 2016a).

The most recent development in skin care is called "nutricosmetics," which blends food, cosmetics, and medications. However, not all dermatologists and consumers are aware of this trend. They are taking dietary and nutritional supplements to enhance skin health and provide cosmetic benefits. The three main areas of focus for action are well-being, skin health, and hair health. These effects can be attributed to a variety of micronutrients (e.g., vitamin C may function as an antioxidant and may help lower the production of free radicals in skin exposed to UV light). In keeping with these developments, the cosmetics industry is currently working harder to create nutricosmetics, or cosmetics with high concentrations of various substances like ceramide, which is known to maintain the structure and function of skin, hyaluronic acid, and collagen (Taofiq et al. 2016a). Currently, species with patents for their use in cosmeceuticals and nutricosmetics include *Grifola frondosa* and *Agaricus subrufescens* (*A. blazei, A. brasiliensis*), *Maeandriformes choiromyces, Cordyceps sinensis, Ganoderma lucidum* and many others, including *Trametes versicolor, Tremella fuciformis, Hypsizygus ulmarius, Inonotus obliquus, Tuber* spp., *Schizophyllum commune, Lentinula edodes, Polyporus* spp., and many others.

12.5 Beneficial Components of Mushrooms

Several studies have been conducted to determine the components of mushrooms that are beneficial to the skin. Numerous bioactive metabolites, including lectins, polysaccharides, phenolics and polyphenolics, terpenoids, ergosterols, and volatile organic compounds, have been shown to be present in mushrooms. Currently, mushroom polysaccharides like lentian, schizophyllan, polysaccharide kerstin (PSK), and polysaccharide peptide (PSP) are sold by the pharmaceutical industry. Certain mushrooms belonging to the higher order of Basidiomycetes have fruiting bodies, cultivated mycelia, and broth that contains biologically active polysaccharides (Lu et al. 2020). The presence of polyphenols, particularly phenolic acids, which are derived from benzoic and cinnamic acids, is linked to the medicinal mushrooms' antioxidant qualities. It has been reported that the main phenolic compounds found in mushrooms are gallic, caffeic, and p-coumaric acids. Mushrooms contain phenolic compounds that have significant antioxidant properties. They prevent metals from oxidizing materials, limit the degradation of peroxide, scavenge reactive oxygen species, and inhibit free radicals. As a result, they prevent changes in cellular DNA and postpone the onset of cancer. Analysis of Chaga ethanol extracts revealed 111 distinct phenolic compounds, including phenolic acids, flavonoids, coumarins, quinones, and styrylpyrones. Medicinal mushrooms contain flavonoids such as myricetin, rutin, naringenin, quercetin, morin, and hesperetin. Among the various medicinal mushrooms, chaga (Inonotus obliquus (Ach. Ex Pers.) Pilát) had the highest polyphenol content and antioxidant activity. This mushroom has a total phenolic content of 97μmol GAE/mg; in comparison, the concentrations of phenolic compounds in shiitake, reishi

(*Ganoderma lucidum (Curtis) P. Karst.*), and turkey tail (*Trametes versicolor (L.) Lloyd*) are 21, 13, and 0.1μmol GAE/mg, respectively. Chaga, or Inonotus obliquus (Ach. Ex Pers.) Pilát, is a type of mushroom that exhibits water-ethanol extract that has antioxidant activity against DPPH that is approximately five times higher than that of other mushrooms. Human health benefits of mushroom polyphenols include anti-aging, hypoglycemic, anticancer, and antioxidant properties (Łysakowska et al. 2023).

Terpenoids are a diverse group of organic compounds made up of five-carbon isoprene units. Ganoderma spp. is the species of mushroom that contains the majority of known terpenoids in terms of terpenes and terpene derivatives. These consist of lanostane, lucidone, ganodermanontriol, ganodermic and ganoderic acids, ganoderals, ganoderols, and ganodermanontriol. These substances are all capable of influencing immune reactions and preventing infection. Carotenoids are organic pigments that belong to the xanthophyll or carotenoid class and are found in a wide variety of mushrooms. Particularly, Sunlotions commonly use these pigments in their products. In mammals, various forms of selenium (Se)-based proteins and/or enzymes depend on selenium (Se), an ultra-trace element. They are inhabited by two different species of mushrooms: *Lentinus edodes* and *Agaricus bisporus*. Strong antitumor agents like γ-glutamyl-Se-methylselenocysteine and Se-methylselenocysteine (MeSeCys), which are known to strengthen teeth, hair, and nails and improve bone health, are found in garlic, which is high in selenium. Shampoos frequently contain selenium (Haneke and Baran 2010). Homoglycans or heteroglycans make up the majority of polysaccharides in mushrooms, depending on the type of monosaccharide that is truly present. These polysaccharides can form polysaccharide–protein complexes or peptidoglycans when they combine with other proteins. Many fungi, including *Pleurotus* species, are rich in polysaccharides, particularly β-(1-3)-D-glucan, which has anticancer properties and is used in cosmetics. In Japan during the late 1960s, the fruiting bodies of Lentinus edodes were the source of the triple-helical β-(1-3)-D-glucan. The active ingredient in creams, leddinan, has the power to boost immunity and fortify the body's resistance to illness and infection. Several of these polysaccharides have been included in cosmetic products because they are all immuno modulatory (Martinez-Medina et al. 2021).

12.6 Cosmetics: Category and Progress

Cosmetic products are any substance with soft action on several parts of the human body that is either scrubbed or sprinkled with the aim to beautify, clean, improve the appeal, safeguard, and keep them in good condition (de Castro 2011). Cosmetics can be divided into two categories: cosmetics and nutricosmetics. Cosmeceuticals can be referred to as products that apply topically only, including lotions, creams, and ointments, while nutricosmetics are concerned with products that are ingested only; therefore, they are referred to as inner beauty cosmetics. Recently, there has been progress in the cosmetics industry in fulfilling the requirements of cosmetic products, such as safety, lack of side effects, and positive effects on the skin. There have been attempts to incorporate natural ingredients, such as fungus metabolites, into nutricosmetics and cosmeceuticals (Hyde et al. 2010). For example, shiitake mushrooms promote faster skin regeneration, have a variety of anti-inflammatory effects, and increase skin elasticity as the skin becomes more radiant. Furthermore, due to the anti-aging properties of mushroom extracts, which can be formulated into capsules and oils, these extracts are used topically, especially in anti-aging skincare products like Reishi mushrooms, which have been around since the 1980s (Wu et al. 2016).

12.6.1 Cosmeceuticals

The term "cosmeceuticals," which describes products that combine pharmaceutical and cosmetic ingredients, was coined by Albert Kligman.

They are applied topically in the form of creams, lotions, and ointments containing biologically active ingredients to improve the appearance, texture, brightness, and anti-aging activity of the skin while also providing essential nutrients for healthy skin (Hyde et al. 2010; Sharma 2011). Conversely, pharmaceutical agents possess the capacity to modify or protect the skin from aberrant or pathological circumstances. Moreover, common ingredients in cosmetics include moisturizers, retinoids, antioxidants, and depigmentation agents. Cosmeceuticals are similar to topical treatments, but they differ in that they are composed of potent ingredients that can change the biological activity of the skin and supply nutrients to support healthy skin (Amer and Maged 2009).

Natural skin care ingredients like dairy products, microbial metabolites, and phytonutrients have gained popularity in recent years (Haruta-Ono et al. 2012; Keller et al. 2014), minerals and components of animal protein (Rodrigues et al. 2015). Fungi, particularly products made from them, are regarded as raw materials that have the potential to treat a variety of skin-related illnesses (Hyde et al. 2010). Human cosmeceuticals' main product categories are anti-aging and moisturizing, with a small number of anti-acne or mushroom-related products and extracts included (Choi and Berson 2006).

12.6.2 Nutricosmetics

Products known as "nutricosmetics" combine food, cosmetics, and pharmaceutical supplements to serve as nutritional supplements (de Castro 2011). It is said that nutricosmetics are "oral cosmetics," "beauty pills," and "beauty from within" (Wu et al. 2016). Due to its acceptance in contemporary culture, nutricosmetics is one of the newest trends in the beauty industry that is receiving a lot of attention. Customers are concerned about the foods they put in their bodies, especially when it comes to natural products that can enhance the look of their skin, nails, and hair without posing any harmful side effects (Dini and Laneri 2019). Nutricosmetics can reduce wrinkles and have anti-aging effects by reducing free radicals (like ROS) produced by solar radiation (Anunciato and da Rocha Filho 2012). Combinations of different natural ingredients, such as proteins, peptides, carotenoids, polyphenols, and extracts from mushrooms (like polyphenolic isoflavones) along with different vitamins (like vitamin C, which has an antioxidant effect and reduces the production of reactive oxygen species to the skin when exposed to UV light) form the basis of the mechanisms underlying the actions of nutricosmetics. Following these components, more research has been conducted to develop natural products based on micronutrients. For example, oral consumption of glycopolyglycans, amino acids, polyunsaturated fatty acids (like omega-3 fatty acids), collagen, hyaluronic acid, elastin, and ceramide may improve the skin's ability to hold water and its barrier function (Haruta-Ono et al. 2012; Dini and Laneri 2019; Barel et al. 2014). Green tea is one of the easily accessible and reasonably priced nutricosmetics. Enzymes such as tyrosinase, collagenase, and elastase are significantly inhibited by various phenolic compounds present in green tea. This preserves structural stability, elasticity, and inhibits melanogenesis, in that order (Hong et al. 2014).

12.7 Moisturizing Effect

The integrity and function of the skin are greatly influenced by the lipids on the skin's surface, the water content of the stratum corneum, and the outermost layer of the epidermis (Sator et al. 2003). Additionally, the skin will appear softer, smoother, and more moisturized the more the stratum corneum is hydrated (Baumann 2007). Moisturizing agents minimize the loss of skin hydration and minimize fine wrinkles. In conventional cosmetic formulations, moisturizers contain active ingredients like pantothenic acid (vitamin B5), 6-palmitoyl-L-ascorbic acid, and hyaluronic acid. The active ingredients from fungi exerted a significant effect on the formulations and showed better results. The polysaccharide that naturally occurs in the extract of *Tremella fuciformis* mushroom can be used either with or without the addition of other chemicals as a moisturizing agent. It was found that applying an extract of Tremella mushroom without adding any chemicals to the skin could inhibit melanin formation with light skin spots (Yang et al. 2006; Zhang et al. 2013). Furthermore, the traditional use of 0.02% hyaluronic acid can be enhanced by adding 0.05% Tremella polysaccharides to cosmetics (Liu and He 2012). The Tremella fungal extract absorbed water more quickly in 5 h than HA did, but the rate of absorption did not slow down over time. After 96 h, the water-insoluble crude carboxymethylated polysaccharide (CATP) of *Tremella fuciformis* reached its moisture capacity at a retention rate of 65.7% (Wang et al. 2015). Additionally, galactosylceramides (GalCers) and glucosylceramides (GlcCer) obtained from filamentous fungi of *Aspergillus* spp. would be sources of moisturizing ingredients in cosmetics (Jiang et al. 2021). Another study revealed that *Auricularia fuscosuccinea* (Montagne) Farlow showed the same effect as sodium hyaluronate, nevertheless lower than that of *Tremella fuciformis* extract. Thus, the addition of polysaccharides extracted from fungi can enhance the moisturizing effect in cosmetics (Liao et al. 2014).

12.8 Effect of Anti-aging (Lifting and Firming)

The skin is thought to be the body's strongest barrier against chemical, biological, and physical attacks. It is the largest organ in the body, making up about 15% of the total weight (Kendall and Nicolaou 2013). Aging leads to a progressive malfunctioning of body organs, including the skin. Internal and external factors are the main contributors to skin aging. One example of an intrinsic factor, or cellular or natural process, that is dependent on psychological and biological interactions is dysfunctional hormone secretion. Along with collagen breakdown, hormone disruption causes the skin to become dry, wrinkled, and lose its elastic fiber networks. The extrinsic mechanism is caused by environmental factors, such as extended exposure to solar radiation, or "photoaging" (Papakonstantinou et al. 2012). There is a production of reactive oxygen species (ROS), which cause photoaging by causing damage to DNA and other cell components. As such, it might affect the expression of genes associated with healthy skin. Increased collagen breakdown results from the overexpression of matrix metalloproteinases (MMP-1, MMP-3, and MMP-9). This is because, among other things, overexpression of MMPs induces transcription factor activator protein-1 (AP-1), which in turn upregulates MMPs to cause collagen degradation (Leem 2015). Moreover, transforming growth factor β (TGF-β) is downregulated in response to sun exposure. TGF-β is a cytokine that increases the accumulation of pro-collagen and, in turn, reduces collagen synthesis. Three main components of skin health are hyaluronic acid, elastin, and collagen. Anti-aging cosmetic ingredients block the actions of elastase, hyaluronidase, tyrosinase, and MMP-1 to delay the aging process of the skin. Anti-aging compounds that serve as antioxidants and ROS scavengers are present in a wide range of fungi (Papakonstantinou et al. 2012; Soto et al. 2015). The antioxidant capacity of mushrooms and their potential to postpone skin aging have been demonstrated by numerous studies (Kosmadaki and Gilchrest 2004). Mushrooms contain phenolic compounds, which have been shown to have this pharmacological effect in grapefruit and wine (Landrault et al. 2001). Furthermore, extracts of naturally occurring *Lentinula edodes* and *Volvariella volvacea* have free radical scavenging abilities (Cheung and Cheung 2005; Puttaraju et al. 2006). These qualities stem from their capacity to activate the body's antioxidant system's two enzymes, glutathione peroxidase (GPx) and superoxidase dismutase (SOD), which protect the skin from oxidative stress and convert myofibroblasts back into fibroblasts (Cheung et al. 2003). L-ergothioneine, a histidine thiourea derivative, was found to reduce mitochondrial membrane oxidation. It was extracted from *Grifola frondosa*, *Pleurotus ostreatus*,

Pleurotus eryngii, Lentinus edodes, and portabellas (brown *Agaricus bisporus*) (Dubost et al. 2006). Additionally, *Pleurotus florida* var. APK2 and *Calocybe indica* var. hybrid mushrooms were added to an alkaline solution for the glucan extraction process. Glucan demonstrated its antioxidant properties and increased immune cell activity (Maity et al. 2011; Sun et al. 2004). Likewise, a lymphocyte proliferation assay showed the immune and antioxidant actions of glucan from *Phellinus ribis* (Yuan et al. 2009). *Tremella fuciformis*, a widely distributed and healthful mushroom in China, is a valuable source of polysaccharides that can be extracted with hot water. There was mannose and 10.77% (w/w) uronic acid. Their anti-photoaging effects were evaluated in an animal model exposed to UV light for 30 days. The results showed that T. fuciformis polysaccharides (TP) could reduce high glycosaminoglycan levels, retain skin hydration, and slow down the skin's loss of collagen. Additionally, by preserving the type I/III collagen ratio, TP prevented endogenous collagen degradation and restored UV-exposed skin, according to a histopathological analysis. Additionally, the study found increased levels of antioxidant enzyme systems such as SOD, GSH-Px, and catalase in comparison to the irradiated control group that did not receive treatment (Wen et al. 2016). Wrinkles are one of the main signs of aging and are brought on by the breakdown of type I collagen, a structural protein in the dermal layer of the skin. Skin aging is accelerated by elastin degradation, which is mediated by metalloproteinase elastase. Similarly, MMPs play a role in the breakdown of collagen; therefore, blocking these enzymes may lessen the appearance of wrinkles on the skin (Lupo and Cole 2007). Numerous investigations into the potential of fungal extracts as anti-wrinkling agents have demonstrated their pharmacological action through the inhibition of these enzymes' activities. The results of the *Tricholoma matsutake* extract indicated that it could, in a dose-dependent way, reduce the activity of MMPs and elastase (Kim et al. 2014). Estimates were made of the naturally occurring methanol's anticholinesterase activity in the fruiting body of the veiled lady mushroom, *Dictyophora indusiate*. Moreover, it showed anti-wrinkling and anti-melanogenesis properties (Nguyen et al. 2013). The pro-inflammatory function of cyclo-oxygenase-2 (COX-2) has an impact on aging. Prostaglandin E2 (PGE2), which is crucial for vasodilation and produces inflammatory mediators in the skin, is one of the prostaglandins (PGs) that COX-2 regulates (Cao and Prescott 2002). *Grifola frondosa* mycelia hexane extract contained three ergosterol compounds that demonstrated COX-2 inhibitory and antioxidant qualities (Zhang et al. 2002). Mouse macrophage RAW 264.7 were used to test the COX-2 inhibitory activity of syringaldehyde and syringic acid, which were extracted from the fruiting bodies of *Elaphomyces granulatus* (deer truffle). As a result, several active components in fungal extracts can reduce the appearance of aging skin (Stanikunaite et al. 2009).

12.9 Skin Whitening Effect

Skin whitening is a popular approach to cosmetics and beauty in many parts of the world. The main pigment that gives skin its color is called melanin, which is also necessary for shielding the skin from ultraviolet light (Lin et al. 2008; Gupta et al. 2014). Tyrosinase is the main modulator of melanin biosynthesis in a range of bioforms because it operates via two distinct oxidation reactions. Tyrosinase catalyzes the hydroxylation of tyrosine to dihydroxyphenylalanine (DOPA) and the oxidation of DOPA to DOPA quinone. Furthermore, depending on the physiological conditions, it is converted into either eumelanin, a brown-black pigment, or pheomelanin, a yellow-red pigment. Therefore, decreasing the catalytic activity of tyrosinase is necessary to enhance the skin-whitening benefits (Parvez et al. 2007; Chang 2009). Many fungal extracts have shown tyrosinase inhibitory activity at different concentrations. The extracts of several mushrooms, including *Cordyceps militaris, Agaricus brasiliensis, Ganoderma lucidum*, and *Antrodia camphorata,* were tested for their ability to inhibit tyrosinase. However, compared to extracts isolated from other Basidiomycetes, *G. lucidum* extracts showed a noticeably higher level of tyrosinase activity (IC50 value 0.32 mg/ml). Nowadays, *G. lucidum*, a traditional Chinese herb, is found in many cosmetic face mask products (Chien et al. 2008b). Similarly, it has been reported that extracts of *Flammulina velutipes* inhibit tyrosinase activity and melanin biosynthesis in B16 murine melanoma cells (Nagasaka et al. 2015). Furthermore, in B16/F10 melanoma cells, methanol that was extracted from *Phellinus bakmi* fruiting bodies demonstrated dose-dependent anti-tyrosinase activity (Kim et al. 2006). The mycelium of *Antrodia cinnamomea* showed anti-melanogenic properties in ethanolic extracts. 30% less intracellular melanin was found in B16-F10 cells when the 50% ethanol-extracted fraction suppressed tyrosinase activity and downregulated its protein expression (Chen et al. 2021). Moreover, sterol extracts from the mycelial cultures of *Ganoderma applanatum, Laetiporus sulphureus,* and *Trametes* versicolor inhibit tyrosinase activity in a way that is dependent on dosage (Sułkowska-Ziaja et al. 2021; Racz et al. 2000). Extracted from xerophytic plants are two fungal strains from the genus *Comoclathris*: CF-090361 and CF-090766. This strain generated brevianamide, graphostrin B, and comoclathrin, three possible bioactive secondary metabolites. The results showed that the anti-tyrosinase activity of comoclathrin was the strongest (IC50 0.16μM), 90 times greater than that of the positive control, kojic acid (IC50 14.07μM) (Georgousaki et al. 2022).

In MSH-stimulated melanocytes (B16F10 cells), tremella fuciformis polyphenol and flavonoid extracts reduced tyrosinase activity and melanin expression. The conclusion was supported by the fact that the mRNA expression of several melanin formation factors, including tyrosinase-related protein-1 (TRP-1), tyrosinase-related protein-2 (TRP-2), and microphthalmia-associated transcription factor (MITF), was significantly downregulated in MSH-stimulated melanocytes (B16F10 cells) (Lee et al. 2016). Fungal enzymes used for bleaching human eumelanin in skin were extracted from *Sporotrichum pruinosum* in suitable culture conditions. These enzymes do not suppress melanin synthesis, but rather act as whitening agents for the already-formed melanin. This different approach of using a melanin discoloration enzyme could be effective because of the biological origin of the enzyme, which is more acceptable for the cosmetics industry than chemical formulations (Mohorčič et al. 2007).

12.10 Hair Cosmetics

The primary goals of hair cosmetics are to preserve hair structure, prevent bacterial and fungal infections, and safeguard scalp hygiene (Blume-Peytavi et al. 2008). Among the many hair cosmetic formulations available are shampoos, serums, conditioners, styling products (sprays, waxes, gels, and mousses), bleaching agents, straighteners, perms, and hair dyes (Blume-Peytavi et al. 2008). Additionally, some cosmeceuticals such as minoxidil and finasteride play a vital role in increasing scalp hair density (Dias 2015). Shampoo formulas are made to accommodate differences in hair type, quantity, and quality as well as particular issues like oily hair, dandruff, and androgenic alopecia (Nazir et al. 2012). Tremella mushroom extracts yield the hydrophilic polysaccharide glucuronoxylomannan (GXM), a powerful ingredient in hair cosmetics with anti-inflammatory and wound-healing characteristics (De Baets and Vandamme 2001). Moreover, hair cosmeceuticals are made using silk film that has been separated from shiitake mushrooms. By lowering dihydrotestosterone or prostatic hyperplasia levels, ganoderma lucidum (Reishi mushroom) extracts combined with three other plants and zinc may improve hair evolution in a male human (Meehan 2015). Lack of certain nutrients, poor lifestyle choices, chemicals, and recurring infections can all contribute to hair problems. Since mushrooms are rich in essential antioxidants, supplemental vitamin D, and minerals like iron, copper, and selenium, they are good for overall health and for strengthening hair free of dandruff and hair loss. This is why the cosmetics industry is very interested in using mushroom extract (Wu et al. 2016). The edible mushroom *Agaricus bisporus* contains traces of Se in its fruiting bodies (Racz et al. 2000). *A. bisporus* extract can be used as Shampoos to relieve hair issues, such as hair loss, greasy hair, and dandruff (Ogra et al. 2004). In addition, *A. bisporus* extract can be used to clean scalp hair and enhance hair growth (Jeong et al. 2008). It is possible to separate vitamin D and various minerals, such as copper and iron, from *A. bisporus*, which helps avoid hair issues (Wu et al. 2016). Moreover, *Pleurotus ostreatus* strengthens hair elasticity from the roots, thus decreasing hair loss and enhancing hair growth (Shukla 2022). The sclerotium of *Inonotus obliquus* (chaga mushrooms), which are used as a traditional hair shampoo in Mongolia, contains five triterpenes of the lanostane type. Lanosterol, inotodiol, lanost-8,24-diene-3β,21-diol, and trametenolic acid are among the extracts from *I. obliquus*. The results indicated that the extracts had a higher impact on the growth of human follicle dermal papilla cells (HFDPCs) in comparison to the positive control, minoxidil, an anti-alopecia medication. *I. obliquus* might therefore be a good candidate for novel hair care products that promote hair growth (Sagayama et al. 2019).

12.11 Conclusion

The rise of fungal-based cosmetics marks a significant transformation in the beauty sector, emphasizing a move towards sustainable and nature-inspired solutions. These cosmetics embody a blend of innovation, sustainability, and natural beauty by tapping into the diverse biodiversity and distinct biochemical characteristics of fungi. The adaptability of fungal ingredients offers formulators a broad spectrum of choices to develop effective products that meet the increasing demand for environmentally conscious beauty solutions. Delving into the realm of fungal-based cosmetics not only unveils the scientific principles governing their creation, but also underscores the potential of nature's medicinal resources in enhancing skincare and beauty. Ultimately, incorporating fungal-based formulations into the beauty industry reflects a dedication to environmental stewardship, fostering a balanced connection between personal care and sustainable practices.

References

Abendrot M, Płuciennik E, Felczak A, Zawadzka K, Piątczak E, Nowaczyk P et al (2021) Zinc (II) complexes of amino acids are novel active ingredients for anti-acne dermatological preparations. Int J Mol Sci 22(4):1641

Abo Nahas HH, Darwish AMG, Abo Nahas YH, Elsayed MA, Abdel-Azeem MA, Abdel-Azeem AM (2021) Fungi are the gold mines of antioxidants. In: Industrially important fungi for sustainable development: volume 2: bioprospecting for biomolecules. Springer, pp 73–113

Ali SA, Choudhary RK, Naaz I, Ali AS (2015) Understanding the challenges of melanogenesis: key roles of bioactive compounds in the treatment of hyperpigmentation disorders. J Pigment Dis 2:1–9

Alkafaas SS, Abdallah AM, Hussien AM, Bedair H, Abdo M, Ghosh S et al (2023) A study on the effect of natural products on the transmission of B. 1.1. 529 Omicron. Virol J 20(1):191

Alsterholm M, Karami N, Faergemann J (2010) Antimicrobial activity of topical skin pharmaceuticals. An in vitro study. Acta Derm Venereol 90(3):239–245

Alves MJ, Ferreira IC, Dias J, Teixeira V, Martins A, Pintado M (2012) A review of the antimicrobial activity of mushroom (Basidiomycetes) extracts and isolated compounds. Planta Med 8(16):1707–1718

Alves MJ, Ferreira IC, Froufe HJ, Abreu R, Martins A, Pintado M (2013) Antimicrobial activity of phenolic compounds identified in wild mushrooms, SAR analysis and docking studies. J Appl Microbiol 115(2): 346–357

Amer M, Maged M (2009) Cosmeceuticals versus pharmaceuticals. Clin Dermatol 27(5):428–430

Anunciato TP, da Rocha Filho PA (2012) Carotenoids and polyphenols in nutricosmetics, nutraceuticals, and cosmeceuticals. J Cosmet Dermatol 11(1):51–54

Bae JY, Choi JS, Kang SW, Lee YJ, Park J, Kang YH (2010) The dietary compound ellagic acid alleviated skin wrinkles and inflammation induced by UV-B irradiation. Exp Dermatol 19(8):e182–ee90

Barel AO, Paye M, Maibach HI (2014) Handbook of cosmetic science and technology. CRC Press

Baumann L (2007) Skin aging and treatment. J Pathol 211(2):241–251

Blume-Peytavi U, Whiting DA, Trüeb RM (2008) Hair growth and disorders. Springer

Cao Y, Prescott SM (2002) Many actions of cyclooxygenase-2 in cellular dynamics and in cancer. J Cell Physiol 190(3):279–286
Carocho M, Ferreira ICJF (2013) A review on antioxidants, prooxidants, and related controversy: natural and synthetic compounds, screening and analysis methodologies, and future perspectives. Food Chem Toxicol 51:15–25
Chang T-S (2009) An updated review of tyrosinase inhibitors. Int J Mol Sci 10(6):2440–2475
Chang T-S (2012) Natural melanogenesis inhibitors act by downregulating tyrosinase activity. Materials (Basel) 5(9):1661–1685
Chen HY, Cheng KC, Wang HT, Hsieh CW, Lai YJ (2021) Extracts of Antrodia cinnamomea mycelium as a highly potent tyrosinase inhibitor. J Cosmet Dermatol 20(7):2341–2349
Cheung L, Cheung PC (2005) Mushroom extract exhibits antioxidant activity against lipid peroxidation. Food Chem 89(3):403–409
Cheung L, Cheung PC, Ooi VE (2003) Antioxidant activity and total phenolics of edible mushroom extracts. Food Chem 81(2):249–255
Chien C-C, Tsai M-L, Chen C-C, Chang S-J, Tseng C-H (2008a) Effects of *G. lucidum* extracts on tyrosinase activity and related mushrooms. Mycopathologia 166: 117–120
Chien C-C, Tsai M-L, Chen C-C, Chang S-J, Tseng C-H (2008b) Effects on tyrosinase activity by the extracts of Ganoderma lucidum and related mushrooms. Mycopathologia 166:117–120
Choi CM, Berson DS (eds) (2006) Cosmeceuticals. Seminars in cutaneous medicine and surgery. Elsevier
D'Mello SA, Finlay GJ, Baguley BC et al (2016) Signaling pathways in melanogenesis. Int J Mol Sci 17(7): 1144
Dandapat S, Sinha MP (2015) The antioxidant and anti-inflammatory activities of *Pleurotus tuber-regium* (Rumph. ex Fr.) Singer. Adv Biol Res 9(3):140–145
De Baets S, Vandamme E (2001) Extracellular *Tremella polysaccharides*: structure, properties and applications. Biotechnol Lett 23:1361–1366
de Castro ML (2011) Cosmetobolomics as an incipient '-omics' with high analytical involvement. TrAC Trends Anal Chem 30(9):1365–1371
De Silva DD, Rapior S, Hyde KD, Bahkali AH (2012) Medicinal mushrooms for the prevention and control of diabetes mellitus. Fungal Divers 56:1–29
Dias MFRG (2015) Hair cosmetics: an overview. Int J Trichology 7(1):2
Dini I, Laneri S (2019) Nutricosmetics: a brief overview. Phytother Res 33(12):3054–3063
Dubost NJ, Beelman RB, Peterson D et al (2006) Identification and quantification of ergothioneine in cultivated mushrooms using liquid chromatography-mass spectroscopy. Int J Med Med 8(3):215–222
El Enshasy HA, El Marzugi NA, Elsayed EA, Ling OM, Malek RA, Kepli AN et al (2018) Medical and cosmetic applications of fungal nanotechnology: production, characterization, and bioactivity. In: Fungal nanobionics: principles and applications. Springer, pp 21–59
Elsner P (2006) Antimicrobials and the skin physiological and pathological flora. Curr Probl Dermatol 33:35–41
Fernandes MS, Kerkar S (2017) Microorganisms as a source of tyrosinase inhibitors: a review. Ann Microbiol 67(4):343–358
Ferreira ICFR, Vaz JA, Vasconcelos MH, Martins A (2010) Compounds from wild mushrooms with antitumor potential. Anti Cancer Agents Med Chem 10(5):424–436
Georgousaki K, González-Menéndez V, Tormo JR, Tsafantakis N, Mackenzie TA, Martín J et al (2022) Comoclathrin, a novel potent skin-whitening agent produced by endophytic Comoclathris strains associated with Andalusia desert plants. Sci Rep 12(1):1649
Gündoğdu A, Kılıç H, Kutateladze M (2016) Susceptibility of multidrug-resistant pathogens responsible for complicated skin and soft tissue infections in standard bacteriophage cocktails. Mikrobiyol Bul 50(2): 215–223
Gupta S, Gautam A, Kumar S (2014) Natural skin whitening agents: a current status. Adv Biol Res (Rennes) 8(6):257–259
Gupta PL, Rajput M, Oza T, Trivedi U, Sanghvi G (2019) Eminence of microbial products in cosmetic industry. Nat Prod Bioprospect 9(4):267–278
Haneke E, Baran R (2010) Micronutrients for hair and nails. nutrition for healthy skin: strategies for clinical and cosmetic practice. Springer, pp 149–163
Haruta-Ono Y, Setoguchi S, Ueno HM, Higurashi S, Ueda N, Kato K et al (2012) Orally administered sphingomyelin in bovine milk is incorporated into the skin sphingolipids and is involved in the water-holding capacity of hairless mice. J Dermatol Sci 68(1):56–62
Heleno SA, Ferreira RC, Antonio AL, Queiroz M-JRP, Barros L, Ferreira ICFR (2015) Nutritional values, bioactive compounds, and antioxidant properties of three edible mushrooms from Poland. Food Biosci 11:48–55
Hong YH, Jung EY, Noh DO, Suh HJ (2014) Physiological effects of a formulation containing tannase-converted green tea extract on skin care: physical stability, collagenase, elastase, and tyrosinase activities. Integr Med Res 3(1):25–33
Hyde K, Bahkali A, Moslem M (2010) Fungi: an unusual source of cosmetics. Fungal Divers 43:1–9
Ito A (2014) Effects of a hyaluronan lotion with a molecular weight of approximately 50–110 kDa on aged atrophic skin. J Cosmet Dermatol Sci Appl 4:150–155
Jeong YT, Yang BK, Jeong SC, Kim SM, Song CH (2008) Ganoderma applanatum: a promising mushroom for antitumor and immunomodulating activity. Phytother Res 22(5):614–619
Jiang C, Ge J, He B, Zeng B (2021) Glycosphingolipids in filamentous fungi: biological roles and potential

applications in cosmetics and health foods. Front Microbiol 12:690211
Kalra R, Conlan XA, Goel M (2020) Fungi as a potential source of pigments: harnessing filamentous fungi. Front Chem 8:369
Keller S, Le H-Y, Rödiger C, Hipler U-C, Kertscher R, Malarski A et al (2014) Supplementation of a dairy drink enriched with milk phospholipids in patients with atopic dermatitis: a double-blind, placebo-controlled, randomized, crossover study. Clin Nutr 33(6): 1010–1016
Kendall AC, Nicolaou A (2013) Bioactive lipid mediators in skin inflammation and immunity. Prog Lipid Res 52(1):141–164
Kim D, Park J, Kim J, Han C, Yoon J, Kim N et al (2006) Flavonoids as mushroom tyrosinase inhibitors: a fluorescence quenching study. J Agric Food Chem 54(3): 935–941
Kim SY, Go KC, Song YS, Jeong YS, Kim EJ, Kim BJ (2014) Extract of the mycelium of *T. matsutake* inhibit elastase activity and TPA-induced MMP-1 expression in human fibroblasts. Int J Mol Med 34(6):1613–1621
Kosmadaki M, Gilchrest B (2004) Role of telomeres in skin aging and photoaging. Micron 35(3):155–159
Landrault N, Poucheret P, Ravel P, Gasc F, Cros G, Teissedre P-L (2001) Antioxidant capacities and phenolic levels of French wines of different varieties and vintages. J Agric Food Chem 49(7):3341–3348
Lee KH, Park HS, Yoon IJ, Shin YB, Baik YC, Kooh DH et al (2016) Whitening and anti-wrinkle effects of *Tremella fuciformis* extracts. Korean J Medicinal Crop Sci 24(1):38–46
Leem K-H (2015) Effects of Olibanum extracts on collagenase activity and procollagen synthesis in Hs68 human fibroblasts and tyrosinase activity. Int J Biosci Biotechnol 88:172–175
Liao WC, Hsueh C-Y, Chan C-F (2014) Antioxidative activity, moisture retention, film formation, and viscosity stability of *Auricularia fuscosuccinea* white strain water extract. Biosci Biotechnol Biochem 78(6): 1029–1036
Lin J-W, Chiang H-M, Lin Y-C, Wen K-C (2008) Natural products with skin-whitening effects. J Food Drug Anal 16(2):8
Liu H, He L (2012) Comparison of the moisture retention capacity of *Tremella polysaccharides* and hyaluronic acid. J Anhui Agric Sci 40:13093–13094
Lu H, Lou H, Hu J, Liu Z, Chen Q (2020) Macrofungi: a review of cultivation strategies, bioactivity, and application of mushrooms. Compr Rev Food Sci Food Saf 19(5):2333–2356
Lupo MP, Cole AL (2007) Cosmeceutical peptides. Dermatol Ther 20:343–349
Łysakowska P, Sobota A, Wirkijowska A (2023) Medicinal mushrooms: their bioactive components, nutritional value and application in functional food production—a review. Molecules 28(14):5393
Maity K, Kar E, Maity S, Gantait SK, Das D, Maiti S et al (2011) Structural characterization and study of immunoenhancing and antioxidant property of a novel polysaccharide isolated from the aqueous extract of a somatic hybrid mushroom of *Pleurotus florida* and *Calocybe indica* variety APK2. Int J Biol Macromol 48(2):304–310
Martinez-Medina GA, Chávez-González ML, Verma DK, Prado-Barragán LA, Martínez-Hernández JL, Flores-Gallegos AC et al (2021) Bio-functional components in mushrooms, a health opportunity: ergothiononine and huitlacohe as recent trends. J Funct Foods 77: 104326
Martins AM, Marto JM (2023) A sustainable life cycle for cosmetics: from design and development to the post-use phase. Sustain Chem Pharm 35:101178
Masaki H (2010) Role of antioxidants in the skin: anti-aging effects. J Dermatol Sci 58(2):85–90
Mattila P, Könkö K, Eurola M, Pihlava JM, Astola J, Vahteristo L et al (2001) Vitamin, mineral element, and phenolic compound contents in cultivated mushrooms. J Agric Food Chem 49(5):2343–2348
McMullen RL, Dell'Acqua G (2023) History of natural ingredients in cosmetics. Cosmetics 10(3):71
Meehan K (2015) Composition to promote hair growth in humans. Google Patents
Meng T-X, Furuta S, Fukamizu S, Yamamoto R, Ishikawa H et al (2011) Evaluation of biological activities of extracts from the fruiting body of *Pleurotus citrinopileatus* in skin cosmetics. J Wood Sci 57:452–458
Meng T-X, Zhang C-F, Miyamoto T, Ishikawa H, Shimizu K, Ohga S et al (2012) Melanin biosynthesis stimulates compound isolation from the fruiting bodies of *Pleurotus citrinopileatus*. J Cosmet Dermatol Sci Appl 2(03):151
Miyake M, Yamamoto S, Sano O, Fujii M, Kohno K, Ushio S et al (2010) Inhibitory effects of 2-Amino-3 H-phenoxazin-3-one on melanogenesis in murine B16 melanoma cell line. Biosci Biotechnol Biochem 74(4): 753–758
Mohorčič M, Friedrich J, Renimel I, André P, Mandin D, Chaumont J-P (2007) Production of melanin bleaching enzyme of fungal origin and its application in cosmetics. Biotechnol Bioprocess Eng 12:200–206
Nagasaka R, Ishikawa Y, Inada T, Ohshima T (2015) Depigmenting effect of winter medicinal mushroom *Flammulna velutipes* (higher Basidiomycetes) on melanoma cells. Int J Med Mushrooms 17(6):511–520
Nazir H, Wang L, Lian G, Zhu S, Zhang Y, Liu Y et al (2012) Multilayered silicone oil droplets of narrow size distribution: preparation and improved deposition on hair. Colloids Surf B: Biointerfaces 100:42–49
Ndlovu G, Fouche G, Tselanyane M, Cordier W, Steenkamp V (2013) In vitro determination of anti-aging potential of four medicinal plants in Southern Africa. BMC Complement Altern Med 13:1–7
Nesy E, Mathew L (2014) Studies on antimicrobial and antioxidant efficacy of *Thevetia neriifolia*, Juss leaf extracts against human skin pathogens. Int J Pharm Sci Drug Res 6(2):164–168

Nguyen TK, Shin DB, Lee KR, Shin PG, Cheong JC, Yoo YB et al (2013) Antioxidant and anti-inflammatory activities of fruiting bodies of *Dyctiophora indusiata*. J Mushroom 11(4):269–277

Ogra Y, Ishiwata K, Ruiz Encinar J, Łobiński R, Suzuki KT (2004) Speciation of selenium in selenium-enriched shiitake mushroom, *Lentinula edodes*. Anal Bioanal Chem 379:861–866

Onar HC, Yusufoglu A, Turker G, Yanardag R (2012) Elastase, tyrosinase, and lipoxygenase inhibition and antioxidant activity of an aqueous extract of *Epilobium angustifolium* L. leaves. J Med Plant Res 6(5):716–726

Papakonstantinou E, Roth M, Karakiulakis G (2012) Hyaluronic acid: a key molecule in skin aging. Dermatoendocrinology 4(3):253–258

Park HS, Hwang YH, Kim MK, Hong GE, Lee HJ, Nagappan A et al (2015a) Functional polysaccharides from *Grifola frondosa* aqueous extract inhibited atopic dermatitis-like skin lesions in NC/Nga mice. Biosci Biotechnol Biochem 79(1):147–154

Park KM, Kwon KM, Lee SH (2015b) Evaluation of the antioxidant activities and tyrosinase inhibitory properties of mycelium culture extracts. Evid Based Complement Alternat Med 2015:616298

Parvez S, Kang M, Chung HS, Bae H (2007) Naturally occurring tyrosinase inhibitors: mechanism and applications in skin health, cosmetics and agriculture industries. Phytother Res 21(9):805–816

Puttaraju NG, Venkateshaiah SU, Dharmesh SM et al (2006) Antioxidant activity of indigenous edible mushrooms. J Agric Food Chem 54(26):9764–9772

Racz L, Bumbalova A, Harangozo M, Tölgyessy J, Tomeček O (2000) Determination of cesium and selenium in cultivated mushrooms using radionuclide X-ray fluorescence technique. J Radioanal Nucl Chem 245:611–614

Reid T, Kashangura C, Chidewe C, Benhura MA et al (2016) Antibacterial properties of wild edible and non-edible mushrooms found in Zimbabwe. Afr J Microbiol Res 10(26):977–984

Reis FS, Martins A, Barros L, Ferreira IC (2012) Antioxidant properties and phenolic profile of the most widely appreciated cultivated mushrooms: a comparative study between in vivo and in vitro samples. Food Chem Toxicol 50(5):1201–1207

Ribeiro AS, Estanqueiro M, Oliveira MB, Sousa Lobo JM (2015) Main benefits and applicability of plant extracts in skin care products. Cosmetics 2(2):48–65

Rodrigues F, Pimentel FB, Oliveira MBP (2015) Olive by-products: challenging application in the cosmetic industry. Ind Crop Prod 70:116–124

Sagayama K, Tanaka N, Fukumoto T, Kashiwada Y (2019) Lanostane-type triterpenes from the sclerotium of Inonotus obliquus (Chaga mushrooms) as proproliferative agents on human follicle dermal papilla cells. J Nat Med 73:597–601

Salah LA, Faergemann J (2015) Retrospective analysis of skin bacterial colonization, susceptibility, and resistance in patients with atopic dermatitis and impetigo. Acta Derm Venereol 95(5):532–535

Sarkar R, Arora P, Garg K (2013) Cosmeceuticals for hyperpigmentation: what is available? J Cutan Aesthet Surg 6(1):4–11

Sator P-G, Schmidt JB, Hönigsmann H (2003) Comparison of epidermal hydration and skin surface lipids in healthy individuals and in patients with atopic dermatitis. J Am Acad Dermatol 48(3):352–358

Sharma P (2011) Cosmeceuticals: regulatory scenario in the US, Europe, and India. Int J Pharm Technol 3: 1512–1535

Shukla AC (2022) Applied mycology: entrepreneurship with fungi. Springer

Soto ML, Falqué E, Domínguez H (2015) Relevance of natural phenolics from grapes and derivative products in the formulation of cosmetics. Cosmetics 2(3): 259–276

Stanikunaite R, Khan SI, Trappe JM, Ross SA (2009) Cyclooxygenase-2 inhibitory and antioxidant compounds from the truffle *Elaphomyces granulatus*. Phytother Res 23(4):575–578

Sułkowska-Ziaja K, Grabowska K, Apola A, Kryczyk-Poprawa A, Muszyńska B (2021) Mycelial culture extracts of selected wood-decay mushrooms as a source of skin-protecting factors. Biotechnol Lett 43: 1051–1061

Sułkowska-Ziaja K, Trepa M, Olechowska-Jarząb A, Nowak P, Ziaja M, Kała K et al (2023) Natural compounds of fungal origin with antimicrobial activity have potential cosmetic applications. Pharmaceuticals (Basel, Switzerland) 16(9):1200

Sun J, He H, Xie BJ (2004) Novel antioxidant peptides from fermented mushroom *Ganoderma lucidum*. J Agric Food Chem 52(21):6646–6652

Tanuja Y, Mishra S, Das S, Aggarwal S et al (2015) Anticedants and natural prevention of environmental toxicants induced accelerated aging of skin. Environ Toxicol Pharmacol 39(1):384–391

Taofiq O, González-Paramás AM, Martins A, Barreiro MF, Ferreira ICFR (2016a) Mushroom extracts and compounds in cosmetics, cosmeceuticals, and nutricosmetics: a review. Ind Crop Prod 90:38–48

Taofiq O, Martins A, Barreiro MF et al (2016b) Anti-inflammatory potential of mushroom extracts and isolated metabolites. Trends Food Sci Technol 50: 193–210

Thomas NV, Manivasagan P, Kim S-K (2014) Potential matrix metalloproteinase inhibitors from edible marine algae: a review. Environ Toxicol Pharmacol 37(3): 1090–1100

Vachtenheim J, Borovanský J (2010) "Transcription physiology" of pigment formation in melanocytes: central role of MITF. Exp Dermatol 19(7):617–627

Vitale GA, Coppola D, Palma Esposito F, Buonocore C, Ausuri J, Tortorella E et al (2020) Antioxidant molecules from marine fungi: methodologies and perspectives. Antioxidants 9(12):1183

Wang X, Zhang Z, Zhao M (2015) Carboxymethylation of polysaccharides from *Tremella fuciformis* for antioxidant and moisture-preserving activities. Int J Biol Macromol 72:526–530

Wen L, Gao Q, Ma C-w, Ge Y, You L, Liu RH et al (2016) Effect of polysaccharides from *Tremella fuciformis* on UV-induced photoaging. J Funct Foods 20:400–410

Woźniak B, Chudzińska J, Szczyglewska P, Nowak I, Feliczak-Guzik A (2023) Optimization of the composition of a cosmetic formulation containing *Tremella fuciformis* extract (fungi). Cosmetics 10(3):82

Wu G, Li L, Sung GH, Kim TW, Byeon SE, Cho JY et al (2011) Inhibition of 2, 4-dinitrofluorobenzene-induced atopic dermatitis by topical application of the butanol extract of *Cordyceps bassiana* in NC/Nga mice. J Ethnopharmacol 134(2):504–509

Wu Y, Choi M-H, Li J, Yang H, Shin H-J (2016) Mushroom cosmetics: the present and future. Cosmetics 3(3):22

Yahaya YA, Don MM (2012) Evaluation of *Trametes lactinea* extracts on the inhibition of hyaluronidase, lipoxygenase, and xanthine oxidase activity in vitro. J Phys Ther Sci 23:1–15

Yan Z-F, Yang Y, Tian F-H, Mao X-X, Li Y, Li C-T et al (2014) Inhibitory and acceleratory effects of Inonotus obliquus on tyrosinase activity and melanin formation in B16 melanoma cells. Evid Based Complement Alternat Med 2014:259836

Yang S-H, Liu H-I, Tsai SJ (2006) Edible *Tremella polysaccharides* for skin care. Google Patents

Yuan C, Huang X, Cheng L, Bu Y, Liu G, Yi F et al (2009) Evaluation of antioxidant and immune activity of *Phellinus ribis* glucan in mice. Food Chem 115(2): 581–584

Zhang Y, Mills GL, Nair MG (2002) Cyclooxygenase inhibitory and antioxidant compounds from the mycelia of the edible mushroom *Grifola frondosa*. J Agric Food Chem 50(26):7581–7585

Zhang K, Meng X, Sun Y, Guo P (2013) Preparation of *Tremella*, *Speranskiae tuberculatae*, and *Eriocaulon buergerianum* extracts and their performance in cosmetics. Deterg Cosmet 36:28

13 Commercially Available Cosmetics with Fungal-Based Ingredients 2- Haircare Products

Elijah Kolawole Oladipo, Helen N. Onyeaka, and Stephen Feranmi Adeyemo

Abstract

The global cosmetic industry, valued at approximately $430 billion in 2023, is rapidly evolving with a growing emphasis on sustainability and natural ingredients. A significant shift toward eco-friendly products has led to the increased use of fungal-based ingredients in haircare solutions. This chapter explores the utilization of fungal-derived compounds in commercially available haircare products, highlighting their benefits and market presence. Fungal extracts such as beta-glucans, chitosan, and cordycepin are known for their antifungal, antimicrobial, antioxidant, and hair-conditioning properties. These ingredients are sustainably produced through fermentation processes, aligning with the industry's move toward green chemistry. Fungal-based ingredients offer numerous benefits including moisturizing and hydration, improved scalp health, hair strengthening and repair, and a natural appeal. Products like Nizoral Anti-Dandruff Shampoo, ProBliva Fungus Shampoo, and Neutrogena T-Gel Shampoo leverage these properties to address scalp issues such as dandruff, fungal infections, and hair loss. Additionally, ingredients like beta-glucans from mushrooms provide deep hydration and protection, enhancing overall hair health. The chapter also discusses the fungal-based ingredients used in hair care, their benefits, and the commercial products that incorporate these innovative compounds. Exploring fungal-derived components presents an exciting opportunity to develop sustainable haircare solutions that effectively address various hair and scalp issues while being gentle with the environment. As consumer demand for natural and eco-friendly products continues to rise, the integration of fungal-based ingredients in haircare products is poised to play a pivotal role in the future of the cosmetic industry.

Keywords

Fungal-based ingredients · Haircare products · Beta-glucans · Antifungal properties · Sustainable cosmetics · Scalp health · Natural formulations

E. K. Oladipo (✉)
Division of Medical Artificial Intelligence, Helix Biogen Institute, Ogbomoso, Oyo State, Nigeria

School of Chemical Engineering, University of Birmingham, Birmingham, UK

Department of Molecular Biology, Immunology and Bioinformatics, Adeleke University, Ede, Osun State, Nigeria
e-mail: koladipo2k3@helixbiogeninstitute.org

H. N. Onyeaka (✉)
School of Chemical Engineering, University of Birmingham, Birmingham, UK
e-mail: h.onyeaka@bham.ac.uk

S. F. Adeyemo
Division of Medical Artificial Intelligence, Helix Biogen Institute, Ogbomoso, Oyo State, Nigeria

S. Ghosh, H. N. Onyeaka (eds.), *Fungal-Based Cosmetics: Formulation and Usage*, The Mycota 17,
https://doi.org/10.1007/978-3-032-09291-5_13

13.1 Introduction

The cosmetic industry is a large, dynamic, and fast-paced global industry. It encompasses a wide range of products including makeup, fragrances, skincare products, and haircare solutions. The global cosmetic industry generates significant revenue, estimated at around $430 billion globally (Berg et al. 2023). The cosmetic industry has been projected to generate revenues amounting to nearly 129 billion U.S. dollars by the year 2028 (Petruzzi 2024). The industry's largest markets are in the European Union, the United States, and Japan (Ortega-Requena et al. 2024). The cosmetic industry comprises key segments such as color cosmetics, deodorants, and talcum powder, with color cosmetics experiencing the highest growth. This industry is known for its continuous innovation, as it introduces new products and formulations to keep up with changing consumer trends and preferences. Sustainability and environmental impact are also growing concerns that the industry is working to address (Ferreira et al. 2022). Overall, the cosmetic industry is a large, lucrative, and dynamic global market that continues to evolve and grow to meet consumers' diverse needs and expectations worldwide. There is a shift in the cosmetics industry toward sustainability due to an increase in consumer awareness and concern about the environmental and health impacts of synthetic chemicals in cosmetics. This led to the growing consumer demand for natural and eco-friendly products. An increasing number of consumers, surpassing 40%, are now placing greater importance on the use of natural ingredients in their beauty and personal care products (Ameta et al. 2024). The advancements in biotechnology, green chemistry, and extraction methods have enhanced the sustainable production of natural cosmetic ingredients. Some of the examples of new natural ingredients being developed include plant-based emollients, natural preservatives such as extracts from tea, thyme, and cinnamon, microbial-derived compounds, and upcycled botanical extracts (Lavuri et al. 2022). Fungal-derived compounds have emerged as valuable ingredients in cosmetics, including haircare products due to their diverse functional properties. These include antifungal, antimicrobial, antioxidant, and hair-conditioning effects that can benefit hair and scalp health (Sang et al. 2023). Fungal extracts and metabolites such as chitosan, cordycepin, and schizophyllan have been found to possess strong antifungal and antimicrobial properties. Many fungal-derived ingredients can be produced through sustainable fermentation processes, reducing the environmental impact compared to synthetic alternatives (Mukherjee et al. 2024). This, however, aligns with the growing consumer demand for natural, sustainable, and eco-friendly haircare products (Fig. 13.1 and Table 13.1).

13.2 Types of Fungal-Based Ingredients Used in Haircare

The haircare industry is diverse, covering a wide range of products and services designed to improve the health and appearance of hair. This includes a variety of items such as shampoos, conditioners, hair oils, serums, and specialized treatments (Pedro et al. 2023). Consumers are often in search of remedies for issues such as dryness, frizz, dandruff, and hair loss, leading to a growing interest in microbial-based ingredients in haircare products. The types of fungal-based ingredients used in haircare are not limited to polysaccharides (beta-glucans), fermented fungal extracts, and proteins and peptides. Beta-glucans derived from fungi, such as yeast and mushrooms, have emerged as valuable ingredients in haircare products as they contain antifungal, antimicrobial, antioxidant, and hair-conditioning effects that can benefit hair and scalp health (Sousa et al. 2023). They are mostly derived from the cell walls of yeast such as *Saccharomyces cerevisiae* or mushrooms like *Tremella fuciformis* (Snow Fungus) and confer benefits such as moisture, strength, and protection to the hair and scalp (Camelini et al. 2022). Fermented extracts from fungi like *Aspergillus* and *Penicillium* also contain compounds with potent antifungal and antimicrobial activities.

Fig. 13.1 Applications of macrofungal (mushrooms) in different industries including the cosmetic industry (Srivastava et al. 2024)

These properties can help combat common scalp issues like dandruff and fungal infections that can lead to hair loss and irritation. They enhance nutrient absorption, scalp health, and hair growth stimulation (Choi and Choi 2022). Similarly, proteins and peptides sourced from fungi such as *Tricholoma matsutake*, in addition to plants and algae, are increasingly being used as cosmetic ingredients. These fungal-derived proteins and peptides exhibit diverse functional properties including repairing of damaged hair, enhancing elasticity, and providing shine all of which benefit hair and scalp health (Belsito et al. 2017). In conclusion, fungal-derived components utilized in haircare present an exciting opportunity for creating eco-friendly haircare solutions that can effectively target various hair and scalp issues while being sustainable and gentle on the environment (Table 13.2).

13.3 Commercially Available Haircare Products with Fungal-Based Ingredients

There are a variety of products designed to maintain hair health, cleanliness, and aesthetics. Some of these products include shampoo, conditioner, styling agents, hair loss treatments, hair colorants, relaxers, and salon haircare products. Antifungal shampoos often contain ingredients like ketoconazole, zinc pyrithione, and selenium sulfide that target and eliminate fungal infections on the scalp (Hill et al. 2024). Examples include ProBliva Fungus Shampoo which contains natural ingredients like emu oil, coconut oil, and jojoba oil to soothe the scalp, and Nizoral anti-dandruff shampoo containing 1% ketoconazole (Amazon 2024a). Ketoconazole was shown to be ten times more effective at fighting fungus which is the

Table 13.1 Key segments of the cosmetic industry

S/N	Key segment	Description	Market trend	References
1	Color cosmetics	Products that include makeup such as foundation, blush, eyeshadow, and lipstick.	Increased demand for long-lasting and transfer-proof products.	Ortega-Requena et al. (2024)
2	Deodorants	Products designed to mask body odor and reduce sweating.	Growing interest in natural and aluminum-free deodorants.	Ortega-Requena et al. (2024)
3	Talcum powder	Products that absorb moisture and reduce friction are often used for personal hygiene.	Shift toward talc-free formulations due to health concerns.	Ortega-Requena et al. (2024)
4	Skincare products	Includes moisturizers, creams, serums, and lotions aimed at improving skin health and appearance.	Rise in demand for products with natural and organic ingredients, and products with anti-aging benefits.	Ferreira et al. (2022)
5	Haircare solutions	Products such as shampoos, conditioners, hair oils, and treatments designed to improve hair health.	Increased popularity of products with natural and sustainable ingredients, focusing on scalp health.	Pedro et al. (2023)
6	Fragrances	Perfumes and colognes are designed to provide pleasant scents.	A trend toward personalized and niche fragrances.	Berg et al. (2023)

Table 13.2 Types of fungal-based ingredients used in haircare

S/N	Ingredient type	Description	Benefits	References
1	Beta-glucans	Polysaccharides are derived from the cell walls of yeast and mushrooms like *Saccharomyces cerevisiae* and *Tremella fuciformis*.	Provide moisture, strength, and protection to hair and scalp.	Camelini et al. (2022) and Sousa et al. (2023)
2	Fermented fungal extracts	Extracts from fungi such as Aspergillus and Penicillium.	Contains antifungal and antimicrobial compounds that combat scalp issues like dandruff and fungal infections, enhance nutrient absorption, and stimulate hair growth.	Choi and Choi (2022)
3	Proteins and peptides	Sourced from fungi like Tricholoma matsutake.	Repair damaged hair, enhance elasticity, and provide shine.	Belsito et al. (2017)
4	Chitosan	A bioactive compound extracted from Cordyceps fungi.	Offers film-forming, moisturizing, and antimicrobial properties beneficial for hair and scalp health.	Mukherjee et al. (2024)
5	Cordycepin	A bioactive compound extracted from Cordyceps fungi.	Exhibits antifungal, antimicrobial, and anti-inflammatory properties that promote scalp health.	Sang et al. (2023)
6	Schizophyllan	A polysaccharide produced by Schizophyllum commune.	Acts as a conditioning agent, providing moisture and protection to the hair.	Mukherjee et al. (2024)

primary cause of dandruff. Ketoconazole works by disrupting the fungal cell membrane and inhibiting the synthesis of ergosterol, an essential component of the fungal cell wall. This dual mechanism of action makes it difficult for the fungus to develop resistance, improving the long-term efficacy of ketoconazole-based treatments (Camelini et al. 2022). Neutrogena T-Gel shampoos contain a combination of active ingredients like salicylic acid and zinc pyrithione that make them effective for hair loss treatments and also confer special benefits to consumers as they help in treating several scalp conditions (Kim et al. 2022). Furthermore, some brand products contain selenium sulfide which is effective for the treatment of hyperkeratosis and adjunctive therapy for tinea capitis. Selenium sulfide is a highly effective antifungal ingredient used in haircare products to treat severe dandruff and related scalp conditions (Cohen and Anderson 2018). Vichy Dercos offers a range of anti-dandruff shampoos formulated for different hair types, including their anti-dandruff shampoo for normal to oily hair and anti-dandruff shampoo for dry hair. The shampoos contain antifungal ingredients to combat the fungus *Malassezia* that causes dandruff and can also be considered a hair regrowth shampoo (Amazon 2024b). These products have been of good significance to consumers as they serve as alternatives to synthetic products (Table 13.3).

13.4 Benefits of Fungal-Based Ingredients in Haircare

The benefits of fungal-based ingredients in the haircare industry are wide. They confer benefits such as moisturizing and hydration, scalp health, strengthening and repair of hair, and sustainability and natural appeal. Moisture and hydration are important for keeping healthy hair as they can contribute to the elasticity, shine, softness, manageability of the hair, and scalp health (Tian et al. 2023). Fungal polysaccharides and lipids, such as beta-glucans, can act as effective hair moisturizers and conditioners. Hair dryness and brittleness oppose well-moisturized hair, making the hair unhealthy. This dryness can be caused by free radicals present in the hair which can be protected by certain antioxidants like ergothioneine produced by some mushrooms. Fungal-based ingredients also improve scalp health through anti-inflammatory and anti-

Table 13.3 Commercially available haircare products with fungal-based ingredients

S/N	Product name	Fungal-based ingredient	Benefits	References
1	ProBliva Fungus Shampoo	Natural oils (emu oil, coconut oil, jojoba oil)	Soothes the scalp and reduces ringworm and itchy scalp	Amazon (2024a)
2	Nizoral Anti-Dandruff Shampoo	Ketoconazole	Targets and eliminates fungal infections on the scalp, reduces dandruff	Camelini et al. (2022)
3	Neutrogena T-Gel Shampoo	Salicylic acid, zinc pyrithione	Treats scalp conditions like dandruff, and hair loss, and provides additional scalp benefits	Kim et al. (2022)
4	Vichy Dercos Anti-Dandruff Shampoo	Selenium sulfide	Treats severe dandruff and related scalp conditions	Amazon (2024b)
5	Paul Mitchell Tea Tree Special Shampoo	Tea tree oil, antifungal peptides	Soothes and cleanses the scalp, and provides antifungal benefits	Hill et al. (2024)
6	Pura D'or Original Gold Label Anti-Hair Thinning Shampoo	Fermented rice water, biotin	Strengthens hair, reduces hair thinning, and enhances scalp health	Choi and Choi (2022)
7	Klorane Shampoo with Quinine and B Vitamins	Quinine, fermented plant extracts	Strengthens hair, reduces hair loss, and promotes hair growth	Sang et al. (2023)
8	Alterna Caviar Anti-Aging Replenishing Moisture Shampoo	Beta-glucans from fungi (mushrooms)	Provides moisture, improves hair strength, and reduces breakage	Pedro et al. (2023)
9	Briogeo Scalp Revival Charcoal + Tea Tree Scalp Treatment	Tea tree oil, charcoal	Detoxifies scalp reduces dandruff and soothes irritation	Mukherjee et al. (2024)

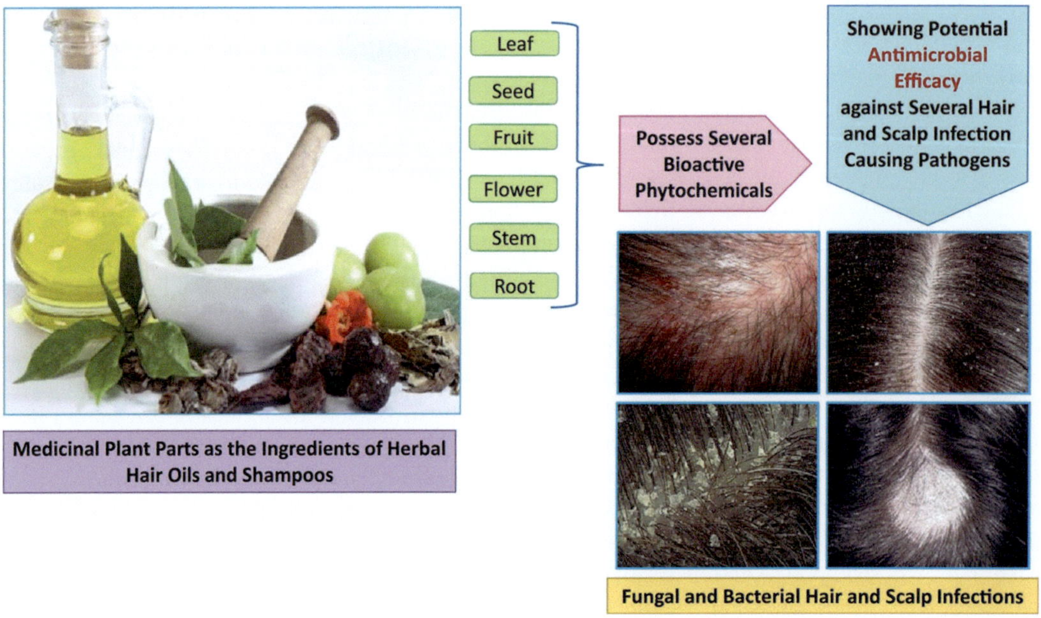

Fig. 13.2 Medicinal plants (including macrofungal) showing potential antimicrobial efficacy against several hair and scalp infections (Rajput and Kumar 2020)

microbial properties contained in them. An unhealthy scalp will give rise to unhealthy hair, having poor growth and quality (Trueb et al. 2018). Folliculitis is a condition that causes infection and inflammation of the hair follicles, leading to bald spots and reducing the overall health of the scalp. However, the use of fungal-based ingredients in haircare products can reduce dandruff and scalp irritation for consumers. Similarly, fungal-based ingredients strengthen and repair hair by enhancing hair strength and reducing breakage (Winters and Mitchell 2024). Hair might experience some weakening and breakdown when not properly taken care of, resulting in weak ends, hair thinning, breakage, and even hair loss. Hair loss can be a result of oxidative stress, and one major organism known to contribute to oxidative damage is *Malassezia* sp. which can be combated by antioxidants like flavonoids, carotenoids, tannins, exopolysaccharides, etc., which can be found in mushrooms (Trueb et al. 2018). Lentinan, a homopolysaccharide produced by shiitake mushroom (*Lentinus edodes*), is capable of reducing hair loss, especially in patients undergoing chemotherapy (Stoica et al. 2023). These ingredients also serve as an eco-friendly and biodegradable alternative to synthetic ingredients. Over time, synthetic and chemical-based hair products have been found to cause more damage to the hair than good, hence the need for more natural, eco-friendly, and biodegradable products (Luengo et al. 2023). The need for this has pushed consumers to use products including plant extracts, essential oils, herbs, shea butter, and fungal-based substances which this work focuses on (Fig. 13.2 and Table 13.4).

13.5 Conclusion

The cosmetic industry is a vast and ever-changing global market that offers significant financial opportunities. It constantly adapts to meet the diverse needs and high expectations of consumers worldwide, evolving to meet the demand for natural, eco-friendly products, with a shift toward sustainability and innovation. The utilization of fungal-derived components in haircare presents an exciting prospect for developing environmentally friendly haircare solutions. These solutions

Table 13.4 Description of the benefits of fungal-based ingredients in haircare

S/N	Benefit	Fungal-based ingredient	Description	References
1	Antifungal properties	Ketoconazole, tea tree oil	Helps in treating and preventing dandruff, scalp infections, and other fungal-related issues	Camelini et al. (2022)
2	Scalp soothing and healing	Tea tree oil, emu oil	Reduces scalp irritation, redness, and itchiness	Hill et al. (2024)
3	Moisturizing	Beta-glucans, fermented rice water	Provides deep hydration to the scalp and hair, preventing dryness and flakiness	Pedro et al. (2023)
4	Strengthening hair	Quinine, beta-glucans	Enhances hair strength, reducing breakage and promoting healthier hair	Sang et al. (2023)
5	Anti-dandruff	Ketoconazole, selenium sulfide	Effectively targets dandruff and other scalp conditions	Kim et al. (2022)
6	Hair growth promotion	Fermented plant extracts, biotin	Stimulates hair follicles, promoting hair growth and reducing hair thinning	Choi and Choi (2022)
7	Detoxifying the scalp	Charcoal, tea tree oil	Removes impurities, excess oil, and buildup from the scalp	Mukherjee et al. (2024)
8	Anti-inflammatory	Tea tree oil, jojoba oil	Reduces inflammation and soothes the scalp	Amazon (2024a)
9	Conditioning	Fermented rice water, jojoba oil	Provides natural conditioning, making hair softer and more manageable	Amazon (2024b)
10	Antioxidant properties	Beta-glucans, mushroom extracts	Protects the scalp and hair from oxidative stress and environmental damage	Hill et al. (2024)

can effectively address various hair and scalp issues while remaining sustainable and gentle on the environment. Fungal-derived ingredients offer advantages such as moisturization, hydration, scalp health improvement, hair strengthening, and repair, as well as a natural and sustainable appeal. Components like beta-glucans, fermented extracts, and proteins sourced from fungi provide antifungal, antimicrobial, and antioxidant effects conducive to hair and scalp wellness. Commercially available products like antifungal shampoos with ingredients such as ketoconazole and selenium sulfide effectively address scalp conditions like dandruff, fungal infections, and hair loss. Fungal-based ingredients contribute to improved hair and scalp health, offering an eco-friendly, natural alternative to synthetic products, aligning with the trend toward sustainable beauty solutions in the cosmetics industry.

References

Amazon (2024a, July 01) ProBliva fungus shampoo, psoriasis shampoo, itchy scalp shampoo for hair & scalp—help to reduce ringworm, itchy scalp—contains natural ingredients coconut oil, jojoba oil, emu oil. https://www.amazon.com/ProBliva-Fungus-Shampoo-Hair-Scalp/dp/B07RZ9B33D

Amazon (2024b, July 03) Vichy dercos shampoo. https://www.amazon.com/vichy-dercos-shampoo/s?k=vichy+dercos+shampoo

Ameta M, Tiwari H, Chandrul KK (2024) Green beauty: the growing demand for organic cosmetics. Eur J Modern Med Pract 4(5):383–389. Retrieved from https://inovatus.es/index.php/ejmmp/article/view/3316

Belsito DV, Hill RA, Klaassen CD, Liebler DC, Marks JG Jr, Shank RC et al (2017) Safety assessment of plant-derived proteins and peptides as used in cosmetics. Int J Toxicol 41(2_suppl):5S–20S

Berg A, Hudson S, Weaver KK, Pacchia ML, Amed I (2023) The beauty market in 2023: a special state of fashion report. E-source: https://www.mckinsey.com/industries/retail/our-insights/the-beauty-market-in-2023-a-special-state-offashion-report

Camelini CM, Rossi MJ, de Sousa FTG, Giachini A (2022) Fungal polysaccharide production for dermatological purposes. In: Oliveira JM, Radhouani H, Reis RL (eds) Polysaccharides of microbial origin. Springer, Cham. https://doi.org/10.1007/978-3-030-42215-8_64

Choi M, Choi E (2022) Usefulness of smilax China leaves fermented product as a cosmetic material for scalp hair protection. Protect Converg 7(2):1–14

Cohen PR, Anderson CA (2018) Topical selenium sulfide for the treatment of hyperkeratosis. Dermatol Ther 8(4):639–646. https://doi.org/10.1007/s13555-018-0259-9

Ferreira M, Matos A, Couras A, Marto J, Ribeiro H (2022) Overview of cosmetic regulatory frameworks around the world. Cosmetics 9(4):72. https://doi.org/10.3390/cosmetics9040072

Hill RC, Caplan AS, Elewski B, Gold JA, Lockhart SR, Smith DJ, Lipner SR (2024) Expert panel review of skin and hair dermatophytoses in an era of antifungal resistance. Am J Clin Dermatol 25(3):359–389. https://doi.org/10.1007/s40257-024-00848-1

Kim HT, Park HS, Kim YM, Lee IC, Lee SJ, Choi JS (2022) Double-blind randomized placebo-controlled study of the efficacy and safety of hair loss prevention shampoo containing salicylic acid, panthenol, and niacinamide in alopecia patients. Toxicol Environ Heal Sci 14(2):173–185

Lavuri R, Jabbour CJC, Grebinevych O, Roubaud D (2022) Green factors stimulating the purchase intention of innovative luxury organic beauty products: implications for sustainable development. J Environ Manag 301:113899. https://doi.org/10.1016/j.jenvman.2021.113899

Luengo GS, Leonforte F, Greaves A, Rubio RG, Guzman E (2023) Physico-chemical challenges on the self-assembly of natural and bio-based ingredients on hair surfaces: towards sustainable haircare formulations. Green Chem 25(20):7863–7882. https://doi.org/10.1039/D3GC02763E

Mukherjee A, Das S, Chakraborty D, Pal N, Das N (2024) Fungi's treasure in cosmeceuticals-a comprehensive chemical approach. S Afr J Bot 166:311–331. https://doi.org/10.1016/j.sajb.2024.01.036

Ortega-Requena S, Montiel C, Máximo F, Gómez M, Murcia MD, Bastida J (2024) Esters in the food and cosmetic industries: an overview of the reactors used in their biocatalytic synthesis. Materials 17(1):268. https://doi.org/10.3390/ma17010268

Pedro S, Diana T, Manuela A, João A, Manuela P, João F (2023) β-Glucan extracts as high-value multifunctional ingredients for skin health. Carbohydr Polym 322(3): 121329. https://doi.org/10.1016/j.carbpol.2023.121329

Petruzzi D (2024, July 02) Cosmetics industry—statistics & facts. https://www.statista.com/topics/3137/cosmetics-industry/#editorsPicks

Rajput M, Kumar N (2020) Medicinal plants: a potential source of novel bioactive compounds showing antimicrobial efficacy against pathogens infecting hair and scalp. Gene Rep 21:100879. https://doi.org/10.1016/j.genrep.2020.100879

Sang H, Akowuah GA, Liew KB, Lee K, Keng W, Lee K, Yon L, Tan CS, Chew L (2023) Natural alternatives from your garden for hair care: revisiting the benefits of tropical herbs. Heliyon 9(11):e21876. https://doi.org/10.1016/j.heliyon.2023.e21876

Sousa P, Tavares-Valente D, Amorim M, Azevedo-Silva J, Pintado M, Fernandes J (2023) β-Glucan extracts as high-value multifunctional ingredients for skin health: a review. Carbohydr Polym 322:121329. https://doi.org/10.1016/j.carbpol.2023.121329

Srivastava M, Kumari M, Karn SK, Bhambri A, Mahale VG, Mahale S (2024) Submerged cultivation and phytochemical analysis of medicinal mushrooms (Trametes sp.). Front Fungal Biol 5:1414349. https://doi.org/10.3389/ffunb.2024.1414349

Stoica RM, Moscovici M, Lakatos ES, Cioca LI (2023) Exopolysaccharides of fungal origin: properties and pharmaceutical applications. PRO 11(2):335. https://doi.org/10.3390/pr11020335

Tian X, Thorne JL, Moore JB (2023) Ergothioneine: an underrecognised dietary micronutrient required for healthy ageing? Br J Nutr 129(1):104–114. https://doi.org/10.1017/S0007114522003592

Trueb RM, Henry JP, Davis MG, Schwartz JR (2018) Scalp condition impacts hair growth and retention via oxidative stress. Int J Trichology 10(6):262–270

Winters RD, Mitchell M (2024, July 04) Folliculitis. In: StatPearls [Internet]. StatPearls Publishing, Treasure Island. Available from https://www.ncbi.nlm.nih.gov/books/NBK547754/

14 Commercially Available Cosmetics with Fungal-Derived Bioactive Compounds: Hand Sanitizer, Washing Soap, and Saponification Products

Tanveen Kaur, Asma Ashraf Khan, and Chandresh Kumari

Abstract

The cosmetic industry has been exploring the use of fungal-derived bioactive compounds in recent years. This chapter provides an overview of commercially available cosmetics that contain fungal-derived bioactive compounds, including hand sanitizers, washing soaps, and saponification products. The chapter highlights the benefits of using these products and their potential applications in the cosmetic industry. Fungal-derived bioactive compounds have been found to have anti-aging, antioxidant, anti-wrinkle, skin lightening, moisturizing, anti-acne, and anti-dandruff effects. The chapter also provides eight ready-to-use recipes of different cosmetic formulations fortified with mushroom extracts for the benefit of entrepreneurs and researchers in the cosmetic industry. The chapter concludes with a discussion of the future prospects of fungal-derived bioactive compounds in the cosmetic industry.

T. Kaur · C. Kumari (✉)
Faculty of Applied Sciences and Biotechnology, Shoolini University of Biotechnology and Management Sciences, Solan, Himachal Pradesh, India

A. A. Khan
School of Bioengineering and Food Technology, Shoolini University, Solan, Himachal Pradesh, India

Keywords

Fungal-derived compounds · Bioactive compounds · Hand sanitizer · Washing soap · Saponification products

14.1 Introduction

14.1.1 Overview of Fungal-Derived Bioactive Compounds in Cosmetics

In recent years, there has been a growing interest in harnessing the potential of natural sources for developing innovative cosmetic products. Fungi, a diverse group of microorganisms, have emerged as a promising source of bioactive compounds with various beneficial properties. Fungal-derived bioactive compounds refer to the molecules synthesized or produced by fungi that possess therapeutic, protective, or cosmetic attributes.

This chapter aims to provide an overview of the utilization of fungal-derived bioactive compounds in commercially available cosmetics, specifically focusing on hand sanitizers, washing soaps, and saponification products. By exploring the potential of these compounds, cosmetic manufacturers can create products that offer not only cosmetic benefits but also additional advantages derived from fungal bioactive components.

S. Ghosh, H. N. Onyeaka (eds.), *Fungal-Based Cosmetics: Formulation and Usage*, The Mycota 17,
https://doi.org/10.1007/978-3-032-09291-5_14

14.1.2 Importance of Hand Sanitizers, Washing Soaps, and Saponification Products in Personal Hygiene

Personal hygiene is of paramount importance in maintaining overall well-being and preventing the transmission of diseases. Practicing good personal hygiene habits, such as regular handwashing and proper body cleansing, is essential for minimizing the risk of infections and promoting a healthy lifestyle (*The Importance of Personal Hygiene in Our Life* 2021).

Hand sanitizers, washing soaps, and saponification products are indispensable tools in daily hygiene routines. They provide effective cleansing and sanitization properties, enabling the removal of dirt, bacteria, and other harmful microorganisms from the skin. Hand sanitizers, in particular, have gained immense popularity due to their convenience and ability to quickly eliminate germs when soap and water are not readily available. Washing soaps and saponification products, on the other hand, offer thorough cleansing by removing impurities and maintaining skin health (Singh et al. 2020).

These personal hygiene products serve as the first line of defense against the spread of infectious diseases. Proper hand hygiene, including the use of hand sanitizers, has been proven to significantly reduce the transmission of pathogens, including bacteria and viruses. By incorporating regular handwashing and the use of sanitizers into daily routines, individuals can help prevent the transmission of illnesses, such as the common cold, influenza, and gastrointestinal infections (Centers for Disease Control and Prevention 2021).

In addition to their crucial role in preventing the spread of diseases, hand sanitizers, washing soaps, and saponification products contribute to overall cleanliness and well-being. Clean and healthy skin is not only aesthetically appealing but also aids in maintaining the skin's natural barrier function, protecting against external aggressors and preventing skin-related issues.

Given the vital role that personal hygiene products play in our lives, it is essential to explore novel approaches to enhance their efficacy and improve their formulations. One promising avenue is the incorporation of natural ingredients, such as fungal-derived bioactive compounds. These compounds offer the potential to enhance the therapeutic and cosmetic properties of hand sanitizers, washing soaps, and saponification products (Atanasov et al. 2021).

Fungal-derived bioactive compounds, with their diverse range of bioactive molecules, present exciting opportunities for improving the effectiveness of personal hygiene products. These compounds can possess antimicrobial, antifungal, antioxidant, anti-inflammatory, and other beneficial properties that can augment the cleansing and protective actions of hand sanitizers, washing soaps, and saponification products (Rai et al. 2021). By incorporating fungal bioactive compounds, it is possible to enhance the overall efficacy and functionality of these essential products, offering additional benefits beyond basic cleanliness.

Furthermore, the use of natural ingredients aligns with the growing consumer preference for sustainable and eco-friendly cosmetic products. Fungal-derived bioactive compounds offer a renewable and environmentally friendly alternative to synthetic additives, catering to the increasing demand for natural and plant-based cosmetic formulations.

14.1.3 Purpose and Scope of the Chapter

The purpose of this chapter is to examine the use of fungal-derived bioactive compounds in hand sanitizers, washing soaps, and saponification products from a commercial perspective. By focusing on these specific product categories, we aim to provide a comprehensive understanding of how fungal bioactive compounds can be utilized to enhance their functionality and offer unique advantages.

The scope of this chapter encompasses a range of commercially available cosmetics that incorporate fungal-derived bioactive compounds. We will explore the scientific background and mechanisms underlying the bioactivity of these compounds, highlighting their potential applications in personal hygiene products. Furthermore, we will discuss the challenges and opportunities associated with incorporating fungal bioactive components into cosmetic formulations, considering factors such as safety, stability, and regulatory considerations.

By the end of this chapter, readers will gain insights into the current landscape of fungal-derived bioactive compounds in hand sanitizers, washing soaps, and saponification products. This knowledge will aid cosmetic formulators, researchers, and consumers in making informed decisions about the usage and benefits of these innovative cosmetic formulations.

14.2 Fungal-Derived Bioactive Compounds in Hand Sanitizers

14.2.1 Definition and Function of Hand Sanitizers

Hand sanitizers are topical formulations that are specifically formulated to kill or inhibit the growth of microorganisms on the hands. They are designed to promote disinfection and maintain hand hygiene. Hand sanitizers are particularly useful in situations where soap and water are not readily accessible or convenient, providing a quick and convenient alternative for sanitizing the hands.

The primary function of hand sanitizers is to effectively eliminate harmful microorganisms, including bacteria, viruses, and fungi, that may be present on the hands. These microorganisms can be acquired from various sources such as contaminated surfaces, objects, or direct contact with infected individuals. Regular use of hand sanitizers can significantly reduce the risk of spreading infectious diseases and contribute to overall public health.

Hand sanitizers typically contain active ingredients that are antimicrobial in nature, such as alcohol (e.g., ethanol or isopropyl alcohol) or other antiseptic agents. These active ingredients work by disrupting the outer membrane or cellular structures of microorganisms, leading to their inactivation or death. The concentration of the active ingredient in hand sanitizers is carefully regulated to ensure effective disinfection while also considering safety and tolerability for regular use on the skin.

One of the significant advantages of hand sanitizers is their convenience and portability. Unlike traditional handwashing with soap and water, which may not be feasible in certain situations (e.g., when traveling, in public spaces, or during emergencies), hand sanitizers provide a portable and accessible solution for hand hygiene. They can be easily carried in pockets, purses, or bags, allowing individuals to maintain hand cleanliness when soap and water are not immediately available.

Furthermore, hand sanitizers are designed to be quick-acting, requiring minimal time for application and drying. This makes them an efficient choice for individuals who need to sanitize their hands on the go, such as healthcare professionals, travelers, or individuals working in environments where frequent hand hygiene is essential.

It is important to note that while hand sanitizers are effective in killing or inhibiting the growth of many types of microorganisms, they may not be as effective against certain types of pathogens (e.g., norovirus or spores) or when the hands are visibly soiled or contaminated with certain substances (e.g., chemicals, oils). In such cases, thorough handwashing with soap and water is recommended.

To sum up, hand sanitizers are crucial for promoting disinfection and upholding hand hygiene. They offer a convenient, portable, and fast-acting solution when traditional handwashing with soap and water is not easily accessible or feasible. By efficiently reducing the presence of harmful microorganisms on the hands, hand sanitizers play a significant role in preventing the spread of infectious diseases and protecting public health.

14.2.2 Types of Fungal-Derived Bioactive Compounds Used in Hand Sanitizers

14.2.2.1 Antifungal Properties of Fungal-Derived Compounds

Fungal-derived bioactive compounds are known for their natural antifungal properties, which make them valuable additions to hand sanitizers. These compounds have the ability to inhibit the growth and activity of various fungal pathogens, thereby reducing the risk of fungal infections (Singh et al. 2019).

Fungi produce a wide range of bioactive compounds as part of their defense mechanisms against other microorganisms or environmental stresses. Many of these compounds have been identified and studied for their antifungal properties. By incorporating these compounds into hand sanitizers, their antifungal efficacy can be harnessed to provide enhanced protection against fungal pathogens (Gehlot and Singh 2018).

14.2.2.2 Examples of Fungal-Derived Bioactive Compounds in Hand Sanitizers

Commercial hand sanitizers often utilize fungal-derived bioactive compounds to augment their antifungal properties summarized in Table 14.1 (Chojnacki et al. 2021). Here are some commonly used examples:

(a) Amphotericin B: Amphotericin B is a polyene macrolide antibiotic produced by *Streptomyces nodosus*. It exhibits broad-spectrum antifungal activity and is effective against various fungi, including *Candida* species, *Aspergillus* species, and *Cryptococcus neoformans*. Amphotericin B works by binding to the fungal cell membrane, forming pores that disrupt its integrity and ultimately leading to cell death (Ellis 2002).

(b) Caspofungin: Caspofungin is a cyclic lipopeptide derived from *Glarea lozoyensis*. It specifically targets the fungal cell wall by inhibiting the synthesis of β-(1,3)-D-glucan, an essential component of the fungal cell wall. By disrupting the integrity of the cell wall, caspofungin prevents fungal growth and replication. It is effective against a broad range of fungal pathogens, including *Candida* species and *Aspergillus* species (Morrison 2005).

(c) Griseofulvin: Griseofulvin is a natural product isolated from *Penicillium griseofulvum*. It has been used for many years as an oral antifungal medication for the treatment of dermatophyte infections, such as ringworm and athlete's foot. Griseofulvin acts by interfering with fungal cell division, specifically targeting the microtubules involved in mitosis. Its mechanism of action prevents fungal growth and leads to the eventual eradication of the infection (Aris et al. 2022).

These examples highlight the diverse mechanisms by which fungal-derived bioactive compounds exert their antifungal effects. By incorporating such compounds into hand sanitizers, their antifungal activity can complement the sanitizing properties of the product, providing an additional layer of protection against fungal pathogens.

14.2.3 Benefits and Potential Applications of Fungal-Derived Compounds in Hand Sanitizers

14.2.3.1 Enhanced Effectiveness

The incorporation of fungal-derived compounds in hand sanitizers can significantly enhance their overall effectiveness, particularly in combating fungal pathogens. Fungal infections can be challenging to treat, and their prevalence in healthcare settings poses a significant risk to patients, especially those with compromised immune systems (Alajlan et al. 2022). By harnessing the potent antifungal properties of fungal-derived compounds, hand sanitizers can provide an additional layer of defense against fungal infections, complementing their primary role as antimicrobial agents. This enhanced effectiveness can

Table 14.1 List of fungal-derived bioactive compounds and their sources

Fungal-derived bioactive compound	Fungal source	Hand sanitizer activity	Mode of action	References
Palmarumycin C 8	*Lophiotrema* sp. F6932	Antiproliferative activity against MIA PaCa-2 and PANC-1 cells	Inducing apoptosis and cell cycle arrest in cancer cells	Singh et al. (2019)
Palmarumycin CP 30	*Lophiotrema* sp. F6932	Antifungal activity against *Aspergillus fumigatus* and *Candida albicans*	Inhibiting ergosterol biosynthesis and disrupting fungal cell membrane	Singh et al. (2019)
Muyocopronol A-C	*Muyocopron laterale* F5912	Antifungal activity against *Aspergillus fumigatus* and *Candida albicans*	Inhibiting ergosterol biosynthesis and disrupting fungal cell membrane	Singh et al. (2019)
Tropicicolide	*Colletotrichum tropicicola* F10154	Antifungal activity against *Aspergillus fumigatus* and *Candida albicans*	Inhibiting ergosterol biosynthesis and disrupting fungal cell membrane	Singh et al. (2019)
Chitosan	Various fungi	Wound healing and anti-inflammatory activity	Stimulating fibroblast proliferation and collagen synthesis, modulating cytokine production and immune response	Rai et al. (2022)
Penicillin	*Penicillium* sp.	Antibiotic activity against Gram-positive bacteria	Inhibiting bacterial cell wall synthesis by binding to penicillin-binding proteins (PBPs)	Keshri et al. (2021)
Huperzine A	*Huperzia serrata* endophytic fungi	Acetylcholinesterase inhibitor and neuroprotective agent	Increasing acetylcholine levels in the brain by reversibly inhibiting acetylcholinesterase enzyme, protecting neurons from oxidative stress and glutamate toxicity	Gupta et al. (2023)
Resveratrol	*Botryosphaeria* sp. and *Phomopsis* sp. endophytic fungi	Antioxidant, anti-inflammatory, and anticancer activity	Scavenging free radicals, modulating inflammatory pathways, regulating cell cycle and apoptosis in cancer cells, inhibiting angiogenesis and metastasis	Gupta et al. (2023)

contribute to improved patient safety and reduce the transmission of fungal pathogens in healthcare environments (Hans et al. 2023).

14.2.3.2 Safety and Tolerability

Fungal-derived compounds are often well-tolerated by human skin, making them suitable for use in hand sanitizers. Unlike some synthetic antifungal agents that may cause skin irritation or sensitization, fungal-derived compounds offer a natural alternative that is generally gentle on the skin. This characteristic is particularly advantageous for individuals with sensitive skin or those who require frequent hand sanitization, such as healthcare workers. The use of fungal-derived compounds in hand sanitizers can provide effective antifungal protection without compromising skin health or causing adverse effects (Aly et al. 2011).

14.2.3.3 Broad-Spectrum Activity

Many fungal-derived compounds exhibit broad-spectrum antifungal activity, meaning they can target a wide range of fungal pathogens. This versatility is advantageous as it allows hand sanitizers containing these compounds to provide protection against various types of fungi, including common dermatophytes, *Candida* species, and molds. By targeting a broad spectrum of fungal pathogens, hand sanitizers with fungal-derived compounds can address diverse fungal infections and contribute to comprehensive hand hygiene practices (Hans et al. 2023).

Furthermore, the broad-spectrum activity of fungal-derived compounds can be particularly beneficial in certain settings, such as public spaces, where individuals may come into contact with different types of fungal pathogens. It ensures that hand sanitizers are effective against a wide array of fungi, providing a robust defense against potential infections.

Hence, the utilization of fungal-derived compounds in hand sanitizers offers several benefits and potential applications. These compounds enhance the overall effectiveness of hand sanitizers by providing potent antifungal properties, which is particularly valuable in healthcare settings. Fungal-derived compounds are well-tolerated by the skin, reducing the risk of adverse effects and providing a natural alternative to synthetic antifungal agents. Additionally, their broad-spectrum activity enables hand sanitizers to offer protection against a wide range of fungal pathogens, ensuring comprehensive hand hygiene. Incorporating fungal-derived compounds in hand sanitizers can thus enhance their efficacy, safety, and versatility, making them valuable tools in the prevention and control of fungal infections (Fendler and Groziak 2002).

14.2.4 Safety Considerations and Regulatory Aspects

14.2.4.1 Safety Concerns Associated with Fungal-Derived Compounds

While fungal-derived compounds generally have a favorable safety profile, it is important to consider potential allergenicity or sensitization in certain individuals. Some people may have allergic reactions or sensitivities to specific fungal compounds, which can manifest as skin irritation, redness, or itching. Therefore, it is essential for manufacturers to conduct thorough testing to identify and address any potential safety concerns associated with the use of fungal-derived compounds in hand sanitizers.

Proper quality control measures, including rigorous testing of raw materials and finished products, are crucial to ensure the safety of hand sanitizers containing fungal-derived compounds. These tests can help identify and eliminate any impurities or contaminants that may pose a risk to consumer health. Additionally, manufacturers should provide clear instructions and warnings on product labels to inform users about potential allergenicity and sensitization risks, allowing individuals to make informed decisions regarding product use.

14.2.4.2 Regulatory Requirements

Regulatory bodies, such as the Food and Drug Administration (FDA) in the United States, have specific guidelines and requirements for the formulation and labeling of hand sanitizers. These regulations aim to ensure the safety, efficacy, and quality of consumer products, including those containing fungal-derived bioactive compounds.

Manufacturers of hand sanitizers with fungal-derived compounds must adhere to these regulatory requirements, which may include providing detailed ingredient lists, demonstrating product effectiveness through testing, and complying with labeling standards. It is crucial for manufacturers to stay updated with the latest regulatory guidelines and adjust their formulations and labeling accordingly to meet the necessary standards.

14.2.4.3 Importance of Adherence to Safety Standards

To maintain the safety and effectiveness of hand sanitizers with fungal-derived bioactive compounds, it is imperative for manufacturers to follow Good Manufacturing Practices (GMP) and implement stringent quality control procedures. These practices involve ensuring proper hygiene and cleanliness during the manufacturing process, as well as conducting regular monitoring and testing to detect any potential issues.

Adherence to safety standards helps minimize the risk of contamination, ensures consistent product quality, and provides consumers with reliable and safe hand sanitizers. It is the responsibility of manufacturers to prioritize safety and quality assurance, implementing robust quality control measures to prevent any compromises in product safety or efficacy.

Therefore, the incorporation of fungal-derived bioactive compounds in hand sanitizers offers an opportunity to enhance their antifungal properties and provide effective and safe solutions for hand hygiene. However, it is crucial for manufacturers to understand and address any potential safety concerns associated with these compounds, such as allergenicity or sensitization. Compliance with regulatory requirements, including thorough testing and proper labeling, is essential to ensure consumer safety and meet the standards set by regulatory bodies. Adherence to safety standards, including GMP and quality control procedures, is crucial for maintaining the safety and effectiveness of hand sanitizers with fungal-derived bioactive compounds and ensuring the delivery of high-quality products to consumers.

14.3 Fungal-Derived Bioactive Compounds in Washing Soaps

14.3.1 Importance of Washing Soaps in Personal Hygiene

Maintaining personal hygiene is of utmost importance for overall well-being and the promotion of good health. Within the realm of personal hygiene, washing soaps hold a crucial role. These soaps are specifically formulated to cleanse the skin by effectively removing dirt, oil, and impurities that accumulate on the skin's surface.

One of the primary functions of washing soaps is to facilitate the removal of microorganisms from the skin. The skin is constantly exposed to various microorganisms, including bacteria, viruses, and fungi, which can potentially cause infections and diseases. Washing soaps work by creating a lather that helps to dislodge and eliminate these microorganisms, reducing the risk of infections and promoting a healthier environment for the skin.

By washing away dirt, debris, and microorganisms, washing soaps contribute significantly to maintaining cleanliness. Cleanliness is vital in preventing the transmission of pathogens from person to person or from surfaces to individuals. Regular and thorough use of washing soaps helps to minimize the presence of harmful microorganisms on the skin, reducing the likelihood of contracting infections.

Furthermore, washing soaps play a role in reducing body odor. Bacteria on the skin can break down sweat and produce unpleasant odors. Washing soaps help to remove sweat, oils, and bacteria from the skin, which can alleviate body odor concerns and contribute to a more pleasant and fresh-smelling individual.

In addition to the immediate benefits of cleanliness and odor reduction, washing soaps also aid in the preservation of skin health. Accumulated dirt, excess oil, and impurities can clog pores and lead to various skin issues, including acne, inflammation, and infections. By effectively cleansing the skin, washing soaps help to maintain its natural balance, prevent pore blockages, and promote healthier skin overall.

Regular use of washing soaps should be considered a fundamental aspect of personal hygiene routines. Incorporating washing soaps into daily hygiene practices ensures that individuals maintain a clean and healthy appearance while minimizing the risk of infections and skin-related problems. It is important to choose washing soaps that are suitable for specific skin types and needs, as different formulations can offer additional benefits such as moisturization or targeted treatment for specific conditions.

In summary, washing soaps are vital for personal hygiene as they cleanse the skin, remove impurities and microorganisms, minimize the risk of infections, reduce body odor, and promote the overall health and well-being of the skin. By incorporating washing soaps into daily hygiene routines, individuals can maintain a clean and healthy appearance, prevent the transmission of pathogens, and preserve the integrity of their skin.

14.3.2 Fungal-Derived Compounds Used in Washing Soaps

14.3.2.1 Fungistatic and Fungicidal Properties of Fungal-Derived Compounds

Fungal-derived compounds offer unique properties that make them highly valuable in the formulation of washing soaps. Certain fungal species have the ability to produce bioactive compounds that exhibit fungistatic properties, which means they inhibit the growth and reproduction of fungi. Additionally, some fungal-derived compounds possess fungicidal properties, meaning they can directly kill fungi. These properties have attracted significant attention and research due to their potential applications in various industries, including personal care.

The discovery of these fungistatic and fungicidal properties in fungal-derived compounds has opened up new avenues for developing effective antimicrobial agents. These compounds have been studied extensively to understand their mechanisms of action and to evaluate their potential in combating fungal infections. By incorporating these compounds into washing soaps, their antimicrobial properties can be harnessed to provide an extra layer of protection against fungal growth on the skin.

14.3.2.2 Examples of Fungal-Derived Bioactive Compounds in Washing Soaps

Commercial washing soaps have begun incorporating fungal-derived bioactive compounds due to their beneficial properties. Among the fungal species that have been extensively studied for their bioactive compounds are Aspergillus, Penicillium, and Trichoderma.

Examples of fungal-derived bioactive compounds commonly found in washing soaps include terpenoids, phenolics, and peptides. Terpenoids are a diverse class of compounds that are naturally synthesized by fungi and have demonstrated antimicrobial activity against various fungi. Phenolics, another class of compounds derived from fungi, exhibit broad-spectrum antimicrobial properties and have been shown to be effective against both dermatophytes (fungal infections of the skin, hair, and nails) and yeast species. Peptides derived from fungi possess antimicrobial properties and can target specific fungal strains, making them valuable in combating fungal infections.

These fungal-derived bioactive compounds have shown promise in inhibiting the growth of fungi and preventing fungal infections. Their incorporation into washing soaps offers a natural and potentially effective alternative to traditional synthetic antimicrobial agents.

Hence, fungal-derived compounds used in washing soaps possess fungistatic and fungicidal properties that make them valuable in combating fungal infections. Compounds such as terpenoids, phenolics, and peptides derived from fungi like Aspergillus, Penicillium, and Trichoderma have demonstrated antimicrobial activity against a wide range of fungi. By incorporating these bioactive compounds into washing soap formulations, their unique properties can be utilized to enhance the antimicrobial effectiveness of the soap, providing an additional layer of protection against fungal growth on the skin.

14.3.3 Efficacy and Advantages of Using Fungal-Derived Compounds in Washing Soaps

The incorporation of fungal-derived compounds in washing soaps brings forth several noteworthy advantages. One of the primary advantages is their enhanced antimicrobial properties, which make them highly effective in combating fungal infections on the skin. Fungal-derived compounds possess fungistatic and fungicidal activities that can inhibit the growth and reproduction of pathogenic fungi. By preventing the proliferation of these fungi, the risk of developing skin diseases caused by fungal overgrowth can be significantly reduced.

Furthermore, the use of fungal-derived compounds in washing soaps offers a natural alternative to synthetic antimicrobial agents. With the growing demand for organic and sustainable personal care products, consumers are increasingly seeking natural alternatives to conventional chemical-based products. Fungal-derived compounds align well with this trend as they are derived from natural sources and can provide effective antimicrobial properties without the need for synthetic additives. This not only caters to the preferences of eco-conscious consumers but also promotes the use of sustainable ingredients in personal care formulations.

In addition to their enhanced antimicrobial properties and natural origin, fungal-derived compounds may offer other benefits when incorporated into washing soaps. For instance, certain compounds derived from fungi have been found to possess antioxidant properties, which can help protect the skin against oxidative stress and free radical damage. Moreover, some fungal-derived compounds exhibit anti-inflammatory properties, which can contribute to soothing and calming the skin.

The efficacy of fungal-derived compounds in washing soaps can be further enhanced through careful formulation and optimization. Researchers and manufacturers can explore different extraction techniques and purification processes to ensure the preservation and potency of these compounds in the final product. This allows for the development of washing soaps that offer maximum efficacy in terms of antimicrobial action while providing the desired sensory attributes and skin compatibility.

Overall, the utilization of fungal-derived compounds in washing soaps provides numerous advantages. Their enhanced antimicrobial properties make them effective in combating fungal infections, reducing the risk of skin diseases. Additionally, the use of natural alternatives aligns with the demand for organic and sustainable personal care products. By incorporating fungal-derived compounds into washing soap formulations, manufacturers can meet the needs of eco-conscious consumers while offering effective and safe hygiene solutions.

14.3.4 Consumer Perception and Market Trends

Consumer perception of fungal-derived compound-based washing soaps is an important aspect to consider. As awareness of the benefits of natural and eco-friendly products continues to grow, consumers are increasingly interested in using personal care items that incorporate fungal-derived compounds. The perception of these compounds as natural, plant-based ingredients can contribute to a positive consumer response.

Analyzing current market trends reveals that the demand for fungal-derived compound-based washing soaps is on the rise. Consumers are seeking products that offer effective cleansing properties while being gentle on the skin and environmentally friendly. This trend has encouraged manufacturers to explore the potential of fungal-derived compounds and develop innovative washing soap formulations that cater to this growing market segment.

In conclusion, the utilization of fungal-derived bioactive compounds in washing soaps provides a promising avenue for improving personal hygiene and addressing the challenges posed by fungal infections. The fungistatic and fungicidal properties of these compounds, coupled with their natural origin, contribute to their efficacy and appeal to consumers. As market trends continue to evolve, it is crucial for researchers, manufacturers, and consumers to recognize the potential benefits and opportunities associated with fungal-derived compound-based washing soaps.

14.4 Fungal-Derived Bioactive Compounds in Saponification Products

14.4.1 Introduction to Saponification Products

Saponification products, such as soaps and cleansers, play a vital role in personal care and hygiene routines. These products are formulated to remove dirt, oil, and impurities from the skin, leaving it clean and refreshed. The process of saponification involves the hydrolysis of fats or oils in the presence of an alkali, resulting in the formation of soap and glycerol. Soap, the primary component of saponification products, possesses excellent cleansing properties and is widely used for various personal care applications.

14.4.2 Role of Fungal-Derived Bioactive Compounds in Saponification

14.4.2.1 Antimicrobial Effects of Fungal-Derived Compounds

Fungal-derived bioactive compounds have gained significant attention due to their antimicrobial properties. Many fungal species produce secondary metabolites with inherent antimicrobial activity, making them valuable for incorporation into saponification products. These compounds can combat a wide range of microorganisms, including bacteria, fungi, and even some viruses, making them effective in maintaining cleanliness and hygiene (Singh et al. 2019).

14.4.2.2 Exploring the Antimicrobial Effects of Fungal-Derived Compounds in Saponification Products and Their Contribution to Product Quality

The inclusion of fungal-derived bioactive compounds in saponification products enhances their antimicrobial efficacy. These compounds can inhibit the growth of harmful microorganisms, reducing the risk of infections and promoting a healthier skin environment. By incorporating these bioactive compounds, saponification products can provide an added layer of protection and hygiene beyond their basic cleansing function.

14.4.2.3 Examples of Fungal-Derived Bioactive Compounds in Saponification Products

There is a wide range of fungal-derived bioactive compounds that are commonly utilized in commercial saponification products mentioned in Table 14.2. For instance, compounds such as triterpenoids, flavonoids, phenolic acids, and alkaloids derived from fungal sources have demonstrated antimicrobial properties. Some notable examples include chitosan derived from fungal cell walls, antifungal peptides derived from filamentous fungi, and fungal-derived essential oils.

14.4.3 Impact of Fungal-Derived Compounds on the Quality and Properties of Saponification Products

Fungal-derived compounds not only contribute to the antimicrobial efficacy of saponification products but also impact their overall quality and properties. These compounds can influence the sensory attributes, stability, and shelf life of the products. For instance, some fungal-derived bioactive compounds may impart specific scents, colors, or textures to the final product, enhancing its aesthetic appeal. Furthermore, these compounds can also contribute to the physical stability of the product, preventing rancidity and improving its shelf life.

14.4.4 Future Directions and Potential Innovations

The use of fungal-derived bioactive compounds in saponification products is a promising area for future research and innovation. Continued exploration of fungal species and their metabolites may uncover novel compounds with enhanced antimicrobial properties or other beneficial characteristics. Furthermore, advancements in extraction techniques, formulation strategies, and delivery systems may facilitate the effective incorporation of fungal-derived bioactive compounds into saponification products. Future innovations could focus on optimizing the concentrations and combinations of these compounds to maximize their antimicrobial efficacy while maintaining the desired product attributes.

To sum up, fungal-derived bioactive compounds have demonstrated significant potential in enhancing the antimicrobial efficacy and overall quality of saponification products. Their

Table 14.2 Bioactive compounds from different fungal sources and their saponification product

Fungal species	Bioactive compound	Saponification product	References
Aspergillus niger	Lovastatin	Soaps, detergents	Javed et al. (2014)
Penicillium chrysogenum	Penicillin	Antibacterial soaps	Fierro et al. (2022)
Trichoderma spp	Chitinase	Exfoliating scrubs	Woo et al. (2023)
Cordyceps sinensis	Cordycepin	Anti-aging creams	Tuli et al. (2014)
Ganoderma lucidum	Triterpenoids	Moisturizing lotions	Dong et al. (2019)
Phoma spp	Phomopsins	Shampoo, hair products	Ding et al. (2016)
Fusarium oxysporum	Enniatins	Facial cleansers	Gautier et al. (2020)
Aspergillus terreus	Terretonin	Antifungal soaps	Vahedi and Lass-Flörl (2019)
Penicillium roqueforti	Roquefortine C	Body wash	Coton et al. (2020)
Candida albicans	Mannan-binding lectin	Hand sanitizers	Hammad et al. (2018)

inclusion in such products opens up new avenues for maintaining personal hygiene, promoting skin health, and preventing the spread of infectious diseases. Continued research and innovation in this field will undoubtedly lead to the development of improved saponification products with enhanced properties and a positive impact on personal care practices.

14.5 Safety, Regulation, and Quality Control

14.5.1 Safety Assessment of Fungal-Derived Bioactive Compounds in Cosmetics

Fungal-derived bioactive compounds have gained significant attention in the cosmetic industry due to their potential beneficial properties. However, before these compounds are incorporated into cosmetic products, it is essential to conduct a thorough safety assessment to ensure consumer safety. This section will discuss the safety evaluation procedures and considerations for fungal-derived bioactive compounds used in cosmetic products.

14.5.1.1 Preclinical Safety Evaluation

Preclinical safety evaluation involves conducting laboratory studies to assess the potential toxicity and adverse effects of fungal-derived bioactive compounds. The following considerations are crucial during this evaluation process:

(a) Cytotoxicity Studies: These studies assess the effects of the bioactive compounds on cell viability and proliferation. Various cell lines representative of human skin, such as keratinocytes and fibroblasts, are commonly used for these studies. The compounds' cytotoxic effects are evaluated through assays like the MTT (3-(4,5-dimethylthiazol-2-yl)-2,5-diphenyltetrazolium bromide) assay.

(b) Skin Irritation and Sensitization Studies: These studies determine the potential of the bioactive compounds to cause skin irritation or sensitization. Tests such as the Draize patch test and the local lymph node assay (LLNA) are conducted to evaluate these effects. Additionally, reconstructed human epidermis models can be used as an alternative to animal testing.

(c) Genotoxicity and Mutagenicity Studies: These studies assess the potential of the compounds to induce genetic damage or mutations. Tests like the Ames test and the chromosomal aberration assay are commonly employed to evaluate genotoxicity.

(d) Phototoxicity Studies: Fungal-derived compounds with photosensitizing potential should be evaluated for their ability to induce phototoxic reactions when exposed to ultraviolet (UV) light. Phototoxicity tests using human skin models or animal models can be conducted to assess this property.

14.5.1.2 Human Safety Evaluation

Following preclinical safety assessment, human safety evaluation studies are conducted to further assess the safety of fungal-derived bioactive compounds. These studies involve human volunteers and focus on the compounds' potential for skin irritation, sensitization, and allergic reactions. Key considerations include:

(a) Patch Testing: Patch testing involves applying the cosmetic product containing the fungal-derived bioactive compound to the skin under occlusion for a specified period. The skin is then observed for any signs of irritation or allergic reactions. This procedure helps identify potential adverse effects in human subjects.
(b) Repeat Insult Patch Testing (RIPT): RIPT is performed to assess the potential of the compound to cause cumulative irritation or sensitization over repeated applications. The compound is applied under occlusion multiple times, and the skin is monitored for any adverse reactions.
(c) Human Skin Compatibility Studies: These studies evaluate the compatibility of the bioactive compounds with human skin. Parameters such as transepidermal water loss (TEWL), skin hydration, and skin barrier function are measured to ensure that the compounds do not disrupt the natural protective barrier of the skin.
(d) Phototoxicity and Photosafety Studies: If the fungal-derived bioactive compounds have phototoxic potential, phototoxicity studies on human volunteers can be conducted to evaluate their effects under UV exposure.

14.5.1.3 Considerations for Specific Groups

Certain population groups, such as pregnant women, infants, and individuals with compromised immune systems, require special attention regarding the safety of cosmetic products. Additional studies may be necessary to ensure the safety of fungal-derived bioactive compounds in these specific populations.

14.5.2 Regulatory Considerations and Guidelines for Commercial Cosmetics

The commercialization and marketing of cosmetics containing fungal-derived compounds are subject to various regulatory frameworks and guidelines. These regulations aim to ensure the safety, efficacy, and quality of cosmetic products. The following are key regulatory considerations and guidelines relevant to the commercialization of cosmetics containing fungal-derived bioactive compounds:

14.5.2.1 Ingredient Safety Evaluation

Regulatory agencies, such as the U.S. Food and Drug Administration (FDA) and the European Union (EU) Cosmetic Ingredient Review (CIR) panel, evaluate the safety of cosmetic ingredients, including fungal-derived bioactive compounds. These agencies assess the available scientific data on the compounds' toxicology, genotoxicity, and other safety aspects before approving their use in cosmetic products.

14.5.2.2 Labeling Requirements

Cosmetic products must comply with labeling requirements to provide consumers with accurate information about the ingredients and potential risks. The International Nomenclature of Cosmetic Ingredients (INCI) system provides a standardized way to list ingredients on product labels, ensuring transparency and clarity.

14.5.2.3 Claims Substantiation

Regulatory authorities require scientific substantiation for any claims made regarding the efficacy or benefits of cosmetic products. Manufacturers of cosmetics containing fungal-derived bioactive compounds must provide evidence supporting their claims through clinical studies or other appropriate methods.

14.5.2.4 Good Manufacturing Practices (GMP)

Compliance with Good Manufacturing Practices is crucial for ensuring the quality, safety, and consistency of cosmetic products. GMP guidelines outline the processes, facilities, and quality control measures that manufacturers should follow to produce cosmetics in a controlled and reliable manner.

14.5.2.5 Notification and Registration

Depending on the jurisdiction, cosmetic products may need to be notified or registered with regulatory authorities before they can be marketed. This process ensures that authorities are aware of the products being sold and can monitor their safety and compliance with regulations.

14.5.3 Quality Control Measures for Fungal-Derived Compound-Based Products

To ensure the consistency, efficacy, and safety of fungal-derived compound-based cosmetic products, rigorous quality control measures must be implemented throughout the manufacturing process. The following outlines key quality control measures and best practices:

14.5.3.1 Raw Material Evaluation

Thorough testing and evaluation of raw materials, including fungal-derived bioactive compounds, are essential. Suppliers should provide detailed specifications and certificates of analysis to ensure the quality and authenticity of the materials. Analytical techniques such as high-performance liquid chromatography (HPLC) can be employed to verify the presence and concentration of bioactive compounds.

14.5.3.2 Stability Testing

Stability testing assesses the physical, chemical, and microbiological stability of cosmetic products over time. This testing helps determine the shelf life and storage conditions required to maintain the quality and efficacy of the products. Accelerated stability studies, real-time stability studies, and compatibility testing with packaging materials should be conducted.

14.5.3.3 Manufacturing Process Controls

Stringent controls should be implemented during the manufacturing process to ensure product consistency and quality. Standard operating procedures (SOPs) should be developed and followed, covering aspects such as ingredient weighing, mixing, temperature control, and sanitation. Adequate personnel training and hygiene practices are also essential.

14.5.3.4 Microbiological Testing

Microbiological testing is crucial to ensure the safety and microbial quality of cosmetic products. Testing should be performed on raw materials, in-process samples, and finished products to detect the presence of harmful microorganisms. Preservative efficacy testing should also be conducted to ensure the effectiveness of preservative systems.

14.5.3.5 Packaging Integrity

The integrity of the packaging is critical for maintaining the quality and safety of fungal-derived compound-based cosmetic products. Quality control measures should be implemented to ensure that the packaging materials are free from contaminants and provide an adequate barrier against moisture, light, and air.

14.5.4 Case Studies and Success Stories

The successful commercialization of cosmetics incorporating fungal-derived bioactive compounds serves as an inspiration for further research and development in this field. The following case studies and success stories highlight the utilization of fungal-derived bioactive compounds in commercially available cosmetics:

14.5.4.1 Case Study: Antifungal Hand Sanitizer

A company developed a hand sanitizer enriched with a fungal-derived bioactive compound known for its antifungal properties. The product underwent rigorous safety evaluation, including preclinical and human safety studies. Regulatory compliance and quality control measures were implemented, resulting in a safe and effective hand sanitizer with added antifungal benefits.

14.5.4.2 Success Story: Anti-aging Cream

A skincare brand introduced an anti-aging cream formulated with a fungal-derived bioactive compound known for its antioxidant and collagen-boosting properties. The product gained popularity due to its proven efficacy in reducing the appearance of wrinkles and improving skin elasticity. Strict quality control measures were implemented to ensure consistent product quality and safety.

14.5.4.3 Case Study: Natural Soap with Antimicrobial Properties

A soap manufacturer incorporated a fungal-derived bioactive compound with natural antimicrobial properties into their soap formulation. The compound demonstrated effectiveness against various pathogens, making the soap an attractive option for consumers seeking natural and effective antimicrobial skincare products. Regulatory compliance and quality control ensured the product's safety and efficacy.

14.5.4.4 Success Story: Hair Growth Serum

A haircare brand developed a hair growth serum containing a fungal-derived bioactive compound known for its ability to promote hair follicle growth. The product underwent safety evaluation and efficacy testing, demonstrating positive results in improving hair thickness and density. Quality control measures ensured consistent product performance and consumer satisfaction.

These case studies and success stories demonstrate the potential of fungal-derived bioactive compounds in creating innovative and effective cosmetic products. They emphasize the importance of safety evaluation, regulatory compliance, and quality control to bring such products to the market successfully.

14.6 Conclusion

14.6.1 Recap of the Chapter's Main Points

Throughout this chapter, we have explored the utilization of fungal-derived bioactive compounds in commercially available cosmetics, focusing specifically on hand sanitizers, washing soaps, and saponification products. The key points discussed can be summarized as follows:

Fungal-derived bioactive compounds: Fungi are a rich source of bioactive compounds with diverse properties, including antimicrobial, antioxidant, and anti-inflammatory activities. These compounds have the potential to enhance the effectiveness of cosmetic products and offer new opportunities for the development of innovative formulations.

Hand sanitizers: Fungal-derived bioactive compounds, such as chitosan and various fungal extracts, have demonstrated promising antimicrobial properties against a wide range of pathogens, including bacteria and viruses. Incorporating these compounds into hand sanitizers can provide improved protection against infectious agents while maintaining skin health.

Washing soaps: Fungal-derived enzymes, such as proteases and lipases, offer advantages in the formulation of washing soaps. These enzymes can effectively break down protein and lipid-based stains, resulting in enhanced cleaning performance. Additionally, fungal extracts containing bioactive compounds can provide additional benefits, such as moisturization and soothing effects.

Saponification products: Fungal-derived bioactive compounds have the potential to enhance the properties of saponification products, such as shampoos and shower gels. These compounds can contribute to improved cleansing efficacy, scalp health, and hair conditioning, making them valuable ingredients in cosmetic formulations.

14.6.2 Future Prospects and Challenges in the Field

Looking ahead, the utilization of fungal-derived bioactive compounds in commercial cosmetics presents promising future prospects but also some challenges that need to be addressed. Some of the key areas for future exploration and potential challenges include:

Identification of novel fungal sources: Further research is needed to identify and explore novel fungal species that may possess unique bioactive compounds with beneficial properties for cosmetics. This requires extensive screening and characterization efforts, considering both known and unexplored fungal biodiversity.

Sustainable sourcing and production: As the demand for fungal-derived bioactive compounds increases, sustainable sourcing and production methods need to be developed and implemented. This includes strategies for efficient cultivation, extraction, and processing of fungal biomass while ensuring minimal environmental impact.

Safety and regulatory considerations: The safety and regulatory aspects surrounding the use of fungal-derived bioactive compounds in cosmetics require careful evaluation. Extensive toxicological studies, stability testing, and regulatory compliance assessments are necessary to ensure consumer safety and regulatory compliance.

14.6.3 Closing Thoughts on the Potential of Fungal-Derived Bioactive Compounds in Commercial Cosmetics

In conclusion, the utilization of fungal-derived bioactive compounds in commercial cosmetics holds immense potential to revolutionize the cosmetic industry and improve personal hygiene products. The diverse range of bioactive compounds produced by fungi offers exciting opportunities for developing innovative formulations that provide enhanced efficacy, safety, and consumer satisfaction.

By harnessing the antimicrobial, antioxidant, and other beneficial properties of fungal-derived compounds, we can create cosmetics that not only enhance personal hygiene but also promote overall skin and hair health. Furthermore, the sustainable production and utilization of these compounds can contribute to a more environmentally friendly and socially responsible cosmetic industry.

As we continue to deepen our understanding of fungal biodiversity and unlock the potential of these bioactive compounds, the future of fungal-derived cosmetics looks bright. By combining scientific knowledge, technological advancements, and a commitment to sustainability, we can usher in a new era of cosmetics that truly harness the power of nature and improve the well-being of consumers worldwide.

References

Alajlan AA, Mukhtar LE, Almussallam AS, Alnuqaydan AM, Albakiri NS, Almutari TF et al (2022) Assessment of disinfectant efficacy in reducing microbial growth. PLoS One 17(6):e0269850. https://doi.org/10.1371/journal.pone.0269850

Aly AH, Debbab A, Proksch P (2011) Fifty years of drug discovery from fungi. Fungal Divers 50:3–19. https://doi.org/10.1007/s13225-011-0116-y

Aris P, Wei Y, Mohamadzadeh M, Xia X (2022) Griseofulvin: an updated overview of old and current knowledge. Molecules 27(20):7034. https://doi.org/10.3390/molecules27207034. PMID: 36296627; PMCID: PMC9610072

Atanasov AG, Zotchev SB, Dirsch VM et al (2021) Natural products in drug discovery: advances and opportunities. Nat Rev Drug Discov 20:200–216. https://doi.org/10.1038/s41573-020-00114-z

Centers for Disease Control and Prevention (2021, August 10) Hand sanitizer use out and about. Centers for Disease Control and Prevention. https://www.cdc.gov/handwashing/hand-sanitizer-use.html

Chojnacki M, Dobrotka C, Osborn R, Johnson W, Young M, Meyer B, Laskey E, Wozniak RAF, Dewhurst S, Dunman PM (2021) Evaluating the antimicrobial properties of commercial hand sanitizers. mSphere 6(2):e00062–e00021. https://doi.org/10.1128/mSphere.00062-21. PMID: 33658274; PMCID: PMC8546682

Coton E, Coton M, Hymery N, Mounier J, Jany JL (2020) *Penicillium roqueforti*: an overview of its genetics,

physiology, metabolism and biotechnological applications. Fungal Biol Rev 34(2):59–73. https://doi.org/10.1016/j.fbr.2020.03.001
Ding W, Liu WQ, Jia Y, Li Y, van der Donk WA, Zhang Q (2016) Biosynthetic investigation of phomopsins reveals a widespread pathway for ribosomal natural products in Ascomycetes. Proc Natl Acad Sci 113(13):3521–3526. https://doi.org/10.1073/pnas.1522907113
Dong Q, He D, Ni X et al (2019) Comparative study on phenolic compounds, triterpenoids, and antioxidant activity of *Ganoderma lucidum* affected by different drying methods. Food Measure 13:3198–3205. https://doi.org/10.1007/s11694-019-00242-0
Ellis D (2002) Amphotericin B: spectrum and resistance. J Antimicrob Chemother 49(Suppl 1):7–10. https://doi.org/10.1093/jac/49.suppl_1.7. PMID: 11801575
Fendler E, Groziak P (2002) Efficacy of alcohol-based hand sanitizers against fungi and viruses. Infect Control Hosp Epidemiol 23(2):61–62. https://doi.org/10.1086/503455
Fierro F, Vaca I, Castillo NI, García-Rico RO, Chávez R (2022) *Penicillium chrysogenum*, a vintage model with a cutting-edge profile in biotechnology. Microorganisms 10(3):573. https://doi.org/10.3390/microorganisms10030573. PMID: 35336148; PMCID: PMC8954384
Gautier C, Pinson-Gadais L, Richard-Forget F (2020) *Fusarium* mycotoxins enniatins: an updated review of their occurrence, the producing *Fusarium* species, and the abiotic determinants of their accumulation in crop harvests. J Agric Food Chem 68(17):4788–4798. https://doi.org/10.1021/acs.jafc.0c00411. Epub 2020 Apr 17. PMID: 32243758
Gehlot P, Singh J (eds) (2018) Fungi and their role in sustainable development: current perspectives. Springer
Gupta A, Meshram V, Gupta M, Goyal S, Qureshi KA, Jaremko M, Shukla KK (2023) Fungal endophytes: microfactories of novel bioactive compounds with therapeutic interventions; a comprehensive review on the biotechnological developments in the field of fungal endophytic biology over the last decade. Biomolecules 13(7):1038. https://doi.org/10.3390/biom13071038
Hammad NM, El Badawy NE, Ghramh HA, Al Kady LM (2018) Mannose-binding lectin: a potential therapeutic candidate against *Candida* infection. Biomed Res Int 2018:2813737. https://doi.org/10.1155/2018/2813737. PMID: 29854737; PMCID: PMC5954966
Hans M, Lugani Y, Chandel AK et al (2023) Production of first- and second-generation ethanol for use in alcohol-based hand sanitizers and disinfectants in India. Biomass Convers Biorefinery 13:7423–7440. https://doi.org/10.1007/s13399-021-01553-3
Javed S, Bukhari SA, Zovia I, Meraj M (2014) Screening of indigenously isolated fungi for lovastatin production and its in vivo evaluation. Curr Pharm Biotechnol 15(4):422–427. https://doi.org/10.2174/1389201015666140528152138. PMID: 24894549
Keshri PK, Rai N, Verma A et al (2021) Biological potential of bioactive metabolites derived from fungal endophytes associated with medicinal plants. Mycol Prog 20:577–594. https://doi.org/10.1007/s11557-021-01695-8
Morrison VA (2005) Caspofungin: an overview. Expert Rev Anti-Infect Ther 3(5):697–705. https://doi.org/10.1586/14787210.3.5.697. PMID: 16207161
Rai N, Kumari Keshri P, Verma A, Kamble SC, Mishra P, Barik S, Kumar Singh S, Gautam V (2021) Plant associated fungal endophytes as a source of natural bioactive compounds. Mycology 12(3):139–159. https://doi.org/10.1080/21501203.2020.1870579
Rai N, Gupta P, Keshri PK et al (2022) Fungal endophytes: an accessible source of bioactive compounds with potential anticancer activity. Appl Biochem Biotechnol 194:3296–3319. https://doi.org/10.1007/s12010-022-03872-1
Singh AK, Rana HK, Pandey AK (2019) Fungal-derived natural product: synthesis, function, and applications. In: Yadav A, Singh S, Mishra S, Gupta A (eds) Recent advancement in white biotechnology through fungi. Fungal biology. Springer, Cham. https://doi.org/10.1007/978-3-030-14846-1_8
Singh P, Potlia I, Malhotra S, Dubey H, Chauhan H (2020) Hand sanitizer an alternative to hand washing—a review of literature. J Adv Oral Res 11(2):137–142. https://doi.org/10.1177/2320206820939403
The Importance of Personal Hygiene in Our Life (2021, April 13) The scientific world—let's have a moment of science. https://www.scientificworldinfo.com/2021/04/importance-of-personal-hygiene.html
Tuli HS, Sandhu SS, Sharma AK (2014) Pharmacological and therapeutic potential of Cordyceps with special reference to Cordycepin. 3 Biotech 4(1):1–12. https://doi.org/10.1007/s13205-013-0121-9. Epub 2013 Feb 19. PMID: 28324458; PMCID: PMC3909570
Vahedi Shahandashti R, Lass-Flörl C (2019) Antifungal resistance in *Aspergillus terreus*: a current scenario. Fungal Genet Biol 131:103247. https://doi.org/10.1016/j.fgb.2019.103247. Epub 2019 Jun 25. PMID: 31247322
Woo SL, Hermosa R, Lorito M et al (2023) *Trichoderma*: a multipurpose, plant-beneficial microorganism for eco-sustainable agriculture. Nat Rev Microbiol 21: 312–326. https://doi.org/10.1038/s41579-022-00819-5

Fungal Carotenoids and Their Cosmeceutical Effects on Human Bodies 15

Haidy Nasser, Asmaa M. El-Makawy, Esraa E. Ammar, Shrouk Elshafey, Sohaila Emad, and Heba Bedair

Abstract

Carotenoids are lipid-soluble natural amalgams, including fungi, assured bacteria, algae, and plants. In fungi, carotenoids are widely distributed across various taxonomic classifications. These pigments have garnered interest due to their antioxidant assets and their role in helping fungi survive in their natural environment. Biotechnological methods offer a more efficient way to produce carotenoids compared to chemical synthesis or plant extraction. This chapter focuses on industrially significant carotenoids in advanced fungal and yeast strains, discussing genetic modification for heightened production and proposing greener extraction methods, the different pathways of β-carotene and carotenoids biosynthesis, and potential applications of carotenoids.

Keywords

β-carotene · Cosmetics · Carotenoid applications · Fungal carotenoids

15.1 Introduction

Carotenoids are characterized as lipophilic terpenoid chemicals that naturally exist in photosynthetic organisms and plants. They are accountable for the prevalent orange, yellow, and red hues noticed in nature (Marizcurrena et al. 2019). They are present in all fungal taxonomic categories. Yeasts are the favored choice for carotenoid synthesis, as compared to plants or other bacteria (Paul et al. 2023). The research community has become intrigued by red oleaginous yeast because of its facility to combine multiple beneficial and environmentally friendly characteristics for the creation of valuable, non-harmful, and medically important products (Singh et al. 2018; Sinha et al. 2022; Villegas-Méndez et al. 2022; Yoo et al. 2016). Fungi do not require carotenoids, unlike plants and algae, resulting in significantly lower levels of carotenoid accumulation (Sandmann et al. 2002; Valadon 1976). Fungal carotenoids exhibit

H. Nasser
Botany and Microbiology Department, Faculty of Science, Kafrelsheikh University, Kafrelsheikh, Egypt

A. M. El-Makawy
Botany and Microbiology Department, Faculty of Science, Alexandria University, Alexandria, Egypt

E. E. Ammar
Plant Ecology, Botany Department, Faculty of Science, Tanta University, Tanta, Egypt

School of Food and Biological Engineering, Jiangsu University, Zhenjiang, People's Republic of China

S. Elshafey · S. Emad
Biotechnology branch, Botany Department, Faculty of Science, Tanta University, Tanta, Egypt

H. Bedair (✉)
Plant Ecology, Botany Department, Faculty of Science, Tanta University, Tanta, Egypt
e-mail: heba.taha@science.tanta.edu.eg

S. Ghosh, H. N. Onyeaka (eds.), *Fungal-Based Cosmetics: Formulation and Usage*, The Mycota 17,
https://doi.org/10.1007/978-3-032-09291-5_15

distinct chemical structures compared to those found in other organisms (Sandmann 2022).

Carotenoids synthesized by fungi are integral to numerous biotic functions, such as shielding against oxidative stress and acting as edifice blocks meant for the creation of other bioactive substances. The mevalonic acid pathway elucidates the production of isopentenyl pyrophosphate (IPP), a fundamental precursor of carotenoids containing five carbon atoms. This metabolic pathway entails a series of reactions facilitated by specific enzymes, including reductases, kinases, and decarboxylases. Geranyl-geranyl pyrophosphate (GGPP), a compound comprising 20 carbon atoms per capita molecule, is synthesized through the addition reactions of three isopentenyl pyrophosphates (IPPs). Phytoene (C40) is formed when two GGPP particles undergo condensation, a process facilitated by the enzyme phytoene synthase. It is a crucial prerequisite for the production of lycopene. Lycopene can be converted into β-carotene, lutein, γ-carotene, torulene, astaxanthin, and zeaxanthin, depending on the specific bacteria used (Mussagy et al. 2019).

Carotenoids have various commercial uses such as being used as food coloring, additives in animal feed, and as a therapy for obesity. As stated by Narsing Rao et al. (2017), these industries, including pharmaceutical, cosmetic, and nutraceutical, currently have extensive utilization of them. Carotenoids have the main advantage of efficiently neutralizing reactive oxygen species (ROS) besides providing protective health benefits when obsessive as per food otherwise supplements. Specific antioxidants, such as β-carotene and, to a slighter extent, β-cryptoxanthin and α-carotene, have the ability to convert into vitamin A, hence their designation as "provitamin A." Carotenoids obtained from microorganisms are suitable for inclusion in sunscreens and other anti-UV products because of their capacity to form stable emulsions, their solubility, their antioxidant activities, and other individualities that make them advantageous for use in cosmetics (Martins et al. 2013). In addition, β-carotene, lycopene, and astaxanthin are carotenoids that have been researched and used as photo-protection agents, due to their antioxidant properties. These properties help combat the injurious possessions of UV radiation, such as the fabrication of free radicals and a resulting erythema over skin (Mendes-Silva et al. 2020).

This chapter provides a swift of the role of carotenoids, which are used in many industries such as medicines, cosmetics, and nutraceuticals. Carotenoids have diverse commercial applications, including food coloring, animal feed additives, and obesity medication. When eaten in food or supplements, they efficiently counteract reactive oxygen species (ROS) and offer health advantages as antioxidants. Specific carotenoids, like β-carotene, devour the facility to transform into vitamin A, which is referred to as "provitamin A." Carotenoids generated from microorganisms are beneficial for sunscreens and anti-UV products because they possess stability, solubility, and antioxidant characteristics. Carotenoids, including lycopene, β-carotene, and astaxanthin, function as photoprotective agents by shielding against UV radiation, neutralizing free radicals, and preventing skin redness (erythema).

15.2 What Are Carotenoids?

Carotenoids mostly consist of a straightforward C40 hydrocarbon backbone with 3–15 conjugated double bonds (Armstrong and Hearst 1996), which greatly contribute to their physiochemical features. Carotenoids, like many others, exhibit a yellow hue consequently the animation of conjugated double bonds. A requisite number of 7 conjugated double bonds is compulsory to generate a vibrant carotenoid pigment. The references cited are Lin and Xu (2020) and Mapelli-Brahm et al. (2020). The citation for this information is from Rapoport et al. (2021).

Carotenoids can be categorized into 2 groups: 1-Oxygen-free carotenes include neurosporene, β-carotene, α-carotene, γ-carotene, δ-carotene, ζ-carotene, phytoene, lycopene, phytofluene, α-zeacarotene, β-zeacarotene, and torulene.

There are several xanthophylls that contain oxygen, including astaxanthin, canthaxanthin, bixin, auroxanthin, capsorubin, capsanthin, crocetin, cryptoxanthin, taraxanthin, lycoxanthin, luteoxanthin, lutein, neoxanthin, rubixanthin, violaxanthin, lycophyll, zeaxanthin, tunaxanthin, zeinoxanthin, and torularhodin (Aziz et al. 2020; Hegde et al. 2022).

15.3 What Type of Organisms Produce Carotenoids?

Except for animals, they are produced in organisms from all taxonomic groups. Fungal carotenoids display distinct structural variety compared to carotenoids found in other sources. Fungi also acquire these elements in very less quantities compared to plants and algae, as they do not rely on them for their growth.

Only a limited number of species have had their specific carotenoid concentrations reckoned. Their measures varied from 100 to 200 g/g dw, with some being significantly lower (Almeida and Cerdá-Olmedo 2008; Davoli et al. 2004; Sandmann et al. 2008; Schrantz and Lemoine 1995). Carotenoids are synthesized by all fungi, including non-carotenogenic species, as a result of inheriting the ability to make carotenoids from Archaea (Valadon 1976).

Monocyclic γ-carotene is the specific carotenoid that builds up in lower fungi, such as Blastocladiomycota and Chytridiomycota. The achievement was accomplished by the use of an altered lycopene cyclase enzyme, which facilitated the production of a single ionone ring as opposed to the usual two rings found in the manufacture of β-carotene. The Mucoraceae family is recognized for possessing bicyclic β-carotene. In addition, it was shown that a small number of Ascomycota and Basidiomycota fungi also produce this compound as an ultimate merchandise. However, the bulk of carotenoids in phyla are spinoffs that have been oxygenated (Sandmann 2022) (Table 15.1). Light regulates carotenoid synthesis in numerous species of fungi (Rau 1985). The study conducted by Armstrong and Hearst in 1996 is widely believed to be the main factor answerable for carotenogenesis. According to the research conducted by (Dzurendova et al. in 2022), it was found that light stimulates the formation of carotenoids, highlighting their significance as antioxidative agents, as indicated by the study conducted by (Schimek and Wöstemeyer in 2009).

15.4 Evolution of Carotenoid Biosynthesis in Fungi

Fungi have evolved the mevalonate route from acetyl-CoA to aid in carotenogenesis as a result of their evolutionary interaction with Archaea, and additional β-carotene plus its precursors oxygenation that progressed specially in fungi (Sandmann 2022). Although it shares the C-40 pathway of Halobacteria to produce carotenoids, it doesn't use deoxyxylulose 5-phosphate pathway that normally pervades in bacteria. crtYB fusion gene is essential for carotenogenesis in fungi (Verdoes et al. 1999b), as it developed from the co-transcribed overlap seen between the crtY and crtB genes of Archaea, which are prescribed in a gene cluster, such as in *Sulfolobus solfaticus* (Hemmi et al. 2003).

15.4.1 Fungal Carotenoid Biosynthesis Pathways

Carotenoid biosynthesis can be likened to an upturned tree, where the bole symbolizes the timely reactions shared by wholly organisms. The branches symbolize the wide range of xanthophylls produced by incorporating oxygen functions into both acyclic and cyclic carotenes, which is unique to each species. These oxygen functionalities, including methoxy, hydroxy, epoxy, oxo, carboxy, and aldehydic groups, contribute significantly to the structural variability observed in carotenoids. The exploration of these biosynthetic pathways has been comprehensive. Molecular biology techniques alongside traditional enzymology have been instrumental in elucidating them. Nonetheless, the mechanisms

Table 15.1 Summarizes the distribution of carotenoids in several fungal species (Sandmann 2022)

Species	Carotenoid type
Ascomycota	
Fusarium species	Neurosporaxanthin
Neurospora crassa	β-Carotene
Protomyces inndates	Torularhodin
Cookeina sulcipes	
Basidiomycota	
Tremella mesenteric	β-Carotene
Cantharellus species	Canthaxanthin
Xanthophyllomyces dendrorhous	Astaxanthin
Rhodotorula glutinis	Torularhodin
Chytridiomycota	
Blastocladiomycota	γ-Carotene
Allomyces arbuscula	γ-Carotene
Mucoromycotina	
Blakeslea trispora	β-Carotene
Phycomyces blakesleanus	β-Carotene

behind certain cis-trans isomerizations within the process remain elusive, highlighting areas where further enquiry is compulsory to abundantly comprehend the regulation of carotenoid biosynthesis (Armstrong and Hearst 1996).

15.4.2 The Fundamental Carotenoid Biosynthesis Pathway

The amalgamation of two molecules of the C20 intermediate geranylgeranyl pyrophosphate (GGPP) through tail-to-tail condensation is facilitated by the GGPP synthase enzyme prearranged by the al-3 gene, which originates from the *N. crassa* fungus (Armstrong 1994), is the very first step in the production of carotenoids. As this condensation leads to forming intermediate prephytoene pyrophosphate (PPPP), which is unstable, then phytoene by phytoene synthase enzyme produced by al-2 gene extracted from *N. crassa* as well (Schmidhauser et al. 1994).

Phytoene is a very common precursor of many C40 acyclic hydrocarbon carotenoids (Armstrong and Hearst 1996). Phytoene undergoes a sequence of desaturation and cyclization steps to form cyclic carotenes, such as β-carotene.

Whether an organism lives via oxygenic photosynthesis determines the functional and structural properties of the enzymes that convert phytoene into acyclic and cyclic carotenes (Armstrong and Hearst 1996). For instance, the CrtI-type phytoene desaturases derived from the al-1 gene in *N. carassa* (Armstrong 1994) produce lycopene or neurosporene by three or four ATP-dependent desaturations.

Nevertheless, in most cyanobacteria, algae, and higher plants, CrtI-type enzymes have been substituted with CrtP-type phytoene desaturases, which exhibit distinct structural and functional characteristics. These desaturases facilitate two desaturations to transform phytoene into ζ-carotene. The conversion is assisted by the enzymes (Bartley et al. 1991; Pecker et al. 1992; Hugueney et al. 1992). Following that, ζ-carotene undergoes two more desaturations catalyzed by the ζ-carotene desaturase (CrtQ), concluding in the transformation into lycopene. This process was described by Armstrong and Hearst in 1996.

15.4.2.1 Biosynthetic Route of β-Carotene in Fungus

In fungi, the production of β-carotene occurs by a process where a single enzyme facilitates four desaturation stages following the combination, consisting of 2 molecules of geranylgeranyl pyrophosphate are required to synthesize phytoene. The enzymes are produced by two genes: a fusion gene called al-2, crtYB, or carRA, which

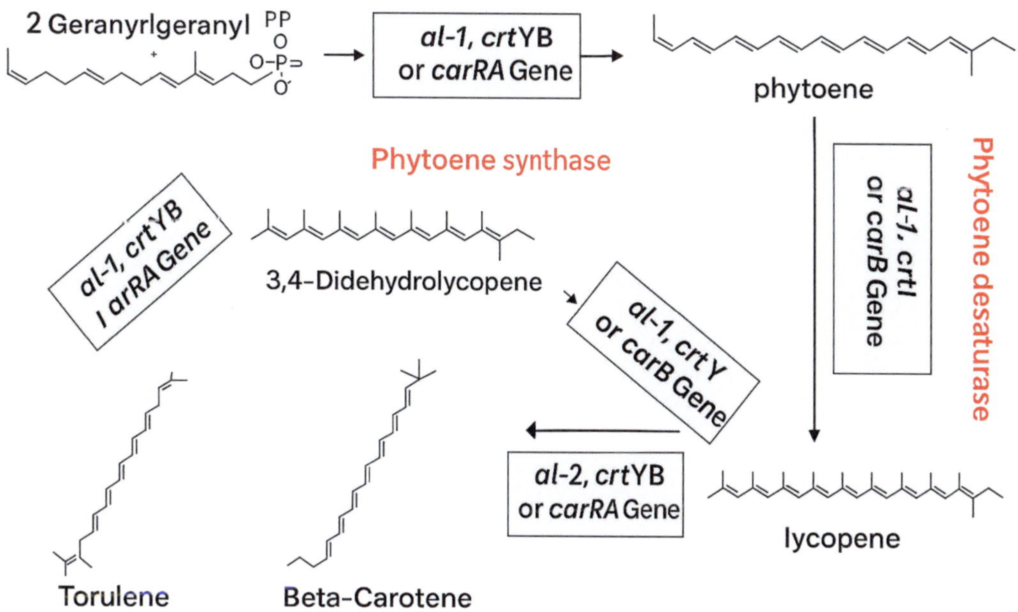

Fig. 15.1 β-carotene biosynthesis pathway in fungi (Sandmann 2022)

combines phytoene synthase and lycopene cyclase, and a separate gene known as al-1, crtI, or carB, which codes for phytoene desaturase (Avalos et al. 2017; Schmidhauser et al. 1994; Verdoes et al. 1999a, b). The functional investigation of the *X. dendrorhous* homolog crtYB demonstrated that the gene product performs two important functions: the generation of phytoene and the cyclization of lycopene. This suggests that the gene sequence contains both functional domains. The expression of both domains in a different organism and the production of antibodies against both proteins were carried out.

The level of specificity exhibited by desaturases in the production of carotenogenic species' products varies. An example of this is the CarB enzyme found in *P. blakesleanus*, which contains a 4-step desaturase that produces lycopene. This has been verified through heterologous expression and the observation of lycopene buildup in the C9 mutant (Ruiz-Hidalgo et al. 1997; Bramley and Davies 1975). This is congruent with the synthesis of β-carotene that is reported in *Mucorales* fungi. Conversely, the al-1 gene merchandise derived from *N. crassa* adds an additional double bond at the C3-C4 position, as reported by Hausmann and Sandmann in 2000. This addition hinders the formation of additional cyclization at the molecule's more unsaturated end, allowing a single cyclization footstep to produce torulene (Fig. 15.1).

15.5 Fungal Carotenoids' Potential Applications (Sunscreen, Whitening, Anti-aging, and Anti-acni Activity)

The demand for skincare products and cosmeceuticals is steadily rising year after year, fungal carotenoids multiplied attention due to their promising benefits and applications in cosmetics and cosmeceuticals. Carotenoids are nutritional organic pigments occurring naturally in various fungi. Researchers linked the carotenoids' health-promoting properties with their bioactivity as free radical searchers, UV protectors, plus antioxidants (Kot et al. 2018; Lin and Xu 2020; Rebelo et al. 2020). Recently, focusing on skincare at the cellular level takes

Table. 15.2 Variation in carotenoids produced by various fungus and their activities benefits/effect on human skin

Fungi	Carotenoid	Activity	References
Blakeslea trispora	β-carotene-lycopene	Antioxidant Provitamin A	Mantzouridou and Tsimidou (2008)
Sporobolomyces roseus (yeast)	β-carotene, torulene, torularhodin	Antioxidant Provitamin A	Davoli et al. (2004)
Rhodotorula glutinis (yeast)	β-carotene, γ-carotene, torulene, and torularhodin	Antioxidant Provitamin A	Elfeky et al. (2019) and Kot et al. (2018)
Phaffia rhodozyma (yeast)	Astaxanthin, β-carotene, canthaxanthin, zeaxanthin	Antioxidant/ photoprotectant/anti-inflammatory	Jinsong et al. (2011)
Sporobolomyces ruberrimus (yeast)	Torularhodine, torulene, β-carotene, and γ-carotene	Antioxidant Provitamin A	Cardoso et al. (2016)
Sporidiobolus pararoseus	Torulene	Antioxidant	Wei et al. (2020)
Saccharomyces cerevisiae (yeast)	Lycopene	Antioxidant	Martínez-Cámara et al. (2021)
Monascus sp.	Canthaxanthin	Antioxidant	
Cantharellus cinnabarinus	Canthaxanthin	Antioxidant	Haxo (1950)
Aspergillus carbonarius	Canthaxanthin	Antioxidant	Barreiro and Barredo (2018)
Cordyceps militaris	Cordycepene, lutein, zeaxanthin, cordyxanthin-I, II, III, and IV	Antioxidant/ photoprotectant/anti-inflammatory	Tang et al. (2019), Yan et al. (2010), and Chen et al. (2013)

Main sources for concentrated carotenoids (Cardoso et al. 2017)

place by combining the nutritional supplements in cream, ointments, lotions, etc. to treat skin damages as dark spots, acne, and wrinkles (Taofiq et al. 2016). Some researchers reported that photoprotection based on topical sunscreen is recommended than that based on dietary supplement (Meléndez-Martínez et al. 2019). Carotenoids remain commonly used in cosmetics exclusively in sun lotions (Hyde et al. 2010). A variety of filamentous fungi, mushroom, and yeast produce various carotenoids used in cosmetics (Sujarit et al. 2021). This chapter will report on some of them (Table 15.2) (Cardoso et al. 2017).

15.5.1 β-Carotene

β-Carotene relics an effective pigment with antiphotodamage activity used against visible and infrared radiation. It also showed effective antioxidant properties (Darvin et al. 2011; Freitas et al. 2019). As a result of its beneficial properties, it is recommended to use β-carotene in mixture of antioxidants due to its instability (Darvin et al. 2011). β-carotene is popularly used in sunscreens (Meléndez-Martínez et al. 2019).

15.5.2 Provitamin a Carotenoids

Provitamin A carotenoids (α-carotene; β-carotene; plus α-cryptoxanthin) be there important for skins by producing retinoic acid, which is used for wound healing, as anti-acne and anti-aging and other several skin disorders (Álvarez et al. 2014; Meléndez-Martínez et al. 2019).

Scientists associated lycopene, which is produced through a biosynthetic pathway of β-carotene in some species of fungus as *Blakeslea trispora* and *Rhodotorula* spp. of yeast, with various positive effects on human health as protecting skin against sunburns (Lin and Xu 2020). It is reported that carotenoids specially lycopene are used in cosmetics such as anti-aging facials, sunblocks, sun lotions, sunscreens, and face creams (Narsing Rao et al. 2017; Sajid and Akbar 2018).

15.5.3 Astaxanthin

Astaxanthin is widely used in cosmeceuticals for its potent antioxidant plus anti-aging properties (Anarjan et al. 2012). Jiang et al. (2017) could produce astaxanthin sunscreen or essence, using *Phaffia rhodozyma,* with elevated DPPH free radical rummaging activity showing good applicability and stability. Cheng et al. (2019) also suggested developing a promising mask using Astaxanthin extracted from *P. rhodozyma* distinguished by its antioxidant properties. A distinguished astaxanthin lipstick was developed through the addition of esterified astaxanthin to the lipstick formula, which indicated superb antioxidant, anti-wrinkling, anti-inflammatory, sunscreen activity, plus coloring function (Cheng et al. 2019). Another researcher could improve that astaxanthin has potential to reduce wrinkling of skin (Davinelli et al. 2018).

15.5.4 Canthaxanthin

Canthaxanthin is an effective anti-aging that rejuvenates and strengthens the skin (Haxo 1950). Also, it is widely used in cosmetics as a natural skin-tanning agent (Sujak 2009). Canthaxanthin exists in the edible mushroom *Cantharellus cinnabarinus* (Haxo 1950).

15.5.5 Lutein and Zeaxanthin

A comparative study demonstrated that the combination use of oral and topical formulations of lutein and zeaxanthin enhanced the elasticity of the skin, more cutaneous hydration (Palombo et al. 2007), and a potent antioxidation effect more than only topical or oral application (Anunciato and da Rocha Filho 2012; Palombo et al. 2007).

15.5.6 Torulene and Torularhodin

Studies demonstrated that torulene and torularhodin are potent antioxidants (Kot et al. 2018). Some researchers showed the ability of torulene and torularhodin to neutralize free radicals (Peterson et al. 1954).

15.5.7 Cordycepene

Cordycepene is another promising novel carotenoid with antioxidative activity and natural cosmetic color extracted from *Cordyceps militaris*, an edible mushroom (Tang et al. 2019).

15.5.8 Antioxidant Activity

Antioxidants are crucial for protecting skin from multiple environmental pro-oxidative factors such as air pollutants, smoke, and ultraviolet radiation (UV) (Kim 2014). Antioxidants such as carotenoids safeguard human skin against the resulting reactive oxygen species (ROS) and free radicals (Lin and Xu 2020) from damaging DNA, proteins, and membrane lipids causing skin damage (Kim 2014).

Researchers revealed that astaxanthin exhibits significant antioxidant action not only through direct free radical scavenging but also by activating the cellular antioxidant defense system via modulation of the Nrf2 pathway (Davinelli et al. 2018). Another study showed that astaxanthin prevents the progression of early burn-wound by alleviating oxidative stress resulting from ROS in a deep-burn model of a rat (Fang et al. 2017).

15.5.9 Anti-photodamage Activity

There are several forms of products that can protect skin against photodamage such as sunscreens, sun lotions, and sun blocks. Carotenoids are extensively used in photoprotection products specially sun lotions such as "Kiehl's Since 1851 Multi-Purpose Facial Formula" and "Methode Jeanne Piaubert-Anti-Ageing Facial Sun" (Hyde et al. 2010).

15.5.10 Skin-Whitening Activity

The increase in melanin production and hyperpigmentation of human skin occur due to the increase in the activity of tyrosinase (Abdelshafy et al. 2022). Inhibiting the activity of tyrosinase enzymes is a crucial component in skin whitening cosmeceuticals (Mago et al. 2023). Carotenoids can regulate melanin synthesis through blocking tyrosinase expression or by regulating melanosomes distribution (Chang 2009). As a result of these beneficial properties, carotenoids are widely used in skin whitening cosmeceuticals.

15.5.11 Anti-aging and Anti-wrinkling Activity

Materialization of reactive oxygen species (ROS) and oxidative stress stays a pivotal mechanism that causes skin aging and dermal damage (Davinelli et al. 2018). Skin aging resulting from oxidation involves DNA damage, the response of inflammation, reduction in antioxidants production, and production of matrix metalloproteinases (MMPs) which degrade elastin and collagen in the dermal layer of skin (Kammeyer and Luiten 2015).

15.5.12 Anti-acne Activity

Researchers on the usage of fungal carotenoids for the treatment of acne are relatively limited but there is evidence that astaxanthin and β-carotene can reduce the symptoms of acne through inhibiting the inflammation and production of sebum.

15.6 Fungal Carotenoids: Properties, Structure, and Types

Carotenoids are a class of secondary metabolites. Carotenoids are hydrocarbons consisting of 40 carbon atoms and 2 terminal rings (Bell et al. 2000). These individuals belong to the isoprenoids subcategory, also known as "terpenoids" in general. The tetraterpenoids (C40) are composed of 8 isoprene units that are arranged in a linear fashion and symmetrical molecule, with the order reversed in the middle. The hydrocarbons consist of eight isoprenoid units, each containing five carbons, which are connected by conjugated double bonds. This molecular structure results in the formation of distinct geometrical isomers, specifically cis/trans molecules. In general, the all-trans arrangement is considered to be stable, as supported by the studies steered by Domonkos et al. in 2013 and Rodriguez-Concepcion et al. in 2018. Carotenoids can be synthesized from the acyclic C40H56 structure through processes such as dehydrogenation, cyclization, hydrogenation, oxidation, or a combination of these methods (Pfander 1992). The structure of carotenoids consists of a long central chain of conjugated double bonds (Fig. 15.2).

Carotenoids are hydrophobic compounds that cannot dissolve in water and have a chromophore consisting of a lengthy sequence of conjugated double bonds (at peak absorption wavelength of 400–500 nm). This chromophore is responsible for the characteristic yellow to orange and reddish dyes of carotenoids (Barredo et al. 2017; Lichtenthaler and Buschmann 2001; Saini and Keum 2017).

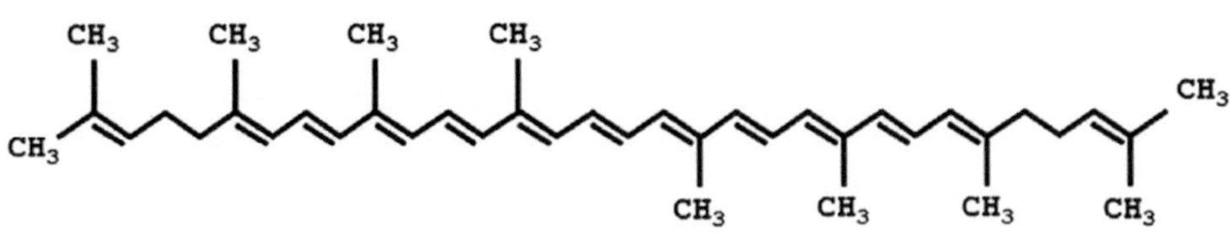

Fig. 15.2 The basic acyclic $C_{40}H_{56}$ carotene model structure from which wholly carotenoids are generated

The carotenoids group comprises α-carotene, β-carotene, β-cryptoxanthin, lycopene, lutein, torularhodin plus torulene. Carotenoids are classified into two chemical categories: carotenes and xanthophylls. Carotenes such as β-carotene, α-carotene, torulene, plus lycopene, are pure hydrocarbons without oxygen atoms. Xanthophylls including lutein, zeaxanthin, violaxanthin, canthaxanthin, astaxanthin, besides neoxanthin, are oxygenated hydrocarbon spin-offs known as oxycarotenoids. These oxycarotenoids contain epoxy, hydroxy, and/or oxo groups.

15.7 Dissemination of Carotenoids in Fungi

Carotenoids stay a rarity among fungi, with only about 200 carotenogenic fungus species identified out of over 330 known species (Goodwin 1992). Table 15.1 provides a list of carotenoids found across the four main fungal subdivisions: Zygomycotina, Ascomycotina, Basidiomycotina, and Deuteromycotina. Fungi capable of producing carotenoids exhibit significant diversity, including species like *Phycomyces blakesleeanus*, *Blakeslea trispora*, *Fusarium sporotrichioides*, plus, *Mucor circinelloides* (Dufossé 2006), *Rhodotorula mucilaginosa* (Aksu and Eren 2005), *Rhodosporium paludigenum*, plus *Rhodotorula glutinis* (Panesar et al. 2015). The lower fungi notably Phycomycetes, Chytridiomycetes, and Blastocladiomycetes, and Ascomycetes mostly accumulate the monocyclic γ-carotene (Goodwin 1992; Sandmann 2022). It was created using a modified lycopene cyclase that catalyzed the creation of only one ionone ring rather than the two molecules, which are involved in the process of β-carotene synthesis. Conversely, the brightly colored Pyrenomycetes and also Discomycetes synthesize more complex carotenoids such as phillipsiaxanthin, aleuriaxanthin, and plectaniaxanthin (Johnson and Schroeder 2006).

The presence of the bicyclic β-carotene is a characteristic feature of plants belonging to the Mucoraceae family. Additionally, it was discovered that a small number of other fungi belonging to the *Ascomycota* and *Basidiomycota phyla* also included this compound as the ultimate consequence. However, the widely held of carotenoids detected in these phyla were modified by the addition of oxygen (Sandmann 2022). Torulene is present in certain basidiomycetous red yeasts and a few ascomycetes, such as Cookeina sulcipes. It is formed through oxidation and contains a C16' carboxylic group known as torularhodin (Sandmann 2022). The carotenoid diversity is greater in basidiomycetes, which is the most evolved phylum of fungi.

Yeasts rarely contain carotenoids, and when they do, they are only found in hetero-basidiomycetes (Kreger-van Rij 1984).

Unfortunately, these bacteria are absent in industrial yeasts such as *Saccharomyces cerevisiae*, *Kluyveromyces*, and *Candida utilis*. Plectaniaxanthin, which is present in Discomycetes, has been identified in the inadequate yeast *Cryptococcus laurentii* (Goodwin 1980). Carotenoids are used as a basis for classifying yeast, according to Kreger-van Rij (1984). While it may seem logical due to the involvement of multiple genes and limited expression in yeasts, carotenoid synthesis can be an unstable characteristic. Phaffia was first identified in the 1960s. In 1976, researchers described its unique physiological characteristics in comparison to other colored yeasts. Additionally, the structure of its pigments was determined in the same year. This discovery has contributed significant diversity to yeast carotenoids (Table 15.3) (Andrewes et al. 1976).

15.8 Fungi-Derived Carotenoids Production Methods

Carotenoids are produced within the cytoplasm and then stored in lipid droplets before being employed as functional compounds in the plasma membrane (Eman 2019). The collection of fungal biomass is carried out to extract carotenoids through a specific fermentation process that is dedicated to the production of carotenoids in either solid or liquid state (Landolfo et al. 2019).

Table 15.3 Representative carotenoids reported in slime molds and fungi

Fungal group		Representative carotenoids
Myxomycota	Myxomycetes (slime molds)	3,4-didehydrolycopene Neurosporaxanthin γ-carotene β-carotene
Mastigomycotina	Chytridiomycetes Oomycetes	γ-carotene Lycopene β-carotene
Zygomycotina	Zygomycetes	β-carotene
Ascomycotina	Plectomycetes	γ-carotene Lycopene β-carotene
	Pyrenomycetes	γ-carotene, lycopene β-carotene Torulene Lycoxanthin Neurosporaxanthin
	Discomycetes	β-carotene γ-carotene Phillipsiaxanthin, 3,4-didehydrolycopene Torulene, Aleuriaxanthin Plectaniaxanthin
Basidiomycotina	Hymenomycetes	γ-carotene β-carotene Cryptoxanthin Canthaxanthin Ataxanthin
	Gasteromycetes	γ-carotene β-carotene
	Teliomycetes	β-zeacarotene γ-carotene β-carotene Torulene Cryoptoxanthin
Deuteromycotina		γ-carotene β-carotene Torulene
Basidiomycetous yeasts		β-carotene HDCO Echinenone Astaxanthin Torulene
Mucoromycotina		Lycopene

Modified from Andrewes et al. (1976)

A cellular biomass, along with an intracellular carotenoid, is isolated from the supernatant by conventional filtration or centrifugation techniques. The cells that have been retrieved are further fragmented to facilitate the release of carotenoids that are present inside the cells (Mussagy et al. 2019). Cell disruption is often necessary to enhance the recovery of intracellular carotenoids (Gong and Bassi 2016). Cellular degradation, involving the disintegration of both cell wall and cell membrane, can be achieved through many methodologies. These approaches might be categorized as mechanical, chemical, or enzymatic. It is essential for each researcher to determine the most effective approach or combination of procedures for extracting a substance. The extraction yield is influenced by the variability of each technique (Vargas-Sinisterra and Ramírez-Castrillón 2021).

To effectively extract intracellular carotenoids, it is typically necessary to disrupt the cells either thoroughly or partially. This step is crucial as it countenances the solvent to pass in the cells and dissolve the carotenoids. It is noteworthy that the processes of cell disruption and extraction can be combined into a single operational unit or else carried out separately using distinct operational units. For example, one can combine chemical pre-treatment of cells utilizing subsequent extraction via Soxhlet. Following this, additional downstream meting out steps can be implemented to achieve saponification and the farewell of specific carotenoids (Mussagy et al. 2019). Commonly used solvents for extracting non-polar carotenoids like lutein include hexane, petroleum ether, plus tetrahydrofuran, while ethanol, acetone, and ethyl acetate are preferred for extracting polar carotenoids such as zeaxanthin. This selection is based on the unique properties of carotenoids, particularly their hydrophobic nature. Extracting carotenoids poses challenges when carotenoids with altered oppositions coexist within the same system. The process of removing polar and nonpolar carotenoids from moist tissues is made easier by the fact that acetone and ethanol can mix with water (Saini and Keum 2017). Volatile organic compounds (VOCs) are substances that are commonly utilized as solubilizing agents in most traditional extraction methods. While VOC-based approaches have the advantage of achieving high extraction yields, their strong potential to solubilize carotenoids and disrupt cells by permeabilizing their walls and membranes poses significant risks to human health besides the environment (Salar-García et al. 2017). Therefore, in latest years, several researchers have been seeking novel and resourceful methods to address these challenges. Their focus has primarily shifted toward (a) substituting volatile organic compounds (VOCs) utilizing more environmentally benign and less deleterious solvents such as supercritical fluids, biosolvents, or ionic liquids (Yara-Varón et al. 2016), and (b) tumbling solvent consumption by integrating chemical extraction with innovative physical methods such as microwave- and ultrasound-assisted extractions.

Traditional techniques, like atmospheric liquid extraction with maceration and Soxhlet extraction, mostly depend on volatile organic compounds (VOCs), with the choice of solvent dictating the method used. These VOCs can also be associated with non-traditional techniques like ultrasound-assisted extraction (UAE) and enzyme-assisted extraction (EAE). Unconventional solvents, like biosolvents and green solvents, are primarily employed in advanced processing methods due to their improved biocompatibility, while others like supercritical fluid extraction (SFE) are chosen for their suitability for specific processes (Saini and Keum 2017).

15.9 Extraction of Carotenoids

15.9.1 Conventional Methods

Extraction of carotenoids can be recognized by two main ways: Conventional methods and Saponification. Carotene is an unbound molecule, while lutein can exist as either an unbound fatty acid or a more stable fatty acid esterified form. The esterified carotenoids are commonly extracted using saponification, a process that releases carotenoids in their ester forms. Typically, KOH or NaOH are employed as saponification agents. The prime influences that touch yield are the parameters of saponification, namely the duration, temperature, and concentration of alkali (Saini and Keum 2018). In a reading conducted by Liu et al. (2021b), it was discovered that the use of a saponifier containing both alcohol and KOH resulted in a higher recovery rate (78.7 ± 2.4%) when separating carotenoids in *Rhodosporidium toruloides*. In order to avert the deterioration of carotenoids, it is commonly carried out at temperatures lower than 60 °C using a solution of 10% potassium hydroxide in either methanol or ethanol (Gong and Bassi 2016).

The alkali is then eliminated from the carotenoid solution through the process of rinsing it with water. Nevertheless, when exposed to alkali and air, carotenoids that contain allylic hydroxyl besides keto-groups, such as neurosporaxanthin, undergo oxidation. Aerobic saponification is not endorsed by these carotenoids. In their publication, Schiedt et al. (1993) present a comprehensive account of this procedure.

15.9.1.1 Soxhlet Extraction

This technique involves using a solvent to dissolve the anticipated substance, where the dirtiness typically remains insoluble while the substance with the desired characteristics exhibits restricted solubility in the same solvent. By employing solvents at both low-slung and boiling temperatures, this method selectively extracts the desired substances. Although time-consuming, Soxhlet extraction yields significant amounts of recoverable carotenoids. However, due to the need for larger quantities of solvent, this procedure is less cost-effective (Macías-Sánchez et al. 2010). Prolonged extraction times and higher temperatures also surge the risk of carotenoid cis-trans isomerization after degradation, then Soxhlet extraction offers advantages such as a continuous manufacturing process, straightforward design, low solvent flow, easy visual process monitoring, and potential for solvent reuse after distillation and stripping. However, a primary limitation of the Soxhlet extraction process is the discharge of potentially toxic compounds through extraction, stemming from the use of incendiary and lethal liquid organic solvents in significant quantities (Ashokkumar et al. 2023).

15.9.1.2 Atmospheric Liquid Extraction Using Maceration

The extraction of carotenoids involves using various volatile organic solvents like methanol, hexane, isopropanol, chloroform, acetone, plus, diethyl ether. Effective extraction of intracellular carotenoids requires a successful cell disruption method, which can be mechanical, chemical, or physical in nature. To dissolve cellular contents effectively, the solvent needs to penetrate the cell. However, the company of a significant amount of water in that sample poses a challenge, necessitating comprehensive treatment to minimize analyte extraction (Ashokkumar et al. 2023; Valduga et al. 2009).

The extraction efficiency of unalike amalgamations of volatile organic compounds (VOCs) (acetone, petroleum ether, methanol, hexane, ethyl acetate, dimethyl sulfoxide (DMSO), plus, chloroform) through liquid nitrogen was assessed using the conventional VOC atmospheric liquid extraction method. This involved successive macerations of cellular biomass from *Sporidiobolus salmoni* color CBS 2636. Notably, the most effective method for carotenoid extraction involved exhausting liquid nitrogen (N2) besides dimethyl sulfoxide (DMSO) to dislocate the cell structure, charted by liquid extraction using a mixture of acetone and methanol with a ratio of 7:3 (volume/volume). This yielded the highest concentration of total carotenoids at 253.8 grams per gram.

The impact of various volatile organic compounds (VOCs) such as DMSO, petroleum ether, acetone, chloroform; plus hexane was investigated on the extraction efficiency of carotenoids from *R. glutinis* cells. Unlike maceration, the authors directly injected each organic solvent mixture into the biomass after initially freeze-drying the yeast cells. Total carotenoids bent by *R. glutinis* cells were successfully recovered through a combination of liquid extraction and lyophilization. Hexane exhibited the lowest extraction capacity, with a measure of 0.19 mg/g, while DMSO and petroleum ether showed the top recovery yields; of values of 0.23 mg/g and 0.24 g/g, in turn (Ashokkumar et al. 2023; Valduga et al. 2009).

15.9.2 Non-conventional Methods

15.9.2.1 Green Solvent Extraction

Most organic solvents are derived from non-renewable sources, posing environmental pollution risks due to their flammability, volatility, and toxicity. Consequently, it is imperative to minimize their usage to mitigate these adverse effects (Liu et al. 2021a). When extracting astaxanthin from *Phaffia rhodozyma* using different solvents, Routray et al. (2019)found that employing the ecologically favorable solvent ethanol produced the highest astaxanthin extraction rates. The ability of surfactant aids to break down cell membrane and remove carotenoids has also been discovered. Ionic liquids (ILs) are an entirely novel type of green solvents that have a variety of benefits, such as the ability to preserve a salt anions and cations composition while being non-flammable and non-hazardous to the environment (Liu et al. 2021a).

15.9.2.2 Enzymatic-Assisted Extraction

Although carotenoids can be extracted using a solvent, the heat generated during the extraction progression can have a sway on the long-term stability of the carotenoids. The researchers utilized enzyme-assisted extraction to enrich the astaxanthin stability, aiming to generate astaxanthin with an extraordinary stability level (Nadar et al. 2018; Saini and Keum 2018). Strati et al. (2015) found that the enzyme pretreatment resulted in a greater output of carotenoids besides lycopene compared to pretreatment with acetone, ethanol, ethyl lactate, hexane, plus, ethyl acetate.

Enzymatic extraction speeds up manufacturing and uses less organic solvents because enzymes are selective, renewable, and friendly to the environment. Enzymes, however, should receive special consideration because they are costly, need optimal performing conditions, and deteriorate quickly. The efficacy of enzyme-based extraction depends on the enzymes' capacity to perform reactions with precise specificity and regional selectivity, while preserving the biological activity of bioactive substances at moderate circumstances. In this method, the integrity of cell walls is compromised by the use of hydrolytic enzymes, exposing intracellular components. Enzyme-assisted extraction (EAE) offers numerous advantages such as operating under mild conditions, achieving high efficiency, maintaining characteristics, and ensuring extraction stability. However, implementing enzyme assays extraction on an industrial scale is not particularly practical due to the limitations imposed by environmental factors on enzymes (Ashokkumar et al. 2023).

15.9.2.3 Microwave-Assisted Extraction

Ionic conduction and dipole rotation result from the direct interaction of microwaves with the molecules during the extraction process. The method is quick, easy, and affordable for removing carotenoids. It can be utilized when a process needs to be completed quickly and with a small amount of solvent. Abridged tackle size, quicker heating to extract bioactive compounds, enhanced extraction, and decreased thermal gradients are all benefits of MAE. The main drawbacks of MAE include the necessity for special equipment, low selectivity, and inevitable reactions at extremely high temperatures (Ashokkumar et al. 2023).

15.9.2.4 Ultrasound-Assisted Extraction

This approach employs cyclical sound pressure at a incidence exceeding 20 kHz to enhance the extraction process by inducing cavitation in liquids. The production of small bubbles through acoustic cavitation leads to the disintegration of the cell, hence improving the transfer of extract and mass. This clues to a rise in both local pressure besides temperature, concluding in enhanced diffusivity, transportation of solvent besides template molecules, penetration, and plus solubility (Singh et al. 2019). The research conducted by Cheng et al. (2019) reported the destruction of *Phaffia rhodozyma* cell clusters using lactic acid. Additionally, they effectively isolated astaxanthin using ultrasonic extraction and anticipated its successful application in food and feed. The advantages of the UAE (Ashokkumar et al. 2023) are minimal economic requirements, quick extraction times, efficient use of solvent volume, and limited instrument demands.

15.9.2.5 Pressure-Assisted Pulsed Electric Field (PEF)-Assisted Extraction

As per the findings of Zhao et al. (2019), the carotenoids' yield was influenced by the extraction pressure. Liu et al. (2019) conducted a study to probe the impression of different wall-breaking techniques on the yeast *S. pararoseus*. The efficacy of acid heating, dimethyl sulfoxide, enzymatic methods, high-pressure homogenization, and cell autolysis in disrupting yeast cell walls was evaluated. The high-pressure homogenization approach was determined to have the least duration, was safe, and caused no pollution, while also achieving the maximum rate of carotenoid extraction (82.5%). The continuous pressured solvent extraction technology demonstrated high efficiency and environmental friendliness, resulting in reduced production costs and less environmental harm.

In a novel non-thermal technique called pulsed electric field technology, biological material enclosed between two electrodes is repeatedly exposed to brief, high-voltage pulsates lasting less than a few milliseconds (Liu et al. 2021a). Electroporation, which increases membrane permeability, allows the extraction of intracellular carotenoid pigments. The magnitude of the applied electric field determines the outcome; electroporation may be either transient (reversible) or permanent (irreversible). PEF-assisted extraction has benefits such as increased extraction, decreased water and energy usage for slow mass transport operations, etc. PEF aided extraction has some drawbacks, such as reduced effectiveness caused by air bubbles, reversibility given by membrane modification, and technique efficiency dependent on electrode gap besides electric field strength (Ashokkumar et al. 2023).

15.9.2.6 Supercritical Fluid Extraction

The supercritical fluid extraction (SFE) technique utilizes a supercritical fluid to extract the specific amalgams of interest. Supercritical carbon dioxide may efficiently remove heat-sensitive chemicals, such as carotenoids, from both solid and liquid substances. The extraction of polar and also non-polar carotenoids requires careful control of flow rate, carbon dioxide density, temperature, plus pressure (Ashokkumar et al. 2023). A supercritical fluid is a distinct form of fluid that possesses the capacity to dissolve substances like a liquid and escape like a gas.

The pharmaceutical and food industries choose to employ supercritical fluid extraction due to its utilization of benign solvents such as water and carbon dioxide (Liu et al. 2021a). By adjusting important variables, such as essential parameters, it is possible to enhance the extraction efficiency by introducing natural oils and hydrocarbons into the system. This addition increases the attraction between CO_2 and decreases its interaction between the substance and the surrounding material leading to a notable enhancement in the effectiveness of extracting the substance (Yousefi et al. 2019). The solvents commonly used in the SFE method include methylene chloride, methanol, acetone, ethyl acetate, and propane. Throughout this extraction method, it is often necessary to carry out drying and grinding steps. By using water-miscible solvents, the aforementioned procedures can be circumvented.

This can obviate the need for sample desiccation, solvent volatilization, and cell lysis. Dimethyl ether is frequently utilized in supercritical fluid extraction as a solvent capable of mixing with water (Goto et al. 2015). Efficiency can be enhanced by integrating various extraction methods. For instance, the process of extracting astaxanthin involved the utilization of microwave pretreatment and supercritical fluid technologies. Astaxanthin extraction is significantly accelerated, with a speed 12 times greater than traditional procedures, while employing diverse solvent systems (with a water content ranging from 0 to 50 vol% of the water-alcohol mixture). The integration of these methods significantly enhanced the proficiency of the extraction progression. The citation "Nunes et al. 2021" refers to a publication by Nunes and colleagues in the year 2021.It serves as an environmentally friendly alternative to traditional carotenoid extraction techniques. Supercritical fluid extraction technology has superior permeability and diffusibility compared to conventional extraction procedures, resulting in faster and more nominal extraction of carotenoids (Yousefi et al. 2019).

15.10 Conclusion

Carotenoids possess a broad spectrum of applications in several industries, including cosmetics, pharmaceuticals, beside nutraceuticals. Liposoluble organic terpenoids, which are naturally found in photosynthetic organisms and plants, and are answerable for the vivid red, yellow, plus, orange colors seen in nature. Yeasts are preferred producers across all fungal taxonomic groupings because they have efficient synthesis compared to plants and other microbes. There has been a substantial proliferation in interest in red oleaginous yeast due to its potential for producing profitable and eco-friendly products.

Carotenoids have various practical uses in industries, such as being employed as natural food dyes, additions in animal feed, and even as a potential treatment for obesity. When ingested as food or supplements, their antioxidant qualities effectively neutralize reactive oxygen species (ROS), which is helpful for human health. Specifically, it is worth mentioning that certain carotenoids, like β-carotene; have the ability to transform into vitamin A, thereby increasing their nutritional significance.

Carotenoids generated from microorganisms possess stability, solubility, plus, antioxidant properties, which make them very suitable for incorporation into sunscreens and other products designed to protect against UV radiation. Furthermore, carotenoids such as lycopene, β-carotene, and also astaxanthin possess photoprotective properties, shielding against the detrimental impacts of UV radiation.

Finally, carotenoids possess a diverse assortment of functions in many industries, providing both visual and health advantages. Their adaptability and capacity for originality emphasize their significance in various domains, offering prospective progress in medicine, cosmetics, and beyond.

References

Abdelshafy AM, Belwal T, Liang Z, Wang L, Li D, Luo Z, Li L (2022) A comprehensive review on phenolic compounds from edible mushrooms: occurrence, biological activity, application and future prospective. Crit Rev Food Sci Nutr 62(22):6204–6224. https://doi.org/10.1080/10408398.2021.1898335

Aksu Z, Eren AT (2005) Carotenoids production by the yeast *Rhodotorula mucilaginosa*: use of agricultural wastes as a carbon source. Process Biochem 40(9): 2985–2991

Almeida ERA, Cerdá-Olmedo E (2008) Gene expression in the regulation of carotene biosynthesis in Phycomyces. Curr Genet 53(3):129–137

Álvarez R, Vaz B, Gronemeyer H, De Lera RA (2014) Functions, therapeutic applications, and synthesis of retinoids and carotenoids. Chem Rev 114(1):1–125. https://doi.org/10.1021/CR400126U

Anarjan N, Tan CP, Nehdi IA, Ling TC (2012) Colloidal astaxanthin: preparation, characterisation and bioavailability evaluation. Food Chem 135(3):1303–1309. https://doi.org/10.1016/j.foodchem.2012.05.091

Andrewes AG, Phaff HJ, Starr MP (1976) Carotenoids of *Phaffia rhodozyma*, a red-pigmented fermenting yeast. Phytochemistry 15(6):1003–1007

Anunciato TP, da Rocha Filho PA (2012) Carotenoids and polyphenols in nutricosmetics, nutraceuticals, and cosmeceuticals. J Cosmet Dermatol 11(1):51–54. https://doi.org/10.1111/J.1473-2165.2011.00600.X

Armstrong GA (1994) Eubacteria show their true colors: genetics of carotenoid pigment biosynthesis from microbes to plants. J Bacteriol 176(16):4795–4802. https://doi.org/10.1128/jb.176.16.4795-4802.1994

Armstrong GA, Hearst JE (1996) Genetics and molecular biology of carotenoid pigment biosynthesis. FASEB J 10(2):228–237. https://doi.org/10.1096/fasebj.10.2.8641556

Ashokkumar V, Flora G, Sevanan M, Sripriya R, Chen WH, Park J-H, Kumar G (2023) Technological advances in the production of carotenoids and their applications—a critical review. Bioresour Technol 367:128215. https://doi.org/10.1016/j.biortech.2022.128215

Avalos J, Pardo-Medina J, Parra-Rivero O, Ruger-Herreros M, Rodríguez-Ortiz R, Hornero-Méndez D, Limón MC (2017) Carotenoid biosynthesis in Fusarium. J Fungi 3(3):39

Aziz E, Batool R, Akhtar W, Rehman S, Shahzad T, Malik A et al (2020) Xanthophyll: health benefits and therapeutic insights. Life Sci 240:117104. https://doi.org/10.1016/j.lfs.2019.117104

Barredo JL, García-Estrada C, Kosalkova K, Barreiro C (2017) Biosynthesis of astaxanthin as a main carotenoid in the heterobasidiomycetous yeast *Xanthophyllomyces dendrorhous*. J Fungi 3(3):44

Barreiro C, Barredo J-L (2018) Carotenoids production: a healthy and profitable industry—microbial carotenoids: methods and protocols. In: Barreiro C, Barredo J-L (eds) Methods in molecular biology. Springer New York, pp 45–55. https://doi.org/10.1007/978-1-4939-8742-9_2

Bartley GE, Viitanen PV, Pecker I, Chamovitz D, Hirschberg J, Scolnik PA (1991) Molecular cloning and expression in photosynthetic bacteria of a soybean cDNA coding for phytoene desaturase, an enzyme of the carotenoid biosynthesis pathway. Proc Natl Acad Sci USA 88(15):6532–6536. https://doi.org/10.1073/pnas.88.15.6532

Bell JG, McEvoy J, Tocher DR, Sargent JR (2000) Depletion of α-tocopherol and astaxanthin in *Atlantic salmon* (Salmo salar) affects autoxidative defense and fatty acid metabolism. J Nutr 130(7):1800–1808

Bramley PM, Davies BH (1975) Carotene biosynthesis by cell extracts of mutants of *Phycomyces blakesleeanus*. Phytochemistry 14(2):463–469

Cardoso LAC, Jäckel S, Karp SG, Framboisier X, Chevalot I, Marc I (2016) Improvement of *Sporobolomyces ruberrimus* carotenoids production by the use of raw glycerol. Bioresour Technol 200: 374–379. https://doi.org/10.1016/J.BIORTECH.2015.09.108

Cardoso LAC, Karp SG, Vendruscolo F, Kanno KYF, Zoz LIC, Carvalho JC, Cardoso LAC, Karp SG, Vendruscolo F, Kanno KYF, Zoz LIC, Carvalho JC (2017) Biotechnological production of carotenoids and their applications in food and pharmaceutical products. Carotenoids. https://doi.org/10.5772/67725

Chang TS (2009) An updated review of tyrosinase inhibitors. Int J Mol Sci 10(6):2440–2475. https://doi.org/10.3390/IJMS10062440

Chen C, Bau T, Bao H (2013) Chemical composition analysis of cultured Cordyceps militaris. Food Sci 34(11):36–40. https://doi.org/10.7506/spkx1002-6630-201311009

Cheng XY, Xiong YJ, Yang MM, Zhu MJ (2019) Preparation of astaxanthin mask from *Phaffia rhodozyma* and its evaluation. Process Biochem 79:195–202. https://doi.org/10.1016/j.procbio.2018.12.027

Darvin ME, Fluhr JW, Meinke MC, Zastrow L, Sterry W, Lademann J (2011) Topical beta-carotene protects against infra-red-light-induced free radicals. Exp Dermatol 20(2):125–129. https://doi.org/10.1111/j.1600-0625.2010.01191.x

Davinelli S, Nielsen ME, Scapagnini G (2018) Astaxanthin in skin health, repair, and disease: a comprehensive review. Nutrients 10(4). https://doi.org/10.3390/NU10040522

Davoli P, Mierau V, Weber RWS (2004) Carotenoids and fatty acids in red yeasts *Sporobolomyces roseus* and *Rhodotorula glutinis*. Appl Biochem Microbiol 40(4): 392–397. https://doi.org/10.1023/B:ABIM.0000033917.57177.f2

Domonkos I, Kis M, Gombos Z, Ughy B (2013) Carotenoids, versatile components of oxygenic photosynthesis. Prog Lipid Res 52(4):539–561

Dufossé L (2006) Microbial production of food grade pigments. Food Technol Biotechnol 44(3):313–323

Dzurendova S, Losada CB, Dupuy-Galet BX, Fjær K, Shapaval V (2022) Mucoromycota fungi as powerful cell factories for modern biorefinery. Appl Microbiol Biotechnol 106(1):101–115. https://doi.org/10.1007/s00253-021-11720-1

Elfeky N, Elmahmoudy M, Zhang Y, Guo JL, Bao Y (2019) Lipid and carotenoid production by *Rhodotorula glutinis* with a combined cultivation mode of nitrogen, sulfur, and aluminium stress. Appl Sci (Switzerland) 9(12):2444. https://doi.org/10.3390/app9122444

Eman M (2019) Fungal and yeast carotenoids. J Yeast Fungal Res 10:30–44. https://doi.org/10.5897/JYFR2019.0192

Fang Q, Guo S, Zhou H, Han R, Wu P, Han C (2017) Astaxanthin protects against early burn-wound progression in rats by attenuating oxidative stress-induced inflammation and mitochondria-related apoptosis. Sci Rep 7(January):1–13. https://doi.org/10.1038/srep41440

Freitas JV, Junqueira HC, Martins WK, Baptista MS, Gaspar LR (2019) Antioxidant role on the protection of melanocytes against visible light-induced photodamage. Free Radic Biol Med 131:399–407. https://doi.org/10.1016/J.FREERADBIOMED.2018.12.028

Gong M, Bassi A (2016) Carotenoids from microalgae: a review of recent developments. Biotechnol Adv 34(8): 1396–1412. https://doi.org/10.1016/j.biotechadv.2016.10.005

Goodwin T (1980) The biochemistry of the carotenoids. Vol. 1: plants. Chapman and Hall, New York

Goodwin T (1992) Distribution of carotenoids. In: Methods in enzymology, vol 213. Elsevier, pp 167–172

Goto M, Kanda H, Wahyudiono, Machmudah S (2015) Extraction of carotenoids and lipids from algae by supercritical CO2 and subcritical dimethyl ether. J Supercrit Fluids 96:245–251. https://doi.org/10.1016/j.supflu.2014.10.003

Hausmann A, Sandmann G (2000) A single five-step desaturase is involved in the carotenoid biosynthesis pathway to β-carotene and torulene in *Neurospora crassa*. Fungal Genet Biol 30(2):147–153. https://doi.org/10.1006/fgbi.2000.1212

Haxo F (1950) Carotenoids of the mushroom *Cantharellus cinnabarinus*. Bot Gaz 112(2):228–232. https://doi.org/10.1086/335653

Hegde P, Bhat Agni M, Rai P, Mohana Kumar B, Gowda KMD (2022) Impact of carotenoids on gut microbiome: implications in human health and disease. J Appl Nat Sci 14:1085–1099. https://doi.org/10.31018/jans.v14i3.3582

Hemmi H, Ikejiri S, Nakayama T, Nishino T (2003) Fusion-type lycopene β-cyclase from a thermoacidophilic archaeon *Sulfolobus solfataricus*. Biochem Biophys Res Commun 305(3):586–591

Hugueney P, Römer S, Kuntz M, Camara B (1992) Characterization and molecular cloning of a flavoprotein catalyzing the synthesis of phytofluene and zeta-carotene in *Capsicum chromoplasts*. Eur J Biochem 209(1):399–407. https://doi.org/10.1111/j.1432-1033.1992.tb17302.x

Hyde KD, Bahkali AH, Moslem MA (2010) Fungi—an unusual source for cosmetics. Fungal Divers 43:1–9. https://doi.org/10.1007/s13225-010-0043-3

Jiang GL, Zhou LY, Wang YT, Zhu MJ (2017) Astaxanthin from Jerusalem artichoke: production by fed-batch fermentation using *Phaffia rhodozyma* and application in cosmetics. Process Biochem 63:16–25. https://doi.org/10.1016/j.procbio.2017.08.013

Jinsong Y, Haisheng T, Rui Y, Xiaohuan S, Hairui Z, Kaimian L (2011) Astaxanthin production by *Phaffia rhodozyma* fermentation of cassava residues substrate. Agric Eng Int CIGR J 13(2) http://cigrjournal.org/index.php/Ejounral/article/view/1847

Johnson EA, Schroeder WA (2006) Microbial carotenoids. In: Downstream processing biosurfactants carotenoids. Springer, pp 119–178

Kammeyer A, Luiten RM (2015) Oxidation events and skin aging. Ageing Res Rev 21:16–29. https://doi.org/10.1016/j.arr.2015.01.001

Kim SK (2014) Marine cosmeceuticals. J Cosmet Dermatol 13(1):56–67. https://doi.org/10.1111/jocd.12057

Kot AM, Błazejak S, Gientka I, Kieliszek M, Bryś J (2018) Torulene and torularhodin: "new" fungal carotenoids for industry? Microb Cell Factories 17(1):1–14. https://doi.org/10.1186/s12934-018-0893-z

Kreger-van Rij N (1984) Yeasts. Elsevier, Amsterdam

Landolfo S, Chessa R, Zara G, Zara S, Budroni M, Mannazzu I (2019) *Rhodotorula mucilaginosa* C2.5t1 modulates carotenoid content and CAR genes transcript levels to counteract the pro-oxidant effect of hydrogen peroxide. Microorganisms 7(9). https://doi.org/10.3390/microorganisms7090316

Lichtenthaler HK, Buschmann C (2001) Chlorophylls and carotenoids: measurement and characterization by UV-VIS spectroscopy. Curr Protocol Food Anal Chem *1*(1):F4. 3.1–F4. 3.8

Lin L, Xu J (2020) Fungal pigments and their roles associated with human health. J Fungi 6(4):1–37. https://doi.org/10.3390/jof6040280

Liu C, Cheng Y, Du C, Lv T, Guo Y, Han M et al (2019) Study on the wall-breaking method of carotenoids producing yeast *Sporidiobolus pararoseus* and the antioxidant effect of four carotenoids on SK-HEP-1 cells. Prep Biochem Biotechnol 49(8):767–774. https://doi.org/10.1080/10826068.2019.1608448

Liu C, Hu B, Cheng Y, Guo Y, Yao W, Qian H (2021a) Carotenoids from fungi and microalgae: a review on their recent production, extraction, and developments. Bioresour Technol 337:125398. https://doi.org/10.1016/j.biortech.2021.125398

Liu Z, van den Berg C, Weusthuis RA, Dragone G, Mussatto SI (2021b) Strategies for an improved extraction and separation of lipids and carotenoids from oleaginous yeast. Sep Purif Technol 257:117946. https://doi.org/10.1016/j.seppur.2020.117946

Macías-Sánchez MD, Fernandez-Sevilla JM, Fernández FGA, García MCC, Grima EM (2010) Supercritical fluid extraction of carotenoids from *Scenedesmus almeriensis*. Food Chem 123(3):928–935. https://doi.org/10.1016/j.foodchem.2010.04.076

Mago P, Sharma R, Hafeez I, Nawaz I, Joshi M, Mehrotra R (2023) Mushroom based cosmeceuticals: an upcoming biotechnology sector. Biosci Biotechnol Res Asia 20(2):381–394

Mantzouridou F, Tsimidou MZ (2008) Lycopene formation in *Blakeslea trispora*. Chemical aspects of a bioprocess. Trends Food Sci Technol 19(7):363–371. https://doi.org/10.1016/j.tifs.2008.01.003

Mapelli-Brahm P, Barba FJ, Remize F, Garcia C, Fessard A, Mousavi Khaneghah A et al (2020) The impact of fermentation processes on the production, retention and bioavailability of carotenoids: an overview. Trends Food Sci Technol 99:389–401. https://doi.org/10.1016/j.tifs.2020.03.013

Marizcurrena JJ, Cerdá MF, Alem D, Castro-Sowinski S (2019) Living with pigments: the colour palette of Antarctic life. In: The ecological role of microorganisms in the Antarctic environment. Springer, Cham, pp 65–82

Martínez-Cámara S, Ibañez A, Rubio S, Barreiro C, Barredo J-L (2021) Main carotenoids produced by microorganisms. Encyclopedia 1(4):1223–1245. https://doi.org/10.3390/encyclopedia1040093

Martins A, Tenreiro T, Andrade G, Gadanho M, Chaves S, Abrantes M, Calado P, Tenreiro R, Vieira H (2013) Photoprotective bioactivity present in a unique marine bacteria collection from Portuguese deep sea hydrothermal vents. Mar Drugs 11(5):1506–1523

Meléndez-Martínez AJ, Stinco CM, Mapelli-Brahm P (2019) Skin carotenoids in public health and nutricosmetics: the emerging roles and applications of the UV radiation-absorbing colourless carotenoids phytoene and phytofluene. Nutrients 11(5). https://doi.org/10.3390/nu11051093

Mendes-Silva TDCD, da Silva Andrade RF, Ootani MA, Mendes PVD, da Silva MRF, Souza KS, dos Santos Correia MT, da Silva MV, de Oliveira MBM (2020) Biotechnological potential of carotenoids produced by extremophilic microorganisms and application prospects for the cosmetics industry. Adv Microbiol 10(8):397–410

Mussagy CU, Winterburn J, Santos-Ebinuma VC, Pereira JFB (2019) Production and extraction of carotenoids produced by microorganisms. Appl Microbiol Biotechnol 103(3):1095–1114. https://doi.org/10.1007/s00253-018-9557-5

Nadar SS, Rao P, Rathod VK (2018) Enzyme assisted extraction of biomolecules as an approach to novel extraction technology: a review. Food Res Int 108:309–330. https://doi.org/10.1016/j.foodres.2018.03.006

Narsing Rao MP, Xiao M, Li WJ (2017) Fungal and bacterial pigments: secondary metabolites with wide applications. Front Microbiol 8(JUN):1113. https://doi.org/10.3389/fmicb.2017.01113

Nunes A, Roda A, Gouveia L, Fernández N, Bronze M, Matias A (2021) Astaxanthin extraction from marine crustacean waste streams: an integrate approach between microwaves and supercritical fluids. ACS Sustain Chem Eng 9. https://doi.org/10.1021/acssuschemeng.0c06534

Palombo P, Fabrizi G, Ruocco V, Ruocco E, Fluhr J, Roberts R, Morganti P (2007) Beneficial long-term effects of combined oral/topical antioxidant treatment with the carotenoids lutein and zeaxanthin on human skin: a double-blind, placebo-controlled study. Skin Pharmacol Physiol 20(4):199–210. https://doi.org/10.1159/000101807

Panesar R, Kaur S, Panesar PS (2015) Production of microbial pigments utilizing agro-industrial waste: a review. Curr Opin Food Sci 1:70–76

Paul D, Kumari PK, Siddiqui N (2023) Yeast carotenoids: cost-effective fermentation strategies for health care applications. Fermentation 9(2):147

Pecker I, Chamovitz D, Linden H, Sandmann G, Hirschberg J (1992) A single polypeptide catalyzing the conversion of phytoene to zeta-carotene is transcriptionally regulated during tomato fruit ripening. Proc Natl Acad Sci USA 89(11):4962–4966. https://doi.org/10.1073/pnas.89.11.4962

Peterson WJ, Bell TA, Etchells JL, Smart WW (1954) A procedure for demonstrating the presence of carotenoid pigments in yeasts. J Bacteriol 67(6):708–713. https://doi.org/10.1128/jb.67.6.708-713.1954

Pfander H (1992) Carotenoids: an overview. Methods Enzymol 213:3–13

Rapoport A, Guzhova I, Bernetti L, Buzzini P, Kieliszek M, Kot AM (2021) Carotenoids and some other pigments from fungi and yeasts. Metabolites 11(2):92

Rau W (1985) Mechanism of photoregulation of carotenoid biosynthesis in plants. Pure Appl Chem 57(5): 777–784

Rebelo BA, Farrona S, Rita Ventura M, Abranches R (2020) Canthaxanthin, a red-hot carotenoid: applications, synthesis, and biosynthetic evolution. Plants 9(8):1–18. https://doi.org/10.3390/plants9081039

Rodriguez-Concepcion M, Avalos J, Bonet ML, Boronat A, Gomez-Gomez L, Hornero-Mendez D, Limon MC, Melendez-Martinez AJ, Olmedilla-Alonso B, Palou A (2018) A global perspective on carotenoids: metabolism, biotechnology, and benefits for nutrition and health. Prog Lipid Res 70:62–93

Routray W, Dave D, Cheema S, Vasudevan Ramakrishnan V, Pohling J (2019) Biorefinery approach and environment-friendly extraction for sustainable production of astaxanthin from marine wastes. Crit Rev Biotechnol 39:1–20. https://doi.org/10.1080/07388551.2019.1573798

Ruiz-Hidalgo MJ, Benito EP, Sandmann G, Eslava AP (1997) The phytoene dehydrogenase gene of Phycomyces: regulation of its expression by blue light and vitamin A. Mol Gen Genet MGG 253(6): 734–744. https://doi.org/10.1007/s004380050378

Saini RK, Keum Y-S (2017) Progress in microbial carotenoids production. Indian J Microbiol 57(1): 129–130. https://doi.org/10.1007/s12088-016-0637-x

Saini RK, Keum Y-S (2018) Carotenoid extraction methods: a review of recent developments. Food Chem 240:90–103. https://doi.org/10.1016/j.foodchem.2017.07.099

Sajid S, Akbar N (2018) Applications of fungal pigments in biotechnology. Pesq Agrop Brasileira 7(2):922–930. https://doi.org/10.19045/bspab.2018.700111

Salar-García MJ, Ortiz-Martínez VM, Hernández-Fernández FJ, de los Ríos AP, Quesada-Medina J (2017) Ionic liquid technology to recover volatile organic compounds (VOCs). J Hazard Mater 321: 484–499. https://doi.org/10.1016/j.jhazmat.2016.09.040

Sandmann G (2022) Carotenoids and their biosynthesis in fungi. Molecules 27(4):1431. https://doi.org/10.3390/molecules27041431

Sandmann G, Karl Esser P, Bennett J, Heinz D (2002) The mycota X industrial applications. In: The mycota X industrial applications. Springer, Osiewacz

Sandmann G, Takaichi S, Fraser PD (2008) C35-apocarotenoids in the yellow mutant *Neurospora crassa* YLO. Phytochemistry 69(17):2886–2890. https://doi.org/10.1016/j.phytochem.2008.09.016

Schiedt K, Bischof S, Glinz E (1993) Metabolism of carotenoids and in Vivo racemization of (3S,3′S)-Astaxanthin in the crustacean Penaeus. In: Methods in enzymology, vol 214. Academic, pp 148–168. https://doi.org/10.1016/0076-6879(93)14062-N

Schimek C, Wöstemeyer J (2009) Carotene derivatives in sexual communication of zygomycete fungi. Phytochemistry 70(15–16):1867–1875

Schmidhauser TJ, Lauter FR, Schumacher M, Zhou W, Russo VE, Yanofsky C (1994) Characterization of al-2, the phytoene synthase gene of *Neurospora crassa*. Cloning, sequence analysis, and photoregulation. J Biol Chem 269(16):12060–12066

Schrantz JP, Lemoine Y (1995) Carotenoid composition of mycelium and apothecia in the discomycete *Scutellinia umbrarum*. Phytochemistry 40(1):33–35. https://doi.org/10.1016/0031-9422(95)00294-H

Singh G, Sinha S, Bandyopadhyay K, Lawrence M, Paul D (2018) Triauxic growth of an oleaginous red yeast *Rhodosporidium toruloides* on waste 'extract' for enhanced and concomitant lipid and β-carotene production. Microb Cell Factories 17:1–10

Singh D, Khattar JIS, Rajput A, Chaudhary R, Singh R (2019) High production of carotenoids by the green microalga *Asterarcys quadricellulare* PUMCC 5.1.1 under optimized culture conditions. PLoS One 14: e0221930. https://doi.org/10.1371/journal.pone.0221930

Sinha S, Das S, Saha B, Paul D, Basu B (2022) Antimicrobial, anti-oxidant, and anti-breast cancer properties unraveled in yeast carotenoids produced via cost-effective fermentation technique utilizing waste hydrolysate. Front Microbiol 13:1088477

Strati IF, Gogou E, Oreopoulou V (2015) Enzyme and high pressure assisted extraction of carotenoids from tomato waste. Food Bioprod Process 94:668–674. https://doi.org/10.1016/j.fbp.2014.09.012

Sujak A (2009) Interactions between canthaxanthin and lipid membranes—possible mechanisms of canthaxanthin toxicity. Cell Mol Biol Lett 14(3):395–410. https://doi.org/10.2478/s11658-009-0010-8

Sujarit K, Suwannarach N, Kumla J, Lomthong T (2021) Mushrooms: splendid gifts for the cosmetic industry. Chiang Mai J Sci 48(3):699–725

Tang H, Chen C, Zou Y, Lou H, Zheng Q, Guo L, Lin J, Ye Z, Yun F (2019) Purification and structural characterization of a novel natural pigment: cordycepene from edible and medicinal mushroom *Cordyceps militaris*. Appl Microbiol Biotechnol 103(19): 7943–7952. https://doi.org/10.1007/s00253-019-10101-z

Taofiq O, González-Paramás AM, Martins A, Barreiro MF, Ferreira ICFR (2016) Mushrooms extracts and compounds in cosmetics, cosmeceuticals and nutricosmetics—a review. Ind Crop Prod 90:38–48. https://doi.org/10.1016/j.indcrop.2016.06.012

Valadon L (1976) Carotenoids as additional taxonomic characters in fungi: a review. Trans Br Mycol Soc 67(1):1–15

Valduga E, Valério A, Tatsch PO, Treichel H, Furigo A, Di Luccio M (2009) Assessment of cell disruption and carotenoids extraction from *Sporidiobolus salmonicolor* (CBS 2636). Food Bioprocess Technol 2(2):234–238. https://doi.org/10.1007/s11947-008-0133-3

Vargas-Sinisterra AF, Ramírez-Castrillón M (2021) Yeast carotenoids: production and activity as antimicrobial biomolecule. Arch Microbiol 203(3):873–888. https://doi.org/10.1007/s00203-020-02111-7

Verdoes JC, Misawa N, van Ooyen AJ (1999a) Cloning and characterization of the astaxanthin biosynthetic gene encoding phytoene desaturase of *Xanthophyllomyces dendrorhous*. In: Biotechnol bioeng, vol 63. Wiley, pp 750–755. https://doi.org/10.1002/(sici)1097-0290(19990620)63:6<750::aid-bit13>3.0.co;2-7

Verdoes J, Krubasik P, Sandmann G, Van Ooyen A (1999b) Isolation and functional characterisation of a novel type of carotenoid biosynthetic gene from *Xanthophyllomyces dendrorhous*. Mol Gen Genet MGG 262:453–461

Villegas-Méndez MÁ, Montañez J, Contreras-Esquivel JC, Salmerón I, Koutinas A, Morales-Oyervides L (2022) Coproduction of microbial oil and carotenoids within the circular bioeconomy concept: a sequential solid-state and submerged fermentation approach. Fermentation 8(6):258

Wei C, Wu T, Ao H, Qian X, Wang Z, Sun J (2020) Increased torulene production by the red yeast, *Sporidiobolus pararoseus*, using citrus juice. Prep Biochem Biotechnol 50(1):66–73. https://doi.org/10.1080/10826068.2019.1663533

Yan X, Bao H, Bau T (2010) Isolation and identification of one natural pigment from cultured *Cordyceps militaris*. Mycosystema 29(9):777–781. https://doi.org/10.13346/j.mycosystema.2010.05.018

Yara-Varón E, Fabiano-Tixier AS, Balcells M, Canela-Garayoa R, Bily A, Chemat F (2016) Is it possible to substitute hexane with green solvents for extraction of carotenoids? A theoretical versus experimental solubility study. RSC Adv 6(33):27750–27759. https://doi.org/10.1039/C6RA03016E

Yoo AY, Alnaeeli M, Park JK (2016) Production control and characterization of antibacterial carotenoids from the yeast *Rhodotorula mucilaginosa* AY-01. Process Biochem 51(4):463–473

Yousefi M, Rahimi-Nasrabadi M, Pourmortazavi SM, Wysokowski M, Jesionowski T, Ehrlich H, Mirsadeghi S (2019) Supercritical fluid extraction of essential oils. TrAC Trends Anal Chem 118:182–193. https://doi.org/10.1016/j.trac.2019.05.038

Zhao T, Yan X, Sun L, Yang T, Hu X, He Z et al (2019) Research progress on extraction, biological activities and delivery systems of natural astaxanthin. Trends Food Sci Technol 91:354–361. https://doi.org/10.1016/j.tifs.2019.07.014

Fungal-Derived Ceramides for the Cosmetics Industry

16

Roshnai Sen, Anuvab De, Moupriya Nag, Dibyajit Lahiri, Debasmita Bhattacharya, and Tanmay Sarkar

Abstract

Cosmetics are an ancient culture that enhances appearance. In the past, it was limited to the queens of higher-classified women. But now it is available to all, and people also started using cosmetics in their daily lives, which is a great revolution for the cosmetics industry. Cosmetics are usually made of a mixture of various organic or inorganic compounds. Now, the cosmetics industries are focusing on premium organic ingredients with relevant bioactive qualities, including the ability to fight inflammation, aging, wrinkles, and pigmentation. They also have antioxidant and anti-collagenase qualities. There are some fungi-derived molecules found in nature that affect cosmetic products' efficiency. This review focuses on a few fungus-based cosmeceuticals, which offer information on their synthesis, mode of action, efficacy, safety, and possible effects on the cosmetics industry.

R. Sen · A. De · M. Nag · D. Lahiri (✉) · D. Bhattacharya
Department of Biotechnology, Institute of Engineering and Management, Kolkata, University of Engineering and Management, Kolkata, West Bengal, India
e-mail: dibyajit.lahiri@uem.edu.in

T. Sarkar
Department of Food Processing Technology, Malda Polytechnic, West Bengal State Council of Technical Education, Government of West Bengal, Malda, West Bengal, India

Keywords

Cosmetics · Ceramides · TEWL (transepidermal water loss) · Bioremediation

16.1 Introduction

Cosmeceuticals are basically a cross between cosmetics and pharmaceuticals that mainly include topical face creams and lotions made with bioactive chemicals that improve skin's look while also having medicinal effects (Mukherjee et al. 2024). These goods are especially appreciated for their capacity to improve and maintain skin-related functions, providing fixes for a range of dermatological problems. The usage of organic materials in cosmeceutical has significantly changed as a result of consumer desire for goods that are safer, more environmentally friendly, and more biocompatible than synthetic equivalents. The change shows a great change in consumer preferences toward natural ingredients (Mukherjee et al. 2024; Hannun and Obeid 2018).

In recent years, the cosmeceutical industry has shown a great shift toward the advantages of using organic and organically derived materials that promise the safety, biocompatibility, and environmental effects of synthetic components (Zouboulis and Katsambas 2009). The cosmetics industry and consumers are becoming more conscious about the substances in skincare products

S. Ghosh, H. N. Onyeaka (eds.), *Fungal-Based Cosmetics: Formulation and Usage*, The Mycota 17,
https://doi.org/10.1007/978-3-032-09291-5_16

and how they can affect their health (Schnuch and Uter 2016). The use of organic components has grown in popularity as safer as and more effective than synthetic chemicals, which cause negative effects on the environment, skin irritation, and hormone disturbance. The cosmetics industry is continually using organic and organically derived materials from natural sources such as plants, algae, and fungi to fulfill the customer demand for safer, sustainable skincare products (Zouboulis and Katsambas 2009; Schnuch and Uter 2016).

In the cosmeceuticals industry the organic and organically derived materials are becoming more and more popular, especially the fungal-derived components. Fungi can produce a diverse range of bioactive compounds that can be used in skincare products due to their great metabolic flux. It is a great achievement to use fungal strains to produce ceramides. Ceramide is a type of lipid molecule that is essential to the formation and function of the skin's barrier. It creates a protective barrier that protects the skin from external factors such as pollutants, viruses, and UV radiation. It also helps in preventing moisture loss and maintains skin hydration, softness, and overall skin health (Schnuch and Uter 2016).

Ceramides have been recognized since many years ago as an organic component that can be used for skincare. The traditional way to extract ceramides is from animal sources such as the tissues, the skin of cows or pigs, etc. These traditional methods of obtaining ceramides from animals have several difficulties and they also hold lots of ethical concerns about sustainability and animal welfare. Also, the extraction method of ceramides from animal tissues sometimes shows variations in quality. It questions the safety across batches as well as possible contamination with allergens or other unwanted chemicals. There were chemically generated ceramides present in the market before. But it had different properties than biologically derived ceramides. It causes several skin problems such as etching, rashes, allergies, etc. (Figueiredo and Monteiro 2021).

The development of fungal-derived ceramides provides a revolutionary solution to these problems. Using the knowledge of applied biotechnology, researchers have developed methods to use specific fungal strain to produce ceramides through regulated fermentation. The biotechnology methods give control over the quality and consistency of the end product, and it is also a more ethical and sustainable process than the traditional methods. Hence it gets accepted worldwide (Fukushima and Kaneko 2018). There are some fungi strains like *Saccharomyces cerevisiae* and *Pichia pastoris* that produce higher yields of ceramides naturally. The production can be increased by further development like genetic modification or altering the metabolic pathway, etc. (Fukushima and Kaneko 2018; Kikuchi and Ohno 2020). These microorganisms are grown in fermentation tanks under carefully monitored conditions, where the metabolic pathways are used that have been optimized for productivity and efficiency to change simple substrates into ceramides (Fukushima and Kaneko 2018).

Fungal-derived ceramides have an extremely high level of purity, which is one of their main benefits. Comparison with ceramides derived from animal tissues, fungal-derived ceramides are highly purified. There are possibilities of the presence of contaminants or undesired by-product in animal-derived ceramides, while in fungal-derived ceramides the risk is reduced. The purity of cosmetic products is very much important, it ensures the safety and liability of the product. Furthermore, it is possible that by adjusting the fermentation process to produce customized types of ceramide, the skincare compositions can be tailored to suit a variety of skin types and treat particular problems like aging, dryness, or irritation (Smith and Patel 2018).

Fungal-derived ceramides also have many advantages in terms of sustainability. Fungi are an effective resource because they can grow quickly in a very short amount of time, require less effort, and are low-cost to grow. Compared to the traditional method of producing ceramides, the fermentation process is very efficient. It required a very small amount of resources, and it also produced a very small amount of waste (Khan and Damsky 2020). In the cosmetics industry, it is also important to aim toward providing products that are both eco-friendly and effective while also minimizing their negative impact on the environment. The traditional source, or animal-derived ceramides, raises lots of ethical issues. Also, there were many vegan customers who were against using animal-derived ceramides. Hence, fungal-derived ceramides solved both of those issues and increased the number of vegan customers (Khan and Damsky 2020; Poucheret et al. 2006).

Fungi are not only a rich source of ceramides but also some other important nutrients and bioactive compounds that can improve the overall efficacy of skincare products. For example, fungi have significant amounts of proteins, vitamins, and minerals that are also good for skin health. There are also other nutrients such as thiamine, β-glucan, selenium, riboflavin, niacin, pantothenic acid, and folic acid present in fungi that help in skin regeneration, hydration, and defense against oxidative stress, etc. These additional nutrients obtained from fungi can be added to cosmeceutical products to improve their therapeutic effects and attract customers who are looking for comprehensive skincare solutions (Ahmad et al. 2013).

The transition of animal-derived ceramides to fungal-derived ceramides can be considered as an overall shift in the cosmeceutical industry toward more ethical and environmentally friendly outcomes. The awareness of health and biocompatible products increased the need for natural, biocompatible ingredients among consumers. Fungal-derived ceramides have a potential in future as an essential component in cosmetics due to their perfect balance of safety, purity, and environmental sustainability. These developments create new opportunities for the creation of personalized, high-performance skincare products (Ahmad et al. 2013).

Fungal-derived ceramides are having significant growth in cosmeceutical industry as they are easily adaptable, morally acceptable, and sustainable compared to the traditional ceramide sources. Through the combination of fermentation techniques and biotechnology during the process of production, these produced ceramides satisfy both terms and quality and environmental sustainability (Ahmad et al. 2013; Madison 2003). Fungal-derived ceramides have the potential to have a huge impact on skincare in the future by providing customers with natural, eco-friendly, and effective skincare solutions with the growth of the cosmetics industry. Fungal-derived ceramides have a significant role in the future into their scientific basis, advantages, and economic potential with the growth of cosmeceutical industries (Madison 2003).

16.2 Fungal-Derived Ceramides

Fungal-derived ceramides are considered as a revolutionary development in cosmeceutical industry. It provides a sustainable and eco-friendly substitute for the traditional ceramides found in most of the cosmetic products. In the past, ceramides have been made chemically or extracted from animal tissues, which raise serious concerns regarding environmental impact, sustainability, and ethical sourcing. Ceramides are very crucial for maintaining skin health as it helps in the skin's barrier function, more specifically it helps in maintaining moisture and defense against outside stresses. Ceramides are in very high demand in the cosmetics industry due to their ability to effectively deal with problems like dryness, irritation, and symptoms of aging. However, traditional chemical production processes are going through more and more research, which contributes to the search for more sustainable and ethical substitutes (Zhao and Geng 2019).

Fungal-derived ceramides apply advanced biotechnological methods to harness the potential of various fungi and have become known as a potential solution to skin problems such as aging, dryness, or irritation. These ceramides are produced by carefully selected and modified fungal strains for high-yield ceramide production. Fungi are especially useful for this because of their wide range of metabolic capacities, which are easily cultivated, and there are very low environmental impacts. Animal-derived ceramides raise concerns about sustainability and animal welfare or are chemically synthesized which may require energy-intensive procedures and produce harmful byproducts. Fungal-derived ceramides provide a more environmentally and ethically acceptable alternative (Mao and Obeid 2008).

Saccharomyces cerevisiae, or baker's yeast, is one of the main fungal species used in the Ceramide synthesis. It is very easy to manipulate the genetic material of this species. That is why this species is considered as the pillar of industrial biology. *Saccharomyces cerevisiae* is a wonderful option for metabolic engineering to biologically modify ceramide production because it will open several ways in research for having well-recognized genetic pathways. Similarly, another yeast species called *Pichia pastoris* is commonly used in the biotechnological ceramide synthesis procedure. Because *Pichia pastoris* can grow to high cell densities in fermentation vessels, it is a very ideal strain for industrial production on a big scale. It is possible to modify both of these yeasts to improve biosynthesis of ceramides, providing an affordable and effective production method (Fukushima and Kaneko 2018; Kikuchi and Ohno 2020; Lee and Park 2019).

Apart from yeast species, other varieties of fungi can also synthesize ceramides. The *Aspergillus oryzae* is a species that has been widely used in traditional Asian fermentation techniques. This is commonly used for the making of miso, sake, and soy sauce (Kim and Kim 2021). Researchers found that this species has inbuilt fermentation capability and it can simply adapt to the industrial environment hence it has shown a great potential to produce ceramides. The species *Ustilago maydis*, which is responsible for smut in corn, is another remarkable species. The species *Ustilago maydis* is mainly known as a plant pathogen, however, it has shown great potential in biotechnology for the synthesis of ceramides. These fungal species can be modified to generate highly pure ceramides in an environmentally friendly and cost-effective process with the combination of modern fermentation procedures (Chen and Zhang 2020).

The use of fungal-derived ceramides represents an important step in the development of healthier and more ethical cosmetics. The consumers are becoming more aware of environmental and ethical issues, and they are putting more and more demand on cosmetics to contain natural and ethically derived components. The fungal-derived ceramides provide products that promote skin health and conform to the ethical sourcing and sustainability standard. Compared to traditional methods, the production process for ceramides obtained from fungi uses very less resources, produces less waste, and requires very less raw materials. The ability to produce fungus on a large scale without destroying natural resources and using renewable substrates improves the sustainability of the ceramides (Figueiredo and Monteiro 2021).

Furthermore, fungal-derived ceramides have several useful advantages for skincare. Compared to the traditional method of producing ceramides, these tend to be more pure, which reduces the possibility of contaminants that could irritate the skin or cause allergic reactions. Since ceramides are produced biotechnologically, the companies have more control over the final product and they can easily create unique formulas customized to specific skincare needs. In the cosmeceutical industry, this personalization is very beneficial in the treatment of certain ailments like eczema, psoriasis, or age-related skin issues. The cosmeceutical industries can produce ceramides with specific qualities that can increase their effectiveness in treating various disorders by manipulating the production process (Figueiredo and Monteiro 2021; Smith and Patel 2018).

Fungal-derived ceramides have a considerably less environmental impact than traditionally derived ceramides, which are obtained from animals or chemically synthesized. Compared to the traditional methods, the cultivation of fungi uses less space, water, and energy, and there are almost zero greenhouse gas emissions throughout the fermentation process. Another way to reduce the environmental impact of the fungal-derived ceramide production is to develop the fungus in such a way that it can be used as waste items or by-products as materials for nutrients. This is a great initiative taken by the cosmeceutical industry to eliminate environmental impact during the production of ceramides (Smith and Patel 2018).

Ceramides produced from fungi are considered a revolutionary development in the cosmeceutical industry. They solve several issues such as the treatment of animals, utilization of resources, and impact on the environment; they provide a sustainable, moral, and eco-friendly substitute for traditionally derived ceramides. A future development in the cosmeceutical industry shows that there are several significant roles for fungal-derived ceramides, as customer demand for sustainable and natural skincare products grows significantly. Their capacity to provide customized, high-quality skincare treatments while reducing ethical and environmental problems places them in a significant position as a component of the cosmeceutical industry. In addition to improving skincare products' sustainability, further development and modification of fungal-derived ceramide production will contribute to the great achievement of developing a more ethical and environmentally friendly cosmetics industry (Smith and Patel 2018; Huang and Liu 2019).

16.3 Ceramides: The Foundations of Skin Obstruction Function

Ceramides are a type of sphingolipids that are essential for preserving the structure and functionality of the skin's barrier. These lipid molecules provide a higher amount of the intercellular lipid matrix. They are mostly found in the outermost layer of epidermis, the stratum corneum. The stratum corneum is made up of dead, flattened skin cells called corneocytes, as well as the complicated structure of lipids that fill the spaces between them. The lipid matrix, including cholesterol, free fatty acids, and ceramides, is responsible for the skin's two primary barrier functions, one is limiting water loss and another is blocking the entry of harmful substances. Hence, ceramides are essential components of this lipid matrix for maintaining the skin's resilience, health, and attractiveness (Elias and Choi 2005).

16.3.1 Barrier Function

Ceramides play a very crucial role in the development and maintenance of the skin's hydrophobic intercellular lipid bilayers. This barrier helps in protecting the skin from external stresses that could otherwise penetrate the epidermis and trigger immunological or inflammatory reactions, such as allergens, irritants, pollutants, and infections. Ceramides also helps in limiting trans-epidermal water loss (TEWL). In this process water evaporates from the skin and into the surrounding environment. Ceramides aid to keep moisture in and keep the skin moisturized, supple, and elastic by creating a semi-permeable barrier. Insufficient ceramide levels can cause the lipid bilayers to break down, which increases TEWL, dryness, and vulnerability to infections and environmental harm (Holleran et al. 2006).

16.3.2 Hydration

Ceramides assist the skin's barrier function, which indirectly but significantly contributes to skin hydration even though they don't directly bind water molecules. Minimizing TEWL is greatly aided by a ceramide-rich, well-organized lipid barrier. Maintaining the moisture balance is essential to maintaining the smooth, moisturized, and free-of-pain epidermis that comes with dry skin (Ishikawa and Nishikawa 2018). On the other hand, the skin's barrier becomes weakened and more water loss occurs when there is a deficiency in ceramides, which can occur from aging, environmental factors, or skin diseases like atopic dermatitis. Dry, rough, and itchy skin is the result of this moisture loss, and it can make psoriasis and eczema worse. Therefore, restoring the skin's general health and moisture levels can be facilitated by replacing ceramides with topical treatments or other interventions (Ishikawa and Nishikawa 2018; Bouwstra and Ponec 2006).

16.3.3 Cell Separation

Ceramides play a significant role in the maturation and upward migration of keratinocytes, the major cells in the epidermis, during the complicated and closely regulated process of keratinocyte differentiation. During this phase, keratinocytes perform some morphological and metabolic changes that prepare them to create the cornified envelope, a strong protective layer that forms on the skin's surface. Ceramides play an important role here. The ceramides modify the structure and composition of the cornified membrane via their interaction with proteins such as involucrin and loricrin, to form the barrier. Ceramides provide an important role in the stratum corneum's lipid makeup, assuring that the cornified cells are surrounded in a strong lipid matrix. This matrix is essential to the barrier's overall performance because it provides mechanical strength as well as hydrophobicity, preventing water loss and the entry of harmful substance (Elias and Choi 2005; Ishikawa and Nishikawa 2018; Bouwstra and Ponec 2006).

16.3.4 Immunomodulation

Ceramides are a kind of variable component of the skin barrier. It constantly modifies the skin's immunological responses. According to current research, certain ceramide species possess anti-inflammatory properties that help to regulate inflammatory processes in the skin. For example, certain ceramides have the ability to inhibit the release of cytokines that promote inflammation, which can reduce irritation and aid in the healing of the skin. On the other hand, abnormal ceramide profiles have been connected to increased inflammation and barrier dysfunction in inflammatory skin disorders such as psoriasis and atopic dermatitis. These imbalances in ceramide composition have also been linked to these conditions. Ceramides' dual function as immune response modulators and barrier integrity defenders highlights how crucial they are to preserving general skin health. Developing knowledge about the interactions between various ceramide species and immune cells as well as signaling pathways creates new opportunities for the development of targeted therapeutics that address the immunological and structural components of skin problems (Nguyen and Eisinger 2018).

16.3.5 Clinical Implication and Skincare Strategies

Ceramides play so many different roles in cosmetics that their significance cannot be overstated. Ceramides and their role in skin health must be thoroughly understood in order to treat skin issues related to the skin barrier. Skin diseases such as ichthyosis, psoriasis, and eczema can become less effective barriers to external stimulation when ceramide levels are low. Skincare professionals can aid impacted individuals by developing products that restore ceramide levels, which will improve barrier function, increase hydration, and reduce inflammation. Additionally, as ceramide levels naturally decline with age, incorporating ceramides into regular skincare routines can benefit both the general public and those who suffer from certain skin conditions (Elias and Choi 2005; Ishikawa and Nishikawa 2018).

Innovative skincare formulations that copy the natural lipid balance of the stratum corneum especially well when they combine ceramides with other lipid components like cholesterol and fatty acids. These formulas can support the skin's barrier function, stop moisture loss, and offer protection from pollutants (Choi and Kim 2017). New methods that take advantage of ceramides' immunomodulatory qualities may surface as research into their many functions develops. These methods would provide more individualized and accurate skincare regimens that target the biological as well as the physical components of skin health (Takeshita and Yamada 2018).

Ceramides are essential for maintaining the skin's moisture, barrier integrity, and overall wellness. Their significance in the prevention and treatment of skin problems is highlighted by their functions in preserving lipid bilayers, promoting keratinocyte differentiation, and regulating immunological responses (Zhang and Wang 2021). The possibility to create more focused and efficient skincare solutions that preserve and repair skin barrier function is becoming more and more appealing as our understanding of ceramides grows. Understanding the complex relationship between ceramides and skin health allows scientists and formulators to keep coming up with new ideas and solutions that not only improve the skin's look but also guarantee its strength and resilience in the face of environmental stresses (Kim and Lee 2019).

16.4 Extraction of Fungal-Derived Ceramides: An Application of Bioprocessing

16.4.1 Cultivation of Fungi

Fungal-derived ceramides are produced by several complicated processes, each of those is necessary to ensure the final product's quality and efficacy when used in cosmeceuticals.

The first step in the procedure is the cultivation of fungus. Some specific fungal strains, like *Aspergillus oryzae*, *Saccharomyces cerevisiae*, or *Mortierella alpina*, are chosen because they possess natural capacity to make ceramides (Fukushima and Kaneko 2018; Kikuchi and Ohno 2020; Liu and Zheng 2019). As a medium, well-balanced mixture of carbon sources, such as glucose, nitrogen sources, such as peptone, and vital minerals and vitamins is used. These nutrients enhance the production of ceramides and increase the fungal growth. The fungi's metabolic activity is maintained by carefully regulating environmental factors such as temperature, pH, aeration, and agitation. This has a direct effect on the quantity and quality of produced ceramides. Depending on the needs of the production, the cultivation can take place in a variety of bioreactor systems, from large-scale continuous bioreactors to small-scale batch reactors (Kikuchi and Ohno 2020; Liu and Zheng 2019).

16.4.2 Cell Harvesting

The process of harvesting cells, which involves separating the fungal biomass from the culture media, comes after the growth has reached its peak. For this, methods like filtering or centrifugation are frequently used. While filtration entails passing the culture through a membrane to retain the cells, centrifugation uses high rotational speeds to pellet the fungal cells. Now that the harvested cells have been condensed into biomass, they can be processed further (Kikuchi and Ohno 2020; Liu and Zheng 2019; Yang and Xu 2021).

16.4.3 Cell Disruption

The fungal cells are disrupted after harvesting to release inner ceramides. Fungi have hard cell walls that must be broken down by mechanical or physical means. Common methods include bead milling, which grinds the cells with tiny beads, ultrasonication, which uses high-frequency sound waves to crack open the cells, and high-pressure homogenization, which forces the cells through a narrow area at high pressure, forcing them to burst. The desired efficiency of ceramide recovery and the hardness of the cell wall determine which disruption approach is best (Song and Liang 2019; Bai and Li 2020).

16.4.4 Lipid Extraction

Following the disruption, the process of lipid extraction starts to separate the ceramides from the cell lysate. In a biphasic extraction technique, organic solvents like methanol and chloroform are usually used for this (Bai and Li 2020; Folch et al. 1957). Ceramides and other lipids are dissolved by the solvents, which makes it possible to extract them from the cell lysate's aqueous components. Repeating the extraction procedure multiple times can optimize the recovery of ceramides. After extraction, a crude lipid mixture is obtained by evaporating the solvent (Silva and Almeida 2019).

16.4.5 Purification

Ceramides are separated from other lipid species by purifying this crude lipid extract. Thin-layer chromatography (TLC) and high-performance liquid chromatography (HPLC) are two popular chromatographic techniques. Ceramides can be separated using HPLC according to the interaction between the stationary phase and the mobile phase. Ceramides can also be found and isolated using TLC, this method divides lipids on a thin plate according to their polarity. The refined ceramides are used for additional examination (Miller and Simon 1985; Sato and Sato 2020).

16.4.6 Characterization and Quantification

To ensure that the purified ceramides achieve the required quality criteria, they must be characterized and quantified. The molecular structure and purity of the ceramides are verified by methods such as mass spectrometry (MS), nuclear magnetic resonance (NMR) spectroscopy, and Fourier-transform infrared (FTIR) spectroscopy (Muir and Yamauchi 2016). Ceramide concentrations can also be measured using quantitative techniques such as gas chromatography (GC) or high-performance liquid chromatography coupled with mass spectrometry (HPLC) and MS. This maintains homogeneity between batches (Hsu and Popp 2018).

16.4.7 Formulation for Cosmeceuticals

At last, the ceramides are prepared to be marketed as cosmetic products. This involves the manipulation of ceramides into various cosmetic products such as emulsions, creams, lotions, and serums. The ceramides' stability, compatibility with other ingredients, and the desired texture and effectiveness of the final product must all be carefully checked during the formulation process. The performance and shelf life of the product may be improved by the addition of stabilizers, emulsifiers, and other additives. The modified ceramides' bioactive qualities, which support the skin barrier, maintain moisture, and improve general skin health, are very much beneficial to the cosmeceutical industry (Kang and Kim 2019).

16.5 The Work Function of Ceramides

16.5.1 Strengthening the Skin Barrier

Ceramides play an important role in strengthening the skin barrier. Ceramides provide a significant amount of the lipid matrix that holds skin cells together in the stratum corneum, the outermost layer of the skin. Corneocytes, or the dead skin cells, are placed in the stratum corneum. This is a lipid matrix that is primarily made up of free fatty acids, cholesterol, and ceramides. Ceramides are the most common among the other lipid molecules; half of the lipids in this layer are made up of ceramides. Because of the lamellar structure of these lipids, a barrier is formed that is crucial for retaining skin hydration, guarding against aggressors in the environment, and blocking the entry of dangerous infections and irritants (Madison 2003; Rawlings and Harding 2004).

The skin barrier works similarly to a wall made of bricks and mortar, with the lipid matrix, especially ceramides, serving as the mortar and the corneocytes as the bricks. By assisting in the sealing of the spaces between skin cells, ceramides aid in stopping excessive transepidermal water loss (TEWL) (Holleran et al. 2006). A moisturized skin barrier is stronger and more resistant to dryness, irritation, and injury. Ceramides help to keep the skin moisturized, which is essential for preserving the skin's suppleness, smoothness, and softness.

They achieve this by reducing TEWL (Rawlings and Harding 2004).

The barrier receives damages when the natural ceramide levels are reduced. As a result of aging, severe environmental circumstances, or using aggressive skincare products can occur. Increased TEWL may result in making the skin rough, dry, and more sensitive to infections, allergies, and irritants. When the skin's barrier is getting damaged, it becomes more sensitive to inflammatory skin diseases such as dermatitis and eczema. Furthermore, as the capacity of the skin to maintain moisture and protect itself from external stresses decreases, a weaker barrier increases the appearance of fine lines, wrinkles, and other signs of aging.

If ceramides are commonly used in skincare products such as serums and moisturizers, they can be applied directly to restore the skin's natural lipid levels and repair the skin's barrier. These ceramides help to maintain the skin barrier by filling in the gaps between corneocytes and reintegrating into the skin's lipid matrix. This improves the skin's defense mechanisms and preservation of moisture capacity, making it less sensitive to irritation and more resistant to external stresses. Regular usage of ceramide-rich products can help in building up more healthier, more moisturized skin with a better texture and fewer signs of aging over time.

Ceramides are essential to the skin's barrier function because they keep the skin hydrated, shield it from the elements, and promote general skin health. They are essential to skincare because of their capacity to strengthen the skin barrier, particularly for people who want to preserve or rebuild the skin's natural defenses.

16.5.2 Enhancing Skin Hydration

Ceramides are very much essential for enhancing skin hydration because they create a barrier that prevents moisture loss and increases water retention in the skin. This is an important function because healthy, durable skin depends on maintaining proper moisture levels. Lipids called ceramides are found naturally in the stratum corneum, the outermost layer of skin. They produce a barrier that keeps moisture in and protects the skin from environmental stressors by collaborating with other lipids like cholesterol and fatty acids (Rawlings and Harding 2004).

The lipids, such as ceramides, serve as the mortar keeping the skin cells (corneocytes), which make up the stratum corneum, together like bricks in a wall. Transepidermal water loss (TEWL), the mechanism by which water evaporates from the skin's surface, is inhibited by this structure. This barrier holds while the skin's ceramide levels are adequate, keeping the skin moisturized, smooth, and soft (Holleran et al. 2006; Rawlings and Harding 2004).

However, several things can reduce the skin's natural ceramide levels, including aging, environmental damage, and harsh skincare treatments. When this occurs, TEWL increases because the barrier is breached. Skin that has lost moisture is flaky, dry, and more prone to irritation. Since moisturized skin is more elastic and resistant to aging symptoms, this dryness can also hasten the development of fine lines and wrinkles (Holleran et al. 2006; Rawlings and Harding 2004; Schmuth and Elias 2012).

Skincare products containing topical ceramides aid in restoring the skin's natural lipid balance and moisture-retaining capacity. Ceramides increase the skin's barrier function by incorporating into its lipid matrix, which lowers TEWL and increases skin moisture. This eventually results in skin that is smoother, and softer and has improved texture and appearance. Furthermore, moisturized skin is less prone to irritation and injury since it is better able to defend against environmental aggressors (Holleran et al. 2006; Rawlings and Harding 2004; Schmuth and Elias 2012).

16.5.3 Repairing Damaged Skin

Through the restoration and reinforcement of the skin's natural barrier, ceramides play a critical role in the process of mending injured skin. The first line of defense against environmental stresses, pathogens, and irritants is the skin barrier, which is mainly made up of ceramides, cholesterol, and fatty acids. The skin loses its capacity to retain moisture and protect itself when this barrier is disrupted, which can happen as a result of aging, severe weather, frequent scrubbing, or skin diseases like eczema. This can cause dryness, irritation, and increased sensitivity (Schmuth and Elias 2012).

Damage to the skin barrier causes a continuous cycle of moisture loss, irritation, and increased skin degradation. The lack of sufficient ceramides destroys the lipid matrix. As a result, the skin cells are destroyed, leading to skin barrier degradation. These gaps enable transepidermal water loss (TEWL), as well as it allows the entry of harmful substances and outside stresses into the skin. As a result, the skin gets dry, flaky, and irritated skin, which is more vulnerable to infections, redness, and other signs of disease (Holleran et al. 2006; Rawlings and Harding 2004; Schmuth and Elias 2012).

Externally applied ceramides are very effective in repairing skin damage and healing injuries. When the ceramides are applied externally to the skin, they close the gaps between skin cells and restore the barrier's integrity by absorbing into the lipid matrix. This method can minimize TEWL and allow the skin to maintain moisture and recover to normal levels of hydration. When the skin's barrier is repaired, inflammation and sensitivity decrease, making it less sensitive to irritants and environmental stresses (Holleran et al. 2006; Rawlings and Harding 2004; Schmuth and Elias 2012).

Furthermore, ceramides help restore the skin barrier, which not only reduces sensations of dryness and irritation but also improves the skin's overall health. Ceramides help the skin's natural healing processes by restoring moisture and protecting it from the environment, hence improving skin regeneration and repair. This eventually results in a skin surface that is more robust, smoother, and has better texture with less aging symptoms (Zhang and Liu 2017).

Ceramides are crucial for healing injured skin by strengthening the skin's barrier, lowering moisture loss, and preventing more damage. Their capacity to reinstate barrier function renders them an indispensable component in skincare formulations intended to remediate, calm, and safeguard vulnerable or irritated skin, culminating in the development of more robust and healthful skin (Zhang and Liu 2017).

16.5.4 Reducing Signs of Aging

By preserving skin hydration, strengthening the skin barrier, and enhancing skin suppleness, ceramides significantly contribute to the reduction of indications of aging. A weakening skin barrier, increased moisture loss, and a reduction in skin resilience are caused by the skin's natural production of ceramides declining with age. These alterations add to the outward manifestations of aging, including wrinkles, drooping skin, fine lines, and rough texture (Rawlings and Harding 2004).

Restoring and preserving the skin's moisture balance is one of the main ways ceramides prevent aging. Well-hydrated skin appears smoother, plumper, and younger because moisture fills in wrinkles and keeps new ones from forming. Ceramides are essential for maintaining the lipid matrix of the skin and aid in stopping trans epidermal water loss (TEWL). This moisture retention makes the skin more youthful-looking and vibrant-looking by enhancing its natural glow and improving its texture and suppleness (Holleran et al. 2006; Rawlings and Harding 2004; Schmuth and Elias 2012).

Ceramides not only hydrate the skin but also fortify its barrier, which deteriorates with age. In order to shield the skin from environmental stressors like pollution, UV radiation, and free radicals—all of which hasten the aging process—a robust barrier is essential. Ceramides help protect the skin from these harmful elements by strengthening the skin barrier. This reduces the risk of premature aging and maintains the overall health and vitality of the skin (Zhang and Liu 2017).

Ceramides help to the skin's overall strength by maintaining the bonds of its cells. As the skin ages, it loses rigidity and elasticity due to the breakdown of collagen and elastin. Ceramides have no direct effect on the production of collagen, but they indirectly contribute to making a stronger and more flexible skin structure by maintaining hydration and barrier function. Ceramides minimize the development of wrinkles and loosen the skin by maintaining the outermost layer of skin smooth and intact, resulting in a younger appearance (Zhang and Liu 2017; Fluhr and Maibach 2007).

Furthermore, the damages due to age can be repaired by using ceramides' medicinal qualities. The skin's natural healing mechanisms weaken with time, resulting in long-term damage that appears as aging symptoms. Ceramides help in the regeneration of healthy skin cells, reduce inflammation, and aid in the healing of microdamage to the skin barrier. These help in smoothing up the skin tone, reducing roughness, and help to hide the wrinkles and other imperfections of the skin (Zhang and Liu 2017).

16.5.5 Protecting Against Environmental Damage

Ceramides are very helpful for maintaining the skin's natural barrier and making it more resistant to outside stresses. This helps in protecting the skin from harm. The skin barrier, made up of ceramides, cholesterol, and fatty acids, provides the body's primary defense against several external stresses, such as pollution, UV radiation, extreme weather, and other pollutants (Fluhr and Maibach 2007).

Environmental factors such as UV radiation, pollution, or other extreme weather may damage the skin barrier. The weakening of the barrier makes it more vulnerable to losses of moisture and less effective at keeping harmful substances away. This can cause increased dryness, irritation, inflammation, and premature skin aging. Furthermore, as external forces might enter the skin layer more easily, the skin may become more vulnerable to infections and allergies (Fluhr and Maibach 2007).

Ceramides help reduce these effects by enhancing the lipid matrix and binding skin cells together. They form a protective barrier on the skin's surface that helps maintain the moisture inside and keeps away the harmful substances out. This barrier function is essential for reducing transepidermal water loss (TEWL) and maintaining ideal hydration levels. Ceramides help to keep the skin hydrated. It is less sensitive to outside forces that cause dryness and irritation by maintaining the moisture (Holleran et al. 2006; Rawlings and Harding 2004; Schmuth and Elias 2012).

Ceramides not only help in staying properly hydrated, but they also serve as a barrier against pollutants and UV rays. Sunlight's UV rays can harm skin cells and induce oxidative stress, which can result in hyperpigmentation, early aging, and a higher chance of developing skin cancer. Ceramides improve the skin's natural repair processes and strengthen the skin's barrier, minimizing the effects of UV exposure. Ceramides, on the other hand, function as a barrier that reduces the entrance of pollutants and environmental toxins into the skin, protecting against them and causing irritation and damage (Fluhr and Maibach 2007).

Ceramides also have anti-inflammatory qualities that aid in calming the skin and lessening redness and inflammation brought on by stresses in the environment. Ceramides contribute to a more durable and balanced skin surface that can resist daily environmental challenges by reducing inflammation and strengthening the skin barrier (Kang and Kim 2019; Rawlings and Harding 2004; Schmuth and Elias 2012; Zhang and Liu 2017; Fluhr and Maibach 2007) (Fig. 16.1).

16.6 Role Bioinformatics and Synthetic Biology

The synthesis of ceramides produced by fungi is being transformed by applied bioinformatics and synthetic biology, which presents huge possibilities for the cosmetics industry (Park et al. 2010). The identification and definition of important genes involved in the ceramide biosynthesis pathways are made possible by the thorough investigation of fungal genomes made possible by bioinformatics. Bioinformatics assists

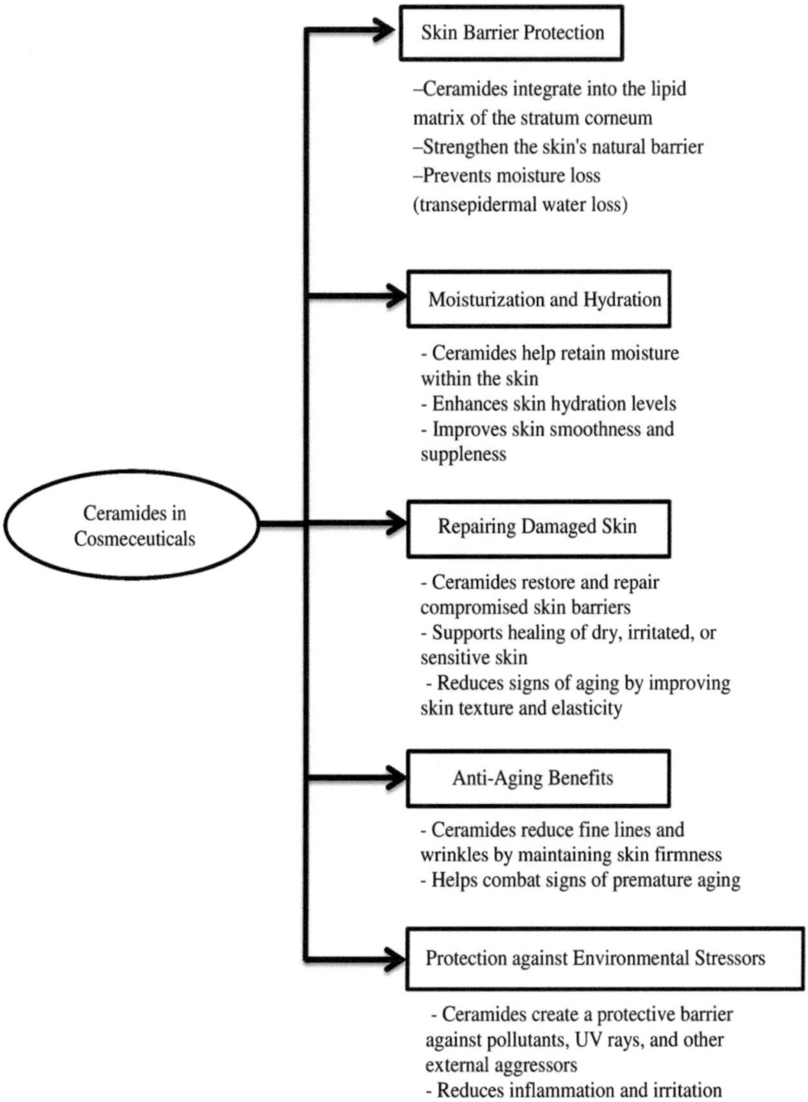

Fig. 16.1 Use of ceramides in cosmeceuticals

in forecasting how changes to these pathways will increase ceramide production, increase efficiency, and decrease byproducts using computer modeling and data analysis (Park et al. 2010; Kaiser and Surade 2019). The design of optimal microbial strains must be guided by this information.

This is further enhanced by synthetic biology, which makes it possible to precisely construct these genetic pathways in fungal hosts or other microbes such as yeast (Kaiser and Surade 2019). Scientists may adjust the lipid profiles to match specific skincare needs, including improved moisturizing or barrier-repair capabilities, and greatly increase ceramide supply by overexpressing or inserting particular genes. They can even create fully synthetic pathways. Furthermore, novel ceramide variations that may provide special advantages not present in naturally occurring ceramides can be created through the use of synthetic biology (Paddon and Keasling 2014; Nielsen and Keasling 2016).

Synthetic biology and bioinformatics work together to create customized, high-performance ceramides that meet the demands of modern, sustainable skincare trends while also increasing manufacturing efficiency (Huang and Liu 2019). These technologies are poised to be crucial in ensuring that fungal-derived ceramides can be produced at scale, with consistent quality and minimal environmental impact, thereby driving innovation and sustainability in the cosmetics industry. This is because the demand for eco-friendly, biotech-driven cosmetics is growing (Czar et al. 2009; Winter et al. 2014).

16.6.1 Bioinformatics

The study of bioinformatics involves the analysis of biological data using computational methods and tools, mainly the genetic and proteomic data of living things. Bioinformatics is important in the context of fungal-derived ceramides in multiple ways:

16.6.1.1 Gene Identification and Sequencing

Fundamental steps in using bioinformatics to produce ceramides from fungi include gene identification and sequencing. These procedures involve the analysis of the enormous amount of genetic data found in fungi using advanced bioinformatics tools and algorithms. Initially, the full genetic code of a fungus species known to make ceramides is decoded by researchers using genome sequencing tools. This has to do with high-throughput sequencing techniques like next-generation sequencing (NGS), which produce massive amounts of data that need to be processed and examined to identify the genes that are important (Tatusova et al. 2015).

The methods of bioinformatics help in sequencing the fungal genome, the annotation process, which involves the discovery of genes and the determination of their functions. The technique of comparative genomics allows researchers to identify similar genes known to be related to ceramide synthesis in several fungal species. This is further strengthened by transcriptome analysis, which examines gene expression patterns under different growth conditions to identify genes that are actively involved in ceramide synthesis (Tatusova et al. 2015).

Bioinformatics makes it possible to predict which genes may encode enzymes that have been identified. These enzymes are important components of metabolic pathways leading to ceramide formation. The structure and function of these enzymes can be predicted by researchers using databases and computational tools, which facilitate the design of studies to confirm their role in ceramide formation. Furthermore, bioinformatics can identify the regulators that control the expression of these genes. It provides information on how to modify or improve ceramide production (Tatusova et al. 2015; Schuster 2008).

Overall, gene identification and sequencing provide comprehensive information to understand the genetic and molecular basis of the synthesis of fungal-derived ceramide. This information is essential for further application in synthetic biology to optimize production procedures, increase ceramide yields, and create modified ceramides. The variations of ceramides with advanced properties will be beneficial for the cosmeceutical industry (Schuster 2008).

16.6.1.2 Metabolic Pathway Analysis

The most important field of bioinformatics is analyzing the metabolism, including identification. It helps in mapping complex biochemical mechanisms that regulate the metabolic pathways that help in the production of fungal-derived ceramides. Starting with the substrate and ending with the final ceramide products, this study shows that all metabolic pathways and enzymes are involved in the synthesis of ceramides. Researchers can recreate the complete ceramide production process and identification of important intermediates, metabolic flux, regulatory mechanisms, etc. along the system by using the methods of bioinformatics and database (Kanehisa et al. 2016).

The connection between the various enzymes in the method and the metabolic product inside the pathway is one of the main benefits of metabolic pathway analysis. This helps scientists to identify the rate-limiting factors that can prevent ceramide synthesis. For example, bioinformatics can help identify genetic changes or enzyme engineering techniques that improve the function of specific enzymes in a pathway that has been identified as a limiting factor due to low levels of activity or expression. This may involve overexpressing the enzyme, modifying the active site to improve catalytic efficiency, or introducing alternative enzymes from other organisms that are more efficient to achieve the same goal (Schuster 2008; Kanehisa et al. 2016).

Furthermore, analyzing metabolic pathways helps researchers to understand the relationship between the production of ceramides and other metabolic pathways in fungal cells. Ceramide synthesis is frequently related to other lipid metabolic pathways that may be identified through bioinformatics methods. This helps scientists to identify alternative competitive pathways that shifts energy or components away from the biosynthetic process of ceramide, this results in the significant decrease of production. Methods for reducing these incompatible pathways could be done by shifting metabolic flux in order to promote ceramide production or by negatively regulating specific genes (Kanehisa et al. 2016).

Furthermore, pathway analysis helps to identify potential feedback mechanisms that regulate ceramide synthesis. For example, bioinformatics may suggest genetic or chemical treatments to reduce feedback inhibition caused by ceramide residue, which inhibits upstream enzymes. Understanding these regulatory networks is necessary for developing fungal strains that are capable of producing huge quantities of ceramide in the face during shifting conditions for development.

Furthermore, metabolic pathway analysis provides the basis for systematic biology approaches, which help simulation and modeling of an organism's entire metabolic system. These models help scientists to predict how environmental changes, genetic mutations, and availability of substrate will affect ceramide synthesis. It makes it easier to produce fungal strains that are specifically suitable for industrial manufacturing. Processes: This type of model of prediction can be extremely helpful in strain engineering (Schuster 2008; Kanehisa et al. 2016; Zengler and Palsson 2012).

Last but not least, the research of metabolic pathways using the methods of bioinformatics helps to understand how the pathways can be modified and optimized while the complete map of ceramide manufacturing process is provided. Researchers can improve the production of ceramides by increasing manufacturing efficiency. They can also produce modified ceramide derivatives with better function for cosmetics industry by studying biochemical interactions and regulatory mechanisms involved in the process (Kanehisa et al. 2016; Zengler and Palsson 2012).

16.6.1.3 Protein Structure Prediction

Protein structure prediction is a key factor of bioinformatics that helps to enhance the knowledge of the enzymes involved in ceramide production. Enzymes, or biological catalysts, catalyze the biochemical pathways in the production of ceramides. The effectiveness of enzymes has a significant impact on both the quantity and quality of ceramides produced. The prediction of the structure of these enzymes is very much crucial for improving their stability and activity because an enzyme's three-dimensional structure is essential to its function (Jones 1999).

The initial phase in the structure prediction of a protein is to determine its amino acid sequence. The prediction can be done using gene sequences identified during the genome research. Based on this sequence, bioinformatics approaches such as homology modeling, ab initio modeling, and threading methods are utilized for determining the enzyme's three-dimensional structure. For example, homology modeling is based on the finding that proteins with similar sequences tend to have similar structures. Researchers can construct a model of the structure of an enzyme by comparing its sequence to databases such as the Protein Data Bank (PDB), which contain information on known protein structures (Jones 1999; Zhang and Skolnick 2004).

After a structural model has been developed, it can be examined to determine significant features such as the active site, substrate binding region, and areas that support the stability of the enzyme. Since the substrate molecules attach and undergo chemical changes in the active site, it is very important for understanding its shape. The production of ceramides may be decreased by an enzyme with poorer catalytic efficiency if the active site is not arranged in the best possible way. By simulating substrate binding to the active site, bioinformatics techniques enable researchers to pinpoint possible problems and make focused changes that increase binding affinity and catalytic cycle (Zhang and Skolnick 2004).

This technique predicts the active site and overall stability of enzymes using protein structure. Traditional enzymatic pathways have been shown they be unsafe for industrial ceramide production, in part because enzymes are sensitive and can easily lose activity in situations such as high temperatures or changing pH over time. Unstable enzymes, such as those not active in industrial conditions like for instance at high temperatures or unstable pHs can make it challenging to synthesize ceramides. These enzymes will denature or even lose their activities. By predicting flexible loops or weak intermolecular interactions, researchers can develop new mutations that lead to improved enzyme stability. This might mean mutating amino acids to more stable alternatives, adding disulfide bonds or improving hydrogen bonding networks (Jones 1999; Zhang and Skolnick 2004).

In addition to this, protein structure prediction allows for the creation of modifications or permutations in enzymes that have some extra features or functions. Scientists can make enzymes more selective for a substrate, resistant to an inhibitor, or function under non-natural conditions; using directed evolution and rational design techniques. The enzymes on which they are based can be developed and added to the production of fungal strains to increase overall ceramide yields while also providing a larger array system that is specific to the environment (Jones 1999; Zhang and Skolnick 2004).

It is also possible that researchers can even predict protein structures and build up derivatives from enzymes with distant or entirely different properties. This allows for well-described simulations of flexibility, conformational changes, and enzyme-substrate/cofactor interactions which can then guide optimization (Jones 1999; Zhang and Skolnick 2004).

Last but not least, protein structure prediction has proven to be an effective method for improving enzymes for ceramide production. Bioinformatics has improved the understanding of enzyme function and structure, helping us to design new enzymes or improve current models through methods like directed evolution or semi-rational design to achieve desired functions and stability. The approach produces faster results, resulting in more efficient production techniques that allow for higher yields with lower costs, as well as increasing the renovations associated with generating new ceramide-based products for the cosmetic industry (Zhang and Skolnick 2004).

16.6.1.4 Data Integration and Modeling

The basic bioinformatics techniques in modeling and data integration serve the aim for us to better interpret the complexity underlying the kinetics of ceramide genes, thus improving their production. The ceramide biosynthesis process can be better understood by using integrative models, in which information from multiple biological databases such as proteomic, metabolomics, and genomic are combined into a single model to depict the whole system. This comprehensive animal-to-silos approach gives researchers a better understanding of how different biological parts are connected and have the potential to influence production efficiency.

16.6.1.4.1 Genomic Data Integration

The basic bioinformatics techniques in modeling and data integration serve the aim for better interpret complexity underlying kinetics of ceramide genes, thus improving their production. The ceramide biosynthesis process can be better understood by using integrative models, in which information from multiple biological databases such as proteomic, metabolomics, and genomic are combined into a single model to depict the whole system. This comprehensive animal-to-silos approach gives researchers a better understanding of how different biological parts are connected and have the potential to influence production efficiency.

16.6.1.4.2 Proteomic Data Integration

Proteomic data is a study of how proteins perform their functions in an organism. It corresponds to the genomic data by identification of various proteins that are expressed based on genes engaged in ceramide synthesis. The researchers use mass spectrometry and other proteomic techniques to identify the presence of protein in fungal cells, determine their quantities, and understand their interactions. The combination of proteomic and genomic data helps not only to identify those proteins that are actively involved in ceramide synthesis but also to confirm the relevance of their functions based on gene activity. This integration also helps identify post-translational modifications that may affect enzyme stability and activity (Jones 1999; Zhang and Skolnick 2004).

16.6.1.4.3 Metabolic Data Integration

These results represent metabolic data profile for substrates, intermediates, and ceramide products, which are principally composed of very small particles that are involved in various cellular metabolic processes. Measurement of the levels of metabolites and metabolic fluxes can give information on how metabolic pathways operate in vivo, which might be informative for the Systems Biology scientist. Using metabolomics data combined with genomic and proteomics will enable the complete mapping of ceramide production across metabolic pathways. This information is essential to further improving process control and achieving higher efficiencies (Mardinoglu and Nielsen 2012; Khatri and Sirota 2012).

16.6.1.4.4 Comprehensive Modeling

Bioinformatics helps to provide these different kinds of information to generate complex models of the metabolic pathways involved in the production of ceramides. These models involve approaches in systemic biology that predict the effects on ceramide synthesis due to changes in metabolite amounts, enzyme activity, and gene expression through advanced biological methods. The computational models can be used to simulate the effects of adding a new enzyme or overexpressing a gene on ceramide production. These models help to design and assist researchers in determining what will happen if certain genes are modified (Khatri and Sirota 2012).

16.6.1.4.5 Optimization and Prediction

Integrated models generate predictions of different modifications, like genetic engineering or fermentation conditions. Similar to the changes that appeared above, researchers can figure out how things such as diet and disease states could be adjusted accordingly so that ceramide production is best optimized. For example, the models could predict how changing the expression levels of specific genes or enzymes affects the overall flux along the ceramide synthesis pathway. Because of this ability to predict, scientists may make decisions based on data and improve business processes (Khatri and Sirota 2012; Bordbar and Palsson 2012).

16.6.1.4.6 Feedback and Iteration

Also by providing researchers to explore the theories and change their plans according to model predictions, the model development will continue to enhance. When researchers compare the model predictions with experimental data, they can verify and improve the accuracy and reliability of their models over time. This iterative process allows for ongoing optimization of ceramide manufacture and improved response to new possibilities or difficulties (Choi and Lee 2016).

Lastly, it can be concluded that the integration of proteomic, metabolomic, and genomic data into comprehensive models was an effective approach since it provides more insight into the production of ceramides. These models predict the effects of modifications, guide the optimization of manufacturing techniques, and provide vital information about the complicated interactions that exist within biosynthetic pathways. Using these comprehensive models, scientists can accelerate advances in ceramide synthesis, resulting in more productive and successful techniques for creating superior ceramides for the cosmetics industry (Choi and Lee 2016; Airoldi and Ma 2014; Borenstein and Gauthier 2008).

16.6.2 Synthetic Biology

In synthetic biology, new biological components, devices, and systems are designed and built, or existing biological systems are redesigned, with specific objectives in mind. Synthetic biology plays the following roles in the synthesis of ceramides obtained from fungi:

16.6.2.1 Stain Engineering

Using advanced synthetic biology methods, strain engineering is used to improve the metabolic pathways involved in ceramide biosynthesis, leading to increased ceramide production in fungi. Choosing an appropriate fungal host with innate ceramide production capability is the first step in this approach. Following identification, particular genes known to increase the production of ceramides from other organisms—typically plants or microbes—are introduced using synthetic biology techniques. These foreign genes have the ability to encode enzymes that catalyze essential ceramide biosynthesis steps, hence boosting the pathway's productivity and efficiency (Zhou et al. 2020; Niu et al. 2021; García et al. 2018).

Apart from introducing new genes, the fungal strain's pre-existing metabolic pathways can be modified. In order to direct more resources toward the synthesis of ceramides, competing processes that use energy or precursors are knocked out or downregulated using methods like RNA interference (RNAi) or CRISPR-Cas9. Additionally, yields can be increased by fine-tuning regulatory components including enhancers, promoters, and ribosome binding sites to maximize the expression of both native and imported genes (Zhou et al. 2020).

Additionally, the design and development of completely new biochemical pathways is made possible by synthetic biology. It is possible to design fungal strains that generate ceramides more efficiently or with modified features, like changed chain lengths or saturation levels, which may be desirable for particular cosmetic applications, by assembling genes and enzymes in novel combinations. Bioinformatics tools, which simulate and analyze the effects of genetic alterations and aid in the identification of the best engineering approaches, frequently support the entire process. In the end, strain engineering pushes the limits of what is possible in cosmeceutical compositions by improving ceramide production and enabling the creation of customized ceramides (Niu et al. 2021; García et al. 2018).

16.6.2.2 Metabolic Engineering

The systematic changing of metabolic pathways in fungi that increase ceramide biosynthesis yield and efficiency is known as metabolic engineering. Usually, this procedure starts with a detailed examination of the fungal metabolic network to identify significant pathways that affect the synthesis of ceramides. Researchers can then modify these pathways to alter the flow of metabolites toward ceramide synthesis, a process known as flux optimization, using synthetic biology approaches. For example, the total flux toward ceramide biosynthesis can be greatly increased by overexpressing genes involved in synthesizing precursor molecules like sphinganine, which is essential for ceramide formation (Pfleger et al. 2017).

Metabolic engineering helps not only to increase the availability of substances but also to reduce or eliminate the competing pathways. These pathways change these substances away from other biosynthetic procedures. Reduced metabolic restrictions provide significantly additional substances for ceramide synthesis. To ensure that these molecules are used preferentially for ceramide production, mechanisms that consume fatty acids, another important component of ceramides, might be blocked (Pfleger et al. 2017).

The improvement of the availability of cofactors is an important part of metabolic engineering. The cofactors are necessary for ceramide synthesis and include NADPH, ATP, and specific ions. It is possible that the efficiency of ceramide production can be further improved by developing methods of increasing cofactor production or enhancing recycling. Also, synthetic biology technologies help in the implementation of regulatory circuits that can adjust metabolic flux in response to current cellular conditions. This ensures that ceramide synthesis remains high even when facing changing environmental stresses (Pfleger et al. 2017; Alper et al. 2020).

Computational modeling is also very helpful in metabolic engineering because this method predicts how genetic changes would affect the overall metabolic pathway. These methods can predict how genetic changes can affect the overall pathways so that scientists can develop more effective methods to maximize ceramide production while minimizing the undesirable outcomes such as the production of harmful intermediates. Lastly, metabolic engineering provides a powerful tool for optimizing fungal strains for ceramide production on a large scale, to open the way for more effective and sustainable manufacturing procedures in the cosmeceutical industry (Pfleger et al. 2017; Alper et al. 2020; Chen et al. 2018).

16.6.2.3 Synthetic Gene Construction

One of the primary methods for increasing the fungal-derived ceramides is the development and assembly of synthetic gene circuits. In basic terms, synthetic gene circuits engineered are a type of network of genes that are linked through transcriptional and post-transcriptional regulation. These circuits are made up of well-designed networks of genes, regulatory components, and promoters that work together to control the expression of target genes required for ceramide production. Scientists achieve more control over the production process by particularly regulating the activity of essential enzymes in the biosynthetic pathway through the addition of synthetic gene constructs (Wang et al. 2019; Ryu et al. 2021).

There is a possible way to design gene circuits with feedback loops that frequently modify a gene's expression in response to changes in ceramide or its precursor levels. For example, if ceramide levels decrease below a certain level, the synthetic circuit may activate genes involved in ceramide synthesis. On the other hand, if ceramide levels are too high, the circuit may reduce synthesis to prevent the development of potentially harmful by-products. Even in an environment with changing external conditions, the level of control assures that ceramide production is consistent and effective (Wang et al. 2019).

There are some other ways to use inducible systems on synthetic gene constructs as well. This system could allow researchers to modulate the amount of ceramide made in response to, for example, light by certain drugs or temperature. The activated regulators that can control the gene expression with respect to its period while in production. Particularly valuable in industrial applications where process requirements or market needs dictate the volume of ceramide to be produced (Ryu et al. 2021).

These can also be applied to synthetic gene circuits that require complementary outputs, as well as factors that affect the strength and stable of the fragments. For example, scientists can develop various paths within the circuit to provide alternative options for ceramide production. If one pathway fails or is obstructed. This stability allows ceramides to be produced consistently at a high-quality level, regardless of changes in the environment or potential influence from genetic factors (Ryu et al. 2021; Petersen et al. 2018).

Many synthetic gene circuits are now constructed with modern bioinformatics tools and machine learning methodologies, which can be used to rationally predict the activity of a circuit in their host organism. By modeling a variety of designs, researchers can select the best-performing circuit combinations for lab experiments. Improved isolation of fungal isolates that perform best in ceramide production could be especially valuable in reducing the requirement for trial and error (Petersen et al. 2018).

Overall, synthetic gene models provide an easy and efficient technique to control ceramide biosynthesis in fungi. It provides affordable and effective output. These arrangements provide the development of fungal strains that meet the extreme quality standards for cosmeceutical applications by consistently producing high-yielding procedures of ceramides with the proper characteristics (Wang et al. 2019; Ryu et al. 2021; Petersen et al. 2018).

16.6.2.4 Optimization of Fermentation Condition

To obtain the highest output and ensure consistent product quality, fermentation conditions for the manufacture of fungal ceramides must be optimized. Because synthetic biology allows for the creation of fungal strains specifically tailored thriving in the proper fermentation settings, it substantially facilitates this process. To do this, the fungus needs to be genetically modified to withstand a range of environmental challenges that can impair their ability to grow and produce. Temperature, pH, oxygen, and nutrition availability variations are some of these stressors (Liu et al. 2020).

A key technique for maximizing fermentation conditions is the genetic modification of fungal strains to boost their resistance to stress factors frequently encountered during large-scale manufacturing. Thus, large-scale industrial fermentation can be supplemented by introducing a heat-shock or chaperone coding genes to improve the overall thermotolerance of fungal strains that experience temperature imbalances. Likewise, altering the expression of pH-regulating genes allows for stable cells to exist at acidic or alkaline conditions. In the meantime, that will keep fungi cold and make ceramide in hostile conditions (Liu et al. 2020; Smith et al. 2021).

The nutritional conditions are key process parameters in the fermented optimization approach. Synthetic biology allows for the exploitation of a wide variety of inexpensive substrates or side-products by expression-optimized and sometimes newly created genes in yeasts capable of growing on non-conventional carbon (and nitrogen) sources. This leads to lower production costs and a broader substrate spectrum applicable for fermentation. Second, one may be able to modulate metabolic pathways in order to reduce the generation of byproducts that could interfere with whether ceramides are formed or not. These engineered strains have a reduced metabolic load which in turn can result in more energy for the ceramide synthesis process—resulting ultimately having higher production (Keller et al. 2019).

Fungal strains with improvements in oxygen usage are also an important part of fermentation optimization. The oxygen concentration in the container can be varied and low levels of free oxygen may reduce ceramide production and metabolism. One solution is to stimulate the expression of the oxygen-uptake and -consumption genes via synthetic biology so that fungi continue producing even under low levels of oxygen. Facilitate the introduction and augmentation of anaerobic pathways that support ceramide biosynthesis in oxygen-limited environments to broaden the spectrum of fermentation conditions at which these fungi can outcompete wild strains (Liu et al. 2020; Smith et al. 2021). Scientists can swiftly ascertain which strains perform best in specific fermentation conditions and enhance the production process by employing these techniques. Ultimately, synthetic biology enables more effective and scalable synthesis of ceramides, satisfying the expectations of the cosmeceutical sector while minimizing costs and environmental impact. This is achieved by developing fungal strains that are durable and highly productive under optimum fermentation conditions (Liu et al. 2020; Smith et al. 2021; Keller et al. 2019).

16.6.2.5 Sustainability and Scalability

The production of fungal-derived ceramides must take sustainability and scalability into consideration, particularly since the cosmetics industry is placing a greater value on economical and environmentally beneficial solutions. By maximizing fungal strains and improving the related production processes, bioinformatics, and synthetic biology are directing efforts to improve the sustainability and efficiency of ceramide production. In addition to reducing production expenses and resource consumption, these developments guarantee a steady and dependable supply of premium ceramides to satisfy the expanding market demands (Luo et al. 2020).

Fungal strains that are designed to use resources more efficiently are one of the main ways that synthetic biology promotes sustainability. Scientists can lower the overall consumption of substrates such as sugar, oils, or other carbon sources by upregulating metabolic pathways that improve how well they convert into ceramides. This increase in resource efficiency directly translates into lower environmental impact and financial savings since lesser input is required to generate the same or higher output. In addition, synthetic biology also has the potential to substitute waste- or renewable-derived ingredients more easily and thereby enhance the sustainability of production (Luo et al. 2020).

Another important consideration is that scaling this up might become necessary, considering the ever-increasing demand for ceramides in the cosmetics industry. Synthetic biology allows us to develop fungal strains that are ideal for industrial-scale production. This entails the creation of strains that could grow in high-density cultures and withstand aspects specific for large bioreactors like oxygen gradients, nutritional barriers, or shear forces, among others. By enhancing the strength and efficiency of these strains, researchers can get higher production, making them effective for large-scale production (Luo et al. 2020).

Bioinformatics integrates with production to make it data-driven, which is essential for ensuring sustainability and scalability. As a result, the researchers might use systems biology and computer modeling tools to simulate the effects of alteration in genes and predict production process blockages, which result in metabolic fluxes. This opens the possibility of engineering large fungal strains and industrial processes to be more productive and scalable, given that we can accurately predict all interactions relevant for protein overload. Among other possibilities, bioinformatics tools could be used to select genes for change—perhaps to increase growth rates and better adapt organisms to industrial fermentation conditions or produce ceramides. This minimizes the time spent on trial and error testing, making it easier to develop production with optimal performance (Luo et al. 2020).

In addition, bioinformatics helps in real-time monitoring and control of industrial processes. Linking process analytics with omics data (transcriptomics, proteomics, and genomes) will allow the development of advanced control strategies to guarantee stable and efficient manufacturing while scaling. Due to the laws and regulations and maintaining safety with variations in consumer perception, the cosmetics industry requires this level of control (Kumar et al. 2021).

Making sustainability more advanced includes systems of closed loops, which work to recycle byproducts and reduce waste. Fungal strains could be developed using synthetic biology to degrade raw materials including dependent waste streams from various industries and produce valuable by-products. The continuous production method helps to add value in the manufacturing process while reducing waste, which also means that enterprises are more commercial and environmentally sustainable (Luo et al. 2020; Kumar et al. 2021).

In summary, the reconstitution of the ceramide pathway in fungi via synthetic biology and bioinformatics represents a solution for scalable sustainable production of these important lipids. These bio-adapted procedures were developed for continuous fermentation using the Fungal System and are designed to assure optimal production performance in terms of strength, high productivity, and resource efficiency. Such a source will support the cosmetics business beyond peak ceramide, and keep a consistent supply at affordable prices for good-quality ceramides, assuring availability due to consumer demand (Johnson et al. 2019).

16.7 Conclusion

In the cosmetics industry, there are several chemical-based products available in the market. However, fungal-derived ceramides signify a great achievement in the cosmetics industry. It offers an ethical, highly effective, and sustainable alternative to the common ceramide sources. The market continues to move toward sustainability and accessibility. Hence, the fungal-derived ceramides have a great potential to be an important component of innovative cosmetics. They give an affordable solution that is compatible with worldwide developments in the environmentally sustainable market. It also reduced the traditional sources such as animals or synthetic sources. The introduction of fungal-derived ceramides into the cosmetics industry demonstrates how biotechnology can impact on the appearance of a person. Hence, the cosmetic industry continues to develop in the future by producing safe and efficient products.

References

Ahmad MF, Ahmad FA, Azad Z, Ahmad A, Alam MI, Ansari JA, Panda BP (2013) Edible mushrooms as health promoting agent. Adv Sci Focus 1:189–196

Airoldi EM, Ma J (2014) Feedback control and systems biology: methods and applications. J Comput Biol 21(8):684–693. https://doi.org/10.1089/cmb.2014.0055

Alper HS et al (2020) Metabolic engineering of fungi for enhanced production of biofuels and chemicals. Trends Biotechnol. https://doi.org/10.1016/j.tibtech.2020.01.002

Bai Y, Li J (2020) Optimization of cell disruption techniques for enhanced recovery of intracellular products from *Aspergillus niger*. Biotechnol Rep 25: e00421. https://doi.org/10.1016/j.btrec.2020.e00421

Bordbar A, Palsson BØ (2012) Constraints-based models: a review of computational approaches for understanding metabolic networks. J Biol Chem 287(1):1–8. https://doi.org/10.1074/jbc.R111.293150

Borenstein E, Gauthier M (2008) Automated analysis of biological pathways and systems: a review of current tools and applications. BMC Syst Biol 2:37. https://doi.org/10.1186/1752-0509-2-37

Bouwstra JA, Ponec M (2006) The skin barrier: old and new concepts. Clin Dermatol 24(1):3–13. https://doi.org/10.1016/j.clindermatol.2005.10.009

Chen X, Zhang T (2020) Organic and natural ingredients in cosmetics: a comprehensive review of their benefits and potential risks. J Dermatol Treat 31(5):453–464. https://doi.org/10.1080/09546634.2020.1778946

Chen X et al (2018) Engineering microbial hosts for efficient biosynthesis of value-added chemicals. Biotechnol J. https://doi.org/10.1002/biot.201700372

Choi HR, Kim SK (2017) Ceramide metabolism and its role in skin inflammatory diseases. Front Immunol 8: 546. https://doi.org/10.3389/fimmu.2017.00546

Choi SK, Lee JH (2016) Systems biology approach to identify and validate potential biomarkers for cancer diagnosis. J Biomed Sci 23(1):50. https://doi.org/10.1186/s12929-016-0225-7

Czar MJ, Anderson JC, McKeown AP (2009) Gene synthesis demystified. Trends Biotechnol 27(2):63–72. https://doi.org/10.1016/j.tibtech.2008.10.007

Elias PM, Choi EH (2005) Interactions among stratum corneum defensive functions. Exp Dermatol 14(10): 719–726. https://doi.org/10.1111/j.1600-0625.2005.00363.x

Figueiredo AM, Monteiro MS (2021) Sustainable and biologically-derived ingredients in cosmetics: an overview of current trends and future directions. Sustainability 13(15):8344. https://doi.org/10.3390/su13158344

Fluhr JW, Maibach HI (2007) Skin barrier function: a review of ceramide and lipid biochemistry. J Dermatol Sci 48(2):89–97. https://doi.org/10.1016/j.jdermsci.2007.08.013

Folch J, Lees M, Stanley GHS (1957) A simple method for the isolation and purification of total lipides from animal tissues. J Biol Chem 226(1):497–509. https://doi.org/10.1016/S0021-9258(18)64849-6

Fukushima K, Kaneko Y (2018) Optimization of yeast fermentation for the production of fungal-derived ceramides. J Ind Microbiol Biotechnol 45(2): 159–168. https://doi.org/10.1007/s10295-017-2007-4

García J et al (2018) Synthetic biology in fungi: engineering metabolic pathways for improved production of value-added compounds. J Ind Microbiol Biotechnol. https://doi.org/10.1007/s10295-018-2093-2

Hannun YA, Obeid LM (2018) Sphingolipids and their metabolism in physiology and disease. Nat Rev Mol Cell Biol 19(3):175–191. https://doi.org/10.1038/nrm.2017.107

Holleran WM et al (2006) Significance of ceramide and other sphingolipids in the epidermis. J Lipid Res 47(2): 203–213. https://doi.org/10.1194/jlr.R500013-JLR200

Hsu CY, Popp HA (2018) Quantification of ceramides and other lipids using high-performance liquid chromatography and mass spectrometry. J Chromatogr B 1083: 26–32. https://doi.org/10.1016/j.jchromb.2018.10.021

Huang X, Liu Z (2019) Sustainable production of fungal-derived ceramides using optimized fermentation processes. J Appl Microbiol 127(3):785–796. https://doi.org/10.1111/jam.14228

Ishikawa J, Nishikawa Y (2018) The role of ceramides in skin hydration and barrier function: clinical implications. J Dermatol Sci 89(3):257–262. https://doi.org/10.1016/j.jdermsci.2017.12.007

Johnson E et al (2019) Sustainability in microbial production systems: from laboratory to industry. Trends Biotechnol. https://doi.org/10.1016/j.tibtech.2019.04.003

Jones DT (1999) Protein secondary structure prediction based on position-specific scoring matrices. J Mol Biol 292(2):195–202. https://doi.org/10.1006/jmbi.1999.3091

Kaiser P, Surade S (2019) Synthetic biology for natural product biosynthesis in fungi. Fungal Biol Rev 33(1): 26–39. https://doi.org/10.1016/j.fbr.2018.12.002

Kanehisa M, Sato Y, Morishima K (2016) BlastKOALA and GhostKOALA: KEGG tools for functional characterization of genome and metagenome sequences. J

Mol Biol 428(4):726–731. https://doi.org/10.1016/j.jmb.2015.11.006
Kang MJ, Kim S (2019) Formulation and evaluation of ceramide-containing lotions and serums: effects on skin hydration and barrier function. J Cosmet Sci 70(4):319–329. https://doi.org/10.1111/jocs.12547
Keller NP et al (2019) Fungal strain engineering for enhanced production of natural products. Front Microbiol. https://doi.org/10.3389/fmicb.2019.01034
Khan MN, Damsky M (2020) Emerging trends in natural and organic cosmeceuticals: focus on sustainability and efficacy. Cosmetics 7(2):35. https://doi.org/10.3390/cosmetics7020035
Khatri P, Sirota M (2012) Ten years of pathway analysis: current approaches and outstanding challenges. Brief Bioinform 13(5):487–500. https://doi.org/10.1093/bib/bbs047
Kikuchi Y, Ohno K (2020) Engineering yeast strains for enhanced production of sphingolipids including ceramides. Appl Microbiol Biotechnol 104(12):5291–5300. https://doi.org/10.1007/s00253-020-10629-5
Kim SY, Kim H (2021) Genetic engineering of *Pichia pastoris* for the production of high-purity ceramides. Fungal Biol 125(4):362–370. https://doi.org/10.1016/j.funbio.2021.01.004
Kim HM, Lee JH (2019) Innovative skincare formulations incorporating ceramides and their impact on skin health. Cosmetics 6(2):25. https://doi.org/10.3390/cosmetics6020025
Kumar A et al (2021) Scalability and sustainability of biotechnological processes for industrial applications. Biochem Eng J. https://doi.org/10.1016/j.bej.2021.107420
Lee YS, Park SW (2019) Enhanced production of ceramides through metabolic engineering of Saccharomyces cerevisiae. J Biotechnol 306:60–67. https://doi.org/10.1016/j.jbiotec.2019.04.014
Liu J, Zheng J (2019) Optimization of fungal fermentation for the production of bioactive lipids, including ceramides: a review. J Ind Microbiol Biotechnol 46(6):775–790. https://doi.org/10.1007/s10295-019-02100-0
Liu Z et al (2020) Optimizing fermentation processes for high-yield production of bio-based chemicals. Appl Microbiol Biotechnol. https://doi.org/10.1007/s00253-020-10651-8
Luo Z et al (2020) Sustainable production of bio-based chemicals through engineered microorganisms. Nat Sustain. https://doi.org/10.1038/s41893-020-00550-8
Madison KC (2003) Barrier function of the skin: "la raison d'être" of the epidermis. J Invest Dermatol 121(2):231–241. https://doi.org/10.1046/j.1523-1747.2003.12359.x
Mao C, Obeid LM (2008) Ceramide and sphingolipid regulation of apoptosis. Biochim Biophys Acta Mol Cell Biol Lipids 1781(9):424–434. https://doi.org/10.1016/j.bbalip.2008.06.002
Mardinoglu A, Nielsen J (2012) Systems biology of metabolism. Annu Rev Biochem 81:425–450. https://doi.org/10.1146/annurev-biochem-061809-093635
Miller EB, Simon M (1985) Thin-layer chromatography of ceramides and other sphingolipids. J Lipid Res 26(4):399–407. https://doi.org/10.1016/S0022-2275(20)32980-4
Muir AR, Yamauchi Y (2016) Characterization of ceramides and their derivatives using mass spectrometry and nuclear magnetic resonance. Anal Chem 88(21):10476–10483. https://doi.org/10.1021/acs.analchem.6b02912
Mukherjee A, Das S, Chakraborty D, Pal N, Das N (2024) Fungi's treasure in cosmeceuticals: a comprehensive chemical approach. S Afr J Bot 154:295–306.s
Nguyen DL, Eisinger M (2018) Ceramides as modulators of skin inflammation and immune responses. J Dermatol Sci 92(3):219–226. https://doi.org/10.1016/j.jdermsci.2018.09.013
Nielsen J, Keasling JD (2016) Engineering cellular metabolism. Cell 164(6):1185–1197. https://doi.org/10.1016/j.cell.2016.02.004
Niu Y et al (2021) Engineering filamentous fungi for high-efficiency production of bioactive compounds. Biotechnol Adv. https://doi.org/10.1016/j.biotechadv.2020.107773
Paddon CJ, Keasling JD (2014) Semi-synthetic artemisinin: a model for the use of synthetic biology in pharmaceutical development. Nat Rev Microbiol 12(5):355–367. https://doi.org/10.1038/nrmicro3240
Park SK et al (2010) Biosynthesis of ceramide in yeast. J Biol Chem 285(36):27010–27021. https://doi.org/10.1074/jbc.M110.141291
Petersen TN et al (2018) Design and implementation of synthetic gene circuits for microbial production. Nat Commun. https://doi.org/10.1038/s41467-018-07400-w
Pfleger BF et al (2017) Metabolic engineering of microorganisms for the production of bio-based chemicals. Curr Opin Biotechnol. https://doi.org/10.1016/j.copbio.2016.10.015
Poucheret P, Fons F, Rapior S (2006) Biological and pharmacological activity of higher fungi: 20-year retrospective analysis. Cryptogam Mycol 27:311–333
Rawlings AV, Harding CR (2004) Moisturization and the skin barrier. Dermatol Ther 17(s1):43–48. https://doi.org/10.1111/j.1529-8019.2004.17305.x
Ryu DD et al (2021) Development of synthetic gene circuits for controlling microbial production. Nat Rev Microbiol. https://doi.org/10.1038/s41579-020-00444-7
Sato H, Sato K (2020) High-performance liquid chromatography for the separation and analysis of ceramides and related lipids. J Chromatogr A 1612:460657. https://doi.org/10.1016/j.chroma.2019.460657
Schmuth M, Elias PM (2012) Ceramide-based therapeutics for dry and sensitive skin: a review. Dermatol Ther

25(5):271–281. https://doi.org/10.1111/j.1529-8019.2012.01446.x

Schnuch A, Uter W (2016) Natural and organic cosmetic ingredients and their impact on skin health: a review. Contact Derm 74(3):165–176. https://doi.org/10.1111/cod.12536

Schuster SC (2008) Next-generation DNA sequencing techniques. Mol Divers 12(1):49–57. https://doi.org/10.1007/s11030-007-9001-5

Silva MF, Almeida JRGS (2019) Optimization of lipid extraction from fungal biomass for the recovery of bioactive compounds: application to *Mortierella alpina*. J Microbiol Methods 165:105–114. https://doi.org/10.1016/j.mimet.2019.105080

Smith A, Patel M (2018) The shift towards organic and natural cosmeceuticals: market trends and consumer preferences. Cosmet Dermatol 17(1):22–29. https://doi.org/10.1111/cod.12345

Smith RL et al (2021) Engineering microbes for efficient fermentation and high product yields. J Biotechnol. https://doi.org/10.1016/j.jbiotec.2021.03.013

Song X, Liang Z (2019) Efficient disruption of fungal cell walls for the extraction of bioactive compounds: a comparative study of physical and mechanical methods. J Biotechnol 299:62–72. https://doi.org/10.1016/j.jbiotec.2019.03.015

Takeshita J, Yamada M (2018) Ceramides in dermatology: basic and clinical perspectives. J Dermatol 45(5):538–549. https://doi.org/10.1111/1346-8138.14230

Tatusova T, Ciufo S, Fedorov B, O'Neill K, Tolstoy I (2015) RefSeq microbial genomes database: new representation and annotation strategies. Nucleic Acids Res 43(D1):D599–D605. https://doi.org/10.1093/nar/gku1065

Wang H et al (2019) Synthetic biology tools for control of gene expression in yeast and filamentous fungi. Fungal Genet Biol. https://doi.org/10.1016/j.fgb.2019.03.005

Winter G et al (2014) Synthetic biology meets natural products. Nat Prod Rep 31(11):1581–1610. https://doi.org/10.1039/C4NP00008A

Yang J, Xu Z (2021) Efficient recovery of fungal biomass from fermentation broths using centrifugation and filtration techniques: a comparative study. J Appl Microbiol 130(3):915–925. https://doi.org/10.1111/jam.14876

Zengler T, Palsson BØ (2012) A road map for the development of industrial microbiology. Nat Rev Microbiol 10(7):470–481. https://doi.org/10.1038/nrmicro2818

Zhang Y, Liu M (2017) Ceramides in skin aging and photoaging: therapeutic strategies. J Dermatol Sci 85(2):85–91. https://doi.org/10.1016/j.jdermsci.2016.10.007

Zhang Y, Skolnick J (2004) Scoring function for automated assessment of protein structure template quality. J Comput Chem 25(1):130–145. https://doi.org/10.1002/jcc.10419

Zhang L, Wang Y (2021) Role of ceramides in the management of skin diseases: insights and clinical applications. Front Dermatol 2:12. https://doi.org/10.3389/fderm.2021.00012

Zhao H, Geng X (2019) Biotechnological production of ceramides from microbial fermentation: recent advances and future prospects. Biotechnol Adv 37(6):107440. https://doi.org/10.1016/j.biotechadv.2019.107440

Zhou S et al (2020) CRISPR/Cas9-based genome editing for improving metabolic pathways in fungi. Fungal Biol Rev. https://doi.org/10.1016/j.fbr.2020.03.002

Zouboulis CC, Katsambas AD (2009) The role of ceramides in the pathogenesis of atopic dermatitis. Dermatology 219(2):97–104. https://doi.org/10.1159/000207339

Polysaccharides Obtained from Fungi and Their Usage in Cosmetics 17

Wan Abd Al Qadr Imad Wan-Mohtar,
Nur Asyiqin Zahia-Azizan, Nurul Aqilah Mohd Zaini,
and Muhamad Hafiz Abd Rahim

Abstract

Polysaccharides, a type of polysaccharide found in fungi, are a potent ally in the fight for natural cosmetic alternatives. These intricate sugar molecules, which are extracted from the cell walls of many species, have an amazing range of characteristics that make them extremely promising for improving our beauty regimens. This article explores the many advantages of fungal polysaccharides for skin and hair care, highlighting their powerful potential. We explore the wide range of applications for these bioactive compounds, demonstrating their superior moisture-retention capabilities and their capacity to prevent wrinkles and encourage hydration. Their calming and anti-inflammatory qualities provide comfort to delicate skin, while their antioxidant capacities protect against damage from free radicals, preserving young radiance. We also examine their potential to support the effectiveness of sunscreens and feed hair follicles, resulting in stronger, healthier strands. Fungal polysaccharides have inherent benefits, but they also align with the current push for sustainability in the cosmetics business. Their organic provenance provides a sustainable substitute for artificial ingredients, catering to the growing environmentally conscious consumer base. To fully realize their potential, handle any allergic responses, and ensure sustainable sourcing techniques, we do understand the need for additional research. In the end, this analysis presents a convincing image of fungal polysaccharides as the upcoming big thing in natural cosmetics. Because of their many uses and innate sustainability, they can be used as revolutionary instruments to shape a future filled with natural beauty and open the door to a brighter, healthier tomorrow.

Keywords

Fungal polysaccharides · Natural anti-aging · Hydration · Anti-inflammatory · Sun protection booster · Cosmetic innovation

17.1 Introduction

People have long benefited from the aesthetic appeal and health-promoting qualities of nature (Wallace and Giusti 2019). The bounty of the

W. A. A. Q. I. Wan-Mohtar (✉) · N. A. Zahia-Azizan
Functional Omics and Bioprocess Development Laboratory, Institute of Biological Sciences, Faculty of Science, Universiti Malaya, Kuala Lumpur, Malaysia
e-mail: qadyr@um.edu.my

N. A. M. Zaini
Department of Food Sciences, Faculty of Science and Technology, Universiti Kebangsaan Malaysia, Bangi, Selangor, Malaysia

M. H. A. Rahim
Faculty of Food Science and Technology, Universiti Putra Malaysia, Serdang, Selangor, Malaysia

S. Ghosh, H. N. Onyeaka (eds.), *Fungal-Based Cosmetics: Formulation and Usage*, The Mycota 17,
https://doi.org/10.1007/978-3-032-09291-5_17

Earth offers a multitude of ingredients, from floral extracts to mineral masks, to enhance our appearance and nourish our skin. However, the kingdom of fungi is a hidden kingdom in nature that is becoming more and more well-known. In addition to being wholesome food sources and significant environmental components, fungi also possess a covert weapon known as polysaccharides (Barbosa and de Carvalho Junior 2020). Inducing anticancer activity, polysaccharides activate pathways of apoptosis, pro-inflammatory substance production, and macrophage. Moreover, the immune system's activation makes vaccine production possible. These complex sugar compounds, with an incredible array of properties that are upending the cosmetics industry, are entwined in the fungal cell walls.

This study delves into the fascinating world of fungal polysaccharides, highlighting their unique makeup, diverse origins, and noteworthy benefits for skin health. The scientific world is aware that polysaccharides and glycoconjugates derived from fungi possess pertinent structural and physicochemical characteristics that make them valuable for usage in pharmaceutical and food applications (Xiao et al. 2020). Few researchers have dedicated themselves to comprehending the roles that polysaccharides play in fungal biology, despite the fact that numerous studies have examined the possibilities of these polymers. Furthermore, how evolution has impacted the growth of increasingly specialized fungus that produce polysaccharides. We think that the missing piece between glycobiology and the creation of new bioinspired technology may be our understanding of the evolution of fungi. We embarked on a journey through the labs of nature, exploring the wide variety of polysaccharides extracted from various fungus and their remarkable characteristics. Every molecule has a fascinating story to tell, ranging from β-glucans, the unsung heroes of hydration, to α-glucans, the champions of antioxidants (Johansson et al. 2006).

But the story is not only about extraction. We need to understand how these natural substances interact with our skin to completely grasp their potential. This study investigates the intricate processes that underlie the advantageous properties of fungus-derived polysaccharides. We examine how they can improve the production of collagen, build barrier layers, and combat free radicals to give you a smoother, more youthful-looking skin. Lastly, we go into the active area of cosmetic formulae, where science and creativity come together to create effective products (Hyde et al. 2010). Fungi are used in skincare products like anti-aging, antioxidants, skin revitalizing, whitening, and hair products. Since the currently used mushrooms have a history of producing medical ingredients, they were the first to be used in cosmetic applications. But there is a ton of other mushroom species with enormous potential for use in the cosmetics industry that have not been studied, described, or are not yet cultivable. We'll examine the sustainable replacements for synthetic ingredients in serums, lotions, and masks that these biodegradable and biocompatible substances can serve. We review the advancements and trends in fungal-based cosmetics and explain the ethical concerns and future possibilities for this rapidly evolving field.

The exploration of potent polysaccharides and the lifestyle of mushrooms is undertaken in the following chapters. The potential of nature's undiscovered treasures is unveiled, opening the door to a brand-new era of radiant, healthy skin, all made possible by the amazing gifts of the fungus kingdom (Mapoung et al. 2021).

17.2 The Rise of Natural Ingredients in Cosmetics

Cosmetics, as defined by the Food, Drug, and Cosmetic Act, include a diverse array of products such as shampoos, makeup, and toothpaste, designed to enhance appearance or cleanliness. The term "cosmeceutical" describes cosmetics that contain active ingredients providing medicinal benefits when applied topically. While skin cosmetics are often associated with makeup, they also encompass products aimed at revitalizing the skin, anti-aging, anti-wrinkle, and skin whitening. Nutricosmetics, which sometimes overlap with nutraceuticals (food), as they are being ingested, offer benefits such as moisturization, hydration, anti-aging effects, melanin production suppression, and skin firming and lifting (Hyde et al. 2010).

The global cosmetics market is substantial, with leading companies like L'Oréal having a market capitalization of nearly 188 billion U.-S. dollars in 2022 (Statista 2022). The beauty market generated approximately 430 billion U.-S. dollars in revenue in 2022 and is expected to reach around 580 billion U.S. dollars by 2027 (McKinsey 2023). Moreover, the European market is particularly interested in new and innovative natural ingredients, which creates opportunities for companies to enter the market with unique offerings (Allied Market Research 2023). The increasing consumer inclination towards natural and active cosmetic ingredients is driving the sector, providing favorable growth opportunities for the market (Grand View Research 2022).

The shift toward "green" products has led to an increase in natural cosmetics, often marketed as "chemical-free" due to their plant-based or organic composition. Although there's no universal definition for natural cosmetics, they generally consist of ingredients sourced from nature and are recognized as such if they contain at least 95% natural substances (Mohd-Setapar 2018). These natural ingredients, including those from plants, animals, and microorganisms, have been used for centuries and are valued for their bioactive properties like antioxidants and vitamins, which contribute to healthier skin (Mohd-Setapar 2018). Natural ingredients, including those derived from fungi in the form of mushroom, present significant opportunities in this market. They align with consumer trends toward sustainability and health-consciousness, which can be leveraged for market growth. For instance, natural ingredients are often associated with fewer side effects and a lower environmental impact compared to synthetic alternatives. Economically, mushrooms can offer significant economic and horticultural advantages over other crops. They have a high profit margin due to their short growth cycle and multiple harvests per year. Their cultivation requires less land, water, and energy, making them suitable for small spaces and urban areas. They contribute to sustainability by utilizing agricultural waste as a growing medium and reducing greenhouse gas emissions associated with food transportation (Thakur 2020).

More importantly, mushrooms are rich in bioactive compounds like terpenoids, selenium, phenolics, and polysaccharides, which offer various skin and hair care benefits (Hyde et al. 2010; Visvanathan et al. 2022) (Table 17.1). These compounds have been found to have anti-aging, antioxidant, anti-wrinkle, skin lightening, moisturizing, anti-acne, and anti-dandruff effects, making them highly valuable in cosmetic formulations. Polysaccharides from fungi, especially beta-glucans, are known for their ability to hydrate and soothe the skin, as well as their anti-aging properties. Polysaccharides can stimulate collagen production, improving skin elasticity and firmness. They also have excellent anti-collagenase or anti-elastase activity, which can delay the skin aging process (Visvanathan et al. 2022). There are numerous examples of using mushroom as the main ingredients for cosmetics, as reviewed by Hyde et al. (2010). For example, Johnson & Johnson's Aveeno brand launched "Positively Ageless," featuring *Ganoderma lucidum* and *Lentinula edodes*, known for immunostimulatory properties. In a recent review by Wu et al. (2016), it was found that out of 24 cosmetics containing mushroom extracts, *Ganoderma lucidum* was the most prevalent, being present in 11 of the products. This was followed by *Lentinula edodes*, which were found in 5 products. The remaining products contained extracts from various other types of mushrooms, such as *Cordyceps sinensis* and *Tremella* species.

Cosmetic formulations often incorporate mushroom extracts rather than individual compounds, owing to the multitude of benefits derived from the thousands of compounds present in the extract (Fig. 17.1). A single compound can have multiple benefits, and even a single mechanism, such as antioxidant activity, can have numerous physiological effects. This makes it challenging to determine the exact impact of

Table 17.1 The general list of ingredients found in mushrooms that make them suitable for cosmetic applications

Compounds	Examples	Cosmetic applications	References
Organic acids	Lactic acid	Hydration, whiteners, and exfoliation	Hyde et al. (2010)
Polysaccharides	Chitin-glucan and chitosan	Moisturizing and anti-aging	Hyde et al. (2010) and Wu et al. (2016)
	Beta-glucan	Antioxidant, anti-inflammatory, immunomodulatory	Kozarski et al. (2023)
	Trehalose	Moisturizing and antioxidant	Hyde et al. (2010)
	Glucuronoxylomannan	Immunomodulatory and wound healing	Wu et al. (2016)
	Lentinan	Immunomodulatory	Hyde et al. (2010)
Fatty acids	Glycosphingolipids	Moisturizing	Jiang et al. (2021)
	Sterols	Antioxidant, immunomodulatory, anti-aging	Taofiq et al. (2019)
	Omega fatty acids	Immunomodulatory, emollients, antioxidants, and cell barriers and regulators	Mohd-Setapar (2018)
Pigments and organic compounds	Phenolics, carotenoids flavonoids	Whiteners, antioxidants, anti-microbial	Kozarski et al. (2023)
	Terpenoids	Immunomodulatory, antimicrobial	Wu et al. (2016)
	Resveratrol	Collagen-protecting, anti-aging	Hyde et al. (2010)
Vitamins	Vitamin A	A micronutrient for skin health	Mohd-Setapar (2018)
	Vitamin D	For hair and skin health	Wu et al. (2016)
Protein	Ergothioneine	Antioxidant	Hyde et al. (2010)
	Fungal immunomodulatory proteins	Antioxidant, inhibition of melanin, antibacterial activity, immunomodulatory	Li et al. (2019)
Others	Tyrosine inhibitors	Whitening	Hyde et al. (2010)
	Melanolytic enzymes	Whitening	Hyde et al. (2010)
	Minerals, e.g., selenium, zinc, iron, copper	Immunomodulatory, hair health	Wu et al. (2016)
	Secondary metabolites	Antimicrobial, immunomodulatory	Sułkowska-Ziaja et al. (2023)

Please note that this list is not exhaustive and only serves as a general guide, given the vast diversity of mushrooms

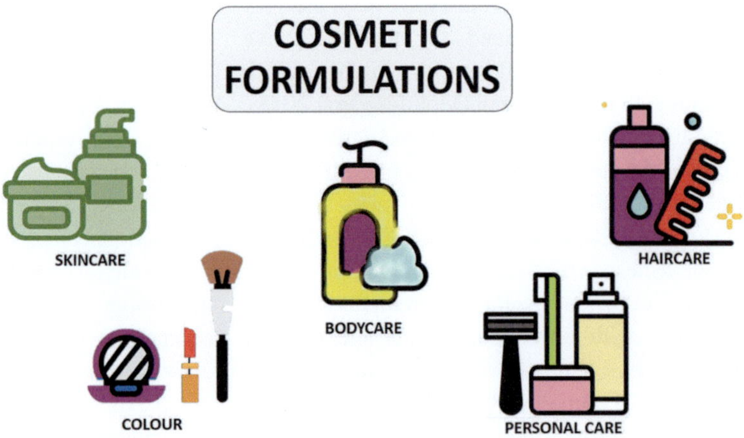

Fig. 17.1 Classification of cosmetic formulations. (Created with BioRender.com)

mushroom extract on cosmetic-related activities. Given the diverse benefits of mushroom extracts, this review will primarily focus on the role of polysaccharides for cosmetic-related applications, which are typically unique and abundant in mushroom species. While the benefits of these extracts generally outweigh the drawbacks, certain compounds, particularly proteins in mushrooms, can cause adverse reactions, such as allergies, in a small population. These potential issues, along with the future direction and challenges associated with the use of mushrooms in cosmetics, will be discussed in more detail toward the end of the review.

17.3 Potential of Fungal Polysaccharides in Cosmetics

For many years, humanity has looked to nature's wealth for solutions to enhance beauty and well-being. A plethora of ingredients for skincare products can be found in nature, from the fragrant rose petals to the nourishing oils of jojoba (Desam and Al-Rajab 2021). However, a new participant has now emerged from the forest floor's depths: the enigmatic world of mushrooms.

Strong chemicals called polysaccharides are tucked away in the cell walls of these lowly creatures. The secret to a new era of skincare lies in these complex sugars, which are expertly weaved into branching chains and offer a multitude of advantages for a glowing and healthy complexion.

Hydration Heroes: Any glowing complexion starts with optimal hydration. Fungal polysaccharides excel in this situation because of their extraordinary ability to bind and cling onto water. Trehalose, a disaccharide found in certain mushrooms, draws, and holds moisture in the skin's deeper layers, acting as a natural humectant (Goel et al. 2024). As a result, wrinkles and fine lines are less noticeable and the appearance is visibly plumper and dewier.

Antioxidant Allies: Free radicals, which are produced by both internal and external activities, can cause havoc on our skin by hastening the ageing process and dulling the complexion. Once more, fungal polysaccharides step up to the plate with their potent antioxidant capabilities. Maitake mushrooms contain beta-glucans, which scavenge these dangerous chemicals to protect skin from oxidative damage and maintain its youthful glow.

Inflammation Interceptors: Prolonged inflammation is frequently the cause of skin issues like redness, sensitivity, and irritation. Fungal polysaccharides help to subdue this inflammatory tempest with a light touch. The herb *Ganoderma lucidum* is highly valued in traditional Chinese medicine because it contains polysaccharides that influence the immune system, lowering inflammation and encouraging skin repair. This makes them perfect for people who have skin issues like rosacea and eczema.

Collagen Champions: The elasticity and firmness of skin are dependent on collagen, a protein that serves as the framework for the skin. Collagen production naturally decreases with age, which leads to sagging and wrinkles in the face area. It has been demonstrated that fungus polysaccharides, such as those present in cordyceps mushrooms, can increase the production of collagen, so preserving the youthful elasticity of the skin and delaying the onset of wrinkles.

Beneath the Skin: Fungal polysaccharides have advantages that go beyond the skin's surface. According to studies, they enable to reach the skin's deeper layers, promoting the growth of new skin cells and the healing of wounds. This makes them useful components for cosmetics that deal with issues like stretch marks and scars.

Using the Magic: Although fungal polysaccharides have enormous potential, meticulous extraction and formulation are necessary to fully realize their promise. We can separate and concentrate polysaccharides with desired effects thanks to advanced technology. Moreover, their incorporation into biocompatible delivery systems guarantees maximum absorption and effectiveness.

A Sustainable Future: The application of fungal polysaccharides corresponds with the expanding market for environmentally and morally conscious cosmetics. Fungi may be cultivated on waste materials like sawdust and agricultural waste, unlike many plant-based substances that can require large amounts of land and water to cultivate. Because of this, they are a green option for consumers who care about the environment (Hanafiah et al. 2024; Mooralitharan et al. 2023).

There seems to be no end to the possibilities of fungal polysaccharides in cosmetics as study into the realm of fungi continues. These multifunctional chemicals, from their antioxidant allies to their hydrating heroes, provide a safe and efficient way to achieve a healthy, glowing complexion. Therefore, think about embracing the wonder of the fungal frontier and stepping beyond your comfort zone the next time you reach for your skincare regimen. Your skin will appreciate it.

To sum up, fungal polysaccharides have a high potential for use in skincare products. Their various functions provide answers for a variety of issues, ranging from collagen stimulation and wound healing to antioxidant protection and hydration. The future of skincare seems to be written in the complicated language of fungal polysaccharides as we continue to uncover the mysteries of these complex molecules and create novel formulations.

It is significant to remember that although research on fungal polysaccharides is encouraging, further investigation is required to completely comprehend their effectiveness and safety over the long run. Before using fungal polysaccharides in your skincare regimen, it is always advised to speak with a dermatologist, just like you do with any new substance.

Fungal frontier research has the potential to completely transform skincare practices with further investigation and advancement. To get a glowing and healthy complexion, embrace the magic of mushrooms and allow the power of polysaccharides to guide you.

17.4 Overview of Fungal Polysaccharides

17.4.1 Potential Side Effects of Fungi Cosmetics

Nowadays, natural cosmetics are becoming more popular among health-conscious users. Compounds primarily sourced from plants are incorporated in the formulation, avoiding the use of synthetic substances that are frequently associated with potential harm. Fungi, such as mushrooms, are renowned for their beneficial properties, including antioxidant, anti-wrinkle, anti-aging, anti-acne, skin-lightening, and moisturizing effects when utilized in appropriate quantities (Visvanathan et al. 2022; Ziemlewska et al. 2022). For example, fungi such as *Lentinula edodes* and *Volvariella volvacea* could serve as sources of antioxidants and enzymes, presenting potential applications in various cosmetic products (Keleş et al. 2011). In addition, chitin-glucan, a copolymer present in the cell walls of various mushrooms, exhibits commendable moisturizing properties. Besides, it shows promise in combating certain indications of skin aging, suggesting its potential utility in formulations designed for skin moisturization and anti-aging (Wu et al. 2016). Thus, numerous cosmetic companies are now integrating various fungi species as functional ingredients with beneficial attributes in the formulation of cosmetic products to increase the values.

However, cosmetic products frequently lead to various side effects, primarily manifesting as hypersensitivity or allergic reactions. The case of allergic reactions especially to cosmetic products is increasingly observed. On the other hand, it is estimated that around 2–6% of the population in developed countries has an allergy to fungi. Sensitivity is most commonly identified with genera such as *Alternaria*, *Cladosporium*, *Aspergillus*, *Penicillium*, and *Fusarium* (Zukiewicz-Sobczak 2013). Therefore, including fungi as an ingredient in cosmetic products would pose a potential risk. Another significant issue associated with cosmetics pertains to the contamination of heavy metals. *Boletus edulis* and *Xerocomus badius*, mushrooms found in the wild, were discovered to contain mercury (Hg), lead (Pb), and cadmium (Cd) (Orywal et al. 2021). Therefore, when the cosmetic products are applied directly on the skin, they will indirectly penetrate through systemic circulation if not properly monitored. Consequently, the adverse impacts of heavy metal contamination can extend to various organs within the body (Almukainzi et al. 2022).

17.4.2 Safety Assessment of Fungi Cosmetic

The research and development of newly formulated cosmetic products depend on customer demand. Whether or not a cosmetic product considers the safety of the consumer is a crucial factor in its formulation. Hence, it is imperative for manufacturers to diligently carry out safety assessments and toxicity studies, adhering to regulatory guidelines and standards, to guarantee the safety and compliance of fungal cosmetics before they are produced and introduced into the market (Fig. 17.2). A toxicity study of fungal cosmetics involves the assessment of potential adverse effects or harm that may arise from the use of cosmetic products containing fungal-derived ingredients. Various tests can be undertaken to specifically assess the toxicity of fungal cosmetics including (Barthe et al. 2021):

- Ingredient Safety Assessment: Examining individual fungal-derived ingredients to identify any known toxic compounds, allergens, or substances that may cause irritation.
- Formulation Evaluation: Investigating the overall formulation of the cosmetic product to ensure that the combination of ingredients does not result in adverse reactions or harmful interactions.
- Skin Irritation Testing: Conducting patch tests or other standardized tests to assess the potential for skin irritation or sensitization when the cosmetic product is applied to the skin.
- Eye Irritation Testing: Assessing the cosmetic product's impact on the eyes to ensure it does not cause irritation or adverse reactions when in contact with ocular tissues.
- Oral Toxicity Testing: If the cosmetic product includes ingredients that may be ingested inadvertently (e.g., lip products), evaluating the oral toxicity to ensure safety in case of accidental ingestion.
- Repeated Dose Toxicity: Examining the effects of prolonged or repeated use of the cosmetic product to identify any cumulative toxicity over time.
- Microbiological Safety: Ensuring that the cosmetic product does not harbor harmful microorganisms, fungi, or bacteria that could pose health risks upon use.
- Dermal Absorption Studies: Assessing the extent to which ingredients are absorbed through the skin to determine if any potentially harmful substances may enter the bloodstream.
- Genotoxicity and Mutagenicity Testing: Evaluating whether the cosmetic product or its ingredients have the potential to cause genetic damage or mutations.
- Long-term Safety Studies: Conducting studies to assess the safety of prolonged use of the cosmetic product over an extended period.

17.4.3 Types of Fungal Polysaccharides

17.4.3.1 Beta-Glucans

Bioactive polysaccharides, such as β-glucans, are crucial constituents of the cell wall of almost all fungi. It is a non-starch polysaccharide made of d-glucose units connected by β-glycoside linkages. β-glucans linked by β-(1,3)-glycosidic bonds, together with different amounts of β-(1,4) and β-(1,6)-glycosidic side chains, have lately shown significant immunoregulatory, anticancer, cholesterol-reducing, hypoglycemic, and anti-infection characteristics (Han et al. 2022). β-Glucans exhibit notable biological efficacy, namely in regulating the immune system. Consequently, they are frequently called biological response modifiers. The chemical β-Glucan has attracted increasing attention due to its recent effects.

Beta-glucans are present in the cellular structure of bacteria, fungi, algae, and some grains including oat and barley (Van Steenwijk et al. 2021). β-glucans derived from various sources exhibit variations in their structure, conformation, physical characteristics, receptor binding affinity,

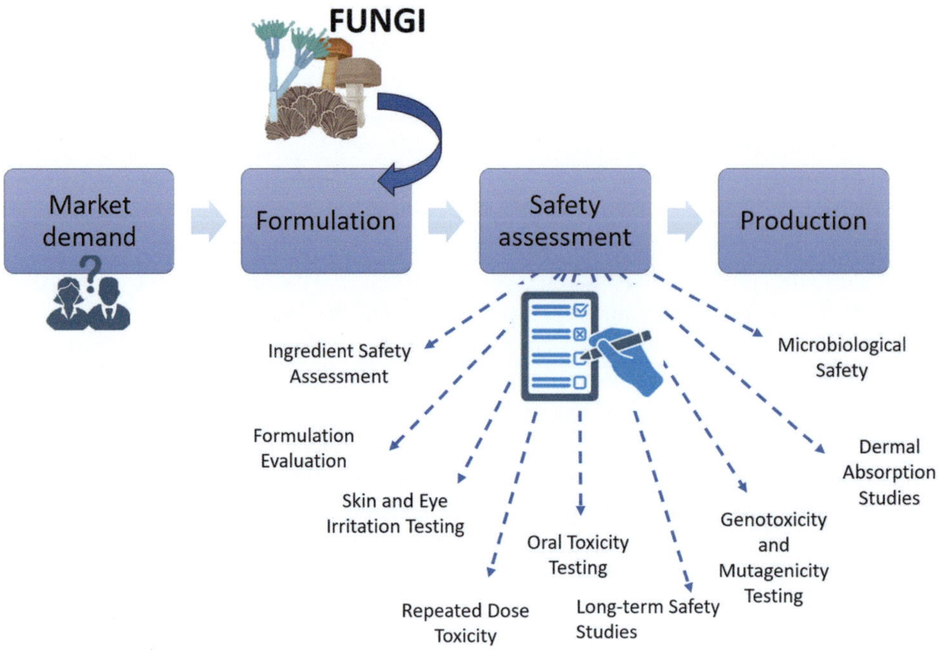

Fig. 17.2 Overview of fungi cosmetics production chain including the potential safety assessment test. (Created with Powerpoint)

and their biological properties (Han et al. 2020). The structure of fungal glucans was first studied using *Saccharomyces cerevisiae* protoplast regeneration glucans. Consist of 65–90% of β-glucans with 1,3 links and 1,6 linkages, *Saccharomyces cerevisiae* yeasts defend their inner compartments with a thick polysaccharide and protein cell wall (Chioru and Chirsanova 2023).

17.4.3.2 Chitin and Chitosan

Henri Braconnot, a chemist, named chitin "fungine" when he first identified it in 1811 after finding it in the cell walls of mushrooms (Iber et al. 2022). In nature, chitin can also be found in many forms in crabs and insects. Though not present in all fungi, chitin is frequently represented in some fungal taxa such as Ascomycota, Zygomycota, and Basidiomycota. Chitin or N-acetyl-glucosamine polymer can be extracted using three methods: chemical extractions (deproteinization, demineralization, and decolorization), enzymatic method which involves proteolytic enzymes, and biological method utilizing microbes (Zainol Abidin et al. 2020). This polysaccharide is characterized as crystalline or semi-crystalline microscopically which functions as a barrier in the cell wall or cuticle, protecting the organism itself (Pusztahelyi 2018).

Chitosan, a prominent variation of chitin, was discovered around 40 years after the identification of chitin. The origins of chitosan can be traced back to Charles Rouget's work in 1859, in which he synthesized the substance by subjecting chitin to heat in an alkaline environment. Traditional chitosan synthesis involves acid and base extraction from fungus cell walls following fermentation. By treating mycelia with 2–4% NaOH at 90–121 °C for 15–120 min, the residual damaged cell wall with proteins and chitosan becomes alkali-insoluble. Following this, the chitin-glucan combination dissociates, releasing chitin, which is deacetylated to chitosan.

17.4.4 Research on Fungal Polysaccharides Species

17.4.4.1 *Ganoderma*

Ganoderma species have been utilized for medicinal applications for centuries, particularly in the Far East countries of China, Japan, and Korea (Kumar 2021). Currently, over 20 species of *Ganoderma* fungi have been examined and a total of more than 600 chemicals have been isolated and identified from the genus *Ganoderma* (Gong et al. 2019). *Ganoderma* contains a variety of chemical substances, including triterpenes, meroterpenoids, steroids, alkaloids, nucleosides, nucleobases, and polysaccharides. Ganoderma polysaccharides consist of linear or branching biopolymers with α or β-linked glucose units (1 → 3), (1 → 6)-β-glucans, and (1 → 3)-α-glucans, and may have side-chains attached at various places. The β-glucans are glucose polymers with a nonbranched (1 → 3)-β-linked backbone or a (1 → 3)-β-linked backbone with (1 → 6)-β-branches. Among other species of genus *Ganoderma*, *Ganoderma lucidum* is often regarded as the most esteemed and venerated medicinal mushroom (Gafforov et al. 2023). It is widely applied in various industries such as to clean sewage wastewater (Hanafiah et al. 2024; Mohd Hanafiah et al. 2019), production of functional fish-feed pellets (Wan-Mohtar et al. 2021), ganodiesel production (Mohamad Jahis et al. 2022), natural food dyes and dye-sensitized solar cells (Ahmad et al. 2022), and act as an antidiabetic treatment to fish (Abdullah et al. 2022).

The discovery of *Ganoderma amboinense* as a novel source of polysaccharides with wound-healing properties has been made. *Ganoderma amboinense* is a prevalent species of *Ganoderma lucidum*, a medicinal fungus that falls within the Polyporaceae family of the Basidiomycota phylum.

Yao and Xu (2022) and Zhao et al. (2021) stated that the polysaccharide derived from the mycelium of *G. amboinense* (GAMPS) has the potential to stimulate cell growth and has antioxidant properties on normal cells. Additionally, GAMPS at a low concentration (0.1 μg/mL) can enhance the movement and healing of scratch injuries in NIH/3T3 cells. Therefore, GAMPS is regarded as possessing favorable wound-healing characteristics. In another study by Mapoung et al. (2021), water-soluble polysaccharide-rich extract from *Auricularia auricula-judae* (AAP) revealed wound healing properties by significantly accelerated wound closure in mice skin wound-healing models on day 9. AAP substantially boosted fibroblast and keratinocyte proliferation, migration, and invasion, and enhanced wound-healing by boosting collagen production and lowering E-cadherin expression ($p < 0.05$).

17.4.4.2 *Poria cocos*

Poria cocos is a prominent member of the *Polyporaceae* family and is widely recognized in traditional Chinese medicine (TCM). It is mainly composed of polysaccharides, triterpenoids, and amino acids and other smaller constituents. The primary constituents of *P. cocos*' dry sclerotia are polysaccharides, which make up over 93% of its composition (Wu et al. 2020). Similar to Ganoderma, *Poria cocos* is an edible and functional medicinal fungus which possesses pharmacological properties such as anticancer, antioxidant, antibacterial, diuretic, and immunomodulatory behaviors (Duan et al. 2023). Remarkably, it is frequently found in the market as both a culinary ingredient and a medicinal supplement. The associated items encompass a range of features, such as cereal, soup, tea, and alcoholic beverages. Polysaccharides obtained from *Poria cocos* can be generated from both naturally occurring fruiting bodies and cultivated mycelia through the process of fermentation (Xu et al. 2022).

According to study by Khan et al. (2018), *Ganoderma lucidum* and *Poria cocos* polysaccharides are proven to modulate gut microbiota composition, making them possess prebiotic functions. Several beneficial bacteria that have been shown to lower obesity, short-chain fatty acid (SCFA) generation, xylan/ polysaccharides breakdown, and lactic acid production were prevalent in polysaccharides-treated mice.

17.4.4.3 *Tremella fuciformis*

Tremella fuciformis is classified in the order *Tremellales* and the family *Tremellaceae* (Gong et al. 2019). It is commonly known as snow fungus, snow ear, silver ear fungus, white jelly mushroom, and white Auricularia. This mushroom is one of the most valuable edible mushrooms in China and acts as medicinal ingredient in traditional Chinese medicine. It contains a low calorie and fat content, but is abundant in protein, polysaccharides, and dietary fiber from a nutritional standpoint. *Tremella fuciformis* also functions as an organic substitute for hyaluronic acid. It promotes cellular hydration retention and enhances the skin's defense mechanism against free radicals, the primary contributors to premature skin ageing (Woźniak et al. 2023).

17.4.4.4 Antarctica Yeast

Antarctica is an ecological niche where microorganisms are compelled to synthesize unusual molecules under pressure to adapt to the region's harsh environmental conditions, which include frigid temperatures, frequent freeze-thaw cycles, strong winds, high sublimation and evaporation rates, and UV radiation (Rusinova-Videva et al. 2022). The most studied yeast genera are *Candida*, *Cryptococcus, Debaryomyces, Mrakia, Rhodotorula,* and *Vishniacozyma* as well as *Leucosporidium.*

17.5 Applications of Fungal Polysaccharides in Cosmetics

The cosmetic business must prioritize the creation and advancement of environmentally friendly goods to meet the expectations of customers and comply with global legislation. This necessitates replacing certain conventional compounds generated from petrochemical sources with alternative ones that possess more environmentally friendly characteristics. Fungal polysaccharides, particularly those obtained from edible mushrooms, are frequently noncytotoxic constituents. This suggests that fungal polysaccharides have significant potential for use in cosmetic and cosmeceutical applications.

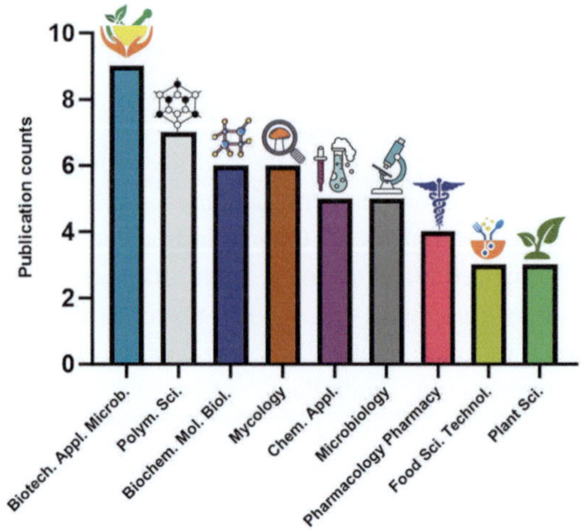

Fig. 17.3 Total relationship findings of fungal polysaccharides and cosmetics in various research fields from 1970 to 2024. (Created with GraphPad Prism 12.0)

Based on Web of Science (2024), the relationship discoveries between fungal polysaccharides and cosmetics in various research fields showed limited studies (Fig. 17.3). Hence, the studies of fungal polysaccharides are vital considering they consist of many natural bioactive compounds contributing to biological properties.

17.5.1 Moisturizers

Besides incorporating in skincare regime, moisturization is a fundamental function of cosmetics that serves to prevent the loss of skin moisture and enhance cellular activity (Pavlou et al. 2021; Zhang et al. 2022). In addition, moisturizers facilitate dermatological treatments for a range of skin conditions and promote a youthful and vibrant appearance (Pavlou et al. 2021). The occlusive elements of moisturizers are principally accountable for their capacity to promptly prevent excessive water loss from the skin. The stratum corneum functions as a protective barrier, preventing the loss of water and other internal fluids while also preventing the entry of external substances into the body. If the water content of the stratum corneum (SC) exceeds 10%, the skin will seem normal and will not exhibit dryness, roughness, or scaliness (Kanlayavattanakul and Lourith 2015); therefore, an optimal moisture level for SC ranges from 20% to 35%. Moisturizers utilize humectants and occlusive substances to replenish the skin's water content and prevent transepidermal water loss (TEWL) (Chirikhina 2022).

Research on the moisturizing properties of certain fungal polysaccharides is limited, with the polysaccharides from *Tremella* being the most extensively investigated. Based on research by Liu and He (2012), a prior investigation showed that *Tremella* polysaccharides (at a concentration of 0.05%) had a greater ability to retain moisture compared to hyaluronic acid at a concentration of 0.02%. This shows that mushroom polysaccharides are natural biological macromolecules with a substantial amount of hydrophilic hydroxyl groups, which give them strong hygroscopicity and excellent film-forming capabilities when used as moisturizers. Similarly, a recent study by Sangthong et al. (2022) reported that twenty volunteers were used in the effectiveness test, and the results revealed that the *Volvariella volvacea* polysaccharides (VVP) gel cream improved the skin's hydration, elasticity, and firmness. In addition, it showed that the skin's melanin level, scaliness, and roughness were all reduced.

17.5.2 Anti-aging Products

Treatment for aging is strategically vital due to its irreversible and complicated degenerative nature. Polysaccharides play crucial roles in combating ageing due to their extensive bioactivities. According to a study by Kozarski et al. (2019), the significant anti-radical activity of *Ganoderma lucidum* extracts proved that the species contains anti-aging property. This property might enhance the skin's barrier function and perhaps inhibit tyrosinase, resulting in a skin-lightening effect.

Fine wrinkles are caused by the loss of structural protein (type I collagen) in the dermal layer of skin, which is a fundamental indicator of skin ageing. Elastase, a metalloproteinase, has a role in breaking down elastin throughout the aging process of the skin. It can either hinder the production of wrinkles or aid in their healing (Freitas-Rodriguez et al. 2017).

17.5.3 Soothing and Anti-inflammatory Products

Mushroom polysaccharides have practical use in creating specialized formulations for individuals with acne-prone or sensitive skin disorders (Dini and Laneri 2021). Their ability to fight against microorganisms may be utilized to successfully control the growth of germs that are linked to the development of acne.

17.5.4 Sun Protection Products

The incorporation of exopolysaccharides into cosmetic creams enhances rheological properties and heat resistance as well as UV protection, with the most significant effects shown when using *Cryptococcus salmonicolor* AL1, which is a psychrophilic yeast strain (Bednarska and Błażejak 2016).

17.5.5 Hair Care Products

Consumers that prioritize health and environmental sustainability are progressively purchasing natural and eco-friendly products. The current endeavor to maintain the health of one's hair has been a difficulty for people in our contemporary fixation. They are becoming aware that continuous exposure to hair chemicals may result in this. With an increasing awareness of the benefits of natural products, more people are opting for safer and healthier alternatives.

Beta-glucan possesses immunomodulatory properties that aid in the alleviation of irritated skin cells in natural shampoo (Panda et al. 2018).

Other than cleansing the hair, this property is important for washing and enhancing the shine, manageability, and oil balance of the hair.

17.6 Challenges, Limitations of Fungal Polysaccharides in Cosmetics

While fungal polysaccharides offer exciting possibilities for the cosmetics industry, there are also challenges and limitations associated with their use. Here are some key hurdles to consider:

17.6.1 Complex Extraction Process

Extracting specific polysaccharides from fungal biomass can be complex and inefficient, requiring multiple purification steps to remove impurities like proteins and lipids. This can be costly and time-consuming (Aggarwal and Kumari 2024; Sarangi et al. 2024).

17.6.2 Variable Purification Composition

Polysaccharide composition can vary depending on the fungal species, growth conditions, and extraction methods. This inconsistency can pose challenges in maintaining product quality and standardization (Bocker and Silva 2024; Bumandalai et al. 2024).

17.6.3 Solubility Properties

Some fungal polysaccharides are poorly soluble in water or solvents commonly used in cosmetics, limiting their formulation options and potential applications (Benalaya et al. 2024; Gericke et al. 2024).

17.6.4 Viscosity and Texture Profiles

Certain polysaccharides can impart a thick, gel-like texture to products, which may not be desirable for all consumers. Finding the right balance between efficacy and user experience can be tricky (Gericke et al. 2024; Zhang et al. 2024).

17.6.5 Stability of Compound

Fungal polysaccharides can be susceptible to degradation by enzymes, microorganisms, and even light. This necessitates the use of stabilizers and appropriate packaging solutions, adding to the cost and complexity (Gericke et al. 2024).

17.6.6 Mycotoxin Contamination Regulation

Some fungi can produce mycotoxins, harmful toxins that can pose safety concerns. Rigorous testing and control measures are needed to ensure the safety of fungal-derived ingredients (Liu et al. 2024).

17.6.7 Limited Data

Compared to plant-based polysaccharides, research on the long-term safety and efficacy of fungal polysaccharides in cosmetics is less extensive. Regulatory clearance and market acceptance might be slower (Chavda et al. 2024).

17.6.8 Sustainability of Polysaccharides

Large-scale cultivation of specific fungi for polysaccharide production may require careful optimization to minimize environmental impact and ensure resource efficiency (Cotas et al. 2024; Riseh et al. 2024).

17.6.9 Consumer Perception

Some consumers may have concerns about using ingredients derived from fungi, even if they are safe and effective. Addressing these concerns through education and transparent labeling is crucial. Despite these challenges, research and development efforts are ongoing to address the limitations of fungal polysaccharides and unlock their full potential in cosmetics. Improvements in extraction, purification, modification, and formulation techniques are key areas of focus (Jiang et al. 2024). With more research and innovation, fungal polysaccharides could become a valuable and sustainable source of functional ingredients for the cosmetics industry.

17.7 Future Developments of Fungal Polysaccharide Cosmetics

The future of fungal polysaccharides in cosmetics is brimming with exciting possibilities, driven by ongoing research and technological advancements. Here are some key areas to watch:

17.7.1 Green and Efficient Techniques Extraction

Development of novel, eco-friendly methods like enzyme-assisted extraction and membrane filtration could reduce processing costs and minimize environmental impact (Maled et al. 2024).

17.7.2 Targeted Isolation

Optimizing extraction processes to isolate specific polysaccharides with desired properties could lead to more potent and tailored cosmetic ingredients (Ren et al. 2024).

17.7.3 Chemical Modification

Modifying polysaccharides through processes like grafting or cross-linking can improve their solubility, texture, and stability, making them more versatile for different cosmetic applications (Angaria et al. 2024).

17.7.4 Encapsulation and Biomimetic Delivery

Encapsulating active ingredients within fungal polysaccharides could enhance their delivery and controlled release within the skin, maximizing their efficacy (Kavur and Yemenicioğlu 2024).

17.7.5 Advanced Research and Development

Exploration of diverse fungal sources: Investigating unique and underexplored fungal species could reveal novel polysaccharides with previously unknown properties and benefits for skin health (Rungjindamai and Jones 2024).

17.7.6 In-Depth Studies on Mechanisms of Action

Understanding the exact mechanisms by which fungal polysaccharides interact with the skin at a molecular level can lead to the development of even more targeted and effective cosmetic formulations (Bharatia et al. 2024).

17.7.7 Computational Modeling and Simulation

Utilizing advanced computational tools could predict the properties and potential applications of polysaccharides before extensive laboratory testing, streamlining the development process (Akhtar and Gupta 2024).

17.7.8 Sustainability and Consumer Acceptance

Sustainable cultivation and bioprocessing: Optimizing fungal cultivation techniques and adopting bio-based solvents for extraction can promote sustainability and reduce the environmental footprint of the industry (Wagh et al. 2024).

17.7.9 Transparency and Education

Educating consumers about the safety, efficacy, and sustainability of fungal polysaccharides can address concerns and build trust, fostering wider adoption. Overall, the future of fungal polysaccharides in cosmetics is bright. Increased research, technological advancements, and a focus on sustainability are paving the way for innovative and effective ingredients that can deliver tangible benefits for skin health and beauty (Bin Abu Sofian et al. 2024). With continued collaboration between researchers, cosmetic manufacturers, and consumers, fungal polysaccharides have the potential to revolutionize the industry and create next-generation products that are both effective and kind to the planet.

17.8 Conclusion

Once a well-kept secret on the forest floor, polysaccharides derived from fungi are becoming increasingly well-known in the cosmetics sector. Their wide range of qualities, which include antioxidant defense and moisturizing magic, reveals a wealth of advantages for the health and appearance of skin. Even if stability and extraction still present obstacles, technological and scientific advancements are shedding light on how to realize their full potential. Ecological extraction and isolation open the door to economical, sustainable production. Like biomimetic delivery techniques, modified polysaccharides offer improved efficacy and deeper penetration. As we learn more about the mechanisms of action, we will be able to precisely customize formulations to address certain skin issues. However, the foundations of this fungal revolution are sustainability and customer trust. Reduced environmental impact can be achieved by using bio-based solutions and optimizing cultivation. The distance between the lab and the bathroom sink will be closed with open communication and honest labeling. In conclusion, fungal polysaccharides hold great promise for the future of cosmetics. This formerly modest resident of a mushroom is ready to change the course by providing practical, environmentally friendly solutions for a prosperous and lovely future. Put parabens and silicones out of business; a new player is emerging that is building from the ground up.

Acknowledgments The authors thank the Ministry of Higher Education (MOHE) under the Fundamental Research Grant Scheme (FRGS:FP089-2023 (FRGS/1/2023/WAB04/UM/02/1) for this project "Antagonistic effects of non-pathogenic mushroom against the Basal Stem Rot disease caused by *Ganoderma boninense*".

References

Abdullah NR, Mohd Nasir MH, Azizan NH, Wan-Mohtar WAAQI, Sharif F (2022) Bioreactor-grown exo-and endo-β-glucan from Malaysian *Ganoderma lucidum*: an in vitro and in vivo study for potential antidiabetic treatment. Front Bioeng Biotechnol 10:960320

Aggarwal S, Kumari A (2024) Role of fungi in biotechnology. In: Entrepreneurship with microorganisms. Elsevier, pp 39–67

Ahmad N, Vunduk J, Klaus A, Dahlan NY, Ghosh S, Muhammad-Sukki F, Dufossé L, Bani NA, Wan-Mohtar WAAQI (2022) Roles of medicinal mushrooms as natural food dyes and dye-sensitised solar cells (DSSC): synergy of zero hunger and affordable energy for sustainable development. Sustainability 14(21):13894

Akhtar ZB, Gupta AD (2024) Advancements within molecular engineering for regenerative medicine and biomedical applications an investigation analysis towards a computing retrospective. J Electronic Electromed Eng Med Inform 6(1):54–72

Allied Market Research (2023) Cosmetics market research 2023. https://www.alliedmarketresearch.com/cosmetics-market

Almukainzi M, Alotaibi L, Abdulwahab A, Albukhary N, El Mahdy AM (2022) Quality and safety investigation of commonly used topical cosmetic preparations. Sci Rep 12(1):18299. https://doi.org/10.1038/s41598-022-21771-7

Angaria N, Saini S, Hussain MS, Sharma S, Singh G, Khurana N, Kumar R (2024) Natural polymer-based hydrogels: versatile biomaterials for biomedical

applications. Int J Polym Mater Polym Biomater 73(17):1–19
Barbosa JR, de Carvalho Junior RN (2020) Occurrence and possible roles of polysaccharides in fungi and their influence on the development of new technologies. Carbohydr Polym 246:116613
Barthe M, Bavoux C, Finot F, Mouche I, Cuceu-Petrenci-C, Forreryd A, Chérouvrier Hansson A, Johansson H, Lemkine GF, Thénot J-P, Osman-Ponchet H (2021) Safety testing of cosmetic products: overview of established methods and new approach methodologies (NAMs). Cosmetics 8(2):50. https://www.mdpi.com/2079-9284/8/2/50
Bednarska AA, Błażejak S (2016) The exopolysaccharides biosynthesis by Candida yeast depends on carbon sources. Electron J Biotechnol 5(1):24
Benalaya I, Alves G, Lopes J, Silva LR (2024) A review of natural polysaccharides: sources, characteristics, properties, food, and pharmaceutical applications. Int J Mol Sci 25(2):1322
Bharatia R, Gupta N, Upadhyay P, Tiwari SK, Gupta S, Sahni K (2024) A review on cosmeceuticals. Int J Pharma Prof Res 15(1):128–141
Bin Abu Sofian ADA, Sun X, Gupta VK, Berenjian A, Xia A, Ma Z, Show PL (2024) Advances, synergy, and perspectives of machine learning and biobased polymers for energy, fuels, and biochemicals for a sustainable future. Energy Fuel 38(3):1593–1617
Bocker R, Silva EK (2024) Pulsed electric field technology as a promising pre-treatment for enhancing orange agro-industrial waste biorefinery. RSC Adv 14(3): 2116–2133
Bumandalai O, Bayliss KL, Moheimani NR (2024) Innovative processes for combating contaminants in fresh Spirulina. Algal Res 78:103397
Chavda VP, Ghali ENHK, Balar PC, Chauhan SC, Tiwari N, Shukla S, Athalye M, Patravale V, Apostolopoulos V, Yallapu MM (2024) Protein subunit vaccines: promising frontiers against COVID-19. J Control Release 366:761–782
Chioru A, Chirsanova A (2023) β-glucans: characterization, extraction methods, and valorization. Food Nutr Sci 14(10):963–983
Chirikhina E (2022) Human face mapping based on TEWL, hydration and ultrasound. South Bank University, London
Cotas J, Lomartire S, Gonçalves AM, Pereira L (2024) From ocean to medicine: harnessing seaweed's potential for drug development. Int J Mol Sci 25(2):797
Desam NR, Al-Rajab AJ (2021) The importance of natural products in cosmetics. In: Bioactive natural products for pharmaceutical applications. Springer, Berlin, pp 643–685
Dini I, Laneri S (2021) The new challenge of green cosmetics: natural food ingredients for cosmetic formulations. Molecules 26(13):3921
Duan Y, Huang J, Sun M, Jiang Y, Wang S, Wang L, Yu N, Peng D, Wang Y, Chen W (2023) Poria cocos polysaccharide improves intestinal barrier function and maintains intestinal homeostasis in mice. Int J Biol Macromol 249:125953
Freitas-Rodriguez S, Folgueras AR, Lopez-Otin C (2017) The role of matrix metalloproteinases in aging: tissue remodeling and beyond. Biochim Biophys Acta, Mol Cell Res 1864(11):2015–2025
Gafforov Y, Umar A, Ghosh S, Tomšovský M, Yamaç M, Rašeta M, Yarasheva M, Wan-Mohtar WAAQI, Rapior S (2023) *Ganoderma adspersum* (Schulzer) Donk; *Ganoderma applanatum* (Pers.) Pat.; *Ganoderma lucidum* (Curtis) P. Karst.; *Ganoderma resinaceum* Boud.-GANODERMATACEAE. In: Ethnobiology of Uzbekistan: ethnomedicinal knowledge of mountain communities. Springer, pp 1135–1169
Gericke M, Amaral AJ, Budtova T, De Wever P, Groth T, Heinze T, Höfte H, Huber A, Ikkala O, Kapuśniak J (2024) The European Polysaccharide Network of Excellence (EPNOE) research roadmap 2040: advanced strategies for exploiting the vast potential of polysaccharides as renewable bioresources. Carbohydr Polym 326:121633
Goel R, Bhardwaj S, Bana S (2024) Pharmaceutical excipients. In: Dosage forms, formulation developments and regulations. Elsevier, pp 311–348
Gong T, Yan R, Kang J, Chen R (2019) Chemical components of Ganoderma. In: Ganoderma and health: biology, chemistry and industry, pp 59–106. https://doi.org/10.1007/978-981-13-9867-4_3
Grand View Research (2022) Organic skin care market [2022 - 2030]. Grand View Research. https://www.grandviewresearch.com/industry-analysis/organic-skin-care-market
Han B, Baruah K, Cox E, Vanrompay D, Bossier P (2020) Structure-functional activity relationship of β-glucans from the perspective of immunomodulation: a mini-review. Front Immunol 11:658
Han X, Luo R, Ye N, Hu Y, Fu C, Gao R, Fu S, Gao F (2022) Research progress on natural β-glucan in intestinal diseases. Int J Biol Macromol 19:1244–1260
Hanafiah ZM, Azmi AR, Wan-Mohtar WAAQI, Olivito F, Golemme G, Ilham Z, Jamaludin AA, Razali N, Halim-Lim SA, Wan Mohtar WHM (2024) Water quality assessment and decolourisation of contaminated ex-mining Lake water using bioreactor dye-eating fungus (BioDeF) system: a real case study. Toxics 12(1): 60
Hyde K, Bahkali A, Moslem M (2010) Fungi—an unusual source for cosmetics. Fungal Divers 43:1–9
Iber BT, Kasan NA, Torsabo D, Omuwa JW (2022) A review of various sources of chitin and chitosan in nature. J Renew Mater 10(4):1097
Jiang C, Ge J, He B, Zeng B (2021) Glycosphingolipids in filamentous fungi: biological roles and potential applications in cosmetics and health foods. Front Microbiol 12:690211
Jiang H, Yao M, Chen J, Zhang M, Hong W (2024) Advances in biomass-based nanofibers prepared by electrospinning for energy storage devices. Fuel 355: 129534

Johansson L, Virkki L, Anttila H, Esselström H, Tuomainen P, Sontag-Strohm T (2006) Hydrolysis of β-glucan. Food Chem 97(1):71–79

Kanlayavattanakul M, Lourith N (2015) Biopolysaccharides for skin hydrating cosmetics. In: Polysaccharides: bioactivity and biotechnology. Springer, New York, pp 1867–1892

Kavur PB, Yemenicioğlu A (2024) An efficient method of improving essential oil retention and sustained release of chitosan films: ultrasound-assisted preparation of chitosan composites with surface active chickpea proteins. Food Hydrocoll 151:109790

Keleş A, Koca I, Gençcelep H (2011) Antioxidant properties of wild edible mushrooms. J Food Process Technol 2(6):2–6

Khan I, Huang G, Li X, Leong W, Xia W, Hsiao WW (2018) Mushroom polysaccharides from *Ganoderma lucidum* and *Poria cocos* reveal prebiotic functions. J Funct Foods 41:191–201

Kozarski M, Klaus A, Jakovljević D, Todorović N, Wan-Mohtar WAAQI, Nikšić M (2019) *Ganoderma lucidum* as a cosmeceutical: antiradical potential and inhibitory effect on hyperpigmentation and skin extracellular matrix degradation enzymes. Arch Biol Sci 71(2):253–264

Kozarski M, Klaus A, van Griensven L, Jakovljevic D, Todorovic N, Wan-Mohtar WAAQI, Vunduk J (2023) Mushroom β-glucan and polyphenol formulations as natural immunity boosters and balancers: nature of the application. Food Sci Human Wellness 12(2):378–396

Kumar A (2021) *Ganoderma lucidum*: a traditional chinese medicine used for curing tumors. Int J Pharm Pharm Sci 13:1–13

Li L-D, Mao P-W, Shao K-D, Bai X-H, Zhou X-W (2019) Ganoderma proteins and their potential applications in cosmetics. Appl Microbiol Biotechnol 103:9239–9250

Liu H, He L (2012) Comparison of the moisture retention capacity of tremella polysaccharides and hyaluronic acid. J Anhui Agric Sci 40:13093–13094

Liu M, Feng J, Yang X, Yu B, Zhuang J, Xu H, Xiang Q, Ma R, Jiao Z (2024) Recent advances in the degradation efficacy and mechanisms of mycotoxins in food by atmospheric cold plasma. Ecotoxicol Environ Saf 270: 115944

Maled SB, Bhat AR, Hegde S, Sivamani Y, Muthuraman A, Elayaperumal S (2024) Enzyme-assisted extraction. In: Bioactive extraction and application in food and nutraceutical industries. Springer, pp 173–200

Mapoung S, Umsumarng S, Semmarath W, Arjsri P, Thippraphan P, Yodkeeree S, Limtrakul P (2021) Skin wound-healing potential of polysaccharides from medicinal mushroom *Auricularia auricula-judae* (Bull.). J Fungi 7(4):247

McKinsey (2023) The beauty market in 2023: new industry trends. McKinsey. https://www.mckinsey.com/industries/retail/our-insights/the-beauty-market-in-2023-a-special-state-of-fashion-report

Mohamad Jahis BM, Ilham Z, Supramani S, Sohedein MNA, Ibrahim MF, Abd-Aziz S, Rowan N, Wan-Mohtar WAAQI (2022) Ganodiesel: a new biodiesel feedstock from biomass of the mushroom *Ganoderma lucidum*. Sustainability 14(17):10764

Mohd Hanafiah Z, Wan Mohtar WHM, Abu Hasan H, Jensen HS, Klaus A, Wan-Mohtar WAAQI (2019) Performance of wild-Serbian *Ganoderma lucidum* mycelium in treating synthetic sewage loading using batch bioreactor. Sci Rep 9(1):16109

Mohd-Setapar H (2018) Natural ingredients in cosmetics from Malaysian plants: a review. Sains Malays 47(5): 951–959

Mooralitharan S, Mohd Hanafiah Z, Abd Manan TSB, Muhammad-Sukki F, Wan-Mohtar WAAQI, Wan Mohtar WHM (2023) Vital conditions to remove pollutants from synthetic wastewater using malaysian *Ganoderma lucidum*. Sustainability 15(4):3819

Orywal K, Socha K, Nowakowski P, Zoń W, Kaczyński P, Mroczko B, Łozowicka B, Perkowski M (2021) Health risk assessment of exposure to toxic elements resulting from consumption of dried wild-grown mushrooms available for sale. PLoS One 16(6):e0252834. https://doi.org/10.1371/journal.pone.0252834

Panda S, Nayak M, Biswas N (2018) Formulation and evaluation of herbal powdered shampoo. J Emerg Technol Innov Res 5:108–113

Pavlou P, Siamidi A, Varvaresou A, Vlachou M (2021) Skin care formulations and lipid carriers as skin moisturizing agents. Cosmetics 8(3):89

Pusztahelyi T (2018) Chitin and chitin-related compounds in plant–fungal interactions. Mycology 9(3):189–201

Ren Y, Jia F, Li D (2024) Ingredients, structure and reconstitution properties of instant powder foods and the potential for healthy product development: a comprehensive review, vol 15. Food Funct, pp 37–61

Riseh RS, Vatankhah M, Hassanisaadi M, Varma RS (2024) A review of chitosan nanoparticles: nature's gift for transforming agriculture through smart and effective delivery mechanisms. Int J Biol Macromol 260(Pt 2):129522

Rungjindamai N, Jones EG (2024) Why are there so few basidiomycota and basal fungi as endophytes? A review. J Fungi 10(1):67

Rusinova-Videva S, Ognyanov M, Georgiev Y, Kambourova M, Adamov A, Krasteva V (2022) Production and chemical characterization of exopolysaccharides by Antarctic yeasts *Vishniacozyma victoriae* and *Tremellomycetes* sp. Appl Sci 12(4):1805

Sangthong S, Pintathong P, Pongsua P, Jirarat A, Chaiwut P (2022) Polysaccharides from *Volvariella volvacea* mushroom: extraction, biological activities and cosmetic efficacy. J Fungi 8(6):572

Sarangi PK, Srivastava RK, Sahoo UK, Singh AK, Parikh J, Bansod S, Parsai G, Luqman M, Shadangi KP, Diwan D (2024) Biotechnological innovations in nanocellulose production from waste biomass with a focus on pineapple waste. Chemosphere 349:140833

Statista (2022) Beauty: leading companies by market cap 2022. https://www.statista.com/statistics/1346623/leading-cosmetics-and-beauty-companies-worldwide-based-on-market-cap/

Sułkowska-Ziaja K, Trepa M, Olechowska-Jarząb A, Nowak P, Ziaja M, Kała K, Muszyńska B (2023) Natural compounds of fungal origin with antimicrobial activity—potential cosmetics applications. Pharmaceuticals 16(9):1200

Taofiq O, Heleno SA, Calhelha RC, Fernandes IP, Alves MJ, Barros L, González-Paramás AM, Ferreira IC, Barreiro MF (2019) Phenolic acids, cinnamic acid, and ergosterol as cosmeceutical ingredients: stabilization by microencapsulation to ensure sustained bioactivity. Microchem J 147:469–477

Thakur M (2020) Advances in mushroom production: key to food, nutritional and employment security: a review. Indian Phytopathol 73:377–395

Van Steenwijk HP, Bast A, De Boer A (2021) Immunomodulating effects of fungal beta-glucans: from traditional use to medicine. Nutrients 13(4):1333

Visvanathan S, Krishnamoorthy R, Sabesan GS (2022) Fungal cosmetics: mushrooms in beauty care and the new age of natural cosmetics. In: Applied mycology: entrepreneurship with fungi. Springer, pp 1–37

Wagh MS, Sowjanya S, Nath PC, Chakraborty A, Amrit R, Mishra B, Mishra AK, Mohanta YK (2024) Valorisation of agro-industrial wastes: circular bioeconomy and biorefinery process—a sustainable symphony. Process Saf Environ Prot 183:708–725

Wallace TC, Giusti MM (2019) Anthocyanins—nature's bold, beautiful, and health-promoting colors. Foods 8: 550. MDPI

Wan-Mohtar WAAQI, Taufek NM, Yerima G, Rahman J, Thiran JP, Subramaniam K, Sabaratnam V (2021) Effect of bioreactor-grown biomass from *Ganoderma lucidum* mycelium on growth performance and physiological response of red hybrid tilapia (Oreochromis sp.) for sustainable aquaculture. Org Agric 11:327–335

Woźniak B, Chudzińska J, Szczyglewska P, Nowak I, Feliczak-Guzik A (2023) Optimization of the composition of a cosmetic formulation containing tremella fuciformis extract (fungi). Cosmetics 10(3):82

Wu Y, Choi M-H, Li J, Yang H, Shin H-J (2016) Mushroom cosmetics: the present and future. Cosmetics 3(3):22

Wu P, Tan H, Zhan J, Wang W, Hu T, Li S (2020) Optimization of bioprocess extraction of Poria cocos polysaccharide (PCP) with *Aspergillus niger* β-glucanase and the evaluation of PCP antioxidant property. Molecules 25(24):5930

Xiao Z, Zhou W, Zhang Y (2020) Fungal polysaccharides. Adv Pharmacol 87:277–299

Xu T, Zhang H, Wang S, Xiang Z, Kong H, Xue Q, He M, Yu X, Li Y, Sun D (2022) A review on the advances in the extraction methods and structure elucidation of Poria cocos polysaccharide and its pharmacological activities and drug carrier applications. Int J Biol Macromol 217:536–551

Yao Y, Xu B (2022) Skin health promoting effects of natural polysaccharides and their potential application in the cosmetic industry. Polysaccharides 3(4): 818–830

Zainol Abidin NA, Kormin F, Zainol Abidin NA, Mohamed Anuar NAF, Abu Bakar MF (2020) The potential of insects as alternative sources of chitin: an overview on the chemical method of extraction from various sources. Int J Mol Sci 21(14):4978

Zhang T, Guo Q, Xin Y, Liu Y (2022) Comprehensive review in moisture retention mechanism of polysaccharides from algae, plants, bacteria and fungus. Arab J Chem 15:104163

Zhang Q, Kong B, Liu H, Du X, Sun F, Xia X (2024) Nanoscale pickering emulsion food preservative films/coatings: compositions, preparations, influencing factors, and applications. Compr Rev Food Sci Food Saf 23(1):1–36

Zhao S, Lei M, Xu H, He H, Suvorov A, Wang J, Qiu J, Zhou Q, Yang J, Chen L (2021) The normal cell proliferation and wound healing effect of polysaccharides from *Ganoderma amboinense*. Food Sci Human Wellness 10(4):508–513

Ziemlewska A, Wójciak M, Mroziak-Lal K, Zagórska-Dziok M, Bujak T, Nizioł-Łukaszewska Z, Szczepanek D, Sowa I (2022) Assessment of cosmetic properties and safety of use of model washing gels with Reishi, Maitake and Lion's Mane extracts. Molecules 27(16). https://doi.org/10.3390/molecules27165090

Zukiewicz-Sobczak WA (2013) The role of fungi in allergic diseases. Postepy Dermatol Alergol 30(1):42–45. https://doi.org/10.5114/pdia.2013.33377

18 Nanocosmetics and Fungal-Based Cosmetics: Similarities and Differences in Formulation and Usage

Suman Bhandary, Yashaswi Dutta Gupta, and Somtirtha Kool Banerjee

Abstract

This chapter explores the similarities and differences in the formulation and usage of nanocosmetics and fungal-based cosmetics. The first section gives a summary of nanocosmetics, including the definition and characteristics of nanocosmetics and the types of nanomaterials used. It also delves into the formulation techniques and applications of nanocosmetics, highlighting their efficacy in skincare, haircare, and makeup products. The second section focuses on fungal-based cosmetics, introducing the concept and sources of fungal-based ingredients. It explores the bioactive substances present in fungus and their possible uses in cosmetics, emphasizing their antibacterial, anti-inflammatory, and antioxidant qualities. The section further discusses the formulation techniques used in fungal-based cosmetics, including the incorporation of fungal extracts and synergistic formulations with natural ingredients. In the third section, the similarities and differences between nanocosmetics and fungal-based cosmetics are examined. This includes a comparative analysis of active ingredients and their mechanisms of action. The formulation and manufacturing techniques utilized in both categories are explored, emphasizing the commonalities and unique challenges faced in each. Additionally, consumer usage and perception of these cosmetic categories are discussed, considering factors such as acceptance, consumer preferences, safety concerns, and market trends. In conclusion, understanding the similarities and differences between nanocosmetics and fungal-based cosmetics is crucial for formulators, researchers, and consumers. This chapter furnishes valuable insights into the formulation, application, and consumer perception of these pioneering cosmetic domains, laying the groundwork for future advancements in the field.

Keywords

Nanocosmetics · Fungal-based cosmetics · Skincare · Haircare · Nanomaterials · Safety · Regulatory guidelines

18.1 Introduction

The cosmetics industry, which includes a diverse array of products, has greatly contributed to human well-being throughout history. In ancient times, it was believed that because products

S. Bhandary (✉) · Y. D. Gupta
Department of Biological Sciences, School of Life Science and Biotechnology, Adamas University, Kolkata, West Bengal, India

S. K. Banerjee
Department of Microbiology, Lady Brabourne College, Kolkata, West Bengal, India

S. Ghosh, H. N. Onyeaka (eds.), *Fungal-Based Cosmetics: Formulation and Usage*, The Mycota 17,
https://doi.org/10.1007/978-3-032-09291-5_18

contained chemicals, they could never be completely natural. The landscape of cosmetics transformed with the advent of industrialization. Later, the emergence of nanotechnology further advanced the development of environmentally friendly cosmetics (Tamrakar and Thakur 2023). Nanotechnology, recognized as the most powerful technology of the twenty-first century, enables the exploration and manipulation of particles within 1–100 nm range, offering a broad spectrum of applications. Its continuous evolution through research and development has significantly advanced innovative solutions, particularly benefiting the cosmetic industry. Historically, personal care relied on household items and preparations, but by the twenty-first century, the advent of advanced technologies revolutionized cosmetic formulations. This shift has caused the production and consumption of cosmeceuticals to rise quickly (Dureja et al. 2005; Saha 2012). In the context of cosmetics, nanotechnology offers innovative solutions to enhance the performance and delivery of cosmetic ingredients. Nanocosmetics leverage nanoparticles to improve product stability, increase efficacy, and provide unique aesthetic benefits (Chauhan and Chauhan 2023; Tamrakar and Thakur 2023). Consumers are expressing a keen interest in adopting comprehensive and technology-driven approaches when purchasing cosmetics, focusing on both beauty and health benefits (Chin et al. 2018; Ghazali et al. 2017; Raj et al. 2012). The primary focus lies in natural products crafted through nanotechnology and bionanotechnology, chosen for their efficacy and safety in promoting well-being and environmental considerations as well (Kaul et al. 2018).

Concerns about the environment, animal welfare, health, and hygiene are driving a growing global interest in green cosmetics. Definitions of these products often highlight aspects such as "green," "natural," "organic," or "bio cosmetics." Additionally, there is an increasing tendency in the use of green cosmetics and a rising consumer interest in online shopping (Barbu et al. 2022; Zhang and Dong 2020). The shift in consumer buying patterns has captured the attention of cosmetic brands, leading them to explore various botanical extracts (Singh et al. 2019). Fungi (mainly mushrooms), in particular, have garnered renewed interest. Certain wild mushrooms, including both their mycelia and fruiting bodies, are rich in proteins, lipids, essential amino acids, glycogens, riboflavin, nicotinic acid, and valuable minerals. They also contain several biologically active compounds such as polysaccharides, lectins, phenolics, terpenoids, ergosterols, and volatile organic chemicals (Ahmad 2018; Mago et al. 2023). These substances have anti-inflammatory, anti-microbial, anti-tyrosinase, and anti-hyaluronidase properties, among other biological advantages (Chien et al. 2008; Meng et al. 2011). Consequently, active ingredients derived from mushroom have become a significant component in a range of cosmetic formulations for hair and skin (Bell et al. 2022; Mukherjee et al. 2024). Since organic materials are safer and more biocompatible than synthetic ones, the cosmetics industry is currently witnessing a global shift in the cosmeceutical sector (Alves et al. 2020). The global market for green and organic cosmetics is experiencing significant growth (Amberg and Fogarassy 2019; Onel 2016). The desire for organic personal care products and a healthy lifestyle has been spurred by rising knowledge of environmental and health issues as well as the potential hazards of synthetic chemicals (Ghazali et al. 2017; Martins and Marto 2023). The global market for cosmetic and personal care surged at an annual compound growth rate of 4.75% from $483 billion in 2020 to $511 billion in 2021. By 2025, it is expected to surpass $716 billion, and by 2027, it will reach $784.6 billion (Ustymenko 2023).

In light of this context, the goal of this chapter is to present an overview of the current state of the art in the fields of cosmeceuticals based on fungi and nanoparticles, offering details on their compositions, modes of action, efficacy, safety, legalities, and possible effects on the cosmetics industry.

18.2 Nanocosmetics

The integration of nanoparticles into dermatological care marks a significant breakthrough, providing a more effective, comfortable, and personalized approach to skincare. This innovative technology enhances the delivery and efficacy of skincare products, allowing for targeted treatment and improved skin health. As advancements in nanotechnology continue to progress, the potential for nanoparticles to revolutionize skincare remains promising. These tiny particles offer the ability for delivering active ingredients more precisely to the desired skin layers, increasing their potency and reducing side effects. It gained momentum with the launch of Christian Dior's Capture line in 1986, and further accelerated with L'Oréal's introduction of Revitalift in 2005 (Ajazzuddin et al. 2015; Battaglia and Ugazio 2019). Since then, the number of cosmetics containing nanomaterials continues to rise. The subsequent sections, encompassing Fig. 18.1 and Table 18.1, provide an in-depth discussion of various submicron-sized innovative drug delivery systems utilized within the cosmetic industry for the effective delivery of active ingredients. These novel delivery systems, which operate at the submicron level, are designed to enhance the efficacy, stability, and absorption of active compounds in cosmetic products. In dermatology, numerous nanocarriers have been generated, offering innovative solutions for topical skin applications. These carriers, including liposomes, micelles, and solid lipid nanoparticles, enhance the delivery of active ingredients, ensuring targeted and effective penetration. Their versatility supports a wide range of medicinal and cosmetic formulations that address concerns such as hydration, aging, and inflammation. By encapsulating beneficial compounds, these nanocarriers are revolutionizing skincare, providing precise and effective treatments for diverse skin issues.

18.2.1 Nanomaterials Used in Cosmetics

Nanocosmetics represent a groundbreaking advancement in the beauty and skincare industry, leveraging the principles of nanotechnology to increase the advantages and efficacy of cosmetic products (Gupta et al. 2022). Nanocosmetics are generally personal care and beauty products containing nanocarriers or nanoparticles (Ajazzuddin et al. 2015). Nanocarriers are categorized into lipid nanocarriers and polymeric nanocarriers. Lipid nanocarriers include solid lipid nanoparticles (SLNs), nanostructured lipid carriers (NLCs), and nanoemulsions. Polymeric nanocarriers encompass nanocapsules, nanospheres, and micelles (Ajazzuddin et al. 2015; Gupta et al. 2022). Recently, the most common application of nanomaterials in cosmetics has been as UV filters in sunscreen products. Nanopowders such as kaolin, iron oxide, zinc oxide (ZnO), and titanium oxide (TiO_2) are utilized in face powders. In addition to enhancing appearance, nanopowders like ZnO also provide sunscreen protection due to their strong light-scattering properties. Incorporating these nanomaterials in cosmetics not only improves the performance and efficacy of the products but also allows for innovative formulations that offer better texture, stability, and sensory experiences for consumers (Fytianos et al. 2020; Sharma et al. 2023; Smijs and Pavel 2011). Below are some types of nanomaterials commonly used in cosmetics, along with their specific applications (Fig. 18.1):

18.2.1.1 Nanometals

Nanosilver and nanogold have gained popularity in the cosmetics industry due to their high efficiency, antimicrobial properties, and unique optical properties. One of the inorganic nanoparticles that is most frequently utilized in sunscreens is titanium dioxide (TiO_2). Because TiO_2 has a

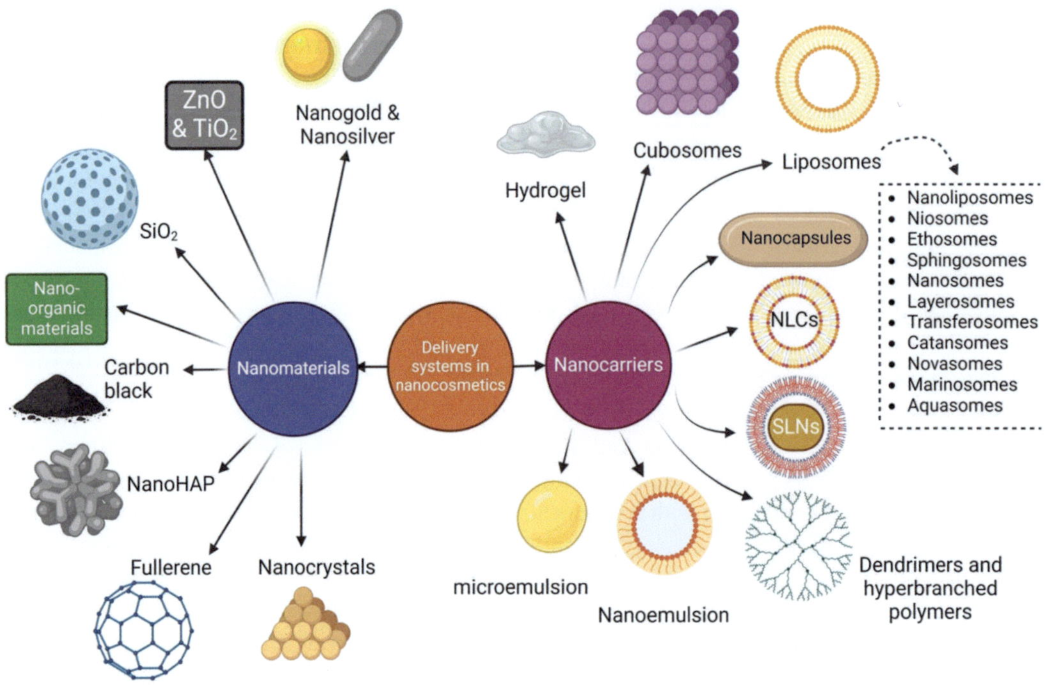

Fig. 18.1 Different nanoformulations utilized in the cosmetic industry. (This figure was created with BioRender (https://biorender.com/))

greater sun protection factor (SPF) at the nanoscale, which makes it more effective and produces superior cosmetic outcomes. Products containing nanoscale TiO_2 or ZnO are often described as "sheer" or "invisible" in the market. Nanoscale TiO_2 and ZnO are reported to have significant advantages over their larger counterparts (Cornier et al. 2019; Fytianos et al. 2020). They are used in formulations of deodorants, toothpastes, anti-aging products, etc. (Shokri 2017). Due to their optical properties, nanogold and nanosilver are successfully used for skin color modulation through blending techniques. Gold nanoparticles have been utilized to create red or blue coloration in the skin, silver nanoparticles to produce a yellowish hue, and gold-silver nanoparticles to achieve an orange tint (Yadwade et al. 2021). A valuable ingredient in cosmeceuticals, gold and silver nanoparticles have antibacterial and antifungal qualities. They are often used in anti-aging lotions, face packs, and deodorant products. It is stated that an ointment with silver nanoparticles has antibacterial properties and can be used to treat wounds and inflammation of the skin (Gupta et al. 2022; Haveli et al. 2012). Fluorescent gold nanoparticles were created inside human hair by French scientist Dr. Philippe Walter and his colleagues. Additionally, nanosilica is employed to improve the effectiveness, texture, and longevity of cosmetic products (Gupta et al. 2022). According to this study, silica nanoparticles can enhance the physical appearance and dispersion of pigments in lipsticks and maintain color saturation (Gupta et al. 2022). In the form of nanodispersions ranging from 5 to 100 nm, these nanoparticles can encapsulate and deliver both hydrophilic and lipophilic substances to their specific targets (Santos et al. 2019). It is anticipated that the use of these nanoparticles in cosmetics will rise. They are frequently found in leave-on and wash-off cosmetic products for hair, skin, lips, face, and nails (Mebert et al. 2017). However, safety considerations are associated with the actual applications of silica-based nanoparticles; size and surface changes should

Table 18.1 A list of some nanocosmetics available on the market

Nanomaterials/ nanocarriers	Type of cosmetics	Manufacturer	Application	Name of the product available in market	References
TiO_2&ZnO nanoparticles	Skin care	Lotus Professionals	Sunscreen	Phytorx UV Defense Sun Block SPF 100	Fytianos et al. (2020)
Nanocrystal	Skin care	NanoInfinity Nanotech Co. Ltd.	Moisturizer	Nano-In Hand and Nail Moisturizing Serum and Foot Moisturizing Serum	Ghasemiyeh and Mohammadi-Samani (2020), Gigliobianco et al. (2018) and Lohani et al. (2014)
NanoSilica (SiO_2)	Lip care	Sugar Cosmetics	Lipstick	Face FWD >> Blush Stick	Fytianos et al. (2020)
Carbon black	Skin care	Carbon BAE	Facemask	Face Masque	Cornils (2020)
Liposomes	Skin care and under eye care	Sesderma	Anti-wrinkle, skin brightening moisturizer& sunscreen	SesdermaC-Vit Liposomal Serum	Dhawan et al. (2020) and Oliveira et al. (2022a, b)
		Dior		Capture Totale Le serum	
		Cristian Diar L'Or'eal		Capture effect du Soleil	
		Payot (Fredinand Muehlens)		Formulae Liposome Gel	
		BiopharmGmbh		Symphatic 2000	
		Avon		Eye Perfector	
Nano-zinc oxide	Lip care	Vortex Health & Beauty Ltd.	Lip care, mostly reduces lip line	Luscious Lips Rejuvenation Duo	Mohammed et al. (2019)
Nanospheres	Lip care	Kara Vita	Lip moisturizer	Lip Tender	Nafisi and Maibach (2017)
	Skin care		Antiwrinkle	Eye Tender	
Dendrimers	Therapeuticskin care agent	Starpharma (Approved for marketing in Europe)	Treatment of bacterial vaginosis	VivaGel® BV	Dave and Krishna Venuganti (2017), Fytianos et al. (2020) and Pentek et al. (2017)
	Skin care		Sunscreen	Topical Resveratrol Formulation	
Nanocapsules	Lip care	Lancome	Antiwrinkle lip treatment	Primordiale Optimum Lip	Dhawan et al. (2020) and Fytianos et al. (2020)
	Skin care	L'Oreal	Antiaging cream	Lancome Soleil Soft-Touch Anti Wrinkle Sun Cream SPF 15	
		Lancome	Antioxidant, hydrating moisturizer	Hydra Flash Bronzer Daily Face moisturizer	
	Haircare	Braziliss	Strengthening hair root	Nano Collagen	
Nano-organic (tris-biphenyl triazine)	Skin care	Allie	Sunscreen	Extra UV Gel	Fytianos et al. (2020)

(continued)

Table 18.1 (continued)

Nanomaterials/ nanocarriers	Type of cosmetics	Manufacturer	Application	Name of the product available in market	References
Nanogold and nanosilver	Skin care	Chantecaille	Anti-aging cream	Nano Gold Firming Treatment	Alaqad and Saleh (2016) and Fytianos et al. (2020)
Nanoliposomes	Skin care	Dior	Antiwrinkle cream	Capture Totale	Fakhravar et al. (2016) and Fytianos et al. (2020)
		Microfluidics	Moisturizer	Dermosome	
Niosomes	Skin care	Loreal, Paris	Anti-aging cream	Lancome®	Kusuma Priya et al. (2020)
		Simply Man Match		Anti-Age Response Cream	
		Laon Cosmetics	Brightening moisturizer	MayuNiosome Base Cream	
Ethosomes	Skin care	Trima, Israel	Moisturizer	Supravir Cream	Verma and Utreja (2019) and Yücel et al. (2019)
Sphingosomes	Skin care	NTT, Israel	Anticellulite cream	Noicellex	Kusuma Priya et al. (2020)
Solid lipid nanoparticles	Skin care and body odor removal	Chanel	Perfume and body cream	Allure Body Cream	Hooda et al. (2018)
	Body moisturizer				
Nanostructured lipid carriers	Skin care	Cutanova	Face spa cream	Dr. Rimpler	Poonia et al. (2016)
Nanocapsules	Skin care	L'Or'eal	Antiwrinkle cream	Primordiale Intense	Blanco-Padilla et al. (2014), Fytianos et al. (2020) and Poonia et al. (2016)
		Braziliss	Haircare	Nano Collagen	
Nano-hydroxyapatite	Dental care	Sangi	Toothpaste	APAGARD M plus	Lee et al. (2019)
		Karex		Kinder Karex Toothpaste	
Nanoemulsions	Skin care	Vitalipid Hanacure	Body lotions, skin creams, balsams, salves, and gels Moisturizer	Cosmeceutical Vitamin A, D, E, K Nano Emulsion Multi-Peptide Moisturizer	Fytianos et al. (2020)
Fullerene (C60)	Skin care	Juva Skincare	Face cream	Brightening Essence	Ghabdian et al. (2021) and Mousavi et al. (2017)
Novasomes	Skin care	Estee lauder	Anti-aging luxury skin care	Renutriv range	Aslam Abdul Rahiman et al. (2021)

be considered when determining toxicity (Mebert et al. 2017; Park et al. 2010). Long-term research is necessary as the practical application of silica nanoparticles in cosmetics is still unknown (Fytianos et al. 2020).

18.2.1.2 Nanocrystals

A nanocrystal refers to a particle composed of atoms arranged either singly or in a polycrystalline structure, typically with at least one dimension smaller than 100 nanometers (nm).

These particles typically range in size from 10 to 400 nm, making them effective for delivering active ingredients that are poorly soluble (Shegokar 2016). In 2007, Juvena launched nanocrystals into the market with their skin-renewing serum, Juvedical. This serum utilized rutin, a flavonoid, as its active ingredient (Pyo et al. 2016). A study discovered that nanocrystal formulation significantly enhances the bioactivity of rutin compared to its water-soluble glucoside form, demonstrating a 500-fold increase (Pyo et al. 2016). Similarly, La Prairie utilizes nanocrystal carriers to incorporate hesperidin, a plant-derived antioxidant glycoside, in its products. Lancôme's anti-wrinkle cream, Renergie Microlift, also leverages nanocrystals in its formulation (Romero et al. 2017; Sadhu et al. 2021).

18.2.1.3 Fullerenes

Fullerene is a carbon-based spherical compound characterized by a three-dimensional structure, featuring carbon rings with an uneven number of atoms (Kroto et al. 1985; Ramsden 2016) and is therefore known as "buckyballs" or buckminsterfullerene. Fullerenes are widely utilized in cosmetics and cosmeceuticals for their antioxidant properties. They are commonly included in skincare formulations aimed at rejuvenation, as they effectively neutralize free radicals, which are linked to UV-induced skin damage such as wrinkles and hyperpigmentation (Lens 2009). Due to their inherent hydrophobic properties, fullerenes have traditionally faced limitations in application. However, their solubility in aqueous solutions has been significantly enhanced by incorporating suitable concentrations of surface-active agents. This improvement has expanded their potential uses in pharmaceutical applications (Lohani et al. 2014). Notably, in 2005, Radical Sponge became the world's first cosmetic product based on fullerenes. In a clinical study by Inui et al. the efficacy of fullerene in treating acne vulgaris was assessed using a Lipo Fullerene gel formulation. The study reported substantial reductions in inflammatory lesions by 23% and 38% after 4 and 8 weeks, respectively (Inui et al. 2011).

18.2.1.4 Carbon Black

Carbon black is a significant component in cosmetics, often used to add color to eye and skin products. It has been approved by the European Union (EU) for use in its nanostructure form and as a colorant, with a maximum allowable concentration of 10%. Research on carbon black nanoparticles indicates they pose higher risks of cytotoxicity, inflammation, and impacts on phagocytosis in human monocytes compared to larger nanoparticles (Sahu et al. 2014). The EU allows its use in cosmetics provided there is no risk of inhalation (Fytianos et al. 2020).

18.2.1.5 Nano-organic Materials

Tris-biphenyl triazine is an innovative and robust UV filter known for its photostability, often employed in sunscreens. In its nano form, it serves as a broad-spectrum UV protector and is widely used in sunscreen products due to its durability. BASF SE markets it in Europe under the brand name TINOSORB® A2B. Another EU-approved UV filter is methylene bis-benzotriazolyl tetramethylbutylphenol (nano), also known as MBBT, which can be used at concentrations of up to 10% w/w in cosmetic products applied to the skin. According to the Scientific Committee on Consumer Safety (SCCS), MBBT does not pose risks when applied to intact skin, although concerns remain regarding potential environmental impacts and potential accumulation in certain tissues (Fytianos et al. 2020).

18.2.1.6 Nano-hydroxyapatite

Nano-hydroxyapatite is utilized in oral care products targeted for specific uses like treating severe dental sensitivity and enhancing tooth enamel remineralization. It has been recognized by the US Food and Drug Administration (USFDA) as a promising and safe option for these applications (Bernauer 2018). These particles are integrated into oral formulations, including toothpastes and mouthwashes, due to their ability to promote enamel remineralization and alleviate sensitivity, offering an alternative to fluoride-based products (Fytianos et al. 2020).

18.2.1.7 Lipid Nanoparticles (SLNs, NLCs)

Solid lipid nanoparticles (SLNs) are nanosized droplets that protect cosmetic ingredients from dehydration. They also impart (i) controlled delivery, (ii) skin penetration, and (iii) skin hydration (Fytianos et al. 2020; Pardeike et al. 2009; Wissing 2003). Nanostructured lipid carriers (NLCs) are designed to accommodate higher concentrations of active ingredients, while solid lipid nanoparticles (SLNs) enhance chemical stability. Both types of carriers improve the ability of active ingredients to penetrate the skin, ensure compatibility with biological systems, and maintain stability, thus serving as effective delivery systems (Kaul et al. 2018; Santos et al. 2019). SLNs and NLCs are commonly used in moisturizing creams and sunscreens, increasing the efficiency of some sunscreen active ingredients (Shokri 2017).

18.2.1.8 Nanoliposomes

Nanoliposomes serve as nanocarriers composed of a hydrophobic tail and a hydrophilic head. These structures act as bilayer vehicles designed to deliver cosmetic components such as vitamins, minerals, antioxidants, and anti-aging agents to skin cells. They achieve this by fusing with the skin's bilayer, thereby enhancing penetration capabilities. Nanoliposomes are known for their biodegradability and biocompatibility, making them highly versatile for cosmetic applications (Rigano and Lionetti 2016). They are utilized to protect active ingredients like vitamins, improve skin permeability, and provide moisturization. Additionally, nanoliposomes are effective for delivering fragrances in products such as antiperspirants, body-spray deodorants, and lipsticks (Santos et al. 2019). Other nanostructures like carbon black, silica, nanohydroxyapatite, nanocrystals, dendrimers, hydrogels, and buckyballs have also been extensively used as delivery vehicles to skin cells (Ramis et al. 2018; Santos et al. 2019; Scientific Committee on Consumer Safety (SCCS) and Chaudhry 2016; Shokri 2017).

18.2.1.9 Ethosomes

Ethosomes are a type of lipid vesicle that includes phospholipids along with high concentrations of alcohol and water. They have gained prominence over the last two decades as carriers for transdermal drug delivery (Chauhan et al. 2022; Touitou et al. 2000). Collaborative efforts between Professor Elka Touitou of Yissum, a research development company at the Hebrew University of Jerusalem, and Trima, a pharmaceutical company, resulted in an ethosomal cream containing acyclovir (Supra-vir) for treating skin and mucous membrane infections caused by herpes simplex virus (Chauhan et al. 2022). However, specific product details are no longer available on the manufacturer's website. Some other products based on ethosomal systems include Nanominox, Cellutight EF, Skin Genuity, and Decorin cream, though comprehensive information on these products or their manufacturers is currently scarce online (Song et al. 2012).

18.2.1.10 Transferosomes

Transferosomes are flexible vesicles with a sophisticated bilayer containing phosphatidylcholine and an edge activator, designed to mimic cellular vesicles (Opatha et al. 2020). Originally proposed by Cevc et al. in 1992, transferosomes have been extensively studied as carriers for transdermal drug delivery (SchÄtzlein and Cevc 1998). These specialized vesicles, designed for transdermal applications, consist of lipids and biocompatible membrane softeners. Compared to traditional liposomes, transferosomes exhibit greater elasticity, making them promising vehicles for enhancing transdermal drug delivery. When a transferosome lipid suspension is applied to the skin, the process utilizes an "osmotic gradient" created by water evaporation, facilitating deeper penetration into the skin layers (Matharoo et al. 2024).

18.2.1.11 Layerosomes

Layerosomes refer to liposomes that have been stabilized by coating them with biocompatible polyelectrolytes, effectively adding one or more layers to their structure (Rao et al. 2019).

18.2.1.12 Sphingosomes

Sphingosomes consist of concentric bilayered vesicles, where an aqueous compartment is completely enclosed by a membrane composed of phospholipids (AlMadalli et al. 2024; Pawar 2022).

18.2.1.13 Catansomes

Catansomes are vesicles characterized by bilayers that consist of an equal mixture of cationic and anionic surfactants, creating an ion-pair amphiphile (IPA) (Sharma et al. 2022; Wu et al. 2007).

18.2.1.14 Dendrimers and Hyperbranched Polymers

Dendrimers are highly structured, symmetrical molecules resembling trees, known for their compact size (2–10 nm diameter) and dense surface functionalization. Companies like L'Oreal, Unilever, and The Dow Chemical Company have secured patents utilizing dendrimers in cosmeceuticals aimed at enhancing skin, hair, and nail health (Bacha et al. 2023). For instance, a patent involving carbo siloxane dendrimers claims benefits such as water and sebum resistance, glossiness, tactile sensation, and adhesive properties for skin and hair (Bacha et al. 2023; Singh et al. 2008). Dendrimers are adept at encapsulating active ingredients internally and on their surfaces, facilitating controlled release mechanisms from their central cores (Chauhan et al. 2020).

18.2.1.15 Niosomes

Niosomes, composed of non-ionic surfactants, represent biodegradable vesicles considered safer than liposomes (Kaur et al. 2024). They enhance stability for encapsulated substances, improve skin penetration, increase bioavailability of poorly absorbed compounds, and facilitate targeted delivery to specific therapeutic sites (Ge et al. 2019; Moammeri et al. 2023). According to Manconi et al. unilamellar niosomes containing Brij®30 were particularly effective in protecting tretinoin against photodegradation (Manconi 2003).

18.2.1.16 Cubosomes

Cubosomes are nanoparticles that consist of a bicontinuous cubic liquid crystalline phase, offering a significantly greater specific surface area compared to the original cubic phase (Sivadasan et al. 2023). Research is exploring their use in skincare, haircare, and antiperspirants (Sivadasan et al. 2023; Yalavarthi et al. 2020).

18.2.1.17 Novasomes

Novasomes represent patented encapsulation techniques characterized by enhanced capabilities such as increased loading capacity, improved stability, aesthetic appeal, and scalability (Fatima et al. 2022). They are derivatives of liposomes or variants of niosomes, which are nonphospholipid paucilamellar vesicles ranging in size from 0.1 to 1 μm (Fatima et al. 2022).

18.2.1.18 Marinosomes

Marinosomes are derived from marine lipid extracts abundant in omega-3 polyunsaturated fatty acids like eicosapentaenoic acid and docosahexaenoic acid (Moussaoui et al. 2002). These lipids are metabolized by enzymes in the epidermis, producing metabolites known for their anti-inflammatory and antiproliferative properties, suggesting marinosomes' potential utility in addressing inflammatory skin conditions (Ahmadi Ashtiani et al. 2016).

18.2.1.19 Aquasomes

Aquasomes are complex structures formed of three layers: a solid nanocrystalline core surrounded by an oligomeric outer layer. These nanoscale entities, often referred to as "bodies of water," mimic water's properties. They possess a hydrophilic component crucial for safeguarding delicate biological molecules, maintaining their structural integrity. Moreover, their large surface area facilitates effective targeting of bioactive substances such as peptides, hormones, antigens, and genes to precise locations within the body (Banerjee and Sen 2018; Damera et al. 2019; Gupta et al. 2021).

18.2.1.20 Nanocapsules

Nanocapsules are excellent nanocarriers of active cosmetic components, typically polymeric materials dispersed in aqueous or oily media (Cardoza et al. 2022). Polymeric nanocapsules play a crucial role in safeguarding active ingredients, minimizing chemical odors, and addressing compatibility challenges among formulation components. They can be directly applied to the skin or integrated into semi-solid formulations, adjusting how ingredients penetrate based on the unique properties of polymers and surfactants (Kaul et al. 2018).

18.2.1.21 Microemulsions

Microemulsions represent stable and transparent mixtures of water and oil, blended with a surfactant, that can be easily formed by combining these components without requiring specialized preparation conditions. These formulations exist in three primary types: oil dispersed in water (o/w), water dispersed in oil (w/o), and bicontinuous structures (Singh et al. 2016). Microemulsions exhibit distinct characteristics based on their composition. In formulations with low oil content, they typically feature o/w droplets, whereas those with a low water fraction tend to have w/o droplets, each ranging in size from 10 to 140 nm. When water and oil are balanced, bicontinuous microemulsions can form. The concept of microemulsions dates back to the early 1940s when Hoar and Schulman developed a clear, single-phase compound by titrating a milky emulsion with a medium-chain alcohol. Schulman and colleagues later coined the term "microemulsion." These versatile systems find application across diverse industries such as fuels, detergents, agrochemicals, food, cosmetics, and pharmaceuticals (Husein and Nassar 2008; Sarkar and Khilar 2009; Subongkot et al. 2022). They have been demonstrated to improve the absorption of cosmetic ingredients like whitening agents, moisturizers, and antioxidants. These substances often face challenges in penetrating the skin barrier, known as the stratum corneum, due to its protective properties (Boonme 2007; Souto et al. 2022).

18.2.1.22 Nanoemulsions

Nanoemulsions are increasingly valued as carriers for delivering active cosmetic agents because of their ability to effectively transport lipophilic molecules compared to liposomes, owing to their lipophilic interior. Their small droplet size and extensive surface area enhance the efficient transfer of active ingredients to the skin (Singh et al. 2016; Souto et al. 2022). Nanoemulsions are well-suited for cosmetic applications as they do not exhibit intrinsic creaming, flocculation, sedimentation, or coalescence, which are common in macroemulsions. Sinerga has pioneered the Nanocream® emulsifier blend, enabling the creation of O/W nanoemulsions. These emulsions are utilized in a wide range of products such as skin lotions, deodorants, shampoos, body lotions, nail enamels, sunscreens, conditioners, and hair serums (Aziz et al. 2019; Bilal and Iqbal 2020). More recently, nanoemulsions have found application in NanoGel systems and emulsion-based wet wipes, expanding into areas like baby care products and makeup cleansers, illustrating their growing versatility and adoption in various consumer goods (Donthi et al. 2023; Singh et al. 2016).

18.2.2 Formulation of Nanocosmetics

Nanomaterials and nanoparticles are becoming more prevalent in the creation of cosmetic products, giving rise to the development of nanocosmetic formulations (Izquierdo et al. 2020). Nanocosmetic compositions often rely on pH adjusters to maintain a stable and functional product. The correct pH level is crucial for the efficacy and safety of the product. Sunscreen agents are frequently added to daytime skincare products to offer protection against harmful UV radiation. Formulations may also include specialty components targeting specific skin concerns, such as brightening agents, acne-fighting compounds, or anti-aging peptides (Izquierdo et al. 2020). Numerous cosmetic companies are making significant investments in products that utilize nanotechnology, attracted by its multitude of benefits (Gupta et al. 2022; Scherer Santos et al. 2023; Tamrakar and Thakur 2023).

Firstly, these tiny ingredients offer several benefits:

(i) *Enhanced entrapment and stability:* Nanoparticles can effectively encapsulate active ingredients, protecting them from degradation and ensuring their stability over time (Raszewska-Famielec and Flieger 2022; Wu and Guy 2009).
(ii) *Improved dispersibility:* The small size of nanoparticles allows for better distribution of ingredients within a formulation, leading to more uniform application (Ganesan and Choi 2016; Raszewska-Famielec and Flieger 2022; Wu and Guy 2009).
(iii) *Enhanced performance:* Nanomaterials can improve the overall effectiveness of cosmetic products by ensuring that active ingredients are delivered efficiently to the targeted areas (Gupta et al. 2022).
(iv) *UV protection:* Nanoparticles can provide superior protection against harmful UV rays, which helps in preventing skin damage (Gupta et al. 2022).
(v) *Improved texture and aesthetic appeal:* The inclusion of nanoparticles can enhance the texture of cosmetic products, making them smoother and more appealing to consumers (Gupta et al. 2022).
(vi) *Better penetration and target specificity:* Due to their small size, nanoparticles can penetrate deeper into the skin, delivering active ingredients precisely where they are needed (Gupta et al. 2022).
(vii) *Prolonged effects:* The sustained release of active ingredients from nanoparticles can lead to longer-lasting effects, reducing the need for frequent reapplication (Gupta et al. 2022).

Secondly, nanocosmetic products are seen as exclusive, influential, and trendy. Their advanced formulations and high-tech appeal attract consumers looking for cutting-edge beauty solutions. The sleek marketing and innovative nature of nanocosmetics make them stand out in the competitive cosmetic industry, appealing to the trend-conscious users (Gupta et al. 2022; Scherer Santos et al. 2023; Tamrakar and Thakur 2023).

Nanocrystal-containing cosmetics are formulated by incorporating nanosuspensions into creams, gels, or ointments, ensuring precise blending of the required amount (Dhawan et al. 2020). It is crucial to avoid adding nanosuspensions at high temperatures because heat significantly impacts solubility, promoting Ostwald ripening (particle growth) during storage, which can diminish the unique properties of nanocrystals. Due to their high effectiveness, only small quantities of nanocrystals, typically ranging from 0.02% to 0.2% (w/w) as active ingredients, are generally needed in the final cosmetic product (Ahmadi Tehrani et al. 2019; Ma et al. 2023). Since nanosuspensions are water-based, they are vulnerable to microbial contamination, necessitating the inclusion of preservatives to maintain high microbial quality.

18.2.3 Applications and Benefits of Nanocosmetics

Nanomaterial-based cosmetics provide distinct benefits compared to their larger-scale counterparts, thanks to their capacity for prolonged efficacy and enhanced stability. The expansive surface area of nanomaterials enables more effective delivery of active ingredients into the skin. Key targets for using nanomaterials in cosmetics include enhancing skin penetration for improved ingredient delivery, introducing new color elements (such as in lipsticks and nail polishes), achieving transparency (especially in sunscreens), and providing long-lasting effects (such as in makeup) (Gupta et al. 2022). Cosmetic industries are increasingly utilizing nanomaterials to precisely deliver active ingredients to targeted body areas while maintaining extended product stability. One prominent application of nanomaterials in cosmetics is within skincare formulations, notably in sunscreens, where they function as effective UV filters (Dhawan et al. 2020; Gupta et al. 2022).

In 1986, Christian Dior launched CaptureTM, an anti-aging cream utilizing liposomes. Since then, numerous cosmetic products have integrated nanomaterials, with several renowned brands adopting these technologies. L'Oréal S.A., a major investor in nanotechnology, ranks among the top patent holders in the United States for nanotechnology innovations. Their formulations often incorporate up to four nano-ingredients including TiO2, ZnO, silica, and carbon black (Ajazzuddin et al. 2015; Dhawan et al. 2020; Gupta et al. 2022).

The cosmeceutical market exhibits significant diversity, featuring products from both major corporations and smaller local businesses globally, which contributes to its rapid expansion. Literature indicates that liposomes are efficient colloidal carriers for delivering therapeutics into the skin. The similarity between liposome components and skin lipids accounts for their safety and efficiency. Nanotechnology-based liposomal cosmetics should be designed and marketed in ways that respect consumer health and the environment. Modified liposomes, such as niosomes, yeast-based liposomes, ethosomes, transferosomes, invasomes, and novasomes, hold promising potential for the cosmetic industry (Shaw et al. 2022; Van Tran et al. 2019).

18.2.4 Consumer Usage and Perception

The adoption and perception of nanotechnology in consumer products, particularly in cosmetics and medicinal products, have been influenced by various factors, including awareness, safety concerns, and the perceived benefits of these products. Nanotechnology has significantly impacted the cosmetics industry, offering products with enhanced properties such as improved UV protection, anti-aging effects, and better skin absorption (Gupta et al. 2022). In 2021, the global market for nanotechnology in cosmetics was valued at USD 4.3 billion. It is projected to grow at a compound annual growth rate (CAGR) of 8.3% from 2022 to 2030, according to a report by Grand View Research. The increasing demand for premium skincare products and the rising consumer awareness of the benefits of nanotechnology are key factors driving this growth. In 2005, an average of 43% of respondents were aware of nanomaterials. This figure rose to 47% in 2010 and further increased to 65% by 2020. These percentages represent the combined responses of "heard something" and "heard a lot" from a representative sample of 5000 individuals across five different EU member states (Tengler et al. 2020). However, among those aware, there is a generally positive perception regarding the benefits of nanotechnology, particularly in cosmetics and healthcare. Despite the growing market, safety concerns remain a significant barrier to widespread acceptance. A survey by the International Journal of Cosmetic Science in 2022 found that 60% of consumers are wary of using products containing nanoparticles due to potential health risks. This apprehension is primarily fueled by a lack of comprehensive information and understanding of the long-term effects of nanoparticles. Consumers who are knowledgeable about nanotechnology tend to appreciate its benefits. Enhanced efficacy, improved product texture, and superior skin absorption are among the top perceived advantages. For instance, a 2021 study published in the Journal of Dermatological Science reported that consumers using nanotechnology-based skincare products experienced noticeable improvements in skin hydration and elasticity compared to those using conventional products. Regulatory bodies play a crucial role in shaping consumer trust in nanotechnology-based products. The stringent regulations imposed by agencies such as the FDA, EU, and SCCS have been instrumental in assuring consumers about the safety of these products. The 2012 and 2013 safety assessments of ZnO and TiO_2 nanoparticles by the SCCS, along with similar endorsements by Canadian authorities, have helped mitigate some consumer concerns (Virmani and Pathak 2021). While the market is growing and consumers are beginning to recognize the benefits, safety concerns and limited awareness continue to challenge widespread acceptance. Efforts to educate consumers and transparent communication about safety regulations are essential to enhance trust and drive the adoption of nanotechnology in everyday products.

18.3 Fungal Cosmetics

Various fungal species, such as *Agaricus subrufescens* (also known as *A. blazei* or *A. brasiliensis*), *Choiromyces maeandriformis*, *Cordyceps sinensis*, *Ganoderma lucidum*, *Grifola frondosa*, *Hypsizygus ulmarium*, *Inonotus obliquus*, *Lentinula edodes*, various Polyporus species, *Trametes versicolor*, *Tremella fuciformis*, Tuber species, and *Schizophyllum commune*, are currently utilized in cosmeceuticals and nutricosmetics (Wu et al. 2016). Mushrooms have ample amount of proteins, vitamins, and minerals and they are also rich source of beta-glycan, selenium, riboflavin, niacin, pantothenic acid, folic acid, etc. (Bell et al. 2022; Mukherjee et al. 2024). Mushrooms have a variety of components like (i) *phenolic and polyphenolic compounds*—Phenolic compounds are grouped into simple phenols, phenolic acids (kojic acid, gallic acid, etc.), Polyphenols (flavonoids, tannins, etc.). Kojic acid from Japanese mushrooms act as skin lightener, age spots corrector, etc. Flavonoids can act as anti-inflammatory, antiproliferative, and anti-oxidant agents. (ii) *terpenoids*—Catotenoids extracted from *Ganoderma* sp. used in sun lotions. (iii) *selenium*—Selenium (*Agaricusbisporus, Lentinus edodes)* strengthens teeth, hair, nails, also applied in shampoo. (iv) *polysaccharides*—Beta (1,3 D-glucan), Beta (1,6 D-glucan) (*Lentinus edodes*), etc. are incorporated in cosmetic products due to antimicrobial properties, antioxidant, anti-inflammatory properties that are also utilized in cosmetic industries to reduce fine lines, wrinkles, uneven skin tone and texture, etc. (Mukherjee et al. 2024; Wu et al. 2016). Well-known brands such as Origins, Shiseido, and Dermalogica are incorporating mushrooms into their formulations for products like face mists, eye creams, and serums. Table 18.2 represents some fungal cosmetic names that are commercially available in the international market and their fungal source.

18.3.1 Bioactive Compounds from Fungi

The kingdom of fungi is diversified into a vast number of species. Different species of fungi are capable of producing different types of bioactive compounds of cosmetic importance (Hyde et al. 2010). This treasure trove of compounds includes bioactive molecules such as polysaccharides, enzymes, etc. offering cosmetic benefits such as moisturization, antioxidants, brightening, UV protection, etc. (Fig. 18.2) (Mukherjee et al. 2024; Wu et al. 2016).

(i) Polysaccharides: Fungi are a rich source of polysaccharides such as chitins, chitosans, and β-glucans which can be incorporated in cosmetic formulations. Fungal polysaccharides have potent moisturizing activity which is an essential aspect of cosmetic products like moisturizers and skin creams (Zhang et al. 2022). Fungal species such as *Tremella fuciformis* (Wang et al. 2021; Xu et al. 2020) and *Cordyceps* (Chen et al. 2014) have been evaluated as sources for obtaining fungal polysaccharides with moisturizing activity. *Cordycepssinensis* is a known source for (1 → 3)-β-d-glucan. β-glucans have also been incorporated into cosmetic formulations as ointments, skin creams for age lines, dermatitis, acne, eczema, etc. (Kady 2019). β-glucans also have potent activity as moisturizing agents and in formulations for wound healing (Majtan and Jesenak 2018; Muthuramalingam et al. 2019). Exopolysaccharides such as galactomannan isolated from *Aspergillus* sp. (Chen et al. 2011; Costa et al. 2019) and glucomannan isolated from *Candida*

Table 18.2 Some fungal cosmetics and their source (Hyde et al. 2010; Wu et al. 2016)

Products name	Fungal source	Product function
1. Surkran Grape Seed Lift Eye Mask	*Tremella polysaccharide*	Eye cream
2. BeautyDiy Aqua Circulation	*Tremella polysaccharide*	Gel moisturizer
Hydrating Gel	*Schizophyllum commune*	Anti-aging cream
3. Alqvimia Eternal Youth Cream	*Ganoderma lucidum Pleurotus*	Skin-tightening cream
Facial Máxima Regeneración	*ostreatus*	Skin lifting and firming cream
4. Tan Ryuk Sang Firming Cream	*Lentinula edodes*	Anti-aging cream
5. Aveeno Positively Ageless	*Ganoderma lucidum*	Moisturizer
6. Yves Saint Laurent Temps Majeur	*Cordyceps sinensis*	Moisturizer
Elixir De Nui	*Tremella fuciformis*	Skin renewal and revitalizing cream
7. Kose Sekkisei Cream	*Agaricus subrufescens*	Anti-aging cream
8. La Prairie Advanced Marine biology	*Ganoderma lucidum*	Ati-aging cream
night solution	*Ganoderma lucidum, Lentinula*	Dirt, oil and makeup remover
9. Vitamega Facial Moisturizing Mask	*edodes, Mucor miehei*	Light weight moisturizer
10. Menard Embellir Refresh Massage	*Lentinula edodes*	Collagen booster,
11. Nano Works Shineluxe	*Lentinula edodes*	Skin elaticizer,
12. Aveeno Positively Ageless Daily	*Ganoderma lucidum*	Improve hydration
Exfoliating Cleanser, USA	*Ganoderma lucidum*	Skin anti-aging cream
13. One Love Organics Vitamin D	*Ganoderma lucidum*	Promote hair growth with sun protection
Moisture Mist, U.K	*Agaricus subrufescens*	and prevent color fading
14. Kat Burki Form Control Marine		Skin renewal and revitalizer cream
Collagen Gel, U.K		
15. Menard Embellir Refresh Massage,		
France		
16. Tela Beauty Organics Encore		
Styling Cream, U.K		
17. Vitamega Facial Moisturizing		
Mask, Brazil		

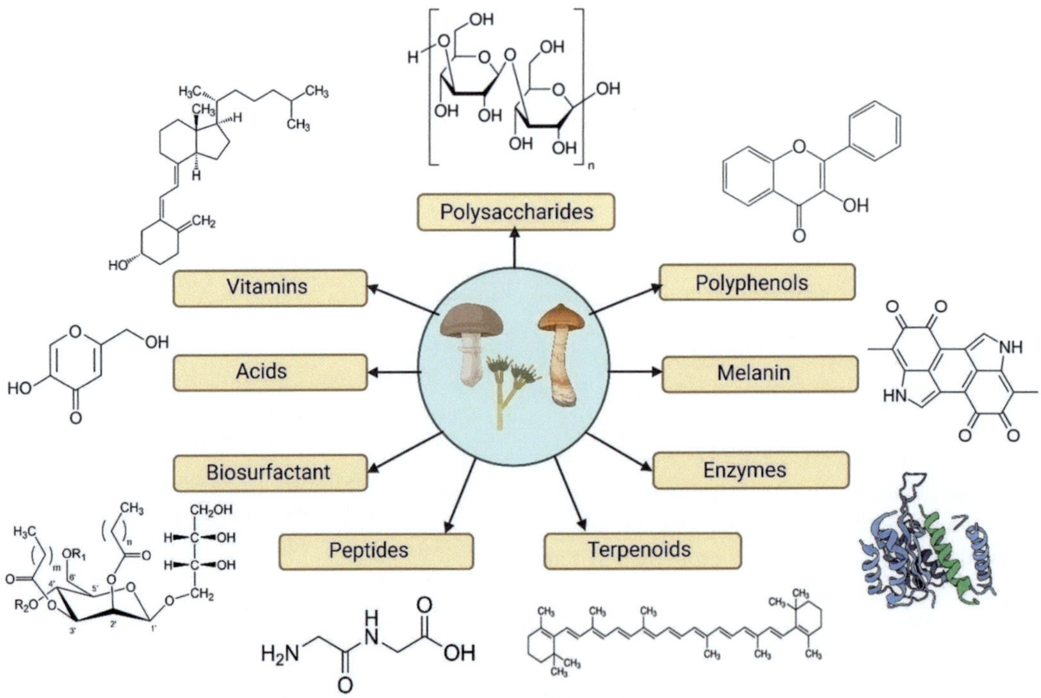

Fig. 18.2 Various bioactive compounds of cosmetic importance obtained from fungi. (This figure was created with BioRender (https://biorender.com/))

sp. (Hájková et al. 2009) have shown good antioxidant activity, which is an essential aspect of skin care products.

(ii) Polyphenols: Fungal polyphenols such as phenolic acids and tannase can be isolated for cosmetic applications. In a study, the authors isolated various polyphenols such as flavonoids and phenolic acids from *Cordyceps militaris* (Pintathong et al. 2021). Polyphenols are an essential component in many cosmetic formulations due to their anti-aging, anti-wrinkle, and antioxidant, etc. properties (Zillich et al. 2015). Studies have reported the extracts from *C. militaris* having great anti-oxidant activity based on DPPH radical scavenging activity, ABTS radical scavenging activity, and FRAP assays. Tannase is another polyphenol of cosmetic importance due to its ability to degrade tannins (Dhiman et al. 2017). Tannase can be isolated from fungal species such as *Aspergillus niger* (Marco 2009), *Aspergillus fumigatus* (Manjit et al. 2008), *Aspergillus nomius* (Farag et al. 2018), *Penicillium purpurogenum* (Reddy 2012), etc.

(iii) Melanin: Melanins are photoprotective agents incorporated in cosmetics such as sunscreens. Melanin isolated from fungus *Amorphotheca resinae* was evaluated for its anti-oxidant activity and its performance as a sunscreen (Oh et al. 2021). The authors reported the fungal-derived melanin to have a broad-spectrum sunscreen activity. Oxygen radical absorbance capacity assays also showed potent antioxidant activity and that the formulation was non-cytotoxic and biocompatible to human cells at a higher concentration. Fungal melanin has been acknowledged as sustainable and biocompatible agents with significant anti-oxidant and photoprotective properties suitable for application in cosmetic industries (Liu et al. 2022).

(iv) Enzymes: Fungal species produce a plethora of enzymes such as proteases, lipases, etc. which have potential in cosmetic applications. Proteases are used in cosmetic industries to eradicate keratin in acne or degrade keratinized skin and treat skin callus (Brandelli et al. 2010). These fungal proteases are also useful in scar reduction and skin smoothening (Chao et al. 2007). Proteases also augment skin renewal processes after removal of dead skin cells and promote keratinolytic activity (Sim et al. 2000). Proteases can be isolated from different fungal species such as *Aspergillus flavus* (Kranthi et al. 2012; Malathi and Chakraborty 1991), *Aspergillus niger* (O'Donnell et al. 2001), *Aspergillus oryzae* Vishwanatha et al. 2009), *Penicillium griseoroseum* (Ikram-Ul-Haq and Mukhtar 2007), *Mucor* sp. (Andrade et al. 2002), etc. Aspartic proteases isolated from fungi also hold applications in the cosmetic industry for skincare (Bernard and Mehul 2009). Fungi are considered to be excellent sources of lipases with sustainable and cost-effective methods of production. Fungal lipases can be extracted from species like Penicillium, Mucor, Fusarium, Alternaria, etc. Lipases are used in cosmetic formulations for oil-removing creams, cleaning and softening skin, aroma enhancement, pearling agent, etc. (Kumar et al. 2023). Mouad et al. also investigated the properties of lipase B isolated from *Candida antarctica* and reported significant potential for uses in cosmetic applications (Mouad et al. 2016). Miguez et al. used lipase isolated from the fungus *Thermomyces lanuginosus* to optimize the enzymatic synthesis of a cosmetic ester-decyloleate (Miguez et al. 2018). Fungal species such as *Aspergillus flavus* also produce cellulytic enzymes that have potent application in cosmetic formulations (Sideney Becker 2013).

(v) Terpenoids: Fungal terpenoids are known to exhibit antioxidant activity making them a suitable contender for applications in cosmetic products (Trepa et al. 2024). Fungal species such as *Ganoderma* have been found to contain over 100 different terpenoids (Chen et al. 2010). Various species of marine fungi have also been attributed to producing different terpenoids exhibiting potent anti-oxidant activity such as β-carotene, g-carotene, neurosporaxanthin, etc. (Ebel 2010).

(vi) Peptides: Fungal extracts are rich in various peptides having applications for cosmetic incorporations. Various dipeptides and tripeptides have been isolated from extracts of *Cordyceps militaris* (Pintathong et al. 2021). Peptides serve as effective components in cosmetics as they exhibit potent anti-oxidant activity, anti-aging, and skin rejuvenating properties (Ferreira et al. 2020).

(vii) Biosurfactants: Biosurfactants are key ingredients in cosmetic formulations due to their skin-hydrating, foaming, emulsifying properties, etc. (Gupta et al. 2019). Biosurfactants are surface active components that can be obtained from filamentous fungi or yeast (Gupta et al. 2019). Mannosylerythritol lipid (MEL) is a commonly used biosurfactant for cosmetic applications obtained from basidiomycetous yeast. Due to its high biocompatibility, MEL is an active ingredient in cosmetic formulations of lipsticks, soaps, shampoos, body oils, anti-wrinkle creams, etc. (Gharaei-Fa 2010).

(viii) Acids: Fungal species secrete various acids such as kojic acid and lactic acid which have potential to be used in cosmetic formulations for skincare (Mukherjee et al. 2024). Fungal species such as *Aspergillus* and *Penicillium* are known producers of kojic acid. Kojic acid is used in the cosmetic industry as a skin-lightening agent and also has potent activity in UV protection and reducing hyperpigmentation (Saeedi et al. 2019). Lactic acid is another potential candidate for incorporation in cosmetic formulations due to its skin whitening properties (Hyde et al. 2010). Fungal species such as *Rhizopus* have been investigated as good producers for lactic acid (Zhang et al. 2007).

(ix) Vitamins: Fungi possess the capacity to perform the synthesis of various vitamins beneficial for the skin, making them suitable for cosmetic applications. Fungi such as *Agaricus bisporus* contain various vitamins such as Vitamin C, D, and B12, which are known to have potent antioxidant activity (Usman et al. 2021). *A. bisporus* also contains vitamin D, which is of high importance in the cosmetic industry due to its cleansing, rejuvenating, and anti-dandruff properties for applications in hair-care products (Wu et al. 2016).

18.3.2 Formulation of Fungal-Based Cosmetics

Formulation of cosmetics with the incorporation of fungal molecules or extracts requires sound knowledge of nature and science. Developing a fungal-based formulation requires several carefully evaluated steps in order to create a safe and effective product. A general outline of the necessary steps required for a fungal-based cosmetic formulation includes (Fig. 18.3):

(i) *Selection of fungal species:* Extensive research and screening must first be conducted in order to select a species of fungi with favorable characteristics to be incorporated in a cosmetic formulation. Different species of fungi have different active biomolecules and extracts each fulfilling different aspects of cosmetic applications. A careful consideration must be made regarding the fungal enzymes, extracts, and biomolecules in order to maximize the overall efficacy of the fungal formulations (Hyde et al. 2010).

(ii) *Cultivation of fungal species:* After selection of suitable fungal species, proper cultivation is required to ensure optimal growth and production of bioactive compounds

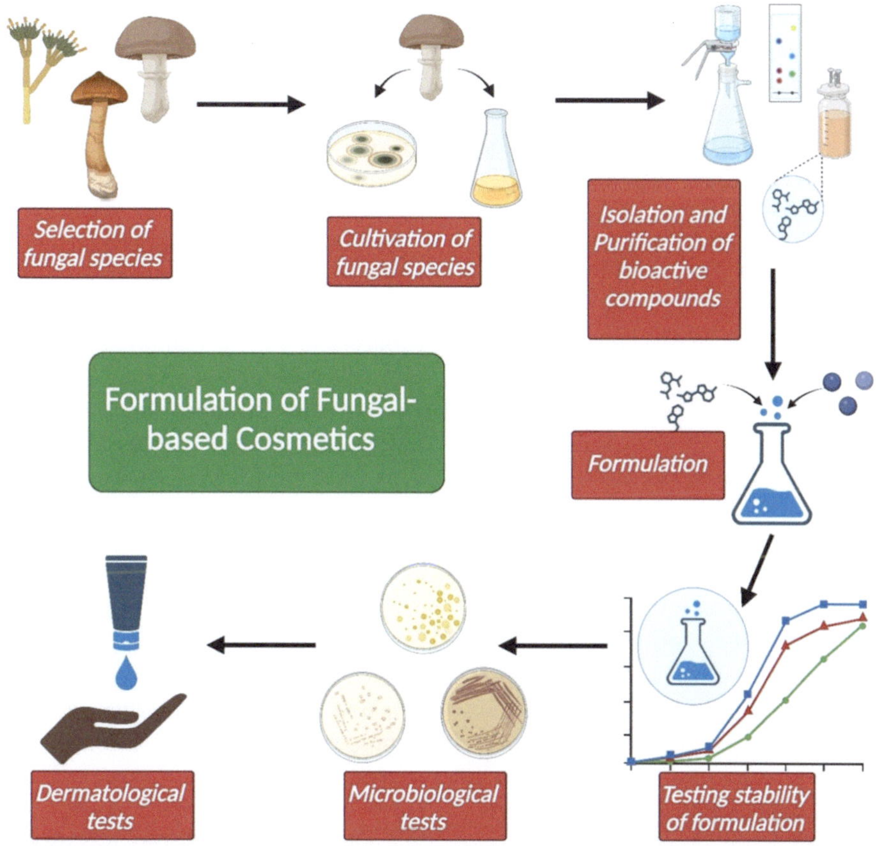

Fig. 18.3 Various steps involved in fungal-based cosmetic formulation. (This figure was created with BioRender (https://biorender.com/))

required for cosmetic formulations (Tanner 2007).

(iii) *Extraction and purification of bioactive compounds:* Fungal biomasses can be harvested and subjected to enzymatic or chemical-assisted extraction techniques and further purification processes such as filtration and chromatography to obtain pure and concentrated bioactive molecules from fungal biomass (McCloud 2010; Miles et al. 1992).

(iv) *Development of formulation:* This step usually involved addition of the fungal extracts to cosmetic ingredients such as emulsifiers, preservatives, etc. For instance, *Tremble fuciformis* extracts were added to a cosmetic emulsion base in order to stabilize and optimize the formulation for cosmetic applications (Woźniak et al. 2023). Addition of other cosmetic ingredients in this step also helps in determining the final texture and stability of the formulation as well as help to preserve the bioactivity of the fungal extracts.

(v) *Testing stability of cosmetic formulation:* At present there is no uniform protocol prescribed by the FDA Code of Federal Regulations or European Commission Regulation to check stability of cosmetic formulations. The stability of the cosmetic formulation can be tested using Turbiscan to observe dispersion stability and detect instabilities such as sedimentation and flocculation (Liu et al. 2011).

(vi) *Microbiological testing:* The formulation must be safety tested against growth of microbiological contaminants such as bacteria, yeast, or molds. These tests must be carried out with strict guidelines as per ISO standards to determine the overall microbial safety of the formulation in order to be considered safe for cosmetic use (Halla et al. 2018; International Organization for Standardization 2016).

(vii) *Dermatological tests:* In the final steps for approval to be used as a safe and effective cosmetic formulation, the product must pass dermatological testing on volunteers such as skin-patch tests and subsequent evaluation by dermatologists to ensure that the cosmetic formulation is safe and effective for commercial use (Kalicińska et al. 2022).

18.3.3 Applications and Benefits of Fungal-Based Cosmetics

The formidable potential held by bioactive fungal molecules has garnered significant amount of attention from the cosmetic industry. In the last few decades, various cosmetic products have appeared which have incorporated the use of these bioactive fungal molecules in their formulations. *KitoZyme* is a company which has developed skincare products using chitin and chitin-glucan from fungal origins. The company's 100% natural cosmetics such as *KiOsmetine*® *CsG* developed using these fungal biopolymers obtained from species such as *Aspergillus niger* are sustainable, provide intensive moisturization to the skin as well as aid in conditioning and repair effects. *Aveeno's Positively Ageless Night Cream* is formulated using different fungal extracts such as lentinan obtained from shitake mushrooms, *Ganoderma lucidum* extracts and *Mucor miehei* extract, attributing to potent antioxidant activity. *Murad's Sleep Reform Serum* consists of Schizophyllan, a water-soluble beta-D-glucan obtained from the fungus *Schizophyllum commune.Lightenex*® cream developed by *PharmaClinix*® contains Japanese mushroom-derived Kojic acid, which is known for its skin whitening properties. Cosmetic products such as *La Prairie Advanced Marine Biology Night Solution, Beauty Diy Aqua Circulation Hydrating Gel,* and *Surkran Grape Seed Lift Eye Mask* have formulations incorporating the use of fungal extracts derived from *Tremella* sp. *Shroom Skincare* has recently launched its first fungal-based cosmetic product—*Mycelium Glow Brightening Serum,* which contains mushroom extracts of *Maitake, Chaga, Reishi,* and *Cordyceps*. A combination of these fungal extracts provides a multi-functional aid to the skin ranging from rejuvenation to protection. *Symrise* have developed fungal-based ingredient *Symhair Thermo*, using extracts from *Polyporus umbellatus* mushroom. This *Symrise* product has been claimed to have a protective effect from heat during hair styling. *Botanical D Moisture Mist* developed by *ONE LOVE ORGANICS*® is a face moisturizer and toner created using fungal extracts of *Lentinus edodes.* The application of fungal extracts and biomolecules in cosmetic formulation has a ton of benefits. Fungal extracts from species such as *Grifola frondosa* (Maitake), *Hericium erinaceus* (Lion's Mane), and *Ganoderma lucidum* (Reishi) have water extracts that are a potent and cost-effective source of hydrating agents for the skin (Ziemlewska et al. 2022) as compared to expensive and hard to obtain hydrating agents such as ceramides (Wu et al. 2016). As compared to traditional chemical ingredients, fungal biomolecules have been established as potent antioxidants based on DPPH radical scavenging activity, ABTS radical scavenging activity, and FRAP assays (Pintathong et al. 2021). Harnessing various properties of different fungal biomolecules has opened up a myriad of options to develop cosmetic formulations for different skin types, conditions, and personalized skincare.

18.3.4 Consumer Usage and Perception

The trend of using and seeking natural materials and additives is currently on the rise, particularly in cosmetic products. This surge in popularity is largely due to the revealed negative effects that synthetic materials have on health and the environment. As a result, marketing trends are increasingly favoring natural solutions for cosmetics, associating them with a healthy lifestyle and linking their usage to healthy eating habits (Gubitosa et al. 2019). Over the years, the demand for sustainable and biological products has paved a route for the cosmetic industry to explore cosmetic formulations involving fungal sources (Gupta et al. 2019). Use of fungal extracts and molecules in cosmetics, or any product of biological origin has drastically reduced the chances of possible side effects as compared to hazardous chemical ingredients (Khan and Abourashed 2009; Mukherjee et al. 2024). However, the use of fungal-based cosmetics still borderlines on a niche market. Since fungal-based cosmetics are of a biological origin, they find most of their consumer base who are looking for 100% natural and sustainable options when it comes to buying cosmetics. The current lines of fungal-based cosmetic products are seemingly priced on the higher end due to their novelty. High prices for cosmetic products definitely appeal to a different set of consumers willing to pay premium prices for luxury cosmetics. Studies conducted on consumer behavior regarding the purchase of cosmetic products are highly influenced by income and awareness regarding the products (Lu and Chen 2017). The acceptance and demand for fungal-based cosmetics may also depend on the geographical location, ethnicity, and cultural factors (De Cianni et al. 2023). Since fungal-based cosmetics are relatively new to the consumers' proper awareness and educating people regarding the benefits, efficacy, and safety of these cosmetic formulations is crucial for shaping their perception and mindset.

18.4 Similarities and Differences: Active Ingredients and Mechanism

Sourcing active ingredients for cosmetic formulations varies for fungal-cosmetics and nanocosmetics. Fungal extracts are 100% natural in origin and can be obtained from the environment or cultured in a laboratory. Nanoparticles for cosmetic applications can be synthesized in a laboratory using chemicals or can also be obtained from plant or microbial origin. Both fungal-based cosmetics and nanocosmetics can harness the properties of bioactive compounds. Fungal-based cosmetics are of fungal origin and as discussed in the previous sections, it is without ambiguity that one can confirm the efficacy of fungi in producing a plethora of bioactive molecules of cosmetic importance. Nanoparticles on the other hand can be used to encapsulate active ingredients for delivery and sustained release to cells. These active ingredients include polysaccharides with moisturizing activity such asβ-glucans, polyphenols with anti-aging, anti-wrinkle, and anti-oxidant properties (Zillich et al. 2015), lipases for oil-removing creams, having properties for cleaning and softening skin (Chandra et al. 2020; Kumar et al. 2023; Mehta et al. 2017), biosurfactants having skin hydrating, foaming, emulsifying properties (Gupta et al. 2019; Kumar et al. 2023), various acids such as kojic acid (Saeedi et al. 2019) as a skin-lightening agent, etc. Some cosmetic formulations may utilize the inherent properties of metal nanoparticles such as silver nanoparticles. Silver nanoparticles have been widely used due to their inherent antimicrobial properties and can be found in sunscreens, haircare, and make-up products (Chakraborty et al. 2024; Gajbhiye and Sakharwade 2016). Titanium oxide nanoparticles are also potent in antimicrobial activity and have been used in cosmetic formulations of skin creams, lip balms, and foundations (Chakraborty et al. 2024; Dréno et al. 2019).

Both fungal-based cosmetics and nanocosmetics strive for deeper penetration into the skin for better efficacy of the formulation. Fungal components derived from the cell wall such as β-glucans and chitosans aim to enhance the permeability of the skin in order to improve absorption and penetration of the bioactive molecules into the skin (Ferreira et al. 2022). Similarly nanocosmetics incorporate the nano-size capabilities of nanoparticles to penetrate into cells and release active ingredients more efficiently, thereby enhancing their effects (Gupta et al. 2022).

18.5 Safety and Regulatory Considerations

Although both nanocosmetics and fungal-based cosmetics show significant potential for the cosmetic industry, their incorporation involves certain risks. Nanoparticles have previously been established as a posing threat to human health due to their toxicity and other attributes such as size, shape, structure, surface properties, etc. which determine their exposure, circulation, and clearance in the human body (Holsapple et al. 2005). The size of the nanoparticles allows them to penetrate into organs which can cause damage or toxicity in them (Li et al. 2003). The small size of nanoparticles also makes them more susceptible to chemical reactivity in the body leading to an increased production of reactive oxygen species (ROS)—a key mechanism responsible for toxicity, oxidative stress, inflammation, DNA damage, etc. (Brown et al. 2002). Even workers involved in manufacturing nanocosmetics are at a risk. Inhalation of nanoparticles such as TiO_2 during the manufacturing process can cause the nanoparticles to travel to the brain and various organs posing a major threat to life. Use of nanoparticles in topical creams such as sunscreens also poses a threat of deep-dermal penetration which can cause health issues (Schulte et al. 2008). Neonatal toxicity is another risk for pregnant women posed by nanoparticles. Nanoparticles may be prone to entering the placenta or fetus and the resulting oxidative damage can cause critical placental damage, hamper neonatal development, and neurotoxicity in infants (Li et al. 2014). In Europe, certain nanomaterials, including Colloidal Silver (nano), Styrene/Acrylates copolymer (nano), Sodium styrene/Acrylates copolymer (nano), and Silica in nanoform, have received inconclusive assessments from the Scientific Committee on Consumer Safety (SCCS). Consequently, the European Commission (EC) has tasked the SCCS with evaluating these substances to identify any potential risks (Apekshaben et al. 2022; Bernauer et al. 2021). Additionally, in 2014, the Food and Drug Administration (FDA) published a document titled "Final Guidance for Industry—Safety of Nanomaterials in Cosmetic Products," which addresses safety considerations and provides regulatory guidance for the cosmetic industry (Center for Food Safety and Applied Nutrition 2014). The FDA has enforced various regulatory guidelines for the use on nanoparticles in cosmetics (Center for Food Safety and Applied Nutrition 2014; Gupta et al. 2022). The FDA recommends several criteria that are required to be fulfilled regarding the safety assessment of nanoparticles to be incorporated in cosmetic formulations. These assessments include tests regarding the physicochemical properties of the nanomaterials, potential for aggregation, and toxicity testing as per the final dosimetry to be used in the final cosmetic formulation, potential routes for nanoparticles exposure, in vitro and in vivo tests for nanoparticle toxicity on dermal cells and other organs (Gupta et al. 2022). These guidelines enhance the security standards for cosmetic products by requiring manufacturers to follow specific guidelines when preparing a product's safety report before it can be marketed. According to regulations, a cosmetic product must ensure human safety during typical or expected usage conditions (Bilia and Costa 2021).

Fungal extracts and biomolecules also require a set of safety assessments before being incorporated into cosmetic formulations. Fungal extracts from species such as *Alternaria, Aspergillus, Candida, Cladosporium, Penicillium, Saccharomyces,* and *Trichophyton* have been previously evaluated as potent allergens with a large subset of population being sensitized to them (Crameri et al. 2014). Tests regarding various fungal extracts on skin on a large cohort of test candidates also revealed skin-sensitive and allergic reactions in many candidates (Mari et al. 2003). More than 10% of the general population test positive for skin-sensitivity toward fungal extracts, and up to 80% of people with respiratory allergies have been reported to be sensitized to fungal allergens (Verhoeff and Burge 1997). Thorough tests regarding potential toxicity, skin irritation, sensitization, etc. must be evaluated before incorporation of fungal extracts in cosmetic formulations. Fungal spores are also potent allergens and pose a threat to workers in the manufacturing units of the fungal cosmetics (Garrett et al. 1998). These allergens may also be present in certain fungal extracts. A comprehensive testing must be carried out to identify potential allergens in the fungal extracts and must also be labeled in the packaging of the final product to ensure consumer safety. Microbiological testing must be thoroughly evaluated to ensure that formulations do not incur the risks of being contaminated by microbial contaminations under various circumstances. Similar to nanocosmetics, fungal-based cosmetics must also abide by the regulatory guidelines such as labeling all ingredients used, submitting safety and cytotoxicity data of the final formulations, etc. in order to satisfy regulatory guidelines and ensure consumer safety. Formulations must be tested for stability under different environmental conditions, shelf-life, etc. to determine efficacy and recommended usage period.

18.6 Targeted Usage and Consumer Preferences

Both nanocosmetics and fungal-based cosmetics are an emerging market in the cosmetic industry. Since a vast majority of the population may not be aware of such formulation, it holds a rather niche group of targeted consumers. Skincare enthusiasts or "cosmetic influencers" willing to test and promote new cosmetic formulations via social media are a first choice for cosmetic companies for promoting these products (Mukherjee et al. 2024; Scherer Santos et al. 2023). As mentioned in previous sections, income also poses as a major influence in purchase of such products (Lu and Chen 2017). Since these types of products are priced toward a higher range, not many would be willing to shelling out premium prices for a cosmetic cream for regular usage. As it is a niche market, a lot of advertisement and marketing would be required to raise awareness regarding the innovation and benefits of nanocosmetics and fungal-based cosmetics. Transparent and accurate product labeling along with proven evidence regarding the benefits and efficacy of these products would also play a key role in swaying the mindset of a vast majority of the population. Consumer appeal would also be enhanced due to fungal-based products being 100% natural or nanocosmetics derived from biological/plant origins as a natural, sustainable, and more skin-friendly as compared to traditional cosmetics made from hazardous chemical ingredients (Migala 2022). Increasingly due to various activists or influences, people would also prefer buying cruelty-free products as they are not tested or sourced from animals. Understanding consumer trends and preferences is crucial for the cosmetic industry in developing nanocosmetics and fungal-based cosmetics that meet consumer needs and expectations.

18.7 Future Prospects

Nanocosmetics and fungal-based cosmetics are a highly innovative trend in advanced skincare for the cosmetic industry and hold tremendous potential in offering promising skincare solutions. Nanocosmetics are expected to witness significant progress in the foreseeable future as the technological advancements for nanotechnology are increasing at a rapid pace. This would lead to development of nanocosmetic formulation with enhanced stability, better safety of the formulation for consumer use, extended shelf life, better bioavailability, and sustained release of encapsulated active ingredients (Morganti et al. 2020). Since the research and exploration of new fungal species is pacing rapidly as well, we can expect to discover a new and vast array of fungal extracts and biomolecules of cosmetic importance. Discovery of new fungal molecules and extracts holds a significant promise for the development of new cosmetics with effective skincare solutions (Wu et al. 2016). Both nanocosmetics and fungal-based cosmetics hold a significant amount of potential in the development of personalized skincare products, i.e., encapsulation and formulation of cosmetics with unique and targeted solutions for personalized skincare. The cosmetic world is slowly ditching orthodox chemical practices and moving towards a 100% natural, eco-friendly, and sustainable approach for the development of cosmetics. The production of nanocosmetics and fungal-based cosmetics from natural sources and environmentally friendly manufacturing and packaging holds a high appeal for prioritizing sustainability and responsible practices for the cosmetic industry (Amberg and Fogarassy 2019). Integration of advanced technologies such as Artificial Intelligence (AI) can be a gamechanger for the industry as it would enable companies to develop cosmetics with better and more effective formulations and the ability to customize unique and targeted cosmetic formulations will revolutionize the field of personalized skincare (Kavyashree et al. 2022). There is a wide scope for a combined and multifunctional aspect since fungal-nanotechnology can be used to harness the potential of nanotechnology and fungi for enhanced and improved skincare solutions (El Enshasy et al. 2018). Future studies in fungal cosmeceuticals should prioritize novel methods and formulations. Exploring tools such as RNA interference (RNAi), CRISPR technology, and omics methodologies can significantly advance our comprehension of marine fungi. Cultivating slower-growing and non-cultivable marine fungi necessitates techniques that mimic natural environments, as highlighted in a study (Owen et al. 2013). Additionally, marine fungi can be fermented using waste materials to lower production expenses, with Solid-State Fermentation (SSF) proving effective in transforming industrial by-products into valuable commodities. The cosmetic industry can benefit from the antioxidant properties of marine fungi and the potential of using epigenetic tools to modulate secondary metabolite production (Vitale et al. 2020). The cosmetic sector's growth will be driven by the exploration of unexplored fungal varieties, which are estimated to be a significant fraction of the total. Molecular cosmetology integrates metabolomics, proteomics, genomics, and systems pharmacology to uncover the molecular underpinnings of medicinal properties (Wu et al. 2016). Utilizing *S. cerevisiae* as a cost-effective source of bioactive peptides supports circular economies and environmental sustainability. These yeast-derived peptides exhibit antioxidant properties and may effectively address signs of aging in skin (Oliveira et al. 2022a, b). Researchers ought to explore new compounds for therapeutic purposes and investigate nanobiotechnology-based methods for targeted drug delivery of cordycepin to enhance its bioavailability. Nanobiotechnology offers exciting prospects in the beauty industry. Chitosan is underutilized in the cosmetics industry despite its potential for use in various fields. Research spanning disciplines such as metabolomics, genomics, and proteomics is revealing the molecular processes behind substance production from previously unexplored fungal species. These insights pave the way for creating biomaterials that find applications in pharmaceuticals, cosmetics, and medicine (Ashraf et al. 2020; Kulka and Sionkowska 2023; Mukherjee et al. 2024). The future holds a promising prospect for the field of nanocosmetics and fungal-based cosmetics and embracing these prospects would surely drive the industry forward and pave the way for a rather sustainable and innovative trend in the cosmetic industry.

18.8 Conclusion

The realm of nanocosmetics and fungal-based cosmetics present an intriguing prospect in an advancing cosmetic landscape. Over the course of this chapter, from the ultramicroscopic realm of nanoparticles to the intricate world of fungi, we have explored two distinct branches of cosmetics and shed light on their features and promising future prospects.

The cutting-edge application of nanotechnology has unveiled nanocosmetics as a new era in the cosmetic world. The capacity to control and engineer materials on a nanoscale has significantly advanced targeted delivery systems, encapsulation techniques, and sustained release methods, thereby improving the bioavailability of active ingredients in cosmetic formulations. The intertwining of nanotechnology in the cosmetic world has not only enhanced product efficacy but also opened the door for a vast array of possibilities for research and development of new nanocosmetics. Fungal-based cosmetics on the other hand have tapped the bountiful world of fungi, harnessing their unprecedented bioactive potential. Attributing to their myriad of active metabolites and vast species, fungi offer a plethora of compounds that possess cosmetic potential such as moisturizing, hydration, rejuvenation, anti-oxidant, and protective properties. Meticulous research and development has led to the infusion of fungi into cosmetic formulations that go beyond the limits of traditional skincare. From the use of ancient medicinal mushrooms of Japan to the untapped potential of fungal metabolites and enzymes, fungal-based cosmetics pompously showcase the harmony of science and nature to provide consumers with a green and sustainable approach to skincare. With the rapidly advancing pace of science and technology, the future prospects of nanocosmetics and fungal-based cosmetics seem promising and boundless. The foreseeable future for nanotechnology seems meticulous in pushing the boundaries of precision and targeted efficacy in skincare. The use of nanoparticles can be envisioned to target intricate pathways to deliver bioactive compounds with precision and accuracy. The entwining of nanotechnology for personalized skincare, boosted by integration of AI can unveil a new era of customized products and tailored formulations for better efficacy. Simultaneously, the discovery of novel metabolites and fungal species is poised to benefit the advancement of fungal-based cosmetics.

In the ever-changing and dynamic landscape of the cosmetic industry, consumer education, raising awareness, and transparency regarding the products will be critical for these cosmetic formulations to gain a strong foothold in the consumer market. Brands must provide accurate and reliable information regarding the science of nanocosmetics and fungal-based cosmetics in a bid to provide consumers a clear communication to empower them in making right skincare choices. In the grand scheme, the use of nanocosmetics and fungal-based cosmetics offers a holistic approach that intertwines the intricacies of nature and science. The foreseeable future holds great promise for a new era of transformative skincare which transcends the boundaries of traditional cosmetics.

References

Ahmad MF (2018) *Ganoderma lucidum*: persuasive biologically active constituents and their health endorsement. Biomed Pharmacother 107:507–519. https://doi.org/10.1016/J.BIOPHA.2018.08.036

Ahmadi Ashtiani HR, Bishe P, Lashgari N-A, Nilforoushzadeh MA, Zare S (2016) Liposomes in cosmetics. J Skin Stem Cell 3(3). https://doi.org/10.5812/jssc.65815

Ahmadi Tehrani A, Omranpoor MM, Vatanara A, Seyedabadi M, Ramezani V (2019) Formation of nanosuspensions in bottom-up approach: theories and optimization. DARU J Pharm Sci 27(1):451–473. https://doi.org/10.1007/s40199-018-00235-2

Ajazzuddin M, Jeswani G, Jha AK (2015) Nano cosmetics: past, present and future trends. Recent Pat Nanomed 5(1):3–11. https://doi.org/10.2174/1877912305666150417232826

Alaqad K, Saleh TA (2016) Gold and silver nanoparticles: synthesis methods, characterization routes and applications towards drugs. J Environ Anal Toxicol 6(4). https://doi.org/10.4172/2161-0525.1000384

AlMadalli HJ, Abdul Rasool BK, Shehab NG, Sala FD, Borzacchiello A (2024) Pomegranate extract-loaded sphingosomes for the treatment of cancer: phytochemical investigations, formulation, and antitumor activity evaluation. PLoS One 19(2):e0293115. https://doi.org/10.1371/journal.pone.0293115

Alves A, Sousa E, Sousa E, Kijjoa A, Pinto M, Pinto M (2020) Marine-derived compounds with potential use as cosmeceuticals and nutricosmetics. Molecules 25(11):2536. https://doi.org/10.3390/MOLECULES25112536

Amberg N, Fogarassy C (2019) Green consumer behavior in the cosmetics market. Resources *8*(3):137. https://doi.org/10.3390/RESOURCES8030137

Andrade VS, Sarubbo LA, Fukushima K, Miyaji M, Nishimura K, de Campos-Takaki GM (2002) Production of extracellular proteases by *Mucor circinelloides* using D-glucose as carbon source/substrate. Braz J Microbiol 33(2). https://doi.org/10.1590/S1517-83822002000200002

Apekshaben PR, Kiran PB, PujaC P, Bhagyashri PA, Damini PM, PracheeKPawar Sunil DP (2022) A review on role of nano technology in cosmeceuticals. Int J Pharm Res Appl 7:2456–4494. https://doi.org/10.35629/7781-070419721993

Ashraf SA, Elkhalifa AEO, Siddiqui AJ, Patel M, Awadelkareem AM, Snoussi M, Ashraf MS, Adnan M, Hadi S (2020) Cordycepin for health and wellbeing: a potent bioactive metabolite of an entomopathogenic medicinal fungus cordyceps with its nutraceutical and therapeutic potential. Molecules 25(12):2735. https://doi.org/10.3390/molecules25122735

Aslam Abdul Rahiman CA, Krishnan K, Sreelekshmi SS, Arjun KK, Nair SC (2021) Novasome: a pioneering advancementin vesicular drug delivery. Int J Appl Pharm:59–64. https://doi.org/10.22159/ijap.2021v13i1.39528

Aziz ZAA, Mohd-Nasir H, Ahmad A, Mohd Setapar SH, Peng WL, Chuo SC, Khatoon A, Umar K, Yaqoob AA, Mohamad Ibrahim MN (2019) Role of nanotechnology for design and development of cosmeceutical: application in makeup and skin care. Front Chem 7. https://doi.org/10.3389/fchem.2019.00739

Bacha K, Chemotti C, Mbakidi J-P, Deleu M, Bouquillon S (2023) Dendrimers: synthesis, encapsulation applications and specific interaction with the stratum corneum—a review. Macromol 3(2):343–370. https://doi.org/10.3390/macromol3020022

Banerjee S, Sen KK (2018) Aquasomes: a novel nanoparticulate drug carrier. J Drug Deliv Sci Technol 43:446–452. https://doi.org/10.1016/j.jddst.2017.11.011

Barbu A, Catană, Ştefan A, Deselnicu DC, Cioca LI, Ioanid A (2022) Factors influencing consumer behavior toward green products: a systematic literature review. Int J Environ Res Public Health 19(24):16568. https://doi.org/10.3390/IJERPH192416568

Battaglia L, Ugazio E (2019) Lipid nano- and microparticles: an overview of patent-related research. J Nanomater 2019(1):2834941. https://doi.org/10.1155/2019/2834941

Bell V, Silva CRPG, Guina J, Fernandes TH (2022) Mushrooms as future generation healthy foods. Front Nutr 9. https://doi.org/10.3389/fnut.2022.1050099

Bernard D, Mehul B (2009) Use of aspartic proteases in cosmetics and therapeutics. United States Patent. http://www.ncbi.nlm

Bernauer U (2018) Opinion of the Scientific Committee on Consumer Safety (SCCS)—revision of the opinion on hydroxyapatite (nano) in cosmetic products. Regul Toxicol Pharmacol 98:274–275. https://doi.org/10.1016/j.yrtph.2018.07.018

Bernauer U, Bodin L, Chaudhry Q, Coenraads PJ, Dusinska M, Gaffet E, Panteri E, Rousselle C, Stepnik M, Wijnhoven S, De Jong WH, Von Götz N (2021) Scientific advice on the safety of nanomaterials in cosmetics (SCCS)—final advice, p 49. https://doi.org/10.2875/125512

Bilal M, Iqbal HMN (2020) New insights on unique features and role of nanostructured materials in cosmetics. Cosmetics 7(2):24. https://doi.org/10.3390/cosmetics7020024

Bilia AR, Costa MdoC (2021) Medicinal plants and their preparations in the European market: why has the harmonization failed? The cases of St. John's wort, valerian, ginkgo, ginseng, and green tea. Phytomedicine 81:153421. https://doi.org/10.1016/j.phymed.2020.153421

Blanco-Padilla A, Soto KM, Hernández Iturriaga M, Mendoza S (2014) Food antimicrobials nanocarriers. Sci World J 2014:1–11. https://doi.org/10.1155/2014/837215

Boonme P (2007) Applications of microemulsions in cosmetics. J Cosmet Dermatol 6(4):223–228. https://doi.org/10.1111/j.1473-2165.2007.00337.x

Brandelli A, Daroit DJ, Riffel A (2010) Biochemical features of microbial keratinases and their production and applications. Appl Microbiol Biotechnol 85(6):1735–1750. https://doi.org/10.1007/s00253-009-2398-5

Brown JS, Zeman KL, Bennett WD (2002) Ultrafine particle deposition and clearance in the healthy and obstructed lung. Am J Respir Crit Care Med 166(9):1240–1247. https://doi.org/10.1164/rccm.200205-399OC

Cardoza C, Nagtode V, Pratap A, Mali SN (2022) Emerging applications of nanotechnology in cosmeceutical health science: latest updates. Health Sci Rev 4:100051. https://doi.org/10.1016/j.hsr.2022.100051

Center for Food Safety and Applied Nutrition (2014) Guidance for industry safety of nanomaterials in cosmetic products. http://www.regulations.gov.

Chakraborty SS, Panja A, Dutta S, Patra P (2024) Advancements in nanoparticles for skin care: a comprehensive review of properties, applications, and future perspectives. Discov Mater 4(1):17. https://doi.org/10.1007/s43939-024-00088-4

Chandra P, Enespa, Singh R, Arora PK (2020) Microbial lipases and their industrial applications: a comprehensive review. Microb Cell Factories 19(1):169. https://doi.org/10.1186/s12934-020-01428-8

Chao Y, Xie F, Yang J, Lu J, Qian S (2007) Screening for a new Streptomyces strain capable of efficient keratin degradation. J Environ Sci 19(9):1125–1128. https://doi.org/10.1016/S1001-0742(07)60183-1

Chauhan A, Chauhan C (2023) Emerging trends of nanotechnology in beauty solutions: a review. Mater Today Proc 81(2):1052–1059. https://doi.org/10.1016/J.MATPR.2021.04.378

Chauhan A, Patil C, Jain P, Kulhari H (2020) Dendrimer-based marketed formulations and miscellaneous applications in cosmetics, veterinary, and agriculture. In: Pharmaceutical applications of dendrimers. Elsevier, pp 325–334. https://doi.org/10.1016/B978-0-12-814527-2.00014-7

Chauhan N, Vasava P, Khan SL, Siddiqui FA, Islam F, Chopra H, Emran TB (2022) Ethosomes: a novel drug carrier. Ann Med Surg 82. https://doi.org/10.1016/j.amsu.2022.104595

Chen Y, Bicker W, Wu J, Xie MY, Lindner W (2010) Ganoderma species discrimination by dual-mode chromatographic fingerprinting: a study on stationary phase effects in hydrophilic interaction chromatography and reduction of sample misclassification rate by additional use of reversed-phase chromatography. J Chromatogr A 1217(8):1255–1265. https://doi.org/10.1016/j.chroma.2009.12.024

Chen Y, Mao W, Tao H, Zhu W, Qi X, Chen Y, Li H, Zhao C, Yang Y, Hou Y, Wang C, Li N (2011) Structural characterization and antioxidant properties of an exopolysaccharide produced by the mangrove endophytic fungus *Aspergillus* sp. Y16. Bioresour Technol 102(17):8179–8184. https://doi.org/10.1016/j.biortech.2011.06.048

Chen X, Siu K-C, Cheung Y-C, Wu J-Y (2014) Structure and properties of a (1→3)-β-d-glucan from ultrasound-degraded exopolysaccharides of a medicinal fungus. Carbohydr Polym 106:270–275. https://doi.org/10.1016/j.carbpol.2014.02.040

Chien CC, Tsai ML, Chen CC, Chang SJ, Tseng CH (2008) Effects on tyrosinase activity by the extracts of *Ganoderma lucidum* and related mushrooms. Mycopathologia 166(2):117–120. https://doi.org/10.1007/S11046-008-9128-X/METRICS

Chin J, Jiang BC, Mufidah I, Persada SF, Noer BA (2018) The investigation of consumers' behavior intention in using green skincare products: a pro-environmental behavior model approach. *Sustainability 10*(11):3922. https://doi.org/10.3390/SU10113922

Cornier J, Keck CM, Van de Voorde M (2019) In: Cornier J, Keck CM, Van de Voorde M (eds) Nano cosmetics. Springer. https://doi.org/10.1007/978-3-030-16573-4

Cornils B (2020) Black carbon. In: Catalysis from A to Z. Wiley. https://doi.org/10.1002/9783527809080.cataz02167

Costa CRLdeM, Menolli RA, Osaku EF, Tramontina R, de Melo RH, do Amaral AE, Duarte PAD, de Carvalho MM, Smiderle FR, Silva JLdaC, Mello RG (2019) Exopolysaccharides from *Aspergillus terreus*: production, chemical elucidation and immunoactivity. Int J Biol Macromol 139:654–664. https://doi.org/10.1016/j.ijbiomac.2019.08.039

Crameri R, Garbani M, Rhyner C, Huitema C (2014) Fungi: the neglected allergenic sources. Allergy 69(2):176–185. https://doi.org/10.1111/all.12325

Damera DP, Kaja S, Janardhanam LSL, Alim S, Venuganti VVK, Nag A (2019) Synthesis, detailed characterization, and dual drug delivery application of BSA loaded aquasomes. ACS Appl Bio Mater 2(10):4471–4484. https://doi.org/10.1021/acsabm.9b00635

Dave K, Krishna Venuganti VV (2017) Dendritic polymers for dermal drug delivery. Ther Deliv 8(12):1077–1096. https://doi.org/10.4155/tde-2017-0091

De Cianni R, Pippinato L, Mancuso T (2023) A systematic review on drivers influencing consumption of edible mushrooms and innovative mushroom-containing products. Appetite 182:106454. https://doi.org/10.1016/j.appet.2023.106454

Dhawan S, Sharma P, Nanda S (2020) Cosmetic nanoformulations and their intended use. In: Nano cosmetics. Elsevier, pp 141–169. https://doi.org/10.1016/B978-0-12-822286-7.00017-6

Dhiman S, Mukherjee G, Kumar A, Mukherjee P, Verekar SA, Deshmukh SK (2017) Fungal tannase: recent advances and industrial applications. In: Developments in fungal biology and applied mycology. Springer Singapore, pp 295–313. https://doi.org/10.1007/978-981-10-4768-8_16

Donthi MR, Munnangi SR, Krishna KV, Saha RN, Singhvi G, Dubey SK (2023) Nanoemulgel: a novel nano carrier as a tool for topical drug delivery. Pharmaceutics 15(1):164. https://doi.org/10.3390/pharmaceutics15010164

Dréno B, Alexis A, Chuberre B, Marinovich M (2019) Safety of titanium dioxide nanoparticles in cosmetics. J Eur Acad Dermatol Venereol 33(S7):34–46. https://doi.org/10.1111/jdv.15943

Dureja H, Kaushik D, Gupta M, Kumar V, Lather V (2005) Cosmeceuticals: an emerging concept. Indian J Pharm 37(3):155. https://doi.org/10.4103/0253-7613.16211

Ebel R (2010) Terpenes from marine-derived fungi. Mar Drugs 8(8):2340–2368. https://doi.org/10.3390/MD8082340

El Enshasy HA, El Marzugi NA, Elsayed EA, Ling OM, Malek RA, Kepli AN, Othman NZ, Ramli S (2018) Medical and cosmetic applications of fungal nanotechnology: production, characterization, and bioactivity. In: Fungal nanobionics: principles and applications. Springer Singapore, pp 21–59. https://doi.org/10.1007/978-981-10-8666-3_2

Fakhravar Z, Ebrahimnejad P, Daraee H, Akbarzadeh A (2016) Nanoliposomes: synthesis methods and applications in cosmetics. J Cosmet Laser Ther 18(3):174–181. https://doi.org/10.3109/14764172.2015.1039040

Farag A, Hassan SW, El-Says A, Ghanem K (2018) Purification, characterization and application of tannase enzyme isolated from marine *Aspergillus nomius* GWA5. J Pure Appl Microbiol 12(4):1939–1949. https://doi.org/10.22207/JPAM.12.4.30

Fatima I, Rasul A, Shah S, Saadullah M, Islam N, Khames A, Salawi A, Ahmed MM, Almoshari Y, Abbas G, Abourehab MAS, Mehmood Khan S, Chauhdary Z, Alshamrani M, Namazi NI, Naguib DM (2022) Novasomes as nano-vesicular carriers to enhance topical delivery of fluconazole: a new approach to treat fungal infections. Molecules 27(9):2936. https://doi.org/10.3390/molecules27092936

Ferreira MS, Magalhães MC, Sousa-Lobo JM, Almeida IF (2020) Trending anti-aging peptides. Cosmetics 7(4):91. https://doi.org/10.3390/cosmetics7040091

Ferreira PG, Ferreira VF, da Silva F de C, Freitas CS, Pereira PR, Paschoalin VMF (2022) Chitosans and nanochitosans: recent advances in skin protection, regeneration, and repair. Pharmaceutics 14(6):1307. https://doi.org/10.3390/pharmaceutics14061307

Fytianos G, Rahdar A, Kyzas GZ (2020) Nanomaterials in cosmetics: recent updates. Nanomaterials 10(5):979. https://doi.org/10.3390/nano10050979

Gajbhiye S, Sakharwade S (2016) Silver nanoparticles in cosmetics. J Cosmet Dermatol Sci Appl 06(01):48–53. https://doi.org/10.4236/jcdsa.2016.61007

Ganesan P, Choi DK (2016) Current application of phytocompound-based nanocosmeceuticals for beauty and skin therapy. Int J Nanomedicine 1987. https://doi.org/10.2147/IJN.S104701

Garrett R, Hooper A, Hooper (1998) Indoor airborne fungal spores, house dampness and associations with environmental factors and respiratory health in children. Clin Exp Allergy 28(4):459–467. https://doi.org/10.1046/j.1365-2222.1998.00255.x

Ge X, Wei M, He S, Yuan W-E (2019) Advances of non-ionic surfactant vesicles (niosomes) and their application in drug delivery. Pharmaceutics 11(2):55. https://doi.org/10.3390/pharmaceutics11020055

Ghabdian Y, Taheri A, Jahanian-Najafabadi A (2021) Development of novel topical formulation from fullerene with antibacterial activity against *Propionibacterium acnes*. Fuller Nanotub Carbon Nanostruct 29(2):163–173. https://doi.org/10.1080/1536383X.2020.1825388

Gharaei-Fa E (2010) Biosurfactants in pharmaceutical industry (a mini-review). Am J Drug Discov Dev 1(1):58–69. https://doi.org/10.3923/ajdd.2011.58.69

Ghasemiyeh P, Mohammadi-Samani S (2020) Potential of nanoparticles as permeation enhancers and targeted delivery options for skin: advantages and disadvantages. Drug Des Devel Ther 14:3271–3289. https://doi.org/10.2147/DDDT.S264648

Ghazali E, Soon PC, Mutum DS, Nguyen B (2017) Health and cosmetics: investigating consumers' values for buying organic personal care products. J Retail Consum Serv 39:154–163. https://doi.org/10.1016/J.JRETCONSER.2017.08.002

Gigliobianco MR, Casadidio C, Censi R, Di Martino P (2018) Nanocrystals of poorly soluble drugs: drug bioavailability and physicochemical stability. Pharmaceutics 10(3):134. https://doi.org/10.3390/pharmaceutics10030134

Gubitosa J, Rizzi V, Fini P, Cosma P (2019) Hair care cosmetics: from traditional shampoo to solid clay and herbal shampoo, a review. Cosmetics 6(1):13. https://doi.org/10.3390/cosmetics6010013

Gupta PL, Rajput M, Oza T, Trivedi U, Sanghvi G (2019) Eminence of microbial products in cosmetic industry. Nat Prod Bioprospect 9(4):267–278. https://doi.org/10.1007/s13659-019-0215-0

Gupta AK, Gupta D, Gupta V (2021) Aquasomes: a self-assembled nano-particulate carrier system. Int J Curr Res Rev 13(04):44–52. https://doi.org/10.31782/IJCRR.2021.13427

Gupta V, Mohapatra S, Mishra H, Farooq U, Kumar K, Ansari M, Aldawsari M, Alalaiwe A, Mirza M, Iqbal Z (2022) Nanotechnology in cosmetics and cosmeceuticals—a review of latest advancements. Gels 8(3):173. https://doi.org/10.3390/gels8030173

Hájková V, Svobodová A, Krejčová D, Číž M, Velebný V, Lojek A, El-Benna J, Kubala L (2009) Soluble glucomannan isolated from *Candida utilis* primes blood phagocytes. Carbohydr Res 344(15): 2036–2041. https://doi.org/10.1016/j.carres.2009.06.034

Halla N, Fernandes IP, Heleno SA, Costa P, Boucherit-Otmani Z, Boucherit K, Rodrigues AE, Ferreira ICFR, Barreiro MF (2018) Cosmetics preservation: a review on present strategies. Molecules 23(7):1571. https://doi.org/10.3390/molecules23071571

Haveli SD, Walter P, Patriarche G, Ayache J, Castaing J, Van Elslande E, Tsoucaris G, Wang P-A, Kagan HB (2012) Hair fiber as a nanoreactor in controlled synthesis of fluorescent gold nanoparticles. Nano Lett 12(12): 6212–6217. https://doi.org/10.1021/nl303107w

Holsapple MP, Farland WH, Landry TD, Monteiro-Riviere NA, Carter JM, Walker NJ, Thomas KV (2005) Research strategies for safety evaluation of nanomaterials, part II: toxicological and safety evaluation of nanomaterials, current challenges and data needs. Toxicol Sci 88(1):12–17. https://doi.org/10.1093/toxsci/kfi293

Hooda A, Sradhanjali M, Popsy. (2018) Formulation and evaluation of novel solid lipid microparticles for the sustained release of ofloxacin. Pharm Nanotechnol 5(4). https://doi.org/10.2174/2211738506666171226121720

Husein M, Nassar N (2008) Nanoparticle preparation using the single microemulsions scheme. Curr Nanosci 4(4):370–380. https://doi.org/10.2174/157341308786306116

Hyde KD, Bahkali AH, Moslem MA (2010) Fungi—an unusual source for cosmetics. Fungal Divers 43(1): 1–9. https://doi.org/10.1007/s13225-010-0043-3

Ikram-Ul-Haq, Mukhtar H (2007) Biosynthesis of acid proteases by *Penicillium griseoroseum* IH-02 in solid-state fermentation. Pak J Bot 39(7):2717–2724

International Organization for Standardization (2016) ISO/TR 19838:2016—microbiology—cosmetics—guidelines for the application of ISO standards on cosmetic microbiology. https://www.iso.org/standard/66336.html?browse=ics

Inui S, Aoshima H, Nishiyama A, Itami S (2011) Improvement of acne vulgaris by topical fullerene application: unique impact on skin care. Nanomedicine 7(2): 238–241. https://doi.org/10.1016/j.nano.2010.09.005

Izquierdo MC, Lillo CR, Bucci P, Gómez GE, Martínez L, del Alonso SV, Calienni MN, Montanari J (2020) Comparative skin penetration profiles of formulations including ultradeformable liposomes as potential nanocosmeceutical carriers. J Cosmet Dermatol 19(11):3127–3137. https://doi.org/10.1111/jocd.13410

Kady EE (2019) Chitin, chitosan and glucan, properties and applications. World J Agric Soil Sci 3(1). https://doi.org/10.33552/WJASS.2019.03.000553

Kalicińska J, Wiśniowska B, Śpiewak R (2022) Irritant patch test reactions to cosmetic ingredients. Alergol Pol 9(1):55–61. https://doi.org/10.5114/pja.2022.112784

Kaul S, Gulati N, Verma D, Mukherjee S, Nagaich U (2018) Role of nanotechnology in cosmeceuticals: a review of recent advances. J Pharm 2018(1):3420204. https://doi.org/10.1155/2018/3420204

Kaur P, Rani R, Singh AP, Singh AP (2024) An overview of niosomes. J Drug Deliv Ther 14(3):137–146. https://doi.org/10.22270/jddt.v14i3.6450

Kavyashree N, Rama Satish KV, Rasal P, Jadhav RJ, Saidireddy M, Kharade KG (2022) Artificial Intelligence based smart cosmetics suggestion system based on skin condition. In: 2022 international conference on automation, computing and renewable systems (ICACRS), pp 797–801. https://doi.org/10.1109/ICACRS55517.2022.10029120

Khan IA, Abourashed EA (2009) Leung's encyclopedia of common natural ingredients used in food, drugs, and cosmetics. https://www.wiley.com/en-gb/Leung's+Encyclopedia+of+Common+Natural+Ingredients%3A+Used+in+Food%2C+Drugs+and+Cosmetics%2C+3rd+Edition-p-9781118213063

Kranthi VS, Rao DM, Jaganmohan P (2012) Production of protease by *Aspergillus flavus* through solid state fermentation using different oil seed cakes. Int J Microbiol Res 3(1):12–15. https://doi.org/10.5829/idosi.ijmr.2012.3.1.613

Kroto HW, Heath JR, O'Brien SC, Curl RF, Smalley RE (1985) C60: Buckminsterfullerene. Nature 318(6042):162–163. https://doi.org/10.1038/318162a0

Kulka K, Sionkowska A (2023) Chitosan based materials in cosmetic applications: a review. Molecules 28(4):1817. https://doi.org/10.3390/molecules28041817

Kumar A, Verma V, Dubey VK, Srivastava A, Garg SK, Singh VP, Arora PK (2023) Industrial applications of fungal lipases: a review. Front Microbiol 14. https://doi.org/10.3389/fmicb.2023.1142536

Kusuma Priya MD, Kumar V, Damini VK, Eswar K, Reddy KR, Brito Raj S, Sucharitha P (2020) Somes: a review on composition, formulation methods and evaluations of different types of "somes" drug delivery system. Int J Appl Pharm:7–18. https://doi.org/10.22159/ijap.2020v12i6.38996

Lee H-S, Byun S-H, Cho S-W, Yang B-E (2019) Past, present, and future of regeneration therapy in oral and periodontal tissue: a review. Appl Sci 9(6):1046. https://doi.org/10.3390/app9061046

Lens M (2009) Use of fullerenes in cosmetics. Recent Pat Biotechnol 3(2):118–123. https://doi.org/10.2174/187220809788700166

Li N, Sioutas C, Cho A, Schmitz D, Misra C, Sempf J, Wang M, Oberley T, Froines J, Nel A (2003) Ultrafine particulate pollutants induce oxidative stress and mitochondrial damage. Environ Health Perspect 111(4):455–460. https://doi.org/10.1289/ehp.6000

Li Y, Zhang Y, Yan B (2014) Nanotoxicity overview: nano-threat to susceptible populations. Int J Mol Sci 15(3):3671–3697. https://doi.org/10.3390/ijms15033671

Liu J, Huang X, Lu L, Li M, Xu J, Deng H (2011) Turbiscan Lab® Expert analysis of the biological demulsification of a water-in-oil emulsion by two biodemulsifiers. J Hazard Mater 190(1–3):214–221. https://doi.org/10.1016/j.jhazmat.2011.03.028

Liu R, Meng X, Mo C, Wei X, Ma A (2022) Melanin of fungi: from classification to application. World J Microbiol Biotechnol 38(12):228. https://doi.org/10.1007/s11274-022-03415-0

Lohani A, Verma A, Joshi H, Yadav N, Karki N (2014) Nanotechnology-based cosmeceuticals. ISRN Dermatol 2014:1–14. https://doi.org/10.1155/2014/843687

Lu Y-C, Chen K-N (2017) Consumer knowledge, brand image, openness to experience and involvement: a case in cosmetic consumption. J Cosmet Dermatol Sci Appl 07(04):349–361. https://doi.org/10.4236/jcdsa.2017.74031

Ma Y, Cong Z, Gao P, Wang Y (2023) Nanosuspensions technology as a master key for nature products drug delivery and in vivo fate. Eur J Pharm Sci 185:106425. https://doi.org/10.1016/j.ejps.2023.106425

Mago P, Sharma R, Hafeez I, Nawaz I, Joshi M, Mehrotra R (2023) Mushroom based cosmeceuticals: an upcoming biotechnology sector. Biosci Biotechnol Res Asia 20(2):381–394. https://doi.org/10.13005/BBRA/3097

Majtan J, Jesenak M (2018) β-glucans: multi-functional modulator of wound healing. Molecules 23(4):806. https://doi.org/10.3390/molecules23040806

Malathi S, Chakraborty R (1991) Production of alkaline protease by a new *Aspergillus flavus* isolate under solid-substrate fermentation conditions for use as a depilation agent. Appl Environ Microbiol 57(3):712–716. https://doi.org/10.1128/aem.57.3.712-716.1991

Manconi M (2003) Niosomes as carriers for tretinoin II. Influence of vesicular incorporation on tretinoin photostability. Int J Pharm 260(2):261–272. https://doi.org/10.1016/S0378-5173(03)00268-0

Manjit, Yadav A, Aggarwal NK, Kumar K, Kumar A (2008) Tannase production by *Aspergillus fumigatus* MA under solid-state fermentation. World J Microbiol Biotechnol 24(12):3023–3030. https://doi.org/10.1007/s11274-008-9847-7

Marco M-G (2009) A novel tannase from the xerophilic fungus *Aspergillus niger* GH1. J Microbiol Biotechnol 19(9):987–996. https://doi.org/10.4014/jmb.0811.615

Mari A, Schneider P, Wally V, Breitenbach M, Simon-Nobbe B (2003) Sensitization to fungi: epidemiology, comparative skin tests, and IgE reactivity of fungal extracts. Clin Exp Allergy 33(10):1429–1438. https://doi.org/10.1046/j.1365-2222.2003.01783.x

Martins AM, Marto JM (2023) A sustainable life cycle for cosmetics: from design and development to post-use phase. Sustain Chem Pharm 35:101178. https://doi.org/10.1016/J.SCP.2023.101178

Matharoo N, Mohd H, Michniak-Kohn B (2024) Transferosomes as a transdermal drug delivery system: dermal kinetics and recent developments. WIREs Nanomed Nanobiotechnol 16(1). https://doi.org/10.1002/wnan.1918

McCloud TG (2010) High throughput extraction of plant, marine and fungal specimens for preservation of biologically active molecules. Molecules 15(7): 4526–4563. https://doi.org/10.3390/molecules15074526

Mebert AM, Baglole CJ, Desimone MF, Maysinger D (2017) Nanoengineered silica: properties, applications and toxicity. Food Chem Toxicol 109:753–770. https://doi.org/10.1016/j.fct.2017.05.054

Mehta A, Bodh U, Gupta R (2017) Fungal lipases: a review. J Biotech Res. https://openurl.ebsco.com/EPDB%3Agcd%3A1%3A19391633/detailv2?sid=ebsco%3Aplink%3Ascholar&id=ebsco%3Agcd%3A123881491&crl=c

Meng TX, Furuta S, Fukamizu S, Yamamoto R, Ishikawa H, Arung ET, Shimizu K, Ohga S, Kondo R (2011) Evaluation of biological activities of extracts from the fruiting body of *Pleurotus citrinopileatus* for skin cosmetics. J Wood Sci 57(5):452–458. https://doi.org/10.1007/S10086-011-1192-Z/METRICS

Migala J (2022) What are "natural" skin-care products, and are they better for you? https://www.everydayhealth.com/skin-beauty/what-are-natural-skin-care-products-are-they-better-you/

Miguez JP, Gama RS, Bolina ICA, de Melo CC, Cordeiro MR, Hirata DB, Mendes AA (2018) Enzymatic synthesis optimization of a cosmetic ester catalyzed by a homemade biocatalyst prepared via physical adsorption of lipase on amino-functionalized rice husk silica. Chem Eng Res Des 139:296–308. https://doi.org/10.1016/j.cherd.2018.09.037

Miles CO, Munday SC, Wilkins AL (1992) Improved method for purification of epidithiodioxopiperazines from fungal extracts: purification of sporidesmin A. J Agric Food Chem 40(12):2458–2460. https://doi.org/10.1021/jf00024a023

Moammeri A, Chegeni MM, Sahrayi H, Ghafelehbashi R, Memarzadeh F, Mansouri A, Akbarzadeh I, Abtahi MS, Hejabi F, Ren Q (2023) Current advances in niosomes applications for drug delivery and cancer treatment. Mater Today Biol 23:100837. https://doi.org/10.1016/j.mtbio.2023.100837

Mohammed YH, Holmes A, Haridass IN, Sanchez WY, Studier H, Grice JE, Benson HAE, Roberts MS (2019) Support for the safe use of zinc oxide nanoparticle sunscreens: lack of skin penetration or cellular toxicity after repeated application in volunteers. J Invest Dermatol 139(2):308–315. https://doi.org/10.1016/j.jid.2018.08.024

Morganti P, Chen H-D, Morganti G (2020) Nano cosmetics: future perspective. In: Nano cosmetics. Elsevier, pp 455–481. https://doi.org/10.1016/B978-0-12-822286-7.00020-6

Mouad AM, Taupin D, Lehr L, Yvergnaux F, Porto ALM (2016) Aminolysis of linoleic and salicylic acid derivatives with *Candida antarctica* lipase B: a solvent-free process to obtain amphiphilic amides for cosmetic application. J Mol Catal B Enzym 126:64–68. https://doi.org/10.1016/j.molcatb.2016.01.002

Mousavi SZ, Nafisi S, Maibach HI (2017) Fullerene nanoparticle in dermatological and cosmetic applications. Nanomedicine 13(3):1071–1087. https://doi.org/10.1016/j.nano.2016.10.002

Moussaoui N, Cansell M, Denizot A (2002) Marinosomes®, marine lipid-based liposomes: physical characterization and potential application in cosmetics. Int J Pharm 242(1–2):361–365. https://doi.org/10.1016/S0378-5173(02)00217-X

Mukherjee A, Das S, Chakraborty D, Pal N, Das N (2024) Fungi's treasure in cosmeceuticals-a comprehensive chemical approach. S Afr J Bot 166:311–331. https://doi.org/10.1016/j.sajb.2024.01.036

Muthuramalingam K, Choi SI, Hyun C, Kim YM, Cho M (2019) β-glucan-based wet dressing for cutaneous wound healing. Adv Wound Care 8(4):125–135. https://doi.org/10.1089/wound.2018.0843

Nafisi S, Maibach HI (2017) Nanotechnology in cosmetics. In: Cosmetic science and technology. Elsevier, pp 337–369. https://doi.org/10.1016/B978-0-12-802005-0.00022-7

O'Donnell D, Wang L, Xu J, Ridgway D, Gu T, Moo-Young M (2001) Enhanced heterologous protein production in *Aspergillus niger* through pH control of extracellular protease activity. Biochem Eng J 8(3): 187–193. https://doi.org/10.1016/S1369-703X(01)00102-4

Oh J-J, Kim JY, Son SH, Jung W-J, Kim DH, Seo J-W, Kim G-H (2021) Fungal melanin as a biocompatible broad-spectrum sunscreen with high antioxidant activity. RSC Adv 11(32):19682–19689. https://doi.org/10.1039/D1RA02583J

Oliveira AS, Ferreira C, Pereira JO, Pintado ME, Carvalho AP (2022a) Spent brewer's yeast (*Saccharomyces cerevisiae*) as a potential source of bioactive peptides: an overview. Int J Biol Macromol 208:1116–1126. https://doi.org/10.1016/j.ijbiomac.2022.03.094

Oliveira C, Coelho C, Teixeira JA, Ferreira-Santos P, Botelho CM (2022b) Nanocarriers as active ingredients enhancers in the cosmetic industry—the European and North America regulation challenges. Molecules 27(5):1669. https://doi.org/10.3390/molecules27051669

Onel N (2016) Pro-environmental purchasing behavior of consumers. Soc Mark Q *23*(2):103–121. https://doi.org/10.1177/1524500416672440

Opatha SAT, Titapiwatanakun V, Chutoprapat R (2020) Transfersomes: a promising nanoencapsulation technique for transdermal drug delivery. Pharmaceutics 12(9):855. https://doi.org/10.3390/pharmaceutics12090855

Owen JG, Reddy BVB, Ternei MA, Charlop-Powers Z, Calle PY, Kim JH, Brady SF (2013) Mapping gene clusters within arrayed metagenomic libraries to expand the structural diversity of biomedically relevant natural products. Proc Natl Acad Sci 110(29): 11797–11802. https://doi.org/10.1073/pnas.1222159110

Pardeike J, Hommoss A, Müller RH (2009) Lipid nanoparticles (SLN, NLC) in cosmetic and pharmaceutical dermal products. Int J Pharm 366(1–2):170–184. https://doi.org/10.1016/j.ijpharm.2008.10.003

Park Y-H, Kim JN, Jeong SH, Choi JE, Lee S-H, Choi BH, Lee JP, Sohn KH, Park KL, Kim M-K, Son SW (2010) Assessment of dermal toxicity of nanosilica using cultured keratinocytes, a human skin equivalent model and an in vivo model. Toxicology 267(1–3): 178–181. https://doi.org/10.1016/j.tox.2009.10.011

Pawar G (2022) Sphingosomes: highlights of the progressive journey and their application perspectives in modern drug delivery. Int J Med Pharm Sci 12(01). https://doi.org/10.31782/IJMPS.2022.12101

Pentek T, Newenhouse E, O'Brien B, Chauhan A (2017) Development of a topical resveratrol formulation for commercial applications using dendrimer nanotechnology. Molecules 22(1):137. https://doi.org/10.3390/molecules22010137

Pintathong P, Chomnunti P, Sangthong S, Jirarat A, Chaiwut P (2021) The feasibility of utilizing cultured *Cordyceps militaris* residues in cosmetics: biological activity assessment of their crude extracts. J Fungi 7(11):973. https://doi.org/10.3390/jof7110973

Poonia N, Kharb R, Lather V, Pandita D (2016) Nanostructured lipid carriers: versatile oral delivery vehicle. Future Sci OA 2(3). https://doi.org/10.4155/fsoa-2016-0030

Pyo S, Meinke M, Keck C, Müller R (2016) Rutin—increased antioxidant activity and skin penetration by nanocrystal technology (smartCrystals). Cosmetics 3(1):9. https://doi.org/10.3390/cosmetics3010009

Raj S, Jose S, Sumod US, Sabitha M (2012) Nanotechnology in cosmetics: opportunities and challenges. J Pharm Bioallied Sci 4(3):186–193. https://doi.org/10.4103/0975-7406.99016

Ramis JM, Coelho CC, Córdoba A, Quadros PA, Monjo M (2018) Safety assessment of nano-hydroxyapatite as an oral care ingredient according to the EU cosmetics regulation. Cosmetics 5(3):53. https://doi.org/10.3390/cosmetics5030053

Ramsden JJ (2016) Carbon-based nanomaterials and devices. In: Nanotechnology. Elsevier, pp 231–244. https://doi.org/10.1016/B978-0-323-39311-9.00015-7

Rao BN, Reddy KR, Mounika B, Fathima SR, Tejaswini A (2019) Vesicular drug delivery system: a review. Int J ChemTech Res 12(05):39–53. https://doi.org/10.20902/IJCTR.2019.120505

Raszewska-Famielec M, Flieger J (2022) Nanoparticles for topical application in the treatment of skin dysfunctions—an overview of dermo-cosmetic and dermatological products. Int J Mol Sci 23(24):15980. https://doi.org/10.3390/ijms232415980

Reddy SB (2012) Gallic acid production and tannase activity of *Penicillium purpurogenum* stoll employing agrobased wastes through solid state fermentation: influence of carbon and nitrogen sources. IOSR J Pharmacy (IOSRPHR) 2(1):109–112. https://doi.org/10.9790/3013-0210109112

Rigano L, Lionetti N (2016) Nanobiomaterials in galenic formulations and cosmetics. In: Nanobiomaterials in galenic formulations and cosmetics. Elsevier, pp 121–148. https://doi.org/10.1016/B978-0-323-42868-2.00006-1

Romero GB, Brysch W, Keck CM, Müller RH (2017) Nanocapsule formation by nanocrystals. In: Nanoencapsulation technologies for the food and nutraceutical industries. Elsevier, pp 165–186. https://doi.org/10.1016/B978-0-12-809436-5.00006-9

Sadhu P, Singh PH, Kumari M, Dash DK, Patel S, Shah N, Seth AK (2021) Trends in cosmeceuticals based nanotechnology: up-to-date. J Pharm Res Int:136–149. https://doi.org/10.9734/jpri/2021/v33i37B32033

Saeedi M, Eslamifar M, Khezri K (2019) Kojic acid applications in cosmetic and pharmaceutical preparations. Biomed Pharmacother 110:582–593. https://doi.org/10.1016/j.biopha.2018.12.006

Saha R (2012) Cosmeceuticals and herbal drugs: practical uses. Int J Pharm Sci Res. https://ijpsr.com/bft-article/cosmeceuticals-and-herbal-drugs-practical-uses/?view=fulltext

Sahu D, Kannan GM, Vijayaraghavan R (2014) Carbon black particle exhibits size dependent toxicity in human monocytes. Int J Inflamm 2014:1–10. https://doi.org/10.1155/2014/827019

Santos AC, Morais F, Simões A, Pereira I, Sequeira JAD, Pereira-Silva M, Veiga F, Ribeiro A (2019) Nanotechnology for the development of new cosmetic formulations. Expert Opin Drug Deliv 16(4): 313–330. https://doi.org/10.1080/17425247.2019.1585426

Sarkar D, Khilar KC (2009) Microemulsion mediated synthesis of nanoparticles. In: Recent advances in nanoscience and technology. Bentham Science Publishers, pp 25–37. https://doi.org/10.2174/978160805053610901010025

Schätzlein A, Cevc G (1998) Non-uniform cellular packing of the stratum corneum and permeability barrier function of intact skin: a high-resolution confocal laser scanning microscopy study using highly deformable vesicles (Transfersomes). Br J Dermatol 138(4): 583–592. https://doi.org/10.1046/j.1365-2133.1998.02166.x

Scherer Santos J, Sousa Ponciano C, Nogueira Barradas T, Pascoal Costa M, Diniz Tavares G (2023) Skin care nano cosmetics. In: Cosmetic products and industry—new advances and applications. IntechOpen. https://doi.org/10.5772/intechopen.111617

Schulte P, Geraci C, Zumwalde R, Hoover M, Kuempel E (2008) Occupational risk management of engineered nanoparticles. J Occup Environ Hyg 5(4):239–249. https://doi.org/10.1080/15459620801907840

Scientific Committee on Consumer Safety (SCCS), Chaudhry Q (2016) Opinion of the Scientific Committee on Consumer safety (SCCS)—second revision of the opinion on carbon black, nano-form, in cosmetic products. Regul Toxicol Pharmacol 79:103–104. https://doi.org/10.1016/j.yrtph.2016.02.021

Sharma B, Samperi M, Jain A, Chaudhary GR, Kaur G, Pérez-García L (2022) Gemini surfactant mediated catansomes for enhanced singlet oxygen generation of rose bengal and their phototoxicity against cancer cells. ACS Biomater Sci Eng 8(5):1878–1891. https://doi.org/10.1021/acsbiomaterials.2c00017

Sharma A, Agarwal P, Sebghatollahi Z, Mahato N (2023) Functional nanostructured materials in the cosmetics industry: a review. ChemEngineering 7(4):66. https://doi.org/10.3390/chemengineering7040066

Shaw TK, Paul P, Chatterjee B (2022) Research-based findings on scope of liposome-based cosmeceuticals: an updated review. Futur J Pharm Sci 8(1):46. https://doi.org/10.1186/s43094-022-00435-3

Shegokar R (2016) What nanocrystals can offer to cosmetic and dermal formulations. In: Nanobiomaterials in galenic formulations and cosmetics. Elsevier, pp 69–91. https://doi.org/10.1016/B978-0-323-42868-2.00004-8

Shokri J (2017) Nano cosmetics: benefits and risks. Bioimpacts 7(4):207–208. https://doi.org/10.15171/bi.2017.24

Sideney Becker O (2013) Production of cellulolytic enzymes by *Aspergillus flavus* using solid state fermentation based on sugarcane bagasse. Am J Biochem 3(2):25–28. https://doi.org/10.5923/j.ajb.20130302.01

Sim Y-C, Lee S-G, Lee D-C, Kang B-Y, Park K-M, Lee J-Y, Kim M-S, Chang I-S, Rhee J-S (2000) Stabilization of papain and lysozyme for application to cosmetic products. Biotechnol Lett 22(2):137–140. https://doi.org/10.1023/A:1005670323912

Singh I, Rehni AK, Kalra R, Joshi G, Kumar M (2008) Dendrimers and their pharmaceutical applications—a review. Pharmazie 63(7):491–496

Singh TG, Dhiman S, Jindal M, Sandhu IS, Chitkara M (2016) Nanobiomaterials. In: Fabrication and self-assembly of nanobiomaterials. Elsevier, pp 401–429. https://doi.org/10.1016/B978-0-323-41533-0.00013-1

Singh A, Kapoor R, Misra R (2019) Green cosmetics—changing young consumer preference and reforming cosmetic industry. Int J Recent Technol Eng 8(4): 12932–12939. https://doi.org/10.35940/ijrte.D6927.118419

Sivadasan D, Sultan MH, Alqahtani SS, Javed S (2023) Cubosomes in drug delivery—a comprehensive review on its structural components, preparation techniques and therapeutic applications. Biomedicine 11(4): 1114. https://doi.org/10.3390/biomedicines11041114

Smijs T, Pavel (2011) Titanium dioxide and zinc oxide nanoparticles in sunscreens: focus on their safety and effectiveness. Nanotechnol Sci Appl 95. https://doi.org/10.2147/NSA.S19419

Song CK, Balakrishnan P, Shim C-K, Chung S-J, Chong S, Kim D-D (2012) A novel vesicular carrier, transethosome, for enhanced skin delivery of voriconazole: characterization and in vitro/in vivo evaluation. Colloids Surf B: Biointerfaces 92:299–304. https://doi.org/10.1016/j.colsurfb.2011.12.004

Souto EB, Cano A, Martins-Gomes C, Coutinho TE, Zielińska A, Silva AM (2022) Microemulsions and nanoemulsions in skin drug delivery. Bioengineering 9(4):158. https://doi.org/10.3390/bioengineering9040158

Subongkot T, Charernsriwilaiwat N, Chanasongkram R, Rittem K, Ngawhirunpat T, Opanasopit P (2022) Development and skin penetration pathway evaluation using confocal laser scanning microscopy of microemulsions for dermal delivery enhancement of finasteride. Pharmaceutics 14(12):2784. https://doi.org/10.3390/pharmaceutics14122784

Tamrakar G, Thakur SS (2023) Nanotechnology's application in cosmetics: dermatology and skin care items. Migrat Lett 20(S13):1–18. https://doi.org/10.59670/ml.v20iS13.6262

Tanner RS (2007) Cultivation of bacteria and fungi. In: Manual of environmental microbiology. Wiley, pp 69–78. https://doi.org/10.1128/9781555815882.ch6

Tengler M, Hamanova J, Novotny T, Simona P, Holomek J, Kovacova V, Oliva F, Travnickova J, Basic D, Dibusz K (2020) Understanding public perception of nanomaterials and their safety in the EU final report. EU Publications. https://doi.org/10.2823/82474

Touitou E, Dayan N, Bergelson L, Godin B, Eliaz M (2000) Ethosomes—novel vesicular carriers for enhanced delivery: characterization and skin penetration properties. J Control Release 65(3):403–418. https://doi.org/10.1016/S0168-3659(99)00222-9

Trepa M, Sułkowska-Ziaja K, Kała K, Muszyńska B (2024) Therapeutic potential of fungal terpenes and terpenoids: application in skin diseases. Molecules 29(5):1183. https://doi.org/10.3390/molecules29051183

Usman M, Murtaza G, Ditta A (2021) Nutritional, medicinal, and cosmetic value of bioactive compounds in button mushroom (*Agaricus bisporus*): a review. Appl Sci 11(13):5943. https://doi.org/10.3390/app11135943

Ustymenko R (2023) Trends and innovations in cosmetic marketing. Econ Educ 8(3):12–17. https://doi.org/10.30525/2500-946X/2023-3-2

Van Tran V, Moon J-Y, Lee Y-C (2019) Liposomes for delivery of antioxidants in cosmeceuticals: challenges and development strategies. J Control Release 300: 114–140. https://doi.org/10.1016/j.jconrel.2019.03.003

Verhoeff AP, Burge HA (1997) Health risk assessment of fungi in home environments. Ann Allergy Asthma Immunol 78(6):544–556. https://doi.org/10.1016/S1081-1206(10)63214-0

Verma S, Utreja P (2019) Vesicular nanocarrier based treatment of skin fungal infections: potential and emerging trends in nanoscale pharmacotherapy. Asian J Pharm Sci 14(2):117–129. https://doi.org/10.1016/j.ajps.2018.05.007

Virmani R, Pathak K (2021) Consumer nanoproducts for cosmetics. In: Handbook of consumer nanoproducts.

Springer Singapore, pp 1–32. https://doi.org/10.1007/978-981-15-6453-6_58-1

Vishwanatha K, Appurao A, Singh S (2009) Characterisation of acid protease expressed from *Aspergillus oryzae* MTCC 5341. Food Chem 114(2): 402–407. https://doi.org/10.1016/j.foodchem.2008.09.070

Vitale GA, Coppola D, Palma Esposito F, Buonocore C, Ausuri J, Tortorella E, de Pascale D (2020) Antioxidant molecules from marine fungi: methodologies and perspectives. Antioxidants 9(12):1183. https://doi.org/10.3390/antiox9121183

Wang L, Wu Q, Zhao J, Lan X, Yao K, Jia D (2021) Physicochemical and rheological properties of crude polysaccharides extracted from *Tremella fuciformis* with different methods. CyTA J Food 19(1):247–256. https://doi.org/10.1080/19476337.2021.1884607

Wissing S (2003) Cosmetic applications for solid lipid nanoparticles (SLN). Int J Pharm 254(1):65–68. https://doi.org/10.1016/S0378-5173(02)00684-1

Woźniak B, Chudzińska J, Szczyglewska P, Nowak I, Feliczak-Guzik A (2023) Optimization of the composition of a cosmetic formulation containing *Tremella fuciformis* extract (Fungi). Cosmetics 10(3):82. https://doi.org/10.3390/cosmetics10030082

Wu X, Guy RH (2009) Applications of nanoparticles in topical drug delivery and in cosmetics. J Drug Deliv Sci Technol 19(6):371–384. https://doi.org/10.1016/S1773-2247(09)50080-9

Wu K-C, Huang Z-L, Yang Y-M, Chang C-H, Chou T-H (2007) Enhancement of catansome formation by means of cosolvent effect: semi-spontaneous preparation method. Colloids Surf A Physicochem Eng Asp 302(1–3):599–607. https://doi.org/10.1016/j.colsurfa.2007.03.049

Wu Y, Choi M-H, Li J, Yang H, Shin H-J (2016) Mushroom cosmetics: the present and future. Cosmetics 3(3):22. https://doi.org/10.3390/cosmetics3030022

Xu X, Chen A, Ge X, Li S, Zhang T, Xu H (2020) Chain conformation and physicochemical properties of polysaccharide (glucuronoxylomannan) from fruit bodies of *Tremella fuciformis*. Carbohydr Polym 245: 116354. https://doi.org/10.1016/j.carbpol.2020.116354

Yadwade R, Gharpure S, Ankamwar B (2021) Nanotechnology in cosmetics pros and cons. Nano Express 2(2): 022003. https://doi.org/10.1088/2632-959X/abf46b

Yalavarthi LP, Jonnadula PK, Manthina MV, Addagalla A (2020) Cubosomes: a novel drug delivery system overview. Int J Ayurveda Pharma Res 11(5):198–204. https://doi.org/10.7897/2277-4343.1105169

Yücel Ç, Şeker Karatoprak G, Değim İT (2019) Antiaging formulation of rosmarinic acid-loaded ethosomes and liposomes. J Microencapsul 36(2): 180–191. https://doi.org/10.1080/02652048.2019.1617363

Zhang X, Dong F (2020) Why do consumers make green purchase decisions? Insights from a systematic review. Int J Environ Res Public Health 17(18):6607. https://doi.org/10.3390/IJERPH17186607

Zhang ZY, Jin B, Kelly JM (2007) Production of lactic acid from renewable materials by Rhizopus fungi. Biochem Eng J 35(3):251–263. https://doi.org/10.1016/j.bej.2007.01.028

Zhang T, Guo Q, Xin Y, Liu Y (2022) Comprehensive review in moisture retention mechanism of polysaccharides from algae, plants, bacteria and fungus. Arab J Chem 15(10):104163. https://doi.org/10.1016/j.arabjc.2022.104163

Ziemlewska A, Wójciak M, Mroziak-Lal K, Zagórska-Dziok M, Bujak T, Nizioł-Łukaszewska Z, Szczepanek D, Sowa I (2022) Assessment of cosmetic properties and safety of use of model washing gels with Reishi, Maitake and Lion's Mane extracts. Molecules 27(16):5090. https://doi.org/10.3390/MOLECULES27165090

Zillich OV, Schweiggert-Weisz U, Eisner P, Kerscher M (2015) Polyphenols as active ingredients for cosmetic products. Int J Cosmet Sci 37(5):455–464. https://doi.org/10.1111/ics.12218

Fungal Food Supplements and Nutraceuticals for a Healthier Life: Techniques and Application

19

Yashaswi Dutta Gupta, Parijat Roy, Antarika Chowdhury, Sagar Pramanik, and Suman Bhandary

Abstract

Nutraceuticals and food supplements derived from fungi have attracted a lot of interest nowadays due to their copious uses and possible benefits for health. This chapter summarizes techniques for producing and extracting valuable compounds from fungi and highlights their role in improving human health. Filamentous molds and mushrooms have a rich history in traditional medicine and cuisine. In modern times, advancements in mycology and biotechnology have opened the pathway for their bioactive compounds to be explored for the promotion of health-related purposes. Techniques for cultivating, fermenting, and extracting these compounds have evolved to ensure their safety, efficacy, and sustainability. This chapter discusses the diverse applications of fungal food supplements and nutraceuticals, including their potential in immune system modulation, anti-inflammatory effects, antioxidant properties, and cholesterol regulation. Fungi serve as an important source of bioactive substances, demonstrating potential in anticancer, antimicrobial, and antidiabetic applications. Furthermore, the potential of fungal extracts in addressing lifestyle-related disorders, such as obesity and metabolic syndrome, is explored. These applications extend beyond oral consumption to include the incorporation of fungal ingredients in functional foods, dietary supplements, and pharmaceutical formulations.

Keywords

Fungi · Nutraceuticals · Food supplements · Probiotics · Phytochemicals · Bioprocessing

19.1 Introduction

Food supplements and nutraceuticals are essential alternatives as they provide the most important nutrients that might be lacking in our regular diet, ensuring that our bodies obtain the appropriate amount of vitamins and minerals needed for maintaining a normal and healthy well-being (Domínguez Díaz et al. 2019). They are often taken over natural food due to their convenience, ability to fill nutritional gaps, and precise dosage, which can be more necessarily benefitted by individuals with a more specific dietary requirement or restriction (Dwyer et al. 2018; Espín et al. 2007). Dietary supplements like vitamins, minerals, probiotics, herbs, and nutraceuticals help address nutritional deficiencies (Siddiqui and Moghadasian 2020). For instance, the inclusion of probiotics as food supplements in diet, which have minimal pathogenicity, elevates

Y. D. Gupta · P. Roy · A. Chowdhury · S. Pramanik · S. Bhandary (✉)
Department of Biological Sciences, School of Life Science and Biotechnology, Adamas University, Kolkata, West Bengal, India

S. Ghosh, H. N. Onyeaka (eds.), *Fungal-Based Cosmetics: Formulation and Usage*, The Mycota 17,
https://doi.org/10.1007/978-3-032-09291-5_19

innate basic nutrition (Oyetayo and Oyetayo 2005). Nutraceuticals encompass those compounds which along with serving as a dietary supplement also prevent diseases. The naturally found components such as Vitamin D, lycopene, and selenium are common examples of nutraceuticals (Rajasekaran and Xavier 2008).

These food supplements and nutraceuticals provide an array of alternatives to lead a healthier lifestyle; however, there are certain challenges presented by these non-fungal sources. One major challenge is the identification of the authenticity of the raw material (Raclariu et al. 2018). For instance, the herb ginseng has many varieties. While all these varieties are marketed as ginseng, neither of them belongs to the real ginseng genus (Siddiqui and Moghadasian 2020). The next challenge includes compound purity and presence of various active components (Lockwood 2010). For instance, certain body building or weight loss supplements may possess illegal components that are harmful for the body. Supplements that possess contaminations such as microbes, heavy metals, toxins, etc. are largely detrimental to the normal functioning of the body (Costa et al. 2019). Other challenges may include illegal adulteration of supplements with drugs, poor experimental proof, and low-quality supplements (Wierzejska 2021).

Despite their prevalent usage, limitations in their bioavailability, environmental impact, and effectiveness, these food supplements and nutraceuticals are often subjects of scrutiny. Conversely, the convenient bioavailability, high nutritional profile, and the presence of novel bioactive components make fungal food supplements more advantageous than non-fungal food supplements (Sanchez and Demain 2017; Xu et al. 2011). A higher sustainability of food supplements of fungal-origin reduces ecological footprint allied with supplement production (Anna et al. 2023). Other advantages include species variety, low allergenic, and elevated medicinal properties (Stoica et al. 2023). In addition to having sustainability and novel bioactive compounds such as polyphenols, terpenoids, etc., fungal-derived nutraceuticals possess immunomodulatory, anti-inflammatory, and antioxidant properties that help reduce the risk of infections and fatal diseases (Sanchez and Demain 2017). In this aspect, *Grifola*, *Pisolithus*, *Agaricus*, *Cordyceps*, and *Pleurotus* are prominent examples of certain edible mushrooms that possess significant nutraceutical properties (Rahi and Malik 2016).

This chapter highlights the need for a paradigm shift toward fungal food supplements and nutraceuticals. This chapter also aims to discuss the diverse techniques and applications to explore and harness the properties of these fungal compounds. Overall, we discuss the potential of fungal food supplements and nutraceuticals as pivotal components encompassing a myriad of benefits to promote a healthier lifestyle.

19.2 Understanding Fungal Food Supplements and Nutraceuticals

Dietary supplements, also known as food supplements, are products recommended to be consumed alongside a regular diet to address deficiencies in essential nutrients, including vitamins and minerals. The utilization of food supplements traces back to ancient times where the Egyptians used some form of herbs and used certain techniques to extract the medicinal compounds (Dias et al. 2012). Over time the use of these dietary supplements increased, and various research are ongoing in this field on how to improve, find different sources, and how to ease the extraction methods from them. Recently it has been observed a striking interest among the scientists to consider or use of fungi in the isolation of various bioactive compounds, use them to produce, pharmaceuticals, food supplements, nutraceuticals, cosmetic products, etc. (Hyde et al. 2019). The use of fungi such as mushrooms or other species as components of food supplements can be traced back to centuries where Chinese civilization and Indian Ayurveda have been known to use them in many medicinal purposes such as use of Chaga (*Inonotus obliquus*) is very popular in Ayurveda. Similarly, the use of Reishi (*Ganoderma lucidum*) or

Shiitake (*Lentinula edodes*) in Chinese civilization is used as a support medicine to improve health (Vetvicka et al. 2021). Simultaneously this may be viewed as a valuable source of nutraceuticals where, apart from the nutritional value, it can give some therapeutic properties as well. Fungal nutraceutical or myco-nutraceutical is a group of products obtained from fungi. These contain various bioactive components which could impart many health benefits, such as acting as an antimicrobial agent, immunomodulatory agent, improving cardiovascular health, etc. (Al-Obaidi et al. 2021). Along with fungal food supplements these are a very important group of compounds which when administered have a positive impact on human health with very little or no side effects at all.

Fungal food supplements or nutraceuticals can be broadly classified into 4 categories—mushroom-based, yeast-based, mycelium-based, and single fungus or combination formulas. Medicinal mushrooms exhibit a diverse array of pharmacological properties like anti-allergic, anti-inflammatory, antibacterial, digestive, and neuroprotective functions (Venturella et al. 2021). These activities are the result of many bioactive compounds present in the mycelium, fruiting bodies, etc. Regarding yeast-based supplements, they can provide a good amount of vitamin B complex, which help in the catabolism of the ingested food and provide energy and improve blood circulation (Jach et al. 2015). Mycelium contains various bioactive compounds like proteins, vitamins, minerals, and certain phytochemicals (Rathore et al. 2019). They exhibit certain antioxidant property, immune regulation, and can help in controlling some physiological problems (Ren et al. 2021).

Nutraceuticals also come under these subheadings, but a worth mention would be *Cordyceps*. They produce cordycepin, a bioactive molecule which is believed to work against some serious health problems like cancer, cardiovascular diseases, and diabetes (Ashraf et al. 2020). Table 19.1 presents numerous bioactive molecules synthesized by fungi that play crucial roles.

Although fungal food supplements and nutraceuticals may have common origin, they differ in some respects like food supplements are just a dietary supplement that completes the incomplete nutrition of a person; on the other hand, nutraceuticals go one more step that besides providing nutrition they also help to treat certain health issues. Nutraceuticals are given in a metered dose, and consultation to an expert before taking is advisable. Another difference is that the food supplements are produced in capsules, tablets, syrup form, and specificity that do not matter regarding in which product form to take, but for nutraceuticals specific formulation designs are required for different bioactive compounds for their efficient delivery (Aronson and Aronson 2017).

Mushrooms are a great dietary supplement since they are full of minerals, protein, fiber, and all essential amino acids (Kumar et al. 2022). It also has several bioactive components which have different health benefits, which upon extraction can be used as a nutraceutical. When used in conjunction with chemotherapy some bioactive polysaccharides like lentinan, krestin, schizophyllan from medicinal mushrooms show antitumor activity (Wasser 2002). One important polysaccharide, β-glucan that is found in fungal cell wall has certain immunomodulatory effects (Chan et al. 2009). It is also used in lowering blood cholesterol, used as an antioxidant, and even has neuroprotective activities.

Mushrooms or some specific fungi produce certain proteins which are also of pharmacologic importance. Lectins have antimicrobial properties, ribosome inactivating proteins inhibit the HIV reverse transcriptase activity, laccases are proven to be inhibiting HCV replication, pleurostrin is thought to have antifungal effects (Chu et al. 2005; EL-Fakharany et al. 2010; Ngai and Ng 2003). Certain lipids like ergosterol have antioxidant properties, and a diet which has balanced amount of sterol can help prevent several heart-related diseases (He et al. 2019). Secondary metabolites obtained from fermentation of fungal colonies show the presence of phenolic compounds which have antiatherogenic,

Table 19.1 Different species of mushrooms along with their bioactive compounds and their beneficial roles

Name	Ingredients	Uses	References
Agaricus bisporus	Flavonoids, beta-glucans, ergosterol, ergothioneine, vitamin D	Immune control, glycemic control, anticancer, improves gut health, hypocholesterolemic effects, etc.	Blumfield et al. (2020)
Agaricus blazei	Phenolic compounds, lipophilic molecules such as agarol, beta-glucans	Immunomodulation, cell signaling, antimicrobial activity, anticancer, hepatoprotective effects, antidiabetic, etc.	da Silva de Souza et al. (2017)
Grifolafrondosa	Beta-glucans, heteroglycans, proteins, glycoproteins, sterols, phenolic compounds, vitamin D2	Antitumor, immunomodulation, antimicrobial, antidiabetic, anti-hypertensive, anti-hyperlipidemic, etc.	Wu et al. (2021)
Pleurotusobstreatus	Terpenoids, phenols, alkaloids, steroids, lectins, and some minerals like Fe, Cu, K, Mg, etc.	Improves bone health, maintains water balance, antioxidant, cardiovascular protector, anticholesterolemic, antifungal, antitumor, etc.	Sen et al. (2021)
Ganoderma lucidum	Tetracyclic triterpenes from ganoderic acid, ergosterol, palmitic acids, flavonoids, lignans, ergosterol peroxide, polysaccharides	Anticancer, antioxidant, supports immune system, cosmeceutical property, antimicrobial, etc.	Martínez-Montemayor et al. (2019) and Taofiq et al. (2017)
Lentinula edodes	Polysaccharides, terpenoids, phenols, nucleotides, glycoprotein derivatives	Anti-hypertensive, antidiabetic, reduces hypercholesterolemia, anticancer, antioxidant, etc.	Nam et al. (2021)
Trametes versicolor	Phenolic compounds, amino acids (Leu, Ile, Met, Tyr, Gln, and Asn), nicotinic acid, linoleic acid, palmitic acid, linolenic acid. Stearic acid	Antioxidant, immunomodulating, anticancer, helps in osteoarthritis, inflammatory bowel disease, induction of analgesia, diabetic cardiomyopathy, antimicrobial, etc.	Habtemariam (2020) and Kɂvrak et al. (2020)
Hericium erinaceous	Flavonoids, phenolics	Neuroprotective, anti-obesity, antioxidant, anti-inflammatory, antimicrobial, etc.	Khan et al. (2013)
Antrodiacinnamomea	Triterpenoids, ubiquinone derivatives, maleic and succinic acid derivatives, benzene derivatives	Hepatoprotection, hypoglycemic, antioxidant, antitumor, anticancer, antioxidant, immunomodulation, gut protection	Li et al. (2022)
Cordyceps militaris	Cordycepin, polysaccharides, gamma-aminobutyric acid, ergothioneine, cerebrosides, lectins, cordycepic acids, carotenoids, ergosterol, vitamins, and minerals	Regulates sleep, memory, learning process, reduces stress, myorelaxant, hepatoprotective, immunomodulation, antioxidant, neuroprotective, improves vision, reduces hypercholesterolemia, reduces development of coronary atherosclerosis, digestive support, protects against oxidative cell stress, etc.	Jędrejko et al. (2021)
Inonotus obliquus	Triterpenoids, polyphenolic compounds, lanosterol, inotodiol, flavonols, beta-glucan	Hepatoprotective, hypoglycemic, anti-inflammatory, antiviral, antioxidant, antitumor, promotes lipid metabolism, modulates cardiac function, etc.	Drenkhan et al. (2022)
Flammulinavelutipes	Polysaccharides, terpenoids, phenolic acids, flavonoids	Antitumor, immunomodulation, antimicrobial, antioxidant, kidney protection, liver protection, neuroprotection, regulation of intestinal flora, improving constipation	Ding et al. (2021)

(continued)

Table 19.1 (continued)

Name	Ingredients	Uses	References
Pleurotus eryngii	Polysaccharides, proteins, cerebrosides, isoflavones, catechols, amines, triacylglycerol, sesquiterpenes, steroids, eryngiolide A, ubiquinone 9, triterpenoids	Anticancer, antiviral, antioxidant, hypolipidemic, immunomodulating, estrogen-like activity	Stajić et al. (2009)
Formitopsis officinalis	Triterpenoids, polysaccharides, phenols, lectins, 3-oxo-dehydrosulfurenic acid, dehydroeburicoic acid, formitopsin C, laricinolic acid, mannofucogalactan	Immunomodulatory, antioxidant, antimicrobial, reduces cholesterol levels, anticancer, anti-inflammatory, neuroprotective, cytotoxic, used in case of rheumatism, excessive night sweating during tuberculosis, etc.	Muszyńska et al. (2020)
Schizophyllum commune	Beta-glucans, tocopherol, hydroxybenzoic acid, hemicellulases, pectinases, schizophyllan	Antitumor, anticancer, antioxidant, immunomodulation, digestion, etc.	Basso et al. (2020)

antithrombotic, and vasodilating functions (Bhambri et al. 2022). Brazilian sun mushroom (*Agaricus blazei*), a worth mention here is believed to be exhibiting antimutagenic and anticarcinogenic properties (Huang et al. 2022). A protein from *Pichia anomala* has also demonstrated inhibitory effects against Influenza A and HIV type 1 virus. A toxin produced by yeast, *P. kudriavzevii*, has shown high inhibitory effects against several disease-causing bacteria in humans (Tadioto et al. 2023). Yeast also produces certain lipids that are important in terms of nutrition like linoleic acid, eicosapentaenoic acid, etc.

Yeast, *Saccharomyces boulardii*, can adapt to high acidic conditions of the stomach and can inhibit pathogenic *Escherichia coli* forms (Pais et al. 2020). *Aspergillus ochraceous* which is a principal toxin producer in food can be controlled by using a volatile bioactive compound, 2-phenylethanol produced by *P. anomala*. *Taxomyces andreanne* yielded taxol upon cultivation in a semi-synthetic liquid medium, which can also fight against cancerous cells (Stierle et al. 1993). Camptothecin is yet another cancer-inhibiting compound originating from the fungus *Entrophosphora infrequens* (Kousar et al. 2022). *Aspergillus fumigatus* while growing produces certain secondary metabolites, one of them is fumagillin which is showing anti-angiogenesis properties (Guruceaga et al. 2019). Certain mycotoxins like ergot alkaloids, which were earlier considered as harmful, but certain studies claim that they can be used in the treatment of angina pectoris, serotonin-related disorder, excessive bleeding after childbirth, etc. (Hodgson 2012). About 20% of biopharmaceuticals come from a single fungus, *S. cerevisiae*. Fungi which were ignored earlier are now becoming an important tool in the modern world to tackle various health problems and related disorders which previously was unachievable or if available, it was so costly that it was out of common man's reach.

19.3 Fungi as a Source of Nutritional Value

Fungi have a great reservoir of nutrients within itself and for that it is preferred as a modern-day food supplement, nutraceutical, biopharmaceutical, etc. Fungal cells themselves are a rich source of varied nutrients which mycophagous animals vastly feed on them for their basic needs of protein, fiber, sugar, fats, minerals, and salts (Elliott et al. 2022). Broadly categorizing the essential components in the fungi, we get polysaccharides, proteins, lipids, polyphenols, terpenoids, etc. (Assemie and Abaya 2022). Polysaccharides share the maximum percentage of dry weight among these. The β-glucans form the most important polysaccharide in the fungi which are useful to humans for developing different types of nutraceuticals and supplements. The major difference between products depends upon these β-glucans' linkage, side chains of monomers, and this difference has led to the production of various bioactive compounds which come under β-glucans family like lentinan, schizophyllan, pleuran, etc. (Wasser 2002). Each bioactive compound has a specific mechanism of action, but, as a whole, it functions to provide antimicrobial, antioxidant, antitumor, prevents cytotoxicity, maintains immune system by activating innate immune system.

One crucial aspect of β-glucans is that their healing properties are improved as their molecular weight increases (Murphy et al. 2020). A research study asserted that β-glucans with a larger molecular weight considerably suppress tumor cell growth, downregulate angiogenesis factors, and induce cell death without negatively impacting normal cells (Ping et al. 2016). β-glucans bind to specific receptors present in the immune cells leading to activation of those and as a result various signaling pathways are upregulated (Lee and Kim 2014). Other immune factors like TNF-α, interleukins, chemokines, and cytokines are released and hence the humoral response and cell-mediated immune response get enhanced. β-glucans from *Pleurotus ostreatus* showed a significant healing effect on children suffering from recurrent respiratory tract infections (Jesenak et al. 2014). The levels of the anti-inflammatory interleukin IL-10 were shown to be increased by β-1, 3-glucans, according to another study (Sarinho et al. 2009). Upon administration of bioactive compound lentinan there was a reduction in immunoglobulin E serum levels (Yamada et al. 2007). A laboratory study demonstrated the clearance of antibiotic-resistant *Staphylococcus aureus* upon exposure to PGG-glucan; along with this the researchers also found an increase in total leucocyte count and oxidative burst activity of neutrophils (Liang et al. 1998). A placebo trial investigating the activity of the antiviral drug didanosine found an upregulation when it was administered with lentinan as a suggested combinatorial therapy (Gordon et al. 1998).

The brewer's yeast is a potent source of B vitamins (Ferreira et al. 2010). Vitamin B complex has some major roles in the body. Cofactors having thiamine origin have been proven to be an essential part in the enzymes of the different metabolic pathways and can even control immunometabolism (Mathis and Shoelson 2011). It also inhibits NF-kB activation due to oxidative stress and prevents discharge of pro-inflammatory cytokines in macrophages (Yadav et al. 2010). On the cell surface, it protects the sulfhydryl groups indirectly by having an antioxidant effect on neutrophils (Jones and Anderson 1983). *Ashbya gossypii* is an ascomycete which is largely used in the production of riboflavin (Aguiar et al. 2015). Riboflavin or vitamin B2 has a very important role in maintaining the NADPH oxidase's phagocytic activity during an infection and helps in generation of superoxide ions (Hashida et al. 2004). It diminishes the expression of TNF-α and nitric oxide by interfering with NF-kB activation; this is achieved upon inhibition of LPS-induced ROS (Qureshi et al. 2011). A research study involving riboflavin demonstrated its ability to arrest melanoma metastasis to lungs (Machado et al. 2013). The risk of lung cancer in females who do not smoke is significantly reduced when they consume riboflavin on a regular basis (Takata et al. 2012). *Grifola frondosa* (maitake mushroom) and *Agaricus bisporus* (portabella mushroom) are excellent sources for vitamin B3 or niacin. Niacin is proposed to be used for patients with fatty liver who do not consume alcohol (Kashyap et al. 2019). Frequent consumption of large doses of niacin has been shown to lower the incidence of hepatocellular cancer (Antwi et al. 2020). Pantothenic acid or vitamin B5, derived from *Lentinula edodes* (Shiitake mushroom), plays a decisive role in the production of coenzyme A. A sodium-dependent multivitamin transporter facilitates pantothenic acid absorption. It helps in the generation of cysteamine which controls the inflammation (Nitto and Onodera 2013). Oyster mushroom (*Pleurotus ostreatus*) can be used as a source of vitamin B7 or biotin. It has several roles in regulating fatty acid metabolism, gluconeogenesis, and lipogenesis. In case of neurodegenerative symptoms, biotin can restore the functionality of mitochondria, and inhibit the release of oxygen free radicals and cellular apoptosis (Sghaier et al. 2019). It also functions in regulating the immunologic effects as the deficiency of biotin is correlated with the expression of NF-κB and SP1/3 (Kuroishi 2015). *Craterellus cornucopioides* (Black trumpet mushroom) and *Cantharellus cibarius* (Golden chanterelles) have high levels of vitamin B12. Its primary function is in the development of myelin sheath and formation of red blood cells and also DNA synthesis. Vitamin B12 also functions in regulating the expression of natural killer cells, CD8+ T-cells, and other immune cells (Tamura et al. 1999).

19.4 Types of Fungal Food Supplements and Nutraceuticals

Vitamin C is an important bioactive compound which is obtained from fungi possessing antioxidant and anti-inflammatory properties. Vitamin C helps in the promotion of collagen synthesis, aids in proliferation, and also helps in migration of the fibroblasts which are necessary for a stabilized epithelial barrier and also greatly improves the adaptive immunity (Jafari et al. 2019). Vitamin C aids in dissemination and segregation of B-cells as well as T-cells and thus increases antibody production (Medoro et al. 2023). Vitamin C in increased concentrations shows greater immunomodulatory properties. The fact that vitamin C is water-soluble enhances its advantages since it can be easily excreted and hence intoxication of vitamin C can never really occur. Deficiency of vitamin C also results in the reduction of humoral and cellular immune responses (Carr and Maggini 2017). Vitamin C also interferes immune cells including dendritic cells and also leads to increased secretion of IL-12 (Mousavi et al. 2019). Another function of vitamin C includes the inhibition of LPS-induced production of proinflammatory cytokines IL-6 and TNF-alpha. The suppression of gene expression induced by LPS in macrophages through NF-κB can be done by vitamin C (Medoro et al. 2023). According to research studies, it was observed that lowered vitamin C concentrations in the mucosa contributed to the onset of Crohn's disease and ulcerative colitis (Mousavi et al. 2019).

Flavonoids derive their importance from the fact that they can function as effective nutraceuticals. Their therapeutic properties include antitumor, antibacterial, antioxidant, anti-inflammatory, cardiovascular protection, etc. Endophytic fungi such as *Ceriporialacerata* can produce flavonoids with anticancer effect (Tang et al. 2020). Another flavonoid that can show anticancer activity is hesperidin. Flavonoids when acting as pro-oxidants may reduce cancer cell proliferation by blocking various signaling pathways such as phosphatidylinositide-3 kinases, epidermal growth factor receptor/ MAPK, NF κ-light-chain-enhancer of B-cells which are active (Kopustinskiene et al. 2020). There are various factors on which the antioxidant properties of flavonoids depend such as the molecular structure, the resonance and conjugation effects, the specific antioxidant mechanism of the compound, etc. (McPhail et al. 2003). Flavonoids which show antioxidant properties and are obtained from fungi include hesperidin, hesperidin, etc. (Wilmsen et al. 2005). Apart from having antioxidant properties they also exhibit anti-microbial, anti-inflammatory, and anticancer effects (Sridevi Sangeetha et al. 2016). Flavonoids can also be used for treating various neurodegenerative diseases such as Alzheimer's disease, dementia, and Parkinson's disease (Ullah et al. 2020).

Fungal terpenes are categorized into four main groups: monoterpenoids, diterpenoids, sesquiterpenoids, and triterpenoids (Amirzakariya and Shakeri 2022). These compounds demonstrate diverse therapeutic properties, such as antioxidant, anticancer, antiviral, antimalarial, anti-inflammatory, and anticholinesterase effects. For example, flammulinolides and flammuninol, two terpenes isolated from *Flammulina velutipes*, have shown cytotoxic effects against tumor cell lines such as KB, HeLa, and HepG2 (Ma et al. 2018). Additionally, terpenes derived from *Pleurotus* species are recognized for their anti-inflammatory properties (Wu et al. 2016). Terpenes like n-butylganoderate H and methyl ganoderate A acetonide, obtained from *Ganoderma lucidum*, exhibit anti-acetylcholinesterase activity, making them promising candidates for treating neurodegenerative diseases (Ma et al. 2018).

19.5 Techniques for Producing Fungal Food Supplements and Nutraceuticals

19.5.1 Extraction and Processing Techniques

Extraction techniques have been in operation for ages and are evolving in accordance with the advancements of the modern world. Initially, herbal medicines were produced using simple extraction methods such as decoctions and infusions and to extract the bioactive compounds, the fungal material was soaked in boiled water which was then consumed either in the form of tea or any other medicinal preparations (Alamgir 2017). Steeping of the fungal component in either alcohol or water is yet another way to extract the bioactive compound (Sopalun et al. 2021). With time, as science developed and demands increased these methods became obsolete and a search for better, faster, more efficient processes was needed. In this regard there was development of solid-liquid extraction, liquid-liquid extraction, etc. As the population continues to grow and resources become scarcer, more advanced methods like Solid Phase Extraction (SPE), Supercritical Fluid Extraction (SFE), and Pressurized Liquid Extraction (PLE) have been developed. These techniques have proven essential in maintaining quality standards and adhering to safety requirements (Vazquez-Roig and Picó 2015).

19.5.1.1 Solid-Phase Extraction (SPE)

The extraction procedure involves usage of solid-phase sorbents which can selectively interact with the target bioactive compounds from a liquid sample including fungi (Fontanals et al. 2019). Kulal et al. perfectly explained the principle of solid phase extraction in which they state, exposing the liquid sample to the sorbent or solid phase leads to selective interaction with the compound of interest. The solid phase is typically contained within cartridges or small tubes. After the sorbent interacts with the specific compound of interest an appropriate solvent is passed to wash away the leftover compounds that did not interact with the sorbent. The compound of interest can then be eluted by the solid phase using an appropriate solvent. The solvent is then analyzed (Kulal et al. 2020).

Sorbents can be categorized into three types: Normal Phase sorbents, Reverse Phase sorbents, and Ion Exchange sorbents (Thurman 2000). For SPE, normal-phase columns use polar substances like florisil, silica, or alumina, while reversed-phase columns make use of non-polar bonded polymers or silica-functionalized materials (Lehotay and Schenck 2000). One of the most frequently used SPE columns consists of nonpolar octadecylsilyl-derivatized silica (C18). SPE has better selectivity and versatility than liquid-liquid extraction. Its other advantages are that it is rapid, requires lesser amount of solvent, smaller sample volumes, prevents the formation of emulsions, and its efficiency is further enhanced by automated SPE systems (Lehotay and Schenck 2000).

The book Solid Phase Extraction has strategically described the general procedure for solid-phase extraction to isolate bioactive compounds and also described the process of solid-phase extraction, which starts from sample preparation (Zwir-Ferenc and Biziuk 2006). This step involves culturing the fungi under optimal conditions and the bioactive compounds are preserved. As discussed earlier, a perfect sorbent is required for proper extraction; selection of sorbent must be done very efficiently. After making the choice the next is conditioning of sorbent. This means that the sorbent must be washed by passing an appropriate solvent to make the functional groups of the sorbent solvate and the packing materials of the sorbent wet. Care must be taken regarding the reconditioning of the sorbent or else occurrence of proper sorption will not take place and the recovery of the bioactive compound will not be proper. Sample loading involves the addition of the fungal sample and sorbent to the column. The interplay between the solid sorbent and the bioactive compound in the sample occurs through Van der Waals interaction, dipole-dipole forces, hydrogen bonding, cation and anion exchange, size exclusion, etc. (Buszewski and Szultka 2012). A strong interaction ensures that the target bioactive compounds are retained along with some matrix compounds. The matrix compounds and various other impurities within the interstitial spaces of the column are removed through the passage of an appropriate solvent retaining the target compound. The solvent also disrupts the interaction between the solid sorbent and the bioactive compound followed by elution.

19.5.1.2 Supercritical Fluid Extraction (SFE)

This extraction method uses an extracting solvent which is called supercritical fluid in order to separate the target bioactive compound from other components (Hedrick et al. 1992). Using this technique bioactive compound can be extracted from solid samples. Some examples of supercritical fluids are ammonia, carbon dioxide, ethane, methanol, water, etc., where carbon dioxide is most commonly utilized as a supercritical fluid.

A substance in a phase exceeding its critical temperature and pressure is described as a supercritical fluid (Hitchen and Dean 1993). Supercritical fluid can either pass through like a gas or dissolve components like a liquid. By changing the temperature and pressure, the properties of supercritical fluid can be changed. The extraction method is mainly based on critical condition, a set of pressure and temperature where the gas and liquid phases of the component are indistinguishable, any compound beyond this critical point is the supercritical fluid (Uwineza and Waśkiewicz 2020).

To separate the bioactive compound, the first step is selection of supercritical fluid; a readily available, non-flammable, non-toxic, and economical fluid is carbon dioxide (CO_2). The production line or instrumentation setup in an industrial setting consists of a modifier cylinder and a CO_2 cylinder. The carbon dioxide cylinder consists of a dip tube through which carbon dioxide is supplied. A jacketed pump head is present for pumping liquefied CO_2 without cavitation. This is usually accomplished using an ethylene glycol mixture. A high-pressure withstanding vessel containing the sample is placed in the oven to obtain the critical temperature. The fungus undergoes several steps before the bioactive compound is obtained from it. In order to extract efficiently, the surface area is increased by drying up the fungal component. Drying also helps in the increased exposure of the supercritical fluid to the fungal components. An optimal temperature and pressure are maintained at which the CO_2 acts as both liquid and gas allowing the penetration of the supercritical CO_2 and solubilizing the bioactive compound of interest. After solubilizing the contents need to be separated and this is done by lowering the pressure. Lowering of pressure converts the supercritical CO_2 into a gaseous form and hence the precipitate is left behind which contains the target bioactive compound (Sánchez-Camargo et al. 2014).

19.5.1.3 Ultrasound-Assisted Extraction (UAE)

Ultrasound-assisted extraction has certain benefits compared to other extraction methods including its low energy consumption and fast processing time. Using this process extraction can be done at low temperature and also the other property that is kept intact is the extract's quality (Chemat et al. 2017). Extraction of bioactive compounds by UAE includes sound waves of high intensity (Fig. 19.1). Due to the application of ultrasound waves the fungal cells are disrupted. During acoustic cavitation, there are physical forces that are developed aiding in the dissolution of bioactive compounds in the solvent where the process is comparatively rapid and enhances the mass transport (Joana Gil-Chávez et al. 2013). Ultrasound technology is widely utilized for extracting bioactive compounds, including polyphenols and polysaccharides, from fungal sources. To properly optimize the process, it is essential to carefully regulate a number of factors

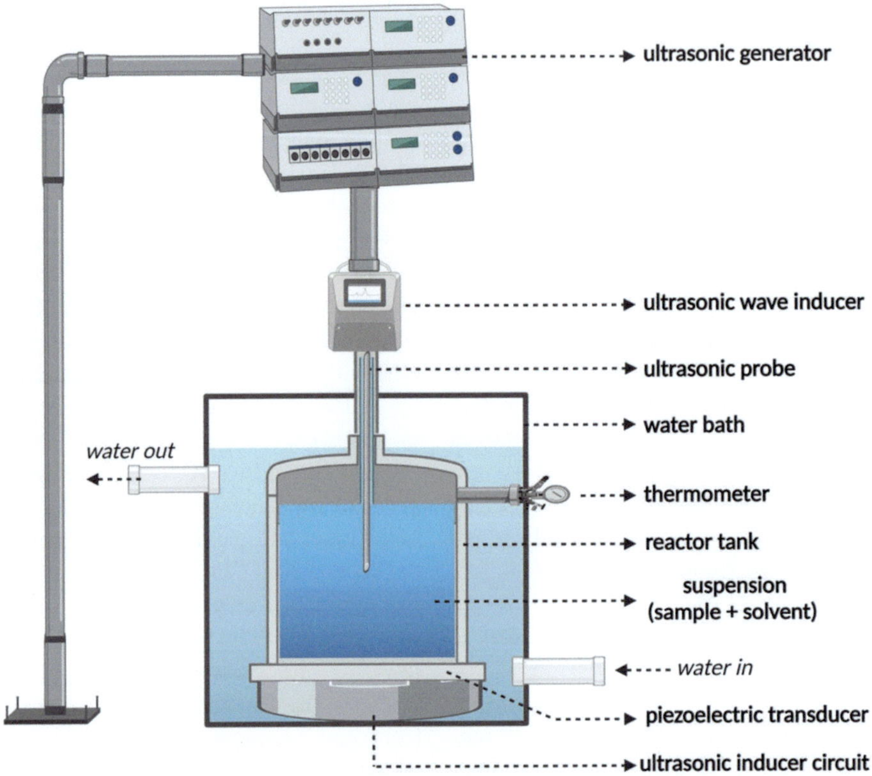

Fig. 19.1 Diagrammatic representation of ultrasound-assisted extraction. (This figure was created with BioRender (https://biorender.com/))

in ultrasonic-assisted extraction (UAE), including power, temperature, pH, frequency, duty cycle, duration, liquid-to-solid ratio, and solvent type (Chemat et al. 2011). The ultrasound waves consist of a series of rarefaction and compression cycles. A series of these waves are penetrated through liquid, gas, or solid medium thus inducing dislodgement and displacement of the original or previous positions. The production of cavitation bubbles is the main step in the process. The cavitation bubbles are created by breaking apart the attractive forces between the molecules which occurs due to the rarefaction and compression cycles. During this process, the temperature and pressure increase to a great extent. Increase in temperature increases the yield of the process by acting on both the solute and the solvent (Kumar et al. 2021).

19.5.1.4 Purification and Concentration Techniques

The target chemicals are purified, identified, and subsequently gathered after the bioactive component of interest has been extracted. The purification techniques aim at isolating and purifying the bioactive compound of interest and removing the unwanted impurities. Various methodologies are employed to isolate and purify bioactive compounds derived from fungi.

19.5.1.5 Chromatography and Fractionation

A range of chromatographic techniques, including thin-layer chromatography (TLC), column chromatography, and high-performance liquid chromatography (HPLC), are widely utilized for the extraction and purification of bioactive substances obtained from fungi (Bele et al. 2011; Kondeti et al. 2014; Majors 1980). These methods separate and purify the bioactive substances according to their specific chemical or physical characteristics, such as size, polarity, affinity, etc.

Fractionation is the process of separating the extract into various fractions. This fractioning of the extract depends upon various properties such as polarity, molecular weight, etc. After fractionation of the extract, it can be further purified using other techniques.

Silica gel column chromatography can be used in the purification of the extracts on silica. Generally in silica gel column the extracts are subjected to chromatography and using a gradient solvent system it can be eluted. Following fractionation of the extract, TLC can be used to analyze the purity of the fractions (Coskun 2016). The fractions having similar R_f values are collected. The characterization of the collected fractions can be done by spectroscopic analysis (Raghava Rao et al. 2017).

Different spectroscopic techniques, including mass spectrometry (MS), nuclear magnetic resonance (NMR), and infrared spectroscopy (IR), are utilized during extract purification to analyze and identify the separated compounds (Yang 2006). With the help of this technique information like chemical structure can also be obtained. To determine the antimicrobial property of the bioactive compound, the antimicrobial compound can be localized on the TLC plate and an agar overlay for bioautographic evaluation (Dewanjee et al. 2015). The antimicrobial compounds may further be subjected to analysis of the compound through GC-MS (Gas Chromatography-Mass Spectrometry) for detailed characterization. For further purification, the active TLC band can be subjected to HPLC (Nishad et al. 2021).

The fractions that are collected after chromatography are concentrated by various techniques: Freeze-drying, known scientifically as lyophilization, is a widely practiced method for post-extraction drying (Gaidhani et al. 2015). During this process, firstly the extracted compound is frozen and then it is subjected to vacuum. The frozen solvent directly sublimes from solid state to the gaseous state leaving behind the concentrated bioactive compound. This process is mostly used for compounds which are heat-sensitive. This process prevents oxidative changes of the metabolites since oxygen is deficient under vacuum (ElNaker et al. 2021). Evaporation entails extracting the solvent from the purified sample that contains the bioactive compound of interest (Pavanetto et al. 1992). Heating the solution under low pressure facilitates a reduction in the solvent's boiling point. This results in the removal of the solvent leaving the concentrated bioactive compound of interest. Rotary evaporators are often used to increase the rate of evaporation. This technique might not always be applicable for heat-sensitive components (ElNaker et al. 2021). Semi-permeable membranes are used in filtration to divide the solvent and bioactive compounds (Martín et al. 2018). The solution is transmitted through the membrane, with the solvent passing through and the compounds being captured. This process can be repeated multiple times to achieve the desired concentration. A semi-permeable membrane is employed in this technique to separate the solvent and bioactive compound based on their size or molecular weight. Ultrafiltration or diafiltration membranes are commonly used for this purpose where the solvent passes through a membrane, leaving behind the concentrated solution of the bioactive compound (Dutré and Trägrdh 1994). Nanofiltration utilizes a semi-permeable membrane with a smaller pore size than ultrafiltration membranes (Abdel-Fatah 2018). They are effective in removing smaller molecules and salts while concentrating the bioactive compounds.

During the processing of fungal supplements and nutraceuticals it is also vital to consider a few other aspects as well. Since the amount of bioactive compound that can be obtained from fungi is usually low, the process of recovery should be very efficient (Joana Gil-Chávez et al. 2013). Another problem can be that many a times the bioactive compound of interest is analogous to other compounds. During separation proper protection strategies must be applied so that instability of the bioactive compound does not affect the processing. These are a few of the many reasons as to why processing must be done in a broad-spectrum manner.

19.5.2 Fermentation and Bioprocessing

Many bioactive compounds that are obtained from fungi are not readily soluble and have complex structures, in order to overcome these problems; the process of fermentation is required. With the aid of microbial metabolism, enzymatic reactions, transformation processes, etc., microbial fermentation can be helpful in altering the structure and characteristics of the bioactive chemical. Through microbial metabolism, enzymatic reactions, and transformation processes, microbial fermentation can enhance or modify the structure and properties of bioactive compounds. Fermentation may benefit from the usage of lactic acid bacteria (LAB). It contributes to the enhancement of nutritional qualities as well as the longevity of the food supplements or nutraceuticals that are obtained from fungi (Samtiya et al. 2021). Microbial fermentation can breakdown complex bioactive compounds into simpler forms (De Villa et al. 2021), hence through the process of fermentation, poorly soluble compounds can be converted to water-soluble forms consequently, the bioavailability of the compound is enhanced, with fermentation products leading to improved absorption of the bioactive supplement in the body (Chen et al. 2020). The minerals obtained from fungi form complexes with non-digestible matter and hence have decreased bioavailability. Through the process of fermentation these complexes can be degraded and hence bioavailability can be increased (Samtiya et al. 2021).

The fungi may contain certain proteins that can act as allergens (Bush and Portnoy 2001). These structures are complex and hence are difficult to degrade. If the food supplement taken contains these types of complexes or allergens, it may cause gastrointestinal problems in the body since these complexes cannot be digested easily (Sun et al. 2022), through fermentation these problems can be easily overcome. According to studies, patulin a mycotoxin synthesized by various fungi like *Penicillium expansum* is considered toxic due to its pronounced reactivity with thiols, which reduces cellular glutathione levels. Fermentation helps in degrading these mycotoxins (Verni et al. 2022). Fermentation has various effects on the nutritional value of the bioactive components. Fermentation leads to increased phenolic compounds, flavonoids, antioxidant activities, etc. (Nkhata et al. 2018).

Microbes secrete a number of enzymes during the fermentation process that can break down complicated molecules, increasing the chemicals' absorption (Chen et al. 2020). Various enzymes including esterase, glycosidases, etc. can lead to increased accessibility of bioactive compounds (Hammed et al. 2013). Known as carboxyl ester hydrolases, esterases assist in the enzymatic hydrolysis and formation of ester bonds (Bulgari et al. 2023). A non-toxic extract with antiangiogenic properties is produced by fermentation and is used as an oral nutraceutical treatment for conditions including age-related macular degeneration. It is marketed under the name Ethnodyne-Visio (Rabhi et al. 2015). Lactobacilli are among the most significant probiotic bacteria, and fermentation with LAB can enhance the content of bioactive compounds, physicochemical characteristics, antioxidant qualities, etc. (Yu et al. 2022).

19.5.2.1 Microbial Fermentation of Fungal Substrates

Through the process of microbial fermentation, the fungal substrates are broken down into fungal biomass which is then converted to several products. Fungi are simple organisms and since they are eukaryotes their biological processes are contingent on the principles, same as that in the higher eukaryotes. Fungi are excellent sources of several compounds such as bioactive components, enzymes, antibiotics, proteins, biofuels, organic acids, etc. (Barrios-González and Tarragó-Castellanos 2017). There are various processes through which microbial fermentation can be done, two such eminent processes are (1) Submerged Fermentation and (2) Solid state Fermentation (Ravichandran 2012).

In case of submerged fermentation, liquid nutrient media is used as substrate to favor the growth of microorganisms (Fazenda et al. 2008). In industrial terms this means that the microbes are grown in closed vessels, termed as bioreactors containing nutrient broths. The bioreactors are designed in a way so that optimal amount of oxygen can be supplied to the aerobic microbes. Also, the bioreactors have the ability to regulate various parameters including temperature, viscosity, pH, substrate utilization, foam formation, desired product formation, etc. (Ouedraogo and Tsang 2021). The fungi used in the submerged fermentation leads to formation of various morphological forms which accordingly affect the process taking place in the bioreactor. These morphological varieties include clumps, pellets, and mycelium which can affect the yield of either upstream processes or downstream processes (Iram et al. 2022). Even the fermentation conditions including mass, rate of transfer of oxygen, etc. vary according to the morphology. The morphology even affects the purification steps. The fungal morphology can also affect the bioactive compound or other metabolites. According to studies, mycelia forms are way more effective fungal microbes to produce enzymes (Adenipekun and Lawal 2012). But choosing an appropriate metabolic form for optimum yield is a challenge. In cases where mycelium forms, it results in the increase in viscosity of the broth resulting in pseudoplastic flow of the broth. Also, certain mass transfer rates are affected negatively. Also, in order to obtain productions in a desired level agitation rate and oxygen input must be regulated well. Sometimes due to formation of pellet, there is increase in the biomass density resulting in greater amount of non-homogeneous pellet structures. The growth of these structures having less porosity continues to increase; lower uptake of nutrient takes place due to the formation of intense hyphal network. Due to this there is lesser amount of substrate and as a result autolysis occurs in the central part of the pellet. Hence in order to obtain optimal products the morphology of the fungi needs to be regulated precisely or else submerged fermentation can become greatly disadvantageous.

One way to avoid this disadvantage even if to a little extent is to add microparticles. Upon addition of microparticle along with filamentous fungi alterations in the cell morphology can be made and hence the productivity can be altered during submerged fermentation. An advantage of this method is that no genetic manipulations of microbes are required. Not only these microparticles can be easily removed by the process of centrifugation or filtration, the microparticles also have no interference with the product. Other advantages of microparticles include enhancing productivity and mass transfer, the results are reproducible and also the overgrowth of fungi is controlled. But the concentration of the microparticle must be precise enough to avoid any adverse effect. Application of various microparticles such as aluminum, nano clay, aluminum oxide, kaolin, talc, etc., can result in better cell growth characteristics such as agitation or aeration rates, nitrogen or carbon concentrations, pH, inoculation ratio, etc. But this method is not efficient enough to control the fungal overgrowth and hence further research is required in this field (Iram et al. 2022).

Certain bacteria are grown on solid, moist, insoluble organic material that acts as a source of nutrients and a physical support in a process known as solid-state fermentation (SSF). This method is characterized by an almost complete water stagnation (Yafetto 2022). A key microbe in the process of solid-state fermentation is fungi. Since fungi can develop hyphal structures, they have the ability to colonize and infiltrate the solid substrate in an enhanced manner (teBiesebeke et al. 2002).

The general steps required in the process of SSF include preparation of inoculum, preparation of solid substrate, sterilization of medium using autoclave, inoculation of microbes, incubation in bioreactors, drying up of the products, followed by extraction, purification, fractionation, and concentration. The efficiency of the process depends upon inoculum type, it should be easy to prepare, resistant, and could be stored for longer times, pH, temperature, and choice of substrate (Gowthaman et al. 2001).

After the process of fermentation is completed, the desired bioactive compound is extracted. The desired compound is then purified and then concentrated. The aim of purification is to purify the desired bioactive compound by removing the impurities from the extracted product.

The purified products are then dried under controlled conditions. Depending on the stability of the desired compound it is then pulverized screened and stored in appropriate containers for further use. The leaching step includes a correct choice of leaching procedure and correct set of solvents, their concentration, physical factors are maintained at an apt so as to make the process easy and economical. Talking of economical processes manufacturers always try to take the simplest approaches and low-cost chemicals. The best and most readily available solvent is water which depending upon the product type can be used to leach out. Some other solvents like salt, glycerin dilute ethanol, or certain buffer solutions can be used as potentially economical solvents. Leaching not only depends upon these factors but also on how much the solute or solvent diffuse, solvent retention, pH, and temperature (Poole and Poole 2010).

This is the last step before the finished product reaches the market after proper labeling including safety regulations. But all these procedures generate large volumes of waste which needs to be safely disposed of. In case of submerged fermentation waste water and in case of SSF extraction and leaching creates waste, which needs to be handled in the same way as in submerged fermentation. The BOD or COD in SSF does not rise much as most of the waste chemicals or matters are present in the spent solid after leaching and this can be treated using simple solid waste treatment techniques (Yazid et al. 2017). After successful treatment, slurry or residue is obtained which can be further sterilized and dried and can be used in various domestic and agricultural purposes.

In terms of enzyme or secondary metabolite production, SSF demonstrates a number of advantages over submerged fermentation, including increased yields. Also, the microbes that grow on the solid substrate in case of solid state fermentation tend to show a different physiology which is not observed in case of submerged fermentation (Barrios-González and Tarragó-Castellanos 2017).

19.6 Health Benefits and Applications of Fungal Food Supplements and Nutraceuticals

Food supplements and nutraceuticals help in stimulating the immune system through different pathways enhancing the immune defenses (Singh et al. 2021). Along with their influences in immune system, food supplements and nutraceuticals also display antitumor, antibacterial, antioxidant, antiviral, and antifungal characteristics. Nutraceuticals have several therapeutic properties which can upset the immune system (Medoro et al. 2023). The utilization of synthetic drugs for disease treatment dates back to ancient times. Numerous antibiotics, antiviral drugs, etc. that are abundant in the market are losing their importance since the patients are developing resistance against the drugs not to mention the side effects that many of these drugs contain (Kunin 1993). This is the reason why biological sources are given more attention to bioactive compounds obtained from fungi is currently an excellent source of interest because of their high nutritional and therapeutic values (Sanchez and Demain 2017). These bioactive compounds contain various nutrients such as protein, fat, carbohydrate, fiber, etc. Bioactive compounds are primarily derived from molecules like polysaccharides, proteins, terpenes, phenolics, and unsaturated fatty acids. These compounds are documented for their effects, including anti-inflammatory, anti-allergic, anticancer, antimicrobial, antidiabetic, and immunomodulatory properties (Chakraborty et al. 2021).

19.6.1 Bioactive Components

19.6.1.1 Polysaccharides

Polysaccharides are among the most frequently extracted bioactive compounds from fungi because of their nutritional and therapeutic importance. They display a diverse array of therapeutic properties such as anti-stimulatory, anti-oxidant, antitumor, properties, etc. (Ślusarczyk et al. 2021). Various polysaccharides are being exploited for their nutritional properties such as lentinan, schizophyllan, pleuran, calocyban, and ganoderan, most of them representing D-glucans (Villares et al. 2012). The polysaccharides not only have nutraceutical properties but are also non-toxic. Typically, the mycelium, fruiting bodies, and fermentation broth of both edible and medicinal species contain polysaccharides generated from fungi (Santa et al. 2014). The actual basis of the antitumor, anticancer, and immunomodulatory properties of the polysaccharides is their structural conformation thereby giving specificity to these properties (Han et al. 2020). The various structural differences in the polysaccharides confer specificity to their functions, for example, variability in their monosaccharide composition. In order to stimulate immune cells including NK cells, cytokines, macrophages, and dendritic cells, polysaccharides derived from fungi interact with cell surface receptors such as toll-like receptors and dectin-1 (Chakraborty et al. 2021). Fungal carbohydrates are recognized by APCs, including macrophages and DCs, along with intestinal epithelial cells, leading to immune cell stimulation (Barreto-Bergter and Figueiredo 2014; McGreal et al. 2005). APCs expressing C-type lectins are specific for galactose, N-acetylgalactosamine, and fucose. When dendritic cells recognize the glycans they get activated and uptake the antigen further stimulating activation of the cells. The immune system is modulated by polysaccharides through the stimulation of T-cells, B-cells, natural killer cells, neutrophils, and macrophages (Santa et al. 2014).

The majority of β-glucans, which are polysaccharides made of D-glucose units, are present in fungal cell walls. They exhibit notable biological properties, including antitumor, antimicrobial, and immunomodulatory effects, which highlight their potential in tumor immunotherapy. While other polysaccharides like cellulose and chitin also play important roles, glucans stand out for their therapeutic potential. D-glucose monomers connected by 1,3-, 1,4-, or 1,6-glycosidic linkages constitute these dietary fibers. The sources of D-glucose monomers are the cause of the differences in β-glucans. The linear backbone of β-glucans in mushrooms is comprised of 1,3-linked glucose units, with branching at the 1,6 connections. The structure, shape, molecular weight, etc., affect their biological activity and specificity (van Steenwijk et al. 2021). The immunomodulatory effect of β-glucan is supported by its binding to certain receptors on immune cells (Santa et al. 2014). During infection, the innate immune system acts as the body's frontline defense, stimulating other immune system components. Innate immune cells help mitigate the infection, paving the way for the adaptive immune response. The pathogens are eliminated through the process of phagocytosis by the innate immune cells. These immune cells identify pathogenic molecules through pattern recognition receptors (PRRs) that bind to pathogen-associated molecular patterns (PAMPs) (Santoni et al. 2015). The β-glucans act as PAMPs thus stimulating the immune system. Since these PAMPs cannot penetrate the cells due to their high molecular weight they require certain immune cell receptors in order for them to bind. Various β-glucan receptors include dectin-1, complement receptor 3, toll-like receptors, scavenger receptors, lactosylceramide, etc. (Chakraborty et al. 2021). β-glucan is connected with the stimulation of immune cells, including T-helper cells, T-lymphocytes, and natural killer cells (Richter et al. 2016; Wasser and Weis 1999). Dectin-1, one of the most common pattern recognition receptors, is highly expressed and recognized by various immune cells including dendritic cells, eosinophils, T-lymphocytes, neutrophils, monocytes, and macrophages (Brown 2005). When β-glucan binds to dectin-1 or complement receptor 3, a series of signaling events is activated, stimulating innate immune cells first, followed by the activation of acquired immune cells. This cascade leads to the generation of proinflammatory cytokines and reactive oxygen and nitrogen species, which combat toxins, carcinogens, and pathogens (Santa et al. 2014) (Fig. 19.2).

Depending upon the conformation, number of branches of β-glucan, and molecular weight, its biological properties vary. Glucose polymers including 1, 3-β-glucans and 1, 6-β-glucans are potent immunomodulatory agents (Thomas et al.

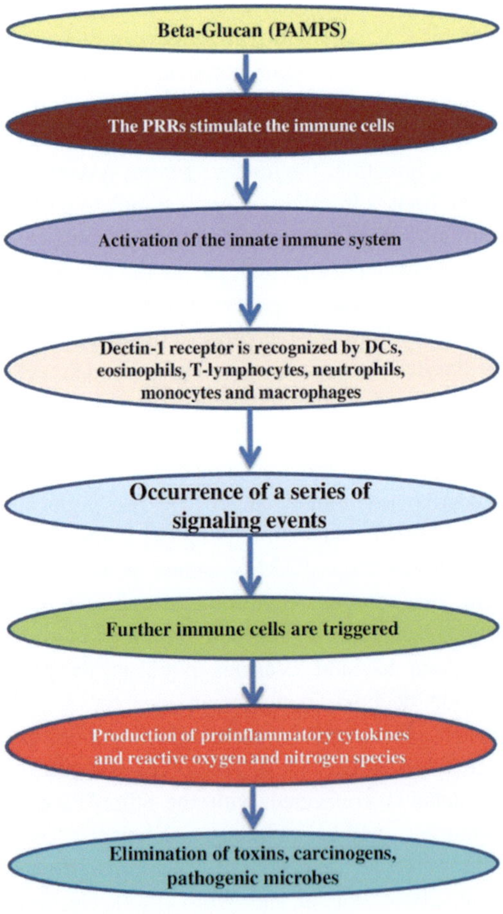

Fig. 19.2 How β-glucans play a role in immunomodulation

2022). The prognosis rate for patients with stomach cancer, particularly individuals with lymph node metastases, can be improved by lentinan because of its potent immunotherapeutic actions (Trivedi et al. 2022). Mycelium extracts from Lentinula edodes resulted in the increased immune activity attributed to an enhanced regulation of antigen-presenting cells (Kajiyama et al. 2023). Additionally, drugs containing lentinan can be used to treat hepatitis, HIV, malignant pleural effusion, lung cancer, colorectal cancer, etc. (Thakur and Singh 2013; Wasser 2005; Zhang et al. 2019).

Beta-glucan when occurs in triple helical conformation has increased antitumor activities. 1, 3-β-D-glucan obtained from *Sclerotinia sclerotiorum* has shown increased antitumor activity (Shinohara et al. 1988). There are several ways by which tumor growth can be inhibited by polysaccharides. Polysaccharides can lead to tumor cell apoptosis and enhance anticancer activity (Wang et al. 2013); other therapeutic potential of polysaccharides include inhibition of tumorigenesis along with tumor metastasis inhibition (Gao et al. 2022). Apoptosis of the cell lines of human breast cancer is conveyed by peptidoglycan which is obtained from Ganoderma lucidum (Cör et al. 2018). This species can also confer anti-inflammatory properties to human carcinoma cells. β-(1,3) and β-(1,6)-glucans obtained from *Pleurotus ostreatus* have immunomodulatory effect in rheumatoid arthritis (Bauerová et al. 2009). Other species of mushrooms conferring immunomodulatory effects include *Grifola frondosa*, *Trametes versicolor,* and *Coriolus versicolor* (Lowenthal et al. 2023).

19.6.1.2 Phenolic Compounds

Phenolic compounds are known for their enhanced antioxidant activity, serving as free radical blockers, metal ion inactivators, oxygen scavengers, and peroxide decomposers (Palacios et al. 2011). These bioactive compounds provide defense against degenerative diseases such as aging, cardiovascular conditions, and brain dysfunction (Ma et al. 2018). The phenolic compounds can have increased inhibition against lipid oxidation. These compounds have enhanced nutritional properties. According to studies, *Ganoderma lucidum* extracts have demonstrated to confer anti-acetylcholinesterase activity (Cör et al. 2017). Phenolic compounds obtained from mushrooms have various therapeutic properties including antimicrobial, antiviral, anti inflammatory, and anticancer effects. Phenolic compounds obtained from Hymenochaetaceae function by inhibiting LPS-stimulated nitric oxide production in the RAW264.7 cell lines (Lee et al. 2017; Taira et al. 2013).

19.6.1.3 Proteins and Peptides

Protein and peptides obtained from fungi have significant nutraceutical properties (Sanchez and Demain 2017). They support optimal digestion and aid in the enhanced absorption of nutrients. Furthermore, proteins and peptides may also function to inhibit particular enzymes and aid in immune system regulation, strengthening the body's defenses against potentially harmful infections. Fungal immunomodulatory proteins (FIPs), laccases, ribosome-inactivating proteins (RIPs), and lectins are examples of such therapeutic proteins and peptides that are produced from fungus (Xu et al. 2011). Lectins show immunomodulatory properties along with antiviral, antifungal, antitumor, and antibacterial properties (Leung et al. 2006). Lectins regulate key recognition events in a range of biological processes, including cellular signaling, host-pathogen interaction, differentiation, and tissue metastasis. Fungal lectins are considered valuable markers for cancer and inflammation due to their ability to specifically bind to epitopes on human glycoconjugates (Naeem et al. 2007). By influencing tumor cell proliferation, they hold promise for targeted drug delivery. Research has shown that GlcNAc-specific lectin (AAL2) exhibits antitumor activity against hepatoma cells, suggesting its potential in cancer diagnosis (Ren et al. 2015).

FIPs can act as adjuvants for treatment of tumor and can also lead to the inactivation of ribosomes by elimination of adenosine residues from rRNA (Barbieri et al. 1997). According to studies, RIPs have also shown significant capabilities in inhibiting the HIV-1 reverse transcriptase activity (Serna-Arbeláez et al. 2021). Ribonucleases obtained from mushrooms can prevent *Staphylococcus aureus*, *Pseudomonas fluorescens*, and *Pseudomonas aeruginosa* from proliferating (Ma et al. 2018). Laccases are useful in the polymerization of phenolics, producing derivatives with increased antioxidant properties (Nagarajan et al. 2020).

19.6.1.4 Poly-Unsaturated Fatty Acids

Poly-unsaturated fatty acids have been known to contribute in reducing serum cholesterol (Ros and Mataix 2006). Ergosterol has been detected as the principle sterol produced by mushrooms. Sterols are required for the prevention of cardiovascular diseases (Shanta Retelny et al. 2008). Tocopherols, a source of effective natural antioxidants, consist of free radical scavenging peroxyl components (Nimse and Pal 2015). These antioxidants have the capability of acting against microbial diseases, degenerative malfunctions, or cardiovascular diseases. Linoleic acids, an important group of fatty acids, reduce inflammation by limiting nitric oxide (NO) production and suppressing pro-inflammatory cytokines such as TNF-α, IL-1β, and IL-6. They also appear to play a role in reducing the impact of Alzheimer's disease (Ma et al. 2018). Omega-3 polyunsaturated fatty acids like eicosapentaenoic acid (EPA) and docosahexaenoic acid (DHA) (Ochi and Tsuchiya 2018) are involved in regulating inflammation. DHA and EPA-derived molecules help exert anti-inflammatory effects (Peng et al. 2020). The oxidized forms of DHA and EPA can better activate peroxisome proliferator-activated receptor gamma which ultimately leads to NF-κB inhibition (Medoro et al. 2023).

19.6.1.5 Cordycepin

Cordycepin, one of the major bioactive compounds obtained from cordyceps, is of utmost importance due to its nutraceutical potential (Fig. 19.3). Another feature of cordycepin is that it is structurally analogous to adenosine enhancing its usefulness (Tuli et al. 2013). The therapeutic properties of cordycepin include antidiabetic,

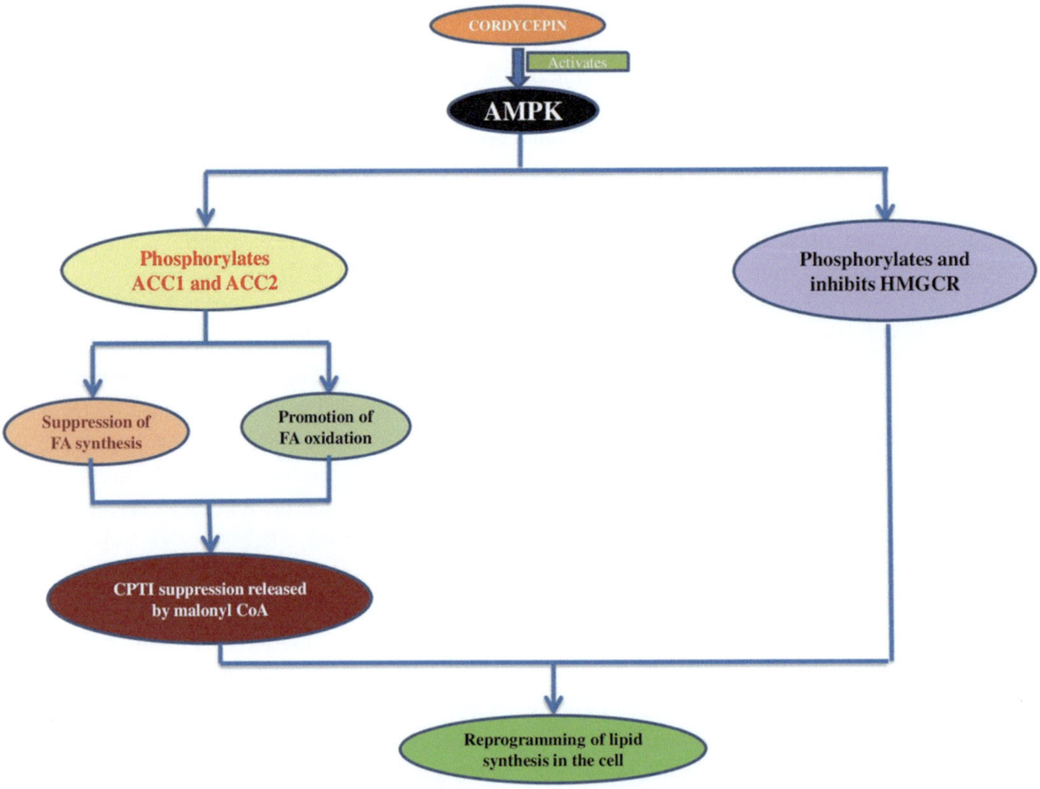

Fig. 19.3 AMPK gets activation upon the action of cordyceps—AMPK pathway

antifungal, anti-hyperlipidemia, immunomodulatory, anti-aging, anti-oxidant, anticancer, antifungal, antiviral, anti-inflammatory etc. (Yadav and Negi 2021). A variety of bioactive components are present in cordyceps which include volatile oils, essential amino acids, phenolic compounds, minerals, carotenoids, flavonoids, carbohydrates, vitamins, etc. (Wang et al. 2015).

Cordycepin has been found to suppress the production of nitric oxide (NO) and pro-inflammatory cytokines, such as TNF-α, IL-1β, and IL-6, in macrophages activated by lipopolysaccharides (LPS). This effect is achieved through the inhibition of key proteins involved in the inflammatory response, leading to a reduced expression of genes associated with type-2 diabetes (Choi et al. 2014). Additionally, higher levels of cordycepin result in a decrease in the expression of co-stimulatory molecules like ICAM-1 and B7-1/2, thereby modulating the immune response (Lee et al. 2020). The compound also inhibits the expression of diabetes-related genes by blocking the NF-κB-driven inflammatory pathway (Ashraf et al. 2020).

Cardiovascular diseases (CVDs) continue to be a major cause of death globally, with high lipid levels being a key factor in their development (Félix-Redondo et al. 2013). Elevated lipid levels are linked to CVDs such as coronary heart disease and atherosclerosis (Kris-Etherton et al. 2023). AMP-activated protein kinase (AMPK) is essential in regulating lipid metabolism. When activated, AMPK reduces fatty acid levels by phosphorylating and inhibiting acetyl-CoA carboxylase (ACC), which decreases fatty acid synthesis and promotes fatty acid breakdown (Li et al. 2009). Additionally, AMPK activation lowers cholesterol and triglyceride levels by inhibiting enzymes such as HMG-CoA reductase and glycerol-3-phosphate acyltransferase (Li et al. 2013; Soto-Acosta et al. 2017). Studies have shown that cordycepin can reduce intracellular lipid accumulation by activating AMPK, particularly its gamma1 subunit (Wu et al. 2014). Thus, targeting AMPK activation through cordycepin may offer a promising therapeutic approach for individuals with hyperlipidemia and obesity-related metabolic disorders.

Allergic asthma which is a chronic inflammatory disease consists of various properties such as airway inflammation, mucus hypersecretion, airway wall remodeling, airway hyperresponsiveness, etc. (Gohy et al. 2016). According to a scientific study, cordycepin possesses therapeutic advantages against allergic asthma through inhibition of immune responses (Lee et al. 2020). Allergic asthma occurs when immune cells such as lymphocytes, neutrophils, leukocytes, eosinophils, etc. are accumulated in the lungs. There is an elevated amount of Th2 cytokines including IL-13, IL-4, IL-5, and also increased amount of IgE levels observed in the lungs (Vroman et al. 2018). These Th2 cells along with cytokines result in the formation of asthma and hence are the initial cause of allergic asthma (Kudo et al. 2013). Research shows that the Th2 cytokine pathway can thus be targeted in order to develop remedy against allergic asthma (León and Ballesteros-Tato 2021). Cordycepin has the ability of delaying IgE, ICAM-1, and eotaxin which are allergen-specific (Lee et al. 2020). The Th2 cytokines causing allergic asthma can also be decreased by cordycepin. The anti-asthmatic abilities of cordycepin include inhibition of Th2-type responses through interfering with MAPKs and NF-κB signaling pathways (Choi et al. 2014).

Cordycepin's ability to reduce oxidative stress and inflammation, major contributors to Parkinson's disease, is another notable benefit (Niranjan 2014; Wei et al. 2018). Thus, cordycepin can confer neuroprotective effects against Parkinson's disease. Neuroprotective effects conferred by cordycepin are associated with the inhibition of the NF-κB signaling pathway (Tan et al. 2021).

19.6.2 Antioxidant and Anti-inflammatory Effects

The substances found in fungi have the power to affect how the body responds to stress and inflammation. Cellular damage and malfunction result from oxidative stress, which happens when the body produces more reactive oxygen species (ROS) than it can neutralize (Sies et al. 2017). Conversely, inflammation is a biological reaction characterized by immune cell activation and the release of pro-inflammatory chemicals such as chemokines, cytokines, and nitric oxide (NO) (Schmid-Schönbein 2006). Inflammation and oxidative stress are important factors in the onset and progression of many diseases, such as diabetes, cancer, heart disease, and neurological problems (Pham-Huy et al. 2008; Rahman 2007). Therefore, natural antioxidants and anti-inflammatory agents found in fungi present a promising avenue for preventing and treating such diseases.

The Pleurotus genus is among the most extensively researched groups of fungi known for their antioxidant and anti-inflammatory capabilities. The bioactive phytochemicals found in these mushrooms, such as phenolic substances, ergosterol, ergothioneine, and mannitol, help them neutralize reactive oxygen species (ROS), chelate metal ions, decrease ferric ions, and increase the activity of antioxidant enzymes (Diaz et al. 2012; Nguyen et al. 2016; Stastny et al. 2022). Furthermore, in a variety of cell-based models, researchers have explored their ability to reduce the generation of inflammatory mediators and to inhibit the expression and activity of inflammatory enzymes (Diaz et al. 2012; Nguyen et al. 2016; Stastny et al. 2022). These effects are thought to be mediated by the modulation of critical inflammation regulators, including nuclear factor-κB (NF-κB) and mitogen-activated protein kinase (MAPK) signaling pathways (Diaz et al. 2012; Nguyen et al. 2016).

Pleurotus flabellatus has been identified as a particularly potent species within the Pleurotus genus, demonstrating remarkable antioxidant and anti-inflammatory properties. Research indicates that the methanol extract of *P. flabellatus* exhibits superior radical scavenging activity and oxygen radical absorbance capacity (ORAC) compared to five other Pleurotus species (Stastny et al. 2022). Additionally, chloroform extracts of this species have shown significant inhibitory effects on cyclooxygenase-2 (COX-2), an enzyme closely linked to inflammatory processes due to its elevated expression during inflammation (Mohsin et al. 2022). Notably, *P. flabellatus* methanol extracts were found to contain the highest levels of ergosterol among the species studied, highlighting its potential role in modulating inflammation and oxidative stress (Stastny et al. 2022). Ergosterol is known for its antioxidant capabilities, particularly its ability to neutralize singlet oxygen and hydroxyl radicals (Paul and Snyder 2010).

Research on various fungi, such as *Marasmius oreades* and *Cyathus striatus*, has revealed their capacity to inhibit the activation of the NF-κB pathway, a crucial regulator of inflammatory and immunological responses (Petrova et al. 2007). In a similar vein, methanol extracts of *Agaricus bisporus* (white button mushroom) have demonstrated an inhibitory effect on NF-κB signaling (Chaturvedi et al. 2018). NF-κB is pivotal in regulating immune responses and the expression of important pro-inflammatory genes, including COX-2, iNOS, TNF-α, IL-1β, and IL-6 (Lawrence 2009). Additionally, fungi are abundant in β-glucans, which are recognized for their immunomodulatory properties. These polysaccharides activate immune cells by interacting with PRRs such as Dectin-1, complement receptor 3 (CR3), scavenger receptors, and lactosylceramide (Han et al. 2020; Camilli et al. 2018). This immune activation contributes to their therapeutic potential in managing inflammation.

Fungi offer a wealth of natural bioactive compounds, including antioxidants and anti-inflammatory agents, which target various molecular pathways associated with oxidative stress and inflammation. Incorporating these fungi into diets as nutraceuticals or functional foods may provide protective effects and support the management of chronic diseases linked to these physiological challenges.

19.6.3 Gut Health and Digestive Support

A wide variety of bacteria known as the gut microbiota can be found in the human gut, which is a complex ecosystem. This microbiota supports crucial physiological processes like digestion, metabolism, and immune system control, all of which are critical for maintaining general health (Rinninella et al. 2019). Various factors, including diet, stress, antibiotic use, and certain health conditions, can disrupt this delicate microbial equilibrium, leading to a state known as dysbiosis, where the balance between beneficial and harmful microbes is compromised. In addition to systemic problems like obesity, diabetes, and allergies, dysbiosis has been linked to gastrointestinal conditions like inflammatory bowel disease (IBD), irritable bowel syndrome (IBS), and colorectal cancer (Rinninella et al. 2019; Thursby and Juge 2017). In order to improve gut health and treat a variety of diseases, methods for reestablishing the diversity and balance of the gut microbiota have drawn interest. Using nutraceuticals and food supplements based on fungi, which contain bioactive substances that might affect the gut microbiota and its interactions with the host, is one such strategy (Cencic and Chingwaru 2010; Jayachandran et al. 2017).

The growth and activity of beneficial bacteria in the gut are greatly aided by prebiotics, which are non-digestible carbohydrates. Fungi-derived prebiotics, including fructo-oligosaccharides (FOS), β-glucans, and chitin, have demonstrated potential in influencing the gut microbiota. FOS, short-chain fructose molecules found in edible mushrooms like Pleurotus spp. and Lentinula edodes, are known to promote the growth of bifidobacteria and lactobacilli, which support gut health and enhance immune function (Sabater-Molina et al. 2009). Likewise, β-glucans, polysaccharides located in fungal cell walls, boost immune responses by activating macrophages, natural killer cells, and dendritic cells, strengthening the body's defense against infections and tumors (Akramienė et al. 2007). Chitin, a polymer of N-acetylglucosamine found in the cell walls of specific fungi and molds, also functions as a prebiotic. It fosters the production of short-chain fatty acids (SCFAs) in the gut, which help regulate intestinal pH, inhibit harmful bacteria, reduce inflammation, and strengthen the integrity of the intestinal barrier (Zheng et al. 2018).

Fungal-derived prebiotics may help maintain digestive health and prevent or treat dysbiosis-related illnesses by altering the gut microbiota and improving host-microbe interactions. Frequent use of these supplements derived from fungi may have major positive effects on gut health and general well-being.

19.6.4 Metabolic Health and Weight Management

Metabolic health is a state of optimal physiological functioning that involves the regulation of factors such as glucose, lipids, as well as the maintenance of hormonal balance and inflammatory response. Numerous factors, such as genetic predisposition, lifestyle decisions, dietary practices, and environmental effects, influence metabolic well-being. Environmental factors, food patterns, choices regarding lifestyle, and genetics all have a role in metabolic health. When the body's metabolic balance is disrupted, it can result in metabolic disorders such as obesity, diabetes, and dyslipidemia. In addition to lowering general health, these diseases increase the chance of cancer, heart disease, and early death (Alberti et al. 2009; Stanaway et al. 2018). Therefore, improving metabolic health and preventing or treating metabolic disorders are important aspects for being a healthy individual. Few fungal compounds have been reported in studies to modulate metabolic pathways and processes connected to metabolic health and weight management, such as glucose utilization, energy regulation, lipid synthesis and oxidation, appetite and satiety, regulation of insulin activity, adipogenesis, lipolysis, etc. (Guillamón et al. 2010; Jayachandran et al. 2017).

The genus *Ganoderma* and *Cordyceps* are among the most studied groups of fungi with metabolic effects. *Ganoderma* sp. contains various phytochemicals, such as polysaccharides, triterpenoids, sterols, peptides, and nucleotides, that can modulate glucose and lipid metabolism (Wachtel-Galor et al. 2011). *Cordyceps* also contain various phytochemicals, such as cordycepin, cordycepic acid, polysaccharides, ergosterol, and adenosine, which have been reported to modulate glucose and lipid metabolism (Bawadekji et al. 2016). Cordycepin has been associated with improvements in hyperlipidemia and is also documented to lower lipid levels by activating AMPK, which inhibits acetyl-CoA carboxylase, reducing cholesterol and triglyceride levels (Borde and Singh 2022). Animal model studies have shown that polysaccharides from *Cordyceps militaris* possess the ability to lower total cholesterol and triglyceride levels (Yu et al. 2021). Flavonoids and polyphenols are abundantly found in fungal species and these compounds could impede lipid synthesis and help lipid oxidation (Shapaval et al. 2023). Adenosine from *Cordyceps sinensis* can improve insulin sensitivity and secretion by enhancing IRS1 phosphorylation and pancreatic β-cell function (Lo et al. 2006). This anti-hyperglycemic activity can help with diabetes.

Fungi from the genera *Ganoderma* and *Cordyceps* are abundant in natural bioactive compounds known to influence metabolic pathways and processes related to metabolic health and weight management. Incorporating these fungi as dietary supplements or nutraceuticals may provide potential benefits in preventing and managing metabolic disorders, including obesity, diabetes, dyslipidemia, and metabolic syndrome.

19.6.4.1 Cognitive Function and Mental Well-Being

Processing information, learning, remembering, reasoning, problem-solving, and decision-making are cognitive functions (Eysenck et al. 1987).

Mental well-being on the other hand is all about feeling good, relaxed, and functioning well in different domains of our lives (Beiser 1974). Both are vital for well-being and living standards of people and proper care is required to mitigate risks of cognitive and mental challenges (Min et al. 2013). Therefore, finding ways to enhance cognitive function and mental well-being is a major challenge for the aging society. Some fungal compounds have demonstrated the ability to modulate cognitive functions and mental well-being by influencing neurotransmission, neuroprotective effects, neuroinflammation, and neuroplasticity (Nelson et al. 2013).

One of the most studied groups of fungi with cognitive and mental effects is the genus *Hericium*, which contain various phytochemicals that can modulate cognitive function and mental well-being in different ways (Chong et al. 2019). For instance, in a research study, polysaccharides from *Hericium erinaceus* have been reported by the authors to demonstrate enhanced cognitive functions by stimulating the synthesis of nerve growth factor and brain-derived neurotrophic factor, which are essential for the regulation of neurons (Lai et al. 2013). Studies involving terpenoids from *Hericium erinaceus* have also been investigated and found to improve mental well-being by reducing depression and anxiety symptoms by modulating serotonin and dopamine levels in the brain (Nagano et al. 2010; Vigna et al. 2019). Phenolic compounds extracted from *Hericium erinaceus* have been shown to exhibit neuroprotective properties, supporting cognitive function by mitigating oxidative stress and preventing mitochondrial dysfunction in neuronal cells (Lee et al. 2014).

Psilocybe fungi also contain phytochemicals, such as psilocybin, that can modulate cognitive function and mental well-being (Cavanna et al. 2022). Research has shown that psilocybin, a compound derived from *Psilocybe cubensis*, can enhance cognitive abilities such as creativity and critical thinking by stimulating serotonin receptors in the prefrontal cortex (Griffiths et al. 2016; Studerus et al. 2011). Additionally, studies have indicated that psilocybin may support mental health by alleviating symptoms of depression and anxiety, an effect attributed to its role in modulating serotonin activity in the amygdala (Carhart-Harris et al. 2016; Kraehenmann et al. 2015). Furthermore, psilocybin has been found to promote mental well-being by inducing feelings of euphoria and relaxation (Carhart-Harris et al. 2016; Griffiths et al. 2016; Studerus et al. 2011). Its mechanism of action primarily involves interactions with serotonin receptors in the brain, particularly the 5-HT2A receptor, which is integral to regulating mood, cognition, and perception (Kometer et al. 2012, 2013).

19.7 Safety and Quality Considerations

Authentication of fungal source is crucial in ensuring the safety and quality of fungal food supplements and nutraceuticals (Zhou et al. 2009). By implementing rigorous authentication protocols, manufacturers must keep a sharp surveillance to guarantee that the intended species are utilized, thereby averting potential risks associated with misidentification (Egbuta et al. 2017; Houbraken et al. 2021). This plays a critical role in upholding consumer trust and assurance in these products. Screening contaminants is a vital process to guarantee the safety and quality of fungal-based products. (Costa et al. 2019). Since fungi are largely obtained from soil-based environments, they are prone to be in conjugation with contaminants such as heavy metals and pesticides (Nasreddine and Parent-Massin 2002; Wong et al. 2020). Also, some species of fungi may possess mycotoxins which confers a significant amount of toxicity (Adeyeye 2016; Moretti et al. 2017). Ensuring that products are thoroughly tested for contaminants and subsequently purified during processing is crucial for maintaining safety and high-quality standards. The application of Good Manufacturing Practices (GMP) and the implementation of Hazard Analysis and Critical Control Points (HACCP) systems are vital for achieving consistent quality in the production of fungal-based nutraceuticals and food supplements (De Oliveira et al. 2016). By implementing stringent control measures at every stage of manufacturing, manufacturers can mitigate risks associated with cross-contamination and ensure product uniformity (Chen et al. 2001; Kirchner et al. 2021). Comprehensive studies are essential for determining the stability of formulations and shelf life of fungal food supplements and nutraceuticals (Sengupta et al. 2018; Wibowo et al. 2018). This can help manufacturers gain critical insights into the effects of factors such as temperature, humidity, and packaging materials on product stability and safety (Siracusa 2012). This would also be crucial for providing accurate storage recommendations and ensuring product safety and efficacy over time (Mercier et al. 2017). Accurate identification and proper labeling of potential allergens in fungal food supplements and nutraceuticals is vital for safeguarding consumers (Mills et al. 2004). By engaging advanced allergen detection approaches, manufacturers can certify transparency and offer consumers with the information they need to make informed decisions about their health (Monaci and Visconti 2009; Poms et al. 2007).

19.8 Safety Evaluation and Regulation

Stringent pre-clinical toxicity studies using animal models are critical for evaluating the safety profile of fungal food supplements and nutraceuticals Atkins et al. (2020). Comprehensive assessments, including acute and chronic toxicity studies, provide critical data on potential adverse effects of fungal supplements, thereby ensuring the safety of these products for human consumption (Heimann and Doughty 2014). Adhering to stringent regulatory frameworks, such as the Dietary Supplement Health and Education Act (DSHEA) in the United States, is essential for ensuring the safety and quality of fungal food supplements and nutraceuticals (Young and Bass 1995). Regular audits and inspections by regulatory bodies help to guarantee ongoing compliance with established safety standards, thereby maintaining consumer trust and confidence in these products (Yapp and Fairman 2006). Development of a strong post-market surveillance system would also allow for real-time monitoring of product safety, enabling quick identification and reporting of any adverse events for timely intervention (Amanor-Boadu 2004; Hepburn et al. 2008). Regulatory bodies such as the Food and Drug Administration (FDA) in the United States and the Food Safety and Standards Authority (FSSAI) in India are also responsible for setting up regulations and enforcing post-market surveillance of commercially launched supplements and nutraceuticals.

19.9 Conclusion and Future Prospects

As outlined in this chapter, fungi are an abundant source of bioactive molecules that have found usage in dietary supplements and nutraceuticals, providing a diverse range of health advantages. When utilized responsibly, these natural resources have the potential to enhance human health and well-being while promoting environmental sustainability, offering a more eco-friendly alternative to conventional sources. Over the last few decades, emerging biotechnology approaches have paved the way for more effective extraction and processing of these fungal bioactive substances. These advancements have not only enhanced the quality and efficacy of the end products but additionally provided novel possibilities for investigating the largely untapped potential of fungi. This has helped create a niche but significant market for fungal food supplements and nutraceuticals.

Current market trends suggest that consumers are increasingly invested in consuming natural and health-enhancing commodities. This tendency is projected to persist, making fungal dietary supplements and nutraceuticals a viable future industry. This booming sector, however, poses regulatory and safety hurdles. Ensuring that new fungal-based products are safe for consumption and comply with regulatory standards is paramount. Addressing these challenges and fostering innovation requires collaboration among researchers, industry experts, and regulatory authorities. These partnerships can expedite the development of novel products and processes by enabling the sharing of knowledge, resources, and best practices.

Consequently, it is essential to disseminate information to the general public regarding the advantages of fungal food supplements and nutraceuticals. As more people become conscious of the advantages of these products, the demand for them will surge parallelly, thus stimulating the growth of the industry. This necessitates a comprehensive approach to education and communication across all parties involved. The potential of fungi as a bioactive compound source for food supplement and nutraceutical products is noteworthy, however, achieving its full potential demands addressing several vital aspects. With the advancement of technology, collaboration, and increased public recognition, the outlook for this sector is optimistic with regards to a healthy and sustainable future.

References

Abdel-Fatah MA (2018) Nanofiltration systems and applications in wastewater treatment: review article. Ain Shams Eng J 9(4):3077–3092. https://doi.org/10.1016/J.ASEJ.2018.08.001

Adenipekun CO, Lawal R (2012) Uses of mushrooms in bioremediation: a review. Biotechnol Mol Biol Rev 7(3):62–68. https://doi.org/10.5897/BMBR12.006

Adeyeye SAO (2016) Fungal mycotoxins in foods: a review. Cogent Food Agric 2(1):1213127. https://doi.org/10.1080/23311932.2016.1213127

Aguiar TQ, Silva R, Domingues L (2015) *Ashbya gossypii* beyond industrial riboflavin production: a historical perspective and emerging biotechnological applications. Biotechnol Adv 33(8):1774–1786. https://doi.org/10.1016/J.BIOTECHADV.2015.10.001

Akramienė D, Akramienė D, Kondrotas A, Didžiapetrienė J, Kėvelaitis E (2007) Effects of b-glucans on the immune system. Medicina (Kaunas) 43(8):597

Alamgir ANM (2017) Herbal drugs: their collection, preservation, and preparation; evaluation, quality control, and standardization of herbal drugs. Prog Drug Res 73: 453–495. https://doi.org/10.1007/978-3-319-63862-1_10/COVER

Alberti KGMM, Eckel RH, Grundy SM, Zimmet PZ, Cleeman JI, Donato KA, Fruchart JC, James WPT, Loria CM, Smith SC (2009) Harmonizing the metabolic syndrome: a joint interim statement of the International Diabetes Federation Task Force on Epidemiology and Prevention; National Heart, Lung, and Blood Institute; American Heart Association; World Heart Federation; International Atherosclerosis Society; and International Association for the Study of Obesity. Circulation 120(16):1640–1645. https://doi.org/10.1161/CIRCULATIONAHA.109.192644

Al-Obaidi JR, Jambari NN, Ahmad-Kamil EI (2021) Mycopharmaceuticals and nutraceuticals: promising agents to improve human well-being and life quality. J Fungi 7(7):503. https://doi.org/10.3390/JOF7070503

Amanor-Boadu V (2004) Post-market surveillance model for potential human health effects of novel foods. Food Policy 29(6):609–620. https://doi.org/10.1016/J.FOODPOL.2004.09.001

Amirzakariya BZ, Shakeri A (2022) Bioactive terpenoids derived from plant endophytic fungi: an updated review (2011–2020). Phytochemistry 197:113130. https://doi.org/10.1016/J.PHYTOCHEM.2022.113130

Anna S, Veronika L, Svetlana T, Alexander O (2023) Patented keratinolytic enzymes for industrial application: an overview. Recent Pat Biotechnol 17(4): 346–363. https://doi.org/10.2174/1872208317666221212122656

Antwi SO, Petrick JL, Campbell PT, Norez DA, Stevens VL, Liao LM, Roberts LR, Patel T, McGlynn KA (2020) One-carbon metabolism-related micronutrients intake and risk for hepatocellular carcinoma: a prospective cohort study. Int J Cancer 147(8): 2075–2090. https://doi.org/10.1002/IJC.33007

Aronson JK, Aronson JK (2017) Defining 'nutraceuticals': neither nutritious nor pharmaceutical. Br J Clin Pharmacol 83(1):8–19. https://doi.org/10.1111/BCP.12935

Ashraf SA, Elkhalifa AEO, Siddiqui AJ, Patel M, Awadelkareem AM, Snoussi M, Ashraf MS, Adnan M, Hadi S (2020) Cordycepin for health and wellbeing: a potent bioactive metabolite of an entomopathogenic medicinal fungus cordyceps with its nutraceutical and therapeutic potential. Molecules 25(12):2735. https://doi.org/10.3390/MOLECULES25122735

Assemie A, Abaya G (2022) The effect of edible mushroom on health and their biochemistry. Int J Microbiol 2022:8744788. https://doi.org/10.1155/2022/8744788

Atkins JT, George GC, Hess K, Marcelo-Lewis KL, Yuan Y, Borthakur G, Khozin S, LoRusso P, Hong DS (2020) Pre-clinical animal models are poor predictors of human toxicities in phase 1 oncology clinical trials. British Journal of Cancer 123(10):1496–1501. https://doi.org/10.1038/s41416-020-01033-x

Barbieri L, Valbonesi P, Bonora E, Gorini P, Bolognesi A, Stirpe F (1997) Polynucleotide: adenosine glycosidase activity of ribosome-inactivating proteins: effect on DNA, RNA and Poly(A). Nucleic Acids Res 25(3): 518–522. https://doi.org/10.1093/NAR/25.3.518

Barreto-Bergter E, Figueiredo RT (2014) Fungal glycans and the innate immune recognition. Front Cell Infect Microbiol 4(OCT):109046. https://doi.org/10.3389/FCIMB.2014.00145/BIBTEX

Barrios-González J, Tarragó-Castellanos MR (2017) Solid-state fermentation: special physiology of fungi. In: Fungal metabolites, pp 319–347. https://doi.org/10.1007/978-3-319-25001-4_6

Basso V, Schiavenin C, Mendonça S, de Siqueira FG, Salvador M, Camassola M (2020) Chemical features and antioxidant profile by *Schizophyllum commune* produced on different agroindustrial wastes and byproducts of biodiesel production. Food Chem 329: 127089. https://doi.org/10.1016/J.FOODCHEM.2020.127089

Bauerová K, Paulovičová E, Mihalová D, Švík K, Poništ S (2009) Study of new ways of supplementary and combinatory therapy of rheumatoid arthritis with immunomodulators. Glucomannan and Imunoglukán® in adjuvant arthritis. Toxicol Ind Health 25(5):329–335. https://doi.org/10.1177/0748233709102945

Bawadekji A, Al Ali K, Al Ali M (2016) A review of the bioactive compound and medicinal value of cordyceps militaris. J North Basic Appl Sci 1(1):69–76. https://doi.org/10.12816/0021378

Beiser M (1974) Components and correlates of mental well-being. J Health Soc Behav 15(4):320–327. https://doi.org/10.2307/2137092

Bele AA, Khale A, Archana M, Bele A (2011) An overview on thin layer chromatography. IJPSR 2(2): 256–267. www.ijpsr.com

Bhambri A, Srivastava M, Mahale VG, Mahale S, Karn SK (2022) Mushrooms as potential sources of active metabolites and medicines. Front Microbiol 13: 837266. https://doi.org/10.3389/FMICB.2022.837266/BIBTEX

Blumfield M, Abbott K, Duve E, Cassettari T, Marshall S, Fayet-Moore F (2020) Examining the health effects and bioactive components in *Agaricus bisporus* mushrooms: a scoping review. J Nutr Biochem 84: 108453. https://doi.org/10.1016/J.JNUTBIO.2020.108453

Borde M, Singh SK (2022) Prospects of cordycepin and polysaccharides produced by cordyceps. In: Fungal diversity, ecology and control management, pp 93–107. https://doi.org/10.1007/978-981-16-8877-5_6

Brown GD (2005) Dectin-1: a signalling non-TLR pattern-recognition receptor. Nat Rev Immunol 6(1):33–43. https://doi.org/10.1038/nri1745

Bulgari D, Renzetti S, Messgo-Moumene S, Monti E, Gobbi E (2023) Optimization of esterase production in solid-state fermentation of agricultural digestate. Fermentation 9(6):524. https://doi.org/10.3390/FERMENTATION9060524/S1

Bush RK, Portnoy JM (2001) The role and abatement of fungal allergens in allergic diseases. J Allergy Clin Immunol 107(3):S430–S440. https://doi.org/10.1067/MAI.2001.113669

Buszewski B, Szultka M (2012) Past, present, and future of solid phase extraction: a review. Crit Rev Anal Chem 42(3):198–213. https://doi.org/10.1080/07373937.2011.645413

Camilli G, Tabouret G, Quintin J (2018) The complexity of fungal β-glucan in health and disease: effects on the mononuclear phagocyte system. Front Immunol 9 (APR):343771. https://doi.org/10.3389/FIMMU.2018.00673/BIBTEX

Carhart-Harris RL, Bolstridge M, Rucker J, Day CMJ, Erritzoe D, Kaelen M, Bloomfield M, Rickard JA, Forbes B, Feilding A, Taylor D, Pilling S, Curran VH, Nutt DJ (2016) Psilocybin with psychological support for treatment-resistant depression: an open-label feasibility study. Lancet Psychiatry 3(7): 619–627. https://doi.org/10.1016/S2215-0366(16)30065-7

Carr AC, Maggini S (2017) Vitamin C and immune function. Nutrients 9(11):1211. https://doi.org/10.3390/NU9111211

Cavanna F, Muller S, de la Fuente LA, Zamberlan F, Palmucci M, Janeckova L, Kuchar M, Pallavicini C, Tagliazucchi E (2022) Microdosing with psilocybin mushrooms: a double-blind placebo-controlled study. Transl Psychiatry 12(1):1–11. https://doi.org/10.1038/s41398-022-02039-0

Cencic A, Chingwaru W (2010) The role of functional foods, nutraceuticals, and food supplements in intestinal health. Nutrients 2(6):611. https://doi.org/10.3390/NU2060611

Chakraborty N, Banerjee A, Sarkar A, Ghosh S, Acharya K (2021) Mushroom polysaccharides: a potent immune-modulator. Review. Biointerface Res Appl Chem 11(2):8915–8930. https://doi.org/10.33263/BRIAC112.89158930

Chan GCF, Chan WK, Sze DMY (2009) The effects of beta-glucan on human immune and cancer cells. J Hematol Oncol 2(1):25. https://doi.org/10.1186/1756-8722-2-25/TABLES/2

Chaturvedi VK, Agarwal S, Gupta KK, Ramteke PW, Singh MP (2018) Medicinal mushroom: boon for therapeutic applications. 3 Biotech 8(8):1–20. https://doi.org/10.1007/S13205-018-1358-0/METRICS

Chemat F, Zill-E-Huma, Khan MK (2011) Applications of ultrasound in food technology: processing, preservation and extraction. Ultrason Sonochem 18(4): 813–835. https://doi.org/10.1016/J.ULTSONCH.2010.11.023

Chemat F, Rombaut N, Sicaire AG, Meullemiestre A, Fabiano-Tixier AS, Abert-Vian M (2017) Ultrasound assisted extraction of food and natural products. Mechanisms, techniques, combinations, protocols and applications. A review. Ultrason Sonochem 34:540–560. https://doi.org/10.1016/J.ULTSONCH.2016.06.035

Chen Y, Jackson KM, Chea FP, Schaffner DW (2001) Quantification and variability analysis of bacterial cross-contamination rates in common food service tasks. J Food Prot 64(1):72–80. https://doi.org/10.4315/0362-028X-64.1.72

Chen X, Zhai R, Li Y, Yuan X, Liu ZH, Jin M (2020) Understanding the structural characteristics of water-soluble phenolic compounds from four pretreatments of corn stover and their inhibitory effects on enzymatic hydrolysis and fermentation. Biotechnol Biofuels 13(1):1–13. https://doi.org/10.1186/S13068-020-01686-Z/FIGURES/7

Choi YH, Kim GY, Lee HH (2014) Anti-inflammatory effects of cordycepin lipopolysaccharide-stimulated RAW 264.7 macrophages through Toll-like receptor 4-mediated suppression of mitogen-activated protein kinases and NF-kappaB signaling pathways. Drug Des Devel Ther 8:1941–1953. https://doi.org/10.2147/DDDT.S71957

Chong PS, Fung ML, Wong KH, Lim LW (2019) Therapeutic potential of hericiumerinaceus for depressive disorder. Int J Mol Sci 21(1):163. https://doi.org/10.3390/IJMS21010163

Chu KT, Xia L, Ng TB (2005) Pleurostrin, an antifungal peptide from the oyster mushroom. Peptides 26(11): 2098–2103. https://doi.org/10.1016/J.PEPTIDES.2005.04.010

Cör D, Botić T, Knez Ž, Gregori A, Pohleven F (2017) The effects of different solvents on bioactive metabolites and in vitro; antioxidant and anti-acetylcholinesterase activity of *Ganoderma lucidum* fruiting body and primordia extracts. Maced J Chem Chem Eng 36(1): 129–142. https://doi.org/10.20450/MJCCE.2017.1054

Cör D, Knez Ž, Hrnčič MK (2018) Antitumour, antimicrobial, antioxidant and antiacetylcholinesterase effect of *Ganoderma lucidum* terpenoids and polysaccharides: a review. Molecules 23(3):649. https://doi.org/10.3390/MOLECULES23030649

Coskun O (2016) Separation techniques: chromatography. North Clin Istanbul 3(2):156–160. https://doi.org/10.14744/nci.2016.32757

Costa JG, Vidovic B, Saraiva N, do Céu Costa M, Del Favero G, Marko D, Oliveira NG, Fernandes AS (2019) Contaminants: a dark side of food supplements? Free Radic Res 53(sup1):1113–1135. https://doi.org/10.1080/10715762.2019.1636045

da Silva de Souza AC, Correa VG, de Goncalves GA, Soares AA, Bracht A, Peralta RM (2017) *Agaricus blazei* bioactive compounds and their effects on human health: benefits and controversies. Curr Pharm Des 23(19):2807–2834. https://doi.org/10.2174/1381612823666170119093719

De Oliveira CAF, Da Cruz AG, Tavolaro P, Corassin CH (2016) Food safety: good manufacturing practices (GMP), sanitation standard operating procedures (SSOP), hazard analysis and critical control point (HACCP). In: Antimicrobial food packaging, pp 129–139. https://doi.org/10.1016/B978-0-12-800723-5.00010-3

De Villa R, Roasa J, Mine Y, Tsao R (2021) Impact of solid-state fermentation on factors and mechanisms influencing the bioactive compounds of grains and processing by-products. Crit Rev Food Sci Nutr 63(21):5388–5413. https://doi.org/10.1080/10408398.2021.2018989

Dewanjee S, Gangopadhyay M, Bhattacharya N, Khanra R, Dua TK (2015) Bioautography and its scope in the field of natural product chemistry. J Pharm Anal 5(2):75–84. https://doi.org/10.1016/J.JPHA.2014.06.002

Dias DA, Urban S, Roessner U (2012) A historical overview of natural products in drug discovery. Metabolites 2(2):303–336. https://doi.org/10.3390/METABO2020303

Diaz P, Jeong SC, Lee S, Khoo C, Koyyalamudi SR (2012) Antioxidant and anti-inflammatory activities of selected medicinal plants and fungi containing phenolic and flavonoid compounds. Chin Med 7:26. https://doi.org/10.1186/1749-8546-7-26

Ding Y, Chen S, Wang H, Li S, Ma C, Wang J, Cui L (2021) Identification of secondary metabolites in *Flammulina velutipes* by UPLC-Q-exactive-orbitrap MS. J Food Qual 2021:4103952. https://doi.org/10.1155/2021/4103952

Domínguez Díaz L, Fernández-Ruiz V, Cámara M (2019) The frontier between nutrition and pharma: the international regulatory framework of functional foods, food supplements and nutraceuticals. Crit Rev Food Sci Nutr 60(10):1738–1746. https://doi.org/10.1080/10408398.2019.1592107

Drenkhan R, Kaldmäe H, Silm M, Adamson K, Bleive U, Aluvee A, Erik M, Raal A (2022) Comparative analyses of bioactive compounds in *Inonotus obliquus* conks growing on alnus and betula. Biomolecules 12(9):1178. https://doi.org/10.3390/BIOM12091178

Dutré B, Trägrdh G (1994) Macrosolute-microsolute separation by ultrafiltration: a review of diafiltration processes and applications. Desalination 95(3):227–267. https://doi.org/10.1016/0011-9164(94)00063-8

Dwyer JT, Coates PM, Smith MJ (2018) Dietary supplements: regulatory challenges and research resources. Nutrients 10(1):41. https://doi.org/10.3390/NU10010041

Egbuta MA, Mwanza M, Babalola OO (2017) Health risks associated with exposure to filamentous fungi. Int J Environ Res Public Health 14(7):719. https://doi.org/10.3390/IJERPH14070719

EL-Fakharany EM, Haroun BM, Ng TB, Redwan E-R (2010) Oyster mushroom laccase inhibits hepatitis C virus entry into peripheral blood cells and hepatoma cells. Protein Pept Lett 17(8):1031–1039. https://doi.org/10.2174/092986610791498948

Elliott TF, Truong C, Jackson SM, Zuniga CL, Trappe JM, Vernes K (2022) Mammalian mycophagy: a global review of ecosystem interactions between mammals and fungi. Fungal Syst Evol 9(1):99–159. https://doi.org/10.3114/FUSE.2022.09.07

ElNaker NA, Daou M, Ochsenkühn MA, Amin SA, Yousef AF, Yousef LF (2021) A metabolomics approach to evaluate the effect of lyophilization versus oven drying on the chemical composition of plant extracts. Sci Rep 11(1):1–11. https://doi.org/10.1038/s41598-021-02158-6

Espín JC, García-Conesa MT, Tomás-Barberán FA (2007) Nutraceuticals: facts and fiction. Phytochemistry

68(22–24):2986–3008. https://doi.org/10.1016/J.PHYTOCHEM.2007.09.014

Eysenck MW, MacLeod C, Mathews A (1987) Cognitive functioning and anxiety. Psychol Res 49(2–3):189–195. https://doi.org/10.1007/BF00308686/METRICS

Fazenda ML, Seviour R, McNeil B, Harvey LM (2008) Submerged culture fermentation of "higher fungi": the macrofungi. Adv Appl Microbiol 63.33–103. https://doi.org/10.1016/S0065-2164(07)00002-0

Félix-Redondo FJ, Grau M, Fernández-Bergés D (2013) Cholesterol and cardiovascular disease in the elderly. Facts and gaps. Aging Dis 4(3):154. /pmc/articles/PMC3660125/

Ferreira IMPLVO, Pinho O, Vieira E, Tavarela JG (2010) Brewer's Saccharomyces yeast biomass: characteristics and potential applications. Trends Food Sci Technol 21(2):77–84. https://doi.org/10.1016/J.TIFS.2009.10.008

Fontanals N, Marcé RM, Borrull F (2019) Materials for solid-phase extraction of organic compounds. Separations 6(4):56. https://doi.org/10.3390/SEPARATIONS6040056

Gaidhani KA, Harwalkar M, Bhambere D, Nirgude PS (2015) Lyophilization/freeze drying—a review. World J Pharm Res *4*(8):517. www.wjpr.net

Gao S, Jiang X, Wang L, Jiang S, Luo H, Chen Y, Peng C (2022) The pathogenesis of liver cancer and the therapeutic potential of bioactive substances. Front Pharmacol 13:1029601. https://doi.org/10.3389/FPHAR.2022.1029601/BIBTEX

Gohy ST, Hupin C, Pilette C, Ladjemi MZ (2016) Chronic inflammatory airway diseases: the central role of the epithelium revisited. Clin Exp Allergy 46(4):529–542. https://doi.org/10.1111/CEA.12712

Gordon M, Bihari B, Goosby E, Gorter R, Greco M, Guralnik M, Mimura T, Rudinicki V, Wong R, Kaneko Y (1998) A placebo-controlled trial of the immune modulator, lentinan, in HIV-positive patients: a phase I/II trial. J Med 29(5–6):305–330

Gowthaman MK, Krishna C, Moo-Young M (2001) Fungal solid state fermentation—an overview. Appl Mycol Biotechnol 1(C):305–352. https://doi.org/10.1016/S1874-5334(01)80014-9

Griffiths RR, Johnson MW, Carducci MA, Umbricht A, Richards WA, Richards BD, Cosimano MP, Klinedinst MA (2016) Psilocybin produces substantial and sustained decreases in depression and anxiety in patients with life-threatening cancer: a randomized double-blind trial. J Psychopharmacol (Oxford, England) 30(12):1181–1197. https://doi.org/10.1177/0269881116675513

Guillamón E, García-Lafuente A, Lozano M, D́arrigo M, Rostagno MA, Villares A, Martínez JA (2010) Edible mushrooms: role in the prevention of cardiovascular diseases. Fitoterapia 81(7):715–723. https://doi.org/10.1016/J.FITOTE.2010.06.005

Guruceaga X, Perez-Cuesta U, De Cerio AAD, Gonzalez O, Alonso RM, Hernando FL, Ramirez-Garcia A, Rementeria A (2019) Fumagillin, a mycotoxin of *Aspergillus fumigatus*: biosynthesis, biological activities, detection, and applications. Toxins 12(1):7. https://doi.org/10.3390/TOXINS12010007

Habtemariam S (2020) *Trametes versicolor* (Synn. Coriolus versicolor) polysaccharides in cancer therapy: targets and efficacy. Biomedicine 8(5):135. https://doi.org/10.3390/BIOMEDICINES8050135

Hammed AM, Jaswir I, Amid A, Alam Z, Asiyanbi-H TT, Ramli N (2013) Enzymatic hydrolysis of plants and algae for extraction of bioactive compounds. Food Rev Int 29(4):352–370. https://doi.org/10.1080/87559129.2013.818012

Han B, Baruah K, Cox E, Vanrompay D, Bossier P (2020) Structure-functional activity relationship of β-glucans from the perspective of immunomodulation: a mini-review. Front Immunol 11:521871. https://doi.org/10.3389/FIMMU.2020.00658/BIBTEX

Hashida S, Yuzawa S, Suzuki NN, Fujioka Y, Takikawa T, Sumimoto H, Inagaki F, Fujii H (2004) Binding of FAD to cytochrome b558 is facilitated during activation of the phagocyte NADPH oxidase, leading to superoxide production. J Biol Chem 279(25):26378–26386. https://doi.org/10.1074/JBC.M309724200

He WS, Cui D, Li L, Tong LT, Rui J, Li H, Zhang H, Liu X (2019) Cholesterol-reducing effect of ergosterol is modulated via inhibition of cholesterol absorption and promotion of cholesterol excretion. J Funct Foods 57:488–496. https://doi.org/10.1016/J.JFF.2019.04.042

Hedrick JL, Mulcahey LJ, Taylor LT (1992) Supercritical fluid extraction. Mikrochim Acta 108(3–6):115–132. https://doi.org/10.1007/BF01242421/METRICS

Heimann KG, Doughty K (2014) Examination of acute and chronic toxicity. In: Regulatory toxicology, pp 81–88. https://doi.org/10.1007/978-3-642-35374-1_31/COVER

Hepburn P, Howlett J, Boeing H, Cockburn A, Constable A, Davi A, de Jong N, Moseley B, Oberdörfer R, Robertson C, Wal JM, Samuels F (2008) The application of post-market monitoring to novel foods. Food Chem Toxicol 46(1):9–33. https://doi.org/10.1016/J.FCT.2007.09.008

Hitchen SM, Dean JR (1993) Properties of supercritical fluids. In: Applications of supercritical fluids in industrial analysis, pp 1–11. https://doi.org/10.1007/978-94-011-2146-0_1

Hodgson E (2012) Toxins and venoms. Prog Mol Biol Transl Sci 112:373–415. https://doi.org/10.1016/B978-0-12-415813-9.00014-3

Houbraken J, Visagie CM, Frisvad JC (2021) Recommendations to prevent taxonomic misidentification of genome-sequenced fungal strains. Microbiol Resourc Announc 10(48):e0107420. https://doi.org/10.1128/MRA.01074-20

Huang K, El-Seedi HR, Xu B (2022) Critical review on chemical compositions and health-promoting effects of mushroom *Agaricus blazei* Murill. Curr Res Food Sci

5:2190–2203. https://doi.org/10.1016/J.CRFS.2022.10.029

Hyde KD, Xu J, Rapior S, Jeewon R, Lumyong S, Niego AGT, Abeywickrama PD, Aluthmuhandiram JVS, Brahamanage RS, Brooks S, Chaiyasen A, Chethana KWT, Chomnunti P, Chepkirui C, Chuankid B, de Silva NI, Doilom M, Faulds C, Gentekaki E et al (2019) The amazing potential of fungi: 50 ways we can exploit fungi industrially. Fungal Divers 97(1): 1–136. https://doi.org/10.1007/S13225-019-00430-9

Iram A, Özcan A, Yatmaz E, Turhan İ, Demirci A (2022) Effect of microparticles on fungal fermentation for fermentation-based product productions. PRO 10(12): 2681. https://doi.org/10.3390/PR10122681

Jach M, Serefko A, Sajnaga E, Kozak E, Poleszak E, Malm A (2015) Dietary supplements based on the yeast biomass. Curr Top Nutraceutical Res 13:83–88

Jafari D, Esmaeilzadeh A, Mohammadi-Kordkhayli M, Rezaei N (2019) Vitamin C and the immune system. In: Nutrition and immunity, pp 81–102. https://doi.org/10.1007/978-3-030-16073-9_5

Jayachandran M, Xiao J, Xu B (2017) A critical review on health promoting benefits of edible mushrooms through gut microbiota. Int J Mol Sci 18(9). https://doi.org/10.3390/IJMS18091934

Jędrejko KJ, Lazur J, Muszyńska B (2021) *Cordyceps militaris*: an overview of its chemical constituents in relation to biological activity. Foods 10(11):2634. https://doi.org/10.3390/FOODS10112634

Jesenak M, Hrubisko M, Majtan J, Rennerova Z, Banovcin P (2014) Anti-allergic effect of pleuran (β-glucan from pleurotusostreatus) in children with recurrent respiratory tract infections. Phytother Res 28(3):471–474. https://doi.org/10.1002/PTR.5020

Joana Gil-Chávez G, Villa JA, Fernando Ayala-Zavala J, Basilio Heredia J, Sepulveda D, Yahia EM, González-Aguilar GA (2013) Technologies for extraction and production of bioactive compounds to be used as nutraceuticals and food ingredients: an overview. Compr Rev Food Sci Food Saf 12(1):5–23. https://doi.org/10.1111/1541-4337.12005

Jones PT, Anderson R (1983) Oxidative inhibition of polymorphonuclear leukocyte motility mediated by the peroxidase/H2O2/halide system: studies on the reversible nature of the inhibition and mechanism of protection of migratory responsiveness by ascorbate, levamisole, thiamine and cysteine. Int J Immunopharmacol 5(5):377–389. https://doi.org/10.1016/0192-0561(83)90012-7

Kajiyama S, Nagatake T, Ishikawa S, Hosomi K, Shimada Y, Matsui Y, Kunisawa J (2023) *Lentinula Edodes* Mycelia extract regulates the function of antigen-presenting cells to activate immune cells and prevent tumor-induced deterioration of immune function. BMC Complement Med Ther 23(1):281. https://doi.org/10.1186/S12906-023-04106-5/FIGURES/5

Kashyap ML, Ganji S, Nakra NK, Kamanna VS (2019) Niacin for treatment of nonalcoholic fatty liver disease (NAFLD): novel use for an old drug? J Clin Lipidol 13(6):873–879. https://doi.org/10.1016/J.JACL.2019.10.006

Khan MA, Tania M, Liu R, Rahman MM (2013) Hericiumerinaceus: an edible mushroom with medicinal values. J Complement Integr Med 10(1):253–258. https://doi.org/10.1515/JCIM-2013-0001/MACHINEREADABLECITATION/RIS

Kirchner M, Goulter RM, Chapman BJ, Clayton J, Jaykus LA (2021) Cross-contamination on atypical surfaces and venues in food service environments. J Food Prot 84(7):1239–1251. https://doi.org/10.4315/JFP-20-314

Kıvrak İ, Kıvrak Ş, Karababa E (2020) Assessment of bioactive compounds and antioxidant activity of Turkey tail medicinal mushroom *Trametes versicolor* (Agaricomycetes). Int J Med Mushrooms 22(6): 559–571. https://doi.org/10.1615/INTJMEDMUSHROOMS.2020035027

Kometer M, Schmidt A, Bachmann R, Studerus E, Seifritz E, Vollenweider FX (2012) Psilocybin biases facial recognition, goal-directed behavior, and mood state toward positive relative to negative emotions through different serotonergic subreceptors. Biol Psychiatry 72(11):898–906. https://doi.org/10.1016/J.BIOPSYCH.2012.04.005

Kometer M, Schmidt A, Jäncke L, Vollenweider FX (2013) Activation of serotonin 2A receptors underlies the psilocybin-induced effects on α oscillations, N170 visual-evoked potentials, and visual hallucinations. J Neurosci 33(25):10544–10551. https://doi.org/10.1523/JNEUROSCI.3007-12.2013

Kondeti RR, Mulpuri KS, Meruga B (2014) Advancements in column chromatography: a review. World J Pharm Pharm Sci:1375–1383. https://wjpsonline.com/index.php/wjps/article/view/advancements-column-chromatography-review

Kopustinskiene DM, Jakstas V, Savickas A, Bernatoniene J (2020) Flavonoids as anticancer agents. Nutrients 12(2):457. https://doi.org/10.3390/NU12020457

Kousar R, Naeem M, Jamaludin MI, Arshad A, Shamsuri AN, Ansari N, Akhtar S, Hazafa A, Uddin J, Khan A, Al-Harrasi A (2022) Exploring the anticancer activities of novel bioactive compounds derived from endophytic fungi: mechanisms of action, current challenges and future perspectives. Am J Cancer Res 12(7):2897. /pmc/articles/PMC9360238/

Kraehenmann R, Preller KH, Scheidegger M, Pokorny T, Bosch OG, Seifritz E, Vollenweider FX (2015) Psilocybin-induced decrease in amygdala reactivity correlates with enhanced positive mood in healthy volunteers. Biol Psychiatry 78(8):572–581. https://doi.org/10.1016/J.BIOPSYCH.2014.04.010

Kris-Etherton PM, Sanders L, Lawler O, Riley T, Maki K (2023) Hyperlipidemia. In: Encyclopedia of human nutrition: volume 1–4, fourth edition, vol 1–4, pp 361–379. https://doi.org/10.1016/B978-0-12-821848-8.00175-X

Kudo M, Ishigatsubo Y, Aoki I (2013) Pathology of asthma. Front Microbiol 4(SEP):64115. https://doi.org/10.3389/FMICB.2013.00263/BIBTEX

Kulal DK, Loni PC, Dcosta C, Some S, Kalambate PK (2020) Cyanobacteria: as a promising candidate for heavy-metals removal. In: Advances in cyanobacterial biology, pp 291–300. https://doi.org/10.1016/B978-0-12-819311-2.00019-X

Kumar K, Srivastav S, Sharanagat VS (2021) Ultrasound assisted extraction (UAE) of bioactive compounds from fruit and vegetable processing by-products: a review. Ultrason Sonochem 70:105325. https://doi.org/10.1016/J.ULTSONCH.2020.105325

Kumar H, Bhardwaj K, Kuča K, Sharifi-Rad J, Verma R, Machado M, Kumar D, Cruz-Martins N (2022) Edible mushrooms' enrichment in food and feed: a mini review. Int J Food Sci Technol 57(3):1386–1398. https://doi.org/10.1111/IJFS.15546

Kunin CM (1993) Resistance to antimicrobial drugs—a worldwide calamity. Ann Intern Med 118(7):557–561. https://doi.org/10.7326/0003-4819-118-7-199304010-00011

Kuroishi T (2015) Regulation of immunological and inflammatory functions by biotin. Can J Physiol Pharmacol 93(12):1091–1096. https://doi.org/10.1139/CJPP-2014-0460

Lai PL, Naidu M, Sabaratnam V, Wong KH, David RP, Kuppusamy UR, Abdullah N, Malek SNA (2013) Neurotrophic properties of the Lion's mane medicinal mushroom, *Hericium erinaceus* (Higher Basidiomycetes) from Malaysia. Int J Med Mushrooms 15(6):539–554. https://doi.org/10.1615/INTJMEDMUSHR.V15.I6.30

Lawrence T (2009) The nuclear factor NF-kappaB pathway in inflammation. Cold Spring Harb Perspect Biol 1(6):a001651. https://doi.org/10.1101/CSHPERSPECT.A001651

Lee DH, Kim HW (2014) Innate immunity induced by fungal β-glucans via dectin-1 signaling pathway. Int J Med Mushrooms 16(1):1–16. https://doi.org/10.1615/INTJMEDMUSHR.V16.I1.10

Lee S, Lee D, Jang TS, Kang KS, Nam JW, Lee HJ, Kim KH (2017) Anti-inflammatory phenolic metabolites from the edible fungus *Phellinus baumii* in LPS-stimulated RAW264.7 cells. Molecules 22(10):1583. https://doi.org/10.3390/MOLECULES22101583

Lee CT, Huang KS, Shaw JF, Chen JR, Kuo WS, Shen G, Grumezescu AM, Holban AM, Wang YT, Wang JS, Hsiang YP, Lin YM, Hsu HH, Yang CH (2020) Trends in the immunomodulatory effects of *Cordyceps militaris*: total extracts, polysaccharides and cordycepin. Front Pharmacol 11:575704. https://doi.org/10.3389/FPHAR.2020.575704/BIBTEX

Lehotay SJ, Schenck FJ (2000) Multiresidue methods: extraction. In: Encyclopedia of separation science, pp 3409–3415. https://doi.org/10.1016/B0-12-226770-2/05621-0

León B, Ballesteros-Tato A (2021) Modulating Th2 cell immunity for the treatment of asthma. Front Immunol 12:637948. https://doi.org/10.3389/FIMMU.2021.637948/BIBTEX

Leung MYK, Liu C, Koon JCM, Fung KP (2006) Polysaccharide biological response modifiers. Immunol Lett 105(2):101–114. https://doi.org/10.1016/J.IMLET.2006.01.009

Li XN, Song J, Zhang L, LeMaire SA, Hou X, Zhang C, Coselli JS, Chen L, Xing LW, Zhang Y, Shen YH (2009) Activation of the AMPK-FOXO3 pathway reduces fatty acid–induced increase in intracellular reactive oxygen species by upregulating thioredoxin. Diabetes 58(10):2246–2257. https://doi.org/10.2337/DB08-1512

Li YY, Yu LF, Zhang LN, Qiu BY, Su MB, Wu F, Chen DK, Pang T, Gu M, Zhang W, Ma WP, Jiang HW, Li JY, Nan FJ, Li J (2013) Novel small-molecule AMPK activator orally exerts beneficial effects on diabetic db/db mice. Toxicol Appl Pharmacol 273(2):325–334. https://doi.org/10.1016/J.TAAP.2013.09.006

Li H-X, Wang J-J, Lu C-L, Gao Y-J, Gao L, Yang Z-Q, Zang M, Li H-X, Wang J-J, Lu C-L, Gao Y-J, Gao L, Yang Z-Q (2022) Review of bioactivity, isolation, and identification of active compounds from antrodiacinnamomea. Bioengineering 9(10):494. https://doi.org/10.3390/BIOENGINEERING9100494

Liang J, Melican D, Cafro L, Palace G, Fisette L, Armstrong R, Patchen ML (1998) Enhanced clearance of a multiple antibiotic resistant *Staphylococcus aureus* in rats treated with PGG-glucan is associated with increased leukocyte counts and increased neutrophil oxidative burst activity. Int J Immunopharmacol 20(11):595–614. https://doi.org/10.1016/S0192-0561(98)00007-1

Lo HC, Hsu TH, Te Tu S, Lin KC (2006) Anti-hyperglycemic activity of natural and fermented *Cordyceps sinensis* in rats with diabetes induced by nicotinamide and streptozotocin. Am J Chin Med 34(5):819–832. https://doi.org/10.1142/S0192415X06004314

Lockwood GB (2010) The quality of commercially available nutraceutical supplements and food sources. J Pharm Pharmacol 63(1):3–10. https://doi.org/10.1111/J.2042-7158.2010.01159.X

Lowenthal R, Taylor M, Gidden JA, Heflin B, Lay JO, Avaritt N, Tackett AJ, Urbaniak A (2023) The mycelium of the Trametes versicolor synn. *Coriolus versicolor* (Turkey tail mushroom) exhibit anti-melanoma activity in vitro. Biomed Pharmacother 161:114424. https://doi.org/10.1016/J.BIOPHA.2023.114424

Ma G, Yang W, Zhao L, Pei F, Fang D, Hu Q (2018) A critical review on the health promoting effects of mushrooms nutraceuticals. Food Sci Human Wellness 7(2):125–133. https://doi.org/10.1016/J.FSHW.2018.05.002

Machado D, Shishido SM, Queiroz KCS, Oliveira DN, Faria ALC, Catharino RR, Spek CA, Ferreira CV (2013) Irradiated riboflavin diminishes the aggressiveness of melanoma in vitro and in vivo.

PLoS One 8(1):e54269. https://doi.org/10.1371/JOURNAL.PONE.0054269

Majors RE (1980) Multidimensional high performance liquid chromatography. J Chromatogr Sci 18(10): 571–579. https://doi.org/10.1093/CHROMSCI/18.10.571

Martín J, Díaz-Montaña EJ, Asuero AG (2018) Recovery of anthocyanins using membrane technologies: a review. Crit Rev Anal Chem 48(3):143–175. https://doi.org/10.1080/10408347.2017.1411249

Martínez-Montemayor MM, Ling T, Suárez-Arroyo IJ, Ortiz-Soto G, Santiago-Negrón CL, Lacourt-Ventura MY, Valentín-Acevedo A, Lang WH, Rivas F (2019) Identification of biologically active *Ganoderma lucidum* compounds and synthesis of improved derivatives that confer anti-cancer activities in vitro. Front Pharmacol 10(FEB):432597. https://doi.org/10.3389/FPHAR.2019.00115/BIBTEX

Mathis D, Shoelson SE (2011) Immunometabolism: an emerging frontier. Nat Rev Immunol 11(2):81–83. https://doi.org/10.1038/NRI2922

McGreal EP, Miller JL, Gordon S (2005) Ligand recognition by antigen-presenting cell C-type lectin receptors. Curr Opin Immunol 17(1):18–24. https://doi.org/10.1016/J.COI.2004.12.001

McPhail DB, Hartley RC, Gardner PT, Duthie GG (2003) Kinetic and stoichiometric assessment of the antioxidant activity of flavonoids by electron spin resonance spectroscopy. J Agric Food Chem 51(6):1684–1690. https://doi.org/10.1021/JF025922V

Medoro A, Davinelli S, Colletti A, Di Micoli V, Grandi E, Fogacci F, Scapagnini G, Cicero AFG (2023) Nutraceuticals as modulators of immune function: a review of potential therapeutic effects. Prev Nutr Food Sci 28(2):89–107. https://doi.org/10.3746/PNF.2023.28.2.89

Mercier S, Villeneuve S, Mondor M, Uysal I (2017) Time–temperature management along the food cold chain: a review of recent developments. Compr Rev Food Sci Food Saf 16(4):647–667. https://doi.org/10.1111/1541-4337.12269

Mills ENC, Valovirta E, Madsen C, Taylor SL, Vieths S, Anklam E, Baumgartner S, Koch P, Crevel RWR, Frewer L (2004) Information provision for allergic consumers—where are we going with food allergen labelling? Allergy 59(12):1262–1268. https://doi.org/10.1111/J.1398-9995.2004.00720.X

Min JA, Lee CU, Lee C (2013) Mental health promotion and illness prevention: a challenge for psychiatrists. Psychiatry Investig 10(4):307. https://doi.org/10.4306/PI.2013.10.4.307

Mohsin N ul A, Aslam S, Ahmad M, Irfan M, Al-Hussain SA, Zaki MEA (2022) Cyclooxygenase-2 (COX-2) as a target of anticancer agents: a review of novel synthesized scaffolds having anticancer and COX-2 inhibitory potentialities. Pharmaceuticals 15(12): 1471. https://doi.org/10.3390/PH15121471

Monaci L, Visconti A (2009) Mass spectrometry-based proteomics methods for analysis of food allergens. TrAC Trends Anal Chem 28(5):581–591. https://doi.org/10.1016/J.TRAC.2009.02.013

Moretti A, Logrieco AF, Susca A (2017) Mycotoxins: an underhand food problem. Methods Mol Biol 1542:3–12. https://doi.org/10.1007/978-1-4939-6707-0_1/COVER

Mousavi S, Bereswill S, Heimesaat MM (2019) Immunomodulatory and antimicrobial effects of vitamin C. Eur J Microbiol Immunol 9(3):73–79. https://doi.org/10.1556/1886.2019.00016

Murphy EJ, Rezoagli E, Major I, Rowan NJ, Laffey JG (2020) β-glucan metabolic and immunomodulatory properties and potential for clinical application. J Fungi 6(4):356. https://doi.org/10.3390/JOF6040356

Muszyńska B, Fijałkowska A, Sułkowska-Ziaja K, Włodarczyk A, Kaczmarczyk P, Nogaj E, Piętka J (2020) *Fomitopsis officinalis*: a species of arboreal mushroom with promising biological and medicinal properties. Chem Biodivers 17(6):e2000213. https://doi.org/10.1002/CBDV.202000213

Naeem A, Saleemuddin M, Hasan Khan R (2007) Glycoprotein targeting and other applications of lectins in biotechnology. Curr Protein Pept Sci 8(3):261–271

Nagano M, Shimizu K, Kondo R, Hayashi C, Sato D, Kitagawa K, Ohnuki K (2010) Reduction of depression and anxiety by 4 weeks *Hericium erinaceus* intake. Biomed Res (Tokyo, Japan) 31(4):231–237. https://doi.org/10.2220/BIOMEDRES.31.231

Nagarajan S, Nagarajan R, Kumar J, Salemme A, Togna AR, Saso L, Bruno F (2020) Antioxidant activity of synthetic polymers of phenolic compounds. Polymers 12(8):1646. https://doi.org/10.3390/POLYM12081646

Nam M, Choi JY, Kim MS (2021) Metabolic profiles, bioactive compounds, and antioxidant capacity in *Lentinula edodes* cultivated on log versus sawdust substrates. Biomolecules 11(11):1654. https://doi.org/10.3390/BIOM11111654/S1

Nasreddine L, Parent-Massin D (2002) Food contamination by metals and pesticides in the European Union. Should we worry? Toxicol Lett 127(1–3):29–41. https://doi.org/10.1016/S0378-4274(01)00480-5

Nelson ED, Ramberg JE, Best T, Sinnott RA (2013) Neurologic effects of exogenous saccharides: a review of controlled human, animal, and in vitro studies. Nutr Neurosci 15(4):149–162. https://doi.org/10.1179/1476830512Y.0000000004

Ngai PHK, Ng TB (2003) Lentin, a novel and potent antifungal protein from shitake mushroom with inhibitory effects on activity of human immunodeficiency virus-1 reverse transcriptase and proliferation of leukemia cells. Life Sci 73(26):3363–3374. https://doi.org/10.1016/J.LFS.2003.06.023

Nguyen TK, Im KH, Choi J, Shin PG, Lee TS (2016) Evaluation of antioxidant, anti-cholinesterase, and anti-inflammatory effects of culinary mushroom *Pleurotus pulmonarius*. Mycobiology 44(4):291. https://doi.org/10.5941/MYCO.2016.44.4.291

Nimse SB, Pal D (2015) Free radicals, natural antioxidants, and their reaction mechanisms. RSC Adv 5(35):27986–28006. https://doi.org/10.1039/C4RA13315C

Niranjan R (2014) The role of inflammatory and oxidative stress mechanisms in the pathogenesis of Parkinson's disease: focus on astrocytes. Mol Neurobiol 49(1): 28–38. https://doi.org/10.1007/S12035-013-8483-X/METRICS

Nishad JH, Singh A, Gautam VS, Kumari P, Kumar J, Yadav M, Kharwar RN (2021) Bioactive potential evaluation and purification of compounds from an endophytic fungus *Diaporthe longicolla*, a resident of *Saraca asoca* (Roxb.) Willd. Arch Microbiol 203(7): 4179–4188. https://doi.org/10.1007/S00203-021-02390-8/METRICS

Nitto T, Onodera K (2013) Linkage between coenzyme a metabolism and inflammation: roles of pantetheinase. J Pharmacol Sci 123(1):1–8. https://doi.org/10.1254/JPHS.13R01CP

Nkhata SG, Ayua E, Kamau EH, Shingiro JB (2018) Fermentation and germination improve nutritional value of cereals and legumes through activation of endogenous enzymes. Food Sci Nutr 6(8):2446–2458. https://doi.org/10.1002/FSN3.846

Ochi E, Tsuchiya Y (2018) Eicosapentaenoic acid (EPA) and docosahexaneoic acid (DHA) in muscle damage and function. Nutrients 10(5):552. https://doi.org/10.3390/NU10050552

Ouedraogo JP, Tsang A (2021) Production of native and recombinant enzymes by fungi for industrial applications. In: Encyclopedia of mycology, pp 222–232. https://doi.org/10.1016/B978-0-12-819990-9.00046-9

Oyetayo VO, Oyetayo FL (2005) Potential of probiotics as biotherapeutic agents targeting the innate immune system. Afr J Biotechnol 4(2):123–127. https://doi.org/10.4314/AJB.V4I2.15065

Pais P, Almeida V, Yılmaz M, Teixeira MC (2020) *Saccharomyces boulardii*: what makes it tick as successful probiotic? J Fungi 6(2):78. https://doi.org/10.3390/JOF6020078

Palacios I, Lozano M, Moro C, D'Arrigo M, Rostagno MA, Martínez JA, García-Lafuente A, Guillamón E, Villares A (2011) Antioxidant properties of phenolic compounds occurring in edible mushrooms. Food Chem 128(3):674–678. https://doi.org/10.1016/J.FOODCHEM.2011.03.085

Paul BD, Snyder SH (2010) The unusual amino acid L-ergothioneine is a physiologic cytoprotectant. Cell Death Differ 17(7):1134–1140. https://doi.org/10.1038/CDD.2009.163

Pavanetto F, Conti B, Genta I, Giunchedi P (1992) Solvent evaporation, solvent extraction and spray drying for polylactide microsphere preparation. Int J Pharm 84(2):151–159. https://doi.org/10.1016/0378-5173(92)90055-7

Peng Z, Zhang C, Yan L, Zhang Y, Yang Z, Wang J, Song C (2020) EPA is more effective than DHA to improve depression-like behavior, glia cell dysfunction and hippcampal apoptosis signaling in a chronic stress-induced rat model of depression. Int J Mol Sci 21(5): 1769. https://doi.org/10.3390/IJMS21051769

Petrova RD, Mahajna J, Reznick AZ, Wasser SP, Denchev CM, Nevo E (2007) Fungal substances as modulators of NF-κB activation pathway. Mol Biol Rep 34(3): 145–154. https://doi.org/10.1007/S11033-006-9027-5/METRICS

Pham-Huy LA, He H, Pham-Huy C (2008) Free radicals, antioxidants in disease and health. Int J Biomed Sci 4(2):89. /pmc/articles/PMC3614697/

Ping Z, Xu H, Liu T, Huang J, Meng Y, Xu X, Li W, Zhang L (2016) Anti-hepatoma activity of the stiff branched β-D-glucan and effects of molecular weight. J Mater Chem B 4(26):4565–4573. https://doi.org/10.1039/C6TB01299J

Poms RE, Klein CL, Anklam E (2007) Methods for allergen analysis in food: a review. Food Addit Contam 21(1):1–31. https://doi.org/10.1080/02652030310001620423

Poole CF, Poole SK (2010) Extraction of organic compounds with room temperature ionic liquids. J Chromatogr A 1217(16):2268–2286. https://doi.org/10.1016/J.CHROMA.2009.09.011

Qureshi AA, Tan X, Reis JC, Badr MZ, Papasian CJ, Morrison DC, Qureshi N (2011) Suppression of nitric oxide induction and pro-inflammatory cytokines by novel proteasome inhibitors in various experimental models. Lipids Health Dis 10(1):1–25. https://doi.org/10.1186/1476-511X-10-177/FIGURES/14

Rabhi C, Arcile G, Cariel L, Lenoir C, Bignon J, Wdzieczak-Bakala J, Ouazzani J (2015) Antiangiogenic-like properties of fermented extracts of ayurvedic medicinal plants. J Med Food 18(9): 1065–1072. https://doi.org/10.1089/jmf.2014.0128

Raclariu AC, Heinrich M, Ichim MC, de Boer H (2018) Benefits and limitations of DNA barcoding and metabarcoding in herbal product authentication. Phytochem Anal 29(2):123–128. https://doi.org/10.1002/PCA.2732

Raghava Rao KV, Mani P, Satyanarayana B, Raghava Rao T (2017) Purification and structural elucidation of three bioactive compounds isolated from *Streptomyces coelicoflavus* BC 01 and their biological activity. 3 Biotech 7(1):1–12. https://doi.org/10.1007/S13205-016-0581-9/TABLES/5

Rahi DK, Malik D (2016) Diversity of mushrooms and their metabolites of nutraceutical and therapeutic significance. J Mycol 2016:1–18. https://doi.org/10.1155/2016/7654123

Rahman K (2007) Studies on free radicals, antioxidants, and co-factors. Clin Interv Aging 2(2):219. /pmc/articles/PMC2684512/

Rajasekaran A, Xavier R (2008) Conventional and microwave assisted synthesis of pyrazolone Mannich bases possessing anti-inflammatory, analgesic, ulcerogenic effect and antimicrobial properties. Bioorg Med Chem Lett. https://www.researchgate.net/publication/268427982

Rathore H, Prasad S, Kapri M, Tiwari A, Sharma S (2019) Medicinal importance of mushroom mycelium: mechanisms and applications. J Funct Foods 56:182–193. https://doi.org/10.1016/J.JFF.2019.03.016

Ravichandran S (2012) Solid state and submerged fermentation for the production of bioactive substances: a comparative study. Int J Sci Nat 3(3):480–486. https://doi.org/10.1016/j.jddst.2017.10.025

Ren XM, Li DF, Jiang S, Lan XQ, Hu Y, Sun H, Wang DC (2015) Structural basis of specific recognition of non-reducing terminal N-acetylglucosamine by an *Agrocybe aegerita* Lectin. PLoS One 10(6): e0129608. https://doi.org/10.1371/JOURNAL.PONE.0129608

Ren L, Zhang J, Zhang T (2021) Immunomodulatory activities of polysaccharides from Ganoderma on immune effector cells. Food Chem 340:127933. https://doi.org/10.1016/J.FOODCHEM.2020.127933

Richter J, Kral V, Pohorska J, RajnohovaDobiasova L, Vetvicka V (2016) Effects of β-glucan on natural killer cells in patients recovering from cancer treatment: clinical trial. Int J Clin Exp Med Sci 2(2):26–30. https://doi.org/10.11648/j.ijcems.20160202.12

Rinninella E, Raoul P, Cintoni M, Franceschi F, Miggiano GAD, Gasbarrini A, Mele MC (2019) What is the healthy gut microbiota composition? A changing ecosystem across age, environment, diet, and diseases. Microorganisms 7(1):14. https://doi.org/10.3390/MICROORGANISMS7010014

Ros E, Mataix J (2006) Fatty acid composition of nuts—implications for cardiovascular health. Br J Nutr 96 (S2):S29–S35. https://doi.org/10.1017/BJN20061861

Sabater-Molina M, Larqué E, Torrella F, Zamora S (2009) Dietary fructooligosaccharides and potential benefits on health. J Physiol Biochem 65(3):315–328. https://doi.org/10.1007/BF03180584

Samtiya M, Aluko RE, Puniya AK, Dhewa T (2021) Enhancing micronutrients bioavailability through fermentation of plant-based foods: a concise review. Fermentation 7(2):63. https://doi.org/10.3390/FERMENTATION7020063

Sanchez S, Demain AL (2017) Bioactive products from fungi. Food Bioactives 59. https://doi.org/10.1007/978-3-319-51639-4_3

Sánchez-Camargo AP, Mendiola JA, Ibáñez E, Herrero M (2014) Supercritical fluid extraction. In: Reference module in chemistry, molecular sciences and chemical engineering. https://doi.org/10.1016/B978-0-12-409547-2.10753-X

Santa HSD, Romão PRT, Sovrani V, Oliveira FR, Peres A, Monteiro MC (2014) Dietary polysaccharides and immune modulation. In: Polysaccharides, pp 1–24. https://doi.org/10.1007/978-3-319-03751-6_6-1

Santoni G, Cardinali C, Morelli BB, Santoni M, Nabissi M, Amantini C (2015) Danger- and pathogen-associated molecular patterns recognition by pattern-recognition receptors and ion channels of the transient receptor potential family triggers the inflammasome activation in immune cells and sensory neurons. J Neuroinflammation 12(1):1–10. https://doi.org/10.1186/S12974-015-0239-2/TABLES/1

Sarinho E, Medeiros D, Schor D, Rego Silva A, Sales V, Motta ME, Costa A, Azoubel A, Rizzo JA (2009) Production of interleukin-10 in asthmatic children after Beta-1-3-glucan. Allergol Immunopathol 37(4): 188–192. https://doi.org/10.1016/J.ALLER.2009.02.005

Schmid-Schönbein GW (2006) Analysis of inflammation. Annu Rev Biomed Eng 8:93–151. https://doi.org/10.1146/ANNUREV.BIOENG.8.061505.095708

Sen P, Kosre A, Koreti D, Chandrawansi NK, Jadhav SK (2021) Nutrients and bioactive compounds of Pleurotus ostreatus mushroom. NewBioWorld 3(2): 8–12. https://doi.org/10.52228/NBW-JAAB.2021-3-2-2

Sengupta P, Chatterjee B, Tekade RK (2018) Current regulatory requirements and practical approaches for stability analysis of pharmaceutical products: a comprehensive review. Int J Pharm 543(1–2):328–344. https://doi.org/10.1016/J.IJPHARM.2018.04.007

Serna-Arbeláez MS, Florez-Sampedro L, Orozco LP, Ramírez K, Galeano E, Zapata W (2021) Natural products with inhibitory activity against human immunodeficiency virus type 1. Adv Virol. https://doi.org/10.1155/2021/5552088

Sghaier R, Zarrouk A, Nury T, Badreddine I, O'Brien N, Mackrill JJ, Vejux A, Samadi M, Nasser B, Caccia C, Leoni V, Moreau T, Cherkaoui-Malki M, Salhedine Masmoudi A, Lizard G (2019) Biotin attenuation of oxidative stress, mitochondrial dysfunction, lipid metabolism alteration and 7β-hydroxycholesterol-induced cell death in 158N murine oligodendrocytes. Free Radic Res 53(5):535–561. https://doi.org/10.1080/10715762.2019.1612891/SUPPL_FILE/IFRA_A_1612891_SM8602.ZIP

Shanta Retelny V, Neuendorf A, Roth JL (2008) Nutrition protocols for the prevention of cardiovascular disease. Nutr Clin Pract 23(5):468–476. https://doi.org/10.1177/0884533608323425

Shapaval V, Deniset-Besseau A, Dubava D, Dzurendova S, Heitmann Solheim J, Kohler A (2023) Multiscale spectroscopic analysis of lipids in dimorphic and oleaginous *Mucor circinelloides* accommodate sustainable targeted lipid production. Fungal Biol Biotechnol 10(1):1–18. https://doi.org/10.1186/S40694-023-00148-Z/FIGURES/10

Shinohara H, Ohno N, Yadomae T (1988) Antitumor activity and structural characterization of a (1→3)-β-D-glucan extracted with cold alkali from sclerotia of *Sclerotinia sclerotiorum* IFO 9395. Chem Pharm Bull 36(2):819–823. https://doi.org/10.1248/CPB.36.819

Siddiqui RA, Moghadasian MH (2020) Nutraceuticals and nutrition supplements: challenges and opportunities. Nutrients 12(6):1593. https://doi.org/10.3390/NU12061593

Sies H, Berndt C, Jones DP (2017) Oxidative stress. Annu Rev Biochem 86:715–748. https://doi.org/10.1146/ANNUREV-BIOCHEM-061516-045037

Singh S, Kola P, Kaur D, Singla G, Mishra V, Panesar PS, Mallikarjunan K, Krishania M (2021) Therapeutic potential of nutraceuticals and dietary supplements in the prevention of viral diseases: a review. Front Nutr 8: 679312. https://doi.org/10.3389/FNUT.2021.679312/BIBTEX

Siracusa V (2012) Food packaging permeability behaviour: a report. Int J Polym Sci 2012:1–11. https://doi.org/10.1155/2012/302029

Ślusarczyk J, Adamska E, Czerwik-Marcinkowska J (2021) Fungi and algae as sources of medicinal and other biologically active compounds: a review. Nutrients 13(9):3178. https://doi.org/10.3390/NU13093178

Sopalun K, Laosripaiboon W, Wachirachaikarn A, Iamtham S (2021) Biological potential and chemical composition of bioactive compounds from endophytic fungi associated with thai mangrove plants. S Afr J Bot 141:66–76. https://doi.org/10.1016/J.SAJB.2021.04.031

Soto-Acosta R, Bautista-Carbajal P, Cervantes-Salazar M, Angel-Ambrocio AH, del Angel RM (2017) DENV up-regulates the HMG-CoA reductase activity through the impairment of AMPK phosphorylation: a potential antiviral target. PLoS Pathog 13(4):e1006257. https://doi.org/10.1371/JOURNAL.PPAT.1006257

Sridevi Sangeetha KS, Umamaheswari S, Uma C, Reddy M, Kalkura SN (2016) Flavonoids: therapeutic potential of natural pharmacological agents. Int J Pharm Sci Res 7(10):3924. https://doi.org/10.13040/IJPSR.0975-8232.7(10).3924-30

Stajić M, Vukojević J, Duletić-Lauević S (2009) Biology of *Pleurotus eryngii* and role in biotechnological processes: a review. Crit Rev Biotechnol 29(1):55–66. https://doi.org/10.1080/07388550802688821

Stanaway JD, Afshin A, Gakidou E, Lim SS, Abate D, Abate KH, Abbafati C, Abbasi N, Abbastabar H, Abd-Allah F, Abdela J, Abdelalim A, Abdollahpour I, Abdulkader RS, Abebe M, Abebe Z, Abera SF, Abil OZ, Abraha HN et al (2018) Global, regional, and national comparative risk assessment of 84 behavioural, environmental and occupational, and metabolic risks or clusters of risks for 195 countries and territories, 1990-2017: a systematic analysis for the Global Burden of Disease Study 2017. Lancet (London, England) 392(10159):1923–1994. https://doi.org/10.1016/S0140-6736(18)32225-6

Stastny J, Marsik P, Tauchen J, Bozik M, Mascellani A, Havlik J, Landa P, Jablonsky I, Treml J, Herczogova P, Bleha R, Synytsya A, Kloucek P (2022) Antioxidant and anti-inflammatory activity of five medicinal mushrooms of the genus *pleurotus*. Antioxidants (Basel, Switzerland) 11(8):1569. https://doi.org/10.3390/ANTIOX11081569

Stierle A, Strobel G, Stierle D (1993) Taxol and taxane production by taxomycesandreanae, an endophytic fungus of pacific yew. Science 260(5105):214–216. https://doi.org/10.1126/SCIENCE.8097061

Stoica RM, Moscovici M, Lakatos ES, Cioca LI (2023) Exopolysaccharides of fungal origin: properties and pharmaceutical applications. Processes 11(2):335. https://doi.org/10.3390/PR11020335

Studerus E, Kometer M, Hasler F, Vollenweider FX (2011) Acute, subacute and long-term subjective effects of psilocybin in healthy humans: a pooled analysis of experimental studies. J Psychopharmacol (Oxford, England) 25(11):1434–1452. https://doi.org/10.1177/0269881110382466

Sun N, Liu Y, Liu K, Wang S, Liu Q, Lin S (2022) Gastrointestinal fate of food allergens and its relationship with allergenicity. Compr Rev Food Sci Food Saf 21(4):3376–3404. https://doi.org/10.1111/1541-4337.12989

Tadioto V, Giehl A, Cadamuro RD, Guterres IZ, dos Santos AA, Bressan SK, Werlang L, Stambuk BU, Fongaro G, Silva IT, Alves SL (2023) Bioactive compounds from and against yeasts in the one health context: a comprehensive review. Fermentation 9(4): 363. https://doi.org/10.3390/FERMENTATION9040363

Taira J, Ohminea W, Nanbub H, Uedab K (2013) Inhibition of LPS-stimulated NO production in RAW264.7 macrophages through iNOS suppression and nitrogen radical scavenging by phenolic compounds from *Agrimonia pilosa* Ledeb. Oxidants Antioxidants Med Sci 2(1):21. https://www.academia.edu/72399185/Inhibition_of_LPS_stimulated_NO_production_in_RAW264_7_macrophages_through_iNOS_suppression_and_nitrogen_radical_scavenging_by_phenolic_compounds_from_Agrimonia_pilosa_Ledeb

Takata Y, Cai Q, Beeghly-Fadiel A, Li H, Shrubsole MJ, Ji BT, Yang G, Chow WH, Gao YT, Zheng W, Shu XO (2012) Dietary B vitamin and methionine intakes and lung cancer risk among female never smokers in China. Cancer Causes Control 23(12):1965–1975. https://doi.org/10.1007/S10552-012-0074-Z/METRICS

Tamura J, Kubota K, Murakami H, Sawamura M, Matsushima T, Tamura T, Saitoh T, Kurabayshi H, Naruse T (1999) Immunomodulation by vitamin B12: augmentation of CD8+ T lymphocytes and natural killer (NK) cell activity in vitamin B12-deficient patients by methyl-B12 treatment. Clin Exp Immunol 116(1):28–32. https://doi.org/10.1046/J.1365-2249.1999.00870.X

Tan L, Song X, Ren Y, Wang M, Guo C, Guo D, Gu Y, Li Y, Cao Z, Deng Y (2021) Anti-inflammatory effects of cordycepin: a review. Phytother Res 35(3): 1284–1297. https://doi.org/10.1002/PTR.6890

Tang Z, Wang Y, Yang J, Xiao Y, Cai Y, Wan Y, Chen H, Yao H, Shan Z, Li C, Wang G (2020) Isolation and identification of flavonoid-producing endophytic fungi from medicinal plant *Conyza blinii* H. Lév that exhibit higher antioxidant and antibacterial activities. PeerJ 2020(4):e8978. https://doi.org/10.7717/PEERJ.8978/SUPP-3

Taofiq O, Heleno SA, Calhelha RC, Alves MJ, Barros L, González-Paramás AM, Barreiro MF, Ferreira ICFR (2017) The potential of *Ganoderma lucidum* extracts as bioactive ingredients in topical formulations, beyond its nutritional benefits. Food Chem Toxicol 108:139–147. https://doi.org/10.1016/J.FCT.2017.07.051

teBiesebeke R, Ruijter G, Rahardjo YSP, Hoogschagen MJ, Heerikhuisen M, Levin A, van Driel KGA, Schutyser MAI, Dijksterhuis J, Zhu Y, Weber FJ, de Vos WM, van den Hondel KAMJJ, Rinzema A, Punt PJ (2002) Aspergillus oryzae in solid-state and submerged fermentations progress report on a multidisciplinary project. FEMS Yeast Res 2(2):245–248. https://doi.org/10.1111/J.1567-1364.2002.TB00089.X

Thakur MP, Singh HK (2013) Mushrooms, their bioactive compounds and medicinal uses: a review. Med Plants 5(1):1–20. https://doi.org/10.5958/J.0975-6892.5.1.004

Thomas S, Rezoagli E, Abidin IZ, Major I, Murray P, Murphy EJ (2022) β-glucans from yeast—immunomodulators from novel waste resources. Appl Sci 12(10):5208. https://doi.org/10.3390/APP12105208

Thurman EM (2000) Sorbent selection for solid-phase extraction. In: Encyclopedia of separation science, pp 4204–4212. https://doi.org/10.1016/B0-12-226770-2/02171-2

Thursby E, Juge N (2017) Introduction to the human gut microbiota. Biochem J 474(11):1823. https://doi.org/10.1042/BCJ20160510

Trivedi S, Patel K, Belgamwar V, Wadher K (2022) Functional polysaccharide lentinan: ROLE in anti-cancer therapies and management of carcinomas. Pharmacol Res Mod Chin Med 2:100045. https://doi.org/10.1016/J.PRMCM.2022.100045

Tuli HS, Sharma AK, Sandhu SS, Kashyap D (2013) Cordycepin: a bioactive metabolite with therapeutic potential. Life Sci 93(23):863–869. https://doi.org/10.1016/J.LFS.2013.09.030

Ullah A, Munir S, Badshah SL, Khan N, Ghani L, Poulson BG, Emwas AH, Jaremko M (2020) Important flavonoids and their role as a therapeutic agent. Molecules 25(22):5243. https://doi.org/10.3390/MOLECULES25225243

Uwineza PA, Waśkiewicz A (2020) Recent advances in supercritical fluid extraction of natural bioactive compounds from natural plant materials. Molecules 25(17):3847. https://doi.org/10.3390/MOLECULES25173847

van Steenwijk HP, Bast A, de Boer A (2021) Immunomodulating effects of fungal beta-glucans: from traditional use to medicine. Nutrients 13(4):1333. https://doi.org/10.3390/NU13041333

Vazquez-Roig P, Picó Y (2015) Pressurized liquid extraction of organic contaminants in environmental and food samples. TrAC Trends Anal Chem 71:55–64. https://doi.org/10.1016/J.TRAC.2015.04.014

Venturella G, Ferraro V, Cirlincione F, Gargano ML (2021) Medicinal mushrooms: bioactive compounds, use, and clinical trials. Int J Mol Sci 22(2):634. https://doi.org/10.3390/IJMS22020634

Verni M, Pontonio E, Montemurro M, Rizzello CG, Verni M, Pontonio E, Montemurro M, Rizzello CG (2022) Fermentation as strategy for improving nutritional, functional, technological, and sensory properties of legumes. In: Legumes research—volume 2. https://doi.org/10.5772/INTECHOPEN.102523

Vetvicka V, Teplyakova TV, Shintyapina AB, Korolenko TA (2021) Effects of medicinal fungi-derived β-glucan on tumor progression. J Fungi 7(4):250. https://doi.org/10.3390/JOF7040250

Vigna L, Morelli F, Agnelli GM, Napolitano F, Ratto D, Occhinegro A, Di Iorio C, Savino E, Girometta C, Brandalise F, Rossi P (2019) Hericiumerinaceus improves mood and sleep disorders in patients affected by overweight or obesity: could circulating pro-BDNF and BDNF be potential biomarkers? Evid Based Complement Alternat Med 2019:7861297. https://doi.org/10.1155/2019/7861297

Villares A, Mateo-Vivaracho L, Guillamón E (2012) Structural features and healthy properties of polysaccharides occurring in mushrooms. Agriculture 2(4):452–471. https://doi.org/10.3390/AGRICULTURE2040452

Vroman H, Bergen IM, van Hulst JAC, van Nimwegen M, van Uden D, Schuijs MJ, Pillai SY, van Loo G, Hammad H, Lambrecht BN, Hendriks RW, Kool M (2018) TNF-α–induced protein 3 levels in lung dendritic cells instruct TH2 or TH17 cell differentiation in eosinophilic or neutrophilic asthma. J Allergy Clin Immunol 141(5):1620–1633.e12. https://doi.org/10.1016/j.jaci.2017.08.012

Wachtel-Galor S, Yuen J, Buswell JA, Benzie IFF (2011) *Ganoderma lucidum* (Lingzhi or Reishi). In: Herbal medicine: biomolecular and clinical aspects: second edition, pp 175–199. https://www.ncbi.nlm.nih.gov/books/NBK92757/

Wang KP, Zhang QL, Liu Y, Wang J, Cheng Y, Zhang Y (2013) Structure and inducing tumor cell apoptosis activity of polysaccharides isolated from *Lentinus edodes*. J Agric Food Chem 61(41):9849–9858. https://doi.org/10.1021/JF403291W/ASSET/IMAGES/MEDIUM/JF-2013-03291W_0008.GIF

Wang J, Kan L, Nie S, Chen H, Cui SW, Phillips AO, Phillips GO, Li Y, Xie M (2015) A comparison of chemical composition, bioactive components and antioxidant activity of natural and cultured *Cordyceps sinensis*. LWT Food Sci Technol 63(1):2–7. https://doi.org/10.1016/J.LWT.2015.03.109

Wasser S (2002) Medicinal mushrooms as a source of antitumor and immunomodulating polysaccharides. Appl Microbiol Biotechnol 60(3):258–274. https://doi.org/10.1007/S00253-002-1076-7/METRICS

Wasser SP (2005) Shiitake (Lentinus edodes). In: Encyclopedia of dietary supplements. CRC Press, pp 653–664. https://doi.org/10.1081/E-EDS-120024880

Wasser SP, Weis AL (1999) Therapeutic effects of substances occurring in higher basidiomycetes mushrooms: a modern perspective. Crit Rev Immunol 19(1):32. https://doi.org/10.1615/CRITREVIMMUNOL.V19.I1.30

Wei Z, Li X, Li X, Liu Q, Cheng Y (2018) Oxidative stress in Parkinson's disease: a systematic review and meta-analysis. Front Mol Neurosci 11:385209. https://doi.org/10.3389/FNMOL.2018.00236/BIBTEX

Wibowo S, Buvé C, Hendrickx M, Van Loey A, Grauwet T (2018) Integrated science-based approach to study quality changes of shelf-stable food products during storage: a proof of concept on orange and mango juices. Trends Food Sci Technol 73:76–86. https://doi.org/10.1016/J.TIFS.2018.01.006

Wierzejska RE (2021) Dietary supplements—for whom? The current state of knowledge about the health effects of selected supplement use. Int J Environ Res Public Health 18(17):8897. https://doi.org/10.3390/IJERPH18178897

Wilmsen PK, Spada DS, Salvador M (2005) Antioxidant activity of the flavonoid hesperidin in chemical and biological systems. J Agric Food Chem 53(12): 4757–4761. https://doi.org/10.1021/JF0502000

Wong TC, Lee FSC, Hu GL, Chang L, Wang X, Fu PP (2020) A survey of heavy metal and organochlorine pesticide contaminations in commercial Lingzhi products. J Food Drug Anal 15(4):16. https://doi.org/10.38212/2224-6614.2401

Wu C, Guo Y, Su Y, Zhang X, Luan H, Zhang X, Zhu H, He H, Wang X, Sun G, Sun X, Guo P, Zhu P (2014) Cordycepin activates AMP-activated protein kinase (AMPK) via interaction with the γ1 subunit. J Cell Mol Med 18(2):293–304. https://doi.org/10.1111/JCMM.12187

Wu Y, Choi MH, Li J, Yang H, Shin HJ (2016) Mushroom cosmetics: the present and future. Cosmetics 3(3):22. https://doi.org/10.3390/COSMETICS3030022

Wu JY, Siu KC, Geng P (2021) Bioactive ingredients and medicinal values of *Grifola frondosa* (Maitake). Foods 10(1):95. https://doi.org/10.3390/FOODS10010095

Xu X, Yan H, Chen J, Zhang X (2011) Bioactive proteins from mushrooms. Biotechnol Adv 29(6):667–674. https://doi.org/10.1016/J.BIOTECHADV.2011.05.003

Yadav D, Negi PS (2021) Bioactive components of mushrooms: processing effects and health benefits. Food Res Int 148:110599. https://doi.org/10.1016/J.FOODRES.2021.110599

Yadav UCS, Kalariya NM, Srivastava SK, Ramana KV (2010) Protective role of benfotiamine, a fat-soluble vitamin B1 analogue, in lipopolysaccharide-induced cytotoxic signals in murine macrophages. Free Radic Biol Med 48(10):1423–1434. https://doi.org/10.1016/J.FREERADBIOMED.2010.02.031

Yafetto L (2022) Application of solid-state fermentation by microbial biotechnology for bioprocessing of agro-industrial wastes from 1970 to 2020: a review and bibliometric analysis. Heliyon 8(3):e09173. https://doi.org/10.1016/J.HELIYON.2022.E09173

Yamada J, Hamuro J, Hatanaka H, Hamabata K, Kinoshita S (2007) Alleviation of seasonal allergic symptoms with superfine beta-1,3-glucan: a randomized study. J Allergy Clin Immunol 119(5):1119–1126. https://doi.org/10.1016/J.JACI.2007.02.005

Yang Z (2006) Online hyphenated liquid chromatography–nuclear magnetic resonance spectroscopy–mass spectrometry for drug metabolite and nature product analysis. J Pharm Biomed Anal 40(3):516–527. https://doi.org/10.1016/J.JPBA.2005.10.002

Yapp C, Fairman R (2006) Factors affecting food safety compliance within small and medium-sized enterprises: implications for regulatory and enforcement strategies. Food Control 17(1):42–51. https://doi.org/10.1016/J.FOODCONT.2004.08.007

Yazid NA, Barrena R, Komilis D, Sánchez A (2017) Solid-state fermentation as a novel paradigm for organic waste valorization: a review. Sustainability 9(2):224. https://doi.org/10.3390/SU9020224

Young AL, Bass IS (1995) The dietary supplement health and education act. Food Drug Law J 50(2):285–292. https://heinonline.org/HOL/Page?handle=hein.journals/foodlj50&id=321&div=&collection=

Yu WQ, Yin F, Shen N, Lin P, Xia B, Li YJ, Guo SD (2021) Polysaccharide CM1 from *Cordyceps militaris* hinders adipocyte differentiation and alleviates hyperlipidemia in LDLR(+/−) hamsters. Lipids Health Dis 20(1):178. https://doi.org/10.1186/S12944-021-01606-6

Yu J, Geng Y, Xia H, Ma D, Liu C, Wu R, Wu J, You S, Bi Y (2022) LAB fermentation improves production of bioactive compounds and antioxidant activity of *Withania somnifera* extract and its metabolic signatures as revealed by LC-MS/MS. J Microbiol Biotechnol 32(4):473–483. https://doi.org/10.4014/JMB.2111.11018

Zhang M, Zhang Y, Zhang L, Tian Q (2019) Mushroom polysaccharide lentinan for treating different types of cancers: a review of 12 years clinical studies in China. Prog Mol Biol Transl Sci 163:297–328. https://doi.org/10.1016/BS.PMBTS.2019.02.013

Zheng J, Cheng G, Li Q, Jiao S, Feng C, Zhao X, Yin H, Du Y, Liu H (2018) Chitin oligosaccharide modulates gut microbiota and attenuates high-fat-diet-induced metabolic syndrome in mice. Mar Drugs 16(2):66. https://doi.org/10.3390/MD16020066

Zhou X, Gong Z, Su Y, Lin J, Tang K (2009) Cordyceps fungi: natural products, pharmacological functions and developmental products. J Pharm Pharmacol 61(3): 279–291. https://doi.org/10.1211/JPP.61.03.0002

Zwir-Ferenc A, Biziuk M (2006) Solid phase extraction technique—trends, opportunities and applications. Pol J Environ Stud 15(5):677–690. http://www.pjoes.com/Solid-Phase-Extraction-Technique-Trends-Opportunities-and-Applications.87920.0.2.html

20 Stability of Fungal-Based Cosmeceutical Formulations

Adedola S. Adeboye, Tosin Olaniran, Oluwatoyin O. Onipe, and Helen N. Onyeaka

Abstract

Cosmeceuticals, combining cosmetic and therapeutic benefits, are a rapidly expanding segment in the personal care industry, driven by growing consumer demand for skincare that enhances both appearance and well-being. This chapter explores the potential and challenges of fungal-based cosmeceuticals as fungal-derived ingredients, such as polysaccharides, beta-glucans, and antioxidants, are gaining popularity in addressing skin concerns like aging, inflammation, and hydration due to their resilience, stability, and effectiveness. The importance of stability in cosmeceutical formulations, with physical, chemical, and microbiological factors affecting product integrity, is addressed. Key techniques for stability assessment, including accelerated and real-time testing, are discussed, along with case studies on fungal ingredients like *Cordyceps sinensis* and *Ganoderma lucidum.* Challenges in formulating stable fungal-based cosmeceuticals, such as microbial contamination and chemical degradation, as well as strategies for improving stability, are also discussed. The chapter concludes by emphasizing the potential of fungal-based cosmeceuticals in meeting consumer demand for natural, effective skincare solutions while underscoring the need for continued research to ensure product safety and sustainability.

Keywords

Cosmeceuticals · Fungal-based cosmeceuticals · Fungal extracts · Stability testing

20.1 Overview of Fungal-Based Ingredients in Cosmeceuticals

Cosmeceuticals arguably is the intersection between cosmetics and pharmaceuticals, and one of the fastest growing categories within the personal care sector. Most cosmeceutical products contain active ingredients offering therapeutic benefits, making it a product line which falls

A. S. Adeboye (✉)
African Food Research Network, Gauteng, Pretoria, South Africa

T. Olaniran
Department of Consumer and Food Sciences, Faculty of Natural and Agricultural Sciences, University of Pretoria, Hatfield, South Africa
e-mail: tosin.olaniran@tuks.co.za

O. O. Onipe
Department of Biotechnology and Food Technology, Faculty of Science, University of Johannesburg, Doornfontein, South Africa
e-mail: oluwatoyino@uj.ac.za

H. N. Onyeaka
School of Chemical Engineering, University of Birmingham, Birmingham, UK
e-mail: h.onyeaka@bham.ac.uk

S. Ghosh, H. N. Onyeaka (eds.), *Fungal-Based Cosmetics: Formulation and Usage*, The Mycota 17,
https://doi.org/10.1007/978-3-032-09291-5_20

between traditional cosmetics and pharmaceutical-grade skincare. In contrast to traditional cosmetics, which mostly only work on the outer appearance of the skin; cosmeceuticals help take care of and maintain long-term endurance, aging issues, hyperpigmentation problems, or acne troubles on the skin (Lee 2016; Alamgir and Alamgir 2017).

The term "cosmeceutical" was first coined in the 1980s by dermatologist Dr. Albert Kligman, who recognized the potential of certain skincare products to deliver both cosmetic and therapeutic benefits (Bharatia et al. 2024). Since then, the cosmeceutical industry has grown exponentially, driven by advances in dermatological science, consumer demand for effective skincare solutions, and a growing awareness of the link between appearance and overall well-being.

In terms of key ingredients, cosmeceuticals often contain active ingredients such as vitamins (e.g., Vitamin C and E), peptides, retinoids, antioxidants, alpha hydroxy acids (AHAs), and plant-based extracts (Bellad et al. 2017). These ingredients are selected for their proven efficacy in promoting skin rejuvenation, collagen production, and protection against environmental stressors. An example is Vitamin C which is known for its brightening effects and ability to reduce the appearance of fine lines and wrinkles (Mohiuddin 2019). Others are retinoids, widely recognized for their role in accelerating cell turnover, reducing acne, and improving skin texture; and peptides which help in stimulating collagen production, which is essential for maintaining skin elasticity and firmness (Bharatia et al. 2024; Mohiuddin 2019).

The rise in the popularity of cosmeceuticals can be attributed to several factors, aging population being prominent among them. As the global population ages, there is an increasing demand for products that can effectively address the signs of aging. Cosmeceuticals offer targeted solutions that go beyond mere moisturization, helping to reduce wrinkles, firm the skin, and restore a youthful appearance (Mohiuddin 2019). Secondly, today's consumers are more knowledgeable and discerning than ever before. With access to vast amounts of information, they seek out products that are not only effective but also backed by scientific research. Cosmeceuticals, with their combination of cosmetic appeal and therapeutic benefits, meet this demand. The growing emphasis on health and wellness has led consumers to prioritize skincare as part of their overall self-care routine. Cosmeceuticals, which promote both aesthetic and health benefits, align perfectly with this trend. The contributory influence of research is significant. Continuous research in dermatology and skin biology such as that of Kaul et al. (2018) has led to the development of new and more effective cosmeceutical formulations. These advancements have expanded the range of treatable conditions and improved the efficacy of these products.

The cosmeceutical market is expected to continue its upward trajectory, driven by innovation and a steady increase in consumer demand (Lee 2016; Alamgir and Alamgir 2017; Bharatia et al. 2024). Key trends include the development of personalized skincare solutions, the incorporation of natural and organic ingredients, and the expansion of product offerings to address specific skin concerns for different demographics (Mohiuddin 2019; Bharatia et al. 2024). In addition, the growing interest in preventative skincare is likely to boost the demand for cosmeceuticals that protect against environmental damage and maintain skin health over time.

Stability in cosmeceutical formulations is critical to ensuring the safety, efficacy, and shelf-life of these products. A stable formulation maintains its intended physical, chemical, and microbiological properties over time, which is essential for delivering consistent therapeutic benefits to the consumer. Unstable formulations can lead to degradation of active ingredients, resulting in reduced potency and effectiveness, which can compromise the product's ability to address specific skin concerns. Additionally, instability can cause changes in texture, color, odor, or pH, potentially leading to skin irritation or adverse reactions.

Furthermore, stability is a key factor in maintaining the product's appeal and consumer trust. A well-formulated cosmeceutical product that remains stable under various storage conditions is more likely to meet regulatory standards and provide the promised benefits throughout its shelf life. Ensuring stability also helps prevent contamination and degradation caused by environmental factors such as light, temperature, and humidity. This is particularly important for cosmeceuticals containing sensitive active ingredients like vitamins, antioxidants, and retinoids, which require careful formulation and packaging to preserve their effectiveness. Overall, stability in cosmeceutical formulations is essential for delivering high-quality, reliable products that meet consumer expectations and regulatory requirements.

Fungal-based ingredients used in cosmeceuticals are often valued for their ability to adapt to and thrive in diverse environments, which translates into their resilience and efficacy in skincare formulations. They are also increasingly popular in natural and clean beauty products, appealing to consumers looking for sustainable and plant-based alternatives. As research into the dermatological applications of fungi continues, these ingredients are likely to play a growing role in the development of innovative cosmeceutical products that offer both therapeutic and cosmetic benefits. The rising interest in mycology and the exploration of fungal biodiversity are paving the way for new discoveries in this field, further expanding the range of effective ingredients available for cosmeceutical formulations. Associated challenges in formulation development and stability of fungal-based cosmeceuticals present a tall order to overcome by the personal care industry.

This chapter provides a special focus on fungal-based ingredients used in cosmeceuticals, drawing importance and attention to the significance of the stability of these ingredients and products derived from them, on shelf stability, quality, and customer satisfaction.

20.2 Fungal-Based Ingredients in Cosmeceuticals

Fungal-based ingredients in cosmeceuticals are gaining attention for their unique properties and potential benefits in skincare. Fungi, including various species of mushrooms and other fungal organisms, are rich in bioactive compounds such as polysaccharides, beta-glucans, antioxidants, and enzymes, which have been shown to promote skin health (Lee 2016; Bharatia et al. 2024). These ingredients offer a range of benefits, including hydration, anti-aging effects, skin barrier enhancement (Alamgir and Alamgir 2017), and anti-inflammatory properties (Taofiq et al. 2016). For instance, beta-glucans derived from fungi are known for their ability to deeply hydrate the skin, stimulate collagen production, and enhance the skin's natural defense mechanisms.

20.2.1 Benefits and Potential Applications of Fungal-Based Derived Components in Skincare

Fungal-derived components significantly impact the cosmetic industry because of their bioactive characteristics that promote skin health, combat aging, and provide protection. The cosmetic industry makes use of the therapeutic components found in the fruiting bodies of certain wild and cultivatable mushrooms like *Hypsizygus ulmarium, Tremella fuciformis, Cordyceps* spp., and *Ganoderma lucidum*. The benefits of fungal extracts in skincare products have been extensively researched, with major brands such as Aveeno and Estee Lauder leading the way in validating their effectiveness using these extracts. These validations, based on rigorous scientific studies, provide confidence in the effectiveness of fungal extracts in skincare products. Recent trends show a consumer-driven growing interest in cosmetic products made from natural, safe ingredients like proteins, vitamins, and antioxidants derived from selected plants (Woźniak et al. 2023). Hence, there is a recent rise in natural-based cosmetics, with consumers experimenting with natural food ingredients in their pantries.

Based on its use in skincare products, fungal derivatives can be classified as follows: extracts, oils, enzymes, polysaccharides, pigments, and acids. This versatility allows for a wide range of skincare products, from moisturizers to serums, to benefit from the unique properties of fungal derivatives. For example, reishi mushrooms contain polysaccharides, proteins, and terpenes, possessing antimicrobial, anti-inflammatory, and immunomodulator compounds, making them a versatile ingredient in skincare products.

20.2.2 Fungal Extracts

(a) *Mushroom Extracts:* Extracts from mushrooms like *Ganoderma lucidum* (Reishi), *Tremella fuciformis* (snow mushroom), and *Cordyceps* are predominantly used. These mushrooms contain nutrients like polysaccharides (beta-glucans) and antioxidants that boost hydration, skin elasticity, and immune function. For instance, *Tremella fuciformis* has exceptional moisturizing characteristics comparable to hyaluronic acid but offers additional antioxidant advantages. In their study, Woźniak et al. (2023) discovered that skincare formulations containing *Tremella fuciformis* extract effectively reduced transepidermal water loss by 12.4% and provided adequate hydration to the skin. Dermatological tests also confirmed that these formulations did not cause any erythema, swelling, or irritation in the test volunteers even after 48 and 72 h of application.

(b) *Fermented Fungal Extracts:* Fermentation enhances the bioavailability and potency of the fungal actives. For example, *Saccharomyces* ferment filtrates to nourish and brighten the skin. Aside from plants, yeasts like *Saccharomyces cerevisiae* and *Pichia pastoris* can synthesize resveratrol as a secondary metabolite. Research has shown that resveratrol affects cellular signaling mechanisms associated with UV-induced photoaging. These mechanisms include MAP kinases, nuclear factor kappa B, and matrix metalloproteinases (Baxter 2008). Using cheese whey as a substrate, *Candida curvata* was used to synthesize microbial oil in a moisturizer formulation (Pérez-Rivero and López-Gómez 2023).

(c) *Fungal-Derived Polysaccharides: Ganoderma lucidum (Curtis) P. Karst*, also known as Reishi mushroom, is an extremely valued mushroom with therapeutic properties. *Ganoderma lucidum*, a prominent species of the Ganoderma genus, has been acknowledged for millennia as an agent in traditional Chinese medicine. Reishi mushrooms contain bioactive molecules that offer multifunctionality as non-toxic natural ingredients in cosmetic products. Rich in beta-glucans, it has anti-inflammatory, moisturizing, and antioxidant properties, helping to reduce redness, fine lines, and wrinkles (Hyde et al. 2010). The polysaccharide extracts of *G. lucidum* at various purification levels effectively protected the skin from aging. Notably, these extracts do not exhibit any harmful effects on HaCaT keratinocytes, ensuring their safety for use. Polysaccharide extracts have proven to be highly effective in inhibiting lipid peroxidation, surpassing the efficacy of commonly used additives found in cosmeceutical formulations. Exposure to chronic ultraviolet rays from the sun or other sources leads to skin damage, accelerated wrinkle formation, photoaging, and disruption of skin barrier function. Choi and Kim (2021) investigated the effects of *Tremella fuciformis* polysaccharide as a functional cosmetic ingredient to combat these dermal issues. The polysaccharide extract derived from the edible fungus *Tremella fuciformis* (TF-Glucan®) demonstrated significant photoprotective effects by inhibiting matrix metalloproteases (MMPs) and promoting collagen synthesis. Moreover, it exhibited anti-inflammatory properties by inhibiting the production of nitric oxide and pro-inflammatory cytokines at concentrations below 1000 μg/ml.

Table 20.1 Overview of fungal-based cosmeceutical ingredients and their benefits

Fungal ingredient	Bioactive compounds	Cosmetic benefits	Example products
Ganoderma lucidum (Reishi)	Beta-glucans, polysaccharides	Anti-aging, anti-inflammatory	Creams, serums
Tremella fuciformis (snow mushroom)	Polysaccharides, antioxidants	Hydration, skin elasticity	Moisturizers
Cordyceps spp.	Polysaccharides, nucleosides	Anti-inflammatory, brightening	Lotions, masks
Saccharomyces cerevisiae (yeast)	Resveratrol, peptides	Anti-aging, UV protection	Serums, creams

Table 20.1 gives an overview of some commonly used fungal ingredients, showcasing their key bioactive components and cosmetic advantages.

20.3 Stability in the Context of Cosmeceuticals

When it comes to cosmeceuticals, stability is the capacity of a product to hold onto its intended physical, chemical, and microbiological characteristics over an extended period when stored under particular conditions (Das and Das 2022). Stability guarantees that the product is safe, functional, and visually appealing for its shelf life. Unstable products may degrade, resulting in decreased effectiveness and perhaps harmful effects. As such, stability is essential for both efficacy and safety (Mittal 2017). A stability study is the first crucial step in developing a cosmetic formulation since it helps forecast any changes in the product's physical, physicochemical, chemical, and microbiological makeup from the time of manufacture until the end of its expiration date (Vieira et al. 2009).

20.3.1 Factors Influencing Stability

Key aspects of stability testing in cosmeceuticals include:

A. *Physical Stability:* To guarantee consistency and repeatability, visual appearance, which is often one of the most important factors influencing consumer decisions, must be quantifiable. This entails assessing the product's color, texture, and overall appearance over time (White and Pointer 2002). Several variables as seen in Fig. 20.1, such as temperature, humidity, and exposure to light, can impact a product's overall physical stability. For instance, exposure to high temperatures during consumption, transportation, or storage may cause phase separation or differences in particle size, which may have an impact on customer satisfaction and perception. Therefore, it involves avoiding problems like crystallization, phase separation, or viscosity variations (Müller et al. 2002).

Physical stability is typically evaluated by long-term testing and visual inspection to determine whether a sample has any instabilities; this process might take several months. In emulsions, sedimentation and creaming tests are essential for determining whether particles stay suspended, sink to the bottom, or separate from the solution (Pal 2019). Comprehending particle size plays a pivotal role in ascertaining and forecasting physical stability, and in conducting rheological analyses to understand the physical and mechanical attributes of the final commodity (Brubaker and Moghtadernejad 2024). The procedure of performing an extensive long-term stability and physical examination might take a lot of time. Proficient analytical equipment can improve the speed at which tests are conducted, offer more impartiality, and enable examination of the product in its original condition. By monitoring variations in the size and distribution of emulsion

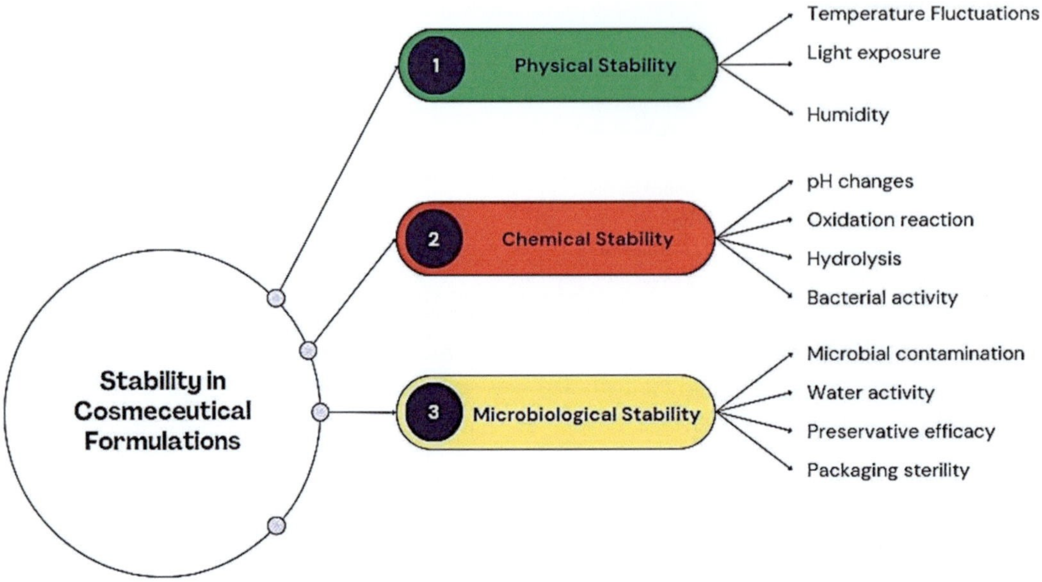

Fig. 20.1 Key factors influencing stability in cosmeceutical formulations

droplets, methods like centrifugation and microscopy can be used to forecast physical instability.

- *Microscopy:* This method aids in determining the distribution and size of emulsion droplets, which can reveal possible instability. Changes in distribution and larger droplet sizes can be used to forecast phase separation and aggregation (Khan et al. 2014).
- *Centrifugation:* This technique, which is meant to quicken the emulsion separation process, can detect early on whether the oil phase in the emulsion tends to separate from the water phase, a sign of insufficient emulsification (Sila-On et al. 2008).
- *Rheology:* The product's viscoelastic behavior and flow characteristics are measured. This aids in figuring out the consistency and forecasting the behavior of the product in various scenarios (Siska et al. 2019).

B. *Chemical Stability:* This evaluates the integrity of active ingredients and excipients. This entails preserving the active components' efficacy and purity. The other ingredients and active substances do not break down or interact in a way that would change the product's safety or effectiveness (Dave et al. 2015). Degradation of cosmeceutical products due to chemical instability can impact their efficacy, safety, and shelf life. This degradation is typically caused by changes in pH, oxidation, hydrolysis processes, or bacterial activity as seen in Fig. 20.1 (Springer and Ziegler 2022). Several medication compounds will react with oxygen and undergo an undesired chemical change, which makes oxygen a major source of problems for chemical stability. Degradation and the creation of contaminants and degradation products are the results of reactions with oxygen. Similar concerns with the synthesis of undesirable new compounds are caused by a process called hydrolysis, which can happen when exposed to water (Snape et al. 2010).

- *Differential Scanning Calorimetry (DSC):* This is one technique that can be used to monitor phase transitions and detect crystallization events, which are signs of chemical changes in the product. DSC monitors phase transitions such as melting and crystallization events. This information is crucial for understanding the thermal behavior and potential degradation of the product (Rafiquzzaman et al. 2017).

- *High-Performance Liquid Chromatography (HPLC):* HPLC is frequently utilized to measure the active component concentration over time, ensuring it remains within specified limits (Wang and Gray 2005).

C. *Microbiological Stability:* It is imperative to guarantee that the product is free from microbiological contamination and does not pose a risk of infection to the user throughout its shelf life. This is especially crucial for products that include water since it can serve as a breeding ground for bacteria and fungi (Alshehrei 2023). Challenge tests, in which the product is injected with germs to assess the effectiveness of the preservative, are a type of stability testing (Russell 2003).
   - *Preservative Efficacy Testing (PET):* This involves challenging the product with specific microorganisms to ensure the preservative system is effective throughout the product's shelf life. It is used across various industries to denote the testing of a product's ability to inhibit microbial growth (Ferreira et al. 2014).
   - *Sterility Testing:* This is essential for products that are supposed to be sterile, ensuring that they remain free from microbial contamination under normal storage conditions. Broadly sensitive techniques are used in sterility testing; samples are typically incubated in growth conditions to encourage the growth of a variety of bacteria where replication is visible (Roesti and Staerk 2023).

20.3.1.1 Importance of Stability Testing and Regulatory Considerations

For cosmeceuticals to have their safety, effectiveness, and quality maintained over their shelf life, stability testing is essential. Setting the variables to be tested, specifying the acceptability standards, and/or identifying the techniques capable of measuring the attribute changes over time are necessary for the stability test to have any significance (Kirkbride et al. 2021). The researcher is in charge of selecting the tests to be run, and criteria and choices are made based on the kind of cosmeceutical product that is related to the scientific literature, guidelines, and Official Compendiums, given that each formulation has unique properties of ingredients and cosmetic forms (Martin and Glaser 2011). The primary features of the product must also be considered in these criteria since they must not change or only be modified in a way that does not negatively impact the product's functionality and appearance. Proper documentation is essential for regulatory compliance. Stability data must be accurately recorded and reported in a manner that meets regulatory standards. This includes detailed descriptions of the testing methods, conditions, and results.

- *Efficacy:* Stability testing ensures that active ingredients maintain their potency and effectiveness over time. For instance, vitamin C and retinoids are known to degrade when exposed to light and air, which can significantly reduce their effectiveness (Temova Rakuša et al. 2021).
- *Safety:* Unstable products can form harmful degradation by-products that may cause skin irritation or other adverse reactions. Stability testing helps in identifying such potential risks before products reach consumers (Panico et al. 2019).
- *Consumer Satisfaction:* Consistency in product texture, color, and fragrance is vital for consumer satisfaction. Stability testing helps ensure that these characteristics remain unchanged over the product's shelf life, maintaining the overall user experience (Rico et al. 2024).
- *Shelf-Life Determination:* Stability testing provides data to support the shelf life claims on product packaging. This is essential for informing consumers about the expected duration of product efficacy and safety (Khan et al. 2018).
- *Regulatory Compliance:* Regulatory bodies require stable data to approve cosmeceutical products for market release. This includes evidence that products will remain stable under various environmental conditions (Ferreira et al. 2022).

20.3.1.2 Stability Testing Methods

Stability testing methods in cosmeceuticals are essential to ensure products remain safe, effective, and aesthetically pleasing throughout their shelf life. These methods encompass a variety of tests that simulate different storage conditions and stress factors to predict how a product will perform over time (Gupta et al. 2022). Stability testing is a common practice used at different phases of product development for cosmeceutical products. Accelerated stability testing, conducted at relatively high temperatures and/or humidity levels, is utilized in the first phases to identify the kinds of degradation products that might be discovered following extended storage (González-González et al. 2022). To find a product's shelf life and expiration date, testing is done under less strict circumstances, such as those advised for long-term shelf storage, at somewhat higher temperatures. Providing a reasonable assurance that the products will maintain an acceptable level of fitness or quality for the duration of their availability in the market and that they will be fit for patients to use up until the patient uses the last unit of the product is the main goal of cosmeceutical stability testing (Isin 2012). The stability test directs the creation of cosmetic formulations, offering details for better preparedness if instability symptoms or chemical incompatibilities arise.

20.3.1.3 Stability Testing Protocols (Accelerated, Real-Time, Stress Testing)

(a) *Accelerated Stability Testing:* Evaluating long-term stability through accelerated stability testing entails keeping items at higher humidity and temperature. A product is kept under high-stress conditions during accelerated stability testing (Haouet et al. 2018). Based on established correlations between the degradation rate and the acceleration factor, as well as the deterioration at every stress level, degradation under suggested storage conditions may be anticipated. A product undergoes stress at many elevated temperatures (above ambient) to ascertain the minimum quantity of heat input necessary to cause product failure. This is done to put the product in an environment that speeds up deterioration. Next, shelf life is estimated using this data, or it can be used to contrast the relative stability of different formulations (Calligaris et al. 2019). This usually reduces the development schedule by giving an early indication of the product's shelf life. During accelerated stability testing, stress factors such as temperature, humidity, light, agitation, gravity, pH, and packing are also applied. The samples are stressed, chilled after stressing, and then simultaneously analyzed (Haouet et al. 2018). Comparing the analysis's brief duration to the stability testing conducted in real-time, there is less chance of instability in the measuring system. Additionally, within the same assay, the unstressed and stressed materials are compared, and the stressed sample recovery is given as a percentage of the unstressed sample recovery.

(b) *Real-Time Stability Testing:* Real-time Stability Testing is conducted under normal storage conditions to confirm these predictions. A product is kept in approved storage conditions and observed for a while during real-time stability tests (Haouet et al. 2018). To allow for significant product degradation under suggested storage conditions, the test period is typically extended. The length of the test depends on the product's stability, which needs to be sufficient to show unequivocally that no detectable deterioration takes place and to allow one to discern degradation from inter-assay variance. Data is gathered throughout testing at a suitable frequency so that trend analysis can differentiate between daily ambiguity and instability. By using a single batch of reference material whose stability properties have already been determined, the reliability of data interpretation can be strengthened. The stability of the reagents and the instrument's consistent performance throughout the stability testing period are also considered aspects of the stability of the reference material. Monitoring is necessary to ensure that system performance and control are not compromised by drift or discontinuity brought on by modifications to the reagents or the instrumentation.

20.3.1.4 Testing Parameters

Common parameters monitored during stability testing include physical appearance, pH, viscosity, microbial load, and active ingredient concentration. This comprehensive approach ensures that all aspects of product stability are evaluated (Kim et al. 2019).

20.3.1.5 Analytical Techniques for Assessing Stability (e.g., Spectroscopy, Chromatography)

Several analytical techniques are essential in the field of cosmeceutical analysis. These techniques are specifically designed to satisfy the need to understand the molecular nature of cosmeceutical products and the intricacy of their formulations as seen in Table 20.2 (Rico et al. 2024). The main focus of this review is on chromatographic methods and spectroscopic methods. Chromatographic techniques that can effectively separate and identify complex combinations seen in cosmeceutical formulations include GC-MS (gas chromatography-mass spectrometry) and LC-MS/MS (liquid chromatography-mass spectrometry/mass spectrometry). Colorimetry is one of the fast and non-invasive spectroscopic techniques that is commonly used for measuring color in cosmeceutical formulations. When combined, these methods offer an extensive toolkit for meta-analysis in the cosmeceutical. They make it easier to comprehend the intricate relationships between the molecular makeup and microstructure of cosmeceutical products and how those relationships affect their macroscopic characteristics.

(a) *Chromatographic Analysis*

One of the most used separation methods in the cosmetics sector is chromatography, which is especially applied to the qualitative and quantitative examination of formulation ingredients. Numerous chromatography techniques exist, including liquid chromatography (LC) and gas chromatography (GC), which are connected to various detectors (UV, DAD, MS, and MS/MS). These techniques are distinguished by their robustness, sensitivity, and reproducibility. Techniques that enable the separation and analysis of mixtures are helpful for formulation and product innovation in complicated matrices like cosmeceutical goods (Coskun 2016).

(b) *Liquid Chromatography*

The principles of liquid chromatography are based on the fact that different analytes in a mixture have different affinities for the adsorbent material in the mobile phase or column. Due to this variance, the components move at different speeds, which eventually leads them to separate (Rusli et al. 2022). Liquid chromatography includes reverse-phase and normal-phase chromatography, among other types of separation mechanisms. There are also published methods using size exclusion chromatography, such as hyaluronic acid analysis. Analytes are separated in the column and then identified by a linked detection device. The most used detectors are mass spectrometry (MS) and diode-array detection (DAD), also known as UV-visible spectroscopy (UV-Vis).

Using a mass spectrometry detector is recommended for both structural identification and trace quantification. However, it is not as widely utilized due to its comparatively high cost. Ultimately, it can be concluded that the liquid chromatography technique is a valuable tool at various stages of the cosmeceutical product development process. It may be applied to reverse engineering and quality control of new products.

Apart from liquid chromatography, gas chromatography is another technique for separation that relies on the polarity and vapor pressure of the constituents. This method uses a carrier gas (He, H, N) and a stationary phase, which can be either liquid

Table 20.2 Analytical techniques applied in cosmeceuticals

Analytical techniques	Principle	Application in cosmeceuticals
Chromatographic analysis	Separation of components based on interactions with a stationary and mobile phase.	Analyzes complex mixtures and supports formulation innovation (Coskun 2016).
Liquid chromatography	Separation of analytes based on differing affinities for adsorbent material in the column.	Used for cosmeceuticals product development, quality control, and reverse engineering (Rusli et al. 2022).
Gas chromatography	Separation of low molecular weight, heat-tolerant chemicals in a gaseous state.	Analyzes fragrance components, essential oils, and perfumes (Stauffer et al. 2008).
Spectroscopic analysis	Quantitative evaluation of color using instruments like spectrophotometers and colorimeters.	Ensures color consistency and meets quality standards (Ashis Kumar 2024; Saleh and Lee 2023).

or solid. These two methods' distinctions limit them to a particular class of chemicals (Stauffer et al. 2008).

(c) *Gas Chromatography*
Gas chromatography (GC) is limited to low molecular weight chemicals that are heat-tolerant since the molecules must enter the column in a gaseous condition. GC is mostly employed in scent analysis in the cosmetics sector; the compounds used in perfumes are typically combinations of hundreds or thousands of analytes, either synthetically or naturally produced. Before mass spectrometric analysis and detection, GC-MS ionizes and fragments analyte molecules using an energetic electron. The compound identity of the analyte will next be ascertained by comparing its fragmentation pattern or molecular fingerprint to the spectra library. For analytes that could be resolved in GC, this approach is specifically designed. Gas chromatography is frequently connected to a mass spectrometer (MS) or flame ionization detector (FID) for these tests because the majority of scent ingredients, like alcohols and essential oils, have low molecular weights and low vapor pressures. Depending precisely on the matrix and detectors employed in the study, the chromatographic techniques utilized for scent analysis vary greatly in terms of column diameters and polarity, temperature ramps, and carrier gases. As a result, a critical step in the process is the quantification of these chemicals before the release of a product. According to reports, the quickest analysis method is gas chromatography because of the physicochemical properties of these substances (Stauffer et al. 2008).

(d) *Spectroscopic Analysis*
There are two methods for evaluating the color of cosmeceutical products: instrumental analysis and visual assessment. The human eye is capable of evaluating chromatic characteristics as well as other geometric elements like texture, shape, etc., qualitatively (Saleh and Lee 2023). The observer evaluates the sample's color in typical lighting conditions in this process, and the assessment is expressed in terms of specific scores after comparison with predetermined color criteria. Such visual assessment is relative, subjective, and contingent upon the observer and the surrounding circumstances. On the other hand, spectrophotometers, spectroradiometers, and colorimeters are only a few of the tools that can be used to quantitatively evaluate colorimetric properties.

The assessment of color has advanced to a more complex degree in modern times. In the past, preventing discrepancies was primarily achieved by simply comparing a pattern to a sample. Consolidating this understanding, color is crucial in the cosmetics sector. It accomplishes several goals, including as enhancing a product's aesthetic appeal and satisfying commercial quality standards. There are various techniques to evaluate color, ranging from the raw ingredients to the finished product. Advanced equipment, software, and regulated lighting conditions can help researchers and producers create better goods more quickly (Ashis Kumar 2024). These instruments allow for the accurate evaluation and control of color properties in a variety of materials, including liquids, powders, semi-solids, and even human skin. As such, they aid in the preservation of uniform quality among various production batches, therefore augmenting adaptability to the demands of the contemporary market.

20.3.2 Case Studies or Examples of Stability Assessment in Fungal-Based Cosmeceuticals

20.3.2.1 *Cordyceps Sinensis* Extract

Cordyceps Sinensis, a medicinal mushroom, is used in cosmeceuticals for its anti-aging and anti-inflammatory properties. Pintathong et al. (2021) evaluated the stability of its extract in a cosmetic formulation. For 6 months, samples were kept at 40 °C and 75% relative humidity. Parameters such as pH, color, and active ingredient concentration (cordycepin) were monitored using HPLC. Preservative efficacy testing was conducted to ensure microbial stability. The researchers found that there were no significant changes in pH or color. The *Cordycepin* remained stable with minimal degradation, the preservative system was effective, and no microbial growth was detected. The *Cordyceps Sinensis* extract demonstrated good stability under accelerated conditions, suggesting a long shelf life for the final product (Pintathong et al. 2021).

20.3.2.2 *Ganoderma lucidum* (Reishi Mushroom) Extract

Ganoderma lucidum, commonly known as Reishi mushroom, is widely used in cosmeceutical products for its anti-aging and antioxidant capabilities. Reishi mushroom extract was incorporated into a cream base. For 6 months, the product was kept in different temperatures (4 °C, 25 °C, and 40 °C) and relative humidity (60% RH). The physical appearance, pH, viscosity, and microbial contamination were regularly checked. The study found that the product remained stable at 4 °C and 25 °C, with minimal changes in pH and viscosity. However, significant degradation of active compounds occurred at 40 °C, suggesting the need for cool storage conditions (Anil Kumar et al. 2023).

20.3.3 Strategies to Enhance Stability

Enhancing the stability of fungal-based cosmeceuticals involves several strategies that address both formulation and packaging aspects. Here are some key approaches:

(a) *Optimizing Formulation Composition*

Use of Stabilizers and Antioxidants: Incorporating stabilizers such as emulsifiers and thickeners can enhance the physical stability of emulsions and gels. Antioxidants, like vitamin E or ascorbic acid, protect sensitive ingredients from oxidation (Ghelichi et al. 2023).

Encapsulation Techniques: Encapsulating active ingredients in liposomes, nanoparticles, or microspheres can protect them from environmental factors such as light, heat, and oxygen. This also allows for controlled release, improving the longevity of the active compounds (Xie et al. 2018).

pH Adjustment: Adjusting the pH of the formulation to the optimal range for the active ingredients can prevent degradation. For example, many fungal extracts are stable at slightly acidic pH levels, which can also inhibit microbial growth (Fernández-López et al. 2023)

(b) *Advanced Preservation Systems*

Natural Preservatives: Using plant extracts or essential oils (like tea tree oil) as natural preservatives with antimicrobial properties can reduce microbial contamination without affecting the product's natural appeal (Herman et al. 2013).

Broad-Spectrum Preservatives: Incorporating broad-spectrum preservatives can protect against a wide range of microorganisms. However, it's crucial to choose preservatives that do not interact negatively with the active ingredients or other components of the formulation (Stroppel et al. 2023).

(c) *Packaging Innovations*
Airless Packaging: Airless pump dispensers and tubes can prevent air from coming into contact with the product, reducing the risk of oxidation and contamination.
Opaque and UV-Protective Containers: Packaging that protects the product from light, especially UV radiation, can significantly enhance the stability of light-sensitive fungal extracts (Gore and Prajapat 2022).
Tamper-Evident Seals: Ensuring the packaging is tamper-evident can protect the product from contamination and degradation (Marsh and Bugusu 2007).

(d) *Environmental Control*
Controlled Storage Conditions: Storing products at stable, moderate temperatures and humidity levels can prevent degradation. This includes recommendations for consumers to avoid storing products in areas with fluctuating temperatures, such as bathrooms.
Refrigeration for Sensitive Ingredients: For extremely sensitive formulations, refrigeration can be recommended, provided that the product's packaging is designed to withstand low temperatures without compromising its integrity.

(e) *Incorporating Chelating Agents*
Metal ions that may catalyze breakdown activities can attach to chelating agents such as EDTA. This helps in maintaining the stability of the formulation, especially in the presence of trace metals that might be present as impurities (Musakhanian et al. 2022).

(f) *Regular Stability Testing and Shelf-Life Studies*
Conducting thorough stability testing under various conditions (temperature, humidity, light) helps in knowing the product's shelf life and the best ways to store it. Regular testing ensures that any changes in formulation or packaging are evaluated for their impact on product stability (Haouet et al. 2018).

20.3.4 Challenges in the Stability of Fungal-Based Cosmeceutical Formulations

There are several challenges in maintaining the stability of fungal-based formulations. Table 20.3 presents an attempt to put these challenges in perspective.

A multidisciplinary strategy involving formulation science, microbiology, chemistry, and regulatory expertise is required to address these issues. Stable and effective cosmeceutical products based on fungi can be successfully developed and introduced to the market.

20.4 Conclusion

The cosmeceutical business remains at the front line of development, overcoming any barrier among beauty care products and drugs with items that upgrade excellence as well as convey restorative advantages. As the interest in successful, deductively upheld skincare arrangements continues to develop, cosmeceuticals are strategically situated to address the issues of a maturing and progressively well-being-conscious populace. The continuous progressions in dermatological exploration and the improvement of steady, excellent plans will be pivotal in keeping up with purchaser trust and guaranteeing the drawn-out progress of this powerful area.

Introducing fungal-based ingredients is a promising new development area for cosmeceuticals. These ingredients, with their unique properties such as [specific properties],

Table 20.3 Challenges and the potential solution in maintaining the stability of fungal-based cosmeceuticals

Challenge	Description	Potential solution
Microbial contamination	Contaminant organisms, particularly bacteria and other fungi, are known to colonize fungal-based ingredients during their production or storage.	Elimination of contamination risk through use of proper preservatives and strict implementation during manufacturing procedures, such as employing aseptic processing condition for making the product into drug form before formulation with good manufacturing practices (GMP).
Chemical stability	Polysaccharides and peptides are known bioactive compounds in mushrooms, but they can easily degrade due to pH optima, temperature sensitivity or light exposure.	Development in formulation techniques like the addition of stabilizers, antioxidants, and encapsulation technology (e.g., liposomes or nanoemulsions) can improve chemical stability on these compounds.
Physical stability	Fungal-based formulations can be too heterogenous, causing the components to separate and eventually settle; this is a challenge not just because it will have lower desired biopesticide activity but also changing in texture which may affect user experience.	Optimizing the formulation's rheology, using emulsifiers and thickeners, and conducting rigorous stability testing can mitigate physical stability issues.
pH sensitivity	Numerous fungal extracts are highly pH-sensitive and can degrade if the "vehicle" is either too high or too low in its own natural precedent.	Buffer and pH modulators can be added to the formulation in order to maintain a level of accessibility for degrading organisms.
Compatibility with other ingredients	Ingredients derived from fungi may interact with other components in the formulation, resulting in instability or decreased efficacy.	Testing and laboratory evaluation prior to mixing the ingredients is essential. This will ensure homogeneity of the ingredients.
Oxidation	Parasitic concentrates, especially those high in phenolic compounds, are prone to oxidation, prompting a variety of changes and diminished movement.	To prevent oxidation, natural extracts or antioxidants like vitamin E can be added to the formulation.
Shelf life	Cosmeceutical products may have a shorter shelf life than synthetic formulations due to the natural origin of fungal ingredients.	The shelf life of the products can be extended through the use of preservatives, stability testing under a variety of conditions, and protective packaging.
Regulatory compliance	Ensuring that contagious-based details fulfill administrative guidelines in various business sectors can be mind-boggling, especially concerning additive use and marking prerequisites.	During product development, compliance with all necessary regulations can be ensured by working closely with regulatory experts.
Consumer perception	Customers might have worries about the Well-being and adequacy of parasitic-based items because of the relationship of growths with diseases or deterioration	Consumer trust can be increased by educating them about the benefits and safety of cosmeceuticals based on fungi through marketing and clear labeling.

meet the changing preferences of customers looking for natural and sustainable skincare options. However, the industry must address the formulation challenges associated with these ingredients, particularly in terms of safety, to fully realize their potential. By overcoming these obstacles, the personal care sector can continue to innovate and provide products that align with the growing emphasis on holistic health and wellness. Looking forward, the future of cosmeceuticals lies in the capacity to join state-of-the-art science with buyer experiences, conveying customized, compelling, and earth-cognizant items. As the business explores these difficulties and potentially opens doors, it will assume an essential part in forming the fate of skincare, offering arrangements that improve the appearance and add to general skin well-being and prosperity.

References

Alamgir ANM, Alamgir ANM (2017) Classification of drugs, nutraceuticals, functional food, and cosmeceuticals; proteins, peptides, and enzymes as drugs. In: Therapeutic use of medicinal plants and their extracts: volume 1: pharmacognosy. Springer, pp 125–175

Alshehrei FM (2023) Isolation and identification of microorganisms associated with high-quality and low-quality cosmetics from different brands in Mecca region-Saudi Arabia. Saudi J Biol Sci 30(12):103852

Anil Kumar NV, Quispe C, Herrera-Bravo J, Herrera Belén L, Loren P, Salazar LA, Silva V, Erdogan Orhan I, Senol Deniz FS, Nemli E, Capanoglu E, Olatunde A, Cristina Cirone Silva N, Zivkovic J, Shorog EM, Calina D, Sharifi-Rad J (2023) Potential of mushrooms bioactive for the treatment of skin diseases and disorders. J Food Biochem 2023(1): 5915769

Ashis Kumar S (2024) Advanced methods and tools for color measuring and matching: for quality check of colored products of textiles and apparel industry. In: Ashis Kumar S (ed) Advances in colorimetry. IntechOpen, Rijeka

Baxter RA (2008) Anti-ageing properties of resveratrol: review and report of a potent new antioxidant skin care formulation. J Cosmet Dermatol 7(1):2–7

Bellad KA, Nanjwade BK, Kamble MS, Srichana T, Idris NF (2017) Development of cosmeceuticals.World. J Pharm Pharm Sci, 6(4):643–691

Bharatia R, Gupta N, Upadhyay P, Tiwari SK, Gupta S, Sahni K (2024) A review on cosmeceuticals. Int J Pharma Prof Res 15(1):128–141

Brubaker J, Moghtadernejad S (2024) A comprehensive review of the rheological properties of powders in pharmaceuticals. Powders 3(2):233–254

Calligaris S, Manzocco L, Anese M, Nicoli MC (2019) 12—accelerated shelf life testing. In: Galanakis CM (ed) Food quality and shelf life. Academic

Choi JH, Kim B (2021) Improvement of skin photoaging by polysaccharide extract derived from *Tremella fuciformis* (white jelly mushroom). Nat Prod Sci 27(4):300–306

Coskun O (2016) Separation techniques: chromatography. North Clin Istanb 3(2):156–160

Das P, Das M (2022) Physical, chemical, and microbiological stability of nanocosmetics. In: Nanocosmeceuticals, pp 139–166. https://doi.org/10.1016/B978-0-323-91077-4.00005-3

Dave V, Haware R, Sangave N, Sayles M, Popielarczyk M (2015) Drug-excipient compatibility studies in formulation development: current trends and techniques. AAPS, FDD Section Newsletter—January 2015

Fernández-López MG, Batista-García RA, Aréchiga-Carvajal ET (2023) Alkaliphilic/alkali-tolerant fungi: molecular, biochemical, and biotechnological aspects. J Fungi (Basel) 9(6):652

Ferreira MRS, Lourenço FR, Ohara MT, Bou-Chacra NA, Pinto TJA (2014) An innovative challenge test for solid cosmetics using freeze-dried microorganisms and electrical methods. J Microbiol Methods 106: 104–109

Ferreira M, Matos A, Couras A, Marto J, Ribeiro H (2022) Overview of cosmetic regulatory frameworks around the world. Cosmetics 9(4):72

Ghelichi S, Hajfathalian M, Yesiltas B, Sørensen A-DM, García-Moreno PJ, Jacobsen C (2023) Oxidation and oxidative stability in emulsions. Compr Rev Food Sci Food Saf 22(3):1864–1901

González-González O, Ramirez IO, Ramirez BI, O'connell P, Ballesteros MP, Torrado JJ, Serrano DR (2022) Drug stability: ICH versus accelerated predictive stability studies. Pharmaceutics 14(11):2324

Gore AH, Prajapat AL (2022) Biopolymer nanocomposites for sustainable UV protective packaging. Front Mater 9:855727

Gupta V, Mohapatra S, Mishra H, Farooq U, Kumar K, Ansari MJ, Aldawsari MF, Alalaiwe AS, Mirza MA, Iqbal Z (2022) Nanotechnology in cosmetics and cosmeceuticals—a review of latest advancements. Gels 8(3):173

Haouet MN, Tommasino M, Mercuri ML, Benedetti F, Bella SD, Framboas M, Pelli S, Altissimi MS (2018) Experimental accelerated shelf life determination of a ready-to-eat processed food. Ital J Food Saf 7(4):6919

Herman A, Herman AP, Domagalska BW, Młynarczyk A (2013) Essential oils and herbal extracts as antimicrobial agents in cosmetic emulsion. Indian J Microbiol 53(2):232–237

Hyde KD, Bahkali AH, Moslem MA (2010) Fungi—an unusual source for cosmetics. Fungal Divers 43:1–9

Isin A (2012) Standard operating procedures (what are they good for?). In: Isin A (ed) Latest research into quality control. IntechOpen, Rijeka

Kaul S, Gulati N, Verma D, Mukherjee S, Nagaich U (2018) Role of nanotechnology in cosmeceuticals: a review of recent advances. J Pharm 2018(1):3420204

Khan M, Sheraz M, Ahmed S, Kazi S, Shahnavi I (2014) Emulsion separation, classification and stability assessment. J Pharm Pharm Sci 2:56–62

Khan MM, Jiang B, Mazzeo A, Huynh-Ba K (2018) Stability challenges not addressed by harmonized guidance—AAPS workshop of the stability focus group, April 3rd-4th, 2017 in Rockville, MD. AAPS Open 4(1):2

Kim JH, Lee K, Jerng UM, Choi G (2019) Global comparison of stability testing parameters and testing methods for finished herbal products. Evid Based Complement Alternat Med 2019:7348929

Kirkbride L, Humphries L, Kozielska P, Curtis H (2021) Designing a suitable stability protocol in the face of a changing retail landscape. Cosmetics 8(3):64

Lee CM (2016) Fifty years of research and development of cosmeceuticals: a contemporary review. J Cosmet Dermatol 15(4):527–539

Marsh K, Bugusu B (2007) Food packaging—roles, materials, and environmental issues. J Food Sci 72(3):R39–R55

Martin KI, Glaser DA (2011) Cosmeceuticals: the new medicine of beauty. Mol Med 108(1):60–63

Mittal B (2017) Chapter 6—analytical considerations. In: Mittal B (ed) How to develop robust solid oral dosage forms from conception to post-approval. Academic

Mohiuddin AK (2019) Skin aging & modern age anti-aging strategies. Int J Clin Dermatol Res 7:209–240

Müller RH, Radtke M, Wissing SA (2002) Solid lipid nanoparticles (SLN) and nanostructured lipid carriers (NLC) in cosmetic and dermatological preparations. Adv Drug Deliv Rev 54(Suppl 1):S131–S155

Musakhanian J, Rodier J-D, Dave M (2022) Oxidative stability in lipid formulations: a review of the mechanisms, drivers, and inhibitors of oxidation. AAPS PharmSciTech 23(5):151

Pal R (2019) Modeling of sedimentation and creaming in suspensions and pickering emulsions. Fluids 4(4):186

Panico A, Serio F, Bagordo F, Grassi T, Idolo A, Giorgi MDE, Guido M, Congedo M, Donno ADE (2019) Skin safety and health prevention: an overview of chemicals in cosmetic products. J Prev Med Hyg 60(1):E50–e57

Pérez-Rivero C, López-Gómez JP (2023) Unlocking the potential of fermentation in cosmetics: a review. Fermentation 9(5):463

Pintathong P, Chomnunti P, Sangthong S, Jirarat A, Chaiwut P (2021) The feasibility of utilizing cultured cordyceps militaris residues in cosmetics: biological activity assessment of their crude extracts. J Fungi (Basel) 7(11):973

Rafiquzzaman SM, Rahman MA, Kong I-S (2017) Chapter 5—ultrasonic-assisted extraction of carrageenan. In: Venkatesan J, Anil S, Kim S-K (eds) Seaweed polysaccharides. Elsevier

Rico F, Mazabel A, Egurrola G, Pulido J, Barrios N, Marquez R, García J (2024) Meta-analysis and analytical methods in cosmetics formulation: a review. Cosmetics 11(1):1

Roesti D, Staerk A (2023) Microbiology. In: Le Brun P, Crauste-Manciet S, Krämer I, Smith J, Woerdenbag H (eds) Practical pharmaceutics: an international guideline for the preparation, care and use of medicinal products. Springer, Cham

Rusli H, Putri RM, Alni A (2022) Recent developments of liquid chromatography stationary phases for compound separation: from proteins to small organic compounds. Molecules (Basel, Switzerland) 27(3):907

Russell AD (2003) Challenge testing: principles and practice. Int J Cosmet Sci 25(3):147–153

Saleh M, Lee Y (2023) Instrumental analysis or human evaluation to measure the appearance, smell, flavor, and physical properties of food. Foods 12(18):3453

Sila-On W, Vardhanabhuti N, Ongpipattanakul B, Kulvanich P (2008) Influence of incorporation methods on partitioning behavior of lipophilic drugs into various phases of a parenteral lipid emulsion. AAPS PharmSciTech 9:684–692

Siska B, Snejdrova E, Machac I, Dolecek P, Martiska J (2019) Contribution to the rheological testing of pharmaceutical semisolids. Pharm Dev Technol 24(1): 80–88

Snape T, Astles A, Davies J (2010) Understanding the chemical basis of drug stability and degradation. Pharm J 285:416–417

Springer A, Ziegler H (2022) The role of preservatives and multifunctionals on the oxidation of cosmetic O/W emulsions. Cosmetics 9(3):59

Stauffer E, Dolan JA, Newman R (2008) Chapter 8—gas chromatography and gas chromatography—mass spectrometry. In: Stauffer E, Dolan JA, Newman R (eds) Fire debris analysis. Academic, Burlington

Stroppel L, Schultz-Fademrecht T, Cebulla M, Blech M, Marhöfer RJ, Selzer PM, Garidel P (2023) Antimicrobial preservatives for protein and peptide formulations: an overview. Pharmaceutics 15(2):563

Taofiq O, Heleno SA, Calhelha RC, Alves MJ, Barros L, Barreiro MF et al (2016) Development of mushroom-based cosmeceutical formulations with anti-inflammatory, anti-tyrosinase, antioxidant, and antibacterial properties. Molecules 21(10):1372

Temova Rakuša Ž, Škufca P, Kristl A, Roškar R (2021) Retinoid stability and degradation kinetics in commercial cosmetic products. J Cosmet Dermatol 20(7): 2350–2358

Vieira R, Fernandes A, Kaneko T, Consiglieri V, Pinto C, Pereira C, Baby A, Velasco M (2009) Physical and physicochemical stability evaluation of cosmetic formulations containing soybean extract fermented by bifidobacterium animalis. Braz J Pharm Sci 45:516

Wang Q, Gray V (2005) 15—HPLC in dissolution testing. In: Ahuja S, Dong MW (eds) Separation science and technology. Academic

White H, Pointer M (2002) Measurement of appearance. Proc SPIE 4421:615–617

Woźniak B, Chudzińska J, Szczyglewska P, Nowak I, Feliczak-Guzik A (2023) Optimization of the composition of a cosmetic formulation containing tremella fuciformis extract (fungi). Cosmetics 10(3):82

Xie L, Sun C, Zheng X, Chen W (2018) Advances in micro and nano-encapsulation of bioactive compounds using biopolymer and lipid-based transporters. Trends Food Sci Technol 78:34–60

Challenges to the Industrial-Scale Production of Fungal-Based Cosmetics

21

Wilgince Apollon

Abstract

The increasing market interest in sustainable beauty products drives researchers to investigate fungal-based cosmetics as alternative solutions. Industrial-scale manufacturing of innovative products encounters multiple major production obstacles. The industrial production of fungal cosmetics encounters three major challenges, which include developing efficient cultivation methods to achieve consistent fungal quality and quantity, improving extraction and formulation techniques to maximize bioactive compound use, and overcoming regulatory challenges for new cosmetic ingredients. The successful entry of fungal-based cosmetics into the mainstream market depends on both consumer acceptance and effective marketing strategies. This chapter identifies and examines the main production barriers for large-scale manufacturing while providing guidance for future industry developments.

W. Apollon (✉)
Centro de Investigación en Ciencia Aplicada y Tecnología Avanzada (CICATA)-Instituto Politécnico Nacional (IPN), Altamira, Mexico

Department of Agricultural and Food Engineering, Faculty of Agronomy, Autonomous University of Nuevo León, San Nicolás de los Garza, Nuevo León, Mexico
e-mail: wilgince.apollon@uanl.edu.mx

Keywords

Cosmetic biotechnology · Bioprocess scalability · Fungal cultivation · Industrial fermentation · Mycotoxin contamination · Solvent extraction methods

21.1 Introduction

The cosmetics industry is undergoing a significant transformation by adopting sustainable natural ingredients instead of conventional synthetic materials. A new market demand for fungal-based cosmetics has emerged as a result of this industry shift (Martins and Marto 2023). Fungi exhibit both abundance and adaptability, providing a variety of bioactive compounds that enhance different cosmetic product formulations, including skincare, haircare, and color cosmetics. These bioactive compounds show great promise as they possess antioxidant, anti-inflammatory, and antimicrobial properties that improve product effectiveness (Wijesekara and Xu 2023; Chatterjee et al. 2019; Paterska et al. 2024). However, the promising potential of fungal ingredients in cosmetics remains unfulfilled, as scaling up production to meet growing market needs presents a significant challenge. The commercial manufacturing of these innovative ingredients faces obstacles due to unresolved issues related to cultivation techniques, extraction processes, and quality consistency (Patel et al. 2020;

S. Ghosh, H. N. Onyeaka (eds.), *Fungal-Based Cosmetics: Formulation and Usage*, The Mycota 17,
https://doi.org/10.1007/978-3-032-09291-5_21

Kumar et al. 2023). The cosmetics industry must navigate both regulatory standards and customer demands regarding sustainable practices and clear product information. The success of mainstream cosmetics with fungal-based products relies on overcoming these critical hurdles.

Research predicts that the cosmetics industry will grow to $ 438.38 billion by 2026 due to the expanding markets for natural and organic products (Fortune Business Insights 2020). The benefits of fungal-based cosmetics include renewable sourcing, enhanced skin compatibility, and superior bioactivity. However, the major production of these products encounters several barriers, including biological and scalability challenges, regulatory issues, safety concerns, as well as economic limitations, market instability, and technological obstacles in fermentation, extraction, and formulation.

This chapter explores the production difficulties and market potential of fungal-based cosmetics during large-scale manufacturing:

- Fungal-based cosmetics face various production hurdles related to biology, scalability, and technology implementation.
- Regulatory frameworks, safety concerns, and labeling requirements.
- Economic viability, market demand, and supply chain management.
- The research examines current solutions alongside upcoming technological advancements and market trends.

21.2 Cultivation Challenges

The commercial production of fungi for cosmetics requires solving multiple essential problems to succeed. The successful cultivation depends on three main factors: developing specialized nutrient solutions for different fungal species and contamination prevention measures and creating large-scale production systems that deliver consistent bioactive compound yields and quality standards. The successful cultivation of fungi requires all these factors to meet market standards and deliver reliable effective outputs for cosmetic formulations.

21.2.1 Nutrient Requirements for Optimal Fungal Growth

Fungi thrive best with particular nutrients, which must be understood to enhance cultivation conditions. The nutritional requirements of fungi differ substantially between different species and between desired bioactive compounds. Fungi's growth depends on receiving carbon sources (Barzee et al. 2021) and nitrogen sources with essential vitamins and minerals (Strong et al. 2022). The primary carbon sources used by fungi include glucose and sucrose, together with other carbohydrates and nitrogen sources derived from organic peptones or inorganic salts (Barzee et al. 2021).

The development of a nutrient medium that optimizes growth at affordable costs represents a major production obstacle. The use of rich media leads to increased growth rates and yields (Jayasekara et al. 2022), but it raises production expenses, which reduces the financial value of the final product. The nutrient composition needs precise balancing to prevent growth inhibition and unwanted by-product formation. Research into customized nutrient media, which optimize yield and bioactivity while reducing costs, plays an essential role in advancing fungal cultivation for cosmetics. The development of custom-designed nutrient solutions remains essential because they control fungal development, biomass accumulation, and bioactive compound synthesis during cosmetic cultivation (Hyde et al. 2019). The development of these media requires consideration of the nutritional needs of target fungal species through a selection of suitable carbon and nitrogen sources along with vital vitamins and minerals to establish beneficial growth conditions. These media designs should integrate safety features to reduce contamination risks because such threats negatively affect the quality and safety of cosmetic products. The nutrient media need to be adaptable for use in both laboratory and industrial production at different scales (Camilleri et al. 2025). The producers can achieve better consistency and bioactivity in fungal extracts by modifying environmental conditions and nutrient compositions, which results in improved fungal-based cosmetic products.

21.2.2 Contamination Risks During Cultivation

The contamination issue affects fungal cultivation widely because it leads to major problems with product quality and safety. Fungi easily become contaminated by bacteria as well as other fungi and elements from their environment (Hernandez and Martinez 2018). The competition between targeted fungal strains and contaminants for essential nutrients leads to reduced yields and decreased bioactive compound production. Pathogenic microorganisms that enter the cultivation process create substantial safety issues (Laezza et al. 2022). The cultivation process needs absolute aseptic methods to prevent contamination risks (Hernandez and Martinez 2018; Herman and Pauwels 2014; Life Sciences 2025). Growth media sterilization combined with sterile equipment use and environment control measures form part of these procedures (Weiskirchen et al. 2023). The maintenance of sterile conditions proves difficult for large-scale production facilities. The controlled environments of bioreactor systems decrease contamination risks (Palladino et al. 2024), although these systems demand substantial funding and specialized knowledge to operate them. The implementation of beneficial microorganisms as biocontrol agents to stop contaminant growth remains an active area of scientific investigation (Hyde et al. 2024). Biocontrol strategies under development hold great promise to boost both safety and reliability in fungal cultivation operations. Manufacturers who implement these strategies will decrease the chances of pathogenic microorganisms and contaminants that threaten fungal product quality. The improved safety measures ensure fungal cultivation meets rigorous health and safety requirements while building trust among consumers about these products. The development of fungal cultivation as an attractive option for cosmetic manufacturers who want natural, sustainable ingredients in their formulations becomes possible because of these advancements (Couceiro et al. 2025). The emerging trend will result in expanded product lines and novel cosmetic industry applications.

21.2.3 Scalability of Production Systems

The process of enlarging fungal cultivation from laboratory or pilot-scale to industrial-scale production creates major difficulties (Camilleri et al. 2025). The transition requires shifting from small, controlled environments to larger systems that need to sustain consistent growth conditions. The optimal growth of fungi requires precise monitoring and control of temperature and pH alongside oxygen levels and nutrient availability (Mukherjee et al. 2024; Bellettini et al. 2019). The main hurdle during expansion involves creating bioreactors that properly support the unique cultivation needs of different fungal species. The design of universal bioreactors becomes complex because different fungal species need varying amounts of oxygen and nutrients (Bellettini et al. 2019). The scale-up process results in unpredictable changes to growth dynamics because it can cause rate modifications and bioactive compound production changes that smaller experiments fail to detect. The economic sustainability of enlarging fungal cultivation operations needs a thorough evaluation to enhance scalability while ensuring cost-effectiveness. The scalability challenges require further investigation to find effective solutions.

21.3 Extraction and Processing Difficulties

This chapter explores the extraction and processing difficulties of fungal cosmetics by explaining extraction methods, standardization requirements, and active ingredient preservation. The extraction of valuable bioactive compounds from fungi involves three main methods: solvent extraction, supercritical fluid extraction, and enzymatic methods, which provide different extraction characteristics and performance levels. The lack of established extraction procedures creates a major challenge because different extraction methods produce variable results in product quality and effectiveness. The preservation of active ingredients throughout extraction and processing steps remains essential because heat, light, and oxygen exposure can reduce their effectiveness. The successful market entry of fungal-based cosmetics depends on solving extraction and processing challenges that meet consumer expectations and regulatory requirements.

21.3.1 Methods of Extracting Bioactive Compounds

Bioactive compound extraction from fungi represents an essential development step for fungal-based cosmetics since these compounds deliver beneficial effects that boost product effectiveness (Taofiq et al. 2020; Zhao et al. 2022). Different extraction methods exist for obtaining bioactive compounds because each method offers distinct benefits and constraints. Three major extraction techniques include solvent extraction as well as supercritical fluid extraction, and microwave-assisted extraction, as described (Bhadange et al. 2024). Solvent extraction stands as a common method that uses ethanol or methanol to extract a wide variety of compounds because of its ease of use and extraction effectiveness, as comprehensively reviewed by Lee et al. (2024). The selection of extraction solvents directly affects both the obtained compound yield and purity, which requires scientists to make precise choices regarding their target bioactive compounds.

The extraction technique known as supercritical fluid extraction (SFE) has become increasingly well-liked in the scientific community in recent years (Manjare and Dhingra 2019). Bioactive compounds can be extracted by using supercritical carbon dioxide as a solvent, which replaces the need for dangerous organic solvents in the process (Goyeneche et al. 2018). The extraction method demonstrates high selectivity because it allows for targeted compound extraction that maintains compound bioactivity. The extraction process becomes adjustable through changes to temperature and pressure conditions that enable the extraction of diverse bioactive compounds from fungi. SFE requires substantial initial funding and operational expenses which creates obstacles for emerging cosmetic companies that produce fungal products at large scales (Herzyk et al. 2024).

Microwave-assisted extraction (MAE) represents an innovative extraction approach that makes use of microwave power to boost extraction rates (Tapia-Quirós et al. 2024). The extraction of bioactive compounds becomes faster through this method because it raises extraction medium temperature and pressure, which results in better mass transfer and greater extraction yields. MAE works efficiently and rapidly since it needs less extraction time than conventional extraction techniques (Zhang et al. 2018). Uniform heating distribution and the possible deterioration of sensitive compounds during heat treatment represent issues that require proper resolution. The selection of extraction methods depends on the target bioactive compounds as well as cost factors and expected purity and yield requirements, thus requiring continuous research to achieve optimal processes for cosmetics manufacturing.

21.3.2 Standardization of Extraction Processes

The development of fungal-based cosmetics necessitates standardized extraction processes since this factor controls the quality, effectiveness, and safety of the final products. Due to the absence of standardized extraction protocols, large discrepancies arise in bioactive compound extraction, which can lead to inconsistent final products. This lack of consistency in fungal-based cosmetics production can undermine consumer trust and hinder market acceptance. Therefore, it is crucial for manufacturers to develop standard extraction methods to produce products that meet customer quality expectations and performance requirements (Sung et al. 2019; Andriani and Yanto 2024). The lack of standardization in extraction methods remains a significant challenge because different fungal species contain unique bioactive compounds. Various fungi require specific extraction techniques to optimize both the quantity and quality of their bioactive constituents. While solvent extraction is suitable for some fungal species, supercritical fluid extraction or microwave-assisted extraction generally yields better results (Chatterjee et al. 2019). A universal extraction protocol cannot be applied across all fungal species, as an appropriate protocol must be tailored to reflect each species (Hyde et al. 2019).

The selection of extraction solvents is critical for standardization purposes. Solvent extraction methods depend on polarity, solubility, and boiling point, as these factors influence both extraction yield and the bioactive compounds extracted (Jayasekara et al. 2022). Systematic studies should be conducted to identify which extraction solvents are most effective with specific fungal species and their targeted bioactive compounds. Optimizing the extraction process should encompass the evaluation of extraction conditions, time, and temperature (Mukherjee et al. 2024). Furthermore, developing quality control measures is essential alongside optimized extraction methods for standardization. Quality control protocols must incorporate assessments of bioactive compound concentrations, purity levels, and contamination analysis. Implementing these measures ensures that extracted compounds meet both regulatory requirements and consumer safety standards (Couceiro et al. 2025). Standardized analytical methods, such as high-performance liquid chromatography (HPLC) and mass spectrometry, help to establish a consistent evaluation of extraction efficiency and product quality (Paterska et al. 2024). Advancing the standardization of extraction processes requires collaboration among researchers, industry stakeholders, and regulatory bodies. Sharing knowledge and resources enables stakeholders to develop best practices and guidelines that promote consistent bioactive compound extraction from fungi. The standardized processes resulting from these collaborative efforts facilitate the industrial application of production methods (Camilleri et al. 2025). In summary, the successful commercialization of fungal-based cosmetics hinges on standardized extraction procedures as a fundamental element. The industry can enhance the reliability and efficacy of fungal-based products through the establishment of specific extraction protocols, quality control measures, and collaboration among stakeholders. This progress will contribute to expanding market acceptance of fungal ingredients in the cosmetics industry.

21.3.3 Preservation of Active Ingredients

The effectiveness of fungal-based cosmetic products depends heavily on maintaining the preservation of fungal-derived active compounds throughout the extraction, processing, and formulation stages. The bioactive molecules, including β-glucans, phenolic compounds, triterpenoids, and peptides, show high sensitivity to environmental degradation because of exposure to heat, oxygen, UV radiation, and pH changes. The study by Paterska et al. (2024) showed that macrofungal extract components that are heat-sensitive experience fast deterioration when subjected to typical extraction temperatures. Chatterjee et al. (2019) found that *Alternaria alternata* extracts lost their antioxidant capacity after exposure to oxidative stress for an extended period, which shows why preservation strategies should be implemented during the early processing stages. The natural instability of fungal bioactives requires scientists to establish specialized preservation methods that match both the chemical characteristics and functional properties of the active compounds.

The preservation of fungal-derived molecules has become possible through encapsulation technologies, which represent one of the most effective solutions. Microencapsulation that utilizes biopolymers like chitosan, alginate, or carrageenan effectively protects active ingredients from environmental stressors while maintaining their chemical stability and delivering controlled release in topical formulations (Mukherjee et al. 2024). The research by Couceiro et al. (2025) revealed that encapsulating fungal antioxidants inside alginate matrices resulted in improved oxidative stability that increased by 30% during a 4-week accelerated shelf-life assessment. Nanoencapsulation techniques have become prominent due to their capability to protect unstable bioactive and deliver them across skin barriers by using liposomes and solid lipid nanoparticles (SLNs) and nanostructured lipid carriers (NLCs) as carriers. The systems are protected by physical capture and steric hindrance, which limits light exposure and oxygen interaction while providing better penetration through the stratum corneum and improved bioavailability (Mukherjee et al. 2024; Paterska et al. 2024). The commercial development of nanoformulations derived from natural sources faces ongoing challenges regarding scalability and cost-effectiveness along with regulatory approval.

The stability of fungal bioactives can be enhanced by formulation optimization along with material compatibility to provide additional preservation beyond encapsulation. The stability of molecules depends heavily on three essential formulation parameters, which include pH value, ionic strength, and antioxidant co-formulants. Jayasekara et al. (2022) established that Xylaria extracts preserved their β-glucan content optimally when kept between pH 5.5 and 6.0 since this range matches the natural skin pH. The combination of ascorbic acid, tocopherol, and rosemary extract stabilizers shows synergistic properties that help protect polyphenolic compounds from oxidative breakdown (Hyde et al. 2019). The main degradation mechanism of chromophoric fungal metabolites, known as photodegradation, can be controlled by using opaque or UV-blocking primary packaging. Active ingredients require emulsifiers and thickeners and preservatives that do not react with them because excipient incompatibilities might cause degradation or bioactive precipitation. The combination of formulation design with analytical stability research serves as an essential method to maintain product activity during its entire shelf life period.

The development of formulations must accompany compliance with stability and product efficacy requirements of regulatory standards. The European Union and other main markets demand that manufacturers show evidence of ingredient stability along with performance throughout the shelf life period specified for cosmetics (Couceiro et al. 2025). Stability testing must be done thoroughly under different storage conditions using HPLC and FTIR and TGA analysis tools to assess compound integrity. Both safety and consumer confidence depend on maintaining batch-to-batch reproducibility and active concentration thresholds. Camilleri et al. (2025) stated that maintaining bioactivity alongside standardized upstream variables, including fungal strain selection, culture conditions, and extraction parameters, is necessary for fungal-based active ingredients. A vertically integrated production model that connects bioprocessing to formulation to regulatory alignment must include effective preservation as an integral component. The commercial potential of fungal-based cosmetics will receive increasing support from forthcoming green stabilization technologies as well as bio-based carriers and smart packaging systems.

21.4 Regulatory and Safety Concerns

Fungal-based cosmetics face specific hurdles in their growing industry, including regulatory and safety matters. The cosmetic industry faces substantial obstacles because fungal ingredients lack approved regulations. Previous research reveals that there exists no dedicated regulatory system for fungal-derived cosmetic ingredients. The absence of standardized testing protocols creates product safety and efficacy discrepancies because manufacturers do not need to follow them

(Ferreira et al. 2022; Vieira et al. 2024). The development of fungus-specific guidelines by regulatory bodies represents an essential requirement to guarantee product dependability and consumer safety. The regulatory framework for fungal-based cosmetics requires safety testing as an essential element. Thorough safety evaluations are essential because fungal materials can trigger allergic reactions in consumers (Crameri et al. 2014; Barthe et al. 2021; Furey et al. 2022; Xing et al. 2022). The research by Furey et al. (2022) brings forward important findings about Fy Protein™, which originates from *Fusarium* strain flavolapis as a secure nutritional replacement for conventional animal protein sources. The amino acid profile of Fy Protein matches animal proteins while delivering high digestibility through its PDCAAS score of 91. The high fiber content, along with low cholesterol and sodium levels in this product, makes it suitable for modern dietary targets that lower cardiovascular risks. Sustainable food technology benefits from Fy Protein because it needs 99% less land and water and produces 94% less greenhouse gas than beef production. The authors conducted a thorough multi-faceted safety evaluation that proved Fy Protein does not trigger genotoxic, mutagenic, allergenic, or toxic effects during laboratory testing. The controlled production of Fy Protein eliminated all harmful secondary metabolites from the *Fusarium* species origin despite its pathogenic and mycotoxin-producing nature. The allergenicity assessment, which included bioinformatics analysis along with literature review and in vitro digestibility studies, revealed that Fy Protein creates an extremely low probability of allergic reactions even among people with mold sensitivity. The research demonstrates that Fy Protein can function as a new food ingredient that provides safe nutritional value for worldwide food security.

The occurrence of toxic compounds within specific fungi groups creates more safety challenges (Milićević et al. 2010). Mycotoxins within particular fungal species create health dangers when used in cosmetic formulations. In one study by Greeff-Laubscher et al. (2019), they provide essential information about the dangers that mycotoxigenic fungi create for aquaculture operations, specifically regarding the security of prepared abalone feed. The researchers analyzed 93 fungal isolates from *Fusarium*, *Aspergillus*, and *Penicillium* genera to demonstrate that abalone feed provides optimal conditions for mycotoxin production, particularly through *Fusarium verticillioides* fumonisins. The results show fumonisin production occurs at higher levels when the temperature reaches 16 °C instead of 26 °C, which goes against previous beliefs that warmer temperatures enhance mycotoxin production (Greeff-Laubscher et al. 2019). The standard storage conditions at South African abalone farms, which feature humid coastal temperatures, may unintentionally increase mycotoxin accumulation, especially during winter months (Novak et al. 2024). The discovery holds substantial importance for aquaculture health and productivity because fumonisins cause growth impairment and cellular damage in aquatic animals. On the other hand, the study introduced a new ecologically significant aspect through its evaluation of fumonisins environmental fate in aquatic systems (Greeff-Laubscher et al. 2019). The fumonisin content in feed exposed to seawater decreased substantially while the concentration in water increased notably, with FB1 detectable after 2 h of submersion. The rapid movement of water-soluble fumonisins from feed into the surrounding aquatic environment occurs through a leaching process. The reduction of direct abalone exposure to contaminants through this process creates environmental concerns about water quality and the contamination potential of adjacent ecosystems. The detected water concentrations of fumonisins remained below critical thresholds yet demonstrate the need for ongoing monitoring and risk mitigation strategies in aquaculture operations, starting from feed formulation through storage to waste management (Greeff-Laubscher et al. 2019).

According to the above, we can argue that consumers require thorough screening methods to identify toxic substances before they become available on the market. The cosmetic industry's focus on natural ingredients makes it crucial to remove potentially dangerous substances from these products. We propose that regulatory agencies should create rigid guidelines to monitor the origin and testing process of fungal materials to avoid toxic contamination. Consumer perceptions regarding fungal-based cosmetics create market acceptance hurdles as well as challenges with regulatory and safety testing. Schiaroli et al. (2024) demonstrate important gaps in consumer behavior toward sustainable clothing disposal because donating to charity and gifting are common practices, but reselling and recycling methods receive limited use because of perceived difficulty and information deficiency. The study demonstrates that consumers need better education about sustainable disposal methods, and these practices hold essential value for environmental protection. The study reveals that sustainable disposal service effectiveness suffers from accessibility problems, which demonstrates the necessity for businesses and municipalities to enhance service availability and consumer participation through incentive programs. The research also demonstrates that targeting environmentally less-conscious consumers matters because their involvement leads to increased adoption of sustainable practices. Future research needs to analyze how digital environments affect consumer behavior while studying how online influencers transform sustainable fashion perceptions to create an expanded sustainable consumption framework for different consumer groups (Li et al. 2024; Vilkaite-Vaitone 2024). Additionally, we can also argue that educational programs should be established to teach consumers about fungal ingredient benefits and safety characteristics because this knowledge will ease their concerns and establish a better market reception. The industry must work on changing how customers perceive fungal-based cosmetics to boost their market acceptance.

To sum up, the successful commercialization of fungal-based cosmetics requires all stakeholders, including researchers, manufacturers, and regulatory bodies, to work together to solve complex regulatory and safety problems. The studies demonstrate that fungi must be integrated into cosmetics through established rules alongside complete safety protocols and consumer education for safe implementation. Successful commercialization of fungal-based cosmetics will depend on addressing current concerns as the industry advances.

21.5 Market Acceptance and Consumer Perception

The acceptance of fungal ingredients in cosmetics has become a topic of interest in recent years, but it still faces market challenges. Previous research reveals that people show mixed feelings toward fungi in personal care products because they know little about the advantages of using fungi. The research conducted by Amberg demonstrates how consumers now choose natural cosmetics because they want products that support their health and environmental values (Amberg 2023). The research shows that natural cosmetics attract 70% of consumers who would consider buying these products, while 56% show interest in exploring new brands. Consumer demand now favors products that combine health-consciousness with environmental friendliness. The research demonstrates that women demonstrate greater interest in buying natural cosmetics than men, according to the study results. Consumers are willing to spend more money on products with natural ingredients rather than chemical alternatives, which demonstrates their commitment to product safety and sustainability. The market data shows that natural product brands have an open market opportunity to develop better marketing strategies that highlight their products' health advantages and environmentally friendly features (Amberg 2023). Additionally, the antibacterial effectiveness of these natural compounds against Gram-positive and Gram-negative bacteria, along with their antifungal properties against yeast-like fungi in cosmetic formulations, was demonstrated by Fajdek-Bieda et al. (2025). The strong inhibition observed against *Neisseria gonorrhoeae* and *Haemophilus influenzae* demonstrates eugenol and isoeugenol as alternative therapeutic agents for treating antibiotic-resistant infections. The study results demonstrate that these compounds can serve as natural preservatives in cosmetics and show promise for treating infections from resistant bacterial strains, which represent a major public health issue. The study demonstrates that eugenol and isoeugenol possess substantial antimicrobial activity against various pathogenic microorganisms, including drug-resistant strains.

Studies revealed that consumers demonstrate an interest in natural, sustainable ingredients yet show resistance to fungi because they equate these organisms with decay and uncleanliness (Sułkowska-Ziaja et al. 2023; Vittayavarakorn and Sornsaruht 2024; Hinčica et al. 2024). The current social perception creates obstacles for brands who want to successfully market fungal-based cosmetics (Amberg and Fogarassy 2019). Fungal-based cosmetics need distinct marketing approaches that deviate from standard product marketing strategies. The conventional cosmetic industry uses established beauty standards and well-known ingredients like oils and botanicals to market their products (Martins and Marto 2023). Marketing fungal ingredients to consumers demands a different method that includes educational content to break through their hesitation.

In summary, the natural cosmetics industry should improve its marketing approach by clearly explaining product value to customers according to these research results. Consumers will trust natural cosmetics more when companies provide open labeling of their ingredients and detailed information about their supply chain operations. The market can be driven through specific advertising efforts that appeal to environmentally minded consumers who are women. Natural cosmetics brands should create educational materials that teach consumers about natural ingredient advantages and sustainable practices as the market keeps growing. Companies that align their products with consumer values will achieve both short-term market demand fulfillment and establish lasting customer loyalty among environmentally conscious consumers who value health. The application of eugenol and isoeugenol in cosmetics and pharmaceuticals requires thorough safety evaluations, which should include toxicity tests and studies of their compatibility with cosmetic formulations. Future studies need to determine the exact molecular processes behind their antimicrobial and anticancer effects while developing methods to achieve maximum effectiveness with minimal harm to normal cells. Research should investigate how eugenol and isoeugenol interact with other natural or synthetic preservatives because this could improve their effectiveness and extend their uses. The cosmetic industry can create sustainable natural product formulations that meet consumer demands by following these research directions, which lead to better product safety and effectiveness.

21.6 Economic Factors

Multiple factors influence the economic landscape of fungal-based cosmetics, including manufacturing costs, financing requirements, and competition with traditional cosmetic ingredients (Grădinaru et al. 2022). Wang et al. (2023) provide an excellent example by utilizing mushroom fermentation to create tastier and more nutritious foods. Their expertise in fungus cultivation applies to the cosmetic industry and its associated economic challenges, even though their focus is not on cosmetics. They investigated the effects of mushroom root fermentation broth (MRFB) on *Flammulina velutipes*' umami flavor and nutrient profile through their study. The quality of the mushrooms improved significantly with the addition of MRFB. The treatment groups receiving MRFB exhibited enhanced levels of crude fiber, crude fat, soluble protein, and soluble sugar compared to the control group. The E2 and E3 groups showed notable increases in monosodium glutamate-like amino acids, essential amino acids, total amino acids, and total 5′--nucleotide content. Equivalent umami concentration (EUC) measurements in these groups were higher than in the others, indicating superior umami flavor. According to the electronic tongue analysis, the E2 group achieved the highest scores for umami flavor along with saltiness and sweetness, though it had lower scores for sourness, astringency, and bitterness. The research indicates that MRFB acts as a nutrient-rich additive that enhances both the umami flavor and nutritional value of *F. velutipes* (Wang et al. 2023).

Zhang et al. (2024) investigated how Morchella crassipes mycelium fermentation affected soybean protein isolate (SPI) through evaluations of functional properties. Protein solubility reached 81.16%, with a 221.66% foaming capacity after 6 days of fermentation, which decreased surface hydrophobicity and increased sulfhydryl content, despite a notable decrease in emulsifying capacity. The analysis of free amino acids revealed that fermentation produced more favorable flavor compounds, especially umami notes, suggesting that fermentation can enhance the sensory quality of SPI. This research demonstrates how mycelium fermentation boosts both the nutritional value and technical functionality of soybean protein for diverse food applications that support sustainable protein production goals. Establishing fermentation facilities entails significant investment, creating entry barriers for new companies wishing to enter the cosmetics industry (Pérez-Rivero and López-Gómez 2023). According to Cardoso et al. (2020), bioreactors are costly, and the fermentation process requires expertise in mycology and fermentation technology (Formenti et al. 2014), which may deter potential investors. Companies should devise a comprehensive strategy that details their financial investment plans and return on investment strategies. The main challenge arises from established consumer acceptance of conventional cosmetic ingredients, which have remained popular for many years (Huong et al. 2024). The profitability of fungal-based cosmetics relies heavily on pricing, according to Halla et al. (2018). To address this issue, companies should implement diverse pricing tiers, offering budget-friendly options alongside premium choices while maintaining brand integrity. The cosmetic market would benefit from such pricing strategies as consumers seek natural and innovative products but require proof of their efficacy before making substantial purchases. A successful pricing strategy for fungal-based cosmetics must be crafted to ensure competitive success in the market.

21.7 Research and Development Barriers

Industrial-scale fungal cosmetics manufacturing is limited by multiple research and development challenges that prevent scientists from fully understanding fungal benefits. Research about fungal extracts in cosmetic formulations remains restricted because there is limited scientific information available regarding their safety and effectiveness in cosmetic products. The unique bioactive compounds found in many fungi have the potential to provide anti-aging effects, moisturizing properties, and antimicrobial protection, but there is a shortage of studies that scientifically prove these advantages (Hashem et al. 2023). Lack of empirical evidence restricts product formulation development while creating safety and regulatory concerns that are essential for the cosmetics industry (Corbu et al. 2023).

Multiple industries need to work together to surpass existing research challenges. Mycology, cosmetic science, and biotechnology create an exceptional chance for cross-disciplinary partnerships that lead to innovation through collaborative efforts. The expertise of mycologists and biotechnologists helps cosmetic companies discover new uses for fungal metabolites in their products (Gehlot and Singh 2018). Companies need to transform their approach by dedicating resources to enduring research projects instead of prioritizing quick product releases. The combined effort of different stakeholders results in finding new fungal ingredients that boost product performance while satisfying customers who want sustainable and environmentally friendly cosmetics.

The fast rate of technological progress creates both hurdles and beneficial prospects for developing cosmetic products from fungi. Modern extraction methods, together with formulation science and product testing, enable the use of fungal ingredients to create standard cosmetic products (Goyal and Jerold 2023). Companies face the task of innovative product development while they need to fulfill consumer needs and maintain compliance with regulations. The task of maintaining stability between product development and consumer requirements becomes particularly challenging for smaller businesses since they typically do not have enough resources to perform thorough research and development. Successful commercialization of fungal-based products depends on solving current research and development barriers as the natural cosmetics market keeps growing.

21.8 Conclusion

The mass manufacturing of fungal cosmetics faces multiple production hurdles because of cultivation and extraction problems, market regulations, consumer acceptance, and economic stability. The obstacles in fungal-based cosmetics production can be overcome through sustained research investment and stakeholder collaboration. Fungal-based cosmetics show promising potential to become a major contributor to sustainable beauty industry development. The successful commercialization of fungal cosmetics requires researchers, manufacturers, and regulatory bodies to collaborate on solving complex safety and regulatory issues and educating consumers about fungal ingredient benefits and safety. The industry must resolve these concerns to boost market acceptance which will create conditions for fungal-based cosmetics to become mainstream.

Acknowledgments W.A. gratefully acknowledges support from the Secretariat of Science, Humanities, Technology, and Innovation (SECIHTI, its Spanish acronym) for the postdoctoral fellowship (Grant No.4500144/Ref. No. 878025), as well as the CICATA, Altamira at the Instituto Politécnico Nacional, for their support.

References

Amberg N (2023) Appearance of natural cosmetics in consumer behavior related to cosmetics in Hungary. Visegr J Bioecon Sustain Dev 12(2):71–74. https://doi.org/10.2478/vjbsd-2023-0014

Amberg N, Fogarassy C (2019) Green consumer behavior in the cosmetics market. Resources 8(3):137. https://doi.org/10.3390/resources8030137

Andriani A, Yanto DHY (2024) Fungal applications in skincare products. In: Biomass-based cosmetics: research trends and future outlook, pp 113–132. https://doi.org/10.1007/978-981-97-1908-2_5

Barthe M, Bavoux C, Finot F, Mouche I, Cuceu-Petrenci-C, Forreryd A et al (2021) Safety testing of cosmetic products: overview of established methods and new approach methodologies (NAMs). Cosmetics 8(2):50. https://doi.org/10.3390/cosmetics8020050

Barzee TJ, Cao L, Pan Z, Zhang R (2021) Fungi for future foods. J Future Foods 1(1):25–37. https://doi.org/10.1016/j.jfutfo.2021.09.002

Bellettini MB, Fiorda FA, Maieves HA, Teixeira GL, Ávila S, Hornung PS et al (2019) Factors affecting mushroom Pleurotus spp. Saudi J Biol Sci 26(4):633–646. https://doi.org/10.1016/j.sjbs.2016.12.005

Bhadange YA, Carpenter J, Saharan VK (2024) A comprehensive review on advanced extraction techniques for retrieving bioactive components from natural sources. ACS Omega 9(29):31274–31297. https://doi.org/10.1021/acsomega.4c02718

Camilleri E, Narayan S, Lingam D, Blundell R (2025) Mycelium-based composites: an updated comprehensive overview. Biotechnol Adv 79:108517. https://doi.org/10.1016/j.biotechadv.2025.108517

Cardoso VM, Campani G, Santos MP, Silva GG, Pires MC, Gonçalves VM et al (2020) Cost analysis based on bioreactor cultivation conditions: production of a soluble recombinant protein using *Escherichia coli* BL21 (DE3). Biotechnol Rep 26:e00441. https://doi.org/10.1016/j.btre.2020.e00441

Chatterjee S, Ghosh R, Mandal NC (2019) Production of bioactive compounds with bactericidal and antioxidant potential by endophytic fungus *Alternaria alternata* AE1 isolated from *Azadirachta indica* A. Juss PLoS One 14(4):e0214744. https://doi.org/10.1371/journal.pone.0214744

Corbu VM, Gheorghe-Barbu I, Dumbravă AŞ, Vrâncianu CO, Şesan TE (2023) Current insights in fungal importance—a comprehensive review. Microorganisms 11(6):1384. https://doi.org/10.3390/microorganisms11061384

Couceiro B, Hameed H, Vieira AC, Singh SK, Dua K, Veiga F et al (2025) Promoting health and sustainability: exploring safer alternatives in cosmetics and regulatory perspectives. Sustain Chem Pharm 43:101901. https://doi.org/10.1016/j.scp.2024.101901

Crameri R, Garbani M, Rhyner C, Huitema C (2014) Fungi: the neglected allergenic sources. Allergy 69(2):176–185. https://doi.org/10.1111/all.12325

Fajdek-Bieda A, Pawlińska J, Wróblewska A, Żwierełło W, Łuś A, Klimowicz A (2025) Antibacterial and preservative potential of eugenol and isoeugenol in cosmetics: a natural solution for product stability. Appl Sci 15(4):2129. https://doi.org/10.3390/app15042129

Ferreira M, Matos A, Couras A, Marto J, Ribeiro H (2022) Overview of cosmetic regulatory frameworks around the world. Cosmetics 9(4):72. https://doi.org/10.3390/cosmetics9040072

Fortune Business Insights (2020) Cosmetics market size to reach USD 438.38 billion by 2026; driven by the increasing number of company collaborations and acquisitions [Press release]. Globe Newswire. https://www.globenewswire.com/news-release/2020/04/17/2017767/0/en/Cosmetics-Market-Size-to-Reach-USD-438-38-Billion-by-2026-Driven-by-the-Increasing-Number-of-Company-Collaborations-and-Acquisitions-says-Fortune-Business-Insights.html?utm_source=chatgpt.com. Accessed 25 Oct 2025

Formenti LR, Nørregaard A, Bolic A, Hernandez DQ, Hagemann T, Heins AL et al (2014) Challenges in industrial fermentation technology research. Biotechnol J 9(6):727–738. https://doi.org/10.1002/biot.201300236

Furey B, Slingerland K, Bauter MR, Dunn C, Goodman RE, Koo S (2022) Safety evaluation of Fy protein™ (nutritional fungi protein), a macroingredient for human consumption. Food Chem Toxicol 166: 113005. https://doi.org/10.1016/j.fct.2022.113005

Gehlot P, Singh J (eds) (2018) Fungi and their role in sustainable development: current perspectives. Springer, Singapore, p 779. https://doi.org/10.1007/978-981-13-0393-7

Goyal N, Jerold F (2023) Biocosmetics: technological advances and future outlook. Environ Sci Pollut Res 30(10):25148–25169. https://doi.org/10.1007/s11356-021-17567-3

Goyeneche R, Fanovich A, Rodrigues CR, Nicolao MC, Di Scala K (2018) Supercritical CO2 extraction of bioactive compounds from radish leaves: yield, antioxidant capacity and cytotoxicity. J Supercrit Fluids 135: 78–83. https://doi.org/10.1016/j.supflu.2018.01.004

Grădinaru C, Obadă DR, Grădinaru IA, Dabija DC (2022) Enhancing sustainable cosmetics brand purchase: a comprehensive approach based on the sor model and the triple bottom line. Sustainability 14(21):14118. https://doi.org/10.3390/su142114118

Greeff-Laubscher MR, Beukes I, Marais GJ, Jacobs K (2019) Mycotoxin production by three different toxigenic fungi genera on formulated abalone feed and the effect of an aquatic environment on fumonisins. Mycology 11(2):105–117. https://doi.org/10.1080/21501203.2019.1604575

Halla N, Fernandes IP, Heleno SA, Costa P, Boucherit-Otmani Z, Boucherit K, Rodrigues AE, Ferreira ICFR, Barreiro MF (2018) Cosmetics preservation: a review on present strategies. Molecules 23(7):1571. https://doi.org/10.3390/molecules23071571

Hashem AH, Attia MS, Kandil EK, Fawzi MM, Abdelrahman AS, Khader MS, Khodaira MA, Emam AE, Goma MA, Abdelaziz AM (2023) Bioactive compounds and biomedical applications of endophytic fungi: a recent review. Microb Cell Factories 22(1): 107. https://doi.org/10.1186/s12934-023-02118-x

Herman P, Pauwels K (2014) Biosafety recommendations on the handling of animal cell cultures. In: Animal cell culture. Springer, Cham, pp 689–716. https://doi.org/10.1007/978-3-319-10320-4_22

Hernandez H, Martinez LR (2018) Relationship of environmental disturbances and the infectious potential of fungi. Microbiology 164(3):233–241. https://doi.org/10.1099/mic.0.000620

Herzyk F, Piłakowska-Pietras D, Korzeniowska M (2024) Supercritical extraction techniques for obtaining biologically active substances from a variety of plant byproducts. Foods 13(11):1713. https://doi.org/10.3390/foods13111713

Hinčica V, Řezanková H, Macias K, Schulzová M (2024) Perception of natural cosmetics among central European consumers. Cent Eur Econ J 11(58):233–251

Huong VTM, Hung NP, Minh NTT, Thuy LK, Duyen LTN, Minh TN (2024) Factors affecting consumers' repurchase intention toward skin care cosmetics: a cross-sectional study in Vietnam. Heliyon 10(11): e32285. https://doi.org/10.1016/j.heliyon.2024.e32285

Hyde KD, Xu J, Rapior S, Jeewon R, Lumyong S, Niego AGT et al (2019) The amazing potential of fungi: 50 ways we can exploit fungi industrially. Fungal Divers 97:1–136. https://doi.org/10.1007/s13225-019-00430-9

Hyde KD, Baldrian P, Chen Y, Thilini Chethana KW, De Hoog S, Doilom M et al (2024) Current trends, limitations and future research in the fungi? Fungal Divers 125(1):1–71. https://doi.org/10.1007/s13225-023-00532-5

Jayasekara LCB, Poonsawad A, Watchaputi K, Wattanachaisaereekul S, Soontorngun N (2022) Media optimization of antimicrobial activity production and beta-glucan content of endophytic fungi Xylaria sp. BCC 1067. Biotechnol Rep 35:e00742. https://doi.org/10.1016/j.btre.2022.e00742

Kumar A, Verma V, Dubey VK, Srivastava A, Garg SK, Singh VP, Arora PK (2023) Industrial applications of fungal lipases: a review. Front Microbiol 14:1142536. https://doi.org/10.3389/fmicb.2023.1142536

Laezza C, Salbitani G, Carfagna S (2022) Fungal contamination in microalgal cultivation: biological and biotechnological aspects of fungi-microalgae interaction. J Fungi 8(10):1099. https://doi.org/10.3390/jof8101099

Lee JE, Jayakody JTM, Kim JI, Jeong JW, Choi KM, Kim TS et al (2024) The influence of solvent choice on the extraction of bioactive compounds from Asteraceae: a comparative review. Foods 13(19):3151. https://doi.org/10.3390/foods13193151

Li M, Choe YH, Gu C (2024) How perceived sustainability influences consumers' clothing preferences. Sci Rep 14(1):28672. https://doi.org/10.1038/s41598-024-80279-4

Life Sciences (2025). Cell culture contamination: understand, manage, and prevent unwelcome microbes. https://www.corning.com/worldwide/en/products/life-sciences/resources/stories/at-the-bench/how-to-control-bacterial-contamination-and-other-unwelcome-micro.html. Accessed 03 May 2025

Manjare SD, Dhingra K (2019) Supercritical fluids in separation and purification: a review. Mater Sci Energy

Technol 2(3):463–484. https://doi.org/10.1016/j.mset.2019.04.005
Martins AM, Marto JM (2023) A sustainable life cycle for cosmetics: from design and development to post-use phase. Sustain Chem Pharm 35:101178. (2023)
Milićević DR, Škrinjar M, Baltić T (2010) Real and perceived risks for mycotoxin contamination in foods and feeds: challenges for food safety control. Toxins 2(4): 572–592. https://doi.org/10.3390/toxins2040572
Mukherjee A, Das S, Chakraborty D, Pal N, Das N (2024) Fungi's treasure in cosmeceuticals-a comprehensive chemical approach. S Afr J Bot 166:311–331. https://doi.org/10.1016/j.sajb.2024.01.036
Novak T, Bridges CR, Naylor M, Yemane D, Auerswald L (2024) Acute hypercapnia at south African abalone farms and its physiological and commercial consequences. Fishes 9(8):313. https://doi.org/10.3390/fishes9080313
Palladino F, Marcelino PRF, Schlogl AE, José ÁHM, Rodrigues RDCLB, Fabrino DL et al (2024) Bioreactors: applications and innovations for a sustainable and healthy future—a critical review. Appl Sci 14(20):9346. https://doi.org/10.3390/app14209346
Patel S et al (2020) Fungal-based cosmetics: challenges and opportunities. J Pharm Cosmet Rev 4(2):1–9
Paterska M, Czerny B, Cielecka-Piontek J (2024) Macrofungal extracts as a source of bioactive compounds for cosmetical anti-aging therapy: a comprehensive review. Nutrients 16(16):2810. https://doi.org/10.3390/nu16162810
Pérez-Rivero C, López-Gómez JP (2023) Unlocking the potential of fermentation in cosmetics: a review. Fermentation 9:463. https://doi.org/10.3390/fermentation9050463
Schiaroli V, Fraccascia L, Dangelico RM (2024) How can consumers behave sustainably in the fashion industry? A systematic literature review of determinants, drivers, and barriers across the consumption phases. J Clean Prod 483:144232. https://doi.org/10.1016/j.jclepro.2024.144232
Strong PJ, Self R, Allikian K, Szewczyk E, Speight R, O'Hara I, Harrison MD (2022) Filamentous fungi for future functional food and feed. Curr Opin Biotechnol 76:102729. https://doi.org/10.1016/j.copbio.2022.102729
Sułkowska-Ziaja K, Trepa M, Olechowska-Jarząb A, Nowak P, Ziaja M, Kała K, Muszyńska B (2023) Natural compounds of fungal origin with antimicrobial activity—potential cosmetics applications. Pharmaceuticals 16(9):1200. https://doi.org/10.3390/ph16091200
Sung HJ, Khan MF, Kim YH (2019) Recombinant lignin peroxidase-catalyzed decolorization of melanin using in-situ generated H_2O_2 for application in whitening cosmetics. Int J Biol Macromol 136:20–26. https://doi.org/10.1016/j.ijbiomac.2019.06.026
Taofiq O, Barreiro MF, Ferreira IC (2020) The role of bioactive compounds and other metabolites from mushrooms against skin disorders—a systematic review assessing their cosmeceutical and nutricosmetic outcomes. Curr Med Chem 27(41):6926–6965. https://doi.org/10.2174/0929867327666200402100157
Tapia-Quirós P, Granados M, Sentellas S, Saurina J (2024) Microwave-assisted extraction with natural deep eutectic solvents for polyphenol recovery from agrifood waste: mature for scaling-up? Sci Total Environ 912: 168716. https://doi.org/10.1016/j.scitotenv.2023.168716
Vieira D, Duarte J, Vieira P, Gonçalves MBS, Figueiras A, Lohani A et al (2024) Regulation and safety of cosmetics: pre-and post-market considerations for adverse events and environmental impacts. Cosmetics 11(6):184. https://doi.org/10.3390/cosmetics11060184
Vilkaite-Vaitone N (2024) From likes to sustainability: how social media influencers are changing the way we consume. Sustainability 16(4):1393. https://doi.org/10.3390/su16041393
Vittayavarakorn N, Sornsaruht P (2024) Word-of-mouth innovation: a product development shortcut for organic skin care entrepreneurs in Thailand? Bus Strategy Dev 7(1):e331. https://doi.org/10.1002/bsd2.331
Wang Z, Bao X, Xia R, Hou Z, Li Y, Feng Y (2023) Effect of mushroom root fermentation broth on the umami taste and nutrients of *Flammulina velutipes*. J Future Foods 3:67–74. https://doi.org/10.1016/j.jfutfo.2022.09.010
Weiskirchen S, Schröder SK, Buhl EM, Weiskirchen R (2023) A beginner's guide to cell culture: practical advice for preventing needless problems. Cells 12(5): 682. https://doi.org/10.3390/cells12050682
Wijesekara T, Xu B (2023) Health-promoting effects of bioactive compounds from plant endophytic fungi. J Fungi (Basel, Switzerland) 9(10):997. https://doi.org/10.3390/jof9100997
Xing H, Wang J, Sun Y, Wang H (2022) Recent advances in the allergic cross-reactivity between fungi and foods. J Immunol Res 2022:7583400. https://doi.org/10.1155/2022/7583400
Zhang QW, Lin LG, Ye WC (2018) Techniques for extraction and isolation of natural products: a comprehensive review. Chin Med 13:1–26. https://doi.org/10.1186/s13020-018-0177-x
Zhang W, Deng Z, Liu T, Liang J, Liu J (2024) Fermentation with edible mushroom mycelia improves flavor characteristics and techno-functionalities of soybean protein. Food Biosci 59:104123. https://doi.org/10.1016/j.fbio.2024.104123
Zhao W, Yang A, Wang J, Huang D, Deng Y, Zhang X et al (2022) Potential application of natural bioactive compounds as skin-whitening agents: a review. J Cosmet Dermatol 21(12):6669–6687. https://doi.org/10.1111/jocd.15437

Index

A
Acceleration factor, 394
Acnes, 6, 7, 11, 24, 25, 35, 48, 50, 63, 64, 88, 139, 173–176, 190, 200, 203, 205, 206, 247, 262, 264, 311, 325, 331, 333, 388
Active ingredient preservation, 406
Agaricus, 28, 30, 32, 41, 56, 99, 352
Agaricus bisporus, 2, 4, 5, 8, 10, 27, 30, 55, 57, 61, 62, 65, 77–78, 89, 90, 97, 99–101, 113, 175, 190, 216, 219, 223, 226, 228, 334, 354, 356, 370
Agaricus blazei, 2, 27, 30, 32, 52, 100, 222, 331, 354, 355
Aging, 6, 10, 24, 25, 27, 41, 50, 54, 60, 61, 64, 88, 97, 99, 121, 131, 132, 139, 145, 146, 151, 152, 154, 155, 157, 174, 175, 179, 191, 203, 204, 215, 220, 225, 226, 278, 279, 281, 285–287, 311, 321, 340, 366, 372, 388
Agro-industrial waste, 16, 40
Aldehydes, 116–118, 123
Algae, 120, 123, 153–154, 156–157, 162, 167, 218, 220, 235, 257, 259, 260, 278, 307
Allergenicity, 11, 246, 247, 409
Allergic reactions, 11, 16, 48, 49, 88, 192, 246, 252, 280, 306, 338, 409
Amphotericin, 244
Analytes, 268, 395, 396
Anti-aging, 3, 51, 88, 130, 154, 170, 186, 200, 218, 236, 251, 262, 302, 322, 368, 389, 412
Anti-aging agents, 2, 174, 326
Anti-aging cosmetics, 31, 36, 225
Anti-aging creams, 3, 10, 65, 136, 155, 171, 176, 186, 190, 191, 251, 254, 324, 330, 332
Antibacterial, 4, 5, 8, 11, 12, 14, 25, 28, 31, 36, 41, 48, 50, 52, 53, 55–59, 74, 75, 79, 97, 98, 101, 163–170, 172–176, 205, 221, 251, 304, 309, 322, 353, 357, 364, 367, 410
Antibacterial agents, 4, 188
Anticancer, 4, 6, 8, 14, 25, 28, 29, 31, 48, 53, 54, 57–60, 74, 79, 80, 134, 135, 163–165, 167, 199, 219, 223, 245, 302, 307, 309, 354, 355, 357, 364–368, 411
Anticancerous agents, 58
Anti-collagenase activities, 151, 153, 155–156
Antidiabetic, 4, 6, 28, 29, 31, 74, 309, 354, 365, 367
Anti-elastase activities, 143–157, 303
Antifungal, 4, 14, 55, 76, 101, 172, 174, 175, 195, 234, 236, 237, 239, 242, 244–247, 250, 251, 253–254, 322, 353, 364, 367, 368
Antifungal properties in cosmetics, 239, 410
Antigenotoxic, 2, 219
Anti hyaluronidase activities, 130–139
Anti-hyperglycemic, 371
Antihypertensive, 28, 75
Anti-inflammatory, 4, 50, 74, 87, 120, 130, 163, 186, 201, 236, 262, 287, 304, 320, 352, 369, 390, 403
Anti-inflammatory fungi, 6, 12, 56, 64, 89, 104, 217, 357, 369, 370
Anti-inflammatory properties, 4–7, 10, 15, 16, 50, 59, 89, 133, 187, 192, 193, 236, 249, 282, 331, 357, 358, 366, 370, 389, 390, 397
Antimicrobials, 5, 6, 8, 10, 15, 24, 26–30, 35, 36, 51, 52, 55–57, 59, 64, 74–78, 80, 81, 98, 101, 120, 135, 137, 161–179, 186–188, 190, 192, 199–201, 205, 216, 219, 221, 234, 236, 238, 239, 242–244, 248–250, 254, 255, 304, 321, 331, 337, 353–356, 361, 365, 367, 390, 397, 403, 411, 412
Antioxidant activities, 26, 31, 35–37, 53, 59, 61, 76–80, 97, 113, 135, 168, 201, 217, 222, 223, 258, 263, 303, 333, 334, 362, 366
Antioxidants, 2, 24, 48, 74, 88, 113, 130, 152, 163, 186, 199, 216, 234, 245, 258, 302, 323, 352, 388, 403
Antioxidative, 28, 29, 53, 56, 97, 100, 201, 259, 263
Anti-tyrosinase, 29, 51–53, 75, 79, 81, 97, 98, 101, 102, 111–123, 201, 207, 217–219, 227, 320
Anti-tyrosinase agents, 51, 80, 122, 164, 173
Antrodia, 31, 56, 60, 76, 201, 226, 227
Applications, 4, 23, 48, 77, 87, 123, 130, 167, 186, 199, 217, 235, 243, 258, 289, 302, 320, 352, 389, 405
Aquasomes, 327
Arbutin, 31, 113, 116, 204, 206, 218
Ascomycota, 3, 27, 74, 206, 259, 260, 265, 308
Aspergillus, 5, 8, 10, 35, 36, 41, 55, 58, 118, 120, 123, 163, 176, 187, 188, 193, 201, 203, 205, 225, 234, 236, 244, 248, 306, 331, 333, 338, 409
Aspergillus niger, 3, 54, 56, 118, 120, 134, 135, 154, 187, 188, 190, 195, 200, 333, 336
Aspergillus oryzae, 35, 36, 54, 55, 57, 58, 168, 174, 205, 280, 283, 333

S. Ghosh, H. N. Onyeaka (eds.), *Fungal-Based Cosmetics: Formulation and Usage*, The Mycota 17, https://doi.org/10.1007/978-3-032-09291-5

Astaxanthin, 8, 59, 258–260, 262–266, 268–271
Auricularia, 27, 29, 32, 41, 61, 75, 78, 80, 97, 98, 100, 101, 200, 225, 309, 310
Azelaic acid (AzA), 10, 55, 113, 116, 118, 120, 176, 204–207

B
Bacteria, 8, 14, 26, 35, 76, 94, 98, 114, 118, 130, 144, 148, 163, 164, 167–169, 172–175, 221, 242, 243, 245, 247, 250, 254, 257–259, 265, 307, 309, 336, 355, 362, 363, 370, 371, 393, 399, 405, 410
Bakuchiol, 51, 205
Basidiomycota, 3, 27, 28, 30, 55, 74, 170, 206, 259, 260, 265, 308, 309
Beauty products, 9, 13, 15, 24, 170, 214, 321, 389
Benefits, 3–7, 9, 13, 14, 24, 29, 30, 37, 39, 41, 50, 58, 64, 66, 74, 99, 103, 132, 134, 135, 137–139, 155, 170, 173, 174, 176, 186, 187, 189, 190, 193, 194, 201–203, 222, 223, 226, 234–239, 241–247, 249, 252, 254, 258, 261, 262, 269, 270, 278, 282, 289, 302, 303, 305, 311, 313, 314, 320, 327, 329–331, 336, 337, 339–341, 352, 353, 359, 362, 364–372, 388–391, 399, 404, 406, 409, 410, 412, 413
Beta-carotene, 8, 79, 80, 258–262, 264–266, 271, 334
Beta-glucans, 2, 4–7, 10, 11, 13, 15, 16, 27, 29, 31, 33, 36, 37, 52, 53, 55, 56, 77, 97, 99, 132–134, 167, 171, 174, 206, 234, 236, 237, 239, 302–305, 307–309, 311, 331, 337, 354–356, 365, 366, 370, 389–391, 407
Bioactive, 6, 23, 48, 75, 87, 122, 130, 152, 162, 188, 199, 214, 236, 242, 255, 258, 277, 303, 327, 353, 389, 407
Bioactive compounds, 5, 6, 11, 14, 15, 23–25, 27, 29–32, 34, 35, 39, 40, 47, 48, 50, 53, 56–58, 60–62, 64–66, 75, 78, 79, 89, 90, 95, 96, 104, 118, 120, 123, 130, 137, 139, 154, 157, 186, 187, 189–191, 193–195, 214, 221, 236, 241–255, 269, 278, 279, 303, 310, 331–335, 337, 341, 352–367, 370, 371, 374, 389, 391, 399, 403–407, 412
Bioactive compounds in haircare, 10, 403
Bioactive fungi, 6, 12, 23–25, 39, 55, 75, 96, 120, 137, 139, 186, 188–190, 194, 195, 236, 244, 254, 255, 278, 279, 331–334, 352, 353, 358, 374, 406, 407
Biodegradable cosmetics, 13, 186, 189
Bioinformatics tools, 289, 293, 294, 296
Biomolecules, 104, 334, 336, 338, 339
Biopolymers, 59, 202, 206, 309, 336, 408
Bioprocessing, 153, 283–284, 313, 362–364, 408
Bioprocess optimization, 177
Bioprocess scalability, 405
Bioreactor design, 363
Biosurfactants, 37, 334, 337
Biosynthesis, 6, 30, 113, 115, 122, 218, 219, 226, 227, 245, 259–261, 280, 287, 291, 293–295
Biotechnological applications of fungi, 9, 279
Biotechnology, 8, 11–13, 15, 16, 53, 58, 66, 104, 139, 194, 196, 234, 278–280, 297, 373
Biotransformation, 4, 15, 60, 74, 187–189
Butyriboletus regius, 26

C
Carbon black, 323, 325, 326, 330
Cardiovascular, 2, 26, 58, 157, 353, 354, 357, 366–368, 409
Carotenoid pathway, 259–261
Carotenoids, 4, 8, 10, 27, 35, 50, 57–59, 76, 89, 153, 157, 162–164, 186, 191, 192, 194, 223, 224, 238, 257–271, 304, 354, 368
Catansomes, 327
Cell disruption, 266–268, 283
Ceramides, 3, 4, 7, 15, 16, 24, 41, 55, 57, 58, 74, 89, 99, 130, 134, 189, 191, 200–203, 207, 222, 224, 278–296, 336
Chaga, 28, 51, 64, 65, 121, 170, 192, 222, 223, 228, 336, 352
Chelating, 78, 97, 117, 118, 120, 398
Chitin, 10, 56, 163, 167, 174, 175, 308, 331, 336, 365, 370, 371
Chitin-glucan, 3, 4, 56, 134, 135, 188, 190, 193, 207, 304, 306, 308, 336
Chitosans, 4, 61, 74, 101, 163, 167, 173, 174, 193, 195, 200, 234, 236, 245, 250, 254, 304, 308, 331, 337, 340, 408
Chromatography, 201, 335, 360–362, 395–396
Chromatography (HPLC, TLC), 78, 95, 253, 284, 360, 361, 393, 397, 407, 408
Circular economy, 36, 52, 56, 138, 340
Clitocybin A, 191
Collagen, 7, 24, 50, 77, 88, 131, 153, 164, 191, 207, 215, 245, 264, 286, 302, 357, 388
Collagenase, 25, 27, 41, 144–155, 216, 224
Collagen degradation, 77, 151, 220, 225, 226
Collagen synthesis, 30, 50, 56, 75, 77, 151, 167, 175, 201, 225, 245, 357, 390
Consumer perception, 12, 249, 296, 312–313, 399, 410–411
Conventional extraction, 54, 270, 406
Cordycepin, 29, 38, 100, 101, 234, 236, 251, 340, 353, 354, 367–369, 371, 397
Cordyceps, 27, 29, 32, 41, 48, 52–54, 56, 58, 64, 65, 89, 99, 134, 136, 171, 222, 236, 251, 303, 305, 331, 332, 336, 352, 353, 367, 368, 371, 389–391, 397
Cordyceps militaris, 8, 29, 38, 53, 89, 98, 100, 101, 173, 226, 262, 263, 333, 334, 354, 371
Cosmeceuticals, 3–5, 9, 15, 27, 29, 30, 48, 50–53, 55, 59, 60, 63–65, 80, 81, 87–90, 97–99, 103, 104, 122, 130, 154–157, 161–164, 166, 173, 176, 179, 205, 215, 216, 219–224, 227, 261, 263, 264, 277–281, 283, 284, 288, 289, 293–295, 302, 310, 320, 322, 324, 325, 327, 330, 331, 340, 354, 387–399
Cosmetic, 2, 23, 47, 73, 89, 111, 130, 152, 164, 191, 199, 216, 234, 251, 258, 278, 302, 319, 352, 387, 407
Cosmetic biotechnology, 412
Cosmetic formulations, 2, 4–6, 8–10, 15, 16, 24, 35, 40, 47–49, 53, 61, 65, 104, 112, 122, 173, 177–178, 199, 200, 202, 204, 207, 224, 227, 242, 243, 254, 303, 304, 313, 320, 321, 331, 333–341, 391, 394, 397, 404, 409–412

Cosmetic ingredients, 25–31, 37, 40, 41, 50–52, 54, 66, 73–81, 121, 163–166, 194, 195, 199, 201, 220, 223, 225, 234, 235, 252, 303, 313, 320, 326, 328, 335, 390, 408, 411, 412
Cosmetic mycology, 412
Cosmetic regulatory frameworks, 40
Cosmetic sustainability, 3, 5
COX-2 inhibition, 101, 226, 370
CRISPR, 340
Cultivable, 23–27, 29, 31, 32, 35, 37, 40, 41, 302
Cultivation, 6, 24, 31–34, 37–41, 53, 99, 139, 164, 195, 199, 255, 280, 283, 303, 312–314, 334, 355, 403–405, 411, 413
Cyclooxygenase-2 (COX-2), 88, 92, 93, 98, 101, 217, 226, 370
Cytokines, 89, 93–95, 149, 170, 217, 225, 245, 282, 356, 357, 365–369, 390

D

Dectin-1 receptor, 365, 370
Degradation rate, 394
Dendrimers, 323, 326, 327
Dermatological, 3, 48, 134, 145, 150, 165, 175, 176, 187, 188, 195, 205, 207, 277, 310, 321, 336, 389, 390, 398
Dermatological science, 330, 388
Dermatology, 49, 112, 145, 169, 187, 190, 204, 205, 321, 388
Dermatophytes, 175, 244, 245, 248
Dermatophyte-targeting ingredients, 175
Desquamation, 200, 202, 203
Dictyophora indusiata, 155
Dietary supplements, 30, 36, 53, 152, 262, 351–353, 371, 373
Dihydroxyphenylalanine (DOPA), 62, 80, 112, 114, 204, 218, 226
Dimethyl sulfoxide (DMSO), 97, 268, 269
Diode-array detection (DAD), 96, 395
2,2-Diphenyl-1-picrylhydrazyl (DPPH), 14, 77–79, 216, 223, 263, 333, 336
Disinfection, 243
Dopaquinone, 25, 62, 80, 114, 218
Drug delivery systems, 104, 321

E

Eco-friendly products, 234, 238, 249, 271, 311
Eczema, 6, 7, 11, 64, 88, 280–282, 285, 286, 305, 331
Edible mushrooms, 29–31, 51, 53, 66, 74, 75, 78, 79, 120, 206, 228, 263, 310, 352, 370
Efficacy, 3, 5, 8, 9, 15, 30, 31, 40, 48–50, 55, 60, 88, 103, 104, 120, 121, 151, 163, 168, 178, 206, 218, 236, 238, 242, 244, 246–250, 252–255, 269, 279, 283, 307, 312–314, 320, 321, 325, 328–330, 334, 337, 339, 341, 373, 388–393, 397, 399, 407, 408, 412
Elastase, 25, 27, 41, 61, 75, 144–155, 191, 216, 220, 224–226, 311
Elastin degradation, 226
Ellagic acid, 48, 116, 134, 190, 218
Emodin, 92
Emollients, 37, 50, 51, 53, 54, 97, 112, 166, 199–208, 214, 234, 304
Emulsion, 9, 101, 214, 215, 258, 284, 324, 328, 335, 358, 391, 392, 397
Environmental impacts, 138, 162, 194, 234, 255, 279, 280, 289, 295, 296, 303, 312–314, 325, 352
Environmental stressors, 134, 285, 286, 388, 408
Environmental sustainability, 279, 311, 340, 373
Enzymatic browning, 112, 217
Enzyme, 5, 24, 50, 75, 89, 112, 130, 143, 165, 189, 203, 216, 245, 258, 289, 304, 327, 356, 389
Enzyme inhibition, 132
Ergosterols, 4, 5, 10, 24, 25, 29–32, 41, 52, 55, 75, 76, 96, 97, 100, 201, 226, 236, 245, 320, 353, 354, 367, 369–371
Ergothioneine, 2, 6, 7, 16, 31, 34, 51, 53, 55, 56, 75, 79, 100, 174, 175, 237, 304, 354, 369
Ethical sourcing, 13, 48, 279, 280
Ethosomes, 324, 326, 330
Exfoliants, 203–204, 207
Exfoliation, 35, 174, 200, 203, 204, 207, 304
Extracellular matrix (ECM), 25, 89, 131, 133, 144–146, 148, 150, 154, 179, 218, 220, 221
Extraction, 8, 36, 48, 95, 137, 164, 194, 200, 220, 234, 249, 267, 283, 302, 335, 353, 403
Extraction methods, 15, 16, 48, 54, 139, 234, 267, 268, 270, 278, 312, 352, 358, 359, 406, 407, 412
Extracts, 2, 25, 48, 74, 89, 112, 134, 151, 162, 191, 199, 214, 234, 254, 267, 278, 302, 320, 358, 388, 406
Extrinsic aging, 145, 149

F

Fermentation, 5, 29, 58, 122, 154, 163, 193, 202, 234, 265, 278, 308, 362, 390, 411
Fermentation extracts, 154
Fermented fungal extracts, 234, 236, 390
Fibroblasts, 9, 29, 36, 56, 57, 61, 99, 144, 151, 175, 220, 225, 245, 251, 309, 357
Filamentous fungi, 9, 10, 24, 35, 36, 54, 65, 66, 187, 191, 202, 205–207, 225, 250, 262, 334, 363
Flavonoids, 10, 29–31, 50, 52, 58, 59, 76, 78–80, 97, 98, 101, 116–117, 152, 162, 202, 222, 227, 238, 250, 304, 325, 331, 333, 354, 357, 362, 368, 371
Formulation, 3, 24, 47, 81, 97, 123, 130, 152, 177, 186, 204, 214, 236, 254, 284, 305, 325, 373, 388, 412
Fourier-transform infrared (FTIR), 284
Free radicals, 2, 4, 6, 10, 27, 31, 35, 50, 56, 59, 61, 65, 76, 77, 79–81, 88, 94, 114, 116, 118, 120, 122, 132, 135, 152, 157, 174–176, 191, 205–207, 215, 217, 222, 224, 225, 237, 245, 249, 258, 261, 263, 286, 302, 305, 310, 325, 357, 366, 367
Fruiting bodies, 4, 29, 30, 32–34, 37, 39–41, 52, 62, 75, 76, 78, 79, 81, 89, 121, 164, 165, 167, 168, 172, 201, 222, 223, 226–228, 309, 320, 353, 365, 389
Fullerenes, 324, 325
Functional cosmetics, 390
Functional foods, 6, 60, 76, 370
Functional fungi, 370

Fungal, 5, 23, 48, 81, 89, 118, 132, 154, 177, 187, 199, 214, 235, 244, 257, 278, 302, 331, 352, 389, 412
Fungal-based cosmetics, 3–4, 6–7, 9, 11–16, 41, 64–66, 111–123, 130–139, 185–196, 199–208, 214, 228, 302, 319–341, 403–413
Fungal bioactives, 24, 40, 54, 60–62, 66, 95, 104, 241–243, 373, 407, 408
Fungal biotechnology, 8–9, 195
Fungal compounds, 6, 7, 14–16, 24, 41, 51, 55, 62–63, 89, 94–96, 132–133, 200, 205–207, 246, 352, 371, 372
Fungal-cosmetics, 25, 135–138, 307, 331–337, 339, 405, 412, 413
Fungal cultivation, 15, 37, 39–41, 66, 313, 404, 405
Fungal-derived ceramides, 277–297
Fungal-derived ingredients, 214, 234, 239, 307, 312
Fungal-derived keratinases, 203
Fungal-derived polysaccharides, 390
Fungal enzymes, 37, 167–168, 174, 227, 334
Fungal extracts, 6, 7, 10, 11, 13, 15, 16, 24, 25, 39, 47, 48, 62, 64, 66, 87, 89, 97–100, 103, 137–139, 143–157, 166, 191–194, 199, 200, 225, 226, 234, 254, 334–339, 389–391, 397–399, 405, 412
Fungal fermentation, 63–64, 139, 179, 202
Fungal melanin, 8, 24, 333
Fungal metabolites, 6, 9, 15, 39–41, 55, 94, 95, 162, 173, 341, 408, 412
Fungal mycelium, 201, 203
Fungal oils, 250
Fungal polysaccharides, 39, 77, 132, 175, 237, 302, 305–314, 331
Fungi, 2, 23, 48, 74, 88, 114, 130, 154, 177, 186, 200, 214, 235, 244, 257, 278, 302, 331, 352, 389, 403
Fungicidal, 248, 249
Fungi-derived ingredients, 192
Fungistatic, 162, 248, 249
Fungus-based cosmetics, 2–16, 214–228

G
Ganoderma, 5, 6, 10, 25, 27, 32, 41, 52, 56, 57, 62, 76, 79, 97–100, 102, 135, 155, 163, 164, 166, 187, 188, 200, 221, 223, 227, 309, 314, 331, 333, 371, 390
Ganoderma lucidum, 2, 4, 7, 10, 13, 26, 27, 48, 52, 57, 61, 62, 64, 65, 76, 78–80, 89, 91, 94, 98, 99, 102, 133, 134, 136, 163, 171, 172, 174, 193, 199, 219, 222, 223, 226, 227, 251, 303, 305, 309, 311, 331, 332, 336, 352, 354, 358, 366, 367, 389–391, 397
Ganoderma lucidum in haircare, 62
Gastrointestinal infections, 242
Gene circuits, 294
Genetic modification, 278, 295
Genotoxicity, 14, 56, 64, 251, 252, 307
Glycerin, 9, 214, 215, 364
Glycosaminoglycan (GAG), 202, 219, 226
Glycosphingolipid (GSL), 36, 57, 201, 304
Gold nanoparticles, 322
Good Manufacturing Practices (GMP), 64, 103, 246, 247, 252–253, 372, 399
Green chemistry, 234
Green cosmetics, 320
Grifola, 41, 56, 97, 352
Grifola frondosa, 4, 27, 30, 33, 48, 55, 61, 75, 79, 89, 95, 97, 157, 175, 201, 219, 222, 225, 226, 331, 336, 356, 366
Growth requirement, 39, 199

H
HaCaT keratinocytes, 390
Hair, 2, 24, 48, 73, 114, 130, 144, 161, 188, 200, 222, 235, 248, 302, 320
Hair-breakage, 24, 25, 41, 55, 63
Hair care, 9, 50, 177, 192–193, 303
Haircare formulations, 403
Hair care products, 177, 228, 311–312, 334
Hair growth, 25, 28, 31, 41, 62, 63, 80, 155, 193, 228, 235–237, 239, 253, 254, 332
Hair growth stimulation (fungal origin), 235
High-performance liquid chromatography (HPLC), 78, 95, 253, 284, 360, 361, 393, 397, 407
Hyaluronic, 132–134
Hyaluronic acid (HA), 9, 24, 61, 63, 88, 89, 98, 131–134, 138, 164, 167, 171, 174, 176, 202, 206, 207, 214, 218–220, 222, 224, 225, 310, 311, 390, 395
Hydrating agents, 24, 41, 74, 336
Hydration, 4, 7, 10, 27, 51, 61, 88, 89, 131–134, 137, 138, 171, 173, 176, 191, 200, 202, 206, 207, 237, 239, 263, 279, 281–282, 286, 287, 302–306, 310, 311, 321, 332, 341, 389–391
Hyperpigmentation, 5, 6, 8, 25, 27, 28, 41, 50, 52, 55, 64, 80, 112, 114, 118, 121–123, 192, 204–206, 217, 218, 264, 287, 325, 334, 388

I
IL-6 suppression, 90, 93
Immunomodulatory effects, 31, 52, 353, 365, 366
Inducible nitric oxide synthase (iNOS), 90, 93, 94, 217
Industrial applications, 294, 407
Industrial fermentation, 295, 296
Inflammation, 2, 7, 27, 30, 50, 51, 53, 59, 64, 65, 78, 79, 88, 89, 98, 121, 133, 134, 139, 149, 152, 170, 173, 186, 187, 190, 192, 204–206, 215, 217, 238, 239, 247, 264, 282, 286, 287, 305, 321, 322, 325, 338, 356, 367, 369–371
Inflammatory, 7, 55, 88, 89, 94, 98, 101, 186, 187, 192, 200, 216, 217, 219–221, 226, 245, 281, 282, 285, 305, 325, 327, 354, 368–371
Ingredients, 2, 24, 47, 74, 87, 112, 130, 152, 162, 186, 199, 214, 234, 243, 279, 302, 320, 355, 387, 403
Inhibitors, 77, 94, 99, 112–120, 123, 151, 155, 204, 206, 217, 220–221, 291
Innovations, 12, 13, 15–16, 48, 66, 123, 138–139, 171, 214, 228, 234, 238, 250–251, 289, 313, 330, 339, 373, 388, 395, 396, 398, 412
Inonotus, 26, 28, 29, 41, 56, 75, 121, 219
Inonotus obliquus, 27, 28, 32, 48, 51, 64, 65, 76, 89, 121, 201, 219, 222, 223, 228, 331, 352, 354
Intrinsic aging, 145, 149, 191
In vitro analysis, 153, 338

K
Keratinocytes, 28, 30, 36, 57, 112, 151, 175, 202, 204, 205, 207, 251, 282, 283, 309
Kojic acids, 8–10, 24, 25, 31, 36, 40, 55, 74, 76, 102, 113, 115, 116, 118–123, 166, 171, 176, 188, 192, 193, 205–207, 218, 227, 331, 334, 336, 337

L
Lactic acid, 3, 4, 15, 24, 35, 41, 74, 130, 134, 190, 191, 200, 202–204, 207, 269, 304, 309, 334, 362
L-ascorbic acid, 116
Layerosomes, 326
Lentinan, 4, 31, 34, 48, 77, 134, 135, 167, 187, 188, 191, 201, 238, 304, 336, 353, 356, 365, 366
Lentinula, 5, 6, 41, 56, 187
Lentinula edodes, 2, 4, 10, 31, 34, 48, 55, 61, 76, 77, 79, 89, 97–99, 134–136, 171, 175, 187, 188, 193, 200, 201, 221, 222, 225, 303, 306, 331, 332, 353, 354, 356, 366, 370
Lignans, 116, 117, 354
Lipid extraction, 283–284
Lipid nanoparticles, 326
Lipid peroxidation, 77, 78, 97, 390
Liposomes, 104, 321, 323, 326–328, 330, 397, 399, 408
Lyophilization, 268, 361

M
Macrofungi, 4, 10, 56, 104, 206
Macromolecules, 200, 311
Makeup applications, 193
MAPK signaling, 95
Marine fungi, 10, 13, 97, 120, 123, 130, 131, 168, 334, 340
Market acceptance, 312, 407, 410–411, 413
Mass spectrometry (MS), 96, 284, 292, 361, 395
Matrix metalloproteinases (MMPs), 61, 75, 77, 99, 145, 146, 148, 150, 155, 215, 217, 219, 220, 225, 264, 390
Medicinal cosmetics, 23–27, 29, 31, 35, 37, 38, 40, 41
Medicinal mushrooms, 4, 5, 11, 27, 79, 222, 309, 341, 353, 397
Medicinal mushrooms in cosmetics, 397
Melanin, 6, 24, 51, 74, 98, 112, 154, 165, 188, 203, 218, 264, 303, 333
Melanin regulation by fungi, 24
Melanocytes, 36, 112–114, 118, 122, 204–206, 227
Melanogenesis, 30, 31, 112–114, 118, 121, 122, 204–206, 217, 218, 224
Metabolic engineering, 280, 293–294
Metabolites, 2, 24, 48, 95, 118, 130, 153, 162, 187, 206, 216, 234, 250, 264, 292, 304, 327, 353, 390, 409
Metalloenzyme, 112, 114
Metalloproteinases, 61, 75, 99, 144, 146, 148, 150, 155, 215, 217, 219, 220, 225, 226, 264, 311, 390
Microbial cosmetics, 2
Microbiological testing, 253, 254, 335, 339
Microbiology, 193, 194, 398
Microemulsions, 328
Microorganisms, 3, 8, 11, 16, 40, 98, 112, 113, 118, 123, 130, 149, 163, 169, 170, 172–176, 186–189, 194, 195, 221, 241–244, 247, 250, 253, 258, 271, 278, 303, 307, 310–312, 363, 393, 398, 405, 411
Microwave-assisted extraction (MAE), 8, 54, 200, 201, 269, 406, 407
Modern techniques, 54
Moisturization, 12, 30, 31, 50, 191, 202, 207, 239, 247, 254, 303, 306, 310, 326, 331, 336, 388
Moisturizers, 3, 10, 24, 35, 55, 61, 65, 80, 112, 171, 176, 185, 187, 188, 190–192, 202, 214, 218, 224, 236, 237, 285, 310–311, 323, 324, 328, 331, 332, 336, 390, 391
Moisturizing, 3–7, 9–11, 24, 25, 30, 48, 50, 51, 53, 56, 59, 61–62, 64–66, 80, 81, 89, 97–99, 103, 133, 134, 163, 167, 174–176, 186, 190–193, 202, 207, 215, 221, 224–225, 236, 237, 239, 251, 288, 303, 304, 306, 323, 326, 331, 332, 337, 341, 390, 412
Moisturizing agents, 15, 24, 36, 60, 80, 154, 167, 224, 225, 331
Monascus purpureus, 54, 154
Mucor miehei, 35, 193, 332, 336
Mushroom, 2, 23, 48, 89, 113, 130, 155, 163, 186, 202, 215, 235, 262, 302, 320, 352, 389, 411
Mushroom-based skincare, 65
Mushroom polysaccharides, 62, 77, 78, 222, 311
Mutagenicity, 64, 251, 307
Mycelia, 7, 33, 37, 40, 52, 58, 75, 78, 79, 89, 201, 222, 226, 308, 309, 320, 363
Mycelium, 7, 28, 37, 54, 65, 76, 80, 81, 113, 121, 155, 227, 309, 353, 363, 365, 366, 412
Mycelium-based cosmetic formulations, 10
Mycology, 187–190, 389, 412
Mycology in cosmetics, 412
Mycotoxin contamination, 312
Mycotoxins, 99, 312, 355, 362, 372, 409

N
Nanocapsules, 321, 323, 324, 328
Nanocarriers, 55, 321, 326, 328
Nanocrystals, 322–326, 329
Nanoemulsions, 94, 100, 321, 324, 328, 399
Nanoencapsulation, 408
Nanogold, 321, 322, 324
Nano-hydroxyapatite, 324–326
Nanoliposomes, 324, 326
Nanometals, 321–324
Nanoparticles, 14, 165, 176, 320–330, 337, 338, 340, 341, 397
Nanosilica, 322, 323
Natural, 3, 23, 48, 74, 89, 111, 130, 143, 162, 186, 199, 215, 236, 241, 263, 278, 302, 320, 351, 388, 404
Natural anti-wrinkle agent, 207
Natural cosmetics, 2, 15, 39, 40, 200, 234, 263, 303, 306, 336, 410, 411, 413
Natural environments, 154, 340
Natural haircare from fungi, 234–239, 311–312
Natural hyaluronidase inhibitors, 98, 131–133

Natural ingredients, 2, 3, 5, 7, 13–15, 49, 103, 157, 207, 218, 223, 224, 234, 235, 242, 277, 302–305, 390, 403, 409–411
Natural pigments, 9, 163, 174
Natural preservatives, 5, 173–174, 234, 397, 411
Natural products, 23, 24, 73, 77, 80, 95, 96, 122, 162, 188, 207, 218, 224, 244, 311, 320, 410, 411
Natural skincare, 281
NF-κB pathway, 217, 370
Niosomes, 324, 327, 330
NO production, 90–93, 98, 101
Novasomes, 324, 327, 330
Nuclear magnetic resonance (NMR), 284, 361
Nutraceuticals, 27, 31, 50, 52, 61, 64, 65, 130, 258, 270, 302, 351–374
Nutricosmetics, 3, 4, 27, 29, 48, 51, 52, 63–65, 130, 157, 162, 215, 221–224, 302, 331

O
Official Compendiums, 393
Omics (genomics, proteomics, metabolomics), 104, 163, 289, 291–292, 296, 340
Organic haircare ingredients, 222, 224
Oxidative stress, 5–7, 25, 27, 30, 31, 36, 54, 61, 79, 88, 89, 94, 97, 100, 132, 133, 135, 144, 146, 154, 174, 176, 179, 187, 191, 206, 216, 217, 225, 238, 239, 245, 249, 258, 263, 279, 287, 338, 356, 369, 370, 372, 407

P
Pathogens, 36, 56, 80, 98, 116, 134, 173, 187, 188, 190, 221, 242–247, 254, 280, 285, 365, 366
Pattern recognition receptors (PRRs), 167, 365
P-Coumaric, 28, 98, 100, 119, 121, 205–207, 221, 222
Penicillium, 3, 5, 10, 26, 35, 39, 41, 54, 55, 58, 91–93, 118, 120, 154, 176, 188, 193, 203–205, 234, 236, 244, 245, 248, 251, 306, 333, 334, 338, 362, 409
Penicillium extracts in conditioners, 234, 236, 333
Personal care industry, 16, 389
Personal care products, 3, 73, 143–157, 185–187, 189, 190, 192–194, 205, 234, 248, 249, 320, 410
Phenolic, 4, 24, 51, 75, 89, 114, 132, 153, 162, 188, 201, 215, 248, 303, 320, 354, 399, 408
Phenolic acids, 24, 25, 27, 56, 75, 76, 97, 98, 102, 116, 117, 134, 162, 177, 200, 201, 221, 222, 250, 331, 333, 354
Phenolic compounds, 26, 29, 31, 51, 53, 55, 59, 74–76, 79, 80, 97, 101, 121, 132, 133, 135, 153, 155, 162, 164, 190, 199, 201, 202, 217, 221, 222, 224, 225, 331, 353, 354, 362, 366–368, 372, 399, 407
Pholiota nameko, 97, 98, 100, 101
Photoaging, 143–146, 150, 151, 176, 225, 390
Photoprotective effects, 41, 60, 390
Phototoxicity, 251, 252
Phytochemicals, 24, 130, 152, 153, 189, 199, 200, 206, 218, 353, 369, 371, 372
Pichia pastoris, 278, 280, 390
Pigment, 8, 25, 54, 76, 112, 137, 165, 204, 218, 258, 304, 322, 389
Pigment stability, 54
Plant-derived compounds, 122
Plants, 2, 14, 23, 32, 35, 36, 38, 39, 48, 50, 51, 54, 57–59, 64, 80, 112–114, 116, 117, 122, 123, 130, 133, 134, 138, 151–153, 155–157, 162, 167, 168, 194, 195, 206, 217, 218, 220–222, 227, 235, 237–239, 257, 259, 260, 265, 271, 278, 280, 293, 303, 306, 337, 339, 389, 390, 397
Pleurotus, 4, 28, 53, 75, 120, 163, 219, 332, 369
Pleurotus ostreatus, 4, 7, 27, 28, 32, 55, 57, 61, 65, 76, 78–80, 98, 100, 166, 216, 225, 228, 332, 356, 366
Pleurotus species applications, 53, 223, 358, 369, 370
Polyphenols, 48, 76, 78, 103, 116–117, 121, 123, 134, 135, 151, 152, 157, 162, 171, 172, 186, 190, 191, 200, 201, 204, 207, 222–224, 227, 331, 333, 337, 352, 355, 359, 371
Polysaccharides, 4, 6, 7, 16, 24, 25, 27–34, 50–53, 56, 60–62, 65, 74, 75, 77–80, 94, 95, 97–101, 103, 132–134, 136, 155, 162–164, 167, 171, 172, 175, 177, 188, 191, 199–202, 215, 217, 222–227, 234, 236, 301–314, 320, 331, 332, 337, 353–356, 359, 364–366, 370–372, 389–391, 399
Poly-unsaturated fatty acids, 367
Prevalence, 244
Probiotics, 27, 64, 351, 362
Product safety, 15, 66, 177, 247, 373, 408, 410, 411
Proliferation, 3, 36, 61, 122, 218, 226, 245, 248, 251, 271, 309, 357, 367
Proteases, 60, 144, 148, 149, 174, 203, 204, 254, 333
Protein and peptides, 367
Protein modeling, 290
Proteins and peptides, 172, 234–236, 367
Proteolytic enzymes, 146, 203, 204, 207, 308
Psoriasis, 48, 98, 152, 221, 280–282
Purification, 8, 99, 204, 249, 284, 312, 313, 335, 360, 361, 363, 364, 390

Q
Quality control protocols, 407

R
Reactive oxygen species (ROS), 60, 77, 79, 89, 97, 113, 143–146, 150, 152, 175, 192, 215–217, 220, 222, 224, 225, 258, 263, 264, 271, 338, 356, 369
Red Pitaya (*Hylocereus* spp.), 151, 152
Regulatory compliance, 12, 103, 177, 254, 255, 393, 399
Regulatory guidelines, 246, 307, 338, 339
Reishi, 4, 5, 27, 52, 64, 65, 89, 134, 170, 171, 174, 192, 193, 200, 206, 222, 223, 227, 336, 352, 390, 391, 397
Reverse phase chromatography, 95
Rheological analysis, 391
Rhizopus, 3, 35, 54, 134, 190, 192, 202, 203, 207, 334

S
Saccharomyces, 5, 41, 163, 207, 338, 355, 390
Saccharomyces cerevisiae, 8, 36–38, 56, 57, 171, 174, 234, 236, 262, 265, 278, 280, 283, 308, 340, 355, 390, 391
Safety evaluation, 11, 12, 251, 252, 254, 373, 409, 411
Saponification, 241–255, 267, 268
Scalp health, 10, 62, 234–237, 239, 254
Scalp treatment with fungal enzymes, 237

Schizophyllan, 2, 4, 31, 34, 56, 77, 89, 134, 136, 167, 189, 190, 222, 234, 236, 336, 353, 355, 356, 365
Schizophyllum, 56, 80, 99, 134, 187
Schizophyllum commune, 2, 31, 34, 80, 89, 98, 136, 167, 187, 188, 190, 222, 236, 331, 332, 336, 355
Selenium, 3, 15, 31, 33, 52, 56, 57, 62, 63, 74, 76–77, 79, 80, 155, 156, 215, 223, 227, 235, 237, 239, 279, 303, 304, 331, 352
Sensitization, 14, 55, 56, 64, 245–247, 251, 252, 307, 339
Serums, 10, 27, 50, 55, 60, 62, 64, 65, 74, 80, 89, 97, 131, 134, 136, 137, 155, 171, 176, 188, 192, 214, 227, 234, 236, 254, 284, 285, 302, 323, 325, 328, 331, 336, 356, 367, 390, 391
Shiitake, 4, 5, 31, 62, 89, 135, 155, 171, 187, 200, 222, 223, 227, 238, 352, 356
Silver nanoparticles (AgNPs), 322, 337
Skin, 2, 24, 48, 73, 87, 112, 130, 143, 161, 192, 200, 220, 236, 242, 258, 287, 302, 320, 388, 404
Skin aging, 7, 27, 28, 77, 80, 99, 131–133, 143, 154, 179, 187, 191, 207, 215, 216, 218, 220, 225, 226, 264, 287, 303, 306
Skin barrier, 7, 35, 36, 50, 57, 97, 99, 171, 174, 177, 191, 200, 201, 252, 282–287, 328, 389, 390, 408
Skin barrier function, 35, 36, 57, 99, 202, 252, 283, 390
Skin barrier repair, 171
Skincare, 2–10, 12, 13, 16, 25, 30, 36, 41, 48–50, 53, 55, 56, 59, 60, 64, 65, 88, 89, 97, 123, 130–135, 137–139, 145, 154–157, 167, 170, 171, 173, 174, 176, 179, 186, 188, 190–193, 200, 205–207, 214, 221–224, 228, 234, 236, 254, 261, 277–283, 285, 286, 288, 289, 302, 305, 306, 310, 321, 323–325, 327–330, 333, 334, 336, 339–341, 388–390, 398, 399, 403
Skin elasticity, 5, 25, 27, 30, 51, 56, 64, 65, 143–145, 149, 152, 154, 155, 167, 175, 191, 223, 254, 303, 388, 390, 391
Skin healing, 97, 99, 187, 188, 190
Skin health, 5–7, 10, 24, 41, 64, 80, 88, 139, 144, 146, 157, 170, 173, 174, 179, 185, 188, 191, 199, 207, 222, 225, 236, 242, 245, 247, 251, 254, 278–280, 282–285, 302, 304, 313, 314, 321, 388, 389
Skin hydration, 5, 9, 10, 24, 25, 28, 56, 61, 63, 65, 99, 100, 134, 144, 176, 177, 201, 202, 206, 207, 214, 224, 226, 252, 278, 281, 284–286, 326, 330
Skin inflammation, 56, 63, 134, 137
Skin integrity, 135, 145, 150
Skin-lightening, 8, 9, 55, 63, 98, 122, 155, 176, 187, 188, 190, 192, 193, 199–208, 303, 306, 311, 334, 337
Skin rejuvenation, 65, 176, 218, 388
Skin texture, 5, 24, 50, 88, 388
Solid-state fermentation (SSF), 37, 38, 340, 363, 364
Solvent extraction methods, 407
Sphingolipids, 58, 200, 201, 281
Stability testing, 252–253, 255, 391, 393–395, 398, 399, 408
Staphylococcus, 26, 55, 56, 164, 169, 200, 221, 356, 367
Sterols, 24, 25, 31, 53, 78, 102, 162, 165, 217, 227, 304, 353, 354, 367, 371
Stilbenes, 76, 116, 117, 162
Strain engineering, 290, 293
Stress-response, 201
Submerged fermentations, 31, 34, 37, 38, 195, 363, 364
Substrates, 9, 29, 32–34, 36–40, 114, 116, 117, 144, 146, 148, 149, 151, 162, 188, 190, 193–195, 202, 204, 206, 215, 220, 278, 280, 289–292, 295, 296, 362–364, 390
Supercritical fluid extraction (SFE), 8, 54, 95, 201, 267, 270, 358, 359, 406, 407
Supply chain management, 404
Sustainability, 3, 5, 12–14, 16, 39, 66, 112, 138, 186, 187, 193–195, 224–225, 228, 234, 237, 238, 255, 278–281, 289, 295–296, 303, 311–314, 340, 352, 373, 405, 410
Sustainable, 3, 24, 56, 138, 153, 171, 194, 199, 214, 234, 242, 278, 302, 333, 374, 389, 403
Sustainable beauty, 14, 186, 195, 239, 413
Sustainable cosmetics, 16, 186, 195
Sustainable fungal ingredients, 138
Sustainable ingredients, 12, 41, 179, 236, 249, 405, 411
Sustainable production, 12, 16, 234, 255, 296, 314
Synthetic, 4, 23, 48, 112, 130, 144, 172, 186, 215, 234, 242, 277, 302, 320, 355, 399, 403
Synthetic alternatives, 48, 187, 234, 303
Synthetic biology, 9, 287–296

T

Terpenoids, 4, 7, 29, 31, 50–54, 57, 75, 76, 79, 89, 94, 98, 103, 131, 155, 163, 164, 174, 199, 201, 215, 222, 223, 248, 257, 258, 264, 270, 303, 304, 320, 331, 333, 334, 352, 354, 355, 372
Therapeutic benefits, 2, 88, 387, 388
Titanium dioxide (TiO_2), 176, 321
Topical applications, 61, 63, 87, 99, 171, 214
Torularhodin, 59, 259, 260, 262, 263, 265
Toxicological, 11, 14, 15, 29, 64, 99, 255
Toxicology, 41, 66, 252
Trametes versicolor, 25, 27, 29, 30, 33, 39, 48, 53, 54, 75, 79, 102, 134, 163, 201, 222, 223, 227, 331, 354, 366
Transepidermal water loss (TEWL), 9, 207, 252, 281, 284–287, 311, 390
Transmission, 242, 245, 247
Trichoderma in cosmetic applications, 174
Tricholoma matsutake, 7, 33, 39, 41, 51, 52, 57, 61, 75, 98, 166, 219–221, 226, 235, 236
Triterpenoids, 5, 10, 32, 52, 55, 57, 76, 78, 89, 97, 98, 101, 133, 134, 165, 192, 250, 251, 309, 354, 355, 357, 371, 407
Tyrosinase, 6, 25, 51, 80, 89, 112, 135, 154, 166, 204, 217, 264, 311
Tyrosinase inhibitors, 36, 51, 98, 99, 112–117, 120, 123, 135, 217

U
Ultrafiltration, 38, 361
Upregulates, 122, 225
Usage, 13, 16, 28, 36, 51, 99, 103, 152, 179, 194, 199, 214–228, 243, 264, 268, 270, 277, 285, 295, 301–314, 319–341, 352, 358, 362, 373
Ustilago maydis, 280
UV damage, 25, 27, 63, 112, 114, 116, 134
UV protection, 9, 56, 163, 167, 174, 176, 311, 329–331, 334, 391
UV radiation, 13, 36, 51, 55, 56, 77, 80, 113, 118, 131, 132, 144–146, 154, 176, 215, 218, 221, 258, 271, 278, 286, 287, 310, 328, 398, 407

V
Vegan skincare, 279
Vitamin B complex, 353, 356
Vitamin C, 56, 65, 75, 77, 79, 88, 152, 204, 222, 224, 334, 357, 388, 393
Vitamin E, 55, 56, 97, 176, 204, 397, 399
Volatile compounds from fungi, 40
Volvariella, 30, 33, 41, 61, 75, 99, 101, 219, 225, 306, 311

W
Well-being, 12, 47, 64, 112, 137, 222, 242, 247, 255, 305, 319, 320, 351, 371–373, 388, 399, 400
Wound healing, 7, 14, 24, 28, 36, 37, 62, 65, 97, 134, 152, 174, 201, 220, 227, 245, 262, 304, 306, 309, 331
Wrinkle prevention, 88, 191, 206

Y
Yeast strains, 311

Z
ZnO nanoparticles, 323

MIX
Papier aus verantwortungsvollen Quellen
Paper from responsible sources
FSC® C105338

If you have any concerns about our products,
you can contact us on
ProductSafety@springernature.com

In case Publisher is established outside the EU,
the EU authorized representative is:
Springer Nature Customer Service Center GmbH
Europaplatz 3, 69115 Heidelberg, Germany

Printed by Libri Plureos GmbH
in Hamburg, Germany